Duality Principles in Nonconvex Systems

Nonconvex Optimization and Its Applications

Volume 39

The titles published in this series are listed at the end of this volume.

Duality Principles in Nonconvex Systems

Theory, Methods and Applications

by

David Yang Gao
Department of Mathematics,
Virginia Polytechnic Institute and State University,
Blacksburg, Virginia, U.S.A.

SPRINGER-SCIENCE+BUSINESS MEDIA, B.V.

A C.I.P. Catalogue record for this book is available from the Library of Congress.

ISBN 978-1-4419-4825-0 ISBN 978-1-4757-3176-7 (eBook)
DOI 10.1007/978-1-4757-3176-7

Printed on acid-free paper

To Rosa

Contents

Preface

The Tao that can be told is not the constant Tao;
The name that can be named is not the real name.
Nameless, the beginning of the universe.
Nameable, the mother of the myriad creatures.

—Lao Chi, 1.1

As far as the laws of mathematics refer to reality, they are not certain;
as far as they are certain, they do not refer to reality

—Albert Einstein

The term *duality* as used in our daily life means the sort of harmony of two opposite or *complementary* parts through which they integrate into a whole. Symmetry and inner beauty in natural phenomena are bound up with duality and, in particular, are significant in art and science. Mathematics lies at the root of duality. The aim of this book is to give, within a unified framework, a self-contained comprehensive presentation of the mathematical theory of duality for general non-convex and non-smooth systems, and the author hopes that it will provide a smooth, stimulating and provocative blend of different flavors of mechanics.

Duality has always been a rich source of inspiration for human knowledge through the centuries. As a philosophy duality had been discussed extensively from the ancient time of great thinkers such as Lao Chi†, Chuang Chi, Plato and Heraclitus to the giants of modern physics such as Mach, Maxwell, Einstein and Bohr. In science, duality and mechanics have been complementary partners since the time of the founding masters, Newton, Euler, Lagrange and Gauss, and their subsequent developers such as Hamilton, Legendre, Riemann, Kelvin, Noether, Poincaré, Hilbert, Cartan, Weyl, and von Neumann. In the present day, the theory of duality has become a vast subject, especially due to the modern work in optimization, game theory, economic science, theoretical physics and chemistry, mathematical programming, variational analysis, nonconvex-nonsmooth analysis and control, critical point theory and in many other areas. However, the duality gap between the two partners is getting larger and larger artificially.

---

† Lao Chi, an ancient, mystic Chinese philosopher in the fourth century BC. His book *Tao De Chin* was the first poetical treatise on the *Ying-Yang duality*. This book, as well as the books by his follower Chuang Chi and by his *duality partner* Confucius (551-479, BC) have influenced Chinese thought throughout the ages out of all proportion to their lengths. According to the *Shih Ji* (Records of the History, the earliest general history of China, written at the beginning of the first century BC by Shi-Ma Chien), Confucius once visited him and asked to be instructed in the "Tao".

Duality in mathematics, roughly speaking, is a fundamental concept that underlies many aspects of *extremum principles* in natural systems. Eigenvectors, geodesics, minimal surfaces, KKT conditions, harmonic maps, subharmonics of Hamiltonian systems and equilibrium states of many field equations are all critical points of certain functionals on some appropriate constraint sets or manifolds. For convex (static or Hamiltonian) systems, the mathematical theory of duality is well established due to the existence of a common symmetric framework for different problems. Several books have already dealt with major theoretical components of the subject. In applications, the so-called *primal-dual algorithms* have emerged as the most important and useful algorithms in linear programming during this decade. However, the nice symmetry is broken for nonconvex systems, where the dual formulation and the accompanying theory for each problem depend mainly on the intrinsic constitutive law. To develop a simple and correct duality theory usually requires certain "physical knowledge" of the system. Many problems in nonconvex systems have remained obscure and open for a long period.

Actually, at the beginning of this century, duality theory and methods for nonconvex problems were studied by engineers and scientists in mechanics. The well-known Hellinger-Reissner complementary energy principle in nonlinear elasticity, proposed by Hellinger in 1914, might be the first dual formulation for nonconvex variational problems. This principle has many important consequences in large deformation theory and computational mechanics. Unfortunately, this very important complementary variational principle is not known by many mathematicians. It turns out that the extremality property of this principle has been an open problem for more than forty years, and this raised many arguments in finite deformation theory and nonconvex mechanics. One of the main results in this book provides the solution of this problem, and leads to an interesting triality theory for nonconvex systems.

It is intended that the present book will fill the (duality) gap between the mathematical and engineering sciences by providing a systematic exposition of duality theory as a self-contained theoretical system with substantial applications in physics, continuum mechanics, large deformation structures, mathematical programming, partial differential equations and geometry. Most parts of this book contain material that is new, both in its manner of presentation and in its research development. The new results arose naturally during the writing of this book, and it turns out that the development of the project has thereby been vastly enriched.

This is a book of *motivated mathematics*, i.e. a book of mathematics motivated by duality in natural phenomena, with particular emphasis on the mechanics rather than a book of mathematical analysis, proofs and applications. The book is therefore written in a dual way: by using two very simple examples, involving an elastostatic string and the dynamics of a particle, as a pair of seeds, the theory is grown continuously from the classical mono-duality of one-dimensional convex static equilibria, the nice bi-duality in dynamical systems, through the interesting tri-duality in nonconvex problems to the complicated multi-duality in general canonical systems. A potentially powerful sequential dual canonical transformation method is developed heuristically and illustrated by use of many interesting simple examples as well as comprehensive applications in three-dimensional finite deformation theory. The book divides naturally into three closely interconnected parts with a total of seven chapters. Each chapter provides some motivation, both at the beginning and throughout, and concludes with substantial applications and commentaries which furnish credits and references.

Duality in nature is amazingly beautiful, for it is the way nature was created. Duality in nature is simply mysterious, for it is the way that nature exists. It is beautiful because all things were originally created in a splendid harmonious world. It is mysterious because different creatures have different patterns of duality. If we are not confused very often about the duality of natural phenomena, we do not really understand what it is. This may be the way that we exist.

Serious study of duality theory or any other subject in science is very difficult (and dangerous). Both Lao Chi and Einstein told us that once a phenomenon has been either named or explained, certain reality has been lost. So, for example, there was a *Copenhagen fog* in quantum mechanics. Contrary to popular opinion, J. P. Aubin (1993) warned us that "mathematics is not simply a *richer* or *more precise Language*". About fifty years ago, Gödel's *incompleteness theorem* destroyed the existing foundations of formal logic and axiomatic systems. The only reason that the author dares to publish this book is that he does not *really* know what duality is. Thus, this is a book of *constructive mathematics*, i.e. a book using formal language and duality methods to model natural phenomena approximately, to construct intrinsic frameworks in different fields and to provide ideas, concepts and certain tools for describing natural phenomena and for solving real problems arising in engineering and science. The functional spaces and operators used in this book are mainly for the convenience and clarity of the descriptions. All these abstract notations are illustrated sufficiently by very simple examples. Many mathematical definitions, theoretical results and numerical methods are explained with either geometrical illustrations or physical applications. With the aid of these potentially useful notations, common mathematical frameworks are constructed for many different systems, independently of their physical content, and the frameworks also apply to wider classes of problems in engineering and science. These common structures are due to certain conservation laws that govern the systems, while the intrinsic symmetries in the abstract frameworks lead to a variety of dualities, each having a wealth of significant and substantial applications.

To the author's best knowledge, there are at the present time several other approaches to duality theory in nonconvex analysis that have proved successful in dealing with the difficulties of infinite-dimensional critical point theory (cf., e.g., Ghoussoub, 1993). However, these nice theories require a more substantial background in abstract analysis, algebraic topology and geometry than is required for the elementary approach adopted in this book. The author apologizes for this and to the many people whose contributions to duality theory are not included or even mentioned in this very limited book. But he sincerely hopes to receive both positive and negative feedback from readers in order to stimulate communication and possible cooperation.

The basic prerequisites for the book are multi-variable calculus and linear algebra. Topics on convex analysis and nonsmooth calculus needed for mathematical duality theories are included in one essentially independent chapter. A three-part appendix provides some necessary background from elementary functional analysis. The whole book or selected chapters can be used as a text for upper-undergraduate or graduate courses in variational methods, global optimization, applied mathematics, general engineering and science, mathematical physics and operations research. It can also be used for an advanced course in theoretical mechanics or to supplement courses on nonlinear optimization and nonconvex analysis. It is the author's hope that this book will also serve as a catalyst for the further development, both in theory and in applications, of this fascinating and fertile scientific field.

September, 1999 DYG

Hence always rid yourself of desires in order to observe its secrets;
But always allow yourself to have desires in order to observe its manifestations.
These two are the same
But diverge in name as they issue forth.
Being the same they are called mysteries,
Mystery upon mystery -
The gateway of the manifold secrets.

—Lao Chi, *Tao De Chi* 1.3

Motivated mathematicians must possess a sound knowledge of another discipline and have an adequate arsenal of mathematical techniques at their fingertips together with the capacity to create new techniques (often similar to those they already know).

—Jean-Pierre Aubin, 1993

Acknowledgments

And now these three remain: faith, hope and love. But the greatest of these is love.

—1 Corinthians, 13.13

The content and style of the book owes much to the influence of many friends and colleagues over the years, but I should like to express my primary gratitude to Professor Hwang Keh-Chih, Tsinghua University, who introduced me to the subject of engineering mechanics when I was a research student (1983-86), to Professor Gilbert Strang at MIT, who led me into the wonderland of applied mathematics and whose constant support changed my life, and to Professor Shing-Tung Yau and his wife Christina at Harvard University, where I spent two years as a post-doctoral fellow (1989-91). Through them I found the spiritual foundation on which to base my life.

Very special thanks go to Professor Ray W. Ogden at the University of Glasgow, my respected scholar and dear friend. It was through his tremendous patience and careful reading of the whole of the manuscript within a short period of time, his very detailed corrections, many important insights, comments and suggestions that the original rough "Chinglish" draft has been molded into a coherent body of work. It was during this period that I was fortunate to learn a great deal about aspects of academic discipline and to understand some of the intrinsic personal character that a real scientist should possess. For all this I owe him a profound debt of gratitude. Since the due date for this book has been seriously passed, I could not let him to have a second look at the whole revised version of the manuscript except Chapter 6. If there are, and I am sure there are many, mistakes, that is due to my own careless personal being.

I must offer my sincere thanks to many nice people for their kind hospitality and important support during the years in which I followed my professional wife while struggling to stay in the academic world. They include Professor T. Wierzbicki at MIT, Professor E.T. Onat at Yale University, Professor Y.K. Cheung at the University of Hong Kong, and Professor Wei H. Yang at the University of Michigan.

I thank Professor R. Olin, Department Head of Mathematics at Virginia Tech, for his understanding and support on this project. Thanks are also due to my colleague, Professor J. Kim of Virginia Tech, for letting me keep several of his important books for a long time, to Professor Y.H. Xu at Grove City College and my students Hyung-Seok Oh and Hoan Kim Nguyen for reading parts of the manuscript and making helpful suggestions. I should of course thank Professor P. Pardalos, the managing editor of this book series and the helpful people at Kluwer Academic Publishers, particularly John Martindale and his assistant Sharon Donovan, for their encouragement and patience in waiting for the delivery of this long-overdue manuscript.

The project of this book was launched about one year after the tragic death of my beloved wife, Professor Rosa Q. Huang (1963-1995), an exceptionally unusual mathematician, an innocent, loving lady with a pure kindhearted personality. It was for her unswervingly passionate love from the time we met till the end of her life that I was willing to sacrifice my own career and to follow her as a semi-professional husband for six years. The wonderful duality in my life was broken, and the beauty of the triality arises from ashes. Although the core results of this book, discovered one year after her death, mean nothing to me compared with her precious life, I do believe that good and honest people can appreciate the intrinsic beauty and value of the triality, as well as the heavy price we paid for it. During the period of writing, another two very special people left me. One was my dearest brother Song Gao (1963-1998), for whom I cared the most during our childhood when our parents were persecuted by the so-called "cultural revolution". I was deeply grieved to learn that he called my name in his last moments while I could not even be with him due to the distance, the difficulties I had to face and the work on this project. The second was Professor P.D. Panagiotopoulos (1950-1998), a distinguished scholar and pioneer in nonsmooth mechanics. It was through his suggestion, encouragement and recommendation that this project started, and that the book is able to be published by Kluwer. I record in this book that my memory of these three very special people and that I wrote with a broken heart and burning spirit. I am very sorry that I could not list here the names of many others who have helped me during this difficult time. I present my sincere thanks to them for their genuine compassion and precious help.

Finally, I affirm my love and gratitude to my parents, and my three wonderful artistic children, Shawn (violinist), Timothy (pianist) and Tonya (a junior cellist who lost her dear mommy at 1.91 years old), for their love, patience, and understanding throughout the most difficult time of writing the book. Without their unfailing support and unconditional cooperation, I don't think I would have been able to finish this project while carrying a heavy teaching load, maintaining my research and bringing three kids up by myself.

I Symmetry in Convex Systems

1 MONO-DUALITY IN STATIC SYSTEMS

In the beginning God created the heavens and the earth. ⋯
And God said, "Let there be light", and there was light.

—Genesis, I

Tao creates one; one begets two; two produce three; three generate all things. Everything carries Ying and embraces Yang and are the harmony of the generative forces of the two.

— Lao Chi

To wit, since the plan of the universe is the most perfect, there can be no doubt that all actions in the world can be determined from the observed phenomena and the causes with the aid of the method of maxima and minima.

—Leonard Euler

What really interests me is whether God has any choice in the creation of the world.

—Albert Einstein

$$P(u) = W(\Lambda u) - F(u)$$

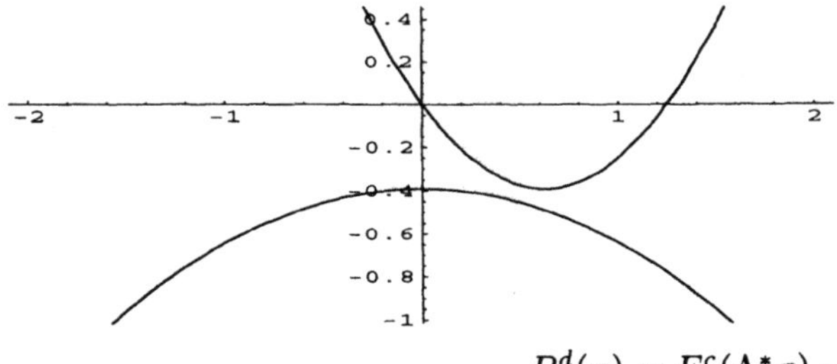

$$P^d(\sigma) = F^c(\Lambda^*\sigma) - W^c(\sigma)$$

Figure 1.1. Min-Max duality in convex optimization (see Example 1.7.1)

The governing equations of equilibrium in nonlinear systems are amazingly beautiful. Their structure is repeated throughout the field equations of mathematical physics and of discrete systems of networks. By introducing abstract notations, we are able to see unifying structures in the different theories. Through pure mathematical analysis, the intrinsic inner beauty in physical nature can be revealed.

We begin with the first problem in the calculus of variations, studied by Euler[1] in the 17th century. Throughout the ensuing discussion of problems in one-dimensional systems, we gradually introduce the ideas, concepts and abstract notations of the subject of duality. These notations and concepts are very powerful in nonlinear analysis, which enables many different problems and systems to be put in a unified framework.

1.1 The First Problem in The Calculus of Variations

1.1.1 Elastic String: Tri-Canonical Forms and Total Potentials

Let us consider an elastic string with an original length $\ell_0 > 0$.The string is fixed at $x = 0$ and is subjected to a distributed load $f(x)$. On the open interval $I = (0, \ell_0) \subset \mathbb{R}$, we denote the displacement by $u(x)$, which is a real-valued function of $x \in I$. The infinitesmall (linear) strain ϵ is defined by

$$\epsilon = \frac{\mathrm{d}}{\mathrm{d}x} u = u_{,x} = u'. \tag{1.1}$$

The stress, which is dual to the strain, is defined physically by the *constitutive law*

$$\sigma = k(\epsilon), \tag{1.2}$$

where the function k is determined from experimental measurements. For linear elasticity, $\sigma = k_0 \epsilon$ is the well-known *Hooke's law* and the material constant k_0 is Hooke's modulus of elasticity. However, for nonlinear elastic materials, the constitutive equation (1.2) is nonlinear in ϵ.

If $f(x)$ is an axial load, the equilibrium equation has the form

$$\frac{\mathrm{d}}{\mathrm{d}x}\sigma + f(x) = 0 \;\; \forall x \in I, \tag{1.3}$$

which holds independently of the equations (1.1) and (1.2).

The elastic energy $U(\epsilon)$ stored in the deformed string is a function of the strain ϵ and is given by

$$U(\epsilon) = \int \sigma(\epsilon) d\epsilon, \tag{1.4}$$

which is called the *stored-energy density*. The constitutive law (1.2) can be written in the general form

$$\sigma = \frac{\partial U(\epsilon)}{\partial \epsilon} = U_{,\epsilon}(\epsilon). \tag{1.5}$$

Let $\mathcal{U}(I)$, or simply $\mathcal{U}$, be a displacement space defined on I, i.e. a space of displacement fields $u(x)$ with domain $x \in I$; $\mathcal{U}^*$ a force space, i.e. a space of force fields $f(x)$. They are placed in duality by the *first bilinear form*

$$\langle *, * \rangle : \mathcal{U} \times \mathcal{U}^* \to \mathbb{R}. \tag{1.6}$$

For example, this bilinear form can be defined simply as

$$\langle u, f \rangle = \int_I u(x) f(x) \, dx \quad \forall (u, f) \in \mathcal{U} \times \mathcal{U}^*.$$

Then, if $\mathcal{U}$ is a *Lebesgue integrable space* $\mathcal{L}^{\alpha}(I)$ with $\alpha \in [1, +\infty]$ (see Appendix A), its dual space is $\mathcal{U}^* = \mathcal{L}^{\alpha^*}(I)$ with $\alpha^* \in [1, +\infty]$, where the constants α, α^* are called *dual numbers*, and satisfying the equation

$$\frac{1}{\alpha} + \frac{1}{\alpha^*} = 1 \quad \forall \alpha, \alpha^* \in [1, \infty]. \tag{1.7}$$

Let $\mathcal{E}$ and $\mathcal{E}^*$ be the corresponding strain and stress spaces, respectively, placed in duality by the *second bilinear form*

$$\langle *; * \rangle : \mathcal{E} \times \mathcal{E}^* \to \mathbb{R},$$

which could, for example, be the same as (1.6), in which case

$$\langle \sigma; \epsilon \rangle = \int_I \epsilon(x) \sigma(x) \, dx \quad \forall (\epsilon, \sigma) \in \mathcal{E} \times \mathcal{E}^*. \tag{1.8}$$

Thus, if $\mathcal{E} = \mathcal{L}^{\beta}(I)$, its dual space is $\mathcal{L}^{\beta^*}(I)$ and β, $\beta^* \in [1, +\infty]$ are dual numbers.

If the string is fixed at both ends, the so-called admissible deformation space $\mathcal{U}_a \subset \mathcal{U}$ can be defined by

$$\mathcal{U}_a = \{u \in \mathcal{U}(I) | \ u(0) = 0, \ u(\ell_0) = 0\}. \tag{1.9}$$

Then, by integration by parts, the identity

$$\int_I \sigma(x) u'(x) \, dI = - \int_I u(x) \sigma'(x) \, dI \tag{1.10}$$

holds for arbitrary differentiable functions $u(x) \in \mathcal{U}_a$ and $\sigma(x) \in C^1(I)$. By introducing the differential operators

$$\Lambda = \frac{d}{dx}, \quad \text{and} \quad \Lambda^* = -\frac{d}{dx},$$

the identity (1.10) can be rewritten as the so-called *Gauss-Green formula*, namely

$$\langle \Lambda u; \sigma \rangle = \langle u, \Lambda^* \sigma \rangle. \tag{1.11}$$

The operator $\Lambda^* : \mathcal{E}^* \to \mathcal{U}^*$ is called the *adjoint operator* of $\Lambda : \mathcal{U} \to \mathcal{E}$. The set of three equations governing this static deformation system can be written in the so-called *tri-canonical forms*

$$\begin{aligned} &(a) \text{ deformation equation:} && \epsilon = \Lambda u = \mathrm{d}u/\mathrm{d}x, \\ &(b) \text{ elastic constitutive law:} && \sigma = DU(\epsilon) = k(\epsilon), \\ &(c) \text{ equilibrium equation:} && f = \Lambda^*\sigma = -\mathrm{d}\sigma/\mathrm{d}x. \end{aligned} \tag{1.12}$$

Let $A : \mathcal{U} \to \mathcal{U}^*$ denote the composition of operators

$$A = \Lambda^* \circ k \circ \Lambda. \tag{1.13}$$

Then, the tri-canonical forms (1.12) can be written jointly in the abstract form

$$A(u) = \Lambda^* k(\Lambda u) = f. \tag{1.14}$$

Since this abstract equilibrium equation appears repeatedly in many systems of mathematical physics, it can reasonably be called the *fundamental equation, of equilibria*. A diagrammatic representation for this static deformation system is shown in Fig. 1.2.

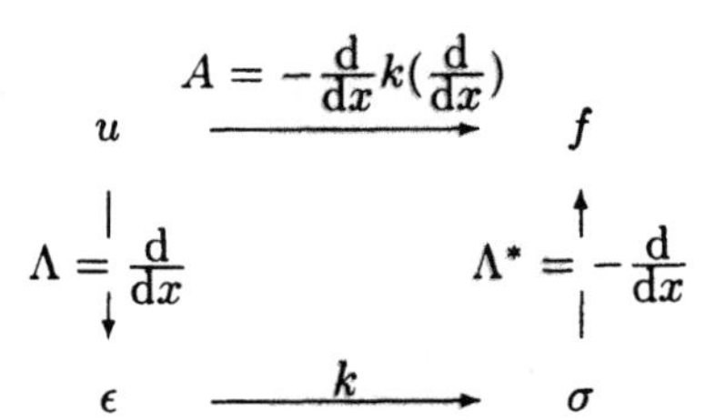

Figure 1.2. Diagrammatic representation in static deformation.

If $f \in \mathcal{U}^*$ is an axial force, the string is in a straight deformation state. For linear elastic string, $k = k_0 > 0$ is a constant, the stored-energy density is simply a quadratic function, $U = \frac{1}{2}k_0\epsilon^2$, and the constitutive law is Hooke's law

$$\sigma = DU(\epsilon) = k_0\epsilon.$$

The total energy stored in the string is a real valued function, W say, of the deformation $\epsilon(x)$ and given by

$$W(\epsilon) = \int_I U(\epsilon)\, \mathrm{d}x = \frac{1}{2}k_0 \int_I \epsilon^2\, \mathrm{d}x.$$

In this case, A is a linear operator and

$$Au = -k_0 u_{,xx} = f. \tag{1.15}$$

The analytic solution $u(x)$ depends uniquely on the external load $f(x)$ and boundary conditions.

Remark 1.1.1 It is easy to check that $A = \Lambda^* k_0 \Lambda = -k_0 \mathrm{d}^2/\mathrm{d}x^2$ is a symmetric operator in the sense that $A : \mathcal{U} \to \mathcal{U}^*$ is *self-adjoint*, so that

$$\langle u, Av\rangle = \langle Au, v\rangle \ \forall u, v \in \mathcal{U}_a.$$

This leads to $A^* = (\Lambda^* k_0 \Lambda)^* = \Lambda^* k_0 \Lambda = A$. ∎

If the string is subjected to a distributed transverse load $f \in \mathcal{U}^*$, the deformation $u(x)$ of the string becomes the deflection, and the string then, takes on a curved deformed state. The square of an element length $\mathrm{d}x$ before the deformation changes to

$$\mathrm{d}\ell^2 = (\mathrm{d}x)^2 + (\mathrm{d}u)^2 = (1 + (u_{,x})^2)(\mathrm{d}x)^2.$$

The coefficient $\xi = 1 + (u_{,x})^2$ is a finite strain. If the constant k_0 is considered as the mass density, then, the integral

$$W(u_{,x}) = \int_\ell k_0 d\ell = \int_I k_0 \sqrt{1 + (u_{,x})^2}\, \mathrm{d}x$$

for this "curved deformed system" represents the total mass of the deformed string. If $k_0 = 1$, W is the total length of the deformed string in classical differential geometry. However, in the context of plasticity theory, if k_0 is considered as the *elastic limit* (denoted by σ_0), then, $W(\epsilon)$ represents the total stored (plastic) energy in structural limit analysis. The stored-energy density U in this case is a nonlinear function of the linear strain $\epsilon = u_{,x}$ given by

$$U(\epsilon) = k_0 \sqrt{1 + \epsilon^2},$$

which leads to the nonlinear constitutive law

$$\sigma = DU(\epsilon) = k(\epsilon) = \frac{k_o}{\sqrt{1 + \epsilon^2}}\epsilon.$$

Clearly, $|\sigma| \le k_0$, and the equality defines the yield condition for plasticity. The fundamental equation in this curved system reads

$$A(u) = -\frac{\mathrm{d}}{\mathrm{d}x}\left(\frac{k_0}{\sqrt{1 + \left(\frac{\mathrm{d}u}{\mathrm{d}x}\right)^2}}\frac{\mathrm{d}}{\mathrm{d}x}u\right) = f(x) \ \forall x \in I,$$

which is a nonlinear ordinary differential equation. It is easy to check that A is not a symmetric operator, i.e. for any given sufficiently smooth functions $u, v \in \mathcal{U}_a$,

$$\langle A(u), v\rangle \ne \langle A(v), u\rangle. \tag{1.16}$$

The broken symmetry in this nonlinear system will be discussed later (see Section 3.1).

For the given external force $f(x)$, the external energy $F(u)$ is given by the integral

$$F(u) = \int_I f(x)u(x)\, dx.$$

A scalar-valued function such as this, whose argument u is itself a function is called a *functional*. The *total potential* of this system is the difference of the internal and external energies, namely

$$P(u) := W(\Lambda u) - F(u) = \int_I U(\Lambda u)\, dx - \int_I fu\, dx,$$

where the notation := stands for "equal by definition". Dually, =: denotes for "defined by". The domain of this functional is a subset of $\mathcal{U}$, called the *kinematically admissible space* . In particular, if $\mathcal{U}_a \subset \mathcal{U}$ is the domain of F, in which the boundary conditions are prescribed, and $\mathcal{E}_a \subset \mathcal{E}$ is the domain of $W(\epsilon)$ on which the constitutive mapping $DU : \mathcal{E}_a \to \mathcal{E}^*$ is well defined, then the kinematically admissible space, denoted $\mathcal{U}_k$, is defined by

$$\mathcal{U}_k = \{u \in \mathcal{U}_a \mid \Lambda u \in \mathcal{E}_a\}. \tag{1.17}$$

For example, if the string is fixed at its two ends, then, $\mathcal{U}_a$ is defined by (1.9). However, if the string is fixed at one end only, then, an external force, the dual variable of the displacement u, has to be given at the other end. For linearly deformed string, $\mathcal{E}_a = \mathcal{E} = \mathcal{L}^2(I)$, which is a self-dual space, i.e. $\mathcal{L}^2 = (\mathcal{L}^2)^*$. On the other hand, for the nonlinear deformation example, $\mathcal{E}_a = \mathcal{E} = \mathcal{L}^1(I)$, which is a *non-reflexive space* (i.e. $(\mathcal{E}^*)^* \neq \mathcal{E}$, see Appendix A.3).

The *minimum potential principle* leads to the minimization problem

$$P(u) \quad \to \min \quad \forall u \in \mathcal{U}_k. \tag{1.18}$$

1.1.2 Variational Problems and Kinematically Admissible Spaces

The *first problem of the calculus of variations* is only a slight generalization of the simple problem (1.18), i.e.

$$P(u) = \int_I \jmath(x, u, \Lambda u)\, dx \quad \to \text{stationary } \forall u \in \mathcal{U}_k, \tag{1.19}$$

where $\jmath$ is a given function of the arguments x, u, and $\Lambda u = u'$. The conditions specified in the definition of $\mathcal{U}_k$ are called *constraints*.

A problem is said to be *homogeneous* if $\jmath$ does not contain the independent variable x explicitly, i.e. $\jmath = \jmath(u, \Lambda u)$.

We shall start with some classical problems, through which the theory of duality evolved and which can easily be generalized for other linear or nonlinear differential operators Λ. This is why we use this abstract notation, instead of the traditional d/ dx.

I. Geodesic Problems with Equality Constraints.

Consider $\jmath(u, \Lambda u) = \sqrt{1+u'^2}$, which corresponds to $k_0 = 1$ in the above nonlinear example, and let the kinematically admissible space be

$$\mathcal{U}_k = \{u \in \mathcal{L}^1(a,b)|\ u' \in \mathcal{L}^1(a,b),\ u(a) = u_a,\ u(b) = u_b\}. \tag{1.20}$$

Then, the graph of the function $u(x)$ is a curve passing through the points $P_a = (a, u_a)$ and $P_b = (b, u_b)$ in $\mathbb{R}^2$. If $\bar{u}$ is the minimizer of P, i.e.

$$P(\bar{u}) \leq \int_a^b \sqrt{1+u'^2}\ \mathrm{d}x \quad \forall u \in \mathcal{U}_k,$$

then, the value of $P(\bar{u})$ identifies minimum length connecting the two end points. Clearly, since $f(x) = 0$ in this case, the solution $u(x)$ for this one-dimensional problem is a linear function, i.e. a straight line connecting the two given points. The analytic solution for nonzero $f(x)$ will be discussed later in this chapter.

In two-dimensional problems, let $\Omega \subset \mathbb{R}^2$ be a simply-connected domain with boundary $\partial\Omega$ and Λ a vector-valued linear partial differential operator defined by

$$\Lambda u = \mathrm{grad}\ u = \nabla u = (u_{,x}, u_{,y}),$$

with

$$(\Lambda u)^2 = |\nabla u|^2 = u_{,x}^2 + u_{,y}^2.$$

For a given continuous function $\phi : \Gamma \to \mathbb{R}$, we define the kinematically admissible space

$$\mathcal{U}_k = \{u \in \mathcal{C}^1(\Omega)|\ \Lambda u \in \mathcal{L}^1(\Omega),\ u(x,y) = \phi(x,y)\ (x,y) \in \partial\Omega\}. \tag{1.21}$$

Then, the graph of $u(x,y) \in \mathcal{U}_k$ is a surface in $\mathbb{R}^3$. The solution $\bar{u}(x)$ of the variational problem

$$P(u) = \int_\Omega \sqrt{1+|\Lambda u|^2}\mathrm{d}\Omega = \int_\Omega \sqrt{1+u_{,x}^2+u_{,y}^2}\mathrm{d}\Omega \quad \to \min \quad \forall u \in \mathcal{U}_k$$

leads to a *minimum graph* $\bar{u} : \Omega \subset \mathbb{R}^2 \to \mathbb{R}$, i.e. a *non-parametric surface* bounded by the space curve $(x, y, \phi(x,y)) \in \partial\Omega \times \mathbb{R}$ with the minimum total surface area $P(\bar{u})$. The duality theory for this interesting geometrical problem will be studied in Chapter 6.

More generally, the *geodesic problem* consists of determining the shortest curve between two given points on a given surface in $\mathbb{R}^n$. This is clearly a variational problem subject to a constraint.

Dual to the geodesic problems, the optimal shape design problem requires determination of the optimal shape under certain constraints. For example, if we let

$$\mathcal{U}_k = \left\{ u \in \mathcal{C}^1(a,b) \,\middle|\, \int_a^b \sqrt{1+u'^2}\ \mathrm{d}x = \ell,\ u(a) = u_a,\ u(b) = u_b \right\}, \tag{1.22}$$

the solution $\bar{u}(x)$ of the maximization problem

$$P(u) = \int_a^b u(x)\, dx \quad \to \max \quad \forall u \in \mathcal{U}_k, \tag{1.23}$$

gives the optimal shape of the string with fixed length ℓ such that the area $P(\bar{u})$ under the graph of $\bar{u}(x)$ is maximal.

II. **Structural Limit Analysis.**

Suppose that the string is fixed at its two ends and is subjected to a distributed transverse load $\varrho f(x)$, where $\varrho > 0$ is the *loading intensity factor.* If the deformation measure $|u_{,x}|$ exceeds a certain limit, μk_0 say, where $\mu \geq 0$ is the so-called plastic flow factor and $k_0 > 0$ is the elastic limit, the string may break. Let

$$\mathcal{U}_k = \left\{ u \in C^1(a,b) \,\middle|\, \int_a^b f(x)u(x)\, dx = 1, \;\; u(a) = u(b) = 0 \right\}. \tag{1.24}$$

Then, the minimization problem

$$\varrho_c = \min \int_a^b k_0 \sqrt{1 + u_{,x}(x)^2}\, dx \;\; \forall u \in \mathcal{U}_k \tag{1.25}$$

gives the *limit loading (or safety) factor* $\varrho_c > 0$. This is a nonlinear variational problem with an integral constraint and boundary conditions. General mathematical results and engineering applications of limit analysis will be given in the last chapter.

In engineering mechanics, for a given structure and distributed load, the main purpose of limit analysis is to find the safety factor ϱ_c under certain constraints, for which the structure reaches its limit. Dually, for a given limit load and certain constraints, the so-called *plastic limit design problem* is to find the optimal shape of the structure such that the structure is in the limit state.

Actually, plastic limit analysis is really a minimum surface type of problem. To see this, let us consider a deformable body in $\mathbb{R}^n$. Before the deformation, the body occupies an open, bounded domain $\Omega \subset \mathbb{R}^n$, with boundary $\partial\Omega$. For a given external force field $f : \Omega \to \mathbb{R}^m$, the deformation of the body is a vector field $u : \Omega \subset \mathbb{R}^n \to \mathbb{R}^m$. For linear deformations, the operator Λ is the gradient of u and Λu is a matrix in $\mathbb{R}^{m\times n}$. Thus,

$$\Lambda u = \operatorname{grad} u = \left(\frac{\partial u_i}{\partial x_j} \right), \quad i = 1, 2, \cdots, m, \;\; j = 1, 2, \cdots, n.$$

Let

$$\mathcal{U}_k = \left\{ u \in \mathcal{L}^2(\Omega; \mathbb{R}^m) \,\middle|\, \Lambda u \in \mathcal{L}^\alpha(\Omega), \;\; \int_\Omega f u\, d\Omega = 1, \quad u(x) = 0 \;\; \forall x \in \partial\Omega \right\}.$$

Then, the one-dimensional problem (1.25) can be generalized to

$$\varrho_c = \int_\Omega k_0 \|\Lambda u\| d\Omega \ \ \forall u \in \mathcal{U}_k,$$

where $\|\cdot\|$ is a certain norm in $\mathbb{R}^{m\times n}$.

For example, in the problem of steady plastic flow in a two-dimensional pipe, the velocity field $u(x,y)$ is a mapping from $\mathbb{R}^2$ to $\mathbb{R}$. The norm is simply the Euclidean norm in $\mathbb{R}^2$, and minimization of the integral

$$\int_\Omega k_0 \sqrt{\left(\frac{\partial u}{\partial x}\right)^2 + \left(\frac{\partial u}{\partial y}\right)^2} d\Omega$$

gives the collapse pressure factor ϱ for plastic flow. This is indeed a minimum surface type problem in differential geometry. More interesting examples in infinitesmall deformation plasticity as well as their dual formulations can be found in the books by Prager and Hodge (1968) and Yang (1991).

III. **Obstacle Problems: Inequality Constraints.**

If the elastic string discussed in Section 1.1 is subjected to a vertical distributed load $f(x)$ and is partially supported on a given concave obstacle with profile $\psi(x)$, then, the constraint included in the kinematically admissible space is an inequality, and we have

$$\mathcal{U}_k = \left\{ u \in \mathcal{L}^2(I) |\ u' \in \mathcal{L}^1(I),\ \ u(x) \geq \psi(x)\ \ \forall x \in I,\ \ u(0) = u(\ell_0) = 0 \right\}.$$

In this case, the equilibrium equation associated with the nonlinear variational problem

$$P(u) = \int_0^{\ell_0} \left(\sqrt{1+u'^2} - fu \right) dx \ \ \to \min \ \ \forall u \in \mathcal{U}_k \tag{1.26}$$

is an inequality

$$A(u(x)) = -\left(\frac{u'(x)}{\sqrt{1+u'(x)^2}} \right)' \geq f(x) \ \ \forall x \in I. \tag{1.27}$$

The contact region

$$I_c = \{x \in I |\ \ u(x) = \psi(x)\},$$

is unknown until the problem is solved. The problem (1.26) is also called a *free boundary value problem.* Clearly, in the contact region the external pressure $-f(x)$ could be bigger than the balance force $-A(u)$; however, in the rest of the region where $u(x) > \psi(x)$, we have $A(u(x)) = f(x)$. Thus, the equality

$$(u(x) - \psi(x))\,(A(u) - f(x)) = 0$$

holds everywhere in the domain I. This equality is the so-called *complementarity condition* in *mathematical programming.* Together with the two inequalities

$$u(x) \geq \psi(x), \quad A(u(x) \geq f(x) \ \ \forall x \in I,$$

these three conditions form the so-called *Karush-Kuhn-Tucker conditions* in *variational inequality*, which will be discussed in Chapter 5.

1.2 Fundamental Lemma and Euler Equation

In this section, we present some basic definitions and results for one-dimensional variational problems. We continue to assume that $\Lambda = d/\, dx$.

1.2.1 Convex Functionals and Gâteaux Variations

Definition 1.2.1 (Convex Set) Let $\mathcal{U}$ be a linear vector space over $\mathbb{R}$. A subset $\mathcal{K} \subset \mathcal{U}$ is said to be *convex* if it contains the line segment between any two of its elements, i.e.

$$\theta u + (1-\theta)v \in \mathcal{K} \quad \forall u,\ v \in \mathcal{K} \tag{1.28}$$

for any given $\theta \in [0, 1]$. ◇

It is easily verified that $\mathcal{K} \subset \mathcal{U}$ is convex if and only if for any given finite subset of elements $u_1, u_2, \cdots, u_n \in \mathcal{K}$,

$$\sum_{i=1}^{n} \theta_i u_i \in \mathcal{K}, \ \ \theta_i > 0, \ \ \sum_{i=1}^{n} \theta_i = 1.$$

The entire space $\mathcal{U}$ is convex. The intersection of convex sets is convex, but the union of convex sets is, in general, not convex. By convention, the empty set $\emptyset$ is convex. The geometrical interpretations of convex and nonconvex sets are shown in Fig. 1.3.

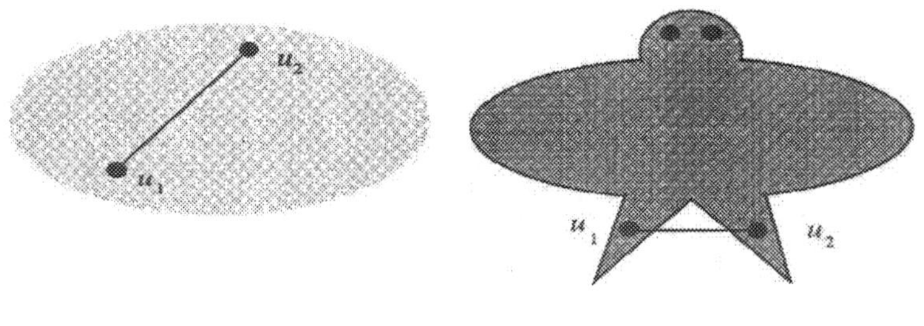

(a) Convex set (b) Nonconvex set

Figure 1.3. Convex and nonconvex sets

Example 1.2.1 The kinematically admissible spaces $\mathcal{U}_k$ defined by (1.20), (1.21) and (1.24) are convex subsets of $\mathcal{U}$, but the space $\mathcal{U}_k$ defined by (1.22) is not a convex subset of $\mathcal{U}$. ♡

Let $P : \mathcal{U} \to \mathbb{R}$ be a real-valued functional. The classical definition of a convex functional is as follows.

Definition 1.2.2 (Convex and Concave Functionals)
(D1) A functional $P : \mathcal{U} \to \mathbb{R}$ is said to be *convex* on $\mathcal{U}$ if

$$P(\theta u_1 + (1-\theta)u_2) \le \theta P(u_1) + (1-\theta)P(u_2) \;\; \forall u_1, u_2 \in \mathcal{U} \tag{1.29}$$

for all $\theta \in [0,1]$, whenever the right-hand side is defined.
(D2) P is said to be *strictly convex* if the strict form of the inequality holds for any $u_1 \neq u_2$.
(D3) P is said to be concave if $-P$ is convex. ◊

Clearly, if $P : \mathcal{U} \to \mathbb{R}$ is linear (i.e. $P(\theta u_1 + u_2) = \theta P(u_1) + P(u_2) \;\; \forall \theta \in \mathbb{R}$), then, P is both convex and concave. If P and G are convex functionals on $\mathcal{U}$, then, for any given $\alpha, \beta \ge 0$, $\alpha P + \beta G$ is also convex on $\mathcal{U}$. If a convex functional $P(u)$ has a local minimum at $u = u_0$, it is the global minimum. Also, if P is convex and differentiable on $\mathcal{K} \subset \mathcal{U}$, the derivative $\partial P(u)/\partial u$ is monotonically increasing with u.

Let $u_\theta = \theta u_1 + (1-\theta)u_2$ and $P_\theta = \theta P(u_1) + (1-\theta)P(u_2)$. The geometrical meaning of convex function is shown in Fig. 1.4.

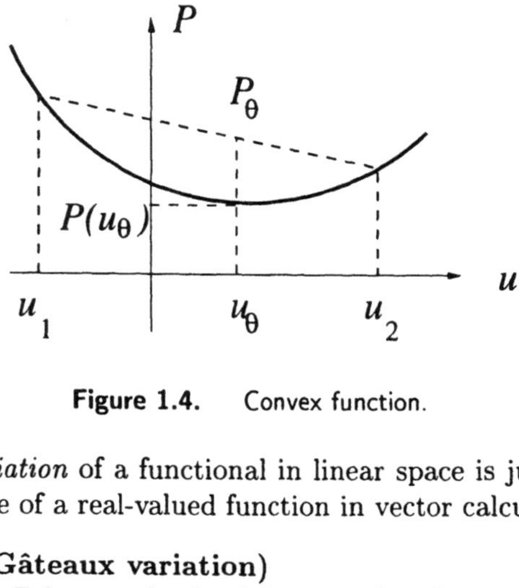

Figure 1.4. Convex function.

The *Gâteaux variation* of a functional in linear space is just the generalized directional derivative of a real-valued function in vector calculus.

Definition 1.2.3 (Gâteaux variation)
(D1) Let $P : \mathcal{U} \to \mathbb{R}$ be a real-valued functional and $\mathcal{U}_a \subset \mathcal{U}$ a subspace. For a given $\bar{u} \in \mathcal{U}_a$, if the limit

$$\delta P(\bar{u}; u) := \lim_{\theta \to 0^+} \frac{P(\bar{u} + \theta u) - P(\bar{u})}{\theta} \;\; \forall u \in \mathcal{U}_a \tag{1.30}$$

exists as $\theta \to 0^+$ (i.e. $\theta \to 0$, $\theta > 0$), then, $\delta P(\bar{u}; u) \in \mathbb{R}$ is called the *Gâteaux variation* of P at $\bar{u}$ in the direction u.
(D2) If the Gâteaux variation is a linear operator in u such that

$$\delta P(\bar{u}; u) = \langle u, DP(\bar{u}) \rangle \;\; \forall u \in \mathcal{U}_a, \tag{1.31}$$

we say that P is *Gâteaux differentiable* at $\bar{u}$. The linear operator $DP(\bar{u}) : \mathcal{U}_a \to \mathcal{U}^*$, which generally depends upon $\bar{u}$, is called the *Gâteaux derivative* of P at $\bar{u}$.

(D3) The functional $P : \mathcal{U} \to \mathbb{R}$ is said to be Gâteaux differentiable on $\mathcal{U}_a$ if it is Gâteaux differentiable at each $u \in \mathcal{U}_a$. ◊

This definition also works for operators. For example, if $\Lambda : \mathcal{U} \to \mathcal{E}$ is a linear operator, say $\Lambda = \text{grad}$, then, $\delta\Lambda(\bar{u}; u) = \text{grad}u$, and the Gâteaux derivative $D\Lambda$ of $\Lambda(u) = \Lambda u$ is simply the operator grad. However, if $\Lambda : \mathcal{U} \to \mathcal{E}$ is nonlinear, say $\Lambda(u) = \frac{1}{2}au_{,x}^2 + bu_{,x} + c$, where a, b and c could be either constants or scalar-valued functions, then, we have $\delta\Lambda(\bar{u}; u) = (a\bar{u}_{,x} + b)u_{,x}$. In this case, the Gâteaux derivative $D\Lambda(\bar{u}) = (a\bar{u}_{,x} + b)\text{d}/\,\text{d}x$ depends on $\bar{u} \in \mathcal{U}$.

For a given Gâteaux differentiable functional $P : \mathcal{U}_k \to \mathbb{R}$, its Gâteaux derivative $DP(\bar{u})$ is an operator from $\mathcal{U}_k$ into $\mathcal{U}^*$. If $DP : \mathcal{U}_k \to \mathcal{U}^*$ is also Gâteaux differentiable on $\mathcal{U}_k$, then, the *second-order Gâteaux variation* of P at $\bar{u}$ in the directions u_1 and u_2 is defined by

$$\begin{aligned} \delta^2 P(\bar{u}; u_1, u_2) &= \lim_{\theta_2 \to 0^+} \frac{DP(\bar{u} + \theta_2 u_2; u_1) - DP(\bar{u}; u_1)}{\theta_2} \\ &= \langle u_1, D^2 P(\bar{u}) u_2 \rangle, \end{aligned} \tag{1.32}$$

where $D^2P(\bar{u})$ stands for the second-order Gâteaux derivative of P at $\bar{u}$.

For example, in the deformation of elastic string system discussed in Section 1.1, if $\bar{f} \in \mathcal{U}_a^* \subset \mathcal{U}^*$ is a given external force, then, the external energy $F(u) = \langle \bar{f}, u \rangle$ is a linear functional of $u \in \mathcal{U}_a$. Its Gâteaux variation at $\bar{u} \in \mathcal{U}_a$ in the direction $\delta u \in \mathcal{U}_a$ (i.e. the virtual displacement) is given simply by

$$\delta F(\bar{u}; \delta u) = \langle \delta u, DF(\bar{u}) \rangle = \langle \delta u, \bar{f} \rangle,$$

which is linearly dependent on δu. Thus, the Gâteaux derivative of the external energy $F : \mathcal{U}_a \to \mathbb{R}$ gives the *duality relation*

$$\bar{f} = DF(\bar{u}) \in \mathcal{U}^*$$

between $\mathcal{U}_a$ and $\mathcal{U}_a^*$. The Gâteaux variation $\delta F(\bar{u}; \delta u)$ represents the *external virtual work*. Since $F(u)$ is a linear functional, its second-order Gâteaux derivative vanishes everywhere on $\mathcal{U}_a$.

Similarly, the Gâteaux variation of the stored-energy $W(\epsilon) = \int_I U(\epsilon)\text{d}I$ at $\bar{\epsilon} \in \mathcal{E}_a \subset \mathcal{E}$ in the direction $\delta\epsilon$, namely

$$\delta W(\bar{\epsilon}; \delta\epsilon) = \left\langle \delta\epsilon; \frac{\partial U}{\partial \epsilon}(\bar{\epsilon}) \right\rangle = \langle \delta\epsilon; \bar{\sigma} \rangle,$$

represents the *internal virtual work*. The Gâteaux derivative of the stored-energy W leads to the duality relation

$$\bar{\sigma} = DW(\bar{\epsilon}) = \frac{\partial U}{\partial \epsilon}(\bar{\epsilon}) \in \mathcal{E}_a^* \subset \mathcal{E}^* \tag{1.33}$$

between $\mathcal{E}_a$ and $\mathcal{E}_a^*$. Thus, the energy conservation law leads to the following universal result.

Theorem 1.2.1 (Virtual Work Principle) *Suppose that for a given system, the external energy $F : \mathcal{U}_a \to \mathbb{R}$, the internal energy $W : \mathcal{E}_a \to \mathbb{R}$ and $\Lambda : \mathcal{U}_a \to \mathcal{E}_a$ are Gâteaux differentiable, and $\bar{u} \in \mathcal{U}_k = \{u \in \mathcal{U}_a(I) | \ \Lambda(u) \in \mathcal{E}_a\}$ is an equilibrium state, then, for any given virtual deformation $\delta u(x) \in \mathcal{U}_k$ which vanish on the boundary ∂I of the domain I, the internal virtual work is equal to the external virtual work, i.e.*

$$\delta W(\epsilon(\Lambda\bar{u}); \delta\epsilon(\bar{u}; \delta u)) = \delta F(\bar{u}; \delta u). \tag{1.34}$$

In particular, if $\Lambda : \mathcal{U}_a \to \mathcal{E}_a$ is linear, Gâteaux differentiable, $\bar{f} = DF(\bar{u})$, and $\bar{\sigma} = DW(\bar{\epsilon}(\bar{u}))$, then

$$\langle \bar{\sigma}; \Lambda \delta u \rangle = \langle \bar{f}, \delta u \rangle \quad \forall \delta u \in \mathcal{U}_k. \tag{1.35}$$

Using *Gauss-Green's formula*, we have

$$\langle \delta u, \Lambda^* \bar{\sigma} - \bar{f} \rangle = 0 \ \ \forall \delta u \in \mathcal{U}_k,$$

which is usually called the *weak form* of the equilibrium equation.

The following lemma, proposed originally by Lagrange[2], is of fundamental importance in the calculus of variations.

Lemma 1.2.1 (Fundamental Lemma of the Calculus of Variations) *If $\psi(x)$ is a continuous function on I and if*

$$\langle \psi, \phi \rangle = 0 \tag{1.36}$$

holds for all functions $\phi(x)$ which vanish on the boundary ∂I and possess a continuous derivative on the closed domain $\bar{I} := I \cup \partial I$, then

$$\psi(x) \equiv 0 \ \ \forall x \in I. \tag{1.37}$$

The first proof of this lemma seems to have been given by L.V. Stegmann in 1854 (cf. e.g., Goldstine, 1980, p. 288). Thus, by using the fundamental lemma, the virtual work principle (1.35) leads to the equilibrium equation

$$\Lambda^* \bar{\sigma} = \bar{f}(x) \ \ \forall x \in I, \tag{1.38}$$

which holds everywhere in I. It is also called the *strong form* of the equilibrium equation.

For Gâteaux differentiable functionals on a convex subset $\mathcal{U}_k \subset \mathcal{U}$, we have an equivalent definition for convex functionals.

Theorem 1.2.2 *If $P : \mathcal{U}_k \subset \mathcal{U} \to \mathbb{R}$ is Gâteaux differentiable, then, the following statements are equivalent to each other.*

$$(S1) \quad P : \mathcal{U}_k \subset \mathcal{U} \to \mathbb{R} \text{ is convex},$$

$$(S2) \quad P(v) - P(u) \geq \langle DP(u), v - u \rangle \ \ \forall v, u \in \mathcal{U}_k, \tag{1.39}$$

$$(S3) \quad \langle v - u, DP(v) - DP(u) \rangle \geq 0 \ \ \forall v, u \in \mathcal{U}_k. \tag{1.40}$$

Proof. (S1) $\Rightarrow$ (S2). From the convexity of P, we have, for $\theta > 0$,

$$\frac{1}{\theta}[P(u+\theta(v-u)) - P(u)] \leq P(v) - P(u) \ \ \forall u, v \in \mathcal{U}_k.$$

Letting $\theta \to 0^+$, we obtain (S2).

(S2) $\Rightarrow$ (S1). From the inequality in (S2), for $\theta \in [0,1]$, we have

$$P(u) \geq P(\theta u + (1-\theta)v) + \langle u - (\theta u + (1-\theta)v), DP(\theta u + (1-\theta)v)\rangle,$$

$$P(v) \geq P(\theta u + (1-\theta)v) + \langle u - (\theta v + (1-\theta)v), DP(\theta u + (1-\theta)v)\rangle.$$

Multiplying the first equation by θ and the second by $(1-\theta)$ and adding them together we obtain the convexity of P.

To prove (S2) $\Rightarrow$ (S3), we use the inequality in (S2). Thus,

$$P(v) \geq P(u) + \langle DP(u), v-u\rangle \ \ \forall u, v \in \mathcal{U}_k,$$

$$P(u) \geq P(v) + \langle DP(v), u-v\rangle \ \ \forall u, v \in \mathcal{U}_k.$$

Combining these two inequalities we obtain

$$\langle v-u, DP(v)\rangle \geq P(v) - P(u) \geq \langle v-u, DP(u)\rangle.$$

This gives (S3).

To show (S3) $\Rightarrow$ (S2), we consider the function

$$h(\theta) = P(u + (\theta(v-u)) \ \ \forall \theta \in [0,1].$$

Since, by (S3),

$$\begin{aligned} h'(\theta) - h'(0) &= \langle v-u, DP(u+\theta(v-u)) - DP(u)\rangle \\ &= \frac{1}{\theta}[\langle u+\theta(v-u) - u, DP(u+\theta(v-u)) - DP(u)\rangle] \geq 0 \end{aligned}$$

integration of this inequality leads to

$$h(\theta) - h(0) \geq \theta h'(0).$$

On setting $w = u + \theta(v-u)$, we obtain

$$P(w) - P(u) \geq \langle w-u, DP(u)\rangle,$$

and the result is obtained. □

Remark 1.2.1 The statement (S3) in this theorem shows that the Gâteaux derivative of a convex function is a *monotone operator* of $\mathcal{U}$ into $\mathcal{U}^*$. Moreover, if $DP : \mathcal{U} \to \mathcal{U}^*$ is Gâteaux differentiable on $\mathcal{U}_k \subset \mathcal{U}$, the symmetrical condition (1.40) implies, by the mean value theorem, that

$$\langle v-u, D^2P(\bar{u})(v-u)\rangle \geq 0 \ \ \forall v, u \in \mathcal{U}_k, \tag{1.41}$$

where $\bar{u} = v + \theta(u-v)$, $\theta \in [0,1]$. Hence, a *sufficient condition* for P to be convex on $\mathcal{U}_k$ is that

$$D^2P(u) \geq 0 \ \ \forall u \in \mathcal{U}_k.$$

■

1.2.2 Primal Variational Problems and Euler Equations

Let $\mathcal{U}$ be a linear space with norm $\| * \|$, and let $\mathcal{D}_r(\bar{u}) \subset \mathcal{U}$ be an open disk centered at $\bar{u}$ with radius r (or, simply, a *neighborhood*), so that

$$\mathcal{D}_r(\bar{u}) = \{u \in \mathcal{U} |\ \ \|u - \bar{u}\| < r, \ \ r \in \mathbb{R}\}.$$

Definition 1.2.4 (Minima, Maxima and Critical Points)
Let $P : \mathcal{U} \to \mathbb{R}$ be a given functional.

(D1) A point $\bar{u} \in \mathcal{U}$ is said to be a *relative minimum* of P on $\mathcal{U}$ if there is an open neighborhood $\mathcal{D}_r(\bar{u}) \subset \mathcal{U}$ such that

$$P(\bar{u}) \leq P(u) \ \ \forall u \in \mathcal{D}_r(\bar{u}).$$

The point $\bar{u}$ is a *strict relative minimum* of P on $\mathcal{U}$ if the inequality is strictly held for all $u \neq \bar{u}$.

(D2) A point $\bar{u} \in \mathcal{U}$ is said to be a *(strict) relative maximum* of P on $\mathcal{U}$ if $\bar{u}$ is a (strict) relative minimum of $-P$ on $\mathcal{U}$.

(D3) A point $\bar{u}$ is called a *critical point* of P if P has a Gâteaux variation at $\bar{u}$ and $\delta P(\bar{u}; u) = 0 \ \ \forall u \in \mathcal{U}$. ◇

The term *extremum* refers to either a maximum or a minimum over any set. A relative extremum is also referred to as a *local extremum*. The subset $\mathcal{U}_k \subset \mathcal{U}$ on which an extremum problem is defined is sometimes called the *feasible set*. A critical point is also called a *stationary point*. The critical point condition

$$\delta P(\bar{u}; u) = 0 \ \ \forall u \in \mathcal{U}_k$$

is the *Euler equation*. The condition $u \in \mathcal{U}_k$ is called a *constraint*.

Three primal variational problems corresponding to these three types of points can be proposed.

Problem 1.2.1 (Primal Variational Problems)
Let $P : \mathcal{U}_k \subset \mathcal{U} \to \mathbb{R}$ be a given functional.

(P1) The *infimum (or inf-) primal variational problem* is to find a global minimizer $\bar{u} \in \mathcal{U}_k$ such that

$$(\mathcal{P}_{\text{inf}}): \quad P(\bar{u}) = \inf P(u) \ \ \forall u \in \mathcal{U}_k. \tag{1.42}$$

(P2) The *supremum (or sup-) primal problem* is to find a global maximizer $\bar{u} \in \mathcal{U}_k$ such that

$$(\mathcal{P}_{\text{sup}}): \quad P(\bar{u}) = \sup P(u) \ \ \forall u \in \mathcal{U}_k. \tag{1.43}$$

(P3) The *stationary (or sta-) primal variational problem* is to find a stationary point $\bar{u} \in \mathcal{U}_k$ such that

$$(\mathcal{P}_{\text{sta}}): \quad P(\bar{u}) = \text{sta}\, P(u) \ \ \forall u \in \mathcal{U}_k. \tag{1.44}$$

♠

The problem $(\mathcal{P}_{\inf})$ is called ***realisable*** if there exists a vector $\bar{u} \in \mathcal{U}_k$ such that the infimum of P is achieved at $\bar{u}$ and is not $+\infty$. This vector $\bar{u}$ is called the *minimizer* of $(\mathcal{P}_{\inf})$ and we write $P(\bar{u}) = \min_u P(u)$. Similarly, a vector $\bar{u} \in \mathcal{U}_k$ is called the *maximizer* of $(\mathcal{P}_{\sup})$ if the supremum is achieved at $\bar{u}$ and is not $-\infty$. We write $P(\bar{u}) = \max_u P(u)$. Both minimizer and maximizer are also called the *optimal solution* to their respective problems. If $P : \mathcal{U}_k \to \mathbb{R}$ is either convex or convex, then, the sup-primal problem can be reduced to $(\mathcal{P}_{\inf})$ by replacing P by to $-P$. In this chapter, we are interested in establishing conditions on P and $\mathcal{U}_k$ sufficient to guarantee that the global minimizers and stationary points exist, this being of great importance in the study of nonlinear differential equations.

Example 1.2.2 The real-valued function $P(x) = e^x$ is convex on $\mathcal{U} = \mathbb{R}$ and

$$\inf P(x) = 0, \quad \sup P(x) = +\infty$$

i.e. P is bounded from below, but is not bounded from above. Thus, on any open domain $\mathcal{U}_k = (a,b) \subset \mathbb{R}$, both the inf-problem and the sup-problem are not realisable. However, on the closed domain $\mathcal{U}_k = [a,b]$ with $-\infty < a < b < +\infty$, the two inf- and sup-problems are realisable and

$$\inf P(x) = \min P(x) = P(a) = e^a$$
$$\sup P(x) = \max P(x) = P(b) = e^b.$$

♡

Now let us consider the first problem in the calculus of variations,

$$P(u) = \int_I \jmath(x, u, \Lambda u)\, dx \quad \to \text{sta} \ \ \forall u \in \mathcal{U}_k. \tag{1.45}$$

Here we assume that the functions $\jmath_{,u}(x,u,\Lambda u)$ and $\jmath_{,\epsilon}(x,u,\epsilon)$ exist and are continuous in each of their variables x, u and $\epsilon = \Lambda u$. For homogeneous boundary conditions, the kinematically admissible space can be defined by

$$\mathcal{U}_k^0 = \{u \in \mathcal{C}^1(0,\ell_0) | \ u(0) = u(\ell_0) = 0\}.$$

Then, the Gâteaux variation of P at $\bar{u} \in \mathcal{U}_k^0$ in the direction $u \in \mathcal{U}_k$ is

$$\begin{aligned}
\delta P(\bar{u}; u) &= \lim_{\theta \to 0^+} \frac{1}{\theta} \int_I [\jmath(x, \bar{u} + \theta u, \Lambda \bar{u} + \theta \Lambda u) - \jmath(x, \bar{u}, \Lambda \bar{u})]\, dx \\
&= \int_I \left[u \frac{\partial \jmath}{\partial u} + \Lambda u \frac{\partial \jmath}{\partial \epsilon} \right] dx = \langle u, \jmath_{,u} \rangle + \langle \Lambda u; \jmath_{,\epsilon} \rangle \\
&= \langle u, \jmath_{,u} + \Lambda^* \jmath_{,\epsilon} \rangle = \langle u, DP(\bar{u}) \rangle.
\end{aligned}$$

Thus, the Gâteaux derivative $DP : \mathcal{U}_k^0 \to \mathcal{U}^*$ is given as

$$DP(\bar{u}) = \jmath_{,u}(x, \bar{u}, \Lambda \bar{u}) + \Lambda^* \jmath_{,\epsilon}(x, \bar{u}, \Lambda \bar{u}).$$

By the fundamental lemma, if $\delta P(\bar{u}; u)$ vanishes for all $u \in \mathcal{U}_k^0$, we have the so-called *Euler equation*

$$\Lambda^* \jmath_{,\epsilon}(x, \bar{u}, \Lambda \bar{u}) + \jmath_{,u}(x, \bar{u}, \Lambda \bar{u}) = 0 \ \ \forall x \in I. \tag{1.46}$$

In the traditional notation $\Lambda = \mathrm{d}/\,\mathrm{d}x$, it reads

$$\frac{\partial \jmath}{\partial u} - \frac{\mathrm{d}}{\mathrm{d}x}\frac{\partial \jmath}{\partial(\mathrm{d}u/\,\mathrm{d}x)} = 0.$$

Lemma 1.2.2 *Let $\mathcal{U}_k$ be an open subset of $\mathcal{U}$. If $P : \mathcal{U}_k \to \mathbb{R}$ is Gâteaux differentiable at $\bar{u}$ and*

$$P(u) \geq P(\bar{u}) \quad \forall u \in \mathcal{U}_k,$$

then, $DP(\bar{u}) = 0$, i.e., $\bar{u}$ is a critical point of P on $\mathcal{U}_k$.

Proof. When $P(u) \geq P(\bar{u})$ $\forall u \in \mathcal{U}_k$ we have

$$P(\bar{u} + \theta u) - P(\bar{u}) \geq \theta\langle 0, u\rangle = 0 \ \ \forall u \in \mathcal{U}_k$$

for all $\theta \in \mathbb{R}$. Divide by $\theta > 0$ and let $\theta \to +0$; then

$$\lim_{\theta \to 0^+} \frac{1}{\theta}(P(\bar{u} + \theta u) - P(\bar{u})) = \langle DP(\bar{u}), u\rangle \geq 0 \ \ \forall u \in \mathcal{U}_k.$$

Thus, $DP(\bar{u}) = 0$ since u is an arbitrary vector in $\mathcal{U}_k$. □

Theorem 1.2.3 (Necessary Conditions)
Let $\mathcal{U}_k$ be an open subset of a Banach space $\mathcal{U}$, and $\jmath \in C^2(I, \mathcal{U}, \mathcal{E})$. Then, every relative extremal point of P is a critical point. Moreover, if P has a relative minimum (resp. maximum) at $\bar{u} \in \mathcal{U}_k$, then, the following hold:
(T1) Euler equation: $DP(\bar{u}) = 0$.
(T2) $\delta^2 P(\bar{u}; u, u) \geq 0$ (resp. ≤ 0) $\forall u \in \mathcal{U}_k$.

This theorem almost has an inverse statement, i.e.

Theorem 1.2.4 (Sufficient Conditions)
Let $\mathcal{U}_k$ be an open subset of a Banach space $\mathcal{U}$, and $\jmath \in C^2(I, \mathcal{U}, \mathcal{E})$. If the following conditions hold for a point $\bar{u} \in \mathcal{U}_k$.
(T1) $DP(\bar{u}) = 0$,
(T2) $\delta^2 P(\bar{u}; u, u) \geq c\|u\|^2$ for all $u \in \mathcal{U}_k$ and a fixed $c > 0$,
then, P has a strict relative minimum at $\bar{u}$.

Proof. If (T1) holds, then, $\bar{u}$ is a critical point of P on the open set $\mathcal{U}_k$. We let $f_u(\theta) := P(\bar{u} + \theta u)$; then,

$$f_u^{(k)}(\theta) = \delta^k P(\bar{u} + \theta u; u^k) \ \ \forall u \in \mathcal{U}_k, \ \ (k = 1, 2).$$

The classical Taylor expansion of f at $\theta = 0$ has the form

$$f_u(\theta) = f_u(0) + \sum_{k=1}^{n} \theta^k \frac{f_u^{(k)}(0)}{k!} + R_n \ \ (n \geq 2)$$

where R_n is the n-th order remainder:

$$R_n = \frac{\theta^n(f_u^{(n)}(\alpha\theta) - f_u^{(n)}(0))}{n!},$$

and the number $\alpha \in (0,1)$ depends on θ and u. Let $\theta = 1$; then, for $n = 2$, we have

$$P(\bar{u}+u) - P(\bar{u}) = f_u(1) - f_u(0) = \frac{f_u^{(2)}(\alpha)}{2}, \quad 0 < \alpha < 1.$$

Since $\jmath \in \mathcal{C}^2(I,\mathcal{U},\mathcal{E})$, P is twice Gâteaux differentiable in a neighborhood of $\bar{u}$ and D^2P is continuous at $\bar{u}$, and, for sufficiently small $\omega > 0$, there exists an $\eta(\omega) > 0$ such that

$$\left|\delta^2 P(v;u^2) - \delta^2 P(\bar{u};u^2)\right| \geq \omega \|u\|^2 \quad \forall u \in \mathcal{U}_k,$$

for all $v \in \mathcal{U}_k$ such that $\|v - \bar{u}\| < \eta(\omega)$. By (T1) and (T2) with $\omega = c/2$, for u with $\|u\| < \eta(c/2)/\alpha$, we have

$$P(\bar{u}+u) - P(\bar{u}) = \frac{1}{2}\delta^2 P(\bar{u}+\alpha u; u^2) \geq \frac{c}{4}\|u\|^2 \geq 0.$$

Thus, $\bar{u}$ is a local minimizer. □

In homogeneous static systems, if the total potential is convex and $P(u) = W(\Lambda u) - F(u)$. In this case, the Euler equation reads

$$\Lambda^* DW(\Lambda \bar{u}) = \Lambda^* U_{,\epsilon}(\Lambda u) = DF. \tag{1.47}$$

If the external potential F is linear, the convexity condition $\delta^2 P(\bar{u};u^2) \geq 0 \ \forall u \in \mathcal{U}_k$ is equivalent to the so-called *Legendre condition*

$$\jmath_{,\epsilon\epsilon}(\bar{u}, \Lambda\bar{u}) = \frac{\partial^2 U}{\partial \epsilon^2}(\Lambda \bar{u}) \geq 0 \ \forall x \in I. \tag{1.48}$$

Thus, if the stored-energy density is convex, the critical points of P are global minimizers. P has at most one minimizer if U is strictly convex.

1.3 Linear Operators and Bilinear Forms

Let $\mathcal{U}^*$ and $\mathcal{E}^*$ be another two spaces, in duality respectively with $\mathcal{U}$ and $\mathcal{E}$ through the bilinear forms

$$\langle *, * \rangle : \mathcal{U} \times \mathcal{U}^* \to \mathbb{R}, \quad \text{and } \langle *; * \rangle : \mathcal{E} \times \mathcal{E}^* \to \mathbb{R}.$$

A *linear operator* $\Lambda : \mathcal{U} \to \mathcal{E}$ is a transformation from one space $\mathcal{U}$ to another space $\mathcal{E}$ satisfying

$$\Lambda(\theta_1 u + \theta_2 \sigma) = \theta_1 \Lambda u + \theta_2 \Lambda \sigma \ \ \forall u, v \in \mathcal{U},$$

where $\theta_1, \theta_2 \in \mathbb{R}$ are arbitrary constants. The *adjoint operator* $\Lambda^* : \mathcal{E}^* \to \mathcal{U}^*$ of Λ is a transformation from $\mathcal{E}^*$ into $\mathcal{U}^*$, defined by

$$\langle \Lambda u; \sigma \rangle = \langle u, \Lambda^* \sigma \rangle \ \ \forall u \in \mathcal{U}, \ \ \sigma \in \mathcal{E}^*. \tag{1.49}$$

Fig. 1.5 gives the diagrammatic representation of the paired spaces and operators. Generally speaking, by defining suitable differential operators and associated bilinear forms, many different problems can be put in this unified framework. In this section, we only discuss the ordinary differential operators and adjoints in one-dimensional systems.

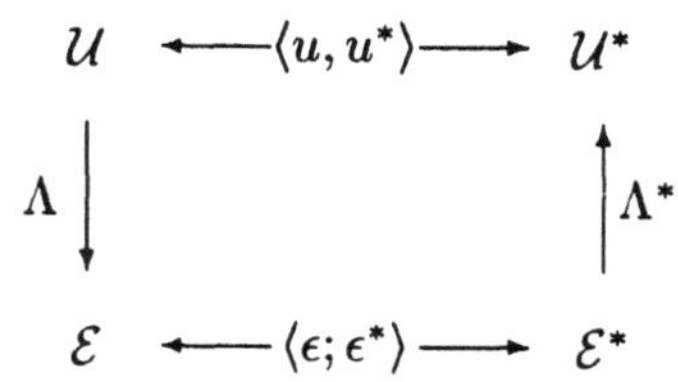

Figure 1.5. Diagrammatic representation for dual spaces and linear operators

1.3.1 Lagrange Identity and Weighted Bilinear Forms

Let $I = (a, b) \subset \mathbb{R}$ be an open interval. Recall the integration by parts identity

$$\int_a^b \sigma(x)u'(x) \, \mathrm{d}x = -\int_a^b u(x)\sigma'(x) \, \mathrm{d}x + u(b)\sigma(b) - u(a)\sigma(a) \tag{1.50}$$

for arbitrary functions $u(x), \sigma(x) \in C^1(I)$. Let $\Lambda = \frac{\mathrm{d}}{\mathrm{d}x}$, $\Lambda^* = -\Lambda$, then, (1.50) can be rewritten as the so-called *Green formula*

$$\int_a^b \sigma \Lambda u \, \mathrm{d}x = \int_a^b u \Lambda^* \sigma \, \mathrm{d}x + J(u, \sigma)|_{x=a}^{x=b}. \tag{1.51}$$

The operator $\Lambda^* = -\mathrm{d}/\mathrm{d}x$ is known as the *formal adjoint* of Λ, and the bilinear form $J(u, \sigma) = u(x)\sigma(x)$ is called the *conjunct* of u and σ. Since the identity is valid for any upper limit b, we differentiate with respect to b and then, set $b = x$ to obtain

$$\sigma \Lambda u - u \Lambda^* \sigma = \frac{\mathrm{d}}{\mathrm{d}x} J(u, \sigma). \tag{1.52}$$

This is the so-called *Lagrange identity.*

In general, let Λ be an ordinary differential operator of order m, i.e.

$$\Lambda = a_m(x)D^m + \cdots + a_1(x)D + a_0(x), \quad D = \frac{\mathrm{d}}{\mathrm{d}x}, \tag{1.53}$$

where the coefficients are in $C^p(I)$. Since, for any $u, \sigma \in C^p(I)$,

$$\begin{aligned} & D[\sigma D^{m-1}u - \sigma' D^{m-2}u + \cdots + (-1)^{m-1}(D^{m-1}\sigma)u] \\ = \; & \sigma D^m u + (-1)^{m-1} u D^m \sigma, \end{aligned}$$

we have

$$\sigma D^m u = (-1)^m u D^m \sigma + D \sum_{j+k=m-1} (-1)^k (D^k \sigma)(D^j u),$$

where the sum ranges over all $j \geq 0$ and $k \geq 0$ satisfying $j + k = m - 1$. If $m = 0$ the summation disappears.

Let $\mathcal{U} = \mathcal{L}^2(I) = \mathcal{U}^*$ and $\mathcal{E} = \mathcal{L}^2(I) = \mathcal{E}^*$ be Hilbert spaces with the simplest bilinear forms

$$\langle u, u^* \rangle = \int_I u(x)u^*(x)\, \mathrm{d}x, \quad \langle \epsilon; \sigma \rangle = \int_I \epsilon(x)\sigma(x)\, \mathrm{d}x.$$

Let $\partial I = \{x = a,\ x = b\}$ be the boundary of the open set $I = (a, b)$. Then, the formal adjoint of Λ can be defined by

$$\Lambda^*\sigma = \sum_{m=0}^{p} (-1)^m D^m(a_m\sigma). \tag{1.54}$$

The conjunct $J(u, \sigma)$ associated with this operator is

$$J(u,\sigma) = \sum_{m=1}^{p} \sum_{j+k=m-1} (-1)^k D^k(a_m\sigma)D^j u(x). \tag{1.55}$$

The corresponding Lagrange identity is

$$\sigma\Lambda u - u\Lambda^*\sigma = \frac{\mathrm{d}}{\mathrm{d}x} J(u, \sigma). \tag{1.56}$$

Observe that the conjunct J contains only derivatives of order up to $p - 1$.

If $\Lambda^* = \Lambda$, we say that Λ is *formally self-adjoint.* This is possible if and only if Λ is of even order and can be written in the form

$$D^r(b_r D^r) + D^{r-1}(b_{r-1}D^{r-1}) + \cdots + D(b_1 D) + b_0, \tag{1.57}$$

where $r = m/2$, and the $b_i (i = 0, 1, \cdots, r)$ are arbitrary functions in $\mathcal{C}^m$.

It is important to realize that there is some degree of flexibility in the definition of the bilinear forms. The choice is guided by the requirements of each individual problem. The pair of adjoint operators Λ and Λ^* depends on the choice of the bilinear forms. For example, *weighted bilinear forms* can be defined by

$$\langle u, u^* \rangle = \int_I w(x)u(x)u^*(x)\, \mathrm{d}x, \quad \langle \epsilon; \sigma \rangle = \int_I \omega(x)\epsilon(x)\sigma(x)\, \mathrm{d}x, \tag{1.58}$$

where $w(x),\ \omega(x) \geq 0\ \ \forall x \in I$ are weighting factors. Weighted bilinear forms play an important role in nonconservative systems.

Example 1.3.1 Let us consider the nonlinear Föppl-Hencky equation in elastic membrane theory (cf., e.g., Arthurs (1980))

$$\frac{\mathrm{d}^2 u}{\mathrm{d}x^2} + \frac{3}{x}\frac{\mathrm{d}u}{\mathrm{d}x} = -\frac{k}{u^2} \quad \forall x \in (0, 1), \ \ k > 0, \tag{1.59}$$

where u is actually the dimensionless stress in a simply supported circular membrane subjected to a constant normal pressure. To put this nonlinear equation in our framework, we let $\mathcal{U} = \mathcal{U}^*$ and $\mathcal{E} = \mathcal{E}^*$ be Hilbert spaces with inner products

$$\langle u, u^* \rangle = \int_0^1 x^3 u u^*\, \mathrm{d}x, \quad \langle \epsilon; \sigma \rangle = \int_0^1 x^3 \epsilon\sigma\, \mathrm{d}x.$$

We choose $\Lambda = \mathrm{d}/\mathrm{d}x$. Then, Gauss-Green's formula (1.51) leads to the formal adjoint

$$\Lambda^* = -\frac{1}{x^3}\frac{\mathrm{d}}{\mathrm{d}x}\left(x^3(\cdot)\right) = -\frac{\mathrm{d}}{\mathrm{d}x} - \frac{3}{x}$$

and the conjunct of u and σ is $J(u,\sigma) = x^3 u\sigma$. Thus, the fundamental equation for this nonlinear problem reads

$$A(u) = -\frac{1}{x^3}\frac{\mathrm{d}}{\mathrm{d}x}\left(x^3\frac{\mathrm{d}}{\mathrm{d}x}u\right) = \frac{k}{u^2}.$$

♡

1.3.2 Boundary Operators and Conditions

The bilinear form and linear operator can be extended in such a way as to include boundary values. Let us illustrate this approach in the one-dimensional case by re-designing the bilinear form to include boundary values. We let $\mathcal{U} = C^1(a,b)$ with inner product $\langle v,u\rangle = \int_a^b v(x)u(x)\,\mathrm{d}x$, and let $\partial\mathcal{U} \subset \mathcal{U}$ denote the space of functions defined on the boundary of $(a,\ b)$. The inner product on $\partial\mathcal{U}$ is defined by

$$\langle v,u\rangle_{\partial I} = u(b)v(b) + v(a)u(a).$$

Let $\bar{\mathcal{U}} = \mathcal{U} \times \partial\mathcal{U}$. By setting

$$\mathbf{u} = \begin{pmatrix} u(x) \\ \gamma_o u \end{pmatrix}, \quad \text{with} \quad \gamma_o u = \begin{pmatrix} u(b) \\ u(a) \end{pmatrix},$$

where $\gamma_o : \mathcal{U} \to \partial\mathcal{U}$ is a boundary operator. Then, we can define a particular bilinear form on $\bar{\mathcal{U}}$, namely,

$$\langle \mathbf{v},\mathbf{u}\rangle_{\bar{\mathcal{U}}} = \langle v,u\rangle + \langle v,u\rangle_{\partial I} = \int_a^b v(x)u(x)\,\mathrm{d}x + (\gamma_o v)^T(\gamma_o u),$$

such that $\{\bar{\mathcal{U}}, \langle *,*\rangle_{\bar{\mathcal{U}}}\}$ is a Hilbert space. The norm of $\mathbf{u}$ induced by this bilinear form is given by

$$\|\mathbf{u}\| = \langle \mathbf{u},\mathbf{u}\rangle_{\bar{\mathcal{U}}}^{1/2} = \left(\int_a^b u^2(x)\,\mathrm{d}x + u^2(b) + u^2(a)\right)^{1/2}.$$

We now introduce the linear operator

$$\bar{\Lambda}_o = \begin{pmatrix} \frac{\mathrm{d}}{\mathrm{d}x} & 0 \\ 0 & B_o \end{pmatrix}, \quad \text{with} \quad B_o = \begin{pmatrix} 0 & 0 \\ 0 & 1 \end{pmatrix}, \tag{1.60}$$

so that

$$\langle \mathbf{v},\bar{\Lambda}_o\mathbf{u}\rangle_{\bar{\mathcal{U}}} \quad = \quad \int_a^b v\frac{\mathrm{d}u}{\mathrm{d}x}\,\mathrm{d}x + (\gamma_o v)^T(B_o\gamma_o u)$$

$$
\begin{aligned}
&= -\int_a^b \frac{dv}{dx} u \, dx + (\gamma_o u)^T \begin{pmatrix} 1 & 0 \\ 0 & -1 \end{pmatrix} (\gamma_o v) + (B_o \gamma_o u)^T (\gamma_o v) \\
&= -\int_a^b \frac{dv}{dx} u \, dx + (B_o^* \gamma_o v)^T (\gamma_o u) \\
&= \langle \Lambda^* v, u \rangle + \langle B^* v, u \rangle_{\partial I} = \langle \bar{\Lambda}_o^* \mathbf{v}, \mathbf{u} \rangle_{\bar{\mathcal{U}}},
\end{aligned}
$$

where

$$
\bar{\Lambda}_o^* = \begin{pmatrix} -\frac{d}{dx} & 0 \\ 0 & B_o^* \end{pmatrix}, \quad \text{with } B_o^* = \begin{pmatrix} 1 & 0 \\ 0 & 0 \end{pmatrix}. \tag{1.61}
$$

Let B and B^* be the boundary operators, defined by

$$
B = -B_o \gamma_o, \quad B^* = B_o^* \gamma_o.
$$

On the Hilbert spaces $\{\mathcal{U}, \langle *.* \rangle\}$ and $\{\partial \mathcal{U}, \langle *, * \rangle_{\partial I}\}$, Gauss-Green's formula can be written as

$$
\langle v, \Lambda u \rangle = \langle \Lambda^* v, u \rangle + J_{\partial I}(u, v),
$$

where $J_{\partial I}(u, v)$ is the so-called *bilinear concomitant*, defined by

$$
\begin{aligned}
J_{\partial I}(u, v) &= J(u, v) n|_{x \in \partial I} \\
&= \langle v, Bu \rangle_{\partial I} + \langle B^* v, u \rangle_{\partial I},
\end{aligned}
$$

where n is an outward *unit vector* normal to the boundary $\partial I = \{x = a, \ x = b\}$, i.e.

$$
n = \begin{cases} +1 & (x = b), \\ -1 & (x = a). \end{cases} \tag{1.62}
$$

In this one-dimensional problem, the relation between the bilinear concomitant $J_{\partial I}(u, v)$ and the conjunct $J(u, v)$ is simply given by

$$
J_{\partial I}(u, v) = J(u, v)|_{x=a}^{x=b} = v(b)u(b) - v(a)u(a).
$$

For general boundary value problems, in addition to the paired function spaces $\{(\mathcal{U}, \mathcal{U}^*), \langle *, * \rangle\}$ and $\{(\mathcal{E}, \mathcal{E}^*), \langle *; * \rangle\}$, we need to introduce certain appropriate boundary bilinear forms:

$$
\begin{aligned}
\langle *, * \rangle_{\partial I} &: \partial \mathcal{U} \times \partial \mathcal{U}^* \to \mathbb{R}, \\
\langle *; * \rangle_{\partial I} &: \partial \mathcal{E} \times \partial \mathcal{E}^* \to \mathbb{R}.
\end{aligned}
$$

Thus, for a given linear operator $\Lambda : \mathcal{U} \to \mathcal{E}$, the formal adjoint Λ^* and the associated boundary operator B may be expressed by the relation

$$
\langle \Lambda u; \sigma \rangle = \langle u, \Lambda^* \sigma \rangle + \langle Bu; \sigma \rangle_{\partial I}. \tag{1.63}
$$

The adjoint B^* of B is defined by

$$
\langle Bu; \sigma \rangle_{\partial I} = \langle u, B^* \sigma \rangle_{\partial I}. \tag{1.64}
$$

In mixed boundary value problems, the boundary ∂I can usually be split into two parts, ∂I_u and ∂I_f say, such that

$$\partial I_u \cup \partial I_f = \partial I, \quad \partial I_u \cap \partial I_f = \emptyset. \tag{1.65}$$

On ∂I_u, the displacement $\bar{u}$ is given, while on its complementary part ∂I_f, the external force $\bar{u}^*$ is prescribed. Then, the bilinear concomitant can be written as

$$J_{\partial I}(u,\sigma) = \langle Bu;\sigma\rangle_{\partial I} = \langle Bu;\sigma\rangle_{\partial I_u} + \langle u, B^*\sigma\rangle_{\partial I_f}.$$

Furthermore, let

$$\bar{\Lambda} = \begin{cases} \Lambda & \text{in } I, \\ -B & \text{on } \partial I, \end{cases} \qquad \bar{\Lambda}^* = \begin{cases} \Lambda^* & \text{in } I, \\ B^* & \text{on } \partial I, \end{cases}$$

and by defining the bilinear forms

$$\begin{aligned} \langle \bar{\Lambda} u;\sigma\rangle_{\bar{I}_u} &= \langle \Lambda u;\sigma\rangle + \langle -Bu;\sigma\rangle_{\partial I_u}, \\ \langle u, \bar{\Lambda}^*\sigma\rangle_{\bar{I}_f} &= \langle u, \Lambda^*\sigma\rangle + \langle u, B^*\sigma\rangle_{\partial I_f}, \end{aligned}$$

Gauss-Green's formula can be written in the more symmetrical form

$$\langle \bar{\Lambda} u;\sigma\rangle_{\bar{I}_u} = \langle u, \bar{\Lambda}^*\sigma\rangle_{\bar{I}_f}. \tag{1.66}$$

Let $J_\sigma(u)$ denote the external boundary energy on ∂I_f, $J_u(\sigma)$ the external complementary boundary energy on ∂I_u such that

$$J_\sigma(u) = \langle u, \bar{u}^*\rangle_{\partial I_f}, \tag{1.67}$$
$$J_u(\sigma) = \langle \bar{u};\ \sigma\rangle_{\partial I_u}. \tag{1.68}$$

Then the boundary conditions can be written in the abstract forms:

$$Bu = DJ_u(\sigma) = \bar{u} \quad \text{on } \partial I_u, \tag{1.69}$$
$$B^*\sigma = DJ_\sigma(u) = \bar{u}^* \quad \text{on } \partial I_f. \tag{1.70}$$

By the Legendre duality theorem, the Legendre transformation on the boundary ∂I can be written as

$$J_\sigma(u) + J_u(\sigma) = J_{\partial I}(u,\sigma). \tag{1.71}$$

Example 1.3.2 For the linear operator $\Lambda = \mathrm{d}/\,\mathrm{d}x$ and the bilinear forms

$$\langle u, u^*\rangle = \int_a^b uu^*\ \mathrm{d}x, \quad \langle \epsilon;\sigma\rangle = \int_a^b \epsilon\sigma\ \mathrm{d}x,$$

$$\langle u, u^*\rangle_{\partial I} = u(a)u^*(a) + u(b)u^*(b), \quad \langle \epsilon;\sigma\rangle_{\partial I} = \epsilon(a)\sigma(a) + \epsilon(b)\sigma(b),$$

three types of boundary conditions can be identified, as follows.

I. *Mixed Boundary Conditions*:

$$u(a) = u_a \quad \text{on } \partial I_u = a, \quad \sigma(b) = \sigma_b \quad \text{on } \partial I_f = b.$$

In this case, we let $\bar{u} = -u_a$, and $\bar{u}^* = \sigma_b$. The boundary energies

$$J_\sigma(u) = \langle u, \bar{u}^* \rangle_{\partial I_f} = \sigma_b u(b), \quad J_u(\sigma) = \langle \bar{u}; \sigma \rangle_{\partial I_u} = -u_a \sigma(a)$$

are linear. The bilinear concomitant is

$$J_{\partial I}(u, \sigma) = \langle Bu; \sigma \rangle_{\partial I_u} + \langle u, B^* \sigma \rangle_{\partial I_f} = Bu(a)\sigma(a) + u(b) B^* \sigma(b).$$

Thus, $B = n = -1$ at $x = a$, and $B^* = n = 1$ at $x = b$. The abstract boundary conditions are

$$\begin{array}{ll} Bu = -u(a) = DJ_u(\sigma) = \bar{u} = -u_a, & x \in \partial I_u = a, \\ B^* \sigma = \sigma(b) = DJ_\sigma(u) = \bar{u}^* = \sigma_b & x \in \partial I_f = b. \end{array} \tag{1.72}$$

II. *Dirichlet Boundary Conditions:*

$$u(a) = u_a, \quad u(b) = u_b, \quad \partial I_u = \{x = a,\ x = b\} = \partial I, \quad \partial I_f = \emptyset.$$

Let $\bar{u} = \{u_b\ (x = b),\ \ -u_a\ (x = a)\}$. Then

$$J_u(\sigma) = \langle Bu; \sigma \rangle_{\partial I_u} = u_b \sigma(b) - u_a \sigma(a).$$

However, $J_\sigma(u) = 0$. Gauss-Green's formula gives

$$Bu = un|_{\partial I} = \{u(b)\ (x = b),\ \ -u(a)\ (x = a)\},$$

and the Dirichlet boundary conditions can be written in an abstract form

$$Bu = DJ_u(\sigma) = \bar{u} \ \ x \in \partial I.$$

III. *Neumann Boundary Conditions:*

$$\sigma(a) = \sigma_a, \quad \sigma(b) = \sigma_b, \quad \partial I_u = \emptyset, \quad \partial I_f = \{x = a,\ x = b\} = \partial I.$$

In this case, we have $J_u(\sigma) = 0$, and

$$J_\sigma(u) = \sigma_b u(b) - \sigma_a u(a).$$

Let $\bar{u}^* = \{\sigma_b\ (x = b),\ -\sigma_a\ (x = a)\}$; the adjoint boundary operator is $B^* = n$ on ∂I and the abstract Neumann condition can be written as

$$B^* \sigma(x) = \sigma(x) n|_{x \in \partial I} = DJ_\sigma(u) = \bar{u}^* \ \ x \in \partial I.$$

Boundary operators for higher order differential operator Λ can be designed in a similar way. Details can be found in Stakgold (1979) and Sewell (1987). ♡

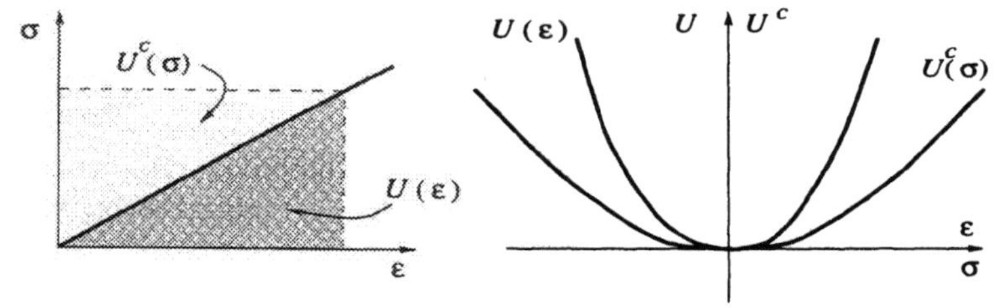

Figure 1.6. Constitutive law and stored-energy densities for a linear elastic spring

1.4 Legendre Transformation and Duality

In continuum mechanics, for a given stored-energy density $U(\epsilon)$ such that the strain-stress relation $\sigma = \partial U / \partial \epsilon$ is invertible, then the so-called *complementary energy density* $U^c(\sigma)$ of U is defined by

$$U^c(\sigma) = \sigma\epsilon(\sigma) - U(\epsilon(\sigma)). \tag{1.73}$$

In mathematics, this the well-known *Legendre transformation*. In the linear deformation elastic spring system, the stored-energy density $U(\epsilon) = \frac{1}{2}k_0\epsilon^2$ is a quadratic function. Hooke's law $\sigma = k_0\epsilon$ is linear and invertible, and the complementary energy density $U^c(\sigma) = \frac{1}{2}\sigma^2/k_0$ is also a quadratic function (see Fig. 1.6).

Generally speaking, the classical Legendre transformation can be viewed as a conversion of one continuous real-valued function into another one. If the transformation is reversible, then we say that each function is the *dual* of the other. The reversible Legendre transformation is also called the *Legendre conjugate transformation*, or simply the *Legendre transformation* (see Sewell (1987)). In this section, we give a basic account of the Legendre transformation. We work in finite dimensions only, with particular attention give to quadratic functions. Detailed study of this subject can be found in Sewell (1987).

1.4.1 Legendre Duality in Euclidean Geometry

Let $\mathcal{E} = \mathbb{R}^n = \mathcal{E}^*$. Then the elements $\boldsymbol{\epsilon} = \{\epsilon_i\} \in \mathcal{E}$ and $\boldsymbol{\sigma} = \{\sigma_i\} \in \mathcal{E}^*$, $(i = 1, \cdots, n)$, are n-dimensional vectors. The bilinear form $\langle *; * \rangle : \mathcal{E} \times \mathcal{E}^* \to \mathbb{R}$ is then the inner product on $\mathbb{R}^n$ defined by

$$\langle \boldsymbol{\sigma}; \boldsymbol{\epsilon} \rangle = \boldsymbol{\sigma}^T \boldsymbol{\epsilon} = \boldsymbol{\sigma} \cdot \boldsymbol{\epsilon} = \sum_{i=1}^{n} \epsilon_i \sigma_i = \sigma_i \epsilon_i.$$

The summation convention over the repeated suffix is used in this section. Let $U : \mathcal{E} \to \mathbb{R}$ be a real-valued function. Its *graph*

$$\mathcal{G}_U = \{(\boldsymbol{\epsilon}, X) \in \mathbb{R}^{n+1} \mid \ X = U(\boldsymbol{\epsilon})\}$$

is a *manifold (or hypersurface)* in the space $\mathbb{R}^{n+1}$. Let any particular point $(\sigma, Y) \in \mathbb{R}^{n+1}$ be called the *pole*. Then the linear function

$$X(\epsilon) = \sigma^T \epsilon - Y \tag{1.74}$$

is called the *polar*, which is a plane in $\mathbb{R}^{n+1}$. Thus, given a pole at a finite point, the polar is well-defined by (1.74). Conversely, given a polar of finite slope, a finite pole can be read off from the equation (1.74). This correspondence is called the *duality* between points and planes.

If U is the quadratic function

$$U(\epsilon) = \frac{1}{2}\epsilon^T \epsilon = \frac{1}{2}\epsilon_i \epsilon_i, \tag{1.75}$$

then the graph $\mathcal{G}_U$ of U is a paraboloid in $\mathbb{R}^{n+1}$. The poles and polars satisfy a nice duality relationship (see Fig. 1.7(a)).

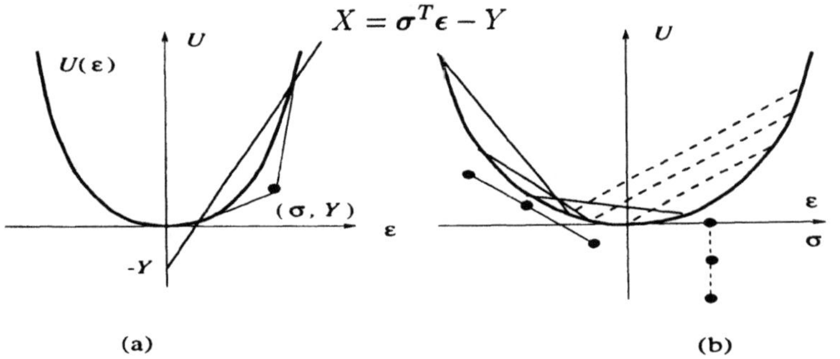

Figure 1.7. Duality between the pole and polar.

Theorem 1.4.1 (Duality between the Pole and the Polar)

(T1) If the pole is outside the paraboloid, the points of contact of all tangents drawn from the pole to the paraboloid lie on the polar.

(T2) If the pole is inside of the paraboloid, the polar lies outside it.

Proof. If the pole is outside the paraboloid, the tangent vector from a tangency point (ϵ, X) to the pole is $\mathbf{t} = (\sigma - \epsilon, Y - X) \in \mathbb{R}^{n+1}$. The vector normal to the level surface $X - \frac{1}{2}\epsilon^T\epsilon = 0$ at the tangency point is $\mathbf{n} = (-\epsilon, 1) \in \mathbb{R}^{n+1}$. Thus,

$$\mathbf{t}^T \mathbf{n} = -(\sigma - \epsilon)^T \epsilon + Y - X = 0.$$

Substituting (1.75) into this leads to (1.74).

If the pole (σ, Y) is inside the paraboloid, we have $U^c > \frac{1}{2}\sigma^T\sigma$. Suppose that the polar is inside or on the paraboloid; then $U \geq \frac{1}{2}\epsilon^T\epsilon$ for any point on the polar. By (1.74), the sum of these inequalities implies $0 > (\sigma - \epsilon)^T(\sigma - \epsilon)$. This contradiction shows that the polar must lie outside the paraboloid. □

It is very interesting to note that for given finite $\mathbf{c} \in \mathbb{R}^n$ and $C \in \mathbb{R}$, when the pole moves on any finite segment of the plane

$$Y = \mathbf{c}^T \boldsymbol{\sigma} - C,$$

the polars all pass through the single point $(\boldsymbol{\epsilon}, X) = (\mathbf{c}, C)$ (see solid lines in Fig. 1.7(b)). However, if the pole is any finite point on the plane $\mathbf{c}^T \boldsymbol{\sigma} - C = 0$, each polar contains a line which is parallel to the fixed direction $(\mathbf{c}, C)$ indicated by the dashed lines in Fig. 1.7(b).

Now we let $U^c : \mathbb{R}^n \to \mathbb{R}$ be a given continuous function such that the graph

$$\mathcal{G}_{U^c} = \{(\boldsymbol{\sigma}, Y) \in \mathbb{R}^{n+1} \mid Y = U^c(\boldsymbol{\sigma}), \ \ \boldsymbol{\sigma} \in \mathbb{R}^n\}$$

of U^c is a continuous surface in $\mathbb{R}^{n+1}$. When the pole $(\boldsymbol{\sigma}, Y)$ moves on this surface, the polar plane changes position and, in general, it will envelop another continuous surface $X = U(\boldsymbol{\epsilon})$, described by another function $U : \mathbb{R}^n \to \mathbb{R}$. The correspondence between the *functions* $U(\boldsymbol{\epsilon})$ and $U^c(\boldsymbol{\sigma})$ is called a Legendre transformation.

Definition 1.4.1 (Regular Point, Domain and Manifold)

Let $U : \mathcal{E} \to \mathbb{R}$ be a piecewise $\mathcal{C}^2$ function.

(D1) A *regular point* of the function $U(\boldsymbol{\epsilon})$ is a point $\boldsymbol{\epsilon} \in \mathcal{E}$ where the determinant of the Hessian matrix $D^2 U = \{\partial^2 U / \partial \epsilon_i \partial \epsilon_j\}$ satisfies

$$\det D^2 U(\boldsymbol{\epsilon}) \neq 0 \ \text{ or } \ \pm \infty. \tag{1.76}$$

(D2) A *regular domain*, denoted by $\mathcal{E}_a$, is a continuous subset of regular points.

(D3) A *regular manifold (or branch)* is a graph of a $\mathcal{C}^2$ function over a regular domain. $\diamond$

A regular point is also called a *non-degenerate point*. A *singular point*, or *degenerate point*, of a $\mathcal{C}^2$ function is a point which is not regular. If U has singular points on $\mathcal{E}$, then $U : \mathcal{E} \to \mathbb{R}$ is usually nonconvex, and this will be discussed in the context of nonconvex analysis. For regular points, we have the following important result.

Theorem 1.4.2 (Legendre Duality Theorem)

Let $U(\boldsymbol{\epsilon})$ be a piecewise $\mathcal{C}^2$ function on $\mathcal{E}$. If $\mathcal{E}_a \subset \mathcal{E}$ is an open, finite subset of the regular domain of U and $\mathcal{E}_a^* \in \mathbb{R}^n$ is the range of the mapping $DU : \mathcal{E}_a \to \mathcal{E}^*$, then on $\mathcal{E}_a^*$ there exists a unique $\mathcal{C}^2$ function U^c, which is dual to U on $\mathcal{E}_a$ in the sense that the Legendre duality relations

$$U(\boldsymbol{\epsilon}) + U^c(\boldsymbol{\sigma}) = \boldsymbol{\sigma}^T \boldsymbol{\epsilon} \ \Leftrightarrow \ \boldsymbol{\sigma} = DU(\boldsymbol{\epsilon}) \ \Leftrightarrow \ \boldsymbol{\epsilon} = DU^c(\boldsymbol{\sigma}) \tag{1.77}$$

hold. Moreover, for each pair $(\boldsymbol{\epsilon}, \boldsymbol{\sigma}) \in \mathcal{E}_a \times \mathcal{E}_a^*$ satisfying (1.77),

$$U_{,ik}(\boldsymbol{\epsilon}) U^c_{,kj}(\boldsymbol{\sigma}) = \delta_{ij} := \begin{cases} 1 & (i = j) \\ 0 & (i \neq j), \end{cases} \tag{1.78}$$

where $U_{,ik} = \partial^2 U/\partial\epsilon_i\partial\epsilon_k$*, and* δ_{ij} *is the Kronecker delta.*

This theorem can be proved easily by the inverse function theorem. A geometrical version of this theorem can be stated as follows.

Geometrical Version of the Legendre Duality. *When a pole moves on a regular manifold of a* $\mathcal{C}^2$ *function over a finite domain, its polar envelops a regular manifold of another function over another finite domain, and vice versa.*

1.4.2 Legendre Conjugate Transformation

Definition 1.4.2 (Legendre Duality Pair and Conjugate Functions)
Let U and U^c be two $\mathcal{C}^2$ functions on the regular domain $\mathcal{E}_a$ and $\mathcal{E}_a^*$, respectively. If, for any $(\epsilon, \sigma) \in \mathcal{E}_a \times \mathcal{E}_a^*$, the Legendre duality relation (1.77) holds, then

(D1) (ϵ, σ) is called the *Legendre duality pair* on $\mathcal{E}_a \times \mathcal{E}_a^*$;

(D2) $U : \mathcal{E}_a \to \mathbb{R}$ and $U^c : \mathcal{E}_a^* \to \mathbb{R}$ are called the *Legendre conjugate functions*;

(D3) $\mathcal{E}_a \times \mathcal{E}_a^*$ is called *Legendre duality domain*, and $\mathcal{G}_U \times \mathcal{G}_U^c$ the *Legendre duality manifold* . ♢

From this definition, it follows that if U and U^c are Legendre dual functions over the duality domain $\mathcal{E}_a \times \mathcal{E}_a^*$, then the duality mappings $DU : \mathcal{E}_a \to \mathcal{E}_a^*$ and $DU^c : \mathcal{E}_a^* \to \mathcal{E}_a$ are bijective.

Let $U : \mathbb{R}^n \to \mathbb{R}$ be a quadratic function defined by

$$U(\epsilon) = \frac{1}{2}\epsilon^T \mathbf{K}\epsilon = \frac{1}{2}K_{ij}\epsilon_i\epsilon_j,$$

where $\mathbf{K} = (K_{ij}) \in \mathbb{R}^{n\times n}$ is a non-singular matrix, i.e. $\det(K_{ij}) \neq 0$. Then U is $\mathcal{C}^2$ on $\mathbb{R}^n$ and

$$\begin{aligned} &U : \mathbb{R}^n \to \mathbb{R} \text{ is strictly convex if } K_{ij} \text{ is positive definite,} \\ &U : \mathbb{R}^n \to \mathbb{R} \text{ is strictly concave if } K_{ij} \text{ is negative definite,} \\ &U : \mathbb{R}^n \to \mathbb{R} \text{ is saddle-shaped if } K_{ij} \text{ is indefinite.} \end{aligned} \tag{1.79}$$

Since $\sigma_i = K_{ij}\epsilon_j$ is invertible on $\mathbb{R}^n$, $\epsilon_i = K_{ij}^{-1}\sigma_j$, the dual function can be given simply by the Legendre conjugate transformation

$$U^c(\sigma) = \epsilon_i\sigma_j - U(\epsilon_i) = \frac{1}{2}K_{ij}^{-1}\sigma_i\sigma_j.$$

Since the properties of definiteness of K_{ij}^{-1} are the same as of K_{ij}, (1.79) holds also for $U^c : \mathbb{R}^n \to \mathbb{R}$. This leads to an important theorem.

Theorem 1.4.3 (Duality between the regular manifolds)
Let U *and* U^c *be Legendre dual functions over the duality domain* $\mathcal{E}_a$ *and* $\mathcal{E}_a^*$*, respectively.*

(1) If U is convex on $\mathcal{E}_a$, then U^c is convex on $\mathcal{E}_a^*$ and

$$U^c(\sigma) = \max_{\epsilon \in \mathcal{E}_a}\{\sigma^T \epsilon - U(\epsilon)\}. \tag{1.80}$$

(2) If U is concave on $\mathcal{E}_a$, then U^c is concave on $\mathcal{E}_a^*$ and

$$U^c(\sigma) = \min_{\epsilon \in \mathcal{E}_a}\{\sigma^T \epsilon - U(\epsilon)\}. \tag{1.81}$$

(3) If U is a saddle-shaped function on $\mathcal{E}_a$, then U^c is also a saddle-shaped function on $\mathcal{E}_a^*$ and

$$U^c(\sigma) = \text{sta}\{\sigma^T \epsilon - U(\epsilon) | \ \forall \epsilon \in \mathcal{E}_a\}. \tag{1.82}$$

Proof. By the inverse function theorem, $U^c_{,ij}(\epsilon) = U^{-1}_{,ij}(\sigma)$ holds for each Legendre duality pair $(\epsilon, \sigma) \in \mathcal{E}_a \times \mathcal{E}_a^*$. Then, for an arbitrarily given vector $\mathbf{u} = \{u_i\} \in \mathcal{E}$, we can construct another vector $\mathbf{v} = \{v_i\} \in \mathcal{E}^*$ such that

$$v_i = U_{,ij}(\epsilon)u_j, \quad u_i = U^c_{,ij}(\sigma)v_j.$$

Therefore, $u_i v_i = U_{ij}(\epsilon)u_i u_j = U^c_{ij}(\sigma)v_i v_j$. This is sufficient to show that the Hessian matrices U_{ij} and U^c_{ij} are either (a) both positive definite, (b) both negative definite or (c) both indefinite.

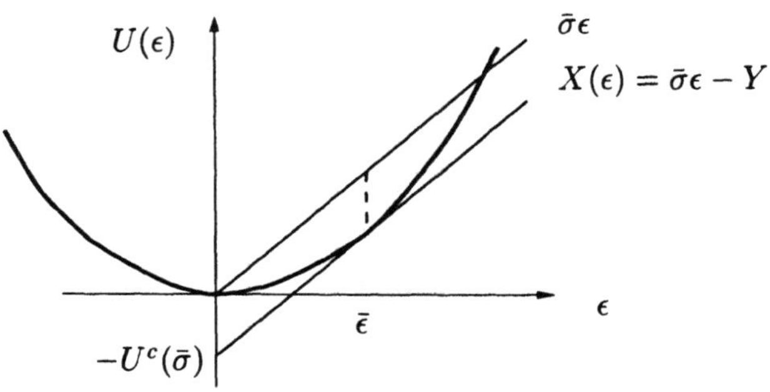

Figure 1.8. Geometry of the Legendre-Fenchel conjugate

The remainder of the proof is given geometrically by using the quadratic function $U(\epsilon) = \frac{1}{2}k_0\epsilon^2$ in one-dimensional space with $k_0 > 0$. As shown in Fig. 1.8, for a given $\bar{\epsilon}$, the associated stress $\bar{\sigma} = k_0\bar{\epsilon}$ is the slope of the polar (line) $X = \bar{\sigma}\epsilon - Y$ tangent to the graph of U at $\bar{\epsilon}$. Since $X = U(\bar{\epsilon})$, the value Y can be determined uniquely by $Y = \bar{\sigma}\bar{\epsilon} - U(\bar{\epsilon}) = U^c(\bar{\sigma})$. However, for any given ϵ, the line $X = \sigma\epsilon - Y$ is below the parabola if

$$Y \geq y(\epsilon) = \bar{\sigma}\epsilon - U(\epsilon) \ \forall \epsilon. \tag{1.83}$$

Since $U(\epsilon)$ is convex, $y(\epsilon)$ is then concave, and it takes its maximum value at $\bar{\epsilon}$ if and only if $y'(\bar{\epsilon}) = 0$. This gives $\bar{\epsilon} = k_0^{-1}\bar{\sigma}$. At this point, the line is tangent to the graph of $U(\epsilon)$, and

$$U^c(\bar{\sigma}) = \bar{\sigma}\bar{\epsilon} - U(\bar{\epsilon}) = \max_{\epsilon}\{\bar{\sigma}\epsilon - U(\epsilon)\}.$$

Similarly, for a concave function, where $k_0 < 0$. The saddle-shaped function is the combination of a convex function with a concave function. □

This theorem will play an important role in the classical Lagrange duality theory in optimization and variational problems.

Remark 1.4.1 (Legendre-Fenchel Transformation)
In infinite-dimensional space $\mathcal{E}$, equation (1.80) is also called *Legendre-Fenchel transformation.* More generally, if $U : \mathcal{E} \to \mathbb{R} \cup \{+\infty\}$ is any given convex function, then

$$U^*(\sigma) = \sup_{\epsilon \in \mathcal{E}} \{\sigma\epsilon - U(\epsilon)\}, \tag{1.84}$$

is called the *Fenchel transformation.* The function $U^* : \mathcal{E}^* \to \mathbb{R} \cup \{+\infty\}$ is called the *Fenchel-conjugate function,* which is also convex. For any given $(\epsilon, \sigma) \in \mathcal{E} \times \mathcal{E}^*$, we have the well-known *Fenchel-Young inequality*

$$U^*(\sigma) \geq \sigma\epsilon - U(\epsilon). \tag{1.85}$$

If U is differentiable on $\mathcal{E}$ and the maximizer of the problem (1.84) exists, then $U^* = U^c$. Dually, if $U : \mathcal{U} \to \mathbb{R} \cup \{-\infty\}$ is concave, its Fenchel-conjugate is defined by

$$U^*(\sigma) = \inf_{\epsilon \in \mathcal{E}} \{\sigma\epsilon - U(\epsilon)\}, \tag{1.86}$$

which is concave on $\mathcal{E}^*$. Detailed study on Fenchel transformation will be given in Chapter 5. ■

Very often, the function U carries some parameters $\boldsymbol{\mu} \in \mathbb{R}^n$, so that $U = U(\boldsymbol{\epsilon}, \boldsymbol{\mu})$. Such parameters also appear in $U^c = U^c(\boldsymbol{\sigma}, \boldsymbol{\mu})$, and it is easy to show that

$$\frac{\partial U}{\partial \boldsymbol{\mu}} + \frac{\partial U^c}{\partial \boldsymbol{\mu}} = 0. \tag{1.87}$$

For example, if $U(\epsilon, \mu) = \frac{1}{2}k_o(\epsilon - \mu)^2$, then $\sigma = U_{,\epsilon} = k_o(\epsilon - \mu)$, $U^c(\sigma, \mu) = \frac{1}{2}k_o^{-1}\sigma^2 + \mu\sigma$. Thus,

$$U_{,\mu} + U^c_{,\mu} = -k_o(\epsilon - \mu) + \sigma = 0.$$

For the linear deformation of Hooke's elastic string, the strain-energy density $U(\epsilon) = \frac{1}{2}k_0\epsilon^2$ is strictly convex on the infinite-dimensional vector space $\mathcal{E} = \mathcal{L}^2$. Its Legendre dual function $U^c(\sigma) = \sigma^2/2k_0$ is also a C^2 convex function on $\mathcal{E}^* = \mathcal{L}^2$, and the nice symmetric relation

$$\sigma = DU(\epsilon) \;\Leftrightarrow\; \epsilon = U^c(\sigma) \;\Leftrightarrow\; U(\epsilon) + U^c(\sigma) = \epsilon\sigma$$

holds on $\mathcal{L}^2 \times \mathcal{L}^2$. Thus, (ϵ, σ) is a Legendre dual pair on the *product space* $\mathcal{L}^2 \times \mathcal{L}^2$.

For the finite deformation of an elastic string discussed in Section 1.1, the stored-energy

$$W(\epsilon) = \int_I U(\epsilon)\, \mathrm{d}x = \int_I k_0\sqrt{1 + \epsilon^2}\, \mathrm{d}x$$

is also a strictly convex functional on $\mathcal{E} = \mathcal{L}^1(I)$, and the constitutive law

$$\sigma = DW(\epsilon) = \frac{\partial U}{\partial \epsilon} = \frac{k_0 \epsilon}{\sqrt{1+\epsilon^2}}$$

is well-defined on the whole space $\mathcal{L}^1(I)$. However, the inverted constitutive relation

$$\epsilon = \frac{k_0}{\sqrt{k_0^2 - \sigma^2}} \sigma, \tag{1.88}$$

is well-definded only on the subset $\mathcal{E}_a^*$ of $\mathcal{E}^* = \mathcal{L}^\infty(I)$, defined by

$$\mathcal{E}_a^* = \{\sigma \in \mathcal{E}^* | \; 0 \leq \sigma(x) < k_0 \quad \forall x \in I\}, \tag{1.89}$$

in which the inequality constraint

$$Y^*(\sigma) = \sigma(x) - k_0 < 0, \;\; \forall x \in I, \tag{1.90}$$

is called the *elastic limit condition* in solid mechanics. The string will be broken if $\sigma \geq k_0$. On $\mathcal{E}_a^*$, the Legendre dual function of U can be written as

$$U^c(\sigma) = -\sqrt{k_0^2 - \sigma^2} \quad \forall \sigma \in \mathcal{E}_a^*. \tag{1.91}$$

The inequality constraint in $\mathcal{E}_a^*$ ensures that the complementary energy

$$W^c(\sigma) = \int_I -\sqrt{k_0^2 - \sigma^2}\, dx$$

is Lebesgue integrable on $\mathcal{E}_a^* \subset \mathcal{L}^\infty(I)$. Also $\mathcal{L}(I)$ is not a *reflexive space*, i.e. $(\mathcal{L}^\infty)^* \neq \mathcal{L}$. Thus, the nice symmetry of the Legendre duality relation is broken for this finite deformation system, and the dual constitutive relation is subjected to the inequality constraint

$$\epsilon = DU^c(\sigma) \quad s.t. \;\; \sigma < k_0.$$

Detailed discussion of the duality symmetry breaking will be given in Chapter 3.

1.5 Alternative Variational Problems: Lagrange Equations and Multipliers

In homogeneous static systems, the total potential can usually be written in the form $P(u) = W(\Lambda u) - F(u)$. For infinitesimal deformation boundary value problems, the internal potential $W(\epsilon)$ is usually convex and the external potential $F(u)$ is simply a linear functional. In this section, we introduce some classical results for complementary variational principles in general boundary value problems. We make the following assumption.

Assumption 1.5.1 Let $\{(\mathcal{U}, \mathcal{U}^*), \langle *, * \rangle\}$ and $\{(\mathcal{E}, \mathcal{E}^*), \langle *; * \rangle\}$ be two duality pairings.

(A1) $\Lambda : \mathcal{U} \to \mathcal{E}$ is a general linear continuous operator.

(A2) $U(\epsilon)$ is a piecewise $\mathcal{C}^2$ function on $\mathcal{E}_a \subset \mathcal{E}$ and $U_{,\epsilon\epsilon} > 0 \;\; \forall \epsilon \in \mathcal{E}_a$. ♣

1.5.1 Boundary-Value Problems and Minimum Potential Energy Principles

Let us consider a static system defined on an open subset $I \subset \mathbb{R}$ with boundary $\partial I = \partial I_u \cup \partial I_f$. In the domain I, the system is subjected to a distributed input (or external load) $\bar{f}$ in the domain $I \subset \mathbb{R}$, and a boundary source (or boundary force) $\bar{t}$ on the boundary ∂I_f. The following general boundary-value problem (($\mathcal{P}_{\text{bv}}$) for short) can then be proposed.

Problem 1.5.1 For a given source $\bar{u}^* = \{\bar{f}(x)\ (x \in I);\ \ \bar{t}\ (\text{on } \partial I_f)\}$, and boundary data $\bar{u}$ on ∂I_u, find u, and the associated ϵ and σ such that

$$(\mathcal{P}_{\text{bv}}): \quad \begin{cases} \epsilon = \Lambda u & \text{in } I, \quad Bu = \bar{u} \quad \text{on } \partial I_u, \\ \sigma = DU(\epsilon) & \text{in } I, \\ \bar{f} = \Lambda^* \sigma & \text{in } I, \quad B^*\sigma = \bar{t} \text{ on } \partial I_f. \end{cases} \tag{1.92}$$

♠

The admissible displacement space $\mathcal{U}_a \subset \mathcal{U}$ for this mixed boundary-value problem is

$$\mathcal{U}_a = \{u \in \mathcal{U} |\ Bu = \bar{u} \quad \text{on } \partial I_u\}.$$

On $\mathcal{U}_a$, the external energy is a linear functional defined by

$$F(u) = \langle \bar{f}, u \rangle + \langle \bar{t}, u \rangle_{\partial I_f} = \int_I \bar{f} u \, \mathrm{d}x + J_\sigma(u), \tag{1.93}$$

where $J_\sigma(u) = \langle \bar{t}, u \rangle_{\partial I_f}$. The duality relation between u and $\bar{u}^*$ is then clearly given by

$$\bar{u}^* = DF(u) = \begin{cases} \bar{f}(x) & \forall x \in I, \\ \bar{t}(x) & \forall x \in \partial I_f. \end{cases}$$

On $\mathcal{E}_a$, the internal energy $W(\epsilon) = \int_I U(\epsilon) \mathrm{d}I$ is Gâteaux differentiable. The kinematically admissible space for this mixed boundary-value problem can be defined by

$$\mathcal{U}_k = \{u \in \mathcal{U} |\ \Lambda u \in \mathcal{E}_a, \ \ Bu = \bar{u} \text{ on } \partial I_u\}.$$

Thus, the total potential energy $P : \mathcal{U}_k \to \mathbb{R}$ of the system is

$$P(u) = W(\Lambda u) - F(u) = \int_I U(\Lambda u) \, \mathrm{d}x - \int_I \bar{f} u \, \mathrm{d}x - J_\sigma(u).$$

The potential variational problem associated with this ($\mathcal{P}_{\text{bv}}$) is

$$(\mathcal{P}_{\text{sta}}): \quad P(u) \to \text{sta} \ \forall u \in \mathcal{U}_k. \tag{1.94}$$

The Theorem of Necessary and Sufficient Conditions leads to the following result in ($\mathcal{P}_{\text{bv}}$).

Theorem 1.5.1 (Minimum Potential Variational Principle) *Suppose that the conditions in Assumption 1.5.1 hold and $\mathcal{U}_k$ is an open subspace of $\mathcal{U}_a$. Then,* *among all kinematically admissible field $u \in \mathcal{U}_k$, the actual equilibrium state $\bar{u}$ is a unique solution of ($\mathcal{P}_{\text{bv}}$) if and only if it is a global minimizer of the total potential P, i.e.*

$$(\mathcal{P}_{\text{inf}}): \quad P(\bar{u}) = \inf P(u) \ \forall u \in \mathcal{U}_k. \tag{1.95}$$

Proof. It is easy to check that the critical condition

$$\delta P(\bar{u}; u) = 0 \ \forall u \in \mathcal{U}_k$$

leads to the Euler equation

$$\Lambda^* DU(\Lambda \bar{u}) = \bar{f} \quad \text{in } I,$$

and the natural boundary condition

$$B^* DU(\Lambda \bar{u}) = DJ_\sigma(u) = \bar{t} \quad \text{on } \partial I_f.$$

This shows that the critical points of P solve the $\mathcal{P}_{\text{bv}}$) and vice versa.

Moreover, since $U_{,\epsilon\epsilon} > 0$ on $\mathcal{E}_a$, the total potential $P : \mathcal{U}_k \to \mathbb{R}$ is strictly convex. Thus, the problem ($\mathcal{P}_{\text{inf}}$) has at most one global minimizer, which is the critical point. □

1.5.2 Complementary Energy Principles and Statically Admissible Spaces

Let $\mathcal{E}_a^* \subset \mathcal{E}^*$ be the range of the constitutive mapping $DU : \mathcal{E} \to \mathcal{E}^*$, i.e.

$$\mathcal{E}_a^* = \{\sigma \in \mathcal{E}^* | \ \sigma \in DU(\mathcal{E}_a)\}. \tag{1.96}$$

On $\mathcal{E}_a^*$, for any given $\epsilon \in \mathcal{E}_a$ such that $\sigma = DU(\epsilon)$, the Legendre duality theorem leads to

$$U^c(\sigma) = \sigma\epsilon - U(\epsilon).$$

Dual to $\mathcal{U}_k$, the *statically admissible space* $\mathcal{S}_s$ can be defined by

$$\mathcal{S}_s = \{\sigma \in \mathcal{E}_a^* | \ \Lambda^*\sigma = \bar{f} \quad \text{in } I, \ B^*\sigma = \bar{t} \quad \text{on } \partial I_f\}. \tag{1.97}$$

For any $u \in \mathcal{U}_a$ and the associated $\sigma \in \mathcal{S}_s$, the external complementary energy F^c is obtained by the Legendre transformation:

$$F^c(\bar{\Lambda}^*\sigma) = \langle \Lambda u; \sigma \rangle - F(u) = J_u(\sigma). \tag{1.98}$$

Thus, the total complementary energy of the system can be defined by

$$P^c(\sigma) = W^c(\sigma) - F^c(\bar{\Lambda}^*\sigma) = \int_I U^c(\sigma) \, dx - J_u(\sigma), \tag{1.99}$$

and the *complementary energy variational problem* associated with ($\mathcal{P}_{\text{bv}}$) is

$$(\mathcal{P}_{\text{sta}}^c): \quad P^c(\sigma) \ \to \text{sta} \ \forall \sigma \in \mathcal{S}_s. \tag{1.100}$$

This is a constrained variational problem. Parallel to the minimum potential principle, we have the following result.

Theorem 1.5.2 (Minimum Complementary Energy Principle)
Suppose that, in addition of Assumption 1.5.1, $U^c : \mathcal{E}^*_a \to \mathbb{R}$ is the Legendre dual function of U and $\mathcal{S}_s$ is an open subset of $\mathcal{E}^*_a$. Then, among all statically admissible $\sigma \in \mathcal{S}_s$, the actual equilibrium state $\bar{\sigma}$ is a unique solution of $(\mathcal{P}_{\text{bv}})$ if and only if it is a global minimizer of the total complementary energy P^c, i.e.

$$(\mathcal{P}^c_{\text{inf}}) : \quad P^c(\bar{\sigma}) = \inf P^c(\sigma) \ \forall \sigma \in \mathcal{S}_s. \tag{1.101}$$

Proof: Introducing the admissible Lagrange multiplier $\lambda \in \mathcal{U}_a$ to relax the equilibrium constraint in $\mathcal{S}_s$, the classical Lagrangian associated with the constrained problem (1.100) can be given by

$$\begin{aligned} L_{hr}(\lambda, \sigma) &= P^c(\sigma) - \langle \bar{\Lambda}^* \sigma - \bar{u}^*, \lambda \rangle_{I_f} \\ &= \int_I U^c(\sigma) \, \mathrm{d}x - J_u(\sigma) - \int_I (\Lambda^* \sigma - \bar{f}) \lambda \, \mathrm{d}x - \langle B^* \sigma - \bar{t}, \lambda \rangle_{\partial I_f}. \end{aligned}$$

The critical condition $\delta L_{hr}(\bar{\lambda}, \bar{\sigma}; \lambda, \sigma) = 0 \quad \forall (\lambda, \sigma) \in \mathcal{U}_a \times \mathcal{E}^*_a$ leads to the *Lagrange canonical equations*

$$\Lambda \bar{\lambda} = DW^c(\bar{\sigma}), \quad \Lambda^* \bar{\sigma} = \bar{f}, \tag{1.102}$$

in the domain I and the boundary condition $B^* \bar{\sigma} = \bar{t}$ on the boundary ∂I_f. By the Legendre duality theorem, the inverted duality relation $\Lambda \bar{\lambda} = DW^c(\bar{\sigma})$ is equivalent to $\bar{\sigma} = DW(\Lambda \bar{\lambda})$. Hence, the critical point $(\bar{\lambda}, \bar{\sigma})$ of L_{hr} solves the $\mathcal{P}_{\text{bv}})$ and the Lagrange multiplier $\bar{\lambda} = \bar{u}$. Moreover, by the Legendre duality theorem, if $U_{,\epsilon\epsilon} > 0$ on $\mathcal{E}_a$, then $U^c_{,\sigma\sigma} > 0$ on $\mathcal{E}^*_a$, and $P^c : \mathcal{S}_s \to \mathbb{R}$ is strictly convex. Therefore, the problem $(\mathcal{P}^c_{\text{inf}})$ has at most one global minimizer on $\mathcal{S}_s$. □

The proof of this theorem shows the following important fact.

Lagrange multiplier's law: *The Lagrange multiplier for the equilibrium constraint in the complementary energy variational problem $(\mathcal{P}^c_{\text{sta}})$ is a solution of the potential variational problem $(\mathcal{P}_{\text{sta}})$.*

Thus, if the external force field $\bar{f}$ is a potential field, i.e. $\bar{f} = DF(u)$, the Lagrange equations in (1.102) can be written in the canonical form:

$$\Lambda u = DW^c(\sigma), \quad \Lambda^* \sigma = DF(u). \tag{1.103}$$

For the quadratic stored-energy density $U(\epsilon; \mu) = \frac{1}{2} k_0 (\epsilon - \mu)^2$ carrying the parameter $\mu \in \mathcal{E}$, the Legendre dual function is

$$U^c(\sigma; \mu) = \frac{1}{2} k^c_0 \sigma^2 + \mu \sigma, \quad k^c_0 = k_0^{-1}.$$

The parameter μ can be considered as a residual strain in an elastic string system. In this case, the *Lagrange canonical equations* (1.103) takes the following forms

$$k_0^c\sigma = \Lambda u - \mu, \quad \Lambda^*\sigma = \bar{f}, \tag{1.104}$$

or, in matrix form,

$$\begin{pmatrix} 0 & \Lambda^* \\ \Lambda & -k_0^c \end{pmatrix}\begin{pmatrix} u \\ \sigma \end{pmatrix} = \begin{pmatrix} \bar{f} \\ \mu \end{pmatrix}. \tag{1.105}$$

By symmetry, the following dual fact also holds generally for a linear operator $\Lambda : \mathcal{U} \to \mathcal{E}$.

Dual Lagrange multiplier's law: *The Lagrange multiplier for the geometrical constraint in the potential energy variational problem* $(\mathcal{P}_{\text{sta}})$ *is a solution of the complementary energy variational problem* $(\mathcal{P}^c_{\text{sta}})$.

Thus, the Lagrangian associated with $(\mathcal{P}_{\text{sta}})$ may be written

$$\begin{aligned} L_{hw}(u, \epsilon, \sigma) &= W(\epsilon) - F(u) - \langle \sigma, \epsilon - \Lambda u \rangle - \langle \sigma, \bar{u} - Bu \rangle_{\partial I_u} \\ &= \int_I (U(\epsilon) - \sigma(\epsilon - \Lambda u)) \, \mathrm{d}x - \int_I \bar{f} u \, \mathrm{d}x + \langle \sigma, Bu - \bar{u} \rangle_{\partial I_u}. \end{aligned} \tag{1.106}$$

It is easy to check that the critical condition $DL_{hw}(u, \epsilon, \sigma) = 0$ leads to the tri-canonical forms

$$\epsilon = \Lambda u, \quad \sigma = DW(\epsilon), \quad \bar{f} = \Lambda^*\sigma.$$

In continuum mechanics, $L_{hr}(u, \sigma)$ is known as the *Hellinger-Reissner generalized energy*, while L_{hw} is the well-known *Hu-Washizu potential energy*. They are linked by the Legendre duality relations

$$U(\epsilon) + U^c(\sigma) = \sigma\epsilon, \quad J_\sigma(u) + J_u(\sigma) = J_{\partial I}(u, \sigma).$$

A diagrammatic representation for the canonical boundary-value problem $(\mathcal{P}_{\text{bv}})$ is shown in Fig. 1.9.

$$\begin{array}{ccccc}
 & F : \mathcal{U}_a \to \mathbb{R} & & F^c : \mathcal{U}_a^* \to \mathbb{R} & \\
 & \mathcal{U}_a \subset \mathcal{U} & \longleftarrow \langle u, u^* \rangle \longrightarrow & \mathcal{U}^* \supset \mathcal{U}_a^* & \\
\mathcal{U}_k = \{u \in \mathcal{U}_a \mid \Lambda u \in \mathcal{E}_a\} & \Lambda \downarrow & & \uparrow \Lambda^* & \mathcal{S}_s = \{\sigma \in \mathcal{E}_a^* \mid \Lambda^*\sigma \in \mathcal{U}_a^*\} \\
 & \mathcal{E}_a \subset \mathcal{E} & \longleftarrow \langle \epsilon; \sigma \rangle \longrightarrow & \mathcal{E}^* \supset \mathcal{E}_a^* & \\
 & W : \mathcal{E}_a \to \mathbb{R} & & W^c : \mathcal{E}_a^* \to \mathbb{R} &
\end{array}$$

Figure 1.9. Diagrammatic representation of the canonical boundary-value problems

1.6 Saddle Lagrange Duality Theory

The classical saddle Lagrange duality theory plays a fundamental role in mathematical optimization and variational problems, and has been studied extensively for both finite and infinite dimensional systems (cf., e.g., Walk (1989) and Ekeland & Temam (1976)). In this section, we introduce some basic idea and results for convex static systems.

1.6.1 Saddle-Lagrangian and Minimax Theorems

Let $L(u,\sigma)$ be a given arbitrary real-valued functional defined on $\mathcal{U} \times \mathcal{E}^*$. Let $\mathcal{U}_a \times \mathcal{E}_a^* \subset \mathcal{U} \times \mathcal{E}^*$ on which L is finite and Gâteaux differentiable.

Definition 1.6.1 (Saddle Points and Saddle-Lagrangian)
(D1) A pair $(\bar{u},\bar{\sigma})$ is called a ***right saddle point*** of L on $\mathcal{U}_a \times \mathcal{E}_a^*$ if

$$L(u,\bar{\sigma}) \geq L(\bar{u},\bar{\sigma}) \geq L(\bar{u},\sigma) \quad \forall (u,\sigma) \in \mathcal{U}_a \times \mathcal{E}_a^*. \tag{1.107}$$

A pair $(\bar{u},\bar{\sigma})$ is called a ***left saddle point*** of L on $\mathcal{U}_a \times \mathcal{E}_a^*$ if

$$L(u,\bar{\sigma}) \leq L(\bar{u},\bar{\sigma}) \leq L(\bar{u},\sigma) \quad \forall (u,\sigma) \in \mathcal{U}_a \times \mathcal{E}_a^*. \tag{1.108}$$

(D2) A pair $(\bar{u},\bar{\sigma})$ is called a ***critical point*** of L if L is partially Gâteaux differentiable at $(\bar{u},\bar{\sigma})$ and

$$D_u L(\bar{u},\bar{\sigma}) = 0, \quad D_\sigma L(\bar{u},\bar{\sigma}) = 0,$$

where $D_u L$ and $D_\sigma L$ denote respectively the partial Gâteaux derivatives of L with respect to u and σ.

(D3) A real-valued functional $L(u,\sigma)$ defined on the product space $\mathcal{U} \times \mathcal{E}^*$ is said to be a ***saddle functional*** if it is convex in one variable and concave in the other one. In particular, $L : \mathcal{U} \times \mathcal{E}^* \to \mathbb{R}$ is said to be a ***right saddle functional*** if L is convex on $\mathcal{U}$ and concave on $\mathcal{E}^*$, and L is said to be a ***left saddle functional*** if $-L$ is a right saddle functional. ◊

Clearly, if $(\bar{u},\bar{\sigma})$ is a right saddle point of L, then it is a left saddle point of $-L$.

Remark 1.6.1 It is a convention that a right saddle functional in convex analysis is simply called a saddle functional. ■

In general, we have the following connection between the saddle-point and the critical points of L.

Theorem 1.6.1 (Saddle-Critical Point Theorem)
Let $L : \mathcal{U}_a \times \mathcal{E}_a^* \to \mathbb{R}$ be a given arbitrary functional. If $(\bar{u},\bar{\sigma})$ is a saddle point of L on an open subset $\mathcal{U}_a \times \mathcal{E}_a^*$ and L is partially Gâteaux differentiable at $(\bar{u},\bar{\sigma})$, then $(\bar{u},\bar{\sigma})$ is a critical point of L, i.e. $DL(\bar{u},\bar{\sigma}) = 0$.

Proof. If $(\bar{u},\bar{\sigma})$ is a saddle-point of L on $\mathcal{U}_a \times \mathcal{E}_a^*$, we have

$$L(\bar{u},\bar{\sigma}) \leq L(u,\bar{\sigma}) \;\; \forall u \in \mathcal{U}_a$$
$$L(\bar{u},\bar{\sigma}) \geq L(\bar{u},\sigma) \;\; \forall \sigma \in \mathcal{E}_a^*.$$

Since $\mathcal{U}_a \times \mathcal{E}_a^*$ is an open subset, Lemma 1.2.2 implies that

$$D_u L(\bar{u},\bar{\sigma}) = 0, \quad D_\sigma L(\bar{u},\bar{\sigma}) = 0.$$

Thus, $(\bar{u},\bar{\sigma})$ is a critical point of L. □

Any critical point of a Gâteaux differentiable saddle functional is a saddle point. However, if $(\bar{u},\bar{\sigma})$ is a saddle point of L it does not follow that L is a saddle functional. In the variational analysis of convex systems, the following statements are of important theoretical value.

Definition 1.6.2 (Saddle-Minimax and Saddle-Point Theorems)
(S1) Under certain necessary and sufficient conditions, suppose that

$$\inf_{u\in\mathcal{U}_a} \sup_{\sigma\in\mathcal{E}_a^*} L(u,\sigma) = \sup_{\sigma\in\mathcal{E}_a^*} \inf_{u\in\mathcal{U}_a} L(u,\sigma). \tag{1.109}$$

A statement of this type is called a *saddle-minimax theorem* and the pair $(\bar{u},\bar{\sigma})$ is called a *saddle-minimax point* of L on $\mathcal{U}_a \times \mathcal{E}_a^*$.

(S2) Under certain conditions, suppose that a pair $(\bar{u},\bar{\sigma}) \in \mathcal{U}_a \times \mathcal{E}_a^*$ exists such that

$$L(u,\bar{\sigma}) \geq L(\bar{u},\bar{\sigma}) \geq L(\bar{u},\sigma) \quad \forall (u,\sigma) \in \mathcal{U}_a \times \mathcal{E}_a^*. \tag{1.110}$$

A statement of this type is called a *right saddle-point theorem.* ◊

Let $\mathcal{U}_k$ be a subset of $\mathcal{U}_a$ such that $\mathcal{U}_k$ contains all point $u \in \mathcal{U}_a$ for which the supremum $\sup_\sigma L(u,\sigma)$ is finite, i.e.

$$\sup_{\sigma\in\mathcal{E}_a^*} L(u,\sigma) < +\infty \quad \forall u \in \mathcal{U}_k.$$

Dually, let $\mathcal{S}_s$ be a subset of $\mathcal{E}_a^*$ such that $\mathcal{S}_s$ contains all points $\sigma \in \mathcal{E}_a^*$ for which the infimum $\inf_u L(u,\sigma)$ is finite, i.e.

$$\inf_{u\in\mathcal{U}_a} L(u,\sigma) > -\infty \quad \forall \sigma \in \mathcal{S}_s.$$

The sets $\mathcal{U}_k$ and $\mathcal{S}_s$ may be either empty or $\mathcal{U}_k = \mathcal{U}_a$ and $\mathcal{S}_s = \mathcal{E}_a^*$. The connection between the minimax theorem and the saddle-point theorem is given by the following general result.

Theorem 1.6.2 (Saddle-Minimax Theorem)
(T1) Let $L : \mathcal{U}_a \times \mathcal{E}_a^* \to \mathbb{R}$ be a given arbitrary functional. If there exists a saddle-minimax point $(\bar{u},\bar{\sigma}) \in \mathcal{U}_a \times \mathcal{E}_a^*$ such that

$$\min_{u\in\mathcal{U}_a} \max_{\sigma\in\mathcal{E}_a^*} L(u,\sigma) = L(\bar{u},\bar{\sigma}) = \max_{\sigma\in\mathcal{E}_a^*} \min_{u\in\mathcal{U}_a} L(u,\sigma), \tag{1.111}$$

then $(\bar{u}, \bar{\sigma})$ *is a saddle point of* L *on* $\mathcal{U}_a \times \mathcal{E}_a^*$.

(T2) Conversely, if $L(u, \sigma)$ *possesses a saddle point* $(\bar{u}, \bar{\sigma})$ *on* $\mathcal{U}_a \times \mathcal{E}_a^*$, *then the saddle-minimax theorem in the form*

$$\min_{u \in \mathcal{U}_k} \max_{\sigma \in \mathcal{E}_a^*} L(u, \sigma) = L(\bar{u}, \bar{\sigma}) = \max_{\sigma \in \mathcal{S}_s} \min_{u \in \mathcal{U}_a} L(u, \sigma) \tag{1.112}$$

holds.

Proof. If $(\bar{u}, \bar{\sigma})$ is a saddle-minimax point of L on $\mathcal{U}_a \times \mathcal{E}_a^*$, then

$$\min_{u \in \mathcal{U}_a} L(u, \bar{\sigma}) = L(\bar{u}, \bar{\sigma}) = \max_{\sigma \in \mathcal{E}_a^*} L(\bar{u}, \sigma),$$

and hence

$$L(u, \bar{\sigma}) \geq L(\bar{u}, \bar{\sigma}) \geq L(\bar{u}, \sigma) \;\; \forall (u, \sigma) \in \mathcal{U}_a \times \mathcal{E}_a^*. \tag{1.113}$$

Thus, $(\bar{u}, \bar{\sigma})$ is a saddle point.

Conversely, if $(\bar{u}, \bar{\sigma})$ is a saddle point, the inequality (1.113) leads back to

$$\min_{u \in \mathcal{U}_a} L(u, \bar{\sigma}) = L(\bar{u}, \bar{\sigma}) = \max_{\sigma \in \mathcal{E}_a^*} L(\bar{u}, \sigma).$$

Thus, the sets $\mathcal{U}_k$ and $\mathcal{S}_s$ are not empty because $\bar{u} \in \mathcal{U}_k$ and $\bar{\sigma} \in \mathcal{S}_s$. Since

$$L(u, \bar{\sigma}) = \sup_{\sigma \in \mathcal{E}_a^*} L(u, \sigma) \geq \inf_{u \in \mathcal{U}_a} L(u, \sigma) = L(\bar{u}, \sigma) \;\; \forall (u, \sigma) \in \mathcal{U}_k \times \mathcal{S}_s,$$

and in particular,

$$L(\bar{u}, \bar{\sigma}) \leq L(u, \bar{\sigma}) \;\; \forall u \in \mathcal{U}_k,$$
$$L(\bar{u}, \bar{\sigma}) \geq L(\bar{u}, \sigma) \;\; \forall \sigma \in \mathcal{S}_s.$$

From this, we obtain

$$\begin{aligned} \min_{u \in \mathcal{U}_k} L(u, \bar{\sigma}) &= \min_{u \in \mathcal{U}_k} \max_{\sigma \in \mathcal{E}_a^*} L(u, \sigma) = L(\bar{u}, \bar{\sigma}) \\ &= \max_{\sigma \in \mathcal{S}_s} L(\bar{u}, \sigma) = \max_{\sigma \in \mathcal{S}_s} \min_{u \in \mathcal{U}_a} L(u, \sigma). \end{aligned}$$

This proves the theorem. □

This theorem shows that the existence of a saddle point implies the existence of a minimax point. However, the inverse statement holds only on $\mathcal{U}_k \times \mathcal{S}_s$. This is because $\max_{\mathcal{E}_a^*} L(u, \sigma)$ may not necessarily exist for all $u \in \mathcal{U}_a$ and also $\min_{\mathcal{U}_a} L(u, \sigma)$ may not necessarily exist for all $\sigma \in \mathcal{E}_a^*$. Thus, the saddle-minimax theorem in the form (1.109) is not equivalent to the saddle-point theorem of the form (1.107). We need a more general result.

Theorem 1.6.3 (Generalized Saddle-Minimax Theorem) *Let* $L : \mathcal{U}_a \times \mathcal{E}_a^* \to \mathbb{R}$ *be an arbitrarily given functional. The pair* $(\bar{u}, \bar{\sigma})$ *is a saddle point* *of L on $\mathcal{U}_k \times \mathcal{S}_s \subseteq \mathcal{U}_a \times \mathcal{E}_a^*$ if and only if it is a saddle-minimax point of L in the form*

$$\inf_{u\in\mathcal{U}_k} \sup_{\sigma\in\mathcal{E}_a^*} L(u,\sigma) = L(\bar{u},\bar{\sigma}) = \sup_{\sigma\in\mathcal{S}_s} \inf_{u\in\mathcal{U}_a} L(u,\sigma). \tag{1.114}$$

Proof. Since for some given $u \in \mathcal{U}_a$, the value of $\sup\{L(u,\sigma)| \ \forall\sigma \in \mathcal{E}_a^*\}$ may be $+\infty$ (see examples in the later of this chapter). Dually, for some $\sigma \in \mathcal{E}_a^*$, the problem $\inf\{L(u,\sigma)| \ \forall u \in \mathcal{U}_a\}$ could be $-\infty$ (it is convention that $+\infty - \infty = \infty$, while $(-\infty) + \infty = -\infty$). Thus,

$$\inf_{u\in\mathcal{U}_a} \sup_{\sigma\in\mathcal{E}_a^*} L(u,\sigma) - \sup_{\sigma\in\mathcal{E}_a^*} \inf_{u\in\mathcal{U}_a} L(u,\sigma) = \inf_{u\in\mathcal{U}_a} \sup_{\sigma\in\mathcal{E}_a^*} (L(u,\sigma) - L(u,\sigma)) \geq 0,$$

and the inequality

$$\inf_{u\in\mathcal{U}_a} \sup_{\sigma\in\mathcal{E}_a^*} L(u,\sigma) \geq \sup_{\sigma\in\mathcal{E}_a^*} \inf_{u\in\mathcal{U}_a} L(u,\sigma)$$

is always true. In particular, for a given point $(\bar{u},\bar{\sigma}) \in \mathcal{U}_k \times \mathcal{S}_s$ such that the supremum $\sup_{\mathcal{E}_a^*} L(\bar{u},\sigma)$ and the infimum $\inf_{\mathcal{U}_a} L(u,\bar{\sigma})$ are finite, then

$$\inf_{u\in\mathcal{U}_k} L(u,\bar{\sigma}) \leq \sup_{\sigma\in\mathcal{S}_s} L(\bar{u},\sigma). \tag{1.115}$$

When $(\bar{u},\bar{\sigma})$ is a saddle point of L on $\mathcal{U}_k \times \mathcal{S}_s$, one has, by the definition,

$$\sup_{\sigma\in\mathcal{E}_a^*} L(u,\sigma) = L(u,\bar{\sigma}) \geq L(\bar{u},\bar{\sigma}) \geq L(\bar{u},\sigma) = \inf_{u\in\mathcal{U}_a} L(u,\sigma) \ \ \forall (u,\sigma) \in \mathcal{U}_k \times \mathcal{S}_s.$$

This leads to

$$\inf_{u\in\mathcal{U}_k} \sup_{\sigma\in\mathcal{E}_a^*} L(u,\sigma) = \inf_{u\in\mathcal{U}_k} L(u,\bar{\sigma}) \geq L(\bar{u},\bar{\sigma}) \geq \sup_{\sigma\in\mathcal{S}_s} L(\bar{u},\sigma) = \sup_{\sigma\in\mathcal{S}_s} \inf_{u\in\mathcal{U}_a} L(u,\sigma).$$

Combining this with (1.115) leads to the minimax theorem in the form (1.114).

Conversely, if $(\bar{u},\bar{\sigma})$ is a minimax point of L in the form (1.114), then

$$\sup_{\sigma\in\mathcal{S}_s} L(\bar{u},\sigma) = \inf_{u\in\mathcal{U}_k} L(u,\bar{\sigma}) \leq L(\bar{u},\bar{\sigma}).$$

Thus,

$$L(\bar{u},\sigma) \leq L(\bar{u},\bar{\sigma}) \ \ \forall\sigma \in \mathcal{S}_s.$$

Similarly,

$$\inf_{u\in\mathcal{U}_k} L(u,\bar{\sigma}) = \sup_{\sigma\in\mathcal{S}_s} L(\bar{u},\sigma) \geq L(\bar{u},\bar{\sigma}),$$

and hence

$$L(u,\bar{\sigma}) \geq L(\bar{u},\bar{\sigma}) \ \ \forall u \in \mathcal{U}_k.$$

This shows that $(\bar{u},\bar{\sigma})$ is a saddle-point of L on $\mathcal{U}_k \times \mathcal{S}_s$. □

Under a certain generalization of the idea of a saddle point, a more general result can be proposed as follows.

Theorem 1.6.4 *Let $L : \mathcal{U}_a \times \mathcal{E}_a^* \to \mathbb{R}$ be a given arbitrary functional, such that*

$$\varrho = \inf_{u \in \mathcal{U}_a} \sup_{\sigma \in \mathcal{E}_a^*} L(u, \sigma), \quad \varrho^* = \sup_{\sigma \in \mathcal{E}_a^*} \inf_{u \in \mathcal{U}_a} L(u, \sigma)$$

are finite. If $\varrho = \varrho^*$, then for any real number $\delta > 0$, there exists a pair $(u^\delta, \sigma^\delta) \in \mathcal{U}_a \times \mathcal{E}_a^*$ such that

$$L(u^\delta, \sigma) - \delta < L(u^\delta, \sigma^\delta) < L(u, \sigma^\delta) + \delta \ \forall (u, \sigma) \in \mathcal{U}_a \times \mathcal{E}_a^*. \tag{1.116}$$

Proof. By the assumption, we have

$$\inf_{u \in \mathcal{U}_a} L(u, \sigma) \le \varrho \ \forall \sigma \in \mathcal{E}_a^*,$$
$$\sup_{\sigma \in \mathcal{E}_a^*} L(u, \sigma) \ge \varrho^* \ \forall u \in \mathcal{U}_a.$$

From the definition of infimum and supremum it follows that for any $\delta > 0$, there exists elements $u^\delta \in \mathcal{U}_a$ and $\sigma^\delta \in \mathcal{E}_a^*$ such that

$$\varrho < \inf_{u \in \mathcal{U}_a} L(u, \sigma^\delta) + \frac{\delta}{2} \le L(u, \sigma^\delta) + \frac{\delta}{2} \ \forall u \in \mathcal{U}_a,$$
$$\varrho^* > \sup_{\sigma \in \mathcal{E}_a^*} L(u^\delta, \sigma) - \frac{\delta}{2} \ge L(u^\delta, \sigma) - \frac{\delta}{2} \ \forall \sigma \in \mathcal{E}_a^*.$$

This leads to

$$L(u, \sigma^\delta) + \frac{\delta}{2} > \varrho = \varrho^* > L(u^\delta, \sigma) - \frac{\delta}{2} \ \forall (u, \sigma) \in \mathcal{U}_a \times \mathcal{E}_a^*. \tag{1.117}$$

Specifically, for $(u^\delta, \sigma^\delta) \in \mathcal{U}_a \times \mathcal{E}_a^*$, we get, from (1.117),

$$\varrho - \frac{\delta}{2} < L(u^\delta, \sigma^\delta) < \varrho^* + \frac{\delta}{2}.$$

and then substitution of (1.117) into this last inequality leads to (1.116). □

1.6.2 Weak and Strong Duality Theorems

Let $P(u)$ and $P^d(\sigma)$ be two real-valued functionals, finite and Gâteaux differentiable on the subsets $\mathcal{U}_k \subset \mathcal{U}$ and $\mathcal{S}_s \subset \mathcal{E}^*$, respectively. If

$$P(u) \ge P^d(\sigma) \ \forall (u, \sigma) \in \mathcal{U}_k \times \mathcal{S}_s,$$

then an infimum of P and a supremum of P^d exist and

$$\inf_{u \in \mathcal{U}_k} P(u) \ge \sup_{\sigma \in \mathcal{S}_s} P^d(\sigma).$$

Definition 1.6.3 (Weak and Strong Duality)

(D1) Under certain conditions, $P : \mathcal{U}_k \to \mathbb{R}$ and $P^d : \mathcal{S}_s \to \mathbb{R}$ are well-defined and

$$P(u) \geq P^d(\sigma) \ \ \forall (u, \sigma) \in \mathcal{U}_k \times \mathcal{S}_s. \tag{1.118}$$

A statement of this type is called a *weak duality theorem.*

(D2) Under certain necessary and sufficient conditions,

$$\inf_{u \in \mathcal{U}_k} P(u) = \sup_{\sigma \in \mathcal{S}_s} P^d(\sigma). \tag{1.119}$$

A statement of this type is called a *strong minimax duality theorem.* ◇

Let $L : \mathcal{U} \times \mathcal{E}^* \to \mathbb{R}$ be a given arbitrary functional which is Gâteaux differentiable on $\mathcal{U}_a \times \mathcal{E}^*_a \subset \mathcal{U} \times \mathcal{E}^*$. For any fixed $u \in \mathcal{U}_a$, we define

$$P(u) = \sup_{\sigma \in \mathcal{E}^*_a} L(u, \sigma). \tag{1.120}$$

Dually, for any fixed $\sigma \in \mathcal{E}^*_a$, we define

$$P^d(\sigma) = \inf_{u \in \mathcal{U}_a} L(u, \sigma). \tag{1.121}$$

Let $\mathcal{U}_k \subseteq \mathcal{U}_a$ be the domain of the functional P defined by (1.120), and let $\mathcal{S}_s \subseteq \mathcal{E}^*_a$ be the domain of P^d defined by (1.121). This means that $\mathcal{U}_k$ contains all point $u \in \mathcal{U}_a$ for which P is finite, and $\mathcal{S}_s$ contains all points $\sigma \in \mathcal{E}^*_a$ for which P^d is finite. The sets $\mathcal{U}_k$ and $\mathcal{S}_s$ may be either empty or $\mathcal{U}_k = \mathcal{U}_a$ and $\mathcal{S}_s = \mathcal{E}^*_a$.

If the functional L is a Lagrangian form associated with a boundary value problem, then the inf-primal variational problem

$$(\mathcal{P}_{\inf}) : \quad P(u) \quad \to \quad \min \ \ \forall u \in \mathcal{U}_k \tag{1.122}$$

is the well-known minimum potential energy problem. Dually, the problem

$$(\mathcal{P}^d_{\sup}) : \quad P^d(\sigma) \quad \to \max \ \ \forall \sigma \in \mathcal{S}_s \tag{1.123}$$

is called the *sup-dual variational problem* . The problems $(\mathcal{P}_{\inf})$ and $(\mathcal{P}^d_{\sup})$ are realisable if $\mathcal{U}_k$ and $\mathcal{S}_s$ are not empty, i.e., there exists a pair $(\bar{u}, \bar{\sigma}) \in \mathcal{U}_k \times \mathcal{S}_s$ such that

$$P(\bar{u}) = \min_{u \in \mathcal{U}_k} P(u) = \inf_{u \in \mathcal{U}_k} P(u),$$

$$P^d(\bar{\sigma}) = \max_{\sigma \in \mathcal{S}_s} P^d(\sigma) = \sup_{\sigma \in \mathcal{S}_s} P^d(\sigma).$$

Theorem 1.6.5 (Complementary Critical Point Theorem)

Let $L : \mathcal{U}_a \times \mathcal{E}^*_a \to \mathbb{R}$ be a given arbitrary functional such that P and P^d are well-defined by (1.120) and (1.121) on the open subsets $\mathcal{U}_k$ and $\mathcal{S}_s$, respectively.

If $(\bar{u}, \bar{\sigma})$ is a saddle point of L on $\mathcal{U}_k \times \mathcal{S}_s$, P is Gâteaux differentiable at $\bar{u}$, and P^d is Gâteaux differentiable at $\bar{\sigma}$, then $DP(\bar{u}) = 0$, $DP^d(\bar{\sigma}) = 0$, and

$$P(\bar{u}) = L(\bar{u}, \bar{\sigma}) = P^d(\bar{\sigma}). \tag{1.124}$$

Proof. By Theorem 1.6.2, if $(\bar{u}, \bar{\sigma})$ is a saddle point of L on the open set $\mathcal{U}_k \times \mathcal{S}_s$, it must be a critical point of L. By (1.107) and the definitions of P and P^d,

$$P(u) = \sup_{\sigma \in \mathcal{E}_a^*} L(u, \sigma) = L(u, \bar{\sigma}) \geq L(\bar{u}, \bar{\sigma}) = P(\bar{u}) \ \ \forall u \in \mathcal{U}_k,$$

$$P^d(\sigma) = \inf_{u \in \mathcal{U}_a} L(u, \sigma) = L(\bar{u}, \sigma) \leq L(\bar{u}, \bar{\sigma}) = P^d(\bar{\sigma}) \ \ \forall \sigma \in \mathcal{S}_s.$$

Thus, Lemma 1.2.2 implies $DP(\bar{u}) = 0$ and $DP^d(\bar{\sigma}) = 0$. Moreover, the definitions of P and P^d lead to

$$P(\bar{u}) = \sup_{\sigma \in \mathcal{S}_s} L(\bar{u}, \sigma) = L(\bar{u}, \bar{\sigma}) = \inf_{u \in \mathcal{U}_k} L(u, \bar{\sigma}) = P^d(\bar{\sigma}).$$

This proves (1.124). □

Theorem 1.6.6 (Weak and Strong Duality Theorems)

Let $L : \mathcal{U}_a \times \mathcal{E}_a^* \to \mathbb{R}$ be a given arbitrary functional such that P and P^d are well-defined by (1.120) and (1.121) on $\mathcal{U}_k$ and $\mathcal{S}_s$, respectively.

(T1) Weak Duality Theorem:

$$P(u) \geq P^d(\sigma) \ \ \forall (u, \sigma) \in \mathcal{U}_k \times \mathcal{S}_s. \tag{1.125}$$

(T2) Strong Duality Theorem: $(\bar{u}, \bar{\sigma})$ is a saddle-point of L on $\mathcal{U}_k \times \mathcal{S}_s \subseteq \mathcal{U}_a \times \mathcal{E}_a^*$ if and only if the strong duality

$$P(\bar{u}) = \inf_{u \in \mathcal{U}_k} P(u) = \sup_{\sigma \in \mathcal{S}_s} P^d(\sigma) = P^d(\bar{\sigma}) \tag{1.126}$$

holds.

Proof. By the definitions (1.120) and (1.121) of P and P^d respectively, it is clear that

$$P(u) \geq L(u, \sigma) \ \ \forall (u, \sigma) \in \mathcal{U}_a \times \mathcal{E}_a^*,$$
$$P^d(\sigma) \leq L(u, \sigma) \ \ \forall (u, \sigma) \in \mathcal{U}_a \times \mathcal{E}_a^*.$$

It follows that

$$P(u) \geq L(u, \sigma) \geq P^d(\sigma) \ \ \forall (u, \sigma) \in \mathcal{U}_a \times \mathcal{E}_a^*. \tag{1.127}$$

Since P and P^d are well-defined on $\mathcal{U}_k$ and $\mathcal{S}_s$, respectively, the inequality (1.127) shows that a supremum of P^d and an infimum of P exist and

$$\inf_{u \in \mathcal{U}_k} P(u) \geq \sup_{\sigma \in \mathcal{S}_s} P^d(\sigma).$$

This shows that the weak duality theorem holds. The strong duality theorem follows directly from Theorem 1.6.3. □

This theorem shows that for a given arbitrary functional $L : \mathcal{U}_a \times \mathcal{E}_a^* \to \mathbb{R}$, if P and P^d are defined respectively by (1.120) and (1.121) but the problems $(\mathcal{P}_{\text{inf}})$ and $(\mathcal{P}_{\text{sup}}^d)$ are not realisable, then the duality theorem

$$\inf_{u \in \mathcal{U}_k} P(u) = \sup_{\sigma \in \mathcal{S}_s} P^d(\sigma)$$

is equivalent to the minimax theorem of the form

$$\inf_{u \in \mathcal{U}_k} \sup_{\sigma \in \mathcal{E}_a^*} L(u, \sigma) = \sup_{\sigma \in \mathcal{S}_s} \inf_{u \in \mathcal{U}_a} L(u, \sigma).$$

However, the existence of a saddle point of L implies that the minimax theorem of the form (1.112) is equivalent to the strong duality theorem of the form

$$\min_{u \in \mathcal{U}_k} P(u) = \max_{\sigma \in \mathcal{S}_s} P^d(\sigma). \tag{1.128}$$

Let us now consider the dual variational principles associated with the boundary value problem $(\mathcal{P}_{\text{bv}})$ given in Problem 1.5.1. The external potential $F : \mathcal{U}_a \to \mathbb{R}$ is assumed to be a linear Gâteaux differentiable functional such that on $\mathcal{U}_a \times \mathcal{U}_a^*$ the Legendre duality relations

$$\bar{u}^* = DF(u) \;\Leftrightarrow\; \bar{u} = DF^c(u^*)$$

holds. Moreover, we assume that the stored-energy $W : \mathcal{E}_a \to \mathbb{R}$ and its Legendre conjugate $W^c : \mathcal{E}_a^* \to \mathbb{R}$ are Gâteaux differentiable and satisfy the Legendre duality relations

$$\sigma = DW(\epsilon) \;\Leftrightarrow\; \epsilon = DW^c(\sigma)$$

on $\mathcal{E}_a \times \mathcal{E}_a^*$. The *Lagrange form* $L : \mathcal{U}_a \times \mathcal{E}_a^* \to \mathbb{R}$ associated with the $(\mathcal{P}_{\text{bv}})$ is defined by

$$\begin{aligned} L(u,\sigma) &= -L_{hr}(u,\sigma) = \langle \Lambda u; \sigma \rangle - W^c(\sigma) - F(u) \\ &= \int_I [\sigma \Lambda u - U^c(\sigma)]\, dx - \int_I \bar{f} u \, dx - J_\sigma(u). \end{aligned} \tag{1.129}$$

Clearly, L is finite and Gâteaux differentiable on $\mathcal{U}_a \times \mathcal{E}_a^*$. The critical condition

$$\delta L(\bar{u}, \bar{\sigma}; u, \sigma) = 0 \;\; \forall (u, \sigma) \in \mathcal{U}_a \times \mathcal{E}_a^* \tag{1.130}$$

leads to the *Lagrange canonical equations*

$$\begin{aligned} D_\sigma L(\bar{u}, \bar{\sigma}) = 0 \;\Rightarrow\; & \Lambda \bar{u} = DW^c(\bar{\sigma}), \\ D_u L(\bar{u}, \bar{\sigma}) = 0 \;\Rightarrow\; & \Lambda^* \bar{\sigma} = DF(\bar{u}). \end{aligned} \tag{1.131}$$

Since $\mathcal{E}_a \times \mathcal{E}_a^*$ is the Legendre duality domain, the Legendre duality theorem shows that $\epsilon = DW^c(\sigma)$ is equivalent to $\sigma = DW(\epsilon)$. Thus, the critical points

$(\bar{u}, \bar{\sigma})$ of L satisfy the tri-canonical form and all boundary conditions, and are solutions of the $(\mathcal{P}_{\rm bv})$.

The Lagrangian L defined by (1.129) plays a fundamental role in classical duality theory. Since $F : \mathcal{U}_a \to \mathbb{R}$ is linear and $\Lambda : \mathcal{U} \to \mathcal{E}$ is a linear operator, then, for any $\sigma \in \mathcal{E}_a^*$, $L(\cdot, \sigma) : \mathcal{U}_a \to \mathbb{R}$ is linear, i.e. both convex and concave. However, for any $u \in \mathcal{U}_a$, the convexity of $L(u, \cdot) : \mathcal{E}_a^* \to \mathbb{R}$ will depend on the stored-energy density. Thus,

$W : \mathcal{E}_a \to \mathbb{R}$ is convex (resp. concave) $\Rightarrow$

$L : \mathcal{U}_a \times \mathcal{E}_a^* \to \mathbb{R}$ is a right (resp. left) saddle functional.

By the Legendre duality theorem, if $W : \mathcal{E}_a \to \mathbb{R}$ is convex and if $\epsilon = \Lambda u \in \mathcal{E}_a$ then

$$\sup_{\sigma \in \mathcal{E}_a^*} \{ \langle \Lambda u ; \sigma \rangle - W^c(\sigma) \} = W(\Lambda u).$$

Thus, on the kinematically admissible space $\mathcal{U}_k = \{ u \in \mathcal{U}_a | \ \Lambda u \in \mathcal{E}_a \}$,

$$P(u) = \sup_{\sigma \in \mathcal{E}_a^*} L(u, \sigma) = W(\Lambda u) - F(u) \tag{1.132}$$

is finite and Gâteaux differentiable. Dually, for a given $\sigma \in \mathcal{E}_a^*$,

$$\begin{aligned} \inf_{u \in \mathcal{U}_a} L(u, \sigma) &= \inf_{u \in \mathcal{U}_a} \{ \langle \Lambda u ; \sigma \rangle - F(u) \} - W^c(\sigma) \\ &= \inf_{u \in \mathcal{U}_a} \{ \langle \bar{\Lambda}^* \sigma - \bar{u}^*, u \rangle_{\Gamma_t} + J_u(\sigma) \} - W^c(\sigma) \\ &= \begin{cases} F^c(\bar{\Lambda}^* \sigma) - W^c(\sigma) & \text{if } \bar{\Lambda}^* \sigma = \bar{u}^* \\ -\infty & \text{otherwise ,} \end{cases} \end{aligned}$$

where the linear functional $F^c(\bar{\Lambda}^* \sigma) = J_u(\sigma)$ is the external complementary energy defined by (1.98). Thus, on the statically admissible space

$$\mathcal{S}_s = \{ \sigma \in \mathcal{E}_a^* | \ \bar{\Lambda}^* \sigma = \bar{u}^* \},$$

$\inf_{\mathcal{U}_a} L(u, \sigma)$ is finite and the functional dual to P is given by

$$P^d(\sigma) = \inf_{u \in \mathcal{U}_a} L(u, \sigma) = F^c(\bar{\Lambda}^* \sigma) - W^c(\sigma) \ \ \forall \sigma \in \mathcal{S}_s, \tag{1.133}$$

which is called the *dual potential energy.* Clearly, $P^d : \mathcal{S}_s \to \mathbb{R}$ is concave, and

$$P^d(\sigma) = -P^c(\sigma) \ \ \forall \sigma \in \mathcal{S}_s, \tag{1.134}$$

where $P^c(\sigma)$ is defined by (1.99). Dual to inf-primal problem $(\mathcal{P}_{\rm inf})$, the *sup-dual problem* associated with the $(\mathcal{P}_{\rm bv})$ is to find $\bar{\sigma} \in \mathcal{S}_s$ such that

$$(\mathcal{P}^d_{\rm sup}) : \quad P^d(\bar{\sigma}) = \sup_{\sigma \in \mathcal{S}_s} P^d(\sigma). \tag{1.135}$$

Theorem 1.6.7 (Complementary Extremum Theorem)

Let $L : \mathcal{U}_a \times \mathcal{E}_a^* \to \mathbb{R}$ be a Lagrangian form defined by (1.129) such that P and P^d are defined by (1.132) and (1.133), respectively. Suppose that $W : \mathcal{E}_a \to \mathbb{R}$ is *convex and* $\mathcal{U}_k \times \mathcal{S}_s$ *is an open, nonempty subset of* $\mathcal{U}_a \times \mathcal{E}_a^*$. *Then, the following statements are equivalent:*

(S1) $(\bar{u}, \bar{\sigma}) \in \mathcal{U}_k \times \mathcal{S}_s$ *is a solution of* $(\mathcal{P}_{\rm bv})$;

(S2) $\bar{u}$ *is a global minimizer of* P *on* $\mathcal{U}_k$ *and* $\bar{\sigma}$ *is a global maximizer of* P^d *on* $\mathcal{S}_s$.

(S3) The complementary extremum theorem holds, i.e.

$$P(u) \geq P(\bar{u}) = L(\bar{u}, \bar{\sigma}) = P^d(\bar{\sigma}) \geq P^d(\sigma) \ \ \forall (u, \sigma) \in \mathcal{K}_s \times \mathcal{S}_s. \tag{1.136}$$

Proof. The point $(\bar{u}, \bar{\sigma})$ is a critical point of L on the open set $\mathcal{U}_k \times \mathcal{S}_s$ if and only if the Lagrangian equations

$$\Lambda \bar{u} = DW^c(\bar{\sigma}), \ \ \bar{\Lambda}^* \bar{\sigma} = DF(\bar{u})$$

holds. By the Legendre duality theorem, $\Lambda \bar{u} = DW^c(\bar{\sigma})$ is equivalent to $\bar{\sigma} = DW(\Lambda \bar{u})$. This leads to

$$\bar{\Lambda}^* DW(\Lambda \bar{u}) = DF(\bar{u}).$$

Thus, $\bar{u}$ is a critical point of P. Replacing $W^c(\bar{\sigma})$ by $\langle \bar{\sigma}; \Lambda \bar{u} \rangle - W(\Lambda \bar{u}) = W(\Lambda \bar{u})$, we have $L(\bar{u}, \bar{\sigma}) = P(\bar{\sigma})$.

Since L is a Lagrangian form for the constrained dual variational problem $\delta P^d(\bar{\sigma}; \sigma) = 0 \ \ \forall \sigma \in \mathcal{S}_s$ and $\bar{u}$ is the Lagrange multiplier, it follows that $\bar{\sigma}$ is a critical point of P^d. Substitution of the equilibrium equation $\bar{\Lambda}^* \bar{\sigma} = DF(\bar{u})$ into L leads to $P^d(\bar{\sigma})$.

When $W : \mathcal{E}_a \to \mathbb{R}$ is convex, $L : \mathcal{U}_a \times \mathcal{E}_a^* \to \mathbb{R}$ is a saddle functional, $P : \mathcal{U}_k \to \mathbb{R}$ defined by (1.132) is convex and $P^d : \mathcal{S}_s \to \mathbb{R}$ defined by (1.133) is concave. Thus, $(\bar{u}, \bar{\sigma}) \in \mathcal{U}_k \times \mathcal{S}_s$ is a critical point of L if and only if $\bar{u}$ is a global minimizer of P on $\mathcal{U}_k$ and $\bar{\sigma}$ is a global maximizer of P^d on $\mathcal{S}_s$. □

This theorem shows that if W is convex on $\mathcal{E}$, any critical point $(\bar{u}, \bar{\sigma})$ of the Lagrangian L solves $(\mathcal{P}_{\rm bv})$, and the sup-dual problem is equivalent to the inf-primal problem in the sense that they have the same solution set. Moreover, the following statements hold.

(S1) The weak duality theorem:

$$P(u) \geq P^d(\sigma) \ \ \forall (u, \sigma) \in \mathcal{U}_k \times \mathcal{S}_s. \tag{1.137}$$

(S2) The strong duality theorem:

$$\inf_{u \in \mathcal{U}_k} P(u) = \sup_{\sigma \in \mathcal{S}_s} P^d(\sigma). \tag{1.138}$$

(S3) The saddle point theorem:

$$L(u, \bar{\sigma}) \geq L(\bar{u}, \bar{\sigma}) \geq L(\bar{u}, \sigma) \ \ \forall (u, \sigma) \in \mathcal{U}_k \times \mathcal{S}_s. \tag{1.139}$$

Moreover, by the properties of the Lagrange multipliers appearing in constrained complementary variational problems, we have the following general result.

Theorem 1.6.8 (Lagrange Multiplier Theorem)
Let $L : \mathcal{U}_a \times \mathcal{E}_a^* \to \mathbb{R}$ be a Lagrangian form defined by (1.129) such that P and P^d are defined by (1.132) and (1.133), respectively.

(T1) The Lagrange multiplier for the inf-primal problem $(\mathcal{P}_{\inf})$ is a solution of the sup-dual problem $(\mathcal{P}^d_{\sup})$.

(T2) Dually, the Lagrange multiplier for the sup-dual problem is a solution of the inf-primal problem.

1.7 Applications and Commentary

To aid digestion of the theory introduced above, this "sabbatical" section gives a list of some simple examples in convex optimization and one-dimensional nonlinear boundary-value problems. Some historical developments in saddle Lagrangian duality theory are discussed briefly in the commentary section. Please enjoy and relax.

1.7.1 Applications in Convex Optimization and Boundary-Value Problems

Example 1.7.1 Let us consider the very simple one-dimensional convex optimization problem with the strict constraints $a < x < b$

$$(\mathcal{P}_{\min}) : \quad P(x) = \frac{1}{2}kx^2 - fx \quad \to \min \;\; \forall x \in I = (a, b), \tag{1.140}$$

where $k > 0$ is a given constant. We assume that $a \leq 0 \leq b$. The critical condition $DP(\bar{x}) = 0$ gives $\bar{x} = f/k$. Clearly, if the external load $f \in \mathcal{U}^* = \mathbb{R}$ satisfies

$$f \in \mathcal{U}_a^* = (ak, bk) \subset \mathbb{R},$$

Then, $\bar{x} = f/k$ is a global minimizer of P on I and

$$\inf_{x \in (a,b)} P(x) = \min_{x \in (a,b)} P(x) = -\frac{1}{2k} f^2.$$

Otherwise, the solution of $(P_{\min})$ does not exists in I, although the infimum of P on I is finite:

$$\inf P(x) = \min\{P(a), P(b)\} > -\infty.$$

To put this problem within our theoretical framework, we need only set $\mathcal{U} = \mathbb{R}$,

$$\mathcal{U}_a = I = (a, b), \quad \Lambda = 1, \quad \epsilon = \Lambda x = x.$$

Thus, $\mathcal{E} = \mathbb{R}$, and the function

$$W(\epsilon) = \frac{1}{2}k\epsilon^2$$

is continuous and differentiable on $\mathcal{E}_a = \mathcal{E}$. The dual variable of ϵ is $\sigma = DW(\epsilon) = k\epsilon$, which is well defined on $\mathcal{E}^* = \mathcal{E}_a^* = \mathbb{R}$. Thus, (ϵ, σ) is a Legendre duality pair on $\mathbb{R} \times \mathbb{R}$ and the complementary energy W^c is also a quadratic function:

$$W^c(\sigma) = \sigma\epsilon(\sigma) - W(\epsilon(\sigma)) = \frac{1}{2k}\sigma^2.$$

On $\mathcal{U}_a \times \mathcal{E}_a^* = I \times \mathbb{R}$, the Lagrangian associated with $(P_{\min})$ is well defined by

$$L(x,\sigma) = x\sigma - \frac{1}{2k}\sigma^2 - fx. \tag{1.141}$$

It is clear that if $x \in \mathcal{U}_k = \mathcal{U}_a = I$,

$$P(x) = \sup_{\sigma \in \mathcal{E}_a^*} L(x,\sigma).$$

However, for any given $\sigma \in \mathcal{E}_a^* = \mathbb{R}$, the dual function is given by

$$\begin{aligned} P^d(\sigma) &= \inf_{x \in \mathcal{U}_a} L(x,\sigma) = \inf_{x \in \mathcal{U}_a} \{x\sigma - fx\} - \frac{1}{2k}\sigma^2 \\ &= \begin{cases} (\sigma - f)b - \frac{1}{2k}\sigma^2 & \text{if } \sigma - f < 0, \\ -\frac{1}{2k}\sigma^2 & \text{if } \sigma - f = 0, \\ (\sigma - f)a - \frac{1}{2k}\sigma^2 & \text{if } \sigma - f > 0. \end{cases} \end{aligned} \tag{1.142}$$

The statically admissible space $\mathcal{S}_s$ for this optimization problem is simply an open interval in $\mathbb{R}$, namely,

$$\mathcal{S}_s = \{\sigma \in \mathcal{E}_a^* = \mathbb{R} |\ \Lambda^*\sigma = \sigma \in \mathcal{U}_a^*\} = (ak, bk).$$

Then, on $\mathcal{S}_s$, the dual problem reads

$$(\mathcal{P}^d_{\max}): \ \ P^d(\sigma) \ \ \to \max \ \ \forall \sigma \in \mathcal{S}_s.$$

Clearly, if $f \in \mathcal{S}_s$, the solution of $(\mathcal{P}^d_{\max})$ is simply $\bar{\sigma} = f \in \mathcal{S}_s$, and

$$P^d(\bar{\sigma}) = \sup_{\sigma \in \mathcal{S}_s} P^d(\sigma) = \max_{\sigma \in \mathcal{S}_s} P^d(\sigma) = -\frac{1}{2k}f^2.$$

In this case, let $y = \sigma - f$. Then the problem

$$P^d_f(y) := -\frac{1}{2k}y^2 - \frac{1}{2k}f^2 \ \ \to \max \ \ \forall y \in \mathbb{R}$$

is equivalent to the dual problem $(\mathcal{P}^d_{\max})$ in the sense that

$$P^d(f) = \max_{\sigma \in \mathcal{S}_s} P^d(\sigma) = \max_{y \in \mathbb{R}} P^d_f(y) = P^d_f(0) = -\frac{1}{2k}f^2.$$

Graphical illustrations of $P(x)$ and $P^d_f(y)$ are shown in Fig. 1.10 with $k = 0.5$, $a = -1$, $b = 2$.

Otherwise, if $f \notin \mathcal{S}_s$, the solution of the dual problem does not exist in $\mathcal{S}_s$, but

$$\sup_{\sigma \notin \mathcal{S}_s} P^d(\sigma) = \max\{P^d(ak), P^d(bk)\} < \infty.$$

Examples of $P(x)$ and $P_f^d(y)$ in this case are shown graphically in Fig. 1.11.

Clearly, in either case, the weak and strong duality theorems hold, i.e.

$$P(x) \geq P^d(\sigma) \quad \forall (x, \sigma) \in (a, b) \times (ak, bk).$$

and

$$\inf_{x \in (a,b)} P(x) = \sup_{\sigma \in (ak,bk)} P^d(\sigma).$$

This very simple example has very important applications in n-dimensional constrained mathematical programming. ♡

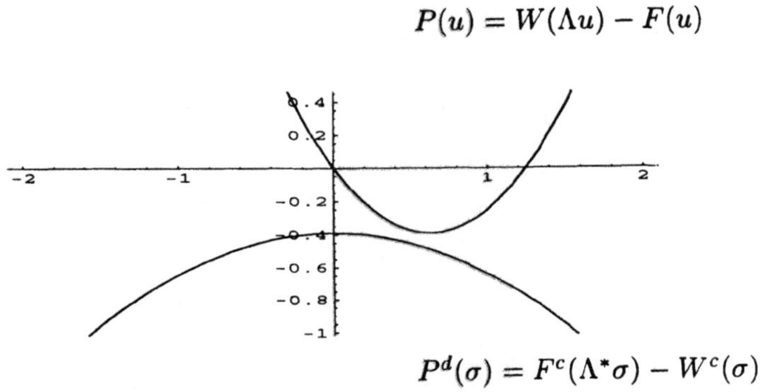

Figure 1.10. Graphs of $P(x)$ and $P_f^d(y)$ for $f \in (ak, bk)$

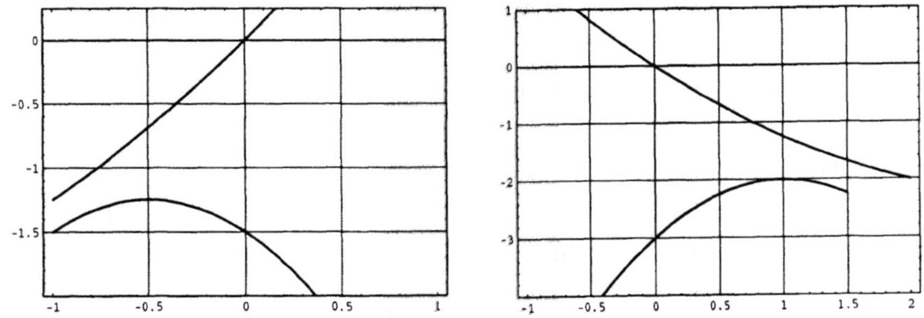

(a) $f = -1.5 < ak$ (b) $f = 1.5 > bk$

Figure 1.11. Graphs of $P(x)$ and $P_f^d(y)$ for $f \notin (ak, bk)$

Example 1.7.2 (Finite Deformation of an Elastic String)

If the string is fixed at its two ends and is subjected to the transverse load $f(x)$ on the undeformed domain $I = (0, 1)$, the primal problem is to find the deflection $u(x)$ such that

$$(\mathcal{P}_{\text{inf}}): \quad P(u) = \int_I k_0\sqrt{1+u'(x)^2}\,\mathrm{d}x - \int_I f(x)u(x)\,\mathrm{d}x \to \min,$$
$$s.t. \quad u(0) = u(1) = 0.$$

This is a constrained nonlinear variational problem. The criticality condition $DP(u) = 0$ leads to a nonlinear differential equation with Dirichlet boundary conditions.

To set this problem precisely within our framework, we let

$$\mathcal{U} = \mathcal{L}^2(I), \quad \mathcal{E} = \mathcal{L}^1(I), \quad \Lambda = \frac{\mathrm{d}}{\mathrm{d}x}.$$

With the standard bilinear forms

$$\langle u, u^*\rangle = \int_I u(x)u^*(x)\,\mathrm{d}x, \quad \langle \epsilon; \sigma\rangle = \int_I \epsilon(x)\sigma(x)\,\mathrm{d}x,$$

the dual spaces of $\mathcal{U}$ and $\mathcal{E}$ are $\mathcal{U}^* = \mathcal{L}^2$ and $\mathcal{E}^* = \mathcal{L}^\infty$, respectively. Let

$$\mathcal{U}_a = \{u \in \mathcal{U} |\ u(0) = u(1) = 0\}.$$

Then, Gauss-Green's formula

$$\langle \Lambda u; \sigma\rangle = \langle u, \Lambda^*\sigma\rangle \ \ \forall (u,\sigma) \in \mathcal{U}_a \times \mathcal{E}_a^*$$

gives the adjoint operator $\Lambda^* = -\Lambda = -\mathrm{d}/\,\mathrm{d}x$. The external functional $F: \mathcal{U}_a \to \mathbb{R}$ is defined by

$$F(u) = \langle u, f\rangle = \int_I f(x)u(x)\,\mathrm{d}x,$$

and, on $\mathcal{E}$, the internal energy is

$$W(\epsilon) = \int_I U(\epsilon)\,\mathrm{d}x = \int_I \sqrt{1+\epsilon^2}\,\mathrm{d}x.$$

The kinematically admissible space is then,

$$\mathcal{U}_k = \{u \in \mathcal{L}^2(I) |\ u'(x) \in \mathcal{L}^1(I), \ \ u(0) = u(1) = 0\}.$$

Thus, the inf-primal variational problem is to find $\bar{u} \in \mathcal{U}_k$ such that

$$(\mathcal{P}_{\text{inf}}): \quad P(\bar{u}) = \inf P(u) \ \ \forall u \in \mathcal{U}_k. \tag{1.143}$$

Since $W(\epsilon)$ is strictly convex on $\mathcal{E}$, the duality relation $\sigma = DW(\epsilon)$ is invertible and, by the Legendre transformation, it is easy to show that

$$W^c(\sigma) = -\int_I \sqrt{k_0^2 - \sigma^2}\,\mathrm{d}x$$

which is well-defined on

$$\mathcal{E}_a^* = \{\sigma \in \mathcal{L}^\infty(I) | \ |\sigma(x)| < k_0 \ \ \forall x \in I\}.$$

Since both ends are fixed, the admissible force space $\mathcal{U}_a^*$ is simply

$$\mathcal{U}_a^* = \{u^* \in \mathcal{L}^2(I) | \ u^*(x) = f(x) \ \ \forall x \in I\}.$$

Since $F(u)$ is a linear functional on $\mathcal{U}_a$, for any given $u^* = DF(u) \in \mathcal{U}_a^*$, its Legendre conjugate is

$$F^c(u^*) = \langle u, u^* \rangle - F(u) = 0 \ \ \forall u \in \mathcal{U}_a.$$

The Lagrange form associated with this problem is a saddle functional on $\mathcal{U}_a \times \mathcal{E}_a^*$, i.e.

$$L(u, \sigma) = \int_I \left(u'\sigma + \sqrt{k_0^2 - \sigma^2} - fu \right) \mathrm{d}I.$$

Clearly, for any given $u \in \mathcal{U}_k$,

$$\sup_{\sigma \in \mathcal{E}_a^*} L(u, \sigma) = P(u).$$

Dually, for any given σ in the statically admissible space defined by

$$\mathcal{S}_s = \left\{ \sigma \in \mathcal{L}^\infty(I) \ \middle| \ -\frac{\mathrm{d}}{\mathrm{d}x}\sigma = f(x) \in \mathcal{L}^2(I), \ \ |\sigma(x)| < k_0 \ \ \forall x \in I \right\},$$

the dual energy $P^d : \mathcal{S}_s \to \mathbb{R}$ is given by

$$\begin{aligned} P^d(\sigma) &= \inf_{u \in \mathcal{U}_a} L(u, \sigma) = F^c(\Lambda^* \sigma) - W^c(\sigma) \\ &= \int_I \sqrt{k_0^2 - \sigma(x)^2} \ \mathrm{d}x. \end{aligned} \tag{1.144}$$

Thus, the dual problem for this finite deformation string is to find $\bar{\sigma} \in \mathcal{S}_s$ such that

$$P^d(\bar{\sigma}) = \sup_{\sigma \in \mathcal{S}_s} P^d(\sigma). \tag{1.145}$$

This is a nonlinear variational problem with both equality and inequality constraints. Solving the equilibrium equation $\Lambda^* \sigma = f$, we have

$$\sigma(x) = -\int_0^x f(t) \mathrm{d}t + c = g(x) + c,$$

where $g(x) = -\int f(x) \ \mathrm{d}x$, and $c \in \mathbb{R}$ is an integral constant. For a given $f \in \mathcal{L}^2$ and the constant k_0 such that the set

$$\mathcal{S}_a^c = \{c \in \mathbb{R} | \ |g(x) + c| < k_0 \ \ \forall x \in I\}$$

is not empty, the dual variational problem is reduced to an optimization problem:

$$P^d_\sigma(c) := P^d(\sigma(c)) = \int_I \sqrt{k_0^2 - (g(x) + c)^2}\, \mathrm{d}x \quad \to \text{ max } \forall c \in \mathcal{S}^c_a.$$

This is a strictly convex dual optimization problem over a convex subset of $\mathbb{R}$, which is much easier than the primal variational problem over the function space $\mathcal{U}_k$. If the maximizer $\bar{c}$ of P^d_σ is in $\mathcal{S}^c_a$, it must be a critical point of $P^d_\sigma(c)$ and the criticality condition gives

$$\int_0^1 \frac{g(x) + \bar{c}}{\sqrt{k_0^2 - (g(x) + \bar{c})^2}}\, \mathrm{d}x = 0. \tag{1.146}$$

The solution $\bar{c}$ of this dual algebraic equation leads to $\bar{\sigma} = g(x) + \bar{c}$. From the duality relation $u' = DW^c(\sigma)$, the integral

$$\bar{u}(x) = \int_0^x \frac{g(t) + \bar{c}}{\sqrt{k_0^2 - (g(t) + \bar{c})^2}} \mathrm{d}t, \tag{1.147}$$

is a solution of the primal problem. Clearly, $\bar{u}(1) = 0$ is just the dual Euler-Lagrange equation (1.146). Thus, by the duality theory, the inf-primal problem is solved completely and we have the following duality theorem.

Theorem 1.7.1 *Suppose that for any external load $f \in \mathcal{L}^2(I)$, the set $\mathcal{S}^c_s$ is not empty. Then, for any solution $\bar{c} \in \mathbb{R}$ of the dual optimization problem (P^d_σ), the solution $\bar{u}$ of the inf-primal problem (1.143) is uniquely defined by (1.147), and the strong duality theorem*

$$P(\bar{u}) = \inf_{u \in \mathcal{U}_k} P(u) = \sup_{c \in \mathcal{S}^c_a} P^d_\sigma(c) = P^d_\sigma(\bar{c})$$

holds.

If the string is subjected to a unit load $f(x) = -1$, Then,

$$\sigma(x) = -\int_0^x f(x)\, \mathrm{d}x + c = x + c.$$

The solution of the dual problem is $\bar{c} = -0.5$, and then

$$\bar{u}(x) = \int_0^x \frac{x - 0.5}{\sqrt{k_0^2 - (x - 0.5)^2}}\, \mathrm{d}x = \sqrt{k_0^2 - 0.5^2} - \sqrt{k_0^2 - (x - 0.5)^2}.$$

The graph of $u(x)$ is shown in Fig. 1.12. Let $u(x) = a\bar{u}(x)$, then the total potential $P_{\bar{u}}(a) = P(a\bar{u})$ is an elliptic function of $a \in \mathbb{R}$, i.e.

$$P_{\bar{u}}(a) = -0.09a + 2.0EllipticE[0.52, 1 - a^2],$$

where *EllipticE* is a *Mathematica* notation for elliptic function. The graph of $P_{\bar{u}}$ is very flat (see Fig. 1.13). However, the dual function $P^d_\sigma(c)$ is an ordinary function

$$P^d_\sigma(c) = \frac{1}{2}\left(-c\sqrt{1 - c2} - \sin^{-1}(c)\right) + \frac{1}{2}\left((1 + c)\sqrt{-c(2 + c)} + \sin^{-1}(1 + c)\right).$$

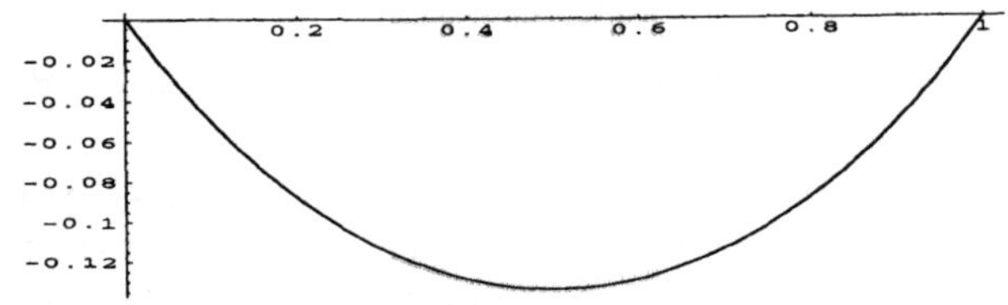

Figure 1.12. Solution $\bar{u}(x)$ for finite deformation elastic string with $k_0 = 1, f = -1$

It is easy to check that

$$P_{\bar{u}}(1) = P(\bar{u}) = P^d(\bar{\sigma}) = P^d_\sigma(-\frac{1}{2}) = 0.95661.$$

Dual to $P_{\bar{u}}$, the graph of P^d_σ is very steep. This fact will be a very important advantage in n-dimensional numerical computations. ♡

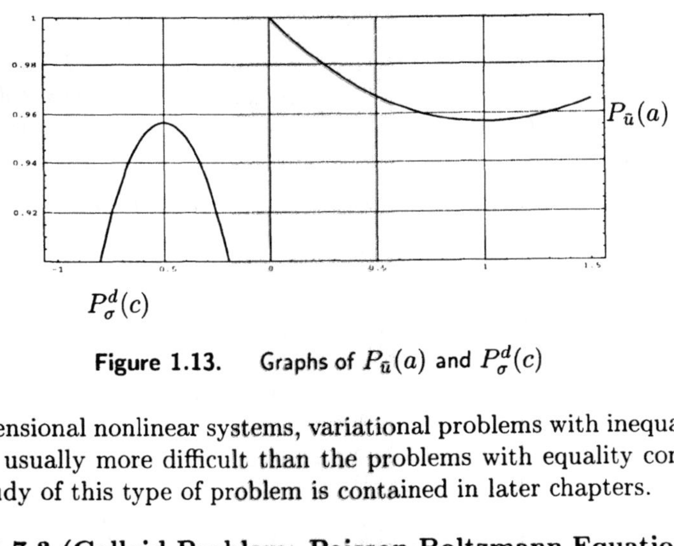

Figure 1.13. Graphs of $P_{\bar{u}}(a)$ and $P^d_\sigma(c)$

In n-dimensional nonlinear systems, variational problems with inequality constraints are usually more difficult than the problems with equality constraints. Detailed study of this type of problem is contained in later chapters.

Example 1.7.3 (Colloid Problem: Poisson-Boltzmann Equation)
For the given numbers u_a, $u_b \in \mathbb{R}$ and the interval $I = (a,b) \in \mathbb{R}$, the boundary-value problem for the Poisson-Boltzmann equation is

$$\frac{d^2}{dx^2}u(x) = 2\sinh u(x), \quad \forall x \in I, \tag{1.148}$$

$$u(a) = u_a, \quad u(b) = u_b. \tag{1.149}$$

This nonlinear boundary-value problem arises, for example, in the theory of colloids with $a = 0, b = \infty$ and $u_a > 0$, $u_b = 0$ (cf., e.g., Alexander and Johnson, 1949; Feynman, Leighton and Sands, 1964) and in the theory of plasmas with

$a, b < \infty$ and $u_a = 0, u_b > 0$ (cf., e.g., Arthurs, (1980)). In these examples, u is the dimensionless quantity $q\phi/kT$, where q is the proton charge, ϕ is the electric potential, kT is the Boltzmann energy and x is measured in units of the Debye length $L_D = \{kT/4\pi q^2 n_0\}^{\frac{1}{2}}$, with n_0 a number density.

Let $\mathcal{U} = \mathcal{U}^*$ and $\mathcal{E} = \mathcal{E}^*$ be two Hilbert spaces with the standard bilinear forms

$$\langle u, u^* \rangle = \int_I u(x) u^*(x) \, \mathrm{d}x, \quad \langle \epsilon; \sigma \rangle = \int_I \epsilon(x) \sigma(x) \, \mathrm{d}x.$$

In abstract notation, this Dirichlet boundary-value problem can be written in the fundamental form

$$\Lambda^* \Lambda u = f(u), \quad \forall x \in I, \tag{1.150}$$

$$Bu = \bar{u} \quad \text{on } \partial I_u, \tag{1.151}$$

with

$$\Lambda = \frac{\mathrm{d}}{\mathrm{d}x}, \quad \Lambda^* = -\Lambda, \quad J_t(u) = 0, \quad J_u(\sigma) = \sigma(b) u_b - \sigma(a) u_a,$$

$$\partial I = \partial I_u = [a, b], \quad \partial I_f = \emptyset, \quad Bu = (u(b), \ -u(a))^T, \quad \bar{u} = (u_b, \ -u_a)^T.$$

The force field in this problem is a potential field: $f(u) = -2 \sinh u = DF(u)$ and $F : \mathcal{U} \to \mathbb{R}$ is a strictly concave functional, namely

$$F(u) = -\int_I 2 \cosh u \, \mathrm{d}x.$$

Let $\mathcal{U}_a = \{u \in \mathcal{U} | \ u(a) = u_a, \ u(b) = u_b\}$, and $\mathcal{E}_a^* = \mathcal{E}^* = \mathcal{L}^2(I)$. The Lagrange form associated with this Dirichlet boundary-value problem is the saddle functional

$$L(u, \sigma) = \int_I \left(u' \sigma - \frac{1}{2} \sigma^2 + 2 \cosh u \right) \, \mathrm{d}x.$$

For any $u \in \mathcal{U}_k = \{u \in \mathcal{U}_a | \ u' \in \mathcal{L}^2(I)\}$, we have

$$P(u) = \sup_{\sigma \in \mathcal{L}^2} L(u, \sigma) = \int_I \left(\frac{1}{2} u'^2 + 2 \cosh u \right) \, \mathrm{d}x. \tag{1.152}$$

However, for any statically admissible $\sigma \in \mathcal{S}_s = \{\sigma \in \mathcal{L}^2(I) | \ \sigma' \in \mathcal{U}^*\}$, the dual functional is given by

$$\begin{aligned} P^d(\sigma) &= \inf_{u \in \mathcal{U}_a} L(u, \sigma) \\ &= \int_I \left[2 \cosh \left(\sinh^{-1} (\frac{1}{2} \sigma') \right) - \sigma' \sinh^{-1} (\frac{1}{2} \sigma') - \frac{1}{2} \sigma^2 \right] \, \mathrm{d}x + J_u(\sigma). \end{aligned}$$

Since the total potential is a strictly convex functional, the solution of the problem is unique and the critical point $(\bar{u}, \bar{\sigma})$ of L satisfies the complementary bounding relation

$$P(u) \geq P(\bar{u}) = L(\bar{u}, \bar{\sigma}) = P^d(\bar{\sigma}) \geq P^d(\sigma) \ \forall (u, \sigma) \in \mathcal{U}_k \times \mathcal{S}_s.$$

Detailed study on the Debye-Húckel theory of colloids was given by Anderson and Arthurs (1968) (see Arthurs, 1980).

♡

1.7.2 Historical Notes and Commentary

The classical (saddle) Lagrangian duality theory in convex static systems has appeared in the literature in various forms for a long time. The earliest examples of the Lagrangian dual variational formulations are probably provided by the complementary energy principles in continuum mechanics and in the theory of structural mechanics (see, for examples, Hellinger, 1914; Trefftz, 1928). There are several methods, closely related, by which the dual problems can be derived. In particular, a heuristic presentation was given by Courant and Hilbert (1953) employing the Legendre transformation. The dual problems can be written down for several variational problems but it is not always possible to establish the relationship between the primal and the dual by this method (see Ekeland and Temam, 1976). Another method, due to Synge (1957), employs the so-called *hypercircle approach*, which is applicable only to certain linear problems. This method is actually a geometrical version of the saddle Lagrangian duality theory in linear systems (see Arthurs, 1980). A systematic approach to complementary variational principles for boundary-value problems was given by Arthurs (1980).

According to Oravas and McLean (1966), the "Legendre" transformation was used by Euler in 1770, 17 years before Legendre himself first used it in 1787. This elegant transformation plays a fundamental role in duality theory and in mathematical physics. However, as indicated in Sewell (1987), it is not as widely appreciated by applied mathematicians as it deserves to be in spite of its long history.

By using the abstract operator Λ and bilinear forms, a unified framework in mathematical physics has been presented heuristically in this chapter through one-dimensional boundary-value problems, for which, the elegant minimax complementary variational principles on their effective domains are established. Based on the Legendre-Fenchel transformation and the unified tri-canonical forms, the duality presented in this chapter gives a clear picture of the structure of primal-dual problems and their connection to boundary-value problems. This theory enable us to obtain more information about the boundary-value problems: the extremality relations, uniqueness and the constraints on the variational problems.

The universe is built on a plan the profound symmetry of which is somehow present in the inner structure of our intellect.

—Paul Valery

Legendre transformation is a very useful mathematical tool; it transforms functions on a vector space to function on dual space. Legendre transformations are related to projective duality and tangential coordinates in algebraic geometry and the construction of dual Banach spaces in analysis. They are often encountered in physics (for example, in the definition of thermodynamic quantities.

—V. I. Arnold, 1978

Of old, these came to be in possession of the One:
Heave in virtue of the One is limpid;
Earth in virtue of One is settled;
Spirits in virtue of One have their potencies;
The valley in virtue of One is full;
The myriad creatures in virtue of One are alive;
Loads and princes in virtue of One become leaders in the empire;
It is the One that makes these what they are.

—Lao Chi, *Tao De Chin*, XXXIX

Read Euler, he is the master of us all.

—Marquis de Pierre Simon Laplace, 1749-1827

The mathematician is perfect only in so far as he is a perfect being, in so far as he perceives the beauty of truth; only then will his work be thorough, transparent, comprehensive, pure, clear, attractive, and even elegant. All this is necessary to resemble Lagrange.

—Johann Wolfgang von Goethe, 1749-1832

Notes

1. Leonard Euler (1707-1783), a great mathematician from Switzerland. During his life he published a total of 886 books and mathematical memoirs, and produced thirteen children (all but five of whom died very young).

2. Joseph Louis Lagrange, 1736-1813, was born in Turin of Italian-French ancestry. His most valuable book *Mechanique Analutique* was published in 1788. In this work, Lagrange unified the science of mechanics, and, as Hamilton said, made of it "a kind of scientific poem".

2 BI-DUALITY IN DYNAMIC SYSTEMS

The Lord God said, "it is not good for the man to be alone, I will make a helper suitable for him." ⋯ Then, the Lord God made a woman from the rib he had taken out of the man, and brought her to the man."

—Genesis, II

Opposition is how the Tao moves; Weakness is the means the Tao employs. All things are born from Something and Something from Nothing.

—Lao Chi

A great truth is a truth whose opposite is also a great truth.

—Niels Bohr

It is a splendid feeling to realize the unity of a complex of phenomena that by physical perception appear to be completely separated.

—Albert Einstein

$$\Pi(q) = W(\Lambda q) - P(q)$$

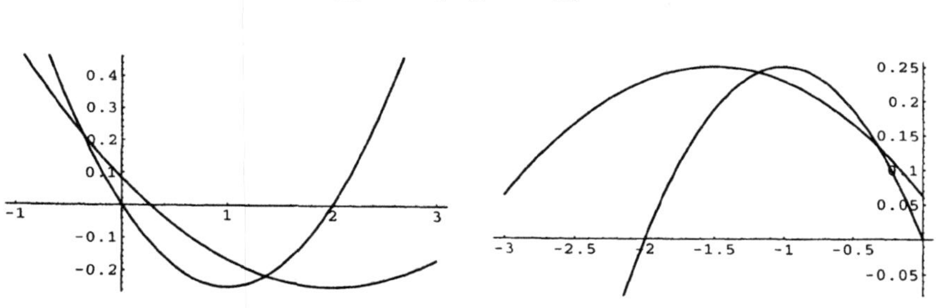

$$\Pi^d(p) = P^c(\Lambda^* p) - W^c(p)$$

Figure 2.1. Double-min and double-max dualities in convex Hamiltonian systems

From static equilibria to dynamical problems, this chapter presents the subject of duality theory for convex Hamiltonian systems. The framework for dynamics is the same as that for static systems, but, the duality theorems are more interesting.

2.1 Particle Dynamics: Newton and Einstein

2.1.1 Newton's Triality Law

Consider a particle of mass m moving in one-dimensional space under a given external force field $f(t) \in \mathbb{R}$, where t is time. The position of the particle is a function of time: $q(t)$. Everyone who has taken high school physics knows that Newton discovered three laws. However, these three laws are really equivalent to just one, i.e. particle motion is governed by our favorite fundamental equation†

$$-\frac{\mathrm{d}}{\mathrm{d}t}\left(m\frac{\mathrm{d}}{\mathrm{d}t}q\right) = f. \tag{2.1}$$

If $f \equiv 0$, this linear equation has an unique solution: $v = q_{,t} = \mathit{constant}$. This is the Newton's first law. If $m = m_0 = \mathit{constant}$, and we let $a = q_{,tt}$, this leads to the popular form of Newton's second law: $F = m_0 a$, where F is the inertial force produced by the acceleration $a = q_{,tt}$. Equation (2.1) then gives $F = -f$, which is the Newton's third law.

We now introduce a pair of new variables: v, the velocity, and p, the momentum. Equation (2.1) can then be split into the tri-canonical equations

$$\begin{array}{lll} (a) \text{ kinematic equation:} & v = \mathrm{d}q/\mathrm{d}t, \\ (b) \text{ momentum-velocity law:} & p = mv, \\ (c) \text{ dynamic equation:} & f = -\mathrm{d}p/\mathrm{d}t. \end{array} \tag{2.2}$$

In this dynamical system, we set $I = (0, t_1) \subset \mathbb{R}$, and let $\mathcal{U}$ be the so-called *configuration space*. The linear operator Λ can be written as

$$\Lambda q = \frac{\mathrm{d}}{\mathrm{d}t}q = q_{,t} = \dot{q}.$$

Thus, Λ maps $\mathcal{U}$ into a space of velocity, $\mathcal{E}$. Its dual space $\mathcal{E}^*$ is the space of momentum. The bilinear forms on the product space $\mathcal{U} \times \mathcal{U}^*$ and $\mathcal{E} \times \mathcal{E}^*$ are still denoted by $\langle *, * \rangle$ and $\langle *; * \rangle$, respectively. For homogeneous time boundary conditions, the admissible space can be defined as

$$\mathcal{U}_a^0 = \{q \in C^1(I) |\ q(0) = 0,\ \ q(t_1) = 0\}.$$

Then, for a given external force $f(t)$ and associated (dynamical) equilibrium state $\bar{q}$, $\bar{p}(\bar{q})$, the virtual work principle

$$\langle \Lambda q; \bar{p}(\bar{q}) \rangle := \int_I \bar{p} \Lambda q \,\mathrm{d}t = \int_I q f \,\mathrm{d}t =: \langle q, f \rangle \ \ \forall q \in \mathcal{U}_k \tag{2.3}$$

† The wonderful negative sign in the fundamental equation (2.1) is anticipated in Lao Chi's observation: "opposition is how the Tao moves" (Tao Te Ching, IX)

leads to Newton's equilibrium equation

$$\Lambda^* \bar{p} = -\frac{\mathrm{d}}{\mathrm{d}t}\bar{p} = f(t) \ \ \forall t \in I. \tag{2.4}$$

Substitution of the physical equation $p = m\Lambda q$ enables the dynamic equation in this dynamical system takes the same symmetrical form

$$A(q) = \Lambda^* m \Lambda q = -\frac{\mathrm{d}}{\mathrm{d}t}(m\frac{\mathrm{d}}{\mathrm{d}t}q) = f. \tag{2.5}$$

When Newton saw an apple falling to the ground, he discovered the great equation (2.4). However, he implicitly assumed the mass of the apple to be unchanged. Thus, for $m = m_0$ the nice fundamental formula (2.5) reduces to the well-known equation

$$m_0 q_{,tt} + f = 0 \tag{2.6}$$

in Newton's mechanical system. The solution of this linear equation is unique for given boundary (or initial) conditions.

2.1.2 Einstein's Relativity

For Einstein, the mass of a moving particle is no longer a constant and is given by

$$m = \frac{m_0}{\sqrt{1 - v^2/c^2}}, \tag{2.7}$$

where m_0 is the mass of the particle at rest, c is the speed of light (the universal speed limit). In this case, the fundamental equation (2.5) gives *Einstein's relativistic equation*

$$A(q) = -\frac{\mathrm{d}}{\mathrm{d}t}\left(\frac{m_0 c}{\sqrt{c^2 - \left(\frac{\mathrm{d}}{\mathrm{d}t}q\right)^2}}\frac{\mathrm{d}}{\mathrm{d}t}q\right) = f. \tag{2.8}$$

Here $A : \mathcal{U} \to \mathcal{U}^*$ is a nonlinear operator. It is easy to check that A is not a self-adjoint operator. Similarly to the finite deformation string system, the symmetry is broken in the Einstein theory. However, the diagrammatic framework for this dynamical system is exactly the analogue of that in Fig. 1.2.

Similarly to the potential energy $U(\epsilon)$ stored in the elastic string, the kinetic energy density stored in this moving particle is defined by

$$K(v) = \int_0^v p(v)dv.$$

For Newton's theory, $m = m_0$, $K(v)$ and its Legendre conjugate $K^c(p)$ are quadratic functions on $\mathcal{E} = \mathcal{E}^* = \mathcal{L}^2(I)$ given by

$$K(v) = \frac{1}{2}m_0 v^2, \ \ K^c(p) = \frac{1}{2m_0}p^2.$$

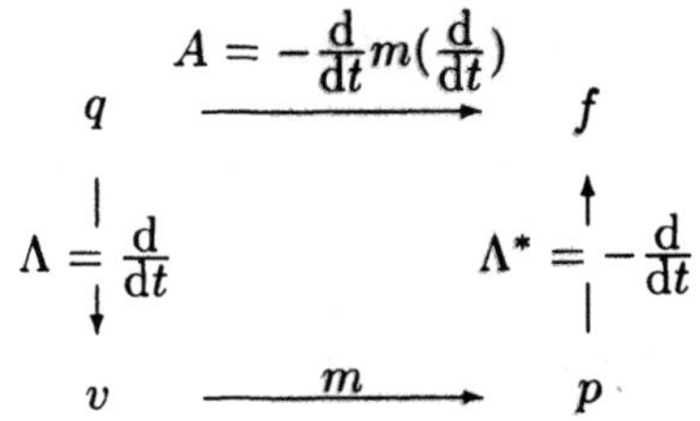

Figure 2.2. Diagrammatic framework representation in particle dynamics

Thus, (v, p) is a canonic duality pair on $\mathcal{E} \times \mathcal{E}^*$, and the Legendre duality relation

$$p = m_0 v \;\Leftrightarrow\; v = p/m_0 \;\Leftrightarrow\; K(v) + K^c(p) = pv \tag{2.9}$$

holds on $\mathcal{L}^2 \times \mathcal{L}^2$. However, for Einstein's theory the kinetic energy density

$$K(v) = -m_0 c\sqrt{c^2 - v^2} \tag{2.10}$$

is nonlinear and defined on a subset of $\mathcal{L}^\infty$, namely

$$\mathcal{E}_a = \{v \in \mathcal{L}^\infty(I) |\; v(t) < c \;\; \forall t \in I\}. \tag{2.11}$$

Although the total kinetic energy

$$W(v) = \int_I -m_0 c\sqrt{c^2 - v^2}\mathrm{d}t$$

is still a strictly convex functional on $\mathcal{E}_a$, the physical equation is subjected to a speed limit constraint:

$$p = \frac{m_0 c}{\sqrt{c^2 - v^2}} v, \quad \text{s.t.} \quad v(t) - c < 0 \;\; \forall t \in I. \tag{2.12}$$

The Legendre-Fenchel conjugate function of K is

$$K^*(p) = \sup_{v \in \mathcal{E}_a} \{pv - K(v)\} = c\sqrt{m_0^2 c^2 + p^2} = K^c(p), \tag{2.13}$$

which is also convex on $\mathcal{E}^* \subset \mathcal{L}^1(I)$. The inverse physical law for Einstein's theory is then

$$v = \frac{cp}{\sqrt{m_0^2 c^2 + p^2}}. \tag{2.14}$$

The symmetry in the Legendre duality relation is broken again for "fast" motion in dynamical systems. It is interesting to note that in static finite deformation theory the dual constitutive law is subjected to the plastic yield condition. However, in this dynamic relativistic theory, the primal constitutive law is subjected to a speed limit constraint. At $v = 0$, we have the best-known Einstein formula

$$E := -K(0) = m_0 c^2.$$

2.2 Convex Hamiltonian Systems

2.2.1 Total Action and Hamilton Principle

The force f is said to be a *conservative field* if there exists a potential function $U(q)$ such that

$$f = DU(q) = \frac{\mathrm{d}U}{\mathrm{d}q}. \tag{2.15}$$

For example, in a mass-spring system, the potential U stored in a linear spring is a quadratic function of the displacement q. Thus,

$$U(q) = \frac{1}{2}k_o q^2,$$

where $k_o > 0$ is the elastic constant. Then, the *action density* is defined by

$$V(q,v) = K(v) - U(q). \tag{2.16}$$

Generally speaking, for any given dynamical system, let $P : \mathcal{U}_a \subset \mathcal{U} \to \mathbb{R}$ be the total potential energy, and $W : \mathcal{E}_a \subset \mathcal{E} \to \mathbb{R}$ the total stored kinetic energy, and let $\Lambda : \mathcal{U} \to \mathcal{E}$ be a linear differential operator. Then, on the *kinematically admissible space*

$$\mathcal{U}_k = \{q \in \mathcal{U}_a | \ \Lambda q \in \mathcal{E}_a\},$$

the total action $\Pi : \mathcal{U}_k \to \mathbb{R}$ of the system is defined by

$$\Pi(q) = W(\Lambda q) - P(q) = \int_I (K(\Lambda q) - U(q)) \ \mathrm{d}t, \tag{2.17}$$

where

$$W(v) = \int_I K(v)\mathrm{d}t, \quad P(q) = \int_I U(q)\mathrm{d}t.$$

Clearly, if $W(v)$ and $P(q)$ are convex, Then, the nonconvex total action Π is a *difference convex functional* (d.c. functional).

The Gâteaux variation of Π at $\bar{q} \in \mathcal{U}_k$ in the direction q is

$$\begin{aligned}
\delta\Pi(\bar{q};q) &= \int_I \left(\frac{\partial K(\Lambda\bar{q})}{\partial v}\frac{\mathrm{d}}{\mathrm{d}t}q - \frac{\partial U(\bar{q})}{\partial q}q \right) \mathrm{d}t \\
&= \langle \Lambda q; DK(\Lambda\bar{q})\rangle - \langle q, DU(\bar{q})\rangle \\
&= \langle q, \Lambda^* DK(\Lambda\bar{q}) - DU(\bar{q})\rangle + DK(\Lambda\bar{q})q(t)|_{t=0}^{t_1}.
\end{aligned}$$

Theorem 2.2.1 (Hamilton's Principle) *Among all kinematically admissible motions $q \in \mathcal{U}_k$, the real motion $\bar{q}$ of the system is a stationary point of Π, i.e.*

$$(\Pi_{\mathrm{sta}}): \quad \delta\Pi(\bar{q};q) = 0 \ \ \forall q \in \mathcal{U}_k. \tag{2.18}$$

This is the form in which the principle was stated in Hamilton's original work (see Hamilton, 1835). By the fundamental lemma, the stationary condition

$$\delta\Pi(\bar{q};q) = 0 \ \ \forall q \in \mathcal{U}_k \tag{2.19}$$

leads to the *Euler-Lagrange equation*

$$\Lambda^* DW(\Lambda\bar{q}) - DP(\bar{q}) = 0. \tag{2.20}$$

The solution $\bar{q}(t)$ of this equation is called the *trajectory* of the system.

Remark 2.2.1 In analytic mechanics, the action density $V(q,v)$ is also called the Lagrangian density and denoted by $L(q,v)$. Thus, Euler's equation $D\Pi(\bar{q}) = 0$ in dynamical systems is also known as the *Euler-Lagrange equation.* Since the notation L is reserved for a general Lagrangian in duality theory, we use V for the Lagrangian density in dynamical systems. ■

2.2.2 Hamiltonian and Canonical Forms

The classical Hamiltonian associated with the nonconvex action Π is defined by

$$H(q,p) := \int_I [K^c(p) + U(q)]\mathrm{d}t, \tag{2.21}$$

which is, however, a convex functional! The *Hamiltonian canonical forms* in this dynamical system are

$$\begin{aligned} \Lambda q &= D_p H(q,p) = DK^c(p), \\ \Lambda^* p &= D_q H(q,p) = DU(q). \end{aligned} \tag{2.22}$$

We let

$$z = \begin{pmatrix} q \\ p \end{pmatrix}, \quad \mathbf{J}^{\oplus} = \begin{pmatrix} 0 & 1 \\ 1 & 0 \end{pmatrix}. \tag{2.23}$$

On the space $\mathcal{Z} := \mathcal{U} \times \mathcal{E}^*$, the Hamiltonian can be written as $H(z) = H(q,p) : \mathcal{Z} \to \mathbb{R}$. We write

$$Dz := \begin{pmatrix} \Lambda q \\ \Lambda^* p \end{pmatrix}, \quad DH(z) := \begin{pmatrix} D_q H(q,p) \\ D_p H(q,p) \end{pmatrix}. \tag{2.24}$$

Then, the Hamilton canonical forms can be written in the more abstract form

$$Dz = \mathbf{J}^{\oplus} DH(z). \tag{2.25}$$

If $\Lambda = \mathrm{d}/\mathrm{d}t$ and $\Lambda^* = -\mathrm{d}/\mathrm{d}t$, this abstract form leads to the *classical Hamiltonian canonical forms*

$$\dot{q} = D_p H(q,p), \quad \dot{p} = -D_q H(q,p). \tag{2.26}$$

In this case, since

$$\frac{\mathrm{d}}{\mathrm{d}t} H(q,p) = D_q H(q,p)\dot{q} + D_p H(q,p)\dot{p} = 0,$$

the Hamiltonian $H(q,p) = c$, a constant, along any trajectory. Moreover, if $\mathcal{U} = \mathbb{R}^n$, Then, $\mathcal{Z} = \mathcal{U} \times \mathcal{E}^* \in \mathbb{R}^{2n}$. Let

$$\dot{z} = \begin{pmatrix} \dot{q} \\ \dot{p} \end{pmatrix}, \quad \mathbf{J}^{\ominus} = \begin{pmatrix} 0 & \mathbf{I} \\ -\mathbf{I} & 0 \end{pmatrix}, \tag{2.27}$$

where $\mathbf{I} \in \mathbb{R}^{n\times n}$ is an identity matrix, Then, the classical Hamiltonian canonical forms can be written as

$$\dot{z} = \mathbf{J}^{\ominus} DH(z). \tag{2.28}$$

We note that $\mathbf{J}^{\ominus} \in \mathbb{R}^{2n\times 2n}$ is the so-called *symplectic matrix*, and satisfies the equations

$$(\mathbf{J}^{\ominus})^T = -\mathbf{J}^{\ominus} = (\mathbf{J}^{\ominus})^{-1}, \quad (\mathbf{J}^{\ominus})^T \mathbf{J}^{\ominus} \mathbf{J}^{\ominus} = \mathbf{J}^{\ominus}.$$

More general (cf., e.g., Arnold, 1978; Ekeland, 1990), a matrix $\mathbf{M} \in \mathbb{R}^{2n\times 2n}$ is said to be symplectic if

$$\mathbf{M}^T \mathbf{J}^{\ominus} \mathbf{M} = \mathbf{J}^{\ominus}.$$

Definition 2.2.1 (Hamiltonian Systems and Phase Space)

(D1) A dynamical system is said to be a *Hamiltonian system* if there exists a functional $H : \mathcal{Z} \to \mathbb{R}$ such that the system is controlled by either (2.28) or (2.25).

(D2) A dynamical system is said to be a *convex Hamiltonian system* if the Hamiltonian $H(z) : \mathcal{Z} \to \mathbb{R}$ is a convex functional.

(D3) The product space $\mathcal{Z} = \mathcal{U} \times \mathcal{E}^*$ in a Hamiltonian system is called the *phase space* in dynamics. ◇

Remark 2.2.2 It is traditional in dynamical systems that a system is said to be *conservative* if it is controlled by the classical canonical form (2.28). As we will see in the next section that by defining suitable bilinear forms, the generalized Hamilton canonical form (2.25) can also be used for certain *dissipative systems*. Thus, we refer those systems which are controlled by (2.25) as the *generalized Hamiltonian systems*. ■

Similarly to the static elastic string problem, for Newton's theory, the duality relation is one-to-one and symmetrical:

$$p = DK(v) \;\Leftrightarrow\; v = DK^*(p) \;\Leftrightarrow\; K(v) + K^*(p) = vp. \tag{2.29}$$

The Euler-Lagrange equation and the Hamiltonian canonical form are equivalent.

But, for Einstein's theory, the primal physical law is subjected to an inequality constraint $v < c$. The classical Hamiltonian system is not equivalent to the primal Euler-Lagrange equation in this case.

2.3 Least Action Principle: Legendre and Jacobian Condition

2.3.1 Euler Equation and Legendre's Condition

Similarly to the first problem for static systems, the generalized total action can be written as

$$\Pi(q) = \int_I \jmath(t, q(t), \Lambda q(t)) \mathrm{d}t, \tag{2.30}$$

where $\jmath : I \times \mathcal{U} \times \mathcal{E} :\to \mathbb{R}$ is an action density. The system is said to be *autonomous* if $\jmath$ does not contain the independent variable t explicitly. For given subspaces $\mathcal{U}_a \subset \mathcal{U}$ and $\mathcal{E}_a \subset \mathcal{E}$, we assume that $\jmath(t, q, v)$ satisfies the standard requirement

$$\jmath \in \mathcal{C}^2(I \times \mathcal{U}_a \times \mathcal{E}_a; \mathbb{R}). \tag{2.31}$$

Let $\mathcal{U}_k \subset \mathcal{U}$ be the kinematically admissible space

$$\mathcal{U}_k = \{q \in \mathcal{U}_a | \ \Lambda q \in \mathcal{E}_a\}. \tag{2.32}$$

In particular, for C^1 problems with homogeneous time boundary conditions, we write

$$\mathcal{U}_k^0 = \{q \in C^1(I) | \ q(t_0) = q(t_1) = 0\}. \tag{2.33}$$

In dynamical systems, since the total action Π is usually a nonconvex functional, the necessary conditions are little different from those in static systems.

Theorem 2.3.1 (Euler's Equation and Legendre's Condition)
Let $\mathcal{U}_k$ be an open subset of a Banach space $\mathcal{U}_a$, $\jmath : I \times \mathcal{U}_a \times \mathcal{E}_a \to \mathbb{R}$ a given action density satisfying the standard assumption (2.31), and assume $D^2\Pi(\bar{q}) \neq 0$. If a trajectory $\bar{q}(t) \in \mathcal{U}_k$ is a local minimum of Π, then, the following conditions hold.

(T1) Euler's equation for a path of stationary action:

$$\jmath_{,q}(t, \bar{q}, \Lambda\bar{q}) + \Lambda^* \jmath_{,v}(t, \bar{q}, \Lambda\bar{q}) = 0 \ \ \forall t \in I. \tag{2.34}$$

(T2) Legendre's condition

$$\jmath_{,vv} = \frac{\partial^2 \jmath}{\partial(\Lambda q)^2}(t, \bar{q}, \Lambda\bar{q}) \geq 0 \ \ \forall t \in I. \tag{2.35}$$

Proof. We already know that if $\bar{q}$ is a local minimum on an open subset $\mathcal{U}_k$, Then, $\bar{q}$ must be a critical point of Π. By definition, $D\Pi(\bar{q}) = 0$ follows the Euler's equation. To prove the Legendre's condition, we need the following Lemma.

Lemma 2.3.1 *Let $f_j : I \to \mathbb{R}$ $(j = 1, 2, 3)$ be continuous functions on I. If the quadratic form*

$$Q(q) = \int_I \left(f_1(t)q(t)^2 + f_2(t)q(t)q'(t) + f_3(t)q'(t)^2\right) dt \geq 0 \ \ \forall q \in \mathcal{U}_k,$$

then, the function $f_3(t)$ is nonnegative on I, i.e. $f_3(t) \geq 0 \ \forall t \in I$.

The proof of this lemma can be found in Blanchard & Brüning (1992, p. 84). Let $f_1 = \jmath_{,qq}$, $f_2 = 2\jmath_{,qv}$ and $f_3 = \jmath_{,vv}$; then $\delta^2\Pi(\bar{q}; q) = Q(q)$. If $\bar{q}$ is a local minimizer of Π, then $\delta^2\Pi(\bar{q}; q) \geq 0 \ \forall q \in \mathcal{U}_k$, i.e. Π is locally convex at $\bar{q}$. Thus, the lemma leads to Legendre's condition (2.35). □

2.3.2 Jacobi Condition and Least Action Principle

According to Theorem 1.2.4, the following conditions are sufficient for the trajectory $\bar{q}(t)$ in $\mathcal{U}_k$ to realize a strict relative minimum of the total action Π.

(a) Euler's equation: $D\Pi(\bar{q}) = 0$,
(b) $\delta^2\Pi(\bar{q}; q, q) > 0 \ \forall q \in \mathcal{U}_k$.

To translate these general conditions into actual requirements on the action density $\jmath$ in dynamical systems, we need to calculate the second-order Gâteaux variation of Π at $\bar{q} \in \mathcal{U}_k$, that is

$$J(q) = \delta^2\Pi(\bar{q}; q^2) = \int_I A_{\bar{q}}(t) q^2 \mathrm{d}t + \int_I B_{\bar{q}}(t)(\Lambda q)^2 \mathrm{d}t, \tag{2.36}$$

where the two functions $A_{\bar{q}}$ and $B_{\bar{q}}$ are calculated by

$$A_{\bar{q}}(t) = \frac{\partial^2 \jmath}{\partial q^2}(t, \bar{q}(t), \Lambda \bar{q}(t)) + \Lambda^* \frac{\partial^2 \jmath}{\partial q \partial v}(t, \bar{q}(t), \Lambda \bar{q}(t)), \tag{2.37}$$

$$B_{\bar{q}}(t) = \frac{\partial^2 \jmath}{\partial v^2}(t, \bar{q}(t), \Lambda \bar{q}(t)) \tag{2.38}$$

For any given functions $A(t)$ and $B(t)$, the convexity of the functional $J(q)$ depends on the following Jacobi condition.

Definition 2.3.1 (Jacobi's Condition)
Let $A \in \mathcal{C}(\bar{I})$ and $B \in \mathcal{C}^1(\bar{I})$ be two real functions on the closed domain $\bar{I} = [t_0, t_1]$. Then, $A(t)$ and $B(t)$ are said to satisfy *Jacobi's condition* if the differential equation

$$\Lambda^* (B(t) \Lambda u) + A(t) u = 0 \tag{2.39}$$

has a nontrivial solution $u(t) \in \mathcal{C}^2(\bar{I})$ such that $u(t) \neq 0 \ \forall t \in \bar{I}$. ◊

Theorem 2.3.2 (Legendre-Jacobi's Convexity Conditions)
(T1) Necessary Condition: suppose two functions $A \in \mathcal{C}(\bar{I})$ and $B \in \mathcal{C}^1(\bar{I})$ satisfy

(C1) The strict Legendre condition: $B(t) > 0 \ \forall t \in \bar{I}$;
(C2) Jacobi's condition.

Then, the functional

$$J(q) = \int_I \left[A(t) q^2(t) + B(t)(\Lambda q(t))^2 \right] dt \tag{2.40}$$

is strictly positive on $\mathcal{U}_k^0$, i.e.

$$J(q) > 0 \ \forall q \in \mathcal{U}_k^0, \ \ q \neq 0.$$

(T2) Sufficient Conditions: conversely, if $A \in \mathcal{C}(\bar{I})$ and $B \in \mathcal{C}^1(\bar{I})$ satisfy

(C1') $B(t) > 0 \ \forall t \in \bar{I}$;
(C2') the functional $J(q)$ is strictly positive on $\mathcal{U}_k^0$.

Then, $A(t)$ and $B(t)$ satisfy Jacobi's condition on $\bar{I}$.

The proof of this theorem can be found in Blanchard & Brüning (1992, p. 90-93).

Corollary 2.3.1 *Suppose two functions $A \in C(\bar{I})$ and $B \in C^1(\bar{I})$ satisfy*
(C1) $B(t) > 0 \ \forall t \in \bar{I}$;
(C2) $J(q)$ is positive on $\mathcal{U}_k^0$, i.e., $J(q) \geq 0 \ \forall q \in \mathcal{U}_k^0$.
Then, the functions $A(t)$ and $B(t)$ satisfy Jacobi's condition on the open interval I.

A particular simplification of Jacobi's condition proposed by Jacobi himself is important in applications.

Theorem 2.3.3 (Jacobi's Equivalent Theorem)
Let $A \in C(\bar{I})$ and $B \in C^1(\bar{I})$ be two real functions on the closed interval $\bar{I} = [t_0, t_1]$. The following statements are equivalent.
(S1) A and B satisfy Jacobi's condition on $\bar{I}$.
(S2) The solution of the initial-value problem

$$(\mathcal{P}_{\text{iv}}): \quad \begin{cases} \Lambda^* (B(t)\Lambda u) + A(t)u = 0 \ \forall t \in I, \\ u(t_0) = 0, \ \ \Lambda u(t_0) = 1 \end{cases} \tag{2.41}$$

has no zero in the semiopen domain $(t_0, t_1]$.

To sum up the results so far, we can formulate the following important theorem in dynamical systems.

Theorem 2.3.4 (Principle of Least Action)
(T1) Necessary Conditions: suppose that $\bar{q} \in \mathcal{U}_k$ is a trajectory for which the action Π attains a local minimum, $D^2\Pi(\bar{q}) \neq 0$ and $A_{\bar{q}}(t)$, $B_{\bar{q}}(t)$ are the associated functions calculated by (2.37) and (2.38), respectively; then, the following conditions hold.

(C1) Euler's equation: $\Lambda^* \jmath_{,v}(t, \bar{q}, \Lambda\bar{q}) + \jmath_{,q}(t, \bar{q}, \Lambda\bar{q}) = 0$;
(C2) Legendre's condition: $B_{\bar{q}}(t) = \jmath_{,vv}(t, \bar{q}(t), \Lambda\bar{q}(t)) \geq 0 \ \forall t \in \bar{I}$;
(C3) Jacobi's condition: whenever $A_{\bar{q}} \in C(\bar{I})$ and $B_{\bar{q}} \in C^1(\bar{I})$, equation (2.39) has a solution $u \in C^2(\bar{I})$, which vanishes nowhere on the (semiopen) domain $(t_0, t_1]$.

(T2) Sufficient Conditions: suppose that a trajectory $\bar{q} \in \mathcal{U}_k$ satisfies

(C1') Euler's equation;
(C2') Legendre's condition;
(C3') $A_{\bar{q}} \in C(\bar{I})$ and $B_{\bar{q}} \in C(\bar{I})$ satisfy Jacobi's condition on $\bar{I}$, and $B_{\bar{q}}(t) > 0 \ \forall t \in \bar{I}$.

Then, the action Π has a strict local minimum at the trajectory $\bar{q}$.

Remark 2.3.1 (Least Action Principle in Periodic Systems)
The least action principle shows that if a trajectory of equilibrium motion minimizes the total action, Jacobi's condition is one of the necessary and sufficient conditions. In periodic dynamical systems with period T, the solution

of the initial-value problem (2.41) usually vanishes within a period. Thus, Jacobi's condition does not hold on the domain $(t_0, t_0 + T)$, and then the least action principle cannot be used alone to study periodic dynamical systems. Its dual principle needs to be introduced. Detailed discussion will be given in later sections. ∎

2.4 Initial-Value Problems and Dissipative Hamiltonian Systems

By choosing suitable bilinear forms, many initial-value problems in dynamical systems can also be set within our unified framework, and the classical Hamiltonian theory can be generalized into nonlinear dissipative systems.

2.4.1 Bilinear Forms in Initial-Value Problems

In order to study initial-value problems, we let

$$\mathcal{U}_a = \{q(t) \in \mathcal{C}^1(0, t_1) | \; q(0) = 0, \; \dot{q}(0) = 0\}. \tag{2.42}$$

On $\mathcal{U}$ we introduce the bilinear form

$$\langle u, v \rangle := \int_0^t u(\tau) v(t - \tau) \mathrm{d}\tau = (u * v)(t) \quad \forall t \in [0, t_1], \tag{2.43}$$

which is called the *convolution* of u with v. Then, the space $\{\mathcal{U}; \langle *, * \rangle\}$ is a Hilbert space. On this space,

$$\begin{aligned} \langle \frac{\mathrm{d}}{\mathrm{d}t} u, v \rangle &= \int_0^t \frac{\mathrm{d}}{\mathrm{d}\tau} u(\tau) v(t - \tau) \mathrm{d}\tau \\ &= \int_0^t -u(\tau) \frac{\mathrm{d}}{\mathrm{d}\tau} v(t - \tau) \mathrm{d}\tau + u(\tau) v(t - \tau)|_{\tau=0}^t \\ &= \int_0^t u(\tau) \frac{\mathrm{d}}{-\mathrm{d}\tau} v(t - \tau) \mathrm{d}\tau \;\; \forall u, v \in \mathcal{U}_a \\ &= \int_0^t u(\tau) \frac{\mathrm{d}}{\mathrm{d}(t - \tau)} v(t - \tau) \mathrm{d}\tau \;\; \forall u, v \in \mathcal{U}_a \\ &= \langle u, \frac{\mathrm{d}}{\mathrm{d}t} v \rangle \quad \forall u, v \in \mathcal{U}_a, \end{aligned} \tag{2.44}$$

so the operator $\Lambda = \frac{\mathrm{d}}{\mathrm{d}t}$ is self-adjoint. It can easily be verified that $\Lambda = \frac{\mathrm{d}^2}{\mathrm{d}t^2}$ is also self-adjoint.

Let us consider the linear initial-value problem

$$(\mathcal{P}_{\mathrm{iv}}) : \quad \begin{cases} m_0 q_{,tt} + k_0 q = f(t) \;\; \forall t \in I = (0, t), \\ q(0) = q_{,t}(0) = 0. \end{cases} \tag{2.45}$$

We let $\mathcal{U} = \mathcal{E}$ be the Hilbert space with the bilinear form (2.43). The admissible space $\mathcal{U}_k$ for this initial-value problem can be defined by

$$\mathcal{U}_k = \{q \in \mathcal{H}(0, t) | \;\; q_{,t} \in \mathcal{H}, \;\; q(0) = 0, \;\; q_{,t}(0) = 0\}.$$

On $\mathcal{U}_k$, the total action is a convex functional of the independent variable t and the dependent variable $q(t)$

$$\Pi(t, q(t)) = \frac{1}{2}m_0\langle(\Lambda q), \Lambda q\rangle + \frac{1}{2}k_0\langle q, q\rangle - \langle f, q\rangle$$

$$= \int_0^t \left(\frac{1}{2}m_0 q_{,\tau}(\tau)q_{,(t-\tau)}(t-\tau) + \frac{1}{2}k_0 q(\tau)q(t-\tau) - f(\tau)q(t-\tau)\right) \mathrm{d}\tau.$$

It is easy to prove that, for any given $t > 0$, the criticality condition $D\Pi(t, \bar{q}; q) = 0 \ \forall q \in \mathcal{U}_k$ leads back to the $(\mathcal{P}_{iv})$. This initial-value problem is represented in Fig. 2.3 within our framework.

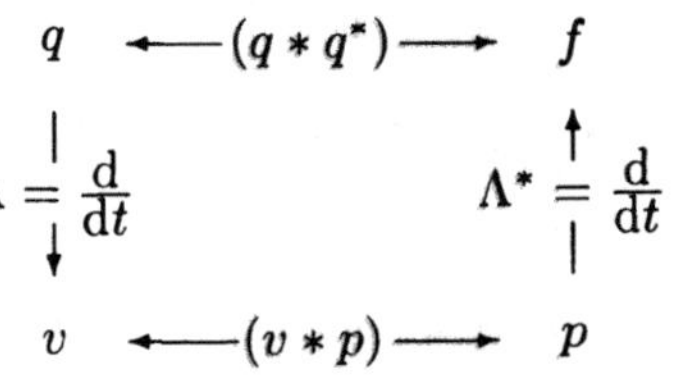

Figure 2.3. Diagrammatic representation for initial-value problem.

2.4.2 Framework for Linear and Nonlinear Dissipative Systems

Generally speaking, in conservative systems, the equilibrium equations are usually of even order. However, in dissipative systems, some odd-order terms may appear in the governing equations. By introducing appropriate bilinear forms, we are still able to put the dissipative systems within the unified framework. We consider the following cases.

I. Linear Dissipative System.

In a one-dimensional mass-spring system with a linear damper, the equation of motion (2.5) for Newton's theory reads

$$m\frac{d^2}{dt^2}q + \mu m\frac{\mathrm{d}}{\mathrm{d}t}q + f = 0, \tag{2.46}$$

where $\mu > 0$ is a constant. If we define the weighted bilinear form

$$\langle p, v\rangle := \int_I e^{\mu t} p(t)v(t)\mathrm{d}t, \quad \mu \in \mathbb{R}, \tag{2.47}$$

then integration by parts leads to

$$\begin{aligned}\langle p, \Lambda q\rangle &= \int_I e^{\mu t}\left(-\frac{\mathrm{d}}{\mathrm{d}t}p - \mu p\right) q\mathrm{d}t + \int_I d\left(e^{\mu t}p(t)q(t)\right) \\ &= \langle \Lambda^* p, q\rangle + e^{\mu t}p(t)q(t)|_{\partial I}.\end{aligned}$$

In this case, the formal adjoint Λ^* of $\Lambda = \frac{\mathrm{d}}{\mathrm{d}t}$ is now the affine operator

$$\Lambda^* = -\frac{\mathrm{d}}{\mathrm{d}t} - \mu.$$

The kinetic energy density in this dissipative system is still a quadratic function $K(v) = \frac{1}{2}mv^2$, and equation (2.46) can still be written in tri-canonical form:

$$\begin{array}{ll} \text{(a) kinetic equation :} & v = \Lambda q = \frac{\mathrm{d}}{\mathrm{d}t}q; \\ \text{(b) physic equation :} & p = DK(v) = mv; \\ \text{(c) balance equation :} & f = \Lambda^* p = -\frac{\mathrm{d}}{\mathrm{d}t}p - \mu p. \end{array}$$

The abstract fundamental equation is then

$$Aq = \Lambda^* m \Lambda q = f.$$

On the open interval $I = (0, t_1)$, the total action in this dissipative problem is

$$\Pi(q) = \int_I e^{\mu t} \left(\frac{1}{2} m (\frac{\mathrm{d}}{\mathrm{d}t} q)^2 - fq \right) \mathrm{d}t. \tag{2.48}$$

The criticality condition for Π leads to the equilibrium equation (2.46). The Hamiltonian for this dissipative problem is still a convex functional, namely

$$H(q,p) = \int_I e^{\mu t} \left(\frac{1}{2m} p^2 + fq \right) \mathrm{d}t, \tag{2.49}$$

and the Hamiltonian canonical forms are

$$Dz = \mathbf{J}^{\oplus} DH(z) \Rightarrow \quad \Lambda q = D_p H(q,p), \quad \Lambda^* p = D_q H(q,p).$$

So, this dissipative system is a convex Hamiltonian system. But the two Hamiltonian canonical forms (2.28) and (2.25) are not equivalent.

II. Dissipative Initial-Value Problem.

Consider the initial-value problem

$$(\mathcal{P}_{\mathrm{iv}}): \quad \left\{ \begin{array}{l} mq'' + \mu m q' + kq = f(t) \ \ \forall t > 0, \\ q(0) = q'(0) = 0. \end{array} \right. \tag{2.50}$$

We let $\mathcal{U} = \mathcal{E}$ be a Hilbert space $\mathcal{H}$ and $\Lambda = \frac{\mathrm{d}}{\mathrm{d}t}$. The bilinear form can be defined by

$$\langle v, p \rangle = \int_0^t e^{-\mu\tau} v(\tau) p(t-\tau) \mathrm{d}\tau. \tag{2.51}$$

On the kinematically admissible space

$$\mathcal{U}_k = \{ q \in \mathcal{H}(0,t) | \ q' \in \mathcal{H}, \ q(0) = 0, \ q'(0) = 0 \},$$

we have

$$\begin{aligned}\langle \Lambda q, v\rangle &= \int_0^t e^{-\mu\tau}\frac{\mathrm{d}}{\mathrm{d}\tau}q(\tau)v(t-\tau)\mathrm{d}\tau \\ &= \int_0^t e^{-\mu\tau}q(\tau)\left(\mu+\frac{\mathrm{d}}{\mathrm{d}(t-\tau)}v(t-\tau)\right)\mathrm{d}\tau \\ &= \langle q, (\mu+\Lambda v)\rangle.\end{aligned}$$

Thus, the adjoint operator Λ^* of Λ associated with this bilinear form is $\Lambda^* = \Lambda + \mu$. Let

$$P(q) = \langle f, q\rangle - \frac{1}{2}k\langle q, q\rangle = \int_0^t e^{-\mu\tau}\left(f(\tau)q(t-\tau) - \frac{1}{2}kq(\tau)q(t-\tau)\right)\mathrm{d}\tau$$

be the total potential. Then, the (*calPiv*) can be written in the fundamental form

$$\Lambda^* m\Lambda q = DP(q).$$

III. **Nonlinear Dissipative System.**

In a dynamical system subject to a quadratic friction, the governing equation is nonlinear:

$$m\frac{d^2}{dt^2}q + m\mu(\frac{\mathrm{d}}{\mathrm{d}t}q)^2 + f = 0. \tag{2.52}$$

In this case, we can introduce a bracket

$$\langle p, v\rangle := \int_I e^{2\mu q}p(t)v(t)\mathrm{d}t, \quad \mu \in \mathbb{R}, \tag{2.53}$$

which is no longer a bilinear form. Let $\Lambda = \mathrm{d}/\mathrm{d}t$; then, integration by parts leads to

$$\langle p, \Lambda q\rangle = \langle \Lambda^* p, q\rangle + e^{2\mu q}pq|_{\partial I},$$

where the formal adjoint Λ^* of Λ is a nonlinear operator depending on $q_{,t}$, namely

$$\Lambda^* = -\frac{\mathrm{d}}{\mathrm{d}t} - 2\mu q_{,t}. \tag{2.54}$$

The nonlinear dissipative equation (2.52) can be written in the same abstract form $A(q) = \Lambda^* \circ m \circ \Lambda(q) = f$. But the total action is defined by

$$\Pi(q) = \int_I \frac{1}{2}e^{2\mu q}\left(m(\frac{\mathrm{d}}{\mathrm{d}t}q)^2 - \frac{1}{\mu}f\right)\mathrm{d}t. \tag{2.55}$$

The Hamiltonian associated with this dissipative system is also a convex functional:

$$H(q, p) = \int_I \frac{1}{2}e^{2\mu q}\left(\frac{1}{m}p^2 + \frac{1}{\mu}f\right)\mathrm{d}t. \tag{2.56}$$

It is easy to check that

$$\delta H(q,p;\delta q,\delta p) = \int_I e^{2\mu q}\left(\frac{1}{m}p^2 + \frac{1}{\mu}f\right)\mu\delta q\mathrm{d}t + \int_I e^{2\mu q}\frac{1}{m}p\delta p\mathrm{d}t$$

Thus,

$$D_qH(q,p) = \left(\frac{1}{m}p^2 + \frac{1}{\mu}f\right)\mu, \quad D_pH(q,p) = \frac{1}{m}p.$$

Hence, the governing equation (2.52) can be written as

$$\dot{z} = \mathbf{J}^{\ominus}DH(z) \;\Rightarrow\; \dot{q} = D_pH(q,p), \quad \dot{p} = -D_qH(q,p),$$

and this nonlinear dissipative system is also a convex Hamiltonian system.

In general, if f is a potential field, $f = DU(q)$, the total action is a nonconvex functional

$$\begin{aligned}\Pi(q) &= \int_I [K(\Lambda q) - U(q))\,\mathrm{d}t \\ &= \int_I \left(\frac{1}{2}me^{2\mu q}(\frac{\mathrm{d}}{\mathrm{d}t}q)^2 - \int^q e^{2\mu x}DU(x)\,\mathrm{d}x\right)\mathrm{d}t. \end{aligned} \quad (2.57)$$

The associated Euler-Lagrange equation is then a nonlinear equation:

$$m\frac{\mathrm{d}^2}{\mathrm{d}t^2}q + m\mu(\frac{\mathrm{d}}{\mathrm{d}t}q)^2 + DU(q) = 0. \quad (2.58)$$

The Lagrangian associated with this general nonlinear dissipative equation can be written as

$$L(q,p) = \int_I e^{2\mu q}\left(pq' - \frac{1}{2m}p^2\right)\mathrm{d}t - \int_I \frac{1}{2}e^{2\mu q}\int^q e^{2\mu x}DU(x)\,\mathrm{d}x\mathrm{d}t. \quad (2.59)$$

It is easy to check that the Euler-Lagrange equation $DL(q,p) = 0$ is equivalent to the governing equation (2.58).

People do not understand how that which is a variance with itself agrees with itself. There is a harmony in the bending back, as in the case of the bow and the lyre.

—Heraclitus

2.5 Complementary Hamiltonian Principles and Euler-Lagrange Equations

As we have seen in dynamical systems, the total action Π is usually nonconvex. Thus, the Hamiltonian H plays an important role. In this section, we present complementary extremum action principles for the autonomous Hamiltonian system

$$Dz = \mathbf{J}^{\oplus} DH(z) \quad \Rightarrow \quad \begin{cases} \Lambda q = D_p H(q,p), \\ \Lambda^* p = D_q H(q,p). \end{cases} \tag{2.60}$$

Two problems associated with this Hamiltonian system can be proposed.

Problem 2.5.1 Let $\{(\mathcal{U},\mathcal{U}^*);\langle *,*\rangle\}$ and $\{(\mathcal{E},\mathcal{E}^*);\langle *;*\rangle\}$ be two pairs of inner product spaces, and let $\Lambda : \mathcal{U} \to \mathcal{E}$ be a linear, continuous operator. The adjoint$\Lambda^* : \mathcal{E}^* \to \mathcal{U}^*$ of Λ is defined by

$$\langle \Lambda q; p\rangle = \langle q, \Lambda^* p\rangle \ \ \forall (q,p) \in \mathcal{U}_a \times \mathcal{E}_a^* \subset \mathcal{U} \times \mathcal{E}^*.$$

Let the Hamiltonian $H : \mathcal{U}_a \times \mathcal{E}_a^* \to \mathbb{R}$ be a given Gâteaux differentiable functional.

(P1) **Initial-Value Problem.** For the given initial data $z_0 = (q_0, p_0) \in \mathcal{Z}_a$, and $t > 0$ such that $I = (0,t)$, find $z(t) = (q(t),p(t)) \in \mathcal{Z}_a$ satisfying the Hamiltonian canonical forms in (2.60) and the initial condition

$$z(0) = (q(0),p(0)) = (q_0,p_0) = z_0. \tag{2.61}$$

(P2) **Periodic Boundary Value Problem.** For a given $T > 0$ such that $I = (0,T)$, find $z(t) = (q(t),p(t)) \in \mathcal{Z}_a$ satisfying the Hamiltonian canonical forms in (2.60) and the periodic boundary condition

$$z(0) = (q(0),p(0)) = (q(T),p(T)) = z(T). \tag{2.62}$$

♠

We need the following assumption.

Assumption 2.5.1 (A1) The Hamiltonian $H : \mathcal{U}_a \times \mathcal{E}_a^* \to \mathbb{R}$ is twice Gâteaux differentiable and can be written as $H(q,p) = W^c(p) + P(q)$.

(A2) On the product spaces $\mathcal{U}_a \times \mathcal{U}_a^*$ and $\mathcal{E}_a \times \mathcal{E}_a^*$, the Legendre duality relations

$$u^* = DP(u) \ \Leftrightarrow \ u = DP^c(u^*) \ \Leftrightarrow \ P(u) + P^c(u^*) = \langle u, u^*\rangle;$$

$$p = DW(v) \ \Leftrightarrow \ v = DW^c(p) \ \Leftrightarrow \ W(v) + W^c(p) = \langle v; p\rangle,$$

respectively hold. ♣

2.5.1 Stabilities and Extremum Action Principles

Let $\mathcal{U}_a \subset \mathcal{U}$ be an admissible configuration space, in which, either the initial condition (2.61) or the periodic boundary condition (2.62) are prescribed. The kinematically admissible space $\mathcal{U}_k$ associated with the dynamical system (2.60) can be defined as

$$\mathcal{U}_k = \{q \in \mathcal{U}_a | \ \Lambda q \in \mathcal{E}_a\}. \tag{2.63}$$

The primal stationary variational problem associated with the Hamiltonian canonical forms (2.62) is

$$(\Pi_{\text{sta}}): \ \Pi(q) = W(\Lambda q) - P(q) \to \text{sta} \ \ \forall q \in \mathcal{U}_k. \tag{2.64}$$

It is easy to check that the criticality condition $\delta\Pi(\bar{q}; q) = 0 \ \ \forall q \in \mathcal{U}_k$ leads to the *Euler-Lagrange equation*

$$\Lambda^* DW(\Lambda q) - DP(q) = 0. \tag{2.65}$$

By the Legendre duality theorem, we know that this Euler-Lagrange equation is equivalent to the control equations in (2.60).

Since $\Pi : \mathcal{U}_k \to \mathbb{R}$ is nonconvex, two primal extremal problems in Hamiltonian systems can be proposed: to find $\bar{q} \in \mathcal{U}_k$ such that

$$(\Pi_{\text{inf}}): \qquad \Pi(\bar{q}) = \inf_{q \in \mathcal{U}_k} \Pi(q), \tag{2.66}$$

$$(\Pi_{\text{sup}}): \qquad \Pi(\bar{q}) = \sup_{q \in \mathcal{U}_k} \Pi(q). \tag{2.67}$$

Clearly, if $\bar{q}$ is a local minimizer of Π, then the kinetic energy $W(\Lambda\bar{q})$ attains its minimum, while the potential energy $P(\bar{q})$ attains its maximum. Dually, if $\bar{q}$ is a local maximizer of Π, then the kinetic energy $W(\Lambda\bar{q})$ attains its maximum, while the potential energy $P(\bar{q})$ attains its minimum (see Fig. 2.4). So we have the following definitions for locally stable states.

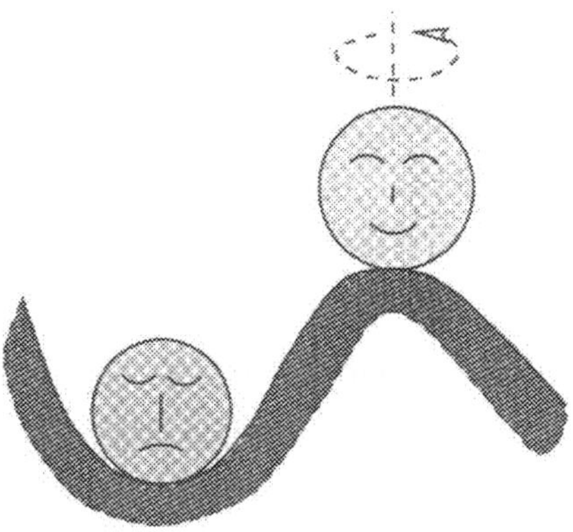

Figure 2.4. Duality in stability: potentially stable and kinematically stable

Definition 2.5.1 (Potential and Kinetic Stabilities)
Let $\mathcal{U}_k$ be an open set and $\bar{q}$ be a critical point of Π on $\mathcal{U}_k$.
(D1) An equilibrium motion $\bar{q}(t) : I \to \mathcal{U}_k$ is called a *potentially stable state* on I if $\bar{q}$ is a local maximizer of Π on $\mathcal{U}_k$.
(D2) An equilibrium motion $\bar{q}(t) : I \to \mathcal{U}_k$ is called a *kinematically stable state* on I if $\bar{q}$ is a local minimizer of Π on $\mathcal{U}_k$. ◊

Let $K(v) \in \mathcal{C}^2(\mathcal{E})$ and $U(q) \in \mathcal{C}^2(\mathcal{U})$ be the kinetic energy density and potential energy density, respectively, and

$$W(v) = \int_I K(v)\mathrm{d}t, \quad P(q) = \int_I U(q)\mathrm{d}t.$$

The least action principle then leads to the following result for Hamiltonian systems.

Theorem 2.5.1 (Locally Extremum Action Theorem)
Let $\mathcal{U}_k$ be an open subset and let $\bar{q}(t) \in \mathcal{U}_k$ be a critical point of Π such that the initial-value problem

$$(\mathcal{P}_{\mathrm{iv}}) : \begin{cases} \Lambda^* \left(K_{,vv}(\Lambda\bar{q})\Lambda u\right) - U_{,qq}(\bar{q})u = 0 \ \forall t \in (t_0, t_1), \\ u(t_0) = 0, \ \ \Lambda u(t_0) = 1, \end{cases} \tag{2.68}$$

has a solution $\bar{u}(t) \in \mathcal{C}^2$, which vanishes nowhere in $I = (t_0, t_1]$. Then,
(T1) if $K_{,vv}(\bar{q}) > 0 \ \forall t \in \bar{I}$, $\bar{q}(t)$ is a kinematically stable state on I;
(T2) if $K_{,vv}(\bar{q}) < 0 \ \forall t \in \bar{I}$, $\bar{q}(t)$ is a potentially stable state on I.

Proof. The functional $J(u)$ associated with $\Pi(q)$ at the equilibrium state $\bar{q}$ is

$$J(u) = \delta^2\Pi(\bar{q}; u^2) = \int_I \left(B_{\bar{q}}(t)(\Lambda u)^2 + A_{\bar{q}}(t)u^2\right) \mathrm{d}t,$$

with

$$A_{\bar{q}}(t) = -U_{,qq}(\bar{q}), \quad B_{\bar{q}}(t) = K_{,vv}(\Lambda\bar{q}).$$

Thus, the Jacobi equation associated with $A_{\bar{q}}$ and $B_{\bar{q}}$, i.e. the Euler equation of the variational problem

$$\delta J(\bar{u}; u) = 0 \ \forall u \in \mathcal{U}_k^0(\bar{I}),$$

is, accordingly,

$$\Lambda^* B_{\bar{q}}(t)\Lambda\bar{u} + A_{\bar{q}}(t)\bar{u} = 0. \tag{2.69}$$

By Theorem 2.3.3, if this equation has a nontrivial solution $\bar{u} \in \mathcal{C}^2(\bar{I})$ with $u(t_0) = 0$ and $\bar{u}(t) \neq 0 \ \forall t \in (t_0, t_1]$, then $A_{\bar{q}}$ and $B_{\bar{q}}$ satisfy Jacobi's condition. By Theorem 2.3.2, if $B_{\bar{q}}(t) = K_{,vv}(\Lambda\bar{q}) > 0$, $J(u)$ is strictly positive on $\mathcal{U}_k^0$. Thus, $\bar{q}$ minimizes Π on $\mathcal{U}_k$, and hence, is a kinematically stable state. If $B_{\bar{q}}(t) = K_{,vv}(\Lambda\bar{q}) < 0$, $J(u)$ is strictly negative on $\mathcal{U}_k^0$, and $\bar{q}$ maximizes Π on $\mathcal{U}_k$, and is therefore a potentially stable state. □

Example 2.5.1 Let $K(v) = \frac{1}{2}m_0 v^2$ and $U(q) = \frac{1}{2}k_0 q^2$, so that the total action for Newton's system is

$$\Pi(q) = \int_0^{t_1} \frac{1}{2}\left(m_0(\Lambda q)^2 - k_0 q^2\right) \mathrm{d}t.$$

The Jacobi equation (2.69) for this simple system is the Euler-Lagrange equation

$$u'' + \lambda_0 u = 0, \quad \lambda_0 = \frac{k_0}{m_0} \in \mathbb{R}$$

for Π. The solution for the initial conditions $u(0) = 0,\ u'(0) = 1$ is

$$u(t) = \begin{cases} \frac{1}{\omega_0}\sin(\omega_0 t) & \text{if } \lambda_0 > 0,\ \omega_0 = \sqrt{\lambda_0}, \\ \frac{1}{\omega_0}\sinh(\omega_0 t) & \text{if } \lambda_0 < 0,\ \omega_0 = \sqrt{-\lambda_0}. \end{cases}$$

For $\lambda_0 < 0$, $u(t)$ vanishes nowhere in $I = (0, t_1]$. If $m_0 = K_{,vv} > 0$, the trajectory $q(t)$ minimizes the action Π on I. If $m_0 < 0$, the trajectory $q(t)$ maximizes Π on I.

On the other hand, for $\lambda_0 > 0$, $u(t)$ is a periodic solution with period $T = 2\pi/\omega_0$. Two cases have to be distinguished:

(a) If $t_1 \in (0, \pi/\omega_0)$, the solution $u(t) \neq 0$ and the trajectory $q(t)$ minimizes the action Π on $(0, t_1]$ for $m_0 > 0$.

(b) If $t_1\omega_0 > \pi$, the corresponding solution $u(t)$ has at least one zero in $(0, t_1]$. In this case, the trajectory $q(t)$ does not minimize the action Π on $(0, t_1)$ for $m_0 > 0$. ♡

Theorem 2.5.1 and this example shows an important fact in dynamical systems: for periodic boundary-value problems, if the solution vanishes in the domain $I = (0, T)$, the actual motion does not minimize the total action.

Mathematically speaking, the "indefinite" functionals may not possess any local minima or maxima other than trivial ones. For example, let $I = (0, \pi)$,

$$\mathcal{U}_k = \{q \in \mathcal{L}^2(I) |\ q' \in \mathcal{L}^2(I),\ q(0) = q(\pi) = 0\},$$

and

$$\Pi(q) = \int_I \frac{1}{2}\left(q'(t)^2 - \frac{1}{2}q(t)^4\right) \mathrm{d}t.$$

It is easy to show that Π is Gâteaux differentiable on $\mathcal{U}_k$ and has $q = 0$ as a local minimum. For any other $q \in \mathcal{U}_k$, and $\theta \in \mathbb{R}$,

$$\Pi(\theta q) = \int_I \frac{1}{2}\left(\theta^2 q'(t)^2 - \frac{1}{2}\theta^4 q(t)^4\right) \mathrm{d}t \to -\infty$$

as $|\theta| \to \infty$, so Π is not bounded from below. Furthermore, for each $\omega \in \mathbb{R}$, $\sin\omega t \in \mathcal{U}_k$, we have

$$\Pi(\sin\omega t) \geq \frac{\pi}{4}\omega^2 - \frac{\pi}{4} \to \infty$$

as $\omega \to \infty$. So Π is not bounded from above either. Thus, it is not obvious, in the sense of the classical direct methods, that Π possesses any critical points other than the trivial one $q = 0$. To overcome this intrinsic difficulty in nonconvex primal problems, we need to study duality theory in dynamical systems. We will see that this nonconvex action Π indeed possesses positive critical values.

2.5.2 Lagrangian and Complementary Actions

The Lagrangian $L : \mathcal{U}_a \times \mathcal{E}_a^* \to \mathbb{R}$ associated with the Hamiltonian $H : \mathcal{U}_a \times \mathcal{E}_a^* \to \mathbb{R}$ is defined by

$$L(q,p) = \langle \Lambda q; p \rangle - H(q,p). \tag{2.70}$$

Clearly, the Lagrange equation $DL(q,p) = 0$ is equivalent to the Hamiltonian canonical forms

$$D_p L(q,p) = 0 \quad \Rightarrow \quad \Lambda q = D_p H(q,p), \tag{2.71}$$
$$D_q L(q,p) = 0 \quad \Rightarrow \quad \Lambda^* p = D_q H(q,p). \tag{2.72}$$

This shows that the critical points of the Lagrangian L are equilibrium states of the the Hamiltonian system.

In a similar way as for the Lagrange equations in static systems, if

$$\begin{aligned} H(q,p) &= W^c(p) + P(q) \\ &= \int_I \left(\frac{1}{2m_0} p^2 + \mu p \right) dt + \int_I \left(\frac{1}{2} k_0 q^2 - f q \right) dt \end{aligned} \tag{2.73}$$

is a quadratic functional, where μ and f are given inputs, the Lagrangian is then

$$L(q,p) = \int_I \left((\Lambda q) p - \frac{1}{2m_0} p^2 - \mu p - \frac{1}{2} k_0 q^2 + f q \right) dt.$$

The Lagrange equations $DL(z) = 0$ for this Newtonian dynamical system can be written in the matrix form:

$$D^2 L(z) z = \begin{pmatrix} -k_0 & \Lambda^* \\ \Lambda & -m_0^{-1} \end{pmatrix} \begin{pmatrix} q \\ p \end{pmatrix} = \begin{pmatrix} -f \\ \mu \end{pmatrix}. \tag{2.74}$$

The Hessian matrix $D^2 L(z)$ is indefinite for $m_0, k_0 > 0$.

Correspondently to the statically admissible space $\mathcal{S}_s$ in static systems, the feasible set associated with the Hamiltonian system (2.60) is defined by

$$\mathcal{S}_s = \{ p \in \mathcal{E}_a^* | \ \Lambda^* p \in \mathcal{U}_a^* \}, \tag{2.75}$$

which is called the *dynamically admissible space.* We use the same notation because the both spaces have the same mathematical structure. For a given $p \in \mathcal{S}_s$ such that $\Lambda^* p = DP(q)$, the Legendre duality theory leads to

$$\begin{aligned} L(q,p) &= \langle \Lambda q; p \rangle - H(q,p) = \langle q, \Lambda^* p \rangle - P(q) - W^c(p) \\ &= P^c(\Lambda^* p) - W^c(p). \end{aligned}$$

Thus, the *dual action* $\Pi^d : \mathcal{S}_s \to \mathbb{R}$ associated with Hamilton's canonical forms (2.60) is obtained naturally via the Lagrangian as

$$\Pi^d(p) = P^c(\Lambda^* p) - W^c(p). \tag{2.76}$$

Clearly, for convex Hamiltonian systems, Π^d is a nonconvex functional. Dual to Theorem 2.2.1 the *dual Hamilton principle* in convex dynamical system can be proposed as follows.

Theorem 2.5.2 (Dual Hamilton principle)
Among all dynamically admissible momenta $p \in \mathcal{S}_s$, the actual one $\bar{p}$ of the system is a stationary point of Π^d, i.e.

$$(\Pi^d_{\text{sta}}): \quad \delta\Pi^d(\bar{p}; p) = 0 \quad \forall p \in \mathcal{S}_s. \tag{2.77}$$

Proof. The criticality condition $\delta\Pi^d(\bar{p}; p) = 0 \quad \forall p \in \mathcal{S}_s$ leads to the dual Euler-Lagrange equation

$$\Lambda DP^c(\Lambda^*\bar{p}) - DW^c(\bar{p}) = 0. \tag{2.78}$$

Let $\bar{q} = DP^c(\Lambda^*\bar{p})$. By the Legendre duality theory, this is equivalent to $\Lambda^*\bar{p} = DP(\bar{q}) = D_qH(\bar{q},\bar{p})$. The dual Euler-Lagrange equation gives $\Lambda\bar{q} = DW^c(\bar{p}) = D_pH(\bar{q},\bar{p})$. Thus, the stationary points of Π^d satisfy the Hamilton canonical forms, and hence is an equilibrium state. □

Similarly to the complementary energy in static systems, the so-called *complementary action* : $\Pi^c : \mathcal{S}_s \to \mathbb{R}$ is defined simply by

$$\Pi^c(p) = -\Pi^d(p) = W^c(p) - P^c(\Lambda^*p). \tag{2.79}$$

At the balance state $(\bar{q}, \bar{p})$, the complementary relation

$$\Pi(\bar{q}) + \Pi^c(\bar{p}) = 0 \tag{2.80}$$

holds. Parallel to the primal extremum problems (Π_{inf}) and (Π_{sup}), the dual extremum problems in convex Hamiltonian systems can be proposed as

$$(\Pi^d_{\text{inf}}): \quad \Pi^d(\bar{p}) = \inf_{p\in\mathcal{S}_s} \Pi^d(p), \tag{2.81}$$

$$(\Pi^d_{\text{sup}}): \quad \Pi^d(\bar{p}) = \sup_{p\in\mathcal{S}_s} \Pi^d(p). \tag{2.82}$$

Dual to Theorem 2.5.1, an extremum dual action theorem also holds for Π^d. A representation of the canonical Hamiltonian system (2.60) is shown in Fig. 2.5.

Example 2.5.2 Consider the dissipative dynamical system

$$\begin{aligned} &m_0q'' + \mu m_0 q' + k_0 q = f(t) \ \ \forall t \in I = (0, t_1), \\ &q(0) = 0, \ \ q'(t_1) = 0. \end{aligned} \tag{2.83}$$

Let $\Lambda = \frac{\mathrm{d}}{\mathrm{d}t}$, and $\mathcal{U} = \mathcal{E} = \mathcal{L}^2(I)$ with weighted bilinear forms

$$\langle u, u^* \rangle = \int_I e^{\mu t} u(t) u^*(t) \mathrm{d}t, \quad \langle v; p \rangle = \int_I e^{\mu t} v(t) p(t) \mathrm{d}t.$$

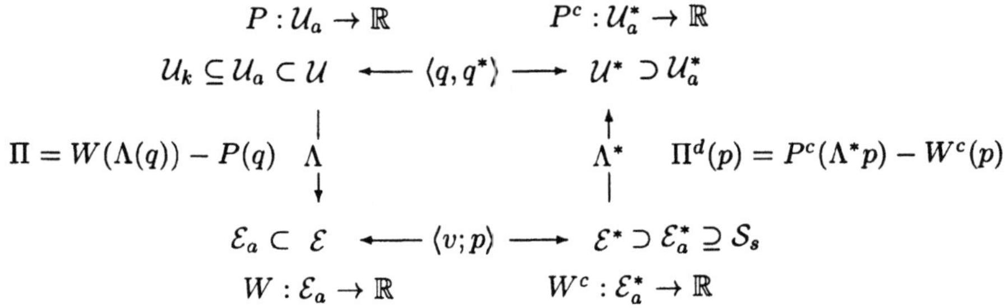

Figure 2.5. Representation of canonical Hamiltonian systems

The admissible spaces $\mathcal{U}_a$ and $\mathcal{E}_a$ are

$$\mathcal{U}_a = \{q \in \mathcal{L}^2(I) |\ q(0) = 0\}, \quad \mathcal{E}_a = \{v \in \mathcal{L}^2(I) |\ v(t_1) = 0\}.$$

On $\mathcal{U}_a$ and $\mathcal{E}_a$, the kinetic and potential energies

$$W(v) = \int_I e^{\mu t} \frac{1}{2} m_0 v^2 \mathrm{d}t, \quad P(q) = \int_I e^{\mu t} \left(\frac{1}{2} k_0 q^2 - f q \right) \mathrm{d}t$$

are convex. From the duality relations

$$p = DW(v) = m_0 v, \quad q^* = DP(q) = k_0 q - f,$$

the dual admissible spaces $\mathcal{U}_a^*$ and $\mathcal{E}_a^*$ can be obtained as

$$\begin{aligned} \mathcal{E}_a^* &= \{p \in \mathcal{E}^* |\ p/m_0 \in \mathcal{E}_a\} = \{p \in \mathcal{L}^2(I) |\ p(t_1) = 0\}, \\ \mathcal{U}_a^* &= \{q^* \in \mathcal{U}^* |\ (q^* + f)/k_0 \in \mathcal{U}_a\} = \{q^* \in \mathcal{L}^2(I) |\ q^*(0) = f(0)\}. \end{aligned}$$

Thus, for any given $(q, p) \in \mathcal{U}_a \times \mathcal{E}_a^*$, the adjoint operator $\Lambda^* : \mathcal{E}_a^* \to \mathcal{U}_a^*$ defined by $\langle \Lambda q; p \rangle = \langle q, \Lambda^* p \rangle$ is

$$\Lambda^* = -\frac{\mathrm{d}}{\mathrm{d}t} - \mu.$$

The Legendre conjugates $P^c : \mathcal{U}_a^* \to \mathbb{R}$ and $W^c : \mathcal{E}_a^* \to \mathbb{R}$ can easily be obtained as

$$P^c(q^*) = \int_I e^{\mu t} \frac{1}{2} k_0^{-1} (q^* + f)^2 \mathrm{d}t, \quad W^c(p) = \int_I e^{\mu t} \frac{1}{2m_0} p^2 \mathrm{d}t,$$

respectively. The Hamiltonian $H : \mathcal{U}_a \times \mathcal{E}_a^* \to \mathbb{R}$

$$H(q, p) = \int_I e^{\mu t} \left(\frac{1}{2m_0} p^2 + \frac{1}{2} k_0 q^2 - f q \right) \mathrm{d}t$$

associated with this dissipative system is convex. The Hamiltonian canonical forms

$$\Lambda q = DW^c(p) = p/m_0, \quad \Lambda^* p = DP(q) = k_0 q - f,$$

are equivalent to (2.83).

The kinematically admissible space $\mathcal{U}_k$ is a subspace of the Sobolev space $\mathcal{W}^{1,2}$, namely,

$$\mathcal{U}_k = \{q \in \mathcal{U}_a | \ \Lambda q \in \mathcal{E}_a\} = \{q \in \mathcal{W}^{1,2}(I) | \ q(0) = 0, \ q'(t_1) = 0\}.$$

The total action associated with this problem is

$$\Pi(q) = W(\Lambda q) - P(q) = \int_I e^{\mu t} \left[\frac{1}{2} m_0 (\Lambda q)^2 - \frac{1}{2} k_0 q^2 + f q\right] dt. \qquad (2.84)$$

The dynamically admissible space reads

$$\mathcal{S}_s = \{p \in \mathcal{E}_a^* | \ \Lambda^* p \in \mathcal{U}_a^*\} = \{p \in \mathcal{L}^2(I) | \ p' \in \mathcal{L}^2(I), \ p(t_1) = 0\}.$$

On $\mathcal{S}_s$, we have

$$q = DP^c(\Lambda^* p) = \frac{1}{k_0}(\Lambda^* p + f).$$

The total complementary action is then

$$\begin{aligned}\Pi^c(p) &= W^c(p) - P^c(\Lambda^* p) \\ &= \int_I e^{\mu t} \frac{1}{2} k_0^{-1} (\Lambda^* p + f)^2 dt - \int_I e^{\mu t} \frac{1}{2} m_0^{-1} p^2 dt.\end{aligned}$$

The dual Euler-Lagrange equation is simply

$$\Lambda k_0^{-1} (\Lambda^* p + f) - m_0^{-1} p = 0,$$

i.e.

$$p'' + \mu p' + \frac{k_0}{m_0} p = f'.$$

Clearly, by replacing p by $p = m_0 q'$ and integrating both sides, this dual Euler-Lagrange equation is reduced to the original equation (2.83).

Moreover, if $\mu = 0, \ f = 0$ and $\omega_0^2 = k_0/m_0 > 0$, the solution for the Jacobi initial-value problem is a periodic function $p(t) = \frac{1}{\omega_0}\sin(\omega_0 t)$. The equilibrium state of the Hamiltonian system minimizes Π^d on the interval $I = (0, \pi/\omega_0)$. ♢

2.6 Super-Lagrangian Duality Theory

This section presents a super-Lagrange duality theory for general convex Hamiltonian systems, by which, the primal and dual actions are linked to each other. This theory reveals an important phenomenon in dynamical systems, i.e. the solution of initial-value problems may either maximize the associated Lagrangian in both variables, or first maximize and then minimize the Lagrangian in each variable. This super-Lagrange duality theory leads to an interesting bi-duality theory for convex Hamiltonian systems.

2.6.1 Super Lagrangian and Super-Critical Points

Recall that in convex static systems, the Lagrangian associated with the primal problems is a saddle functional. However, in convex Hamiltonian systems, the Lagrangian $L : \mathcal{U}_a \times \mathcal{E}_a^* \to \mathbb{R}$ defined by (2.70) is a very interesting functional, in that although $L(q,p)$ is concave for each of its arguments, it is not a concave functional of the vector (q,p). A new definition for the Lagrangian is therefore needed.

Definition 2.6.1 ($\partial^{\pm}$-Critical Points and $\partial^{\pm}$-Lagrangians)
Let $L(q,p)$ be an arbitrary given real-valued functional on $\mathcal{U} \times \mathcal{E}^*$.
(D1$^+$) A point $(\bar{q},\bar{p})$ is said to be a *super-critical (or ∂^+-critical) point* of L on $\mathcal{U} \times \mathcal{E}^*$ if

$$L(\bar{q},p) \le L(\bar{q},\bar{p}) \ge L(q,\bar{p}) \quad \forall (q,p) \in \mathcal{U} \times \mathcal{E}^*. \tag{2.85}$$

(D1$^-$) A point $(\bar{q},\bar{p})$ is said to be a *sub-critical (or ∂^--critical) point* of L on $\mathcal{U} \times \mathcal{E}^*$ if

$$L(\bar{q},p) \ge L(\bar{q},\bar{p}) \le L(q,\bar{p}) \quad \forall (q,p) \in \mathcal{U} \times \mathcal{E}^*. \tag{2.86}$$

(D2$^+$) A functional $L : \mathcal{U} \times \mathcal{E}^* \to \mathbb{R}$ is said to be a *super-critical (or ∂^+) functional* on $\mathcal{U}_a \times \mathcal{E}_a^*$ if it is concave in each of its arguments, i.e.

$$L : \mathcal{U}_a \to \mathbb{R} \text{ is concave } \forall p \in \mathcal{E}_a^*,$$
$$L : \mathcal{E}_a^* \to \mathbb{R} \text{ is concave } \forall q \in \mathcal{U}_a.$$

In particular, if the super-critical functional $L : \mathcal{U} \times \mathcal{E}^* \to \mathbb{R}$ is a Lagrange form, it is called a *super-Lagrangian* or *∂^+-Lagrangian.*
(D2$^-$) A functional $L : \mathcal{U}_a \times \mathcal{E}_a^* \to \mathbb{R}$ is said to be a *sub-critical (or ∂^-) functional* on $\mathcal{U}_a \times \mathcal{E}_a^*$ if $-L$ is a super-critical functional on $\mathcal{U}_a \times \mathcal{E}_a^*$. It is called a *sub-Lagrangian* or *∂^--Lagrangian* if L is a Lagrange form. ◇

Clearly, $(\bar{u},\bar{p})$ is a super-critical point of L on $\mathcal{U} \times \mathcal{E}^*$ if and only if it is a sub-critical point of $-L$ on $\mathcal{U} \times \mathcal{E}^*$.

Remark 2.6.1 If L is a super-critical functional and Gâteaux differentiable on $\mathcal{U}_a \times \mathcal{E}_a^*$, then all its critical points are super-critical points. That L is a super-critical functional does not imply that $L : \mathcal{U}_a \times \mathcal{E}_a^* \to \mathbb{R}$ is concave. ■

For example, let us consider the quadratic function

$$L(x,y) = axy - \frac{1}{2}bx^2 - \frac{1}{2}cy^2.$$

If $b, c > 0$, then L is a super-critical point function and it is concave for each x, and y. But $L(\mathbf{x})$ is not concave on $\mathbf{x} = (x,y)$. The Hessian matrix of L is

$$D^2L(\mathbf{x}) = \begin{pmatrix} -b & a \\ a & -c \end{pmatrix},$$

which is not necessarily to be negative-definite for any $a \in \mathbb{R}$ and $b, c > 0$.

Dually, if $b, c < 0$, then L is a sub-critical function. But L may not be convex on (x, y) for the same reason.

By definition, the Lagrange form $L(q,p)$ associated with any convex Hamiltonian $H(q,p)$ is a super-critical functional, called simply the *super-Lagrangian.* Dually, a Lagrangian associated with any concave Hamiltonian is a sub-critical functional, or the *sub-Lagrangian.* Since L is a sub-Lagrangian if and only if $-L$ is a super-Lagrangian, here we only consider the duality theory for the super-Lagrangian.

Theorem 2.6.1 (Super-Critical Point Theorem)

Let $L(q,p)$ be an arbitrary given functional, partially Gâteaux differentiable on an open subset $\mathcal{U}_a \times \mathcal{E}_a^* \subset \mathcal{U} \times \mathcal{E}^*$. If $(\bar{q}, \bar{p}) \in \mathcal{U}_a \times \mathcal{E}_a^*$ is either a super-critical or sub-critical point of L, then $(\bar{q}, \bar{p})$ is a critical point of L on $\mathcal{U}_a \times \mathcal{E}_a^*$.

Proof. By definition, if $(\bar{q}, \bar{p})$ is a super-critical point of L on $\mathcal{U}_a \times \mathcal{E}_a^*$, then for any $(q,p) \in \mathcal{U}_a \times \mathcal{E}_a^*$, we have

$$L(\bar{q}, p) \leq \sup_{p \in \mathcal{E}_a^*} L(\bar{q}, p) = L(\bar{q}, \bar{p}) = \sup_{q \in \mathcal{U}_a} L(q, \bar{p}) \geq L(q, \bar{p}).$$

The partial Gâteaux differentiability of L on the open subset $\mathcal{U}_a \times \mathcal{E}_a^*$ and Lemma 1.2.2 imply that $(\bar{q}, \bar{p})$ is a critical point. Similarly for a sub-critical point. □

Any critical point of a Gâteaux differentiable super-Lagrangian is a super-critical point. However, if $(\bar{q}, \bar{p})$ is a super-critical point of L, it does not follows that L is a super-Lagrangian. In the variational analysis of convex Hamiltonian systems, the following statements are of important theoretical value.

(S1) Under certain necessary and sufficient conditions we have

$$\inf_{q \in \mathcal{U}_a} \sup_{p \in \mathcal{E}_a^*} L(q,p) = \inf_{p \in \mathcal{E}_a^*} \sup_{q \in \mathcal{U}_a} L(q,p). \tag{2.87}$$

A statement of this type is called a *super-minimax theorem* and the pair $(\bar{q}, \bar{p})$ is called a *super-minimax point* of L on $\mathcal{U}_a \times \mathcal{E}_a^*$.

(S2) Under certain conditions, a pair $(\bar{q}, \bar{p}) \in \mathcal{U}_a \times \mathcal{E}_a^*$ exists such that

$$L(q, \bar{p}) \leq L(\bar{q}, \bar{p}) \geq L(\bar{q}, p) \quad \forall (q,p) \in \mathcal{U}_a \times \mathcal{E}_a^*. \tag{2.88}$$

A statement of this type is called a *super-critical point theorem.*

By the fact that the maxima of $L(q,p)$ can be taken in either order on $\mathcal{U}_a \times \mathcal{E}_a^*$, the equality

$$\sup_{q \in \mathcal{U}_a} \sup_{p \in \mathcal{E}_a^*} L(q,p) = \sup_{p \in \mathcal{E}_a^*} \sup_{q \in \mathcal{U}_a} L(q,p) \tag{2.89}$$

always holds. A pair $(\bar{q}, \bar{p})$ which maximizes L on $\mathcal{U}_a \times \mathcal{E}_a^*$ is called a *super-maximum point* of L on $\mathcal{U}_a \times \mathcal{E}_a^*$.

For a given super-Lagrangian $L : \mathcal{U}_a \times \mathcal{E}_a^* \to \mathbb{R}$, we let $\mathcal{U}_k \subseteq \mathcal{U}_a$ and $\mathcal{S}_s \subseteq \mathcal{E}_a^*$ such that

$$\sup_{p \in \mathcal{E}_a^*} L(q,p) < +\infty \quad \forall q \in \mathcal{U}_k,$$

$$\sup_{q \in \mathcal{U}_a} L(q,p) < +\infty \quad \forall p \in \mathcal{S}_s.$$

Theorem 2.6.2 (Super-Lagrangian Duality Theorem)
Let the Lagrangian $L : \mathcal{U} \times \mathcal{E}^* \to \mathbb{R}$ be an arbitrary given functional.
(T1) If there exists either a super-maximum point $(\bar{q}, \bar{p}) \in \mathcal{U}_a \times \mathcal{E}_a^*$ such that

$$\max_{q \in \mathcal{U}_a} \max_{p \in \mathcal{E}_a^*} L(q,p) = L(\bar{q}, \bar{p}) = \max_{p \in \mathcal{E}_a^*} \max_{q \in \mathcal{U}_a} L(q,p), \tag{2.90}$$

or a super-minimax point $(\bar{q}, \bar{p}) \in \mathcal{U}_a \times \mathcal{E}_a^*$ such that

$$\min_{q \in \mathcal{U}_a} \max_{p \in \mathcal{E}_a^*} L(q,p) = L(\bar{q}, \bar{p}) = \min_{p \in \mathcal{E}_a^*} \max_{q \in \mathcal{U}_a} L(q,p), \tag{2.91}$$

then $(\bar{q}, \bar{p})$ is a super-critical point of L on $\mathcal{U}_a \times \mathcal{E}_a^*$.
(T2) Conversely, if L is partially Gâteaux differentiable on an open subset $\mathcal{U}_a \times \mathcal{E}_a^* \subset \mathcal{U} \times \mathcal{E}^*$, and $(\bar{q}, \bar{p})$ is a super-critical point of L on $\in \mathcal{U}_a \times \mathcal{E}_a^*$, then either the super-maximum theorem in the form

$$\max_{q \in \mathcal{U}_k} \max_{p \in \mathcal{E}_a^*} L(q,p) = L(\bar{q}, \bar{p}) = \max_{p \in \mathcal{S}_s} \max_{q \in \mathcal{U}_a} L(q,p), \tag{2.92}$$

holds, or the super-minimax theorem in the form

$$\min_{q \in \mathcal{U}_k} \max_{p \in \mathcal{E}_a^*} L(q,p) = L(\bar{q}, \bar{p}) = \min_{p \in \mathcal{S}_s} \max_{q \in \mathcal{U}_a} L(q,p) \tag{2.93}$$

holds.

Proof. If $(\bar{q}, \bar{p})$ is either a super-maximum point, or a super-minimax point of L on $\mathcal{U}_a \times \mathcal{E}_a^*$, we have

$$L(q, \bar{p}) \leq \max_{q \in \mathcal{U}_a} L(q, \bar{p}) = L(\bar{q}, \bar{p}) = \max_{p \in \mathcal{E}_a^*} L(\bar{q}, p) \geq L(\bar{q}, p),$$

for any $(q,p) \in \mathcal{U}_a \times \mathcal{E}_a^*$. Thus, $(\bar{q}, \bar{p})$ is a super-critical point of L on $\mathcal{U}_a \times \mathcal{E}_a^*$.

Since $\mathcal{U}_a \times \mathcal{E}_a^*$ is an open set and $(\bar{q}, \bar{p})$ is a super-critical point of L, the super-critical point theorem shows that $(\bar{q}, \bar{p})$ is a critical point of L and

$$\max_{q \in \mathcal{U}_a} L(q, \bar{p}) = L(\bar{q}, \bar{p}) = \max_{p \in \mathcal{E}_a^*} L(\bar{q}, p).$$

Thus, $\mathcal{U}_k$ and $\mathcal{S}_s$ are not empty. Since $\mathcal{U}_a$ is an open subset, for the fixed $\bar{p} \in \mathcal{E}_a^*$, the partial critical point $\bar{q}$ is either an extremum point or a saddle point of $L(q, \bar{p})$. If $\bar{q}$ maximizes $L(q, \bar{p})$ on $\mathcal{U}_k$, then

$$\begin{aligned}
\max_{q \in \mathcal{U}_k} L(q, \bar{p}) &= \max_{q \in \mathcal{U}_k} \max_{p \in \mathcal{E}_a^*} L(q,p) = L(\bar{q}, \bar{p}) \\
&= \max_{p \in \mathcal{S}_s} L(\bar{q}, p) = \max_{p \in \mathcal{S}_s} \max_{q \in \mathcal{U}_a} L(q,p).
\end{aligned}$$

Thus, the super-maximum theorem in the form of (2.92) holds.

If the partial critical point $\bar{q}$ minimizes $L(q,\bar{p})$ on $\mathcal{U}_k$, then

$$L(\bar{q},\bar{p}) = \min_{q\in\mathcal{U}_k} L(q,\bar{p}) = \min_{q\in\mathcal{U}_k}\max_{p\in\mathcal{E}_a^*} L(q,p). \tag{2.94}$$

On the other hand, for fixed $\bar{q}$, $\bar{p}$ is a partial critical point of $L(\bar{q},p)$. On the open set $\mathcal{E}_a^*$, $D_pL(\bar{q},\bar{p}) = 0$. This implies that $\bar{p}$ is either a local extremum point or a saddle point of $L(\bar{q},p)$ on $\mathcal{S}_s$. If $\bar{p}$ maximizes $L(\bar{q},p)$ on $\mathcal{S}_s$, then

$$\begin{aligned} L(\bar{q},\bar{p}) &= \max_{p\in\mathcal{S}_s} L(\bar{q},p) = \max_{p\in\mathcal{S}_s}\max_{q\in\mathcal{U}_a} L(q,p) \\ &= \max_{q\in\mathcal{U}_k}\max_{p\in\mathcal{E}_a^*} L(q,p). \end{aligned}$$

This contradicts to (2.94), i.e. $\bar{p}$ is not a local maximizer of $L(\bar{q},p)$ on $\mathcal{S}_s$. If $\bar{p}$ is a local saddle point of $L(\bar{q},p)$ and it maximizes $L(\bar{q},p)$ in the direction $\bar{p}_0$, then

$$\begin{aligned} L(\bar{q},\bar{p}) &= \max_{\theta\geq 0} L(\bar{q},\bar{p}+\theta\bar{p}_0) = \max_{\theta\geq 0}\max_{q\in\mathcal{U}_a} L(\bar{q},\bar{p}+\theta\bar{p}_0) \\ &= \max_{q\in\mathcal{U}_k}\max_{\theta\geq 0} L(\bar{q},\bar{p}+\theta\bar{p}_0) \end{aligned}$$

This contradicts equation (2.94) again. Thus, $\bar{p}$ is a local minimizer of $L(\bar{q},p)$ for all direction $\bar{p}_0$ on $\mathcal{S}_s$. This proves that if $(\bar{q},\bar{p})$ is a super-critical point, then the super-minimax theorem in the form of (2.93) holds. □

This super-Lagrangian theorem shows a very important fact in Hamiltonian systems, i.e. the critical points of L either maximize L or minimaximize L on $\mathcal{U}_k \times \mathcal{S}_s$ in either order.

Similarly to the generalized super-Lagrange duality theory, we have the following general result for super-Lagrangians.

Theorem 2.6.3 (Generalized Super-Lagrangian Duality Theorem)
Let the Lagrangian $L : \mathcal{U}\times\mathcal{E}^* \to \mathbb{R}$ be an arbitrary given functional, partially Gâteaux differentiable on an open subset $\mathcal{U}_a \times \mathcal{E}_a^* \subset \mathcal{U}\times\mathcal{E}^*$. If $(\bar{q},\bar{p})$ is a super-critical point of L on $\in \mathcal{U}_a\times\mathcal{E}_a^*$, then either the super-maximum theorem in the form

$$\sup_{q\in\mathcal{U}_k}\sup_{p\in\mathcal{E}_a^*} L(q,p) = L(\bar{q},\bar{p}) = \sup_{p\in\mathcal{S}_s}\sup_{q\in\mathcal{U}_a} L(q,p), \tag{2.95}$$

holds, or the super-minimax theorem in the form

$$\inf_{q\in\mathcal{U}_k}\sup_{p\in\mathcal{E}_a^*} L(q,p) = L(\bar{q},\bar{p}) = \inf_{p\in\mathcal{S}_s}\sup_{q\in\mathcal{U}_a} L(q,p) \tag{2.96}$$

holds.

2.6.2 Bi-Duality Theory

Let $L : \mathcal{U}_a \times \mathcal{E}_a^* \to \mathbb{R}$ be an arbitrary given super-critical functional. For any fixed $q \in \mathcal{U}_a$, let

$$\Pi(q) = \sup_{p\in\mathcal{E}_a^*} L(q,p). \tag{2.97}$$

Let $\mathcal{U}_k \subset \mathcal{U}_a$ be the domain of Π such that $\Pi : \mathcal{U}_k \to \mathbb{R}$ is finite and Gâteaux differentiable. Dually, for any fixed $p \in \mathcal{E}_a^*$, let

$$\Pi^d(p) = \sup_{q \in \mathcal{U}_a} L(q,p) \tag{2.98}$$

with domain $\mathcal{S}_s \subset \mathcal{E}_a^*$, on which, $\Pi^d : \mathcal{S}_s \to \mathbb{R}$ is finite and Gâteaux differentiable.

Similarly to the complementary extremum principles in static systems, the following results play important roles for convex Hamiltonian systems.

Theorem 2.6.4 (Super-Critical Point Theorem)

Let $L : \mathcal{U}_a \times \mathcal{E}_a^* \to \mathbb{R}$ be an arbitrary given functional such that Π and Π^d are defined by (2.97) and (2.98), respectively. Suppose that $\mathcal{U}_k \times \mathcal{S}_s$ is an open subset of $\mathcal{U}_a \times \mathcal{E}_a^*$ and $(\bar{q}, \bar{p})$ is a super-critical point of L on $\mathcal{U}_k \times \mathcal{S}_s$, Π is Gâteaux differentiable at $\bar{q}$ and Π^d is Gâteaux differentiable at $\bar{p}$. Then, $\bar{q}$ is a critical point of Π, $\bar{p}$ is a critical point of Π^d and

$$\Pi(\bar{q}) = L(\bar{q}, \bar{p}) = \Pi^d(\bar{p}). \tag{2.99}$$

Proof. By the super-critical point theory, $(\bar{q}, \bar{p})$ is a critical point of L on the open set $\mathcal{U}_k \times \mathcal{S}_s$, and

$$\begin{aligned} \Pi(q) &= \sup_{p \in \mathcal{E}_a^*} L(q,p) = L(q,\bar{p}) \\ &\le L(\bar{q},\bar{p}) = \sup_{p \in \mathcal{E}_a^*} L(\bar{q},p) = \Pi(\bar{q}) \quad \forall q \in \mathcal{U}_k. \end{aligned}$$

Thus, Lemma 1.2.2 implies that $D\Pi(\bar{q}) = 0$, and $\bar{q}$ is a critical point of Π on $\mathcal{U}_k$. Similarly,

$$\begin{aligned} \Pi^d(p) &= \sup_{q \in \mathcal{U}_a} L(q,p) = L(\bar{q},p) \\ &\ge L(\bar{q},\bar{p}) = \sup_{q \in \mathcal{U}_a} L(q,\bar{p}) = \Pi^d(\bar{p}) \quad \forall p \in \mathcal{S}_s, \end{aligned}$$

and hence $D\Pi^d(\bar{p}) = 0$ and $\bar{p}$ is a critical point of Π^d on $\mathcal{S}_s$. □

Theorem 2.6.5 (Double-Min and Double-Max Duality)

Let $L : \mathcal{U}_a \times \mathcal{E}_a^* \to \mathbb{R}$ be an arbitrary given super-critical functional such that $\Pi : \mathcal{U}_k \to \mathbb{R}$ and $\Pi^d : \mathcal{S}_s \to \mathbb{R}$ are defined by (2.97) and (2.98), respectively. If $(\bar{q}, \bar{p})$ is a critical point of L on $\mathcal{U}_k \times \mathcal{S}_s \subseteq \mathcal{U}_a \times \mathcal{E}_a^*$, then

$$\Pi(\bar{q}) = \inf_{q \in \mathcal{U}_k} \Pi(q) \quad \Leftrightarrow \quad \Pi^d(\bar{p}) = \inf_{p \in \mathcal{S}_s} \Pi^d(p); \tag{2.100}$$

$$\Pi(\bar{q}) = \sup_{q \in \mathcal{U}_k} \Pi(q) \quad \Leftrightarrow \quad \Pi^d(\bar{p}) = \sup_{p \in \mathcal{S}_s} \Pi^d(p). \tag{2.101}$$

Proof. This theorem follows from the combination of the super-critical point theorem, the super-Lagrangian theorem and Theorem 2.6.4. □

Remark 2.6.2 (Periodic Bi-Extremality Phenomenon) In convex Hamiltonian systems, the Lagrange form defined by $L(q,p) = \langle \Lambda q; p\rangle - H(q,p)$ is a super-Lagrangian, and its critical points in phase space are the trajectories of the system. For periodic boundary value problems, if the solution $q(t)$ vanishes at least once within a period, then by the local extremum action theorem and the bi-duality theory presented in this chapter, both the primal and the dual actions take periodically the same extremal values within a period. This interesting periodic bi-extremality phenomenon is very important in understanding nonconvex periodic Hamiltonian systems. By this, the classical least action principle is in fact a misnomer. A suitable name might be, for example, the alternative extremum action principle. ■

If the functional $L : \mathcal{U}_a \times \mathcal{E}_a^* \to \mathbb{R}$ is a Lagrange form associated with the Hamiltonian $H(q,p) = W^c(p) + P(q)$, then

$$L(q,p) = \langle \Lambda q; p\rangle - W^c(p) - P(q). \tag{2.102}$$

Clearly, for convex Hamiltonian systems, $W^c(p)$ and $P(q)$ are convex functionals, and the Lagrange form $L : \mathcal{U}_a \times \mathcal{E}_a^*$ is a super-Lagrangian. For any given $q \in \mathcal{U}_a$, the Legendre-Fenchel transformation leads to the sup-variational problem

$$\begin{aligned} \sup_{p\in\mathcal{E}_a^*} L(q,p) &= \sup_{p\in\mathcal{E}_a^*} \{\langle p; \Lambda q\rangle - W^c(p)\} - P(q) \\ &= \begin{cases} W(\Lambda q) - P(q) & \text{if } \Lambda q \in \mathcal{E}_a, \\ +\infty & \text{otherwise.} \end{cases} \end{aligned}$$

Thus, the supremum of $\sup_p L(q,p)$ is finite if $\mathcal{E}_a$ is the range of the dual constitutive mapping $DW^c : \mathcal{E}_a^* \to \mathcal{E}_a$. On the kinematically admissible space

$$\mathcal{U}_k = \{q \in \mathcal{U}_a | \ \Lambda q \in \mathcal{E}_a\},$$

the supremum of $\sup_p L(q,p)$ leads back to the total action

$$\Pi(q) = \sup_{p\in\mathcal{E}_a^*} L(q,p) = L(q,\bar{p}) = W(\Lambda q) - P(q). \tag{2.103}$$

Dually, for any $p \in \mathcal{E}_a^*$, we have

$$L(\bar{q},p) = \sup_{q\in\mathcal{U}_a} L(q,p) = P^*(\Lambda^* p) - W^c(p),$$

in which $P^* : \Lambda^* \circ \mathcal{E}_a^* \to \mathbb{R} \cup \{\infty\}$ is defined by the Legendre-Fenchel transformation

$$\begin{aligned} P^*(\Lambda^* p) &= \sup_{q\in\mathcal{U}_a} \{\langle \Lambda q; p\rangle - P(q)\} = \sup_{q\in\mathcal{U}_a} \{\langle q, \Lambda^* p\rangle - P(q)\} \\ &= \begin{cases} P^c(\Lambda^* p) & \text{if } \Lambda^* p \in \mathcal{U}_a^*, \\ +\infty & \text{otherwise,} \end{cases} \end{aligned}$$

and $\mathcal{U}_a^*$ is the range of the mapping $DP : \mathcal{U}_a \to \mathcal{U}^*$. Thus, on the dynamically admissible space $\mathcal{S}_s$ defined by $\mathcal{S}_s = \{p \in \mathcal{E}_a^* | \ \Lambda^* p \in \mathcal{U}_a^*\}$, we have $P^*(\Lambda^* p) = P^c(\Lambda^* p)$ and the supremum of $\sup_q L(q,p)$ leads to the dual action

$$\Pi^d(p) = \sup_{q \in \mathcal{U}_a} L(q,p) = P^c(\Lambda^* p) - W^c(p). \tag{2.104}$$

For the Newtonian periodic boundary value problem

$$m_0 q''(t) + k_0 q(t) = 0, \quad \forall t \in I = (0,T),$$
$$q(0) = q(T), \quad q'(0) = q'(T),$$

the solution is a periodic function

$$q(t) = R \sin(\omega_0 t - \delta), \quad \text{with } \omega_0^2 = \frac{k_0}{m_0}.$$

The amplitude $R > 0$ and the initial phase $\delta \in \mathbb{R}$ depend on the initial conditions. The primal action and the dual action are quadratic functions, namely

$$\Pi(q) = \int_0^T \frac{1}{2} \left(m_0 q'(t)^2 - k_0 q(t)^2 \right) \mathrm{d}t = \int_0^T \ell(q(t)) \mathrm{d}t$$

and

$$\Pi^d(p) = \int_0^T \frac{1}{2} \left(p'(t)^2 / k_0 - p(t)^2 / m_0 \right) \mathrm{d}t = \int_0^T \ell^d(p(t)) \mathrm{d}t,$$

respectively, where $\ell(q(t))$ and $\ell^d(p(t))$ are the action and dual action densities, respectively, and

$$\ell(q(t)) = \frac{1}{2} k_0 R^2 \left(\cos^2(\omega_0 t - d) - \sin^2(\omega_0 t - d) \right) = -\ell^d(p(t)).$$

They have the same graphs in phase space (see Fig. 2.6).

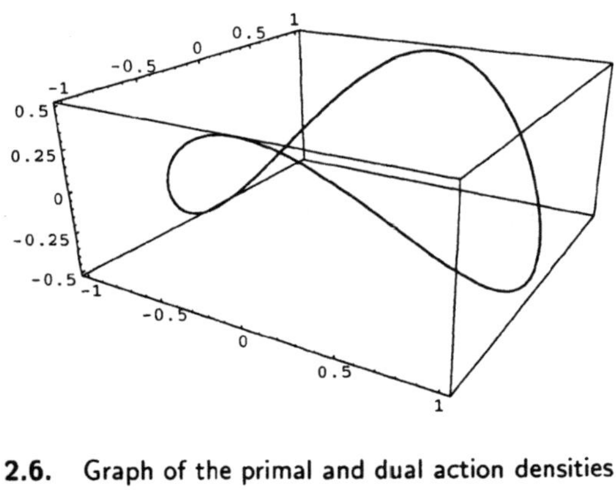

Figure 2.6. Graph of the primal and dual action densities in phase space $\mathcal{U} \times \mathcal{E}^*$

2.7 Applications and Commentary

2.7.1 Nonconvex Optimization and Variational Problem

I. Extremum Problem for Indefinite Optimization.

Consider the quadratic programming problem

$$\Pi(x) = \frac{1}{2}m_0x^2 - \frac{1}{2}k_0x^2 + fx \quad \to \text{sta} \quad \forall x \in \mathbb{R}.$$

Let $\Lambda = 1$, $y = m_0x$. The Lagrangian associated with Π is then

$$L(x,y) = xy - \frac{1}{2m_0}y^2 - \frac{1}{2}k_0x^2 + fx,$$

and the dual action reads

$$\Pi^d(y) = \frac{1}{2k_0}(y-f)^2 - \frac{1}{2m_0}y^2.$$

It is easy to check that the Lagrangian has only one critical point $(\bar{x},\bar{y}) = (\frac{f}{k_0-m_0}, \frac{mf}{k_0-m_0})$, and that

$$\Pi(\bar{x}) = L(\bar{x},\bar{y}) = \Pi^d(\bar{y}).$$

Clearly, if $m_0 > k_0 > 0$, Π and Π^d are convex (see Fig. 2.7(a)), but the Lagrangian is not concave in $(x,y) \in \mathbb{R}^2$ (since $\det(D^2L(x,y)) = k_0/m_0 - 1 < 0$). In this case,

$$\min \Pi(x) = \min_x \max_y L(x,y) = \min_y \max_x L(x,y) = \min_y \Pi^d(y).$$

However, if $0 < m_0 < k_0$, Π and Π^d are concave. In this case, the Lagrangian is concave in $(x,y) \in \mathbb{R}^2$ (see Fig. 2.7(b)), and

$$\max_x \Pi(x) = \max_{x,y} L(x,y) = \max_y \Pi^d(y).$$

II. Constrained Nonconvex Optimization.

Let us consider a very simple one-dimensional optimization problem with constraint

$$(\mathcal{P}_{\text{sup}}): \quad P(x) = \frac{1}{2}kx^2 - fx \quad \to \max \quad \forall x \in \bar{I} = [a,b], \tag{2.105}$$

where $(k > 0)$. We assume that $-\infty < a < 0 < b < \infty$. Since $P(x)$ is strictly convex on I, the maximum is attained only on the boundary of I, i.e.

$$\sup_{x\in[a,b]} P(x) = \max\{P(a), P(b)\} < \infty.$$

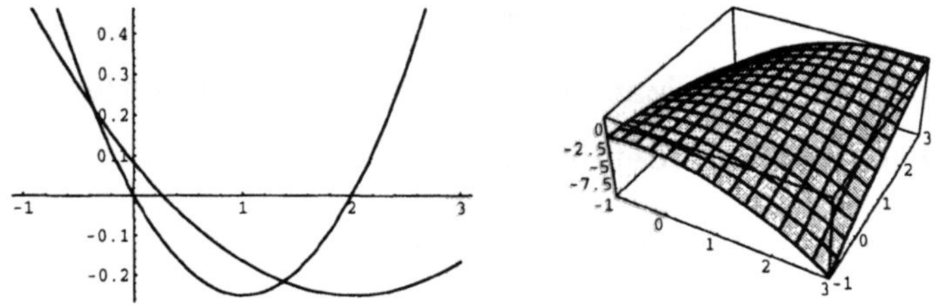

(a) Convex actions Π, Π^d and the associated Lagrangian ($m_0 = 2, k_0 = 1.5$)

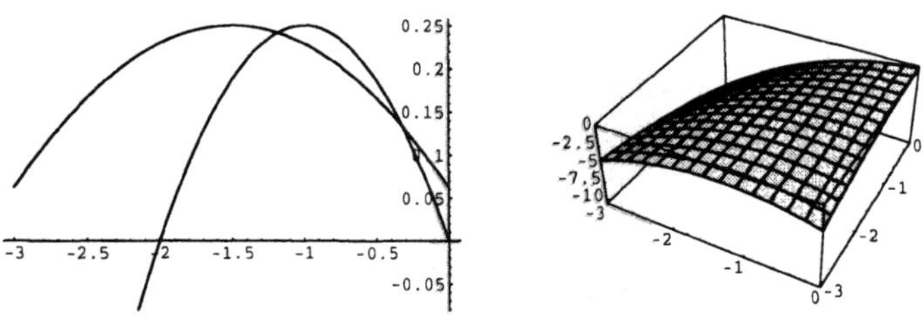

(b) Concave actions Π, Π^d and concave Lagrangian ($m_0 = 1.5, k_0 = 2$)

Figure 2.7. Double minimum (a), double maximum (b) dualities and Lagrangians

The classical Lagrange multiplier method cannot be used for this nonconvex problem. To set this problem within our framework, we need only set $\mathcal{U} = \mathbb{R}$, $\mathcal{U}_a = [a, b]$ and let $\Lambda = 1$, so that

$$\epsilon = \Lambda x = x \in \mathcal{E} = \mathbb{R}.$$

Thus, the range of the mapping $\Lambda : \mathcal{U}_a \to \mathcal{E} = \mathbb{R}$ is $\mathcal{E}_a = [a, b]$. Let

$$W(\epsilon) = \begin{cases} 0 & \text{if } \epsilon \in \mathcal{E}_a, \\ +\infty & \text{if } \epsilon \notin \mathcal{E}_a. \end{cases}$$

It is not difficult to check that $W : \mathcal{E} \to \mathbb{R} \cup \{+\infty\}$ is convex. On $\mathcal{E}_a$, W is finite and differentiable. Thus, the kinematically admissible space can be defined by

$$\mathcal{U}_k = \{x \in \mathcal{U}_a | \ \Lambda x = x \in \mathcal{E}_a\} = [a, b].$$

The constrained maximization problem $(\mathcal{P}_{\sup})$ is then equivalent to the unconstrained nonconvex minimization problem

$$(\Pi_{\inf}) : \quad \Pi(x) = W(\Lambda x) - P(x) \quad \to \min \quad \forall x \in \mathbb{R}. \tag{2.106}$$

Clearly, Π is finite only on $\mathcal{U}_k$. The condition $x \in \mathcal{U}_k$ is called the *implicit constraint* of $(\Pi_{\inf})$.

Since $P(x)$ is strictly convex and differentiable on $\mathcal{U}_a = [a, b]$, and

$$x^* = DP(x) = kx - f \in \mathcal{U}_a^* = [ak - f, bk - f] \subset \mathcal{U}^* = \mathbb{R}$$

is invertible, the Legendre conjugate $P^c : \mathcal{U}_a^* \to \mathbb{R}$ can easily be obtained as

$$P^c(x^*) = \max_{x \in \mathcal{U}_a}\{xx^* - P(x)\} = \frac{1}{2k}(x^* + f)^2.$$

By the Legendre-Fenchel transformation, the conjugate of the nonsmooth function W can be obtained as

$$W^c(y) = \sup_{\epsilon \in \mathcal{E}}\{\epsilon y - W(\epsilon)\} = \max_{\epsilon \in \mathcal{E}_a} \epsilon y = \begin{cases} by & \text{if } y > 0, \\ 0 & \text{if } y = 0, \\ ay & \text{if } y < 0. \end{cases}$$

It is convex and differentiable on $\mathcal{E}_a^* = \mathcal{E}^* = \mathbb{R}$.

On $\mathcal{U}_a \times \mathcal{E}_a^* = [a, b] \times \mathbb{R}$, the Lagrangian associated with the problem (Π_{inf}) is well-defined by

$$\begin{aligned} L(x,y) &= y\Lambda x - W^c(y) - P(x) \\ &= \begin{cases} xy - by - \frac{1}{2}kx^2 + fx & \text{if } y \geq 0, \\ xy - ay - \frac{1}{2}kx^2 + fx & \text{if } y < 0. \end{cases} \end{aligned} \tag{2.107}$$

Since both W^c and P are convex, $L(x, y)$ is a super-critical point function. If $x \in \mathcal{U}_k = [a, b]$, then

$$\Pi(x) = \sup_{y \in \mathcal{E}_a^*} L(x, y).$$

On the other hand, for any y in the dynamically admissible space

$$\mathcal{S}_s = \{y \in \mathcal{E}_a^* = \mathbb{R} | \ \Lambda^* y = y \in \mathcal{U}_a^*\} = [ak - f, bk - f],$$

the dual function is obtained by

$$\begin{aligned} \Pi^d(y) &= \sup_{x \in \mathcal{U}_a} L(x, y) = \sup_{x \in \mathbb{R}}\{\Lambda xy - P(x)\} - W^c(y) \\ &= P^c(\Lambda^* y) - W^c(y), \end{aligned}$$

where

$$\begin{aligned} P^c(\Lambda^* y) &= \sup_{x \in \mathcal{U}_a}\{\Lambda xy - P(x)\} = \sup_{x \in \mathbb{R}}\{x(y + f) - \frac{1}{2}kx^2\} \\ &= \frac{1}{2k}(y + f)^2 = P^c(y). \end{aligned}$$

Thus, the dual action Π^d is well defined on $\mathcal{S}_s$ by

$$\Pi^d(y) = \begin{cases} \frac{1}{2k}(y + f)^2 - by & \text{if } y > 0, \\ \frac{1}{2k}f^2 & \text{if } y = 0, \\ \frac{1}{2k}(y + f)^2 - ay & \text{if } y < 0. \end{cases} \tag{2.108}$$

This is a double-well function on $\mathbb{R}$ (see Fig. 2.8). The dual problem

$$(\Pi_{\text{inf}}^d): \ \ \Pi^d(y) \to \min \quad \forall y \in \mathcal{S}_s$$

is a convex optimization problem on either $\mathcal{S}_s^+ = \{y \in \mathcal{S}_s |\ y > 0\}$ or $\mathcal{S}_s^- = \{y \in \mathcal{S}_s |\ y < 0\}$. In n-dimensional problems, this dual problem is much easier than the primal problem. The criticality condition leads to

$$\bar{y} = \begin{cases} bk - f & \text{if } \bar{y} > 0, \\ 0 & \text{if } \bar{y} = 0 \\ ak - f & \text{if } \bar{y} < 0. \end{cases}$$

It is easy to check that the following duality theorems hold:

$$\min_{x \in \mathcal{U}_k} \Pi(x) = \min_{y \in \mathcal{S}_s} \Pi^d(y),$$

$$\max_{x \in \mathcal{U}_k} \Pi(x) = \max_{y \in \mathcal{S}_s} \Pi^d(y).$$

The graphs of $\Pi(x)$ and $\Pi^d(y)$ are shown in Fig. 2.8.

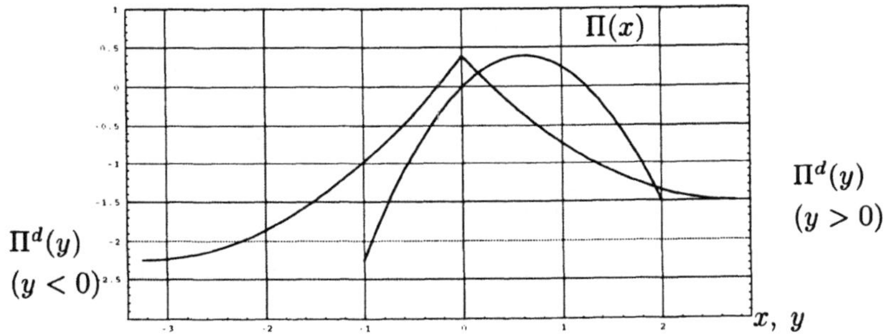

Figure 2.8. Bi-duality in constrained nonconvex optimization

III. **Nonlinear Heavy Rotating Chain** (Toland, 1979)

Let us consider a heavy rotating chain suspended with one end-point fixed, and acted on solely by the the forces of gravity and tension T. Suppose that the chain is lying on a vertical plane and rotating with constant angular speed ω. Let the independent variable s be are-length measured from the free end of the chain, and let the dependent variables $v(s)$ and $T(s)$ be respectively the horizontal displacement and tension of the chain at a point at a distance s from the free end. By using a new variable $u(s) = v(s)T(s)$, the governing equation for this rotating chain is

$$\begin{aligned} & u''(s) + \omega^2 u(s) \left(s^2 + u(s)^2\right)^{-\frac{1}{2}} = 0, \quad \forall s \in I = (0,1), \\ & u(0) = u'(1) = 0. \end{aligned} \tag{2.109}$$

The total "action" associated with this mixed boundary-value problem is the nonlinear functional

$$\Pi(u) = \int_I \left(\frac{1}{2} u'(s)^2 - \omega^2 \sqrt{s^2 + u(s)^2} \right) ds. \tag{2.110}$$

To set this problem within our framework, we let

$$\mathcal{U} = \mathcal{L}(I), \quad \mathcal{E} = \mathcal{L}^2(I), \quad \Lambda = \frac{\mathrm{d}}{\mathrm{d}s}.$$

With the standard bilinear forms

$$\langle u, u^* \rangle = \int_I u(s)u^*(s)\mathrm{d}s, \quad \langle v; v^* \rangle = \int_I v(s)v^*(s)\mathrm{d}s,$$

the dual spaces are $\mathcal{U}^* = \mathcal{L}^\infty$ and $\mathcal{E}^* = \mathcal{L}^2 = \mathcal{E}$. Let

$$\mathcal{U}_a = \{u \in \mathcal{U} |\ u(0) = 0\}, \ \ \mathcal{E}_a^* = \{p \in \mathcal{E}^* |\ p(1) = 0\};$$

then we have

$$\langle \Lambda u; p \rangle = \langle u, \Lambda^* p \rangle \ \ \forall (u, p) \in \mathcal{U}_a \times \mathcal{E}_a^*,$$

and the adjoint operator Λ^* of Λ is $\Lambda^* = -\Lambda$. The functional $P : \mathcal{U}_a \to \mathbb{R}$ is defined by

$$P(u) = \int_I U(u)\mathrm{d}s = \int_I \sqrt{s^2 + u(s)^2}\mathrm{d}s,$$

and, on $\mathcal{E}$,

$$W(v) = \int_I K(v)\mathrm{d}s = \int_I \frac{1}{2\omega^2} v(s)^2 \mathrm{d}s.$$

The kinematically admissible space is then

$$\mathcal{U}_k = \{u \in \mathcal{U} |\ \Lambda u \in \mathcal{L}^2(I), \ \ u(0) = 0\}.$$

On $\mathcal{U}_k$, the primal problem is to find $\bar{u} \in \mathcal{U}_k$ such that

$$\Pi(\bar{u}) = \inf_{u \in \mathcal{U}_k} \Pi(u) = \inf_{u \in \mathcal{U}_k} \{W(\Lambda u) - P(u)\}. \tag{2.111}$$

Since $U(u) = \sqrt{s^2 + u(s)^2}$ is strictly convex on $\mathcal{U}$, the duality relation

$$u^* = \frac{u}{\sqrt{s^2 + u(s)^2}}$$

is invertible and, by the Legendre transformation, it is easy to find that

$$U^*(u^*) = -s\sqrt{1 - u^*(s)^2} \ \ s.t. \ \ |u^*| \leq 1.$$

Incidentally, the Legendre conjugate of $K(v) = v^2/(2\omega^2)$ is simply $K^*(p) = \frac{1}{2}\omega^2 p^2$. Thus, on the dynamically admissible space,

$$\mathcal{S}_s = \{p \in \mathcal{E}_a^* |\ \Lambda^* p \in \mathcal{L}^\infty, \ \ |\Lambda^* p(s)| \leq 1 \ \ \forall s \in I\},$$

and the dual "action" $\Pi^d : \mathcal{S}_s \to \mathbb{R}$ is obtained as

$$\begin{aligned} \Pi^d(p) &= P^c(\Lambda^* p) - W^c(p) \\ &= -\int_I \left(\frac{1}{2}\omega^2 p(s)^2 + s\sqrt{1 - p'(s)^2} \right) \mathrm{d}s. \end{aligned} \tag{2.112}$$

The dual problem for this heavy rotating chain is to find $\bar{p} \in \mathcal{S}_s$ such that

$$\Pi^d(\bar{p}) = \inf_{p \in \mathcal{S}_s} \Pi^d(p). \tag{2.113}$$

It is easy to check that the Euler-Lagrange equation for this variational problem is

$$\frac{\mathrm{d}}{\mathrm{d}s}\left(\frac{sp'(s)}{\sqrt{1-p'(s)^2}}\right) + \omega^2 p = 0. \tag{2.114}$$

By the duality theory developed in this section, we know that

$$\begin{aligned} \inf_{u \in \mathcal{U}_k} \Pi(u) &= \inf_{p \in \mathcal{S}_s} \Pi^d(p), \\ \sup_{u \in \mathcal{U}_k} \Pi(u) &= \sup_{p \in \mathcal{S}_s} \Pi^d(p), \end{aligned}$$

which can be verified directly from the Euler-Lagrange equations. As pointed by Toland (1979), both primal and dual problems have an infinite number of critical points, and there is a one-to-one correspondence between the critical points of Π and those of Π^d, according to Theorem 2.6.4.

2.7.2 Commentary

Similarly to the minimum potential principle in static systems, the least action principle states that the solutions of a dynamic system are extremals of an action over closed curves. This fact was known to Fermat, and Maupertuis put it in the Hamiltonian formalism. However, in spite of its great aesthetic appeal, the least action principle has had little impact in classical Hamiltonian mechanics. This is because the total action of a convex Hamiltonian system is indefinite. The periodic solutions do not minimize the action, are just extremals, i.e. the stationary trajectories. Thus, as an inf-primal variational problem in dynamical systems, the least action principle is in fact a misnomer, and the classical methods of the calculus of variations do not apply to the dynamical systems.

The earliest duality theory of the double-min type is due to Toland (1978, 1979). His work concentrated on the difference of two convex functionals, i.e. the total action can be split into $\Pi(q) = W(\Lambda q) - P(q)$. The generalizations were made by Auchmuty (1983) to general nonconvex functionals with a linear operator Λ. Like Toland, he allowed the nonconvex functionals to be extended-real-valued, so that difficulties such as $\infty - \infty$ have to be handled. It is traditional in the calculus of variations and optimization that the primal problem is always taken to be a minimization problem. This is only for static systems where the sup-primal problems can be converted into inf-primal problems. However, this tradition obscured our view of the general problems. As we have learnt in dynamical systems, the least action principle is somehow misleading since the action density takes minimum and maximum values periodically on the time domains. Both the inf- and the sup-primal problems have to be considered simultaneously in a period.

By the super-Lagrange duality theory presented in this chapter, the interesting bi-duality theory is developed for generalized (both conservative and dissipative) convex Hamiltonian systems. This super-Lagrange duality theory provides a relatively clear picture of the extremal problems in convex Hamiltonian systems. Actually, this super-Lagrange theory was originally discovered by Gao (1997) in general nonconvex finite deformation theory, which will be discussed in the next chapter. Recently while writing this book, the author realized that this super-Lagrange duality theory plays an important role in elementary dynamical systems. Complementary to the double-min duality theorem, the double-max duality is equally important in the study of the periodic dynamics and it cannot be replaced by double-min duality theory.

The most submissive thing in the world can ride roughshod over the hardest in the world-that which is without substance entering that which has no crevices.

—Lao Chi, *Tao De Chin*, XLIII

The most important result of a general character to which the special theory of relativity has led is concerned with the conception of mass. Before the advent of relativity, physics recognized two conservation laws of fundamental importance, namely, the law of conservation of energy and the law of conservation of mass; these two fundamental laws appeared to be quite independent of each other. By means of the theory of relativity they have been united into one law.

—Albert Einstein, Relativity, 1916

II Symmetry Breaking: Triality Theory in Nonconvex Systems

3 TRI-DUALITY IN NONCONVEX SYSTEMS

Hidden harmony is stronger than the explicit one.

—Heraclitus

Symmetry breaking leads to phenomena.

—Pierre Curie

The shortest path between two truths in the real domain passes through the complex domain.

—Jacques Hadamard

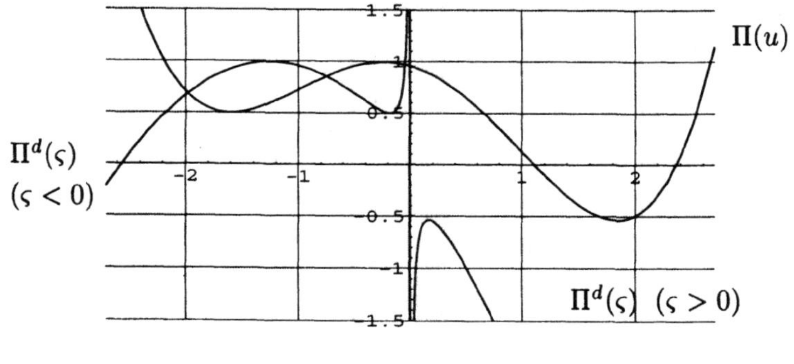

Figure 3.1. Triality in Nonconvex Optimization

From traditional convex systems to general nonlinear systems, we will study, in this chapter, the generalized duality theory and analytic solutions for one-dimensional nonconvex variational problems with applications to phase transitions, post-bifurcation, nonsmooth elastoplasticity and nonconvex dynamical systems. Because of the nonconvexity, the nice simple symmetry in the governing equations is broken and the beautiful one-to-one global duality relation no longer exists. The solutions in these systems are usually not unique. But, by introducing a suitable nonlinear operator Λ, a generalized framework may still be established for many systems. The resulting triality theory reveals interesting local dualities (see Fig. 3.1, the associated problem is given in Example 3.3).

The first two sections discuss, respectively, two kinds of symmetry breaking in one-dimensional problems, i.e.

(1) constitutive symmetry breaking, and

(2) geometrical symmetry breaking.

To recover the broken symmetry, a powerful canonical dual transformation method is proposed, which can be used to solve many nonconvex variational problems. A general analytic solution is obtained for mixed boundary-value problems. Its properties are controlled by the interesting triality theory.

We end up by providing a potentially powerful primal-dual method for solving nonconvex dynamical systems.

3.1 Constitutive Symmetry Breaking in Convex Systems

3.1.1 Legendre Duality Breaking: KKT conditions

Recall the strain-energy density and the kinetic-energy density in linear deformation dynamical systems, i.e.

$$U(\epsilon) = \frac{1}{2}k_0\epsilon^2, \quad K(v) = \frac{1}{2}m_0 v^2.$$

They are strictly convex functions on $\mathcal{E} = \mathcal{L}^2(I)$. Their Legendre conjugates

$$U^c(\sigma) = \frac{1}{2k_0}\sigma^2, \quad K^c(p) = \frac{1}{2m_0}p^2$$

are also strictly convex on the self-dual space $\mathcal{E}^* = \mathcal{L}^2(I)$. The nice symmetrical Legendre duality relations

$$\begin{aligned} \sigma = DU(\epsilon) \ &\Leftrightarrow\ \epsilon = DU^c(\sigma), \\ p = DK(v) \ &\Leftrightarrow\ v = DK^c(p) \end{aligned} \tag{3.1}$$

hold on the product space $\mathcal{L}^2 \times \mathcal{L}^2$. The strain-stress (ϵ, σ) pair for Hooke's material and the velocity-momentum pair (v, p) in Newton's theory are Legendre duality pairs on $\mathcal{L}^2 \times \mathcal{L}^2$.

For the finite deformation system, the stored energy in the geodesic string

$$W(\epsilon) = \int_I U(\epsilon)\,\mathrm{d}x = \int_I k_0\sqrt{1+\epsilon^2}\,\mathrm{d}x$$

is also a strictly convex functional on $\mathcal{E} = \mathcal{L}^1(I)$, and the constitutive law

$$\sigma = DU(\epsilon) = \frac{k_0 \epsilon}{\sqrt{1+\epsilon^2}}$$

is well defined on the whole space $\mathcal{L}^1(I)$. However, the Legendre conjugate of U is defined only on a subset $\mathcal{E}_a^*$ of the dual space $\mathcal{E}^* = \mathcal{L}^\infty(I) \subset \mathcal{L}^1$, i.e.

$$U^c(\sigma) = -\sqrt{k_0^2 - \sigma^2} \quad \sigma \in \mathcal{E}_a^*, \tag{3.2}$$

where $\mathcal{E}_a^* \subset \mathcal{E}^*$ is the so-called elastic yield space, defined by

$$\mathcal{E}_a^* = \{\sigma \in \mathcal{E}^* | \; |\sigma(x)| < k_0 \quad \forall x \in I\}, \tag{3.3}$$

in which the inequality constraint

$$Y^*(\sigma) = |\sigma(x)| - k_0 < 0, \;\; \forall x \in I \tag{3.4}$$

is called the *elastic limit condition* in solid mechanics. This inequality constraint ensures that the complementary energy

$$W^c(\sigma) = -\int_I \sqrt{k_0^2 - \sigma^2}\, dx \;\; s.t. \;\; |\sigma(x)| < k_0$$

is Lebesgue integrable on $\mathcal{E}_a^* \subset \mathcal{L}^\infty(I)$. The string will be broken if $\sigma \geq k_0$.

By the Legendre-Fenchel transformation, the conjugate function

$$W^*(\sigma) = \sup_{\epsilon \in \mathcal{E}}\{\langle \epsilon; \sigma\rangle - W(\epsilon)\} = \begin{cases} W^c(\sigma) & \text{if } \sigma \in \mathcal{E}_a^*, \\ +\infty & \text{if } \sigma \notin \mathcal{E}_a^*. \end{cases}$$

is convex, but nonsmooth on $\mathcal{E}^*$. In nonsmooth analysis, $W^* : \mathcal{E}^* \to \mathbb{R} \cup \{\infty\}$ is the so-called *super-potential* (see Chapter 5).

By the Legendre duality theorem, the inverse constitutive equation is equivalent to the constrained variational problem

$$\max_{\sigma \in \mathcal{E}_a^*}\{\langle \epsilon; \sigma\rangle - W^c(\sigma)\}. \tag{3.5}$$

This is a one-dimensional concave optimization problem with inequality constraint. Introducing a so-called *slate variable* $\delta > 0$, the inequality can be written as the equation

$$\sigma + \delta = k_0.$$

Thus, the classical Lagrangian for this constraint problem is

$$L_\delta(\sigma, \mu) = \int_I \left[\sigma\epsilon + \sqrt{k_0^2 - \sigma^2} - \mu(\sigma + \delta - k_0)\right] dx,$$

where $\mu \in \mathbb{R}$ is the Lagrange multiplier. For a give $\delta > 0$, the criticality condition $DL_\delta = 0$ leads to

$$\epsilon = \frac{k_0}{\sqrt{k_0^2 - \sigma^2}}\sigma + \mu, \quad 0 = \sigma + \delta - k_0. \tag{3.6}$$

Although the inequality constraint has been "relaxed" by the slate variable $\delta > 0$ and the Lagrange multiplier μ, new constrains involving δ and μ have to be satisfied. These are the well-known *Karush-Kuhn-Tucker conditions*, or simply the *KKT conditions*,

$$\delta > 0, \quad \mu \geq 0, \quad \mu\delta = 0. \tag{3.7}$$

General inequality constrained optimization problems and KKT theory will be discussed in later chapters. For the strict inequality constraint (3.4), the KKT conditions hold if and only if $\mu = 0$. Thus, the criticality conditions (3.6) lead back to the constrained inverse constitutive law

$$\epsilon = DU^c(\sigma) = \frac{k_0}{\sqrt{k_0^2 - \sigma^2}}\sigma, \quad s.t. \ \sigma - k_0 < 0.$$

Strictly speaking, $\mathcal{L}^1(I)$ is not a reflexive space. This inverse constitutive mapping is from $\mathcal{E}_a^* \subset \mathcal{L}^\infty$ to $(\mathcal{L}^\infty)^* \neq \mathcal{L}^1$. The nice Legendre duality symmetry (3.1) which exists in the linear deformation system is broken in the finite deformation system. The graphs of the conjugate energy densities and nonlinear constitutive relations are shown in Fig. 3.2.

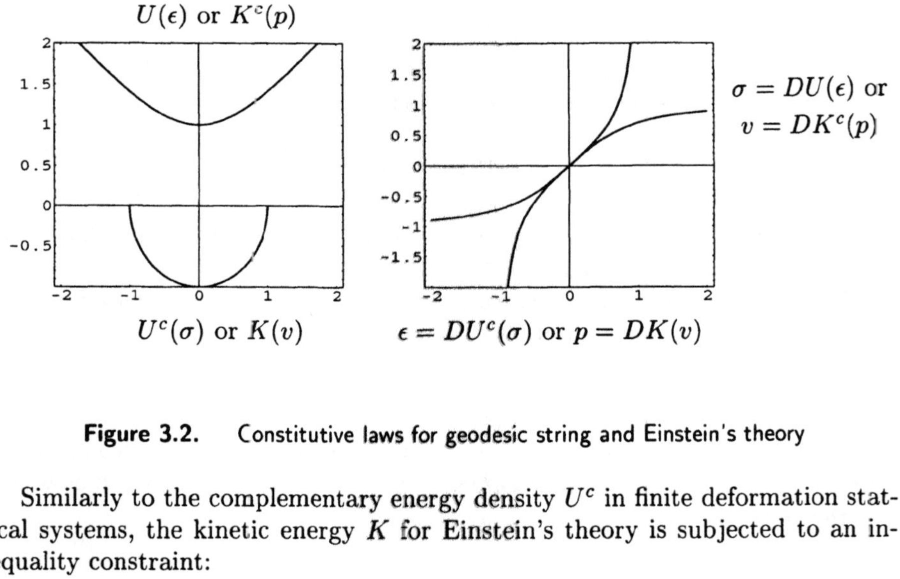

Figure 3.2. Constitutive laws for geodesic string and Einstein's theory

Similarly to the complementary energy density U^c in finite deformation statical systems, the kinetic energy K for Einstein's theory is subjected to an inequality constraint:

$$K(v) = -m_0c^2\sqrt{1 - v^2/c^2}, \quad s.t. \ |v(t)| < c. \tag{3.8}$$

The elastic yield space in this dynamical system becomes the speed limit space:

$$\mathcal{E}_a = \{v \in \mathcal{L}^\infty(I) | \ |v(t)| < c \ \forall t \in I\}.$$

Dually, the complementary kinetic energy density K^c is similar to the stored energy density $U(\epsilon)$ in the statical deformation system:

$$K^c(p) = c\sqrt{m_0^2c^2 + p^2}.$$

The nice symmetrical duality relation in Newton's dynamical system is also broken in the finite deformation dynamical system. This very interesting mirror symmetry between the statical and dynamical systems is a universal phenomena in nature.

3.1.2 Duality Restoration: Finite Deformation Measures

Recall that a pair (ϵ, ϵ^*) is said to be a Legendre duality pair on $\mathcal{E} \times \mathcal{E}^*$ if the Legendre duality relations hold

$$\epsilon^* = \frac{\partial U}{\partial \epsilon} \Leftrightarrow \epsilon = \frac{\partial U^c}{\partial \epsilon^*} \Leftrightarrow \epsilon^* \epsilon = U(\epsilon) + U^c(\epsilon^*) \quad \forall (\epsilon, \epsilon^*) \in \mathcal{E} \times \mathcal{E}^* \tag{3.9}$$

hold, or equivalently, (ϵ, ϵ^*) is a canonical duality pair on $\mathcal{E}_a \times \mathcal{E}_a^* \subset \mathcal{E} \times \mathcal{E}^*$ if and only if $DU : \mathcal{E}_a \to \mathcal{E}_a^*$ is one-to-one and onto.

To recover the broken symmetry in finite deformation systems, we introduce a nonlinear geometrical measure

$$\xi = \Lambda(u) = \frac{1}{2} a u'^2 + \lambda, \tag{3.10}$$

where $\Lambda : \mathcal{U} \to \mathcal{E}$ is a quadratic operator, and a, λ are constants.

For the geodesic string in the finite deformation statical system, $a = 2$, $\lambda = 1$ and

$$\xi = 1 + u_{,x}^2 > 0$$

is a strain measure. Let $\mathcal{E}_a = \{\xi \in \mathcal{L}^2(I) | \ \xi(x) > 0 \ \forall x \in I\}$. In terms of ξ, the canonical stored-energy density $\bar{V} : \mathcal{E}_a \to \mathbb{R}$ can be defined by

$$\bar{V}(\xi) = k_0 \sqrt{\xi} > 0 \tag{3.11}$$

which is a strictly concave function (see Fig. 3.3(a)). The conjugate stress dual to $\xi > 0$ is uniquely defined by the *canonical constitutive law*

$$\varsigma = D\bar{V}(\xi) = \frac{k_o}{2\sqrt{\xi}} > 0. \tag{3.12}$$

This nonlinear constitutive relation is invertible and $\xi = k_o^2/(4\varsigma^2)$. By the Legendre-Fenchel transformation,

$$\bar{V}^*(\varsigma) = \inf_{\xi > 0} \{\xi\varsigma - \bar{V}(\xi)\} = \begin{cases} -\frac{k_o^2}{4\varsigma} & \text{if } \varsigma > 0, \\ -\infty & \text{if } \varsigma \leq 0. \end{cases}$$

On the admissible space $\mathcal{E}_a^* = \{\varsigma \in \mathcal{L}^2(I) | \varsigma(x) > 0 \ \forall x \in I\}$, the complementary energy

$$\bar{V}^c(\varsigma) = \bar{V}^*(\varsigma) = -\frac{k_o^2}{4\varsigma} < 0, \quad \forall \varsigma \in \mathcal{E}_a^*$$

is also a strictly concave function. Thus, by using the finite deformation measure $\Lambda(u)$, the symmetric Legendre duality relation

$$\varsigma = \frac{k_0}{2\sqrt{\xi}} \Leftrightarrow \xi = \frac{k_0^2}{4\varsigma^2} \Leftrightarrow k_0\sqrt{\xi} - \frac{k_0^2}{4\varsigma} = \xi\varsigma \ \forall (\xi, \varsigma) \in \mathcal{E}_a \times \mathcal{E}_a^*$$

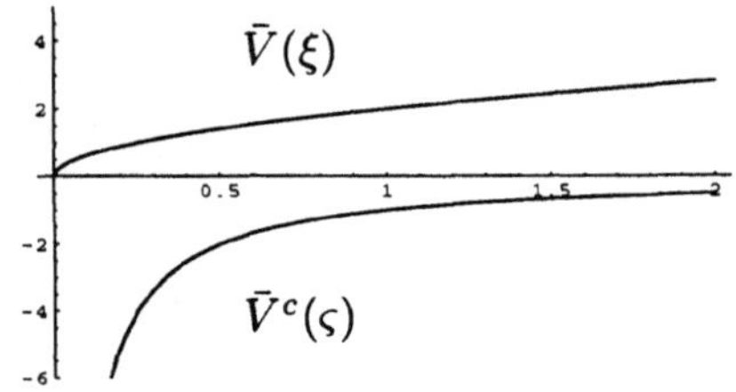

(a) Geodesic string

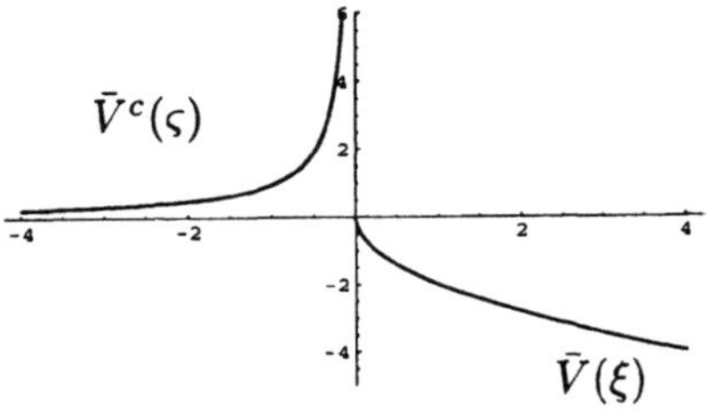

(b) Einstein's theory

Figure 3.3. Mirror constitutive laws: Canonical energies and their conjugates

is recovered. In n-dimensional space, the finite deformation measure ξ becomes the Riemannian metric tensor and its conjugate ς is the so-called *second Piola-Kirchhoff stress* (see Chapter 6).

For Einstein's theory, $a = -2c^2$, $\lambda = 1$, the "finite velocity measure" can be defined by

$$\xi = \Lambda(u) = 1 - u_{,t}^2/c^2 > 0 \quad \forall |u_{,t}| < c.$$

On $\mathcal{E}_a = \{\xi \in \mathcal{L}^2(I) |\ \xi(t) > 0 \ \ \forall t \in I\}$, the kinetic energy density in Einstein's theory is a convex function

$$\bar{V}(\xi) = -m_0 c^2 \sqrt{\xi} < 0 \tag{3.13}$$

of ξ (see Fig. 3.3(b)). The conjugate variable of ξ is given uniquely by the canonical constitutive law

$$\varsigma = D\bar{V}(\xi) = -\frac{m_0 c^2}{2\sqrt{\xi}} < 0. \tag{3.14}$$

By the Legendre-Fenchel transformation, the conjugate of $\bar{V}$ is given by

$$\bar{V}^*(\varsigma) = \sup_{\xi>0}\{\xi\varsigma - \bar{V}(\xi)\} = \begin{cases} -\frac{m_0^2 c^4}{4\varsigma} & \text{if } \varsigma < 0, \\ +\infty & \text{if } \varsigma \geq 0. \end{cases}$$

On the admissible space $\mathcal{E}_a^* = \{\varsigma \in \mathcal{L}^2(I) |\ \varsigma(t) < 0 \ \ \forall t \in I\}$, $\bar{V}^*$ is finite and differentiable. The complementary kinetic-energy density is also a strictly convex function:

$$\bar{V}^c(\varsigma) = -\frac{m_0^2 c^4}{4\varsigma} > 0 \ \ \forall \varsigma < 0,$$

and the inverse canonical constitutive equation for Einstein's theory is

$$\xi = D\bar{V}^c(\varsigma) = \frac{m_0^2 c^4}{4\varsigma^2} > 0.$$

In terms of the nonlinear geometrical measure $\xi = \Lambda(u)$, the broken symmetry of the duality relations is recovered:

$$\varsigma = D\bar{V}(\xi) \;\Leftrightarrow\; \xi = D\bar{V}^*(\varsigma) \;\Leftrightarrow\; \bar{V}(\xi) + \bar{V}^*(\varsigma) = \xi\varsigma.$$

Since $\xi = \Lambda(u)$ is a quadratic function of u', we let $\xi_t = \partial\xi/\partial u' = au'$. Then, the Gâteaux variation of ξ at $\bar{u}$ in the direction u is

$$\delta\xi(\bar{u}; u) = \xi_t(\bar{u})u' = a\bar{u}'u' = \Lambda_t(\bar{u})u,$$

where $\Lambda_t(\bar{u})$ is the Gâteaux derivative of Λ at $\bar{u}$, and

for the geodesic string: $\xi = \Lambda(u) = 1 + u_{,x}^2, \;\Rightarrow\; \Lambda_t(u) = 2u_{,x}\dfrac{\mathrm{d}}{\mathrm{d}x};$

for Einstein's theory: $\xi = \Lambda(u) = 1 - \frac{1}{c^2}u_{,t}^2, \;\Rightarrow\; \Lambda_t(u) = -\frac{2}{c^2}u_{,t}\dfrac{\mathrm{d}}{\mathrm{d}t}.$

For a given external force $\bar{f}$ and the associated equilibrium state $\bar{u}$, if $\bar{\varsigma} = D\bar{V}(\Lambda(\bar{u}))$, the virtual work principle still holds for this nonlinear operator, i.e.

$$\langle \delta\Lambda(\bar{u}; u); \bar{\varsigma}\rangle = \langle \Lambda_t(\bar{u})u; \bar{\varsigma}\rangle = \langle u, \bar{f}\rangle \quad \forall u \in \mathcal{U}_a.$$

Integration by parts leads to

$$\langle \Lambda_t(\bar{u})u; \bar{\varsigma}\rangle = \langle u, \bar{\Lambda}_t^*(\bar{u})\varsigma\rangle,$$

where $\bar{\Lambda}_t^*(u)$ is the adjoint operator of Λ_t, defined by

$$\bar{\Lambda}_t^*(\bar{u}) = \begin{cases} -(a\bar{u}'\varsigma)' & \text{in } I, \\ a\bar{u}'\varsigma n & \text{on } \partial I, \end{cases} \tag{3.15}$$

where n is an outward unit "vector" in $\mathbb{R}$ normal to the boundary ∂I. Then, the fundamental lemma leads to the equilibrium equation

$$\Lambda_t^*(\bar{u})\bar{\varsigma} = \bar{f}. \tag{3.16}$$

The fundamental equation can now be written as

$$A(u) = \Lambda_t^*(u)D\bar{V}(\Lambda(u)) = \bar{f}. \tag{3.17}$$

Clearly, for both geodesic string and Einstein's theory, the nonlinear operator A is not a self-adjoint operator, and the equilibrium operator Λ_t^* associated with nonlinear operator $\Lambda(u)$ depends on the configuration $\bar{u}$. The symmetry in the duality relation is restored, but the symmetry between the geometrical relation and the equilibrium equation is lost. We will see that the duality symmetry is more important then the geometrical symmetry. The diagrammatic framework for this nonlinear operator is shown in Fig. 3.4.

If the force field $\bar{f}$ is a potential field, i.e. if there exists a real-valued functional $F : \mathcal{U}_a \to \mathbb{R}$ such that $\bar{f} = DF(u)$, then in terms of the geometrically nonlinear

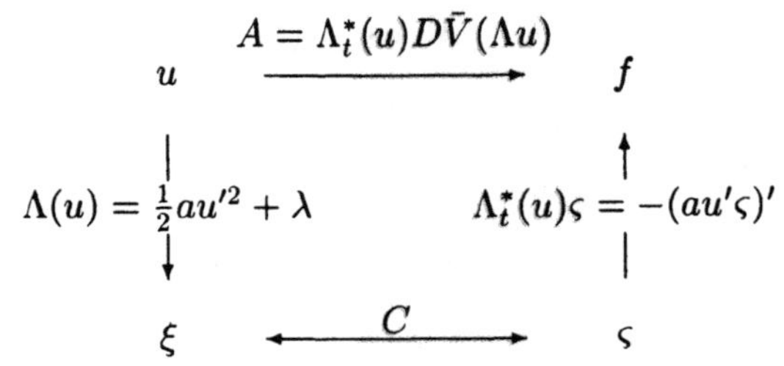

Figure 3.4. Duality representation for the nonlinear operator Λ.

operator Λ, the free energy Π in either the statical and the dynamical system can be written in the unified form

$$\Pi(u) = \int_I \bar{V}(\Lambda(u))\mathrm{d}I - F(u).$$

The kinematically admissible space $\mathcal{U}_k$ is now defined by

$$\mathcal{U}_k = \{u \in \mathcal{U}_a | \;\; \Lambda(u) \in \mathcal{E}_a\}. \tag{3.18}$$

It is easy to check that the Euler-Lagrange equation for the variational problem

$$(\Pi_{\text{sta}}): \quad \Pi(u) \quad \to \text{sta} \quad \forall u \in \mathcal{U}_k$$

is

$$\Lambda_t^*(\bar{u}) D\bar{V}(\Lambda(\bar{u})) = DF(\bar{u}). \tag{3.19}$$

Although this is still a nonlinear differential equation, we will see that by using the canonical dual transformation this problem can be solved completely.

3.1.3 Nonsmooth Constitutive Laws and Mirror Materials

We continue to present some typical one-dimensional examples of the uniaxial elastoplastic laws. These laws play important roles in nonsmooth analysis and mechanics. Theoretical analysis and applications will be given in later chapters.

3.1.3.1 Elastic-perfectly plastic string. When the strain exceeds the elastic limit ϵ_0, plastic deformation begins to take place. For the *elasto-perfectly plastic string* (no hardening effects, see Fig. 3.5(a)), the constitutive law reads

$$\sigma = \begin{cases} k_0\epsilon & \text{if } \epsilon < \epsilon_0, \\ \sigma_0 & \text{if } \epsilon \geq \epsilon_0. \end{cases} \tag{3.20}$$

The stored-energy density is a smooth, convex function of ϵ, given by

$$U(\epsilon) = \int \sigma(\epsilon) d\epsilon = \begin{cases} \frac{1}{2} k_0 \epsilon^2 & \text{if } \epsilon < \epsilon_0, \\ \frac{1}{2} k_0 \epsilon_0^2 + \sigma_0(\epsilon - \epsilon_0) & \text{if } \epsilon \geq \epsilon_0. \end{cases} \tag{3.21}$$

Its Legendre-Fenchel conjugate U^* is a nonsmooth function (see Fig. 3.5(b)), given by

$$U^*(\sigma) = \sup_{\epsilon} \{\sigma\epsilon - U(\epsilon)\} = \begin{cases} \frac{1}{2} k_0^{-1} \sigma^2 & \text{if } \sigma \leq \sigma_0, \\ +\infty & \text{if } \sigma > \sigma_0. \end{cases} \tag{3.22}$$

On the *plastic yield space*

$$\mathcal{S}_a = \{\sigma \in \mathcal{E}^* | \ \sigma \leq \sigma_0\},$$

the complementary energy-density $U^c = U^* = \frac{1}{2} k_0^{-1} \sigma^2$ is convex. By solving the constrained problem

$$\sup_{\sigma} \{\epsilon\sigma - U^c(\sigma)\}, \ \ \forall \sigma \in \mathcal{S}_a,$$

the inverse constitutive law can be obtained in the form

$$\epsilon = \begin{cases} k_0^{-1}\sigma & \text{if } \sigma < \sigma_0, \\ k_0^{-1}\sigma_0 + \mu & \text{if } \sigma = \sigma_0, \ \ \mu \geq 0, \end{cases} \tag{3.23}$$

subject to the *KKT conditions*

$$\mu \geq 0, \ \ \sigma \leq \sigma_0, \ \ \mu(\sigma - \sigma_0) = 0. \tag{3.24}$$

In plasticity theory, the Lagrange multiplier $\mu \geq 0$ is called the *plastic flow factor.*

3.1.3.2 Elastic locking materials. The *universal conservation law* says that if you see a phenomenon in the mirror, then that phenomenon should also occur in nature. The so-called *elastic locking material*, which was first studied by W. Prager (see Fig. 3.5(b)), is completely dual to the elasto-perfactly plastic material. Exchanging the roles of the dual variables ϵ and σ in the constitutive laws discussed above, we have

$$\sigma = \begin{cases} k_0\epsilon & \text{if } \epsilon < \epsilon_0, \\ k_0\epsilon_0 + \mu^* & \text{if } \epsilon = \epsilon_0, \ \ \mu^* \geq 0, \end{cases}$$

subjected to the KKT conditions

$$\mu^* \geq 0, \ \ \epsilon \leq \epsilon_0, \ \ \mu^*(\epsilon - \epsilon_0) = 0$$

in strain space. A possible name for the dual Lagrange multiplier $\mu^* \geq 0$ could be the *locking strength factor.* The admissible space $\mathcal{E}_a$ in this material is

$$\mathcal{E}_a = \{\epsilon \in \mathcal{E} | \ \epsilon \leq \epsilon_0\}. \tag{3.25}$$

Dual to (3.22), the stored potential energy density can be defined by

$$U(\epsilon) = \frac{1}{2} k_0 \epsilon^2 \ \ \ \epsilon \in \mathcal{E}_a, \tag{3.26}$$

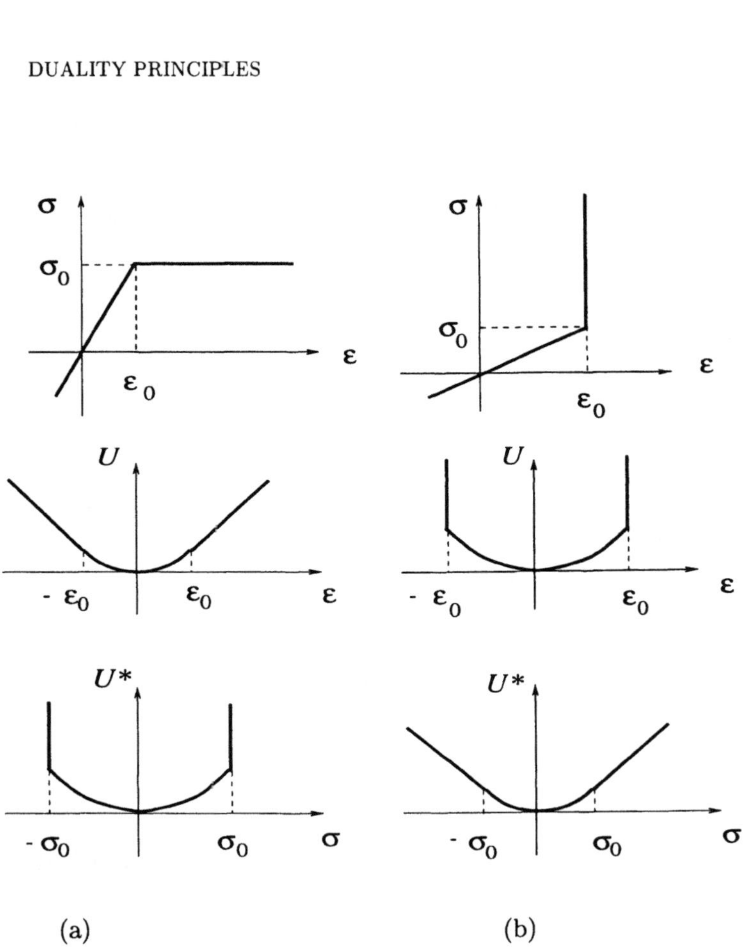

Figure 3.5. Mirror materials: (a) elastic-perfectly plastic and (b) locking material

which is convex and differentiable on $\mathcal{E}_a$. Its Legendre-conjugate function U^* is a convex function on $\mathcal{E}^*$, namely

$$U^c(\sigma) = \max_{\epsilon \in \mathcal{E}_a} \{\sigma\epsilon - U(\epsilon)\} = \begin{cases} \frac{1}{2}k_0^{-1}\sigma^2 & \text{if } \sigma \leq \sigma_0, \\ \frac{1}{2}k_0^{-1}\sigma_0 + \sigma_0(\sigma - \sigma_0) & \text{if } \sigma > \sigma_0. \end{cases}$$

The inverse constitutive law

$$\epsilon = \begin{cases} k_0^{-1}\sigma & \text{if } \sigma < \sigma_0, \\ \sigma_0 & \text{if } \sigma = \sigma_0, \ \ \mu \geq 0, \end{cases} \tag{3.27}$$

for this locking material is continuous on $\mathcal{E}^*$, $\mu \geq 0$ being the plastic flow factor, which is subjected to the KKT conditions (3.24).

3.1.3.3 Nonsmooth materials with linear hardening effects.

The nonsmooth constitutive law for a linear hardening string with stress jump effect can be written as

$$\sigma = \begin{cases} k_0\epsilon & \text{if } \epsilon < \epsilon_0, \\ \sigma_1 + k_1(\epsilon - \epsilon_0) & \text{if } \epsilon \geq \epsilon_0, \end{cases} \tag{3.28}$$

where ϵ_0, k_0, k_1, σ_1 are material constants and $\sigma_0 = k_0\epsilon_0$. The stress-strain graph of this model is shown in Fig. 3.6(a). The stored-energy density is then a nonsmooth, convex function of ϵ (see Fig. 3.6(c)) and is given by

$$U(\epsilon) = \int \sigma(\epsilon)d\epsilon = \begin{cases} \frac{1}{2}k_0\epsilon^2 & \text{if } \epsilon \leq \epsilon_0, \\ \frac{1}{2}k_0\epsilon_0^2 + \sigma_1(\epsilon - \epsilon_0) + \frac{1}{2}k_1(\epsilon - \epsilon_0)^2 & \text{if } \epsilon > \epsilon_0. \end{cases}$$

However, the complementary energy is a smooth function of σ (see Fig. 3.6(d)). This is given by

$$U^c(\sigma) = \begin{cases} \frac{1}{2k_0}\sigma^2 & \text{if } \sigma \leq \sigma_0, \\ \frac{1}{2k_0}\sigma_0^2 + \epsilon_0(\sigma - \sigma_0) & \text{if } \sigma_0 < \sigma \leq \sigma_1, \\ \frac{1}{2k_0}\sigma_0^2 + \epsilon_0(\sigma - \sigma_0) + \frac{1}{2k_1}(\sigma - \sigma_1)^2 & \text{if } \sigma > \sigma_1. \end{cases} \tag{3.29}$$

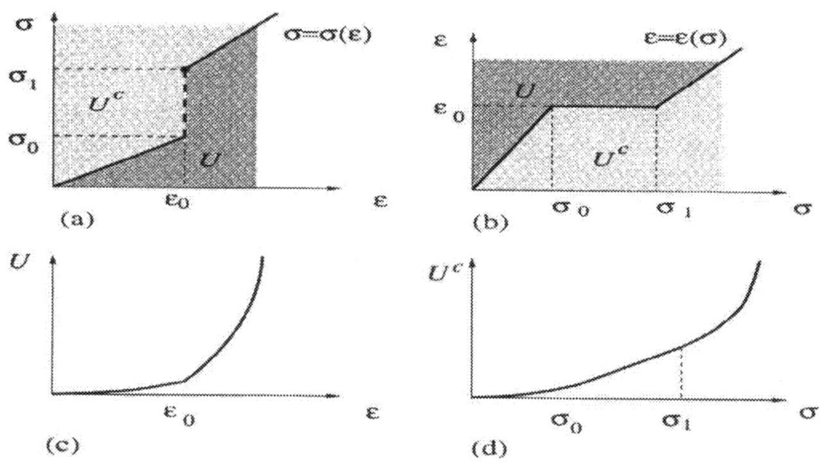

Figure 3.6. Discontinuous constitutive relation: Nonsmooth U and smooth U^c

The inverse constitutive relation $\epsilon = DU^*$ is then a continuous function (see Fig. 3.6(b)), given by

$$\epsilon = \begin{cases} \frac{1}{k_0}\sigma & \text{if } \sigma \leq \sigma_0, \\ \epsilon_0 & \text{if } \sigma_0 < \sigma \leq \sigma_1, \\ \epsilon_0 + \frac{1}{k_1}(\sigma - \sigma_1) & \text{if } \sigma > \sigma_1. \end{cases} \tag{3.30}$$

Similarly, the mirror material is the well-studied elasto-perfect plastic material with linear hardening, defined by

$$\sigma = \begin{cases} k_0\epsilon & \text{if } \epsilon \leq \epsilon_0, \\ \sigma_0 & \text{if } \epsilon_0 < \epsilon \leq \epsilon_1, \\ \sigma_0 + k_1(\epsilon - \epsilon_1) & \text{if } \epsilon \geq \epsilon_1. \end{cases}$$

The stored-energy density in this model is the smooth convex function of ϵ (see Fig. 3.6(d))

$$U(\epsilon) = \int \sigma(\epsilon) d\epsilon = \begin{cases} \frac{1}{2} k_0 \epsilon^2 & \text{if } \epsilon \leq \epsilon_0, \\ \frac{1}{2} k_0 \epsilon_0^2 + \sigma_0(\epsilon - \epsilon_0) & \text{if } \epsilon_0 < \epsilon \leq \epsilon_1, \\ \frac{1}{2} k_0 \epsilon_0^2 + \sigma_0(\epsilon_1 - \epsilon_0) + k_1(\epsilon - \epsilon_0)^2 & \text{if } \epsilon > \epsilon_0. \end{cases}$$

The complementary energy, however, is the nonsmooth function of σ

$$U^c(\sigma) = \begin{cases} \frac{1}{2k_0}\sigma^2 & \text{if } \sigma < \sigma_0, \\ \frac{1}{2k_0}\sigma_0^2 + \epsilon_1(\sigma - \sigma_0) + \frac{1}{2k_1}(\sigma - \sigma_1)^2 & \text{if } \sigma > \sigma_1. \end{cases}$$

The inverse constitutive relation $\epsilon = DU^*$ is then a discontinuous function:

$$\epsilon = \begin{cases} \frac{1}{k_0}\sigma & \text{if } \sigma < \sigma_0, \\ \mu & \text{if } \sigma = \sigma_0, \\ \epsilon_1 + \frac{1}{k_1}(\sigma - \sigma_0) & \text{if } \sigma > \sigma_0. \end{cases}$$

3.2 Geometrical Symmetry Breaking: Framework in Nonconvex Systems

In this section, the symmetry breaking is illustrated by three examples from nonconvex systems. By introducing an appropriate nonlinear operator Λ, the canonical framework is maintained, but the fundamental operator A is no longer symmetric.

3.2.1 Static Bifurcation: van de Waals Double-Well Energy

Let us consider the one-dimensional nonconvex variational problem

$$P(u) = \int_0^1 \frac{1}{2} C_o (\frac{1}{2} u'^2 - \lambda)^2 \, dx - \int_0^1 f u \, dx \quad \to \min \; \forall u \in \mathcal{U}_k, \tag{3.31}$$

where $f(x)$ is a given function; $\lambda > 0$ is a parameter, $C_o > 0$ is a material constant, and the feasible set $\mathcal{U}_k$ can be given simply by

$$\mathcal{U}_k = \{u \in \mathcal{C}^1(0,1) | \, u'^2 \in \mathcal{L}^2(I), \;\; u(0) = 0\}.$$

This nonconvex variational problem appears frequently in association with phase transitions in fluid and solids, and in post-buckling analysis. The Euler equation for this nonconvex variational problem is a nonlinear ordinary differential equation in the domain I with the natural boundary condition at $x = 1$, i.e.

$$\begin{cases} \left(u' C_o (\frac{1}{2} u'^2 - \lambda)\right)' + f(x) = 0 & \forall x \in I = (0,1), \\ u' C_o (\frac{1}{2} u'^2 - \lambda) = 0 & \text{at } x = 1. \end{cases} \tag{3.32}$$

If we choose a linear operator $\Lambda = d/\,dx$, and let $\epsilon = \Lambda u = u'$, the stored-energy density

$$V(\epsilon; \lambda) = \frac{1}{2} C_o (\frac{1}{2} \epsilon^2 - \lambda)^2$$

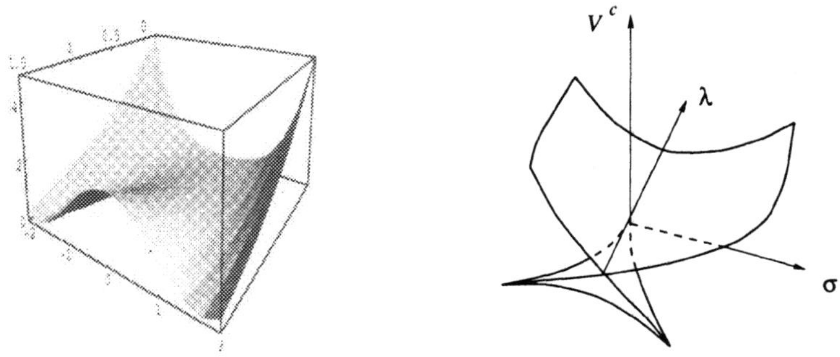

(a) van der Waals' energy (b) Complementary van der Waals' energy

Figure 3.7. van der Waals' energy and its complementary energy

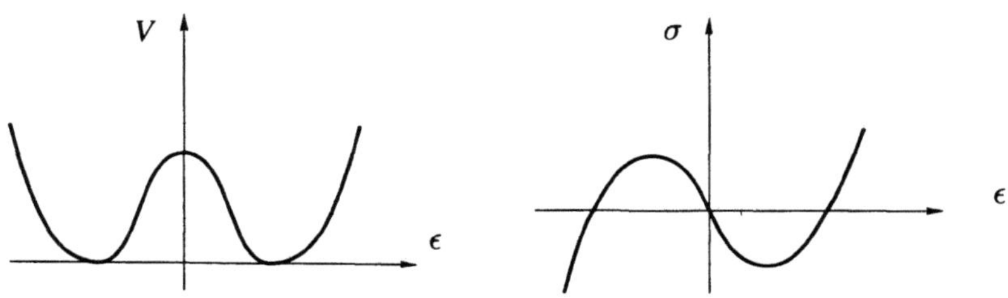

(a) Double-well energy (b) Nonlinear constitutive law

Figure 3.8. Double-well energy and its derivative

is the so-called van der Waals function. It is convex for $\lambda < 0$. However, if $\lambda > 0$, it is a so-called *double-well energy* (see Fig. 3.7(a)). The physical equation is then nonlinearly dependent on ϵ (see Fig. 3.8), i.e.

$$\sigma = \frac{\partial V}{\partial \epsilon} = C_o \epsilon (\frac{1}{2}\epsilon^2 - \lambda) = C(\epsilon),$$

which defines the notation $C(\xi)$. Since $V(\epsilon, \lambda)$ is nonconvex, the inverse constitutive relation $\epsilon = \epsilon(\sigma, \lambda)$ is not unique, and the complementary energy defined by

$$V^c(\sigma; \lambda) = \epsilon(\sigma)\sigma - V(\epsilon(\sigma); \lambda)$$

does not have a simple algebraic form (see Fig. 3.7(b)). Hence, in terms of the linear operator $\Lambda = \mathrm{d}/\,\mathrm{d}x$ and the first Piola-Kirchhoff stress σ, it is difficult to obtain the Hamiltonian for this nonconvex variational problem.

In order to recover the broken symmetry in the duality relation, we introduce a nonlinear geometric measure $\xi = \Lambda(u) = \frac{1}{2}u'^2 - \lambda$, where Λ is a quadratic operator. Thus, the stored energy $\bar{V}(\xi) = \frac{1}{2}C_o\xi^2$ is a strictly convex function of

ξ and the physical equation

$$\varsigma = D\bar{V}(\xi) = C_o \xi$$

is linear. Its inverse is simply equal to $\xi = C_o^{-1}\varsigma$. The complementary energy is then

$$\bar{V}^c(\varsigma) = \frac{1}{2} C_o^{-1} \varsigma^2.$$

Clearly, the following equivalent duality relations hold:

$$\varsigma = D\bar{V}(\xi) \quad \Leftrightarrow \quad \xi = D\bar{V}^c(\varsigma) \quad \Leftrightarrow \quad \bar{V}(\xi) + \bar{V}^c(\varsigma) = \xi\varsigma.$$

The Gâteaux variation of $\xi(u)$ at u in the direction v is then

$$\delta\xi(u; v) = \xi_t(u)v' = u'v' = \Lambda_t(u)v,$$

where $\Lambda_t(u) = u' \frac{\mathrm{d}}{\mathrm{d}x}$ is the Gâteaux derivative of $\xi = \Lambda(u)$ at u. By the chain rule, the relation between σ and ς is given by

$$\sigma = \frac{\partial V(\epsilon)}{\partial \epsilon} = \frac{\partial \bar{V}(\xi)}{\partial \xi} \frac{\partial \xi}{\partial \epsilon} = \varsigma \xi_t. \tag{3.33}$$

Recalling the general virtual work principle Theorem 1.2.1, we note that for the nonlinear operator $\xi = \Lambda(u)$ we have

$$\langle \varsigma; \delta\xi(u; v) \rangle = \int_I \varsigma u'v' \, \mathrm{d}x = \int_I \sigma(u)v' \, \mathrm{d}x = \int_I fv \, \mathrm{d}x = \langle f, v \rangle \quad \forall v \in \mathcal{U}_a.$$

Integration by parts leads to

$$\langle \varsigma; \Lambda_t(u)v \rangle = u'\varsigma vn|_{\partial I} - \int_I (u'\varsigma)' \, \mathrm{d}x = \langle \bar{\Lambda}_t^*(u)\varsigma, v \rangle,$$

where the adjoint operator $\bar{\Lambda}_t^*$ of Λ_t for this mixed boundary-value problem is defined by

$$\bar{\Lambda}_t^*(u)\varsigma = \begin{cases} \Lambda_t^*(u)\varsigma = -(u'\varsigma)' & \forall x \in I = (0, 1), \\ J_{u^*}(u'\varsigma) = u'\varsigma n & \text{on } \partial I, \end{cases}$$

and $\Lambda_t^*(u)$ is the formal adjoint of Λ_t in the domain I. For the given external load $\bar{f}(x)$ in I and f_e on the boundary $x = 1$, let $\bar{u}^* = \{\bar{f}(x) \ (x \in I), \ 0 \ (x = 1)\}$. The equilibrium conditions (3.32) can then be written in the abstract form

$$\bar{\Lambda}_t^*(u)\varsigma = \bar{u}^* \quad \Rightarrow \quad \begin{cases} -(u'\varsigma)' = \bar{f}(x) & \forall x \in (0, 1), \\ u'\varsigma(x) = 0 & \text{at } x = 1. \end{cases}$$

Thus, the tri-canonical forms associated with the nonconvex variational problem (3.31) are

$$\begin{array}{ll} \text{(a) nonlinear deformation eqn:} & \xi = \Lambda(u); \\ \text{(b) linear constitutive eqn:} & \varsigma = C_o\xi; \\ \text{(c) bilinear equilibrium eqn:} & \bar{u}^* = \bar{\Lambda}_t^*(u)\varsigma. \end{array} \tag{3.34}$$

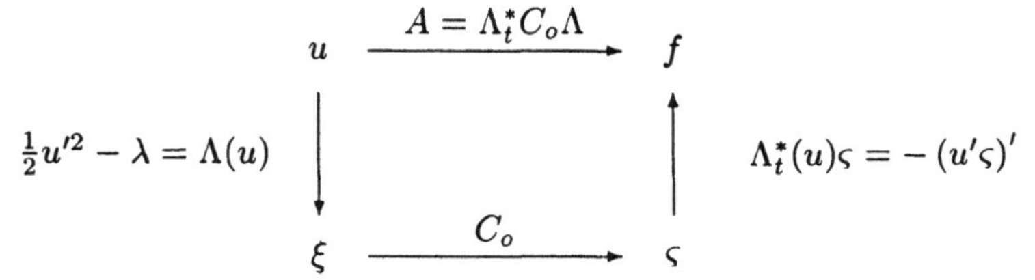

Figure 3.9. Diagrammatic framework for the van der Waals nonconvex system.

The governing equation (3.32) can be written in the abstract form

$$A(u) = \Lambda_t^*(u) C_o \Lambda(u) = \bar{f}(x) \ \ \forall x \in (0,1). \tag{3.35}$$

The nonlinear differential operator $A(u) = \bar{\Lambda}_t^*(u) C_o \Lambda(u)$ is no longer symmetric because of the nonconvexity of the problem.

The generalized Hamiltonian for this nonconvex variational problem is the convex functional

$$H(u, \varsigma) = \int_I \bar{V}^c(\varsigma) \, dx + \int_I \bar{f} u \, dx,$$

but the Hamiltonian canonical forms

$$\Lambda(u) = D_\varsigma H(u, \varsigma), \qquad \Lambda_t^*(u)\varsigma = D_u H(u, \varsigma)$$

are no longer symmetrical. This symmetry breaking is due to the nonlinearity of the finite-deformation operator Λ.

3.2.2 Post-buckling: von Kármán's Paradox and Extended Beam Theory

3.2.2.1 von Kármán's beam model. The classical large deformation thin elastic plate model in the two-dimensional domain $\Omega \subset \mathbb{R}^2$ was first proposed by von Kármán in 1910. Ciarlet's version of this plate model has the form (Ciarlet, 1980)

$$\begin{aligned} &K_0 \Delta\Delta w - h \nabla \cdot (\sigma \cdot \nabla w) = f, \\ &\sigma = C(\epsilon + \tfrac{1}{2} \nabla w \otimes \nabla w), \ \ \epsilon = \tfrac{1}{2}\left(\nabla u + \nabla^T u\right), \\ &\nabla \cdot \sigma = 0, \end{aligned} \tag{3.36}$$

where $w(x)$ is the deflection of the plate, $u(x)$ is the plane displacement vector, $\sigma = \{\sigma_{\alpha\beta}\}$ and $\epsilon = \{\epsilon_{\alpha\beta}\}$ denote, respectively, the stress and strain tensors in the plane of the plate, C is the Hooke's tensor, h is the plate thickness, $K_0 = E_0 h^3/12(1 - \nu^2)$ is the bending rigidity of the plate and E_0 is the modulus of elasticity and $\nu \in (0, \frac{1}{2})$ Poisson's ratio. Clearly, this is a coupled nonlinear partial differential system. The second-order nonlinear term

$$\nabla w \otimes \nabla w = \{w_{,\alpha} w_{,\beta}\}, \quad \alpha, \beta = 1, 2, \tag{3.37}$$

arises from inclusion of moderately large rotation of the plate. The von Kármán model should be considered as a fundamental theory of nonlinear thin elastic plates and the associated boundary-value problems have been studied extensively by both mathematicians and engineers for almost a century.

However, in one-dimensional problems, von Kármán's plate becomes a "beam model" and the equations can be written simply as

$$K_0 w'''' - h(\sigma w')' = f \quad \text{in } I = (0, \ell), \tag{3.38}$$

$$\sigma' = 0 \quad \text{in } I, \tag{3.39}$$

$$\sigma = E_0(u' + \frac{1}{2}w'^2) \quad \text{in } I. \tag{3.40}$$

Here $w'(x)$ is simply the slope of the deformed beam. Equation (3.39) shows that the axial stress σ has to be a constant $\sigma = -\lambda/h$ everywhere in the domain $I = (0, \ell)$. The integral constant $\lambda \in \mathbb{R}$ represents the axial load applied at the end of the beam, determined by the axial boundary condition. Thus, equation (3.38) is really a linear equation:

$$K_0 w'''' + \lambda w'' = f. \tag{3.41}$$

The total potential energy associated with this linear equation is

$$P(w) = \int_I \left(\frac{1}{2} K_0 w''^2 - \lambda w'^2 \right) \, dx - \int_I f w \, dx. \tag{3.42}$$

If the beam is subjected to an axial extension load, then $\lambda \leq 0$. In this case P is strictly convex and the exact solution of this linear problem is uniquely dependent on the external transverse loads $f(x)$ in the domain and the boundary conditions.

However, if the beam is subjected to a compressive load and the magnitude $\lambda > 0$ is not given then the total potential P is nonconvex. If $f = 0$, the homogeneous linear equation (3.41) is equivalent to the well-known eigenvalue problem

$$\lambda_c = \inf_w \frac{\int_I K_0 w''^2 \, dx}{\int_I w'^2 \, dx}, \tag{3.43}$$

and the lowest eigenvalue $\lambda_c > 0$ is the *Euler buckling load.*

In any case, the von Kármán model for one-dimensional problems yields a linear ordinary differential equation, which can only be used in pre-buckling analysis of elastic beam problems. In the history of structural mechanics, the one-dimensional von Kármán model has been considered as a traditional nonlinear beam theory and has been discussed in many well-known textbooks (see, for example, Washizu, 1968). Numerical methods and variational approaches for solving this "nonlinear model" still appear in scientific journals (see Galka and Telega, 1995).

3.2.2.2 Extended Beam Model. The main reason behind the von Kármán paradox is that the seconder order nonlinear term (3.37) is considered for in-plane deformations, but it is ignored in the thickness direction. It may be appropriate for thin plates, but for one-dimensional beam models, this is wrong! It is shown in Gao (1996) that the strain in the thickness direction of the beam is proportional to the second-order term w'^2, and cannot be ignored when the beam is subjected to moderately large rotations. Thus, an extended beam model was proposed such that

$$K_0 w'''' + k_0(\lambda - \frac{1}{2}w'^2)w'' + f = 0 \ \forall x \in I, \tag{3.44}$$

where k_0 is a positive material constant. The total potential energy associated with this nonlinear beam theory is then given by

$$P(w) = \int_I \frac{1}{2}\left(K_0 w''^2 + k_0(\frac{1}{2}w'^2 - \lambda)^2\right) \, \mathrm{d}x - \int_I fw \, \mathrm{d}x \tag{3.45}$$

If the beam is clamped at $x = 0$, simply supported at $x = \ell$, the kinematically admissible space $\mathcal{U}_k$ can be written as

$$\mathcal{U}_k = \{w \in C^2(0,\ell) | \ w(0) = w'(0) = 0, \ \ w(\ell) = w''(\ell) = 0\}.$$

It is clear that for the given Euler buckling load $\lambda_c > 0$

$$\int_I K_0 w''^2 \, \mathrm{d}x \geq \lambda_c \int_I w'^2 \, \mathrm{d}x, \ \ \forall w \in \mathcal{U}_a.$$

Thus, on $\mathcal{U}_k$,

$$\begin{aligned} P(w) \ &\geq \ \int_I \frac{1}{2}\left(\lambda_c w'^2 + k_0(\frac{1}{2}w'^2 - \lambda)^2\right) \, \mathrm{d}x - \int_I fw \, \mathrm{d}x \\ &= \ P_\lambda(w) + \lambda\lambda_c/k_0 - \lambda_c^2/(4k_0^2), \end{aligned}$$

and

$$P_\lambda(w) = \int_I \frac{1}{2}k_0(w'^2 - \lambda_p)^2 \, \mathrm{d}x - \int_I fw \, \mathrm{d}x, \tag{3.46}$$

where $\lambda_p = \lambda - \lambda_c/(2k_0)$. Clearly, when the parameter $\lambda_p > 0$, the beam is in a post-buckled state. The total potential P_λ is then a double-well energy. It has three critical points: two local minimizers, corresponding to two possible stable buckled states, and one local maximizer, corresponding to an unstable buckled state.

It is interesting to note that in the phase transitions of Ericksen's bar subjected to a tensile axial load, the total potential is the same as $P_\lambda(w)$ (see, Ericksen, 1975), while the perturbed problem studied in Carr, Gurtin and Slemrod (1984) has the total potential $P(w)$ with $K_0 = 1/\omega$.

To express this extended beam theory within our framework, we first let

$$\epsilon = \begin{pmatrix} \kappa \\ \epsilon \end{pmatrix} = \Lambda w = \begin{pmatrix} -\dfrac{\mathrm{d}^2}{\mathrm{d}x^2} \\ \dfrac{\mathrm{d}}{\mathrm{d}x} \end{pmatrix} w$$

where $\kappa = -w''$ denotes the bending curvature, and $\epsilon = w'$ is the deformation. Thus, Λ is a vector-valued linear operator. The stored-energy density for this linear deformation operator is then a nonconvex function, namely

$$V(\epsilon, \kappa) = \frac{1}{2} K_0 \kappa^2 + \frac{1}{2} k_0 (\lambda - \frac{1}{2}\epsilon^2)^2.$$

For a given ϵ, the strain energy is a convex function of κ. But, for given bending curvature κ, $V(\epsilon, \kappa)$ is a double-well function of ϵ. The dual variable of ϵ is then defined by

$$\epsilon^* = DV(\epsilon) = \begin{pmatrix} M \\ \sigma \end{pmatrix} \Rightarrow \begin{cases} M = \frac{\partial V}{\partial \kappa} = K_0 \kappa, \\ \sigma = \frac{\partial V}{\partial \epsilon} = k_0 \epsilon (\frac{1}{2}\epsilon^2 - \lambda) \end{cases}$$

where M is the bending moment and σ is the first *Piola-Kirchhoff stress*, which depends nonlinearly on ϵ, and the constitutive relation is not therefore uniquely irreversible.

To find one-to-one constitutive relations, we define the generalized strain vector for this nonconvex problem as

$$\boldsymbol{\xi} = \begin{pmatrix} \kappa \\ \xi \end{pmatrix} = \Lambda(w) = \begin{pmatrix} -w'' \\ \frac{1}{2}w'^2 - \lambda \end{pmatrix},$$

where $\xi = \frac{1}{2}w'^2 - \lambda$ is a Cauchy-Green type axial strain measure, and Λ is a vector-valued quadratic operator. The Gâteaux variation of $\boldsymbol{\xi}$ at $\bar{w}$ is

$$\delta\boldsymbol{\xi}(\bar{w}; w) = (-w'', \ \bar{w}'w')^T.$$

In terms of $\boldsymbol{\xi} = (\kappa, \ \xi)^T$, the stored energy

$$\bar{V}(\boldsymbol{\xi}) = \frac{1}{2} K_0 \kappa^2 + \frac{1}{2} k_0 \xi^2$$

is strictly convex since it is quadratic. Then, the Kirchhoff-type stress, defined by $\boldsymbol{\varsigma} = D\bar{V}(\boldsymbol{\xi})$, is linearly dependent on $\boldsymbol{\xi}$, i.e.

$$\boldsymbol{\varsigma} = \begin{pmatrix} M \\ \varsigma \end{pmatrix} = \begin{pmatrix} K_0 & 0 \\ 0 & k_0 \end{pmatrix} \begin{pmatrix} \kappa \\ \xi \end{pmatrix} = \mathbf{k}_0 \boldsymbol{\xi}. \tag{3.47}$$

The constitutive operator $\mathbf{k}_0 : \mathcal{E} \to \mathcal{E}^*$ is an affine operator, and $\varsigma = k_0(\xi - \lambda)$ is a second Piola-Kirchhoff type axial stress. The parameter λ can be considered either as an axial load, or as a residual strain.

The bilinear form for the vector-valued dual pair$(\boldsymbol{\xi}, \boldsymbol{\varsigma})$ may now be defined by

$$\langle \boldsymbol{\xi}; \boldsymbol{\varsigma} \rangle = \int_I \boldsymbol{\varsigma} \cdot \boldsymbol{\xi} \, \mathrm{d}x = \int_I (M\kappa + \varsigma\xi) \, \mathrm{d}x.$$

Integration by parts leads to the Gauss-Green formula for Λ_t, i.e.

$$\langle \boldsymbol{\varsigma}; \Lambda_t(w)w \rangle = J(\boldsymbol{\varsigma}, w) + \langle \Lambda_t^*(w)\boldsymbol{\varsigma}, w \rangle, \tag{3.48}$$

where the formal adjoint Λ_t^* of Λ_t is defined by

$$\Lambda_t^*(w)\varsigma = -M'' - (w'\varsigma)' \quad \text{in } I, \tag{3.49}$$

and the conjunct $J(\varsigma, w)$ in this beam problem is defined by

$$J(\varsigma, w) = (-Mw' + Mw + \varsigma w'w)\,|_{\partial I}.$$

Thus, the equilibrium equation is

$$\Lambda_t^*(w)\varsigma = -M'' - (w\varsigma)' = f,$$

and, in terms of $\boldsymbol{\xi} = (\kappa, \xi)^T$ and $\varsigma = (M, \varsigma)^T$, the tri-canonical forms in this large deformation beam problem can be written as

$$\begin{aligned}
\boldsymbol{\xi} = \Lambda(w) \quad &\Rightarrow \quad \kappa = -w'', \;\; \xi = \frac{1}{2}w'^2 - \lambda, \\
\varsigma = D\bar{V}(\boldsymbol{\xi}) \quad &\Rightarrow \quad M = K_0\kappa, \;\; \varsigma = k_0\xi \\
f = \Lambda_t^*(w)\varsigma \quad &\Rightarrow \quad M'' + (w'\varsigma)' + f = 0.
\end{aligned}$$

The fundamental governing equation for this extended beam theory is the nonlinear differential equation

$$A(w) = \Lambda_t^*(w)D\bar{V}(\Lambda(w)) = f(x). \tag{3.50}$$

The operator A in this case is not self-adjoint. The framework for this fourth-order nonlinear differential system is depicted in Fig. 3.10.

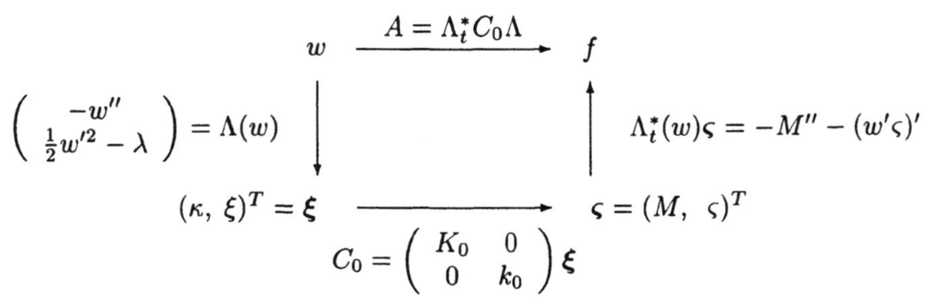

Figure 3.10. Diagrammatic framework in the extended beam theory

3.2.3 Dynamical Bifurcation: Duffing Equation and Its Canonical Form

In the problem of particle dynamics discussed in Chapter 2, if the potential energy $U(q)$ is a double-well energy of the form

$$U(q) = \frac{1}{2}k_0(\frac{1}{2}q^2 - \lambda)^2 + fq,$$

where $k_0 > 0$, $\lambda > 0$ are given parameters, then the total action of the system is

$$\Pi(q) = \int_I \frac{1}{2}\left(m_0 q_{,t}^2 - \frac{1}{2}k_0(\frac{1}{2}q^2 - \lambda)^2\right) dt - \int_I fq dt \quad (3.51)$$

The associated Euler equation is the well-known forced *Duffing equation*

$$m_0 q_{,tt} = k_0 q(\lambda - \frac{1}{2}q^2) + f, \quad (3.52)$$

and the classical Hamiltonian for this system is the nonconvex functional

$$H(q,p) = \int_I \left(\frac{1}{2m_0}p^2 + \frac{1}{2}k_0(\frac{1}{2}q^2 - \lambda)^2\right) dt + \int_I fq dt. \quad (3.53)$$

Hence, the Duffing system is a nonconvex Hamiltonian system. Let $z = (q,\ p)^T$. Then, the Hamiltonian canonical forms for this nonconvex dynamical system leads to the nonlinear initial-value problem

$$\begin{cases} \dot{z}(t) = \mathbf{J}^{\ominus} DH(z), \\ z(t_0) = z_0. \end{cases} \quad (3.54)$$

This nonconvex dynamical system is very sensitive to the numerical methods and the initial condition $z(0) = z_0$. For the given periodic external force $f(t) = 0.1\sin(0.5t)$, the parameter $\lambda = 1.5$, and the fixed initial condition $z_0 = (0,1)^T$, Fig. 3.11 shows that different numerical methods produce very different results. Fig. 3.12 shows that by using the same method, very small change of the initial conditions leads to quite different dynamical states.

Direct methods for solving nonconvex dynamical systems can only provide one-side numerical approaches. The numerical results vary with the methods used. Without comparison with the results of dual methods, numerical results produced by the direct methods cannot be trusted unconditionally.

By introducing the nonlinear geometrical operator Λ such that

$$\boldsymbol{\xi} = \begin{pmatrix} v \\ \xi \end{pmatrix} = \Lambda(q) = \begin{pmatrix} q_{,t} \\ \frac{1}{2}q^2 - \lambda \end{pmatrix},$$

we find that the action density $\bar{V}(v,\xi)$ is a saddle function

$$\bar{V}(v,\xi) = \frac{1}{2}m_0 v^2 - \frac{1}{2}k_0\xi^2. \quad (3.55)$$

The dual variable $\varsigma = D\bar{V}(\epsilon)$

$$\varsigma = \begin{pmatrix} p \\ \varsigma \end{pmatrix} = \begin{pmatrix} m_0 & 0 \\ 0 & -k_0 \end{pmatrix}\begin{pmatrix} v \\ \xi \end{pmatrix} = C_0\boldsymbol{\xi} \quad (3.56)$$

depends linearly on $\boldsymbol{\xi}$. The constitutive matrix C_0 in this canonical dynamical system is an indefinite matrix. The complementary action density $\bar{V}^c$ defined by

$$\bar{V}^c(p,\varsigma) = \varsigma\boldsymbol{\xi} - \bar{V}(\boldsymbol{\xi}) = \frac{1}{2m_0}p^2 - \frac{1}{2k_0}\varsigma^2 \quad (3.57)$$

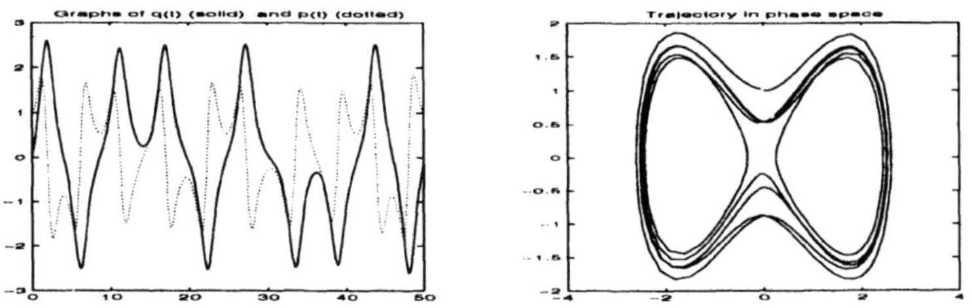

(a) Numerical results computed by 'ode15s'.

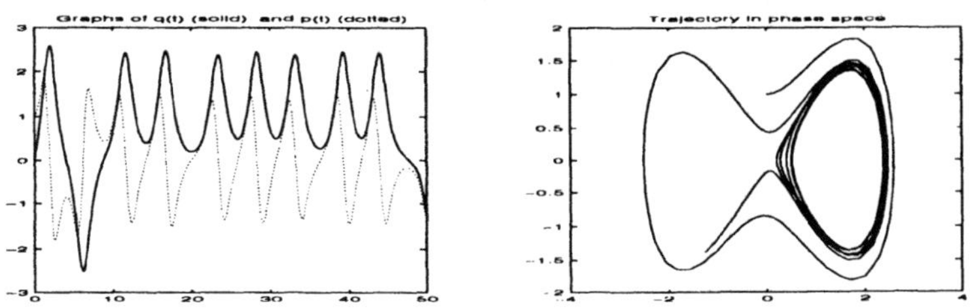

(b) Numerical results computed by 'ode23'.

Figure 3.11. Numerical results produced by two different numerical methods in MATLAB

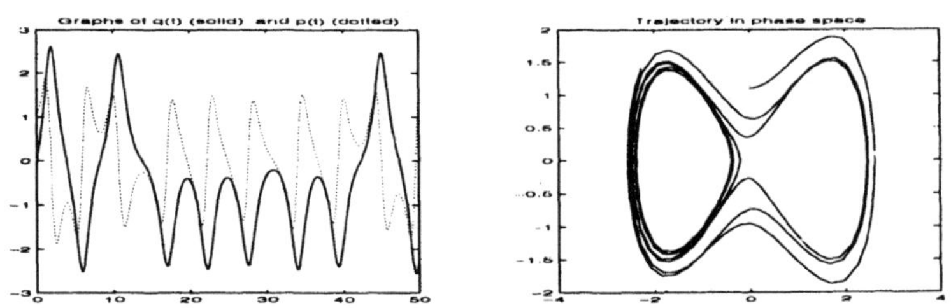

(a) Graphs of $q(t)$ (solid) and $p(t)$ (dashed) (b) Graph in phase space

Figure 3.12. Numerical results computed by 'ode23' with initial condition $z_0 = (0, 1.1)^T$

Period three implies chaos.

—Tien-Yien Li and James A. Yorke, 1975

is also a saddle function. The kinematically admissible space $\mathcal{U}_k$ can be defined by

$$\mathcal{U}_k = \{q \in \mathcal{C}^1(0,t_1) | \ \Lambda(q) \in \mathcal{L}^2(0,t_1), \ \ q(0)=0\}.$$

By defining the bilinear forms

$$\langle \xi ; \varsigma \rangle = \int_I (pv + \varsigma \xi) \mathrm{d}t, \quad \langle q, f \rangle = \int_I fq \mathrm{d}t,$$

the virtual work principle for this nonconvex Hamiltonian system can be written as

$$\langle \delta \xi(\bar{q}; q); \varsigma \rangle = \langle q, \bar{\Lambda}_t^*(\bar{q})\bar{\varsigma} \rangle = \langle q, f \rangle \ \ \forall q \in \mathcal{U}_k, \tag{3.58}$$

where the adjoint operator $\bar{\Lambda}_t^*$ in this nonconvex dynamical system is defined by

$$\bar{\Lambda}_t^*(\bar{q})\varsigma = \begin{cases} \Lambda_t^*(\bar{q})\varsigma = -p_{,t} + \bar{q}\varsigma & \forall t \in I, \\ p & \text{at } t = t_1. \end{cases} \tag{3.59}$$

Thus, the forced Duffing equation can also be written in the general abstract form

$$A(q) = \Lambda_t^*(q) C_0 \Lambda(q) = \bar{f},$$

but, again, the operator A is not a symmetric mapping. The framework for this nonconvex Hamiltonian system is represented in Fig. 3.13.

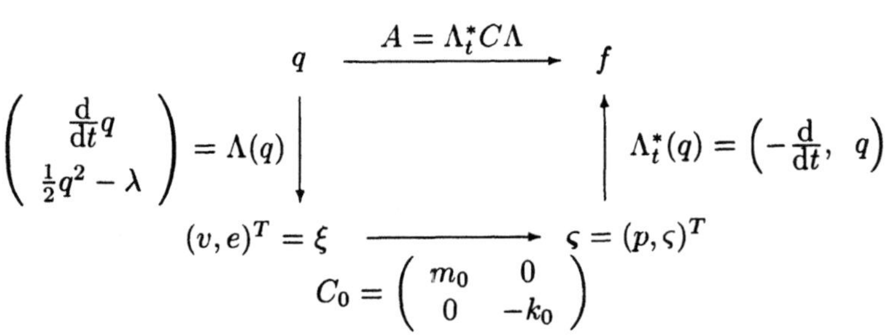

Figure 3.13. Diagrammatic representation for the framework in dynamical bifurcation

Bowed down then preserved; Bent then straight;
Hollow then full; Worn then new;
A little then benefited; A lot then perplexed.
Therefore the sage embraces the One and is a model for the empire.

—Lao Chi, *Tao De Chin*, XXII

3.3 Quadratic Canonical Transformation and the Gap Functional

This section presents a quadratic canonical transformation method for solving some nonconvex variational problems. The key idea of this method is to choose the appropriate operator Λ to ensure that the Lagrange duality relation holds for the canonical constitutive laws. The method and results introduced in this section are of fundamental importance for the rest of book.

3.3.1 Nonconvex Boundary-Value Problems and Duality Gap

Let us consider a nonconvex system defined on an open interval $I = (0, \ell) \subset \mathbb{R}$. At $x = 0$, the configuration of the system $u = u_0$ is given, while at $x = \ell$, the boundary source σ_ℓ is prescribed. Let $V : I \times \mathcal{L}^\gamma(I) \to \mathcal{L}(I)$ be a non-negative differentiable function, i.e.

$$V(x, \epsilon) \geq 0 \ \forall (x, \epsilon) \in I \times \mathcal{L}^\gamma, \quad \gamma \in [1, +\infty],$$

satisfying the generalized *Carathéodory continuity conditions*

$$\begin{array}{l} \text{for almost all } x \in I, \quad V(x,\epsilon) \text{ is piecewise continuous in } \epsilon; \\ \text{for all } \epsilon, \ \ V(x,\epsilon) \text{ is measurable with respect to } x \in I. \end{array} \tag{3.60}$$

We seek the general analytic solution of the following mixed boundary-value problem

Problem 3.3.1 For the given distributed load $\bar{f}(x)$ in the domain I and the boundary data u_0 and σ_ℓ, find $u(x)$ and the associated $\epsilon(x), \sigma(x)$ such that

$$(\mathcal{P}_{\text{bv}}): \quad \left\{ \begin{array}{lll} \epsilon = \dfrac{\mathrm{d}}{\mathrm{d}x} u & x \in I, & u(0) = u_0, \\ \sigma = D_\epsilon V(x, \epsilon) & x \in I, & \\ \bar{f} = -\dfrac{\mathrm{d}}{\mathrm{d}x}\sigma & x \in I, & \sigma(\ell) = \sigma_\ell. \end{array} \right. \tag{3.61}$$

♠

Since the stored energy function $V(x, \epsilon)$ is not convex in $\epsilon = u'$, the constitutive relation $\sigma = D_\epsilon V(x, \epsilon)$ may not be globally invertible. The fundamental equation

$$A(u) = -\left(D_{u'} V(x, u')\right)' = \bar{f}(x) \tag{3.62}$$

is nonlinear and non-symmetric.

Let $\mathcal{U} = \mathcal{L}^\alpha(I)$ be a Lebesgue integrable space with $1 < \alpha < \infty$. Its dual space $\mathcal{U}^* = \mathcal{L}^{\alpha^*}(I)$ is the external force space, and $\alpha^* \in (1, +\infty)$ is the dual number of α, i.e. $\frac{1}{\alpha} + \frac{1}{\alpha^*} = 1$. The bilinear form $\langle *, * \rangle : \mathcal{U} \times \mathcal{U}^* \to \mathbb{R}$ for this boundary-value problem is defined simply by

$$\langle u, u^* \rangle_{\bar{I}} = \int_I u(x) u^*(x) \, \mathrm{d}x + u(x) u^*(x) n|_{x \in \partial I},$$

where $n = \{1(x = \ell), \ \ -1(x = 0)\}$. The admissible configuration space $\mathcal{U}_a$ is defined by

$$\mathcal{U}_a = \{u \in \mathcal{L}^\alpha(I) | \ u(0) = u_0\}. \tag{3.63}$$

For the given external load $\bar{u}^* = \bar{f}(x)$ in the domain and $\bar{u}^* = \sigma_\ell$ at $x = \ell$, the external energy $F : \mathcal{U}_a \to \mathbb{R}$ is then a linear functional:

$$F(u) = \int_I \bar{f} u \, dx + \sigma_\ell u(\ell),$$

and its Gâteaux derivative leads back to the dual variable $\bar{u}^* \in \mathcal{U}^* = \mathcal{L}^{\alpha^*}$ of $u \in \mathcal{U}_a$. Thus,

$$\bar{u}^* = DF(u) = \begin{cases} \bar{f}(x) & \forall x \in I, \\ \sigma_\ell & \text{at } x = \ell. \end{cases}$$

The total potential associated with the nonlinear $(\mathcal{P}_{\text{bv}})$ is the nonconvex functional

$$\begin{aligned} P(u) &= W(u') - F(u) \\ &= \int_I V(x, u') \, dx - \int_I \bar{f} u \, dx - \sigma_\ell u(x_\ell). \end{aligned} \tag{3.64}$$

Since P is generally nonconvex, on the kinematically admissible space $\mathcal{U}_k = \{u \in \mathcal{U}_a | \ u' \in \mathcal{L}^\gamma(I)\}$, the variational problem associated with $(\mathcal{P}_{\text{bv}})$ given in (3.61) is the stationary problem

$$(\mathcal{P}_{\text{sta}}) : \quad P(u) \ \to \text{sta} \ \forall u \in \mathcal{U}_k.$$

For a linear functional $F : \mathcal{U}_a \to \mathbb{R}$, the Legendre conjugate F^c is defined simply by

$$\begin{aligned} F^c(u^*) &= \text{sta}\{\langle u, u^* \rangle_I - F(u) | \ \forall u \in \mathcal{U}_a\} \\ &= -u_0 u^*(0) \ \forall u^* \in \mathcal{U}_a^*, \end{aligned}$$

where $\mathcal{U}_a^* \subset \mathcal{U}^*$ is defined by

$$\mathcal{U}_a^* = \{u^* \in \mathcal{U}^* | \ u^*(x) = \bar{f} \ \forall x \in I, \ u^*(\ell) = \sigma_\ell\}.$$

The Fenchel conjugate of $W(\epsilon)$ is defined by

$$\begin{aligned} W^*(\sigma) &= \sup_\epsilon \{\langle \epsilon; \sigma \rangle - W(\epsilon)\} = \int_I \sup_\epsilon \{\epsilon\sigma - V(x, \epsilon)\} \, dx \\ &= \int_I V^*(x, \sigma) \, dx. \end{aligned}$$

Let $\mathcal{S}_a \subset \mathcal{L}^{\gamma^*}$ be the range of the constitutive mapping $DW : \mathcal{L}^\gamma \to \mathcal{L}^{\gamma^*}$. The statically admissible space is defined by

$$\mathcal{S}_s = \{\sigma \in \mathcal{S}_a | \ \bar{\Lambda}^* \sigma = \bar{u}^*\}.$$

Then, on $\mathcal{S}_s$, the relaxed complementary energy $P^d : \mathcal{S}_s \to \mathbb{R}$ is defined by

$$P^d(\sigma) = F^c(\bar{\Lambda}^* \sigma) - W^*(\sigma).$$

Theorem 3.3.1 (Relaxed Complementary Energy Principle)
Suppose that for a given source $\bar{u}^* \in \mathcal{U}_a^*$, $\mathcal{S}_s$ is no-empty. Then, any critical point $\bar{\sigma}$ of the variational problem

$$(\mathcal{P}^d_{\text{sta}}): \quad \delta P^d(\bar{\sigma};\sigma) = 0 \ \ \forall \sigma \in \mathcal{S}_s \tag{3.65}$$

is a global maximizer of P^d on $\mathcal{S}_s$ and the inequality

$$\inf_{u \in \mathcal{U}_k} P(u) - \sup_{\sigma \in \mathcal{S}_s} P^d(\sigma) \geq 0 \tag{3.66}$$

holds. The equality holds if $W(\epsilon)$ is convex on $\mathcal{E}$. In this case, the critical point $\bar{\sigma}$ of P^d and the associated $\bar{u}$ defined by $\Lambda \bar{u} = DW^*(\bar{\sigma})$ solve the $(\mathcal{P}_{\text{bv}})$ given in Problem 3.3.1. If $W : \mathcal{E} \to \mathbb{R}$ is strictly convex, then the $(\mathcal{P}_{\text{bv}})$ has at most one solution.

Proof. Since $W^* : \mathcal{S}_a \to \mathbb{R}$ is convex, and $F^c : \mathcal{U}_a^* \to \mathbb{R}$ is linear, the relaxed complementary energy $P^d : \mathcal{S}_s \to \mathbb{R}$ is convex. Thus, any critical point on $\mathcal{S}_s$ is a global minimizer. By the *Fenchel-Young inequality*

$$W(\epsilon) + W^*(\sigma) \geq \langle \epsilon; \sigma \rangle \ \ \forall (\epsilon, \sigma) \in \mathcal{L}^\gamma \times \mathcal{L}^{\gamma^*},$$

we have

$$\begin{aligned} P(u) - P^d(\sigma) &= W(\Lambda u) - F(u) + W^*(\sigma) - F^c(\bar{\Lambda}^* \sigma) \\ &\geq \langle u, \bar{\Lambda}^* \sigma - F(u) - F^c(\bar{\Lambda}^* \sigma) \\ &= 0 \ \ \forall (u, \sigma) \in \mathcal{U}_k \times \mathcal{S}_s. \end{aligned}$$

If $W(\epsilon)$ is convex on $\mathcal{L}^\gamma$, then the Fenchel-Young inequality becomes an equality. This leads to $P(u) - P^d(\sigma) = 0$ on $\mathcal{U}_k \times \mathcal{S}_s$. If $W(\epsilon)$ is strictly convex, then $P(u)$ is strictly convex on $\mathcal{U}_k$. If $\mathcal{U}_k$ is non-empty, the primal problem has at most one minimizer, which solves the $(\mathcal{P}_{\text{bv}})$. □

Remark 3.3.1 (Duality Gap) This theorem shows that in variational problem, the primal problem $(\mathcal{P}_{\text{sta}})$ is not equivalent to the relaxed complementary variational problem $(\mathcal{P}^d_{\text{sta}})$. The difference

$$c = \inf_{u \in \mathcal{U}_k} P(u) - \sup_{\sigma \in \mathcal{S}_s} P^d(\sigma) \neq 0$$

is the so-called *duality gap*. ■

In order to eliminate this duality gap, we need to consider a new method.

3.3.2 Canonical Dual Transformation and Operator Decomposition

The key idea of the *canonical dual transformation method* is to introduce a suitable geometric measure $\xi = \Lambda(u)$ to recover the Legendre duality relation. Let

$\mathcal{E} = \mathcal{L}^{\beta}(I)$, $\beta \in (1, +\infty)$, with $\mathcal{E}_a \subset \mathcal{E}$ a subspace, and let $\Lambda : \mathcal{U} \to \mathcal{E}_a \subset \mathcal{E}$ be the quadratic operator

$$\Lambda(x,u) = \frac{1}{2}a(x)u'(x)^2 + b(x)u'(x) + c(x), \quad \text{with} \quad a(x) > 0 \ \forall x \in I. \tag{3.67}$$

Assumption 3.3.1 For a given function $V(x, \epsilon) : I \times \mathcal{L}^{\gamma} \to \mathcal{L}$, assume that the real-valued functions $a(x) > 0$, $b(x)$ and $c(x)$ can be chosen in such a way that there exists a non-negative function $\bar{V}(\xi) : \mathcal{E} \to \mathcal{L}(I)$ satisfying:

(A1) $V(x, u') = \bar{V}(\Lambda(x,u)) \ \forall (x,u) \in I \times \mathcal{C}^1(I)$;

(A2) $\bar{V} : \mathcal{E}_a \subset \mathcal{E} \to \mathcal{L}(I)$ is piecewisely differentiable such that $D\bar{V} : \mathcal{E}_a \to \mathcal{E}_a^* \subset \mathcal{E}^*$ is one-to-one and onto. ♣

Detailed mathematical properties of the nonlinear composition operator $\bar{V} \circ \Lambda : \mathcal{U} \to \mathcal{L}(I)$ are given in the appendix at the end of this chapter. In the following, $\Lambda(x, u)$ is simply written as $\Lambda(u)$.

By this assumption, the canonical constitutive relation $\varsigma = \bar{V}_{,\xi}(\xi)$ is invertible and the Legendre conjugate of $\bar{V}$

$$\bar{V}^c(\varsigma) = \varsigma\xi(\varsigma) - \bar{V}(\xi(\varsigma))$$

is well defined on $\mathcal{E}_a^* \subset \mathcal{E}^*$. The total internal energy $\bar{W} : \mathcal{E} \to \mathbb{R}$ of the system, given by

$$\bar{W}(\xi) = \int_I \bar{V}(\xi) \, dx,$$

is Gâteaux differentiable on $\mathcal{E}_a$. The conjugate variable of $\xi \in \mathcal{E}$ is then defined by

$$\varsigma = D\bar{W}(\xi) = \frac{\partial \bar{V}(\xi)}{\partial \xi} \in \mathcal{E}^*.$$

Let $\mathcal{T} = \mathcal{E}^*$. The spaces $\mathcal{E}$ and $\mathcal{T}$ are placed in duality by the standard bilinear form

$$\langle *; * \rangle : \mathcal{E} \times \mathcal{T} \to \mathbb{R}, \quad \langle \xi; \varsigma \rangle = \int_I \varsigma\xi \, dx.$$

Thus, $\mathcal{T} = \mathcal{L}^{\beta^*}(I)$, and β, $\beta^* \in (1, +\infty)$ are dual numbers defined by (1.7). By the Legendre transformation, the total complementary energy $\bar{W}^c : \mathcal{T} \to \mathbb{R}$ can be obtained as

$$\bar{W}^c(\varsigma) = \text{sta}_{\xi \in \mathcal{E}} \{ \langle \xi; \varsigma \rangle - \bar{W}(\xi) \} = \int_I \bar{V}^c(\varsigma) \, dx.$$

By the assumption, the Legendre duality equivalent relation

$$\varsigma = D\bar{W}(\xi) \ \Leftrightarrow \ \xi = D\bar{W}^c(\varsigma) \ \Leftrightarrow \ \langle \xi; \varsigma \rangle = \bar{W}(\xi) + \bar{W}^c(\varsigma). \tag{3.68}$$

holds on the product space $\mathcal{E}_a \times \mathcal{T}_a$.

Since $\xi = \Lambda(u)$ is a quadratic function of u', its Gâteaux variation at $\bar{u}$ in the direction u is

$$\delta\xi(\bar{u}; u) = \Lambda_t(\bar{u})u = (b + a\bar{u}')u' = \xi_t(\bar{u})u',$$

where $\xi_t(\bar{u}) = \partial\xi/\partial u' = b + a\bar{u}'$, and

$$\Lambda_t(\bar{u}) = (b + a\bar{u}')\frac{\mathrm{d}}{\mathrm{d}x}$$

is the Gâteaux derivative of ξ at $\bar{u}$, which is a differential operator depending on $\bar{u}$. Since $\Lambda(u)$ is Gâteaux differentiable on $\mathcal{U}$, for any $u \in \mathcal{U}$, the *operator decomposition*

$$\Lambda(u) = \Lambda_t(u)u + \Lambda_c(u) \tag{3.69}$$

plays a crucial role in the canonical dual transformation method, where $\Lambda_c : \mathcal{U} \to \mathcal{E}$ is the *complementary operator* of Λ_t, defined by

$$\Lambda_c(u) = \Lambda(u) - \Lambda_t(u)u = c - \frac{1}{2}au'u'.$$

For fixed $\bar{u} \in \mathcal{U}$, the Taylor expansion of Λ at $\bar{u}$ reads

$$\Lambda(\bar{u} + u) = \Lambda(\bar{u}) + \Lambda_t(\bar{u})u - \Lambda_c(u) + c.$$

The decomposition (3.69) of the geometrical operator Λ is of fundamental importance in nonconvex problems. As already seen, the adjoint of Λ_t is the equilibrium operator; on the other hand, Λ_c leads to a so-called *complementary gap function.*

3.3.3 Canonical Boundary-Value Problem and Complementary Gap Functional

For the quadratic operator Λ, the kinematically admissible space in nonconvex problems is defined by

$$\mathcal{U}_k = \{u \in \mathcal{U}_a |\ \Lambda(u) \in \mathcal{E}_a\}. \tag{3.70}$$

Using integration by parts, we obtain

$$\begin{aligned}
\langle \varsigma ; \Lambda_t(\bar{u})u \rangle &= \int_I (b + a\bar{u}')u'\varsigma \, \mathrm{d}x \\
&= (a\bar{u}' + b)\varsigma(x)un|_{\partial I} - \int_I [(b + a\bar{u}')\varsigma]'u \, \mathrm{d}x \\
&= \langle \bar{\Lambda}_t^*(\bar{u})\varsigma, u \rangle_{\bar{I}},
\end{aligned}$$

where the adjoint $\bar{\Lambda}_t^*(\bar{u})$ of $\Lambda_t(\bar{u})$ is defined by

$$\bar{\Lambda}_t^*(\bar{u})\varsigma = \begin{cases} \Lambda_t^*(\bar{u})\varsigma = -[(b + a\bar{u}')\varsigma]' & \forall x \in I, \\ (a\bar{u}' + b)\varsigma n & x \in \partial I, \end{cases} \tag{3.71}$$

and Λ_t^* is the formal adjoint of Λ_t, which depends on $\bar{u}$. The conjunct

$$J(\xi_t(\bar{u})\varsigma, u) = (a\bar{u}' + b)\varsigma u$$

associated with the nonlinear operator $\Lambda(u)$ is also dependent on $\bar{u}$.

Let $\bar{u}^* \in \mathcal{U}^*$ be a given external load, i.e.

$$\bar{u}^* = \begin{cases} \bar{f}(x) & \forall x \in I, \\ \sigma_\ell & \text{at } x = \ell. \end{cases} \tag{3.72}$$

For the nonlinear operator Λ, the virtual work principle can be proposed.

Proposition 3.3.1 (Canonical Virtual Work Principle)
Let $\bar{u}^* \in \mathcal{U}^*$ be a given external source, and $\bar{u}$ and $\bar{\varsigma} = D\bar{W}(\Lambda(\bar{u}))$ the associated equilibrium states. Then, for any given virtual state $u \in \mathcal{U}_k$, which vanishes at $x = 0$,

$$\langle \delta\xi(\bar{u}; u); \bar{\varsigma}\rangle = \langle \bar{u}^*, u\rangle_I \quad \forall u \in \mathcal{U}_k, \tag{3.73}$$

and the equilibrium equation

$$\bar{\Lambda}_t^*(\bar{u})\bar{\varsigma} = \bar{u}^* \tag{3.74}$$

is equivalent to the Euler equation (3.62).

Proof. By the virtual work principle Theorem 1.2.1, for the given external force $\bar{u}^*$ and associated $\bar{u}$ and $\bar{\sigma} = DV(x, \bar{u}')$, we have

$$\langle u'; \bar{\sigma}\rangle = \langle u, \bar{u}^*\rangle_I.$$

By the chain rule, we have

$$\bar{\sigma} = \frac{\partial V(x, \bar{u}')}{\partial \bar{u}'} = \frac{\partial \bar{V}(\xi(\bar{u}'))}{\partial \xi}\frac{\partial \xi(\bar{u}')}{\partial \bar{u}'} = \bar{\varsigma}\xi_t(\bar{u}). \tag{3.75}$$

Thus,

$$\langle u'; \bar{\varsigma}\xi_t(\bar{u})\rangle = \langle \xi_t(\bar{u})u'; \bar{\varsigma}\rangle = \langle \delta\xi(\bar{u}; u); \bar{\varsigma}\rangle = \langle u, \bar{u}^*\rangle_I.$$

Integration by parts leads to

$$\bar{\Lambda}_t^*(\bar{u})\bar{\varsigma} = \left(\frac{\mathrm{d}}{\mathrm{d}x}\right)^* \bar{\varsigma}\xi_t(\bar{u}) = \begin{cases} -\bar{\sigma}' = \bar{f} & \text{in } I, \\ n\bar{\sigma} = \sigma_\ell & \text{at } x = \ell. \end{cases}$$

This is the Euler equation (3.62). □

The virtual work principle for the nonlinear operator $\Lambda(u)$ shows that the value of the bilinear forms $\langle *; *\rangle$ and $\langle *, *\rangle$ are no longer equal to each other at the equilibrium state. By the operator decomposition (3.69), it is clear that

$$\langle \varsigma; \Lambda(u)\rangle = \langle \bar{\Lambda}_t^*(u)\varsigma, u\rangle_I + G(u, \varsigma), \tag{3.76}$$

where G is the so-called *complementary gap functional*, defined by

$$G(u, \varsigma) = \langle \varsigma; -\Lambda_c(u)\rangle = \int_I \left(\frac{1}{2}au'^2 - c\right)\varsigma \,\mathrm{d}x. \tag{3.77}$$

In terms of the quadratic operator Λ, the nonconvex boundary-value problem $(\mathcal{P}_{\rm bv})$ can be written as the so-called *canonical boundary-value problem*, $(\mathcal{P}_{\rm cbv})$ for short, as follows.

Problem 3.3.2 (Canonical Boundary-Value Problem)
For the given boundary data u_0, σ_ℓ and the external load $\bar{f}(x) \in \mathcal{L}^{\alpha^*}(I)$, find $u \in \mathcal{U}_k$ and associated ξ and ς such that

$$(\mathcal{P}_{\rm cbv}): \quad \begin{cases} \xi = \Lambda(u) = \frac{1}{2}au'^2 + bu' + c & \text{in } I, \quad u(0) = u_0, \\ \varsigma = D\bar{V}(\xi) & \text{in } I, \\ \bar{f} = \Lambda_t^*(u)\varsigma = -\left((au' + b)\varsigma\right)' & \text{in } I, \quad (au' + b)\varsigma|_{x=\ell} = \sigma_\ell. \end{cases} \qquad \spadesuit$$

Clearly, this canonical boundary-value problem is equivalent to the $(\mathcal{P}_{\rm bv})$. The diagrammatic representation for this nonlinear canonical boundary-value problem is shown in Fig. 3.14.

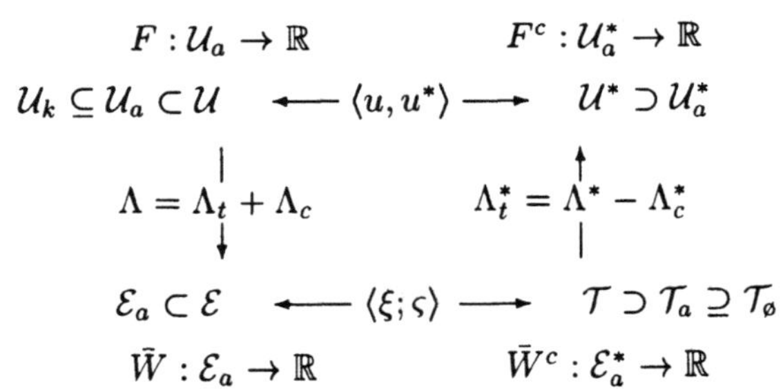

Figure 3.14. Diagrammatic framework for nonconvex canonical boundary-value problems

3.4 Complementary Energy Variational Principle and Analytic Solutions

3.4.1 Dual Problem and Algebraic Euler-Lagrange Equation

In terms of the nonlinear geometrical operator $\Lambda(u)$, the so-called *canonical total potential* associated with the $(\mathcal{P}_{\rm cbv})$ is defined by

$$\Pi : \mathcal{U}_k \to \mathbb{R}, \quad \Pi(u) = \bar{W}(\Lambda(u)) - F(u), \tag{3.78}$$

and the *canonical primal problem* can be proposed as

$$(\Pi_{\rm sta}): \quad \Pi(u) \quad \to \text{sta} \ \ \forall u \in \mathcal{U}_k. \tag{3.79}$$

It is clear that $\Pi(u) = P(u)$ and $(\Pi_{\rm sta})$ is still a nonconvex variational problem. Using the Legendre transformation $\langle \Lambda(u); \varsigma \rangle - \bar{W}^c(\varsigma)$ to replace $\bar{W}(\Lambda(u))$ in Π, the *canonical Lagrangian* $L : \mathcal{U}_a \times \mathcal{T}_a \to \mathbb{R}$ associated with $\mathcal{P}_{\rm cbv})$ can be obtained as

$$L(u, \varsigma) = \langle \Lambda(u); \varsigma \rangle - \bar{W}^c(\varsigma) - F(u). \tag{3.80}$$

Let $\mathcal{Z}_a = \mathcal{U}_a \times \mathcal{T}_a$ denote the admissible canonical phase space. The criticality condition $DL(\bar{u}, \bar{\varsigma}; u, \varsigma) = 0 \quad \forall (u, \varsigma) \in \mathcal{Z}_a$ leads to the *canonical Lagrange equations*

$$\Lambda(\bar{u}) = D\bar{W}^c(\bar{\varsigma}), \quad \Lambda_t^*(\bar{u})\bar{\varsigma} = DF(\bar{u}).$$

Since the equilibrium operator Λ_t^* in nonconvex canonical boundary-value problem depends on the primal variable u, the so-called ***admissible space of equilibrium*** associated with $(\mathcal{P}_{\text{cbv}})$ is a subspace of the admissible phase space $\mathcal{Z}_a$, defined by

$$\mathcal{Z}_e = \{(u, \varsigma) \in \mathcal{U}_a \times \mathcal{T}_a | \; \bar{\Lambda}_t^*(u)\varsigma = DF(u)\}. \tag{3.81}$$

Using the operator decomposition $\Lambda = \Lambda_t + \Lambda_c$ on $\mathcal{Z}_e$, the Lagrangian is reduced to the *pseudo-complementary energy* $\Pi^c : \mathcal{Z}_e \to \mathbb{R}$, given by (see Gao and Strang, 1989a,b)

$$\begin{aligned} \Pi^c(u, \varsigma) &= -L(u, \varsigma) = \bar{W}^c(\varsigma) - F^c(\bar{\Lambda}_t^*(u)\varsigma) + G(u, \varsigma) \\ &= \int_I \bar{V}^c(\varsigma) \, dx + \int_I \left(\frac{1}{2} a u'^2 - c \right) \varsigma \, dx \\ & \quad -(a(0)u'(0) + b(0))\varsigma(0)u_0. \end{aligned} \tag{3.82}$$

Theorem 3.4.1 (Pseudo-Complementary Energy Principle)
For the given boundary data and external load such that the statically admissible space $\mathcal{Z}_e$ is not empty, the solution of the complementary variational problem

$$(\Pi^c_{\text{sta}}): \quad \Pi^c(u, \varsigma) \to \text{sta} \;\; \forall (u, \varsigma) \in \mathcal{Z}_e \tag{3.83}$$

solves the $(\mathcal{P}_{\text{cbv}})$.

To relax the equilibrium constraint in $\mathcal{Z}_e$, we solve the equation $\Lambda_t^*(u)\varsigma = \bar{f}$ for u', i.e.

$$u'(x) = \frac{\sigma(x)}{a(x)\varsigma(x)} - \frac{b(x)}{a(x)} \quad \forall x \in I, \tag{3.84}$$

where $\sigma(x)$ is defined by

$$\sigma(x) = \int_x^\ell \bar{f}(t) dt + \sigma_\ell. \tag{3.85}$$

Substituting (3.84) into Π^c and letting $\Pi^d = -\Pi^c$, we obtain a new functional

$$\begin{aligned} \Pi^d(\varsigma) &= F^c(u^*) - \bar{W}^c(\varsigma) - G(\bar{u}, \varsigma) \\ &= -\sigma(0)u_0 - \int_I \left(\bar{V}^c(\varsigma) + \frac{\sigma^2}{2a\varsigma} + \lambda\varsigma - \frac{b}{a}\sigma \right) dx, \end{aligned} \tag{3.86}$$

which is dual to Π, where the parameter $\lambda : I \to \mathbb{R}$ depends on the coefficients a, b and c and is defined by

$$\lambda(x) = \frac{b(x)^2}{2a(x)} - c(x). \tag{3.87}$$

Let $\mathcal{T}_a \subset \mathcal{T}$ be the range of the constitutive mapping $DW : \mathcal{E}_a \to \mathcal{T}$, i.e.

$$\mathcal{T}_a = \{\varsigma \in \mathcal{T} | \quad \varsigma \in D\bar{W}(\mathcal{E}_a)\}. \tag{3.88}$$

By introducing a subset $\mathcal{T}_\emptyset$ of $\mathcal{T}_a$, defined by

$$\mathcal{T}_\emptyset = \{\varsigma \in \mathcal{T}_a | \ \varsigma(x) \neq 0 \ \ \forall x \in I\}, \tag{3.89}$$

the dual functional $\Pi^d : \mathcal{T}_\emptyset \to \mathbb{R}$ is well-defined and nonconvex. For the given $f \in \mathcal{L}^{\alpha^*}(I)$ and boundary data, $\sigma \in \mathcal{C}^1(I)$ is well-defined by (3.85). The canonical dual variational problem associated with $(\mathcal{P}_{\text{bv}})$ can then be proposed as

$$(\Pi^d_{\text{sta}}) : \qquad \Pi^d(\varsigma) \ \to \text{sta} \ \ \forall \varsigma \in \mathcal{T}_\emptyset. \tag{3.90}$$

The stationary condition $\delta\Pi^d(\bar{\varsigma};\varsigma) = 0 \quad \forall \varsigma \in \mathcal{T}_\emptyset$ leads to the dual Euler-Lagrange equation

$$\bar{\varsigma}^2[D\bar{W}^c(\bar{\varsigma}) + \lambda] = \frac{1}{2a}\sigma^2(x) \ \ \forall \, x \in I. \tag{3.91}$$

The following theorem plays a fundamental role in obtaining the general analytic solution of the $(\mathcal{P}_{\text{cbv}})$.

Theorem 3.4.2 *For any $\bar{f}(x) \in \mathcal{L}^{\alpha^*}(I)$, $\bar{u}$ is a critical point of Π if and only if the associated $\bar{\varsigma} = D_\xi W(\Lambda(\bar{u}))$ is a critical point of Π^d, and then*

$$\Pi(\bar{u}) = \Pi^d(\bar{\varsigma}). \tag{3.92}$$

Proof. Suppose that $\bar{u}$ is a critical point of Π and $\bar{\varsigma} = D_\epsilon \bar{W}(\Lambda(\bar{u}))$ is the associated dual variable. Then, we must have

$$\Lambda_t^*(\bar{u})\bar{\varsigma} = -[(b + a\bar{u}')\bar{\varsigma}]' = \bar{f}(x) \ \ \forall x \in I.$$

Solving for u' gives

$$\bar{u}'(x) = \frac{1}{a}\left[\frac{\sigma(x)}{\bar{\varsigma}(x)} - b\right]. \tag{3.93}$$

Since W is convex in $\epsilon = \Lambda(\bar{u})$, the constitutive law $\bar{\varsigma} = D_\epsilon W_\lambda(\Lambda(\bar{u}))$ is equivalent to

$$\Lambda(\bar{u}) = D\bar{W}^c(\bar{\varsigma}).$$

Substituting (3.93) into this dual constitutive law leads the dual Euler-Lagrange equation (3.91). Thus, $\bar{\varsigma}$ must be a critical point of Π^d.

Conversely, if $\bar{\varsigma}$ is a critical point of Π^d, substituting $\Lambda(\bar{u}) = D\bar{W}^c(\bar{\varsigma})$ into the dual Euler-Lagrange equation (3.91), we obtain

$$\pm(a\bar{u}' + b)\bar{\varsigma} = \sigma(x).$$

Taking the derivative of both sides with respect to x and choosing the correct sign, we obtain the equilibrium equation. The $\bar{u}$ associated with $\bar{\varsigma}$ is therefore a critical point of Π.

Since $\bar{W}(\epsilon)$ is convex, by the equivalent relations in (3.68), we have

$$\bar{W}(\Lambda(\bar{u})) = \langle \Lambda(\bar{u}); \bar{\varsigma} \rangle - \bar{W}^c(\bar{\varsigma}).$$

Substituting this into $\Pi(\bar{u})$ and using the operator decomposition $\Lambda = \Lambda_t + \Lambda_c$, we obtain

$$\begin{aligned} \Pi(\bar{u}) &= \langle \Lambda(\bar{u}), \bar{\varsigma} \rangle - \bar{W}^c(\bar{\varsigma}) - \int_I f \bar{u} \, \mathrm{d}x - \sigma_\ell u(\ell) \\ &= \langle \bar{u}, \Lambda_t^*(\bar{u})\bar{\varsigma} - f \rangle - \bar{W}^c(\bar{\varsigma}) - G(\bar{u}, \bar{\varsigma}) - u_0 \sigma(0). \end{aligned} \tag{3.94}$$

Substituting the equilibrium conditions and (3.93) into (3.94) we then have $\Pi(\bar{u}) = \Pi^d(\bar{\varsigma})$. □

This theorem shows that the critical points of Π and Π^d are solutions of the $(\mathcal{P}_{\mathrm{cbv}})$. Equation (3.92) shows that there is no duality gap between the primal and dual variational problems.

3.4.2 Analytic Solution Theorem

Based on the canonical dual variational problem, the general analytic solution for the $(\mathcal{P}_{\mathrm{bv}})$ can be obtained.

Theorem 3.4.3 *For given boundary data* u_0, σ_ℓ, *and external load* $\bar{f}(x) \in \mathcal{L}^{\alpha^*}(I)$ *such that* $\sigma(x)$ *is defined by* (3.85), *for every solution* $\bar{\varsigma}(x)$ *of the dual Euler-Lagrange equation* (3.91) *there is a unique* $\bar{u}(x)$ *defined by*

$$\bar{u}(x) = \int_0^x \frac{1}{a(t)} \left[\frac{\sigma(t)}{\bar{\varsigma}(t)} - b(t) \right] \mathrm{d}t + u_0, \tag{3.95}$$

and this $\bar{u}$ *solves the mixed boundary-value problem* $(\mathcal{P}_{\mathrm{bv}})$.

Conversely, every solution $\bar{u}$ *of the* $(\mathcal{P}_{\mathrm{bv}})$ *can be written in the form* (3.95) *for some solution* $\bar{\varsigma}$ *of* (3.91).

Moreover, for every positive solution $\bar{\varsigma}(x) > 0 \ \forall x \in I$ *of* (3.91), *the solution* $\bar{u}(x)$ *defined by* (3.95) *is a global minimizer of* Π. *If* $\bar{V}(\xi)$ *is strictly convex and all the algebraic solutions of the dual Euler-Lagrange equation* (3.91) *are positive on* I, *then the solution of the* $(\mathcal{P}_{\mathrm{bv}})$ *is unique.*

Proof. We first prove that the function $u(x)$ defined by (3.95) solves the $(\mathcal{P}_{\mathrm{cbv}})$. For every solution $\bar{\varsigma}(x)$ of (3.91), the derivative of (3.95) leads to

$$a(x)\bar{u}' + b(x) = \frac{\sigma(x)}{\bar{\varsigma}(x)}. \tag{3.96}$$

Substitution of this into (3.91) gives

$$D\bar{V}^c(\varsigma) = \frac{1}{2} a \bar{u}'^2 + b \bar{u}' + c = \Lambda(\bar{u}). \tag{3.97}$$

Since the constitutive relation $\varsigma = D\bar{V}(\xi)$ is invertible, this inverse constitutive equation is equivalent to the geometric-constitutive equation

$$\bar{\varsigma} = D_\xi \bar{V}(\Lambda(\bar{u})). \tag{3.98}$$

Then, we have

$$\begin{aligned} [(a(x)\bar{u}' + b(x))\bar{\varsigma}]' &= [\sigma(x)]' = -\bar{f} && \forall x \in I, \\ (a(x)\bar{u}' + b(x))\bar{\varsigma} &= \sigma(\ell) = \sigma_\ell && \text{at } x = \ell. \end{aligned} \tag{3.99}$$

This shows that for every solution $\bar{\varsigma}$ of (3.91), $\bar{u}(x)$ defined by (3.95) is indeed a solution of the $(\mathcal{P}_{\rm bv})$. Since $D\bar{V}^c(\varsigma)$ is a monotone mapping, the solutions of the algebraic equation (3.91) are well defined for the given $\sigma(x) \in \mathcal{C}^1(I)$. Hence, for each solution $\bar{\varsigma}$ of (3.91), $\bar{u}$ is uniquely defined by (3.95).

Conversely, if $\bar{u}(x)$ is a solution of the $(\mathcal{P}_{\rm bv})$, then $\bar{u}$ has to satisfy the geometric-constitutive equation (3.98) and the equilibrium conditions (3.99). For any given $\varsigma \in \mathcal{T}_\sigma$, solving the equilibrium equation in (3.99), we have

$$a\bar{u}' + b(x) = \sigma(x)/\varsigma(x). \tag{3.100}$$

Substitution of this into the inverse constitutive equation (3.97) leads to the dual Euler-Lagrange equation (3.91). Since the constitutive relation $\varsigma = \varsigma(\xi(u))$ is one-to-one, one of the solutions $\bar{\varsigma}$ of the algebraic equation (3.91) satisfies the inverse constitutive equation (3.97). For this $\bar{\varsigma}$, solving (3.100) for $\bar{u}$ gives the formulation (3.95). Hence, the solution $\bar{u}(x)$ of the $(\mathcal{P}_{\rm bv})$ can be written in the form (3.95).

Now we prove that if $\bar{\varsigma}(x) > 0$ on I, the solution $\bar{u}(x)$ is a global minimizer. Since $\jmath(x, u, u') = V(x, u') - fu = \bar{V}(\Lambda(u)) - fu$, by the chain rule,

$$B(x) = \jmath_{,u'u'}(x, \bar{u}, \bar{u}') = a(x)\bar{\varsigma}(x) + \xi_t^2(\bar{u}')\bar{V}_{,\xi\xi}(\xi).$$

Since $\bar{V}$ is strictly convex, $\bar{V}_{,\xi\xi}(\xi)$ is strictly positive on I, and hence

$$B(x) > 0 \;\; \Leftrightarrow \;\; \bar{\varsigma}(x) > 0 \;\; \forall x \in I.$$

Thus, the Legendre condition $\jmath_{u'u'} > 0$ holds if and only if the solution of the dual Euler-Lagrange equation is strictly positive on I. Since the solution $\bar{u}(x)$ associated with this $\bar{\varsigma}(x) > 0$ is a critical point of Π and Π is strictly convex at $\bar{u}$, then $\bar{u}(x)$ is a global minimizer of Π. □

Remark 3.4.1 In Theorem 3.4.3, it is required that $\varsigma(x) \neq 0$. Actually, $\varsigma(x) = 0$ is a solution of the dual Euler-Lagrange equation (3.91) if and only if $\sigma(x) = 0$. In this case, if $f(x) = 0 = \sigma_\ell$, then

$$D_\xi \bar{V}(\Lambda u(x)) = 0 \;\; \forall x \in I, \quad D_\epsilon W(\Lambda u(x))\,|_{x=L} = 0.$$

Since $D_\xi \bar{V}(\Lambda(u))$ is a one-to-one mapping on $\mathcal{E}$, this homogeneous problem has either a trivial solution or an infinite number of solutions. ■

Example 3.4.1 Analytic Solution of Einstein's Theory

The primal problem in Einstein's relativity theory is the convex variational problem

$$\Pi(q) = -\int_I m_0 c^2 \sqrt{1 - q'(t)^2/c^2}\mathrm{d}t - \int_I f(t)q(t)\mathrm{d}t \quad \rightarrow \min$$

subject to the constraint

$$q \in \mathcal{U}_k = \{q \in \mathcal{L}^2(0,t_1)|\ q' \in \mathcal{L}^\infty(0,1),\ |q'(t)| < c,\ q(0) = q_0,\ q'(0) = v_0\}.$$

In terms of the "finite velocity measure"

$$\xi = \Lambda(q) = 1 - q'(t)^2/c^2 > 0 \quad \text{with } a = -2/c^2,\ \lambda = 1,$$

the stored energy $\bar{W}(\xi) = -m_0 c^2 \sqrt{\xi}$ is strictly convex. The canonical complementary energy density $\bar{V}^c(\varsigma) = -m_0^2 c^4/4\varsigma$ is also strictly convex. The dual Euler-Lagrange equation has a unique solution, namely

$$\varsigma = -\frac{1}{2}c\sqrt{c^2 m_0^2 + \sigma^2}. \tag{3.101}$$

For the given external force field $f(t)$ such that

$$\sigma(t) = -\int f(t)\mathrm{d}t + f_0,$$

where f_0 is an integral constant to be determined by the initial condition $q'(0) = v_0$, the analytic solution for Einstein's theory is

$$q(t) = \int_0^t \frac{c\sigma(t)}{\sqrt{\sigma(t)^2 + m_0^2 c^2}}\mathrm{d}t + q_0. \tag{3.102}$$

If $\sigma(t)/c \approx 0$, then the solution for Einstein's theory is reduced to the solution of Newton's equation, i.e.

$$q(t) \approx \frac{1}{m_0}\int_0^t \sigma(t)\mathrm{d}t + q_0.$$

3.4.3 Quadratic Canonical Energy and Cubic Algebraic Curve

A special case arises when the canonical energy density $\bar{V}$ is a quadratic function, i.e.

$$\bar{V}(\xi) = \frac{1}{2}k_0\xi^2, \quad k_0 > 0. \tag{3.103}$$

The complementary energy density is then also a quadratic function, namely

$$\bar{V}^c(\varsigma) = \varsigma\xi - \bar{V}(\xi) = \frac{1}{2k_0}\varsigma^2.$$

In this case, for the quadratic operator Λ with $a = 1$, i.e.

$$\xi = \Lambda(u) = \frac{1}{2}u'(x)^2 + b(x)u'(x) + c(x), \quad b, c \in \mathcal{C}^0(I),$$

the dual Euler-Lagrange equation (3.91) is the cubic algebraic equation

$$2\varsigma^2(\frac{1}{k_0}\varsigma + \lambda) = \sigma^2(x) \ \ \forall x \in I, \tag{3.104}$$

which has at most three solutions $\varsigma_i(x)$, $i = 1, 2, 3$. For each $\varsigma_i(x)$, the analytic solution of the $(\mathcal{P}_{\rm bv})$ can be written as

$$u_i(x) = \int_0^x \left(\frac{\sigma(t)}{\bar{\varsigma}_i(t)} - b(t) \right) dt + u_0. \tag{3.105}$$

This leads to the following result.

Theorem 3.4.4 (Multi-Solution Criterion) *Suppose that the stored energy $\bar{V}$ is a quadratic function of the quadratic operator Λ with $a = 1$. For given $\bar{f}(x)$ and boundary data u_0 and σ_ℓ, if $\sigma(x) = \int_x^\ell \bar{f}(x)\,dx + \sigma_\ell$ has extremum values $\sigma_{\min}$ and $\sigma_{\max}$ on I such that $\sigma^2_{\min} \le \sigma^2(x) \le \sigma^2_{\max}$ $\forall x \in [0, \ell]$, then there exist two constants*

$$\lambda_{\min} = 1.5\left(\frac{\sigma_{\min}}{k_0}\right)^{2/3}, \qquad \lambda_{\max} = 1.5\left(\frac{\sigma_{\max}}{k_0}\right)^{2/3}. \tag{3.106}$$

If $\lambda(x) = b(x)^2/2 - c(x) \in (-\infty, \lambda_{\min})$, then the cubic equation (3.104) has only one real solution $\bar{\varsigma}(x)$, and the associated $\bar{u}(x)$ is a unique global minimizer of the variational problem $(\Pi_{\rm sta})$ defined by (3.79). If $\lambda \in (\lambda_{\min}, \lambda_{\max})$, then there exists a subdomain $I_s \subset I$ such that on I_s, the problem may have three solutions. If $\lambda \in (\lambda_{\max}, +\infty)$, the problem may have three solutions on the whole domain. If $\varsigma_1(x) > 0 > \varsigma_2(x) > \varsigma_3(x)$ $\forall x \in I$, then $u_1(x)$ is a global minimizer, $u_2(x)$ is a local minimizer and $u_3(x)$ is a local maximizer of the double well energy $\Pi(u)$.

Proof. Here we only prove existence and uniqueness. The extremum properties will be proved using the triality theorem in the next section. Let $h(\varsigma) = 2\varsigma^2(\frac{1}{k_0}\varsigma + \lambda)$. Its graph is shown in Fig. 3.15. If $\lambda \le 0$, the algebraic equation (3.104) has only one real solution for any give $\sigma(x) \ne 0$ (Fig. 3.15(a)).

In this case, the total potential energy Π is strictly convex. If $\lambda > 0$, $h(\varsigma)$ has a local maximum $h_{\max} = 8k_0^2\lambda^3/27$ at $\bar{\varsigma} = -2k_0\lambda/3$, and $h(\varsigma) \le h_{\max}$ $\forall \varsigma \le 0$. Since $\sigma : I \to \mathbb{R}$ is a continuous function, for the given load $\bar{f} \in \mathcal{L}^2(I)$, $\sigma^2(x)$ has a minimum value $\sigma^2_{\min}$ and a maximum value $\sigma^2_{\max}$ on the closed domain $[0, \ell]$. The conditions $h_{\max} = \sigma^2_{\min}$ and $h_{\max} = \sigma^2_{\max}$ give the critical values of $\lambda_{\min}$ and $\lambda_{\max}$ defined in (3.106). Thus, if $\lambda \in (-\infty, \lambda_{\min})$, the algebraic equation (3.104) has only one real solution $\varsigma_1 > 0$. However, if $\lambda \in (\lambda_{\min}, \lambda_{\max})$, there exists a subdomain $I_s \subset I$ such that $\sigma^2(x) < h_{\max}$ $\forall x \in I_s$. In this case, equation (3.104) has three real solutions $\varsigma_i(x)$, $i = 1, 2, 3$, in I_s, and one of them must be positive, and the other two negative (see Fig. 3.15 (b)). If $\lambda > \lambda_{\max}$, the

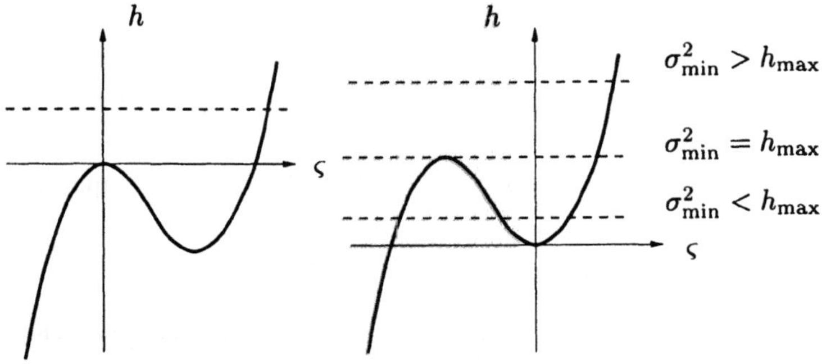

(a) Unique solution: $\lambda < 0$ (b) Multi-solution: $\lambda > 0$

Figure 3.15. Graphs of $h(\varsigma) = 2\varsigma^2(\varsigma/k_0 + \lambda)$ and the criteria of uniqueness

algebraic equation (3.104) has three real roots everywhere in $(0, \ell)$, which give three solutions $u_i (i = 1, 2, 3)$ of the problem. □

In *algebraic geometry*, the dual Euler-Lagrange equation (3.91) is the so-called *singular algebraic curve* in (ς, σ)-space, i.e. $\varsigma = 0$ is on the curve (see Silverman & Tate, 1992, p. 99). With a change of variables, the singular cubic curve (3.104) can be given by the well-known *Weierstrass equation*

$$y^2 = x^3 + \mu x^2 + \nu x + \eta,$$

where $\mu, \nu, \eta \in \mathbb{R}$ are constants. If we let C_{ns} be a set consisting of non-singular points on the curve, then C_{ns} is an Abelian group. This fact in algebraic geometry is very important in understanding the properties of the analytic solution. Since $D\bar{V}^c$ is monotone, then if the continuous function $\sigma(x)$ is one-signed on I, each root $\varsigma(x)$ of (3.104) is also one-signed on I (see Fig. 3.16).

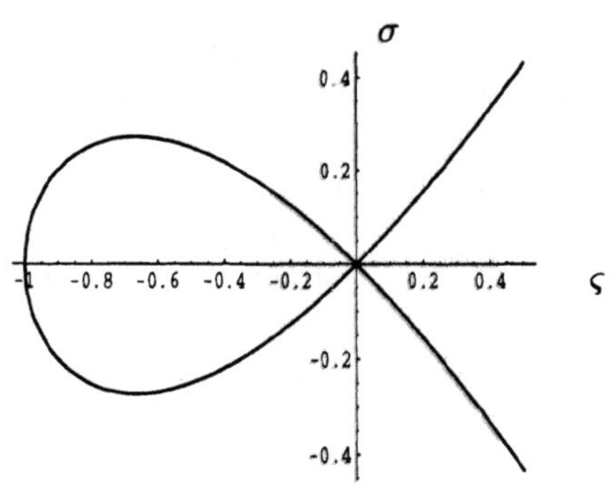

Figure 3.16. Singular algebraic curve for the dual Euler-Lagrange equation

Example 3.4.2 Using Theorem 3.4.4, we are able to solve the static bifurcation problem (3.31), i.e.

$$(\Pi_{\inf}): \qquad \Pi(u) = \int_0^1 \frac{1}{2}k_0(\frac{1}{2}u'^2 - \lambda)^2 \, dx - \int_0^1 fu \, dx \to \min$$
$$s.t. \ u \in \mathcal{U}_k = \{u \in \mathcal{C}^1(0,1) | \ u'^2 \in \mathcal{L}^2(0,1), \ u(0) = 0\}.$$

This is a nonconvex variational problem with double-well energy. Let $\xi = \Lambda(u) = \frac{1}{2}u'^2 - \lambda$ be a nonlinear strain measure, with $a = 1, \ b = 0$, and $c = -\lambda > 0$. For any given $u(x) \in \mathcal{U}_k$, $\xi(x) \geq -\lambda \ \forall x \in I = (0,1)$, the domain for the stored energy $\bar{V}(\xi) = \frac{1}{2}k_0\xi^2$ can be defined by

$$\mathcal{E}_a = \{\xi \in \mathcal{L}^2(I) | \ \xi(x) \geq -\lambda \ \forall x \in I\}.$$

The range of the constitutive mapping $D\bar{V} : \mathcal{E}_a \to \mathcal{T} = \mathcal{L}^2(I)$ is

$$\mathcal{T}_a = \{\varsigma \in \mathcal{L}^2(I) | \quad -k_0\lambda \leq \varsigma(x) \leq +\infty \ \forall x \in I\}.$$

Thus, for any given external force $f(x)$ in the domain I and $\sigma_\ell = 0$ at $x = 1$ such that

$$\sigma(x) = \int_x^1 f(t) dt,$$

the total complementary energy

$$\Pi^d(\varsigma) = -\int_I \left(\frac{1}{2k_0}\varsigma^2 + \lambda\varsigma + \frac{\sigma^2}{2\varsigma} \right) dx$$

is a nonconvex functional on $\mathcal{T}_\emptyset$. For $k_0 = 1, \ f = 1, \ \sigma(x) = 1 - x$, the three roots of the dual cubic algebraic equation $2\varsigma(\varsigma + \lambda) = (x-1)^2$ are

$$\begin{aligned}
\varsigma_1(x) &= -\lambda/3 + 0.839947\lambda^2/e(x) + 0.132283e(x), \\
\varsigma_2(x) &= -\lambda/3 + (-0.41997 - 0.7274i)\lambda^2/e(x) + (-0.06614 + 0.11456i)e(x), \\
\varsigma_3(x) &= -\lambda/3 + (-0.41997 + 0.7274i)\lambda^2/e(x) + (-0.06614 - 0.11456i)e(x),
\end{aligned}$$

where

$$e(x) = [108(x-1)^2 - 16\lambda^3 + \sqrt{-256\lambda^6 + (108(x-1)^2 - 16\lambda^3)^2}]^{1/3}.$$

The graphs of these dual solutions are shown in Fig. 3.17. Since ς_3 does not vanish at $x = 1$, the associated u_3 has to satisfy the boundary condition $u_3'(1) + b(1) = 0$. Since $\sigma^2_{\min} = 0, \ \ \sigma^2_{\max} = 1$, the problem has only one solution for $\lambda < \lambda_{\min} = 0$ (Fig. 3.17 (a)). For $0 < \lambda < \lambda_{\max} = 1.5$, the problem has three solutions in the subdomain $x \in (1 - (2\lambda/3)^{3/2}, 1)$ (Fig. 3.17 (b)). If $\lambda > 1.5$, the problem has three solutions everywhere in $(0,1)$ (Fig. 3.17(c)). ♡

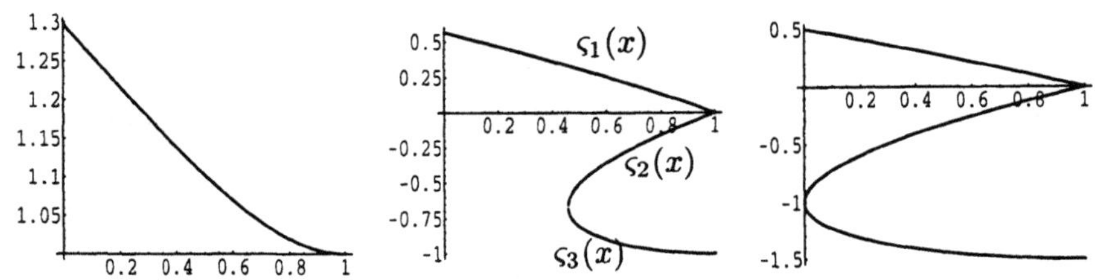

(a) $\lambda < \lambda_{\min}$ (b) $\lambda_{\min} < \lambda < \lambda_{\max}$ (c) $\lambda = \lambda_{\max}$

Figure 3.17. Dual solutions for van der Waals' variational problem $\varsigma_i(x)$ $(i = 1, 2, 3)$

3.5 Tri-Extremum Principles and Triality Theory

We have seen that for convex systems the classical Lagrangian $L : \mathcal{U} \times \mathcal{T} \to \mathbb{R}$ associated with primal variational problems is usually a saddle functional for static problems, or a super-critical functional for convex Hamiltonian systems. The Lagrange duality theory provides a beautiful symmetry between the primal and dual problems.

In non-convex systems, this symmetry is broken. The canonical Lagrangian could have both saddle points and super-critical points. This symmetry breaking yields some very interesting new phenomena. In this section we present the so-called triality theory under the following assumption.

Assumption 3.5.1 Let $\{(\mathcal{U}, \mathcal{U}^*); \langle *, * \rangle\}$ and $\{(\mathcal{E}, \mathcal{T}); \langle *; * \rangle\}$ be two inner product spaces. Assume that

(A1) $\Lambda : I \times \mathcal{U} \to \mathcal{E}$ is a quadratic operator

$$\Lambda(u) = \frac{1}{2} a(x) u'(x)^2 + b(x) u'(x) + c(x), \quad a(x) > 0 \;\; \forall x \in I,$$

where $a, b, c \in C^1(I)$ are given real-valued functions;

(A2) $F : \mathcal{U}_a \subset \mathcal{U} \to \mathbb{R}$ is a linear, Gâteaux differentiable functional and, on $\mathcal{U}_a \times \mathcal{U}_a^* \subset \mathcal{U} \times \mathcal{U}^*$,

$$u^* = DF(u) \;\; \Leftrightarrow \;\; u = DF^c(u^*) \;\; \Leftrightarrow \;\; \langle u, u^* \rangle = F(u) + F^c(u^*);$$

(A3) $\bar{W} : \mathcal{E}_a \subset \mathcal{E} \to \mathbb{R}$ is either convex or concave and on $\mathcal{E}_a \times \mathcal{T}_a \subset \mathcal{E} \times \mathcal{T}$, the Legendre duality relations

$$\varsigma = D\bar{W}(\xi) \;\; \Leftrightarrow \;\; \xi = D\bar{W}^c(\varsigma) \;\; \Leftrightarrow \;\; \langle \xi ; \varsigma \rangle = \bar{W}(\xi) + \bar{W}^c(\varsigma)$$

hold. ♣

Over the kinematically admissible space $\mathcal{U}_k = \{u \in \mathcal{U}_a | \;\; \Lambda(u) \in \mathcal{E}_a\}$, the primal variational problem

$$(\Pi_{\text{sta}}) : \quad \Pi(u) = \bar{W}(\Lambda(u)) - F(u) \;\; \to \text{sta} \;\; \forall u \in \mathcal{U}_k \tag{3.107}$$

is a nonconvex problem.

3.5.1 Nonlinear Lagrangian and Critical Point Theory

Recall the canonical Lagrangian (3.80) associated with the canonical boundary-value problem $(\mathcal{P}_{\rm cbv})$

$$L(u,\varsigma) = \langle \Lambda(u);\varsigma\rangle - \bar{W}^c(\varsigma) - F(u). \tag{3.108}$$

On the canonical phase space $\mathcal{Z}_a = \mathcal{U}_a \times \mathcal{T}_a$, L is well-defined. The criticality condition

$$\delta L(\bar{u},\bar{\varsigma};u,\varsigma) = 0 \;\; \forall (u,\varsigma) \in \mathcal{U}_a \times \mathcal{T}_a$$

leads to the nonlinear Lagrange equations

$$\Lambda(\bar{u}) = D\bar{W}^c(\bar{\varsigma}), \tag{3.109}$$

$$\bar{\Lambda}_t^*(\bar{u})\bar{\varsigma} = DF(\bar{u}). \tag{3.110}$$

By the Legendre duality equivalent relations (3.68) we know that the critical points of L solve the $(\mathcal{P}_{\rm bv})$.

Generally speaking, for any given nonlinear operation $\Lambda(u)$, the dual functional associated with this Lagrangian is defined by

$$\Pi^d(\varsigma) = \mathrm{sta}_{u\in\mathcal{U}_a} L(u,\varsigma) = F_\Lambda^c(\varsigma) - \bar{W}^c(\varsigma), \tag{3.111}$$

where F_Λ^c is called the *generalized Legendre conjugate* of $F(u)$, defined by the *canonical conjugate transformation*

$$F_\Lambda^c(\varsigma) = \underset{u\in\mathcal{U}_a}{\mathrm{sta}} \{\langle\Lambda(u);\varsigma\rangle - F(u)\}. \tag{3.112}$$

For any given $\varsigma \in \mathcal{T}_a$ such that $\mathcal{Z}_e$ is not empty, the solving $\bar{u}$ for this stationary variational problem satisfying the equilibrium equation: $\Lambda_t^*(\bar{u})\varsigma = DF(\bar{u})$.

For the quadratic operator $\Lambda(u) = \frac{1}{2}a(x)u'^2 + b(x)u' + c(x)$, and linear $F(u) = \langle u, \bar{u}^*\rangle$, where $\bar{u}^* \in \mathcal{U}_a^*$ is a given source, we have

$$\Lambda_t^*(\bar{u})\varsigma = \left(\frac{\mathrm{d}}{\mathrm{d}x}\right)^* ((a\bar{u}' + b)\varsigma) = DF(\bar{u}) = \bar{u}^*.$$

Then, $\bar{u}'$ can be solved as

$$\bar{u}' = \frac{\sigma(x)}{a(x)\varsigma(x)} - \frac{b(x)}{a(x)},$$

in which

$$\sigma \in \mathcal{S}_s = \left\{\sigma \in \mathcal{L}^{\beta^*}(I) \,\middle|\, \left(\frac{\mathrm{d}}{\mathrm{d}x}\right)^* \sigma = \bar{u}^*\right\}.$$

Thus, on $\mathcal{T}_\sigma$, the dual functional Π^d is obtained as

$$\Pi^d(\varsigma) = F_\Lambda^c(\varsigma) - \bar{W}^c(\varsigma) = F^c(u^*(\sigma)) - \bar{W}^c(\varsigma) - G^c(\varsigma), \tag{3.113}$$

in which $F^c(u^*(\sigma))$ is the Legendre conjugate of $F(u)$ defined by

$$F^c(u^*(\sigma)) = \mathrm{sta}\{\langle u',\sigma\rangle - F(u) \mid \; \forall u \in \mathcal{U}_a\}, \;\; \forall \sigma \in \mathcal{S}_s,$$

and $G^c : \mathcal{T}_\emptyset \to \mathbb{R}$ is a pure complementary gap functional, given by

$$G^c(\varsigma) = \langle -\Lambda_c(\bar{u}); \varsigma\rangle = \int_I \left[\frac{1}{2}\varsigma \left(\frac{\sigma(x)}{a(x)\varsigma(x)} - \frac{b(x)}{a(x)} \right)^2 - c\varsigma \right] dx. \tag{3.114}$$

Theorem 3.5.1 (Critical Point Classification Theorem)
Suppose that $(\bar{u}, \bar{\varsigma})$ is a critical point of L and $\mathcal{Z}_r = \mathcal{U}_r \times \mathcal{T}_r \subset \mathcal{U}_a \times \mathcal{T}_a$ is a neighborhood of $(\bar{u}, \bar{\varsigma})$.
(T) If $\bar{W} : \mathcal{E} \to \mathcal{L}(I)$ is convex, then
(t1) $(\bar{u}, \bar{\varsigma})$ is a right saddle point of L on $\mathcal{Z}_r$ if and only if $\bar{\varsigma}(x) \geq 0 \ \forall x \in I$.
(t2) $(\bar{u}, \bar{\varsigma})$ is a super-critical point of L on $\mathcal{Z}_r$ if and only if $\bar{\varsigma}(x) < 0 \ \forall x \in I$.

(T') If $\bar{W} : \mathcal{E} \to \mathcal{L}(I)$ is concave, then
(t1') $(\bar{u}, \bar{\varsigma})$ is a left saddle point of L on $\mathcal{Z}_r$ if and only if $\bar{\varsigma}(x) \leq 0 \ \forall x \in I$.
(t2') $(\bar{u}, \bar{\varsigma})$ is a sub-critical point of L on $\mathcal{Z}_r$ if and only if $\bar{\varsigma}(x) > 0 \ \forall x \in I$.

Proof. Since $\Lambda(u)$ is a quadratic operator, for any given $\varsigma \in \mathcal{T}_a$, the Lagrangian $L : \mathcal{U}_a \to \mathbb{R}$ is convex if $\varsigma(x) \geq 0$, concave if $\varsigma(x) \leq 0$. If $(\bar{u}, \bar{\varsigma})$ is a critical point of L, we have

$$\begin{aligned} L(u, \bar{\varsigma}) &\geq L(\bar{u}, \bar{\varsigma}) \ \forall u \in \mathcal{U}_r \quad &\text{iff} \quad \bar{\varsigma} \geq 0 \ \forall x \in I, \\ L(u, \bar{\varsigma}) &\leq L(\bar{u}, \bar{\varsigma}) \ \forall u \in \mathcal{U}_r \quad &\text{iff} \quad \bar{\varsigma} \leq 0 \ \forall x \in I. \end{aligned}$$

If $\bar{W} : \mathcal{E} \to \mathcal{L}(I)$ is convex, then the complementary energy $\bar{W}^c$, defined by

$$\bar{W}^c(\varsigma) = \sup_{\xi \in \mathcal{E}} \left\{ \langle \xi; \varsigma \rangle - \int_I \bar{V}(\xi) \, dx \right\} = \int_I \bar{V}^c(\varsigma) \, dx,$$

is convex. Hence, $(\bar{u}, \bar{\varsigma})$ is a saddle point if and only if $\bar{\varsigma} \geq 0$, and is a super-critical point if and only if $\bar{\varsigma} < 0$.

However, if $\bar{W}(\xi)$ is concave, then the complementary energy $\bar{W}^c$, defined by

$$\bar{W}^c(\varsigma) = \inf_{\xi \in \mathcal{E}} \{ \langle \xi; \varsigma \rangle - \int_I \bar{V}(\xi) \, dx \},$$

is concave. Thus, $(\bar{u}, \bar{\varsigma})$ is a left saddle point of L if and only if $\bar{\varsigma} \leq 0$, and is a sub-critical point if and only if $\bar{\varsigma} > 0$. □

If the admissible space of equilibria $\mathcal{Z}_e$ is not empty, we have the following complementary extremality relations.

Proposition 3.5.1 *Suppose that for any given $(u, \varsigma) \in \mathcal{U}_k \times \mathcal{T}_\emptyset$ the set $\mathcal{Z}_e$ is not empty. Then, for any $u \in \mathcal{U}_k$,*

$$\Pi(u) = \begin{cases} \sup_{\varsigma \in \mathcal{T}_a} L(u, \varsigma) & \text{if } \bar{W} : \mathcal{E}_a \to \mathbb{R} \text{ is convex,} \\ \inf_{\varsigma \in \mathcal{T}_a} L(u, \varsigma) & \text{if } \bar{W} : \mathcal{E}_a \to \mathbb{R} \text{ is concave.} \end{cases} \tag{3.115}$$

Dually, for a given $\varsigma \in \mathcal{T}_\emptyset$,

$$\Pi^d(\varsigma) = \begin{cases} \inf_{u \in \mathcal{U}_a} L(u,\varsigma) & \text{if } \varsigma(x) \geq 0 \ \forall x \in I, \\ \sup_{u \in \mathcal{U}_a} L(u,\varsigma) & \text{if } \varsigma(x) \leq 0 \ \forall x \in I. \end{cases} \tag{3.116}$$

Proof. By the convexity of $\bar{W}(\epsilon)$, we have

$$\sup_{\varsigma \in \mathcal{T}_a} \{\langle \Lambda(u); \varsigma \rangle - \bar{W}^c(\varsigma)\} = \bar{W}(\xi(u)) \ \ \forall u \in \mathcal{U}_k.$$

Thus, $\Pi(u) = \sup_{\varsigma \in \mathcal{T}_a} L(u,\varsigma)$ holds on $\mathcal{U}_k$. However, if $\bar{W}(\epsilon)$ is concave, $\bar{W}(\epsilon) = \inf_\varsigma \{\langle \Lambda(u); \varsigma \rangle - \bar{W}^c(\varsigma)\}$. This leads to $\Pi(u) = \inf_\varsigma L(u,\varsigma)$.

For any $\varsigma \in \mathcal{T}_\emptyset$, if the equilibrium admissible space $\mathcal{Z}_e$ is not empty, then there exists a $\bar{u} \in \mathcal{U}_a$ such that $\delta L(\bar{u},\varsigma; u) = 0 \ \forall u \in \mathcal{U}_a$. If $\varsigma(x) \geq 0$, $L : \mathcal{U} \to \mathbb{R}$ is convex. Then, $\bar{u}$ is the solution of $\inf_u L(u,\varsigma)$ and $\Pi^d(\varsigma) = L(\bar{u},\varsigma) = \inf_u L(u,\varsigma)$. If $\varsigma(x) \leq 0$, then $L : \mathcal{U} \to \mathbb{R}$ is concave. In this case, $\bar{u}$ is the solution of $\sup_u L(u,\varsigma)$ and $\Pi^d(\varsigma) = \sup_u L(u,\varsigma)$. □

This theorem shows the extremality relations among the primal and dual functionals and the nonlinear Lagrangian $L(u,\varsigma)$.

Lemma 3.5.1 (General Critical Point Lemma)

If $(\bar{u},\bar{\varsigma})$ *is any one of the right or left saddle points, the super- or sub-critical points of* L *on the open subset* $\mathcal{U}_k \times \mathcal{T}_\emptyset$, *and* L *is partially Gâteaux differentiable at* $(\bar{u},\bar{\varsigma})$, *then* $(\bar{u},\bar{\varsigma})$ *is a critical point of* L. *If* Π *and* Π^d *are Gâteaux differentiable at* $\bar{u}$ *and* $\bar{\varsigma}$, *respectively, then* $D\Pi(\bar{u}) = 0$, $D\Pi^d(\bar{\varsigma}) = 0$ *and*

$$\Pi(\bar{u}) = L(\bar{u},\bar{\varsigma}) = \Pi^d(\bar{\varsigma}). \tag{3.117}$$

Proof. Suppose that $(\bar{u},\bar{\varsigma})$ is a saddle point of L on $\mathcal{U}_k \times \mathcal{T}_\emptyset$. By Theorem 3.5.1, $\bar{\varsigma}$ is positive for convex $\bar{W}$. From the definition of the saddle point and Proposition 3.5.1, we have

$$\Pi(\bar{u}) = \sup_{\varsigma \in \mathcal{T}_a} L(\bar{u},\varsigma) \leq L(\bar{u},\bar{\varsigma}) \leq \inf_{u \in \mathcal{U}_a} L(u,\bar{\varsigma}) = \Pi^d(\bar{\varsigma}). \tag{3.118}$$

On the other hand, for any $u \in \mathcal{U}_a$ and $\varsigma \in \mathcal{T}_a$,

$$\begin{aligned} \Pi(\bar{u}) - \Pi^d(\bar{\varsigma}) &= \sup_\varsigma L(\bar{u},\varsigma) - \inf_u L(u,\bar{\varsigma}) \\ &= \sup_\varsigma \sup_u \{L(\bar{u},\varsigma) - L(u,\bar{\varsigma})\} \geq 0. \end{aligned}$$

Hence

$$\Pi(\bar{u}) = L(\bar{u},\bar{\varsigma}) = \Pi^d(\bar{\varsigma}).$$

Also, from

$$\Pi(u) = \sup_{\varsigma \in \mathcal{T}_a} L(u,\varsigma) = L(u,\bar{\varsigma}) \geq L(\bar{u},\bar{\varsigma}) = \Pi(\bar{u}) \ \ \forall u \in \mathcal{U}_k,$$

we have $\Pi(\bar{u}) = \inf \Pi(u)$. Since Π is Gâteaux differentiable at $\bar{u} \in \mathcal{U}_k$ and $\mathcal{U}_k$ is open, we have $D\Pi(\bar{u}) = 0$. Similarly, for $D\Pi^d(\bar{\varsigma}) = 0$ and for the left saddle point.

If $(\bar{u}, \bar{\varsigma})$ is a super-critical point of L, then by the definition

$$L(u, \bar{\varsigma}) \leq L(\bar{u}, \bar{\varsigma}) \geq L(\bar{u}, \varsigma) \quad \forall (u, \varsigma) \in \mathcal{U}_a \times \mathcal{T}_a$$

and by Theorem 3.5.1, we know that W is convex and $\bar{\varsigma}$ is negative. Thus, by the fact that $(\bar{u}, \bar{\varsigma}) \in \mathcal{Z}_e$,

$$\Pi(\bar{u}) = \sup_{\varsigma \in \mathcal{T}_a} L(\bar{u}, \varsigma) = L(\bar{u}, \bar{\varsigma}) = \sup_{u \in \mathcal{U}_a} L(u, \bar{\varsigma}) = \Pi^d(\bar{\varsigma}),$$

and hence $\Pi^d(\bar{\varsigma}) = \Pi(\bar{u})$. Moreover, from

$$\Pi^d(\varsigma) = \sup_{u \in \mathcal{U}_a} L(u, \varsigma) = L(\bar{u}, \varsigma) \leq L(\bar{u}, \bar{\varsigma}) = \Pi^d(\bar{\varsigma}),$$

the Gâteaux differentiability of Π^d implies that $\bar{\varsigma}$ is a critical point. Similarly, for $D\Pi(\bar{u}) = 0$ and for the sub-critical point. □

3.5.2 Triality Theorems

Let $\mathcal{L}_c \subset \mathcal{U}_a \times \mathcal{T}_a$ be a critical point set, i.e.

$$\mathcal{L}_c = \{(\bar{u}, \bar{\varsigma}) \in \mathcal{U}_a \times \mathcal{T}_a | \ \delta L(\bar{u}, \bar{\varsigma}; u, \varsigma) = 0 \ \ \forall (u, \varsigma) \in \mathcal{U}_a \times \mathcal{T}_a\}.$$

For any given critical point $(\bar{u}, \bar{\varsigma}) \in \mathcal{L}_c$, we let $\mathcal{U}_r \times \mathcal{T}_r$ be its neighborhood such that, on $\mathcal{U}_r \times \mathcal{T}_r$, $(\bar{u}, \bar{\varsigma})$ is the only critical point of L. The following result is of fundamental importance in nonconvex analysis.

Theorem 3.5.2 (Triality Theorem)
Suppose that $(\bar{u}, \bar{\varsigma}) \in \mathcal{L}_c$, and $\mathcal{U}_r \times \mathcal{T}_r$ is a neighborhood of $(\bar{u}, \bar{\varsigma})$.
(T) Suppose that $\bar{W} : \mathcal{E}_a \to \mathbb{R}$ is convex. If $\bar{\varsigma}(x) \geq 0 \ \forall x \in I$, then

$$\min_{u \in \mathcal{U}_r} \max_{\varsigma \in \mathcal{T}_r} L(u, \varsigma) = L(\bar{u}, \bar{\varsigma}) = \max_{\varsigma \in \mathcal{T}_r} \min_{u \in \mathcal{U}_r} L(u, \varsigma). \tag{3.119}$$

However, if $\bar{\varsigma}(x) < 0 \ \forall x \in I$, then either

$$\min_{u \in \mathcal{U}_r} \max_{\varsigma \in \mathcal{T}_r} L(u, \varsigma) = L(\bar{u}, \bar{\varsigma}) = \min_{\varsigma \in \mathcal{T}_r} \max_{u \in \mathcal{U}_r} L(u, \varsigma), \tag{3.120}$$

or

$$\max_{u \in \mathcal{U}_r} \max_{\varsigma \in \mathcal{T}_r} L(u, \varsigma) = L(\bar{u}, \bar{\varsigma}) = \max_{\varsigma \in \mathcal{T}_r} \max_{u \in \mathcal{U}_r} L(u, \varsigma). \tag{3.121}$$

(T') Suppose that $\bar{W} : \mathcal{E}_a \to \mathbb{R}$ is concave. If $\bar{\varsigma}(x) \leq 0 \ \forall x \in I$, then

$$\max_{u \in \mathcal{U}_r} \min_{\varsigma \in \mathcal{T}_r} L(u, \varsigma) = L(\bar{u}, \bar{\varsigma}) = \min_{\varsigma \in \mathcal{T}_r} \max_{u \in \mathcal{U}_r} L(u, \varsigma). \tag{3.122}$$

However, if $\bar{\varsigma}(x) > 0 \ \forall x \in I$, then either

$$\max_{u \in \mathcal{U}_r} \min_{\varsigma \in \mathcal{T}_r} L(u, \varsigma) = L(\bar{u}, \bar{\varsigma}) = \max_{\varsigma \in \mathcal{T}_r} \min_{u \in \mathcal{U}_r} L(u, \varsigma), \tag{3.123}$$

or

$$\min_{u\in\mathcal{U}_r}\min_{\varsigma\in\mathcal{T}_r} L(u,\varsigma) = L(\bar{u},\bar{\varsigma}) = \min_{\varsigma\in\mathcal{T}_r}\min_{u\in\mathcal{U}_r} L(u,\varsigma). \tag{3.124}$$

Proof. By Theorem 3.5.1, for convex $\bar{W}$, if $\bar{\varsigma}(x) \geq 0 \ \ \forall x \in I$, then $(\bar{u},\bar{\varsigma})$ is a saddle point of L on $\mathcal{U}_r \times \mathcal{T}_r$, and (3.119) follows from the saddle Lagrangian theorem proved in Chapter 1. However, if $\bar{\varsigma}(x) < 0 \ \ \forall x \in I$, then $(\bar{u},\bar{\varsigma})$ is a super-critical point of L on $\mathcal{U}_r \times \mathcal{T}_r$, and, by the super-Lagrangian duality theorem (Theorem 2.6.2), we have either (3.120) or (3.121).

Similarly for concave $\bar{W}$. □

3.5.3 Tri-Duality Theory

Let $\mathcal{U}_c \subset \mathcal{U}_k$ and $\mathcal{T}_c \subset \mathcal{T}_\emptyset$ be the critical point sets of Π and Π^d, respectively:

$$\mathcal{U}_c = \{\bar{u} \in \mathcal{U}_a |\ \ \delta\Pi(\bar{u}; u) = 0 \ \ \forall u \in \mathcal{U}_a\};$$

$$\mathcal{T}_c = \{\bar{\varsigma} \in \mathcal{T}_\emptyset |\ \ \delta\Pi^d(\bar{\varsigma}; \varsigma) = 0 \ \ \forall \varsigma \in \mathcal{T}_\emptyset\}.$$

Theorem 3.5.3 (Tri-Duality Theorem) *Suppose that $\mathcal{U}_c$ and $\mathcal{T}_c$ are not empty, $\bar{u} \in \mathcal{U}_c$ is a critical point of Π and $\mathcal{U}_r \subset \mathcal{U}_k$ is a neighborhood of $\bar{u}$. Let $\bar{\varsigma} = D_\xi \bar{W}(\Lambda(\bar{u}))$ with a neighborhood $\mathcal{T}_r \subset \mathcal{T}_\emptyset$.*

(T) For convex $\bar{W}(\xi)$ on $\mathcal{E}_a$, if $\bar{\varsigma}(x) > 0 \ \forall x \in I$, then

$$\Pi(\bar{u}) = \min_{u\in\mathcal{U}_r} \Pi(u) \quad \textit{if and only if} \quad \Pi^d(\bar{\varsigma}) = \max_{\varsigma\in\mathcal{T}_r} \Pi^d(\varsigma). \tag{3.125}$$

However, if $\bar{\varsigma}(x) < 0 \ \forall x \in I$, then

$$\Pi(\bar{u}) = \min_{u\in\mathcal{U}_r} \Pi(u) \quad \textit{if and only if} \quad \Pi^d(\bar{\varsigma}) = \min_{\varsigma\in\mathcal{T}_r} \Pi^d(\varsigma); \tag{3.126}$$

$$\Pi(\bar{u}) = \max_{u\in\mathcal{U}_r} \Pi(u) \quad \textit{if and only if} \quad \Pi^d(\bar{\varsigma}) = \max_{\varsigma\in\mathcal{T}_r} \Pi^d(\varsigma). \tag{3.127}$$

If $\mathcal{T}_c$ contains only one strictly positive critical point $\bar{\varsigma}$, the primal variational problem $(\Pi_{\rm sta})$ has at most one global minimizer.

(T') For concave $\bar{W}(\xi)$ on $\mathcal{E}_a$, if $\varsigma(x) < 0 \ \forall x \in I$, then

$$\Pi(\bar{u}) = \max_{u\in\mathcal{U}_r} \Pi(u) \quad \textit{if and only if} \quad \Pi^d(\bar{\varsigma}) = \min_{\varsigma\in\mathcal{T}_r} \Pi^d(\varsigma). \tag{3.128}$$

However, if $\bar{\varsigma}(x) > 0 \ \forall x \in I$, then

$$\Pi(\bar{u}) = \min_{u\in\mathcal{U}_r} \Pi(u) \quad \textit{if and only if} \quad \Pi^d(\bar{\varsigma}) = \min_{\varsigma\in\mathcal{T}_r} \Pi^d(\varsigma); \tag{3.129}$$

$$\Pi(\bar{u}) = \max_{u\in\mathcal{U}_r} \Pi(u) \quad \textit{if and only if} \quad \Pi^d(\bar{\varsigma}) = \max_{\varsigma\in\mathcal{T}_r} \Pi^d(\varsigma). \tag{3.130}$$

Moreover, if $\mathcal{T}_c$ contains only one strictly negative critical point $\bar{\varsigma}$, the primal variational problem $(\Pi_{\rm sta})$ has at most one global maximizer.

Proof. From Lemma 3.5.1 we know that $(\bar{u}, \bar{\varsigma})$ must be a critical point of L and $P(\bar{u}) = L(\bar{u}, \bar{\varsigma}) = \Pi^d(\bar{\varsigma})$. If $\bar{\varsigma} > 0$, then, by Theorem 3.5.1, $(\bar{u}, \bar{\varsigma})$ is a saddle point of L on $\mathcal{U}_r \times \mathcal{T}_r$. For any given $u \in \mathcal{U}_a$, $L : \mathcal{T}_a \to \mathbb{R}$ is concave, and

$$\Pi(u) = \max_{\varsigma \in \mathcal{T}_a} L(u, \varsigma) = L(u, \bar{\varsigma}) \geq L(\bar{u}, \bar{\varsigma}) \quad \forall u \in \mathcal{U}_k.$$

By Lemma 3.5.1, $L(\bar{u}, \bar{\varsigma}) = \Pi(\bar{u}) \leq \Pi(u) \quad \forall u \in \mathcal{U}_a$. Thus, the critical point $\bar{u}$ minimizes $\Pi(u)$ on $\mathcal{U}_r$. Since $\bar{\varsigma} > 0$, and $L : \mathcal{U}_r \to \mathbb{R}$ is convex, then, by Proposition 3.5.1,

$$\begin{aligned} \Pi^d(\varsigma) &= \min_{u \in \mathcal{U}_a} L(u, \varsigma) = L(\bar{u}, \varsigma) \leq L(\bar{u}, \bar{\varsigma}) \\ &= \Pi^d(\bar{\varsigma}) \quad \forall \varsigma \in \mathcal{T}_\varnothing. \end{aligned}$$

This shows that the critical point $\bar{\varsigma} \in \mathcal{T}_r$ maximizes Π^d, and Lemma 3.5.1 yields $\Pi(\bar{u}) = \Pi^d(\bar{\varsigma})$.

If $\bar{\varsigma} < 0$, $(\bar{u}, \bar{\varsigma})$ should be a super-critical point of L on $\mathcal{U}_r \times \mathcal{T}_r$, and

$$\max_u L(u, \bar{\varsigma}) = L(\bar{u}, \bar{\varsigma}) = \max_\varsigma L(\bar{u}, \varsigma) \quad \forall (u, \varsigma) \in \mathcal{U}_r \times \mathcal{T}_r. \tag{3.131}$$

If $\bar{u}$ maximizes Π on $\mathcal{U}_r$, then

$$\begin{aligned} \Pi(\bar{u}) &= \max_{u \in \mathcal{U}_r} \Pi(u) = \max_{u \in \mathcal{U}_r} \max_{\varsigma \in \mathcal{T}_r} L(u, \varsigma), \\ &= \max_{\varsigma \in \mathcal{T}_r} \max_{u \in \mathcal{U}_r} L(u, \varsigma) = \max_{\varsigma \in \mathcal{T}_r} P^*(\varsigma) = \Pi^d(\bar{\varsigma}). \end{aligned} \tag{3.132}$$

Thus, $\max \Pi(u) = \max \Pi^d(\varsigma)$ since we can take the maximum in either order on $\mathcal{U}_r \times \mathcal{T}_r$.

If $\bar{u}$ is a local minimizer of Π on $\mathcal{U}_r$, we need to prove that the associated $\bar{\varsigma}$ minimizes Π^d on $\mathcal{T}_r$. Since $\bar{\varsigma}$ is a critical point of Π^d, it must be either a local extremum point or a local saddle point of Π^d. If $\bar{\varsigma}$ is a local maximizer, then by (3.132) the associated $\bar{u}$ should be a local maximizer of Π. This contradicts our assumption. If $\bar{\varsigma}$ is a local saddle point of Π^d and it maximizes Π^d in the direction $\varsigma_o \in \mathcal{T}_r$ such that $\Pi^d(\bar{\varsigma}) = \max_\theta \Pi^d(\varsigma_o + \theta) \ \forall \theta \geq 0$, then the solution θ_o of this problem should satisfy $\bar{\varsigma} = \varsigma_o + \theta_o$. Substituting $\varsigma = \varsigma_o + \theta$ into (3.132) we again arrive at a contradiction. Thus, the critical point $\bar{\varsigma}$ must be a local minimizer of Π^d.

If $\mathcal{T}_c$ contains only one strictly positive critical point $\bar{\varsigma}$, then $L : \mathcal{T}_a \to \mathbb{R}$ is strictly concave. Since $D_\epsilon \bar{W}(\Lambda(u))$ is one-to-one, and $\Lambda(u)$ is quadratic, then $L(u, \varsigma)$ has only one saddle point, and $\Pi : \mathcal{U}_k \to \mathbb{R}$ is strictly convex. The primal problem therefore has at most one solution. □

Example 3.5.1 We now illustrate this interesting tri-duality result with the very simple nonconvex optimization problem

$$\Pi(u) = \frac{1}{2} k_0 (\frac{1}{2} u^2 - \lambda)^2 - \sigma u \quad \to \min \quad \forall u \in \mathbb{R}.$$

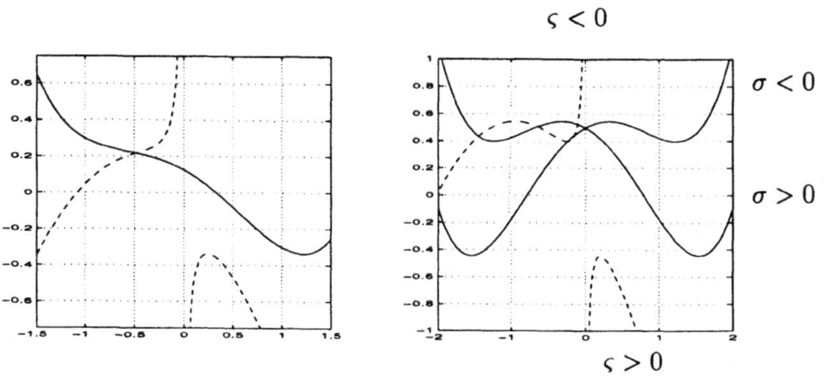

(a) $\lambda < \lambda_{\max}$ (b) $\lambda > \lambda_{\max}$

Figure 3.18. Double-well energy $\Pi(u)$ (solid line) and its dual $\Pi^d(\varsigma)$ (dashed line)

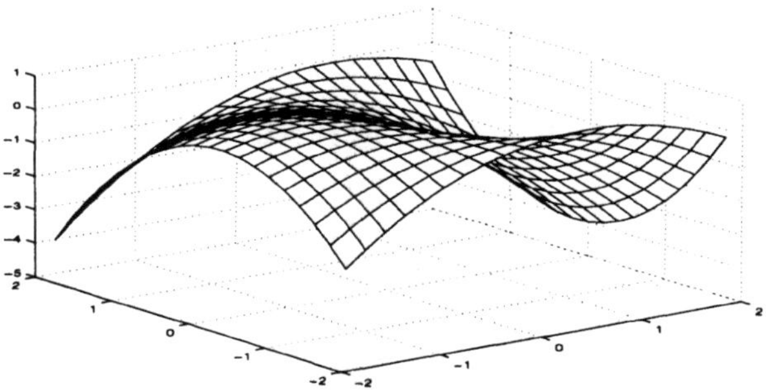

Figure 3.19. Lagrangian for the double-well energy in the Example 3.5.1

Let $\xi = \Lambda(u) = \frac{1}{2}u^2 - \lambda$, $\bar{W}(\xi) = \frac{1}{2}k_0\xi^2$, $\varsigma = D\bar{W}(\xi) = k_0\xi$. Then, the dual function is

$$\Pi^d(\varsigma) = -(\frac{1}{2k_0}\varsigma^2 + \lambda\varsigma) - \frac{\sigma^2}{2\varsigma}.$$

For given data λ and σ, the graphs of P and Π^d are shown in Fig. 3.18. By Theorem 3.4.4 we know that for $\lambda < \lambda_{\min}$ the problem has only one global minimizer (see Fig. 3.18(a)). However, if $\lambda > \lambda_{\min}$, the dual Euler-Lagrange equation (3.104) has three roots $\varsigma_1 > 0 > \varsigma_2 > \varsigma_3$, corresponding to three critical points of Π^d (see Fig. 3.18(b)). Then, ς_1 is a global maximizer of Π^d, $u_1 = \sigma/\varsigma_1$ is a global minimizer of Π, Π^d takes local minimum and local maximum values at ς_2 and ς_3, respectively, $u_2 = \sigma/\varsigma_2$ is a local maximizer of Π, while $u_3 = \sigma/\varsigma_3$ is a local minimizer. ♡

The Lagrangian associated with this double-well energy is

$$L(u, \varsigma) = \frac{1}{2}u^2\varsigma - (\frac{1}{2k_0}\varsigma^2 + \lambda\varsigma) - \sigma u.$$

It is a saddle function for $\varsigma > 0$. If $\varsigma < 0$, it is a super-critical point function (see Fig. 3.19).

Example 3.5.2 (Solution for the Finite Deformation String Problem) Recall the inf-primal problem

$$\Pi(u) = \int_0^1 k_0\sqrt{1 + u'(x)^2}\, dx - \int_0^1 fu\, dx \quad \to \min$$

for the finite deformation elastic string discussed in Chapter 1, subject to the constraint

$$u \in \mathcal{U}_k = \{u \in \mathcal{L}^2(0,1) |\ u' \in \mathcal{L}^1(0,1),\ u(0) = 0\}.$$

In terms of the finite strain measure

$$\xi = \Lambda(u) = u'(x)^2 + 1 > 0,$$

the concave stored energy $\bar{W}(\xi) = k_0\sqrt{\xi}$ is well-defined on

$$\mathcal{E}_a = \{\xi \in \mathcal{L}^2(I) |\ \xi(x) > 0\ \forall x \in I\}.$$

The canonical complementary energy density

$$\bar{W}^c(\varsigma) = \inf_{\xi \in \mathcal{E}_a} \{\langle \xi ; \varsigma \rangle - \bar{W}(\xi)\} = \int_I -\frac{k_0^2}{4\varsigma}\, dx$$

is strictly concave, and the dual feasible space $\mathcal{T}_\emptyset$ is then

$$\mathcal{T}_\emptyset = \{\varsigma \in \mathcal{L}^2(I) |\ \varsigma(x) > 0\ \forall x \in I\}.$$

For any given external load $f(x)$ in the domain I, and σ_ℓ on the boundary $x = 1$, the statically admissible space for this mixed boundary-value problem is

$$\mathcal{S}_s = \{\sigma \in \mathcal{L}^\infty(I) |\ -\sigma'(x) = f(x)\ \forall x \in I,\ \sigma(1) = \sigma_\ell\},$$

which contains a unique element, namely,

$$\sigma(x) = \int_x^1 f(t)dt + \sigma_\ell = g(x) + \sigma_\ell.$$

Thus, on $\mathcal{T}_\emptyset$, the total complementary energy for this curved string is a nonconvex functional

$$\Pi^d(\varsigma) = \sigma_\ell u(1) + \int_I \left(\frac{k_0^2 - \sigma^2}{4\varsigma} - \varsigma\right)\, dx.$$

Its criticality condition leads to the quadratic equation

$$\varsigma^2 = \frac{k_0^2}{4} - \frac{\sigma^2}{4}.$$

For any given $\sigma \in \mathcal{S}_s$ such that $|\sigma(x)| < k_0 \;\; \forall x \in I$, this algebraic equation has the unique solution

$$\bar{\varsigma} = \frac{1}{2}\sqrt{k_o^2 - \sigma^2} \in \mathcal{T}_\varnothing, \tag{3.133}$$

and the analytic solution for this finite deformation mixed boundary-value problem reads

$$\bar{u}(x) = \int_0^x \frac{g(t) + \sigma_\ell}{\sqrt{k_o^2 - (g(t) + \sigma_\ell)^2}} \mathrm{d}t.$$

For Dirichlet boundary conditions $u(0) = u(1) = 0$, the boundary force σ_ℓ should be replaced by the constant $\bar{c}$ defined by (1.146). The solution $\bar{u}$ is then the same as that obtained in Chapter 1.

Since $\bar{W}(\xi)$ is concave and $\varsigma(x) > 0$, the Tri-Duality Theorem gives

$$\Pi(\bar{u}) = \min_{u \in \mathcal{U}_k} \Pi(u) = \min_{\varsigma \in \mathcal{T}_\varnothing} \Pi^d(\varsigma) = \Pi^d(\bar{\varsigma}).$$

To demonstrate this result graphically, we set

$$\begin{aligned} u(x) &= c\bar{u} = c \int_0^x \frac{g(t) + \sigma_\ell}{\sqrt{k_o^2 - (g(t) + \sigma_\ell)^2}} dt \;\; \forall c \in \mathbb{R}, \\ \varsigma(x) &= c^* \bar{\varsigma}(x) = c^* \frac{1}{2}\sqrt{k_o^2 - \sigma^2} \;\; \forall c^* \in \mathbb{R}. \end{aligned}$$

We take $k_0 = 1$, $f(x) = -1$ and $\sigma_\ell = 0.5$. Then $\sigma(x) = x - 0.5$ and the associated functions are

$$\Pi_{\bar{u}}(c) = \Pi(c\bar{u}) = -0.09c + 2.EllipticE[0.52, 1 - c^2],$$
$$\Pi^d_{\bar{\varsigma}}(c^*) = \Pi^d(c^*\bar{\varsigma}) = 0.4783(1 + c^{*2})/c^*.$$

Figure 3.20 shows that

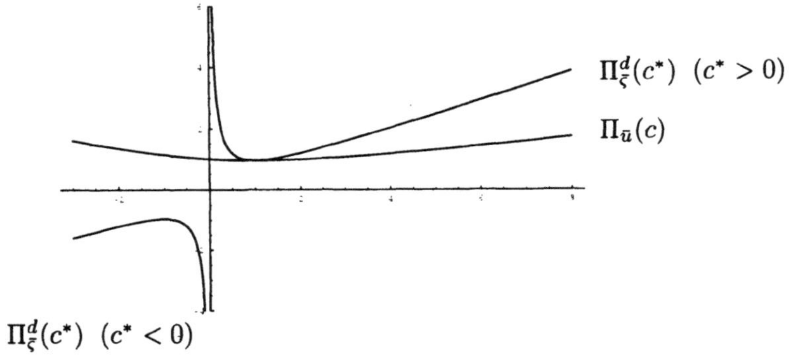

Figure 3.20. Finite deformation string: $\Pi_{\bar{u}}(c)$ and its dual $\Pi^d_{\bar{\varsigma}}(c^*)$

$$\Pi_{\bar{u}}(1) = \min_{c\in\mathbb{R}} \Pi_{\bar{u}}(c) = \min_{c^*>0} \Pi^d_{\varsigma}(c^*) = \Pi^d_{\varsigma}(1) = 0.956611.$$

Since the graph of $\Pi_{\bar{u}}$ is very flat, the numerical method for solving the primal problem needs more iterations than for solving the dual problem. ♡

3.6 Canonical Dual Transformation for Nonconvex Dynamical Systems

This section presents an important applications of the canonical dual transformation method in nonconvex dynamical systems.

Since the stored energy $\bar{W}$ is a saddle functional, the triality theory requires some generalization.

3.6.1 Spacetime geometrical measure and Canonical Dynamical Equations

In dynamical systems, the stored-energy density $V(t,u,u_{,t})$ is usually a nonconvex action function, and may often be written

$$V(t,u,v) = K(t,v) - U(t,u), \tag{3.134}$$

where $K(t,v)$ is the kinetic energy density, and $U(t,u)$ is the stored potential energy. In Newtonian systems, the kinetic energy is usually a quadratic function of the velocity $v = \dot{u} = u_{,t}$. However, the potential $U(t,u)$ is often a nonconvex function of u.

By the canonical dual transformation discussed earlier in Section 3.3.2, the following assumption is essential for this section.

Assumption 3.6.1 For any given action density $V : I \times \mathcal{C} \times \mathcal{C}^1 \to \mathbb{R}$, there exists a vector-valued geometrical mapping $\Lambda = \{\Lambda^{(t)}, \Lambda^{(s)}\} : I \times \mathcal{U} \to \mathcal{E} = \mathcal{E}^t \times \mathcal{E}^s \subset \mathcal{L}^{\beta}(I;\mathbb{R}^2)$ and a real-valued function $\bar{V} : \mathcal{E} \to \mathcal{L}(I)$ such that

(A1) $V(t,u,u') = \bar{V}(\Lambda(u))$ $\forall (t,u) \in I \times \mathcal{C}^1(I)$.

(A2) $\bar{V} : \mathcal{E}_a = \mathcal{E}^t_a \times \mathcal{E}^s_a \subset \mathcal{E} \to \mathcal{L}(I)$ is a Gâteaux differentiable right saddle function. ♠

Thus, in dynamical systems, the geometric equation can be written in the form

$$\boldsymbol{\xi} = \Lambda(t,u) = (\Lambda^{(t)}(t,u),\ \Lambda^{(s)}(t,u))^T = (v(u),\ \xi(u))^T.$$

For the sake of convenience, $\Lambda(t,u)$ is written as $\Lambda(u)$. For each $\boldsymbol{\xi} \in \mathcal{E} = \mathcal{E}^t \times \mathcal{E}^s$, the total internal action

$$\bar{W}(\xi) = \int_I \bar{V}(v,\xi)\mathrm{d}t$$

is also a right saddle functional. Then, the constitutive mapping $D\bar{W}(\boldsymbol{\xi}) \in \mathcal{T} = \mathcal{T}^t \times \mathcal{T}^s \subset \mathcal{L}^{\beta^*}(I)$ is the vector-valued operator

$$\varsigma = \begin{pmatrix} p \\ \varsigma \end{pmatrix} = D\bar{W}(\boldsymbol{\xi}) \Rightarrow \begin{cases} p = D_v\bar{W}(v,\xi), \\ \varsigma = D_{\xi}\bar{W}(v,\xi). \end{cases} \tag{3.135}$$

For the vector-valued geometrical operator $\Lambda : \mathcal{U} \to \mathcal{E}$, the bilinear form on $\mathcal{E} \times \mathcal{T}$ is denoted

$$\langle \boldsymbol{\xi}; \boldsymbol{\varsigma} \rangle = \langle v; p \rangle + \langle \xi; \varsigma \rangle.$$

The complementary action $\bar{W}^c : \mathcal{T}_a \to \mathbb{R}$ is given by

$$\bar{W}^c(\boldsymbol{\xi}) = \text{sta}_{\boldsymbol{\xi} \in \mathcal{E}_a} \{ \langle \boldsymbol{\xi}; \boldsymbol{\varsigma} \rangle - \bar{W}(\boldsymbol{\xi}) \}. \tag{3.136}$$

For given external load $\bar{f}$, let the external action $F : \mathcal{U}_a \subset \mathcal{U} \to \mathbb{R}$ be a linear functional

$$F(u) = \langle \bar{u}^*, u \rangle. \tag{3.137}$$

Thus, on the kinematically admissible space $\mathcal{U}_k = \{ u \in \mathcal{U}_a | \ \Lambda(u) \in \mathcal{E}_a \}$, the primal variational problem in this nonconvex dynamical system can be proposed as

$$(\Pi_{\text{inf}}) : \ \Pi(u) = \bar{W}(\Lambda(u)) - F(u) \quad \to \text{sta} \ \ \forall u \in \mathcal{U}_k. \tag{3.138}$$

The criticality condition leads to the nonlinear partial differential equation

$$\Lambda_t^*(u) D\bar{W}(\Lambda(u)) = DF(u). \tag{3.139}$$

Specially, if

$$\bar{V}(v, \xi) = \bar{K}(v) - \bar{U}(\xi), \tag{3.140}$$

and $\bar{K} : \mathcal{E}^t \to \mathbb{R}$, $\bar{U} : \mathcal{E}^s \to \mathbb{R}$ are strictly convex, then the canonical Euler-Lagrange equations can be written as

$$[\Lambda_t^{(t)}(u)]^* D\bar{K}(\Lambda^{(t)}(u)) - [\Lambda_t^{(s)}(u)]^* D\bar{U}(\Lambda^{(s)}(u)) = DF(u). \tag{3.141}$$

Example 3.6.1 For Newtonian systems, suppose that $\Lambda : \mathcal{U} \to \mathcal{E}$ is the quadratic operator

$$\Lambda(u) = \begin{pmatrix} \Lambda^{(t)} \\ \Lambda^{(s)} \end{pmatrix} = \begin{pmatrix} \frac{\mathrm{d}}{\mathrm{d}t} \\ \frac{1}{2}u^2 + bu + c \end{pmatrix}, \tag{3.142}$$

where $b, c : I \to \mathbb{R}$ are given arbitrary functions. Then,

$$\Lambda(u) = \Lambda_t(u)u + \Lambda_c(u) = \begin{pmatrix} u_{,t} \\ u^2 + bu \end{pmatrix} + \begin{pmatrix} 0 \\ -\frac{1}{2}u^2 + c \end{pmatrix}.$$

Under the standard bilinear form

$$\langle \Lambda(u); \boldsymbol{\varsigma} \rangle = \int_I \left(u_{,t} p + (\frac{1}{2}u^2 + bu + c)\varsigma \right) \mathrm{d}t,$$

the formal adjoints of the operators $\Lambda_t^{(t)}$ and $\Lambda_t^{(s)}$ can be obtained via the virtual work principle as

$$\left(\Lambda^{(t)}\right)^* = -\frac{\mathrm{d}}{\mathrm{d}t}, \quad \left(\Lambda_t^{(s)}(u)\right)^* = u + b.$$

The canonical equation (3.141) reads

$$\frac{\mathrm{d}}{\mathrm{d}t} D\bar{K}(\frac{\mathrm{d}}{\mathrm{d}t}u) + (u+b) D\bar{U}(\frac{1}{2}u^2 + bu + c) = DF(u). \tag{3.143}$$

The complementary gap function is the quadratic functional

$$G(u,\varsigma) = \int_I (\frac{1}{2}u^2 - c)\varsigma \mathrm{d}t$$

of u. Furthermore, if $\bar{K}$ and $\bar{U}$ are quadratic functions, given by

$$\bar{K}(v) = \frac{1}{2}m_0 v^2, \quad \bar{U}(\xi) = \frac{1}{2}k_0 \xi^2,$$

and $F(u) = \int_I fu\mathrm{d}t$ is a linear functional, then the canonical equation (3.141) can be written simply as

$$m_0 u_{,tt} + k_0(u+b)\left(\frac{1}{2}u^2 + bu + c\right) = f. \tag{3.144}$$

If $f = b = 0$ and $c = -\lambda$ is a negative constant, then we have the well-known Duffing equation

$$m_0 u_{,tt} = k_0 u \left(\lambda - \frac{1}{2}u^2\right).$$

3.6.2 Triality Theory in Nonconvex Hamilton Systems

The canonical Lagrangian $L : \mathcal{U}_a \times \mathcal{T}_a \to \mathbb{R}$ in nonconvex dynamical systems is defined by

$$L(u,\varsigma) = \langle \Lambda(u);\varsigma\rangle - \bar{W}^c(\varsigma) - F(u). \tag{3.145}$$

Since $\bar{W}^c : \mathcal{T}_a^t \times \mathcal{T}_a^s \to \mathbb{R}$ is also a right saddle functional, we always have either

$$\inf_{u\in\mathcal{U}_k} \sup_{p\in\mathcal{T}_a^t} \inf_{\varsigma\in\mathcal{T}_a^s} L(u,p,\varsigma) = \inf_{u\in\mathcal{U}_k} \inf_{\varsigma\in\mathcal{T}_a^s} \sup_{p\in\mathcal{T}_a^t} L(u,p,\varsigma) \tag{3.146}$$

or

$$\sup_{u\in\mathcal{U}_k} \inf_{\varsigma\in\mathcal{T}_a^s} \sup_{p\in\mathcal{T}_a^t} L(u,p,\varsigma) = \sup_{u\in\mathcal{U}_k} \sup_{p\in\mathcal{T}_a^t} \inf_{\varsigma\in\mathcal{T}_a^s} L(u,p,\varsigma). \tag{3.147}$$

The triality theory discussed in Section 3.5 can be generalized for nonconvex Newtonian systems.

Theorem 3.6.1 *Suppose that $\Lambda : \mathcal{U} \to \mathcal{E}$ is the quadratic operator*

$$\Lambda(u) = \begin{pmatrix} \frac{\mathrm{d}}{\mathrm{d}t} \\ \frac{1}{2}u^2 + bu + c \end{pmatrix}, \tag{3.148}$$

where $b, c \in \mathcal{C}(I)$ are given arbitrary functions, and $(\bar{u},\bar{p},\bar{\varsigma})$ is a critical point of L on the neighborhood $\mathcal{U}_r \times \mathcal{T}_r^t \times \mathcal{T}_r^s$. If $\bar{\varsigma}(t) \geq 0$ $\forall t \in I$, then, on $\mathcal{U}_r \times \mathcal{T}_r^t \times \mathcal{T}_r^s$,

$$L(\bar{u},\bar{p},\bar{\varsigma}) = \min_u \max_p \min_\varsigma L(u,p,\varsigma) = \max_p \min_{u,\varsigma} L(u,p,\varsigma). \tag{3.149}$$

However, if $\bar{\varsigma}(t) \leq 0 \ \forall t \in I$, then on $\mathcal{U}_r \times \mathcal{T}_r^t \times \mathcal{T}_r^s$, either

$$\begin{aligned} L(\bar{u}, \bar{p}, \bar{\varsigma}) &= \min_u \max_p \min_\varsigma L(u,p,\varsigma) = \min_p \max_u \min_\varsigma L(u,p,\varsigma) \\ &= \min_\varsigma \max_u \min_p L(u,p,\varsigma) = \min_{p,\varsigma} \max_u L(u,p,\varsigma), \end{aligned} \tag{3.150}$$

or

$$\begin{aligned} L(\bar{u}, \bar{p}, \bar{\varsigma}) &= \max_u \min_\varsigma \max_p L(u,p,\varsigma) = \max_p \min_\varsigma \max_u L(u,p,\varsigma) \\ &= \min_\varsigma \max_{u,p} L(u,p,\varsigma). \end{aligned} \tag{3.151}$$

Proof. For the quadratic operator (3.148), the Lagrangian can be written as

$$L(u,p,\varsigma) = \langle \dot{u}; p \rangle + \langle \frac{1}{2}u^2 + bu + c; \varsigma \rangle - \bar{W}^c(p,\varsigma) - F(u). \tag{3.152}$$

If $\bar{\varsigma}(t) \geq 0 \ \ \forall t \in I$, then for any $\varsigma \in \mathcal{T}_r^s$, $L : \mathcal{U}_r \times \mathcal{T}_r^t \to \mathbb{R}$ is a right saddle functional and $\min_u \max_p = \max_p \min_u$.

If $\bar{\varsigma}(t) < 0 \ \ \forall t \in I$, then for any fixed $\varsigma \in \mathcal{T}_r^s$, the Lagrangian L is a super-critical functional on $\mathcal{U}_r \times \mathcal{T}_r^t$, and, for any fixed $p \in \mathcal{T}_r^t$, L is a left-saddle functional on $\mathcal{U}_r \times \mathcal{T}_r^s$. In this case, the statements (3.150) and (3.151) follows directly from the combination of the super-Lagrange duality theory and saddle-Lagrange duality theory. □

For a fixed $\varsigma \in \mathcal{T}_a$, the dual action $\Pi^d : \mathcal{T}_a \to \mathbb{R}$ can be obtained via the Lagrangian through

$$\Pi^d(\varsigma) = \text{sta}\{L(u,\varsigma) | \ \forall u \in \mathcal{U}_a\} = F_\Lambda^c(\varsigma) - \bar{W}^c(\varsigma), \tag{3.153}$$

where the generalized Legendre conjugate F_Λ^c of $F(u)$ is defined by

$$F_\Lambda^c(\varsigma) = \underset{u \in \mathcal{U}_a}{\text{sta}} \ \{\langle \Lambda(u); \varsigma \rangle - F(u) | \ \forall u \in \mathcal{U}_a\}.$$

If the action density $\bar{V}$ is given by (3.140), then the dual action $\Pi^d : \mathcal{T}_a^t \times \mathcal{T}_a^s \to \mathbb{R}$ can be written as

$$\Pi^d(p,\varsigma) = \int_I [\bar{U}^c(\varsigma) - \bar{K}^c(p)] \, \mathrm{d}t + -G^c(p,\varsigma). \tag{3.154}$$

In this nonconvex dynamical system, the pure complementary gap functional is defined by

$$G^c(p,\varsigma) = -\underset{u \in \mathcal{U}_a}{\text{sta}} \int_I \left[p\Lambda^{(t)}(u) + \varsigma\Lambda^{(s)}(u) - fu \right] \, \mathrm{d}t = \langle -\Lambda_c(\bar{u}); \varsigma \rangle \ \ \forall (\bar{u},\varsigma) \in \mathcal{Z}_e,$$

where $\mathcal{Z}_e$ is the admissible space of equilibrium defined by (3.81).

Theorem 3.6.2 *Suppose that $\Lambda : \mathcal{U} \to \mathcal{E}$ is a quadratic operator defined by (3.148), and $(\bar{u}, \bar{p}, \bar{\varsigma})$ is a critical point of L on the neighborhood $\mathcal{U}_r \times \mathcal{T}_r$ ($\mathcal{T}_r = \mathcal{T}_r^t \times \mathcal{T}_r^s$). If $\bar{\varsigma}(t) \geq 0 \ \forall t \in I$, then*

$$\min_{u \in \mathcal{U}_r} \Pi(u) = \max_{p \in \mathcal{T}_r^t} \min_{\varsigma \in \mathcal{T}_r^s} \Pi^d(p,\varsigma). \tag{3.155}$$

If $\bar{\varsigma}(t) < 0 \;\; \forall t \in I$, then either

$$\min_{u \in \mathcal{U}_r} \Pi(u) = \inf_{(p,\varsigma) \in \mathcal{T}_r} \Pi^d(p,\varsigma), \tag{3.156}$$

or

$$\sup_{u \in \mathcal{U}_r} \Pi(u) = \sup_{(p,\varsigma) \in \mathcal{T}_r} \Pi^d(p,\varsigma). \tag{3.157}$$

Example 3.6.2 Let us consider the nonconvex optimization problem

$$\Pi(x) = \frac{1}{2} m x^2 - \frac{1}{2} k (\frac{1}{2} x^2 - \lambda)^2 + f x \to \text{sta} \;\; \forall x \in \mathbb{R}. \tag{3.158}$$

The criticality condition for Π leads to the cubic algebraic equation

$$k x (\frac{1}{2} x^2 - \lambda) - m x = f.$$

Its three real roots $x_i (i = 1, 2, 3)$ are critical points of Π (see Fig. 3.21). In terms of the nonlinear operator defined by

$$\xi = \Lambda(x) = \left(x, \;\; \frac{1}{2} x^2 \right)^T, \tag{3.159}$$

the function

$$\bar{V}(v, \xi) = \frac{1}{2} m v^2 - \frac{1}{2} k (\xi - \lambda)^2$$

is a saddle function. The dual variables (p, ς) are defined by

$$p = D_v \bar{V}(v, \xi) = m v, \quad \varsigma = D_\xi \bar{V}(v, \xi) = -k(\xi - \lambda).$$

Let $\varsigma = -y$, then the Legendre conjugate of $\bar{V}$ is also a saddle function, namely

$$\bar{V}^c(p, y) = \text{sta}\{p v - \xi y - \bar{V}(v, \xi)\} = \frac{1}{2m} p^2 - \frac{1}{2k} y^2 - \lambda y.$$

Thus, the nonlinear Lagrangian for this nonconvex problem is

$$L(x, p, y) = p x - y (\frac{1}{2} x^2 - \lambda) - \frac{1}{2m} p^2 + \frac{1}{2k} y^2 + \lambda y + f x. \tag{3.160}$$

The criticality condition $D_x L(\bar{x}, p, y) = 0$ leads to the equilibrium equation

$$p - \bar{x} y = -f \;\; \Rightarrow \;\; \bar{x} = \frac{p + f}{y}. \tag{3.161}$$

For any given $p \in \mathbb{R}$ and $y \neq 0$, the dual action is obtained as

$$\Pi^d(p, y) = \frac{1}{2k} y^2 + \lambda y + \frac{(p + f)^2}{2y} - \frac{1}{2m} p^2. \tag{3.162}$$

The criticality condition for Π^d leads to the dual Euler-Lagrange equations

$$\frac{p}{m} = \frac{p+f}{y}, \tag{3.163}$$

$$y^2(\frac{1}{k}y + \lambda) = \frac{1}{2}(p+f)^2. \tag{3.164}$$

Solving the first equation for $p = fm/(y-m)$ and substituting it into the second, we obtain

$$(m-y)^2(\frac{1}{k}y + \lambda) = \frac{1}{2}f^2,$$

$$\Pi_p^d(y) := \frac{1}{2k}y^2 + \lambda y + \frac{yf^2}{2(m-y)^2} - \frac{1}{2m}\left(\frac{mf}{m-y}\right)^2.$$

The graphs of $\Pi(x)$ and $\Pi_p^d(y)$ are shown in Fig. 3.21. ♡

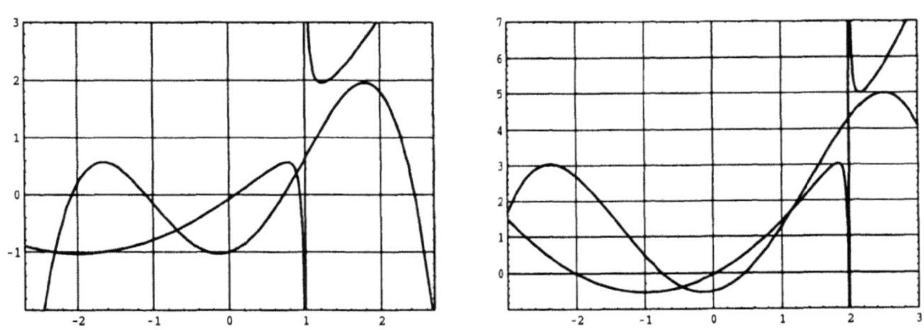

(a) $m > k$ (b) $m < k$

Figure 3.21. Triality for the dynamical system in Example 3.6.2

3.6.3 Duality for Duffing System

As a typical example, let us consider the very simple nonconvex dynamical problem

$$\Pi(u) = \int_I [m_0 u'^2 - \frac{1}{2}a(\frac{1}{2}u^2 - \lambda)^2 - fu]\mathrm{d}t \ \to \min \ \ \forall u \in \mathcal{U}_k. \tag{3.165}$$

The kinematically admissible space $\mathcal{U}_k$ for the initial-value problem of this one-dimensional dynamical system is given simply as

$$\mathcal{U}_k = \{u \in \mathcal{L}^2(0,T) |\ u' \in \mathcal{L}^2(0,T),\ u(0) = u_0,\ u'(0) = v_0\}.$$

The criticality condition for Π leads to the well-known Duffing equation

$$m_0 u'' = au(\lambda - \frac{1}{2}u^2) - f(t), \ \forall t \in I = (0,T),\ u \in \mathcal{U}_k. \tag{3.166}$$

In terms of the nonlinear operator

$$\boldsymbol{\xi} = \Lambda(u) = \left(u', \ \frac{1}{2}u^2 - \lambda\right)^T$$

the action density $\bar{V}(v, \xi)$ and its conjugate $\bar{V}^c(p, \varsigma)$ are saddle functions:

$$\bar{V}(v, \xi) = \frac{1}{2}m_0 v^2 - \frac{1}{2}a\xi^2, \quad \bar{V}^c(p, \varsigma) = \frac{1}{2m_0}p^2 - \frac{1}{2a}\varsigma^2.$$

Let $\tau = -\varsigma$, and introduce the dual space

$$\mathcal{T}_\varnothing = \left\{ (p, \tau) \in \mathcal{C}^1(I) \left| \begin{array}{l} p(0) = m_0 v_0, \quad -\lambda a \leq \tau(t) < +\infty, \\ \tau(t) \neq 0 \ \ \forall t \in I, \ \ \tau(0) = a(\frac{1}{2}u_0^2 - \lambda) \end{array} \right. \right\}.$$

The nonlinear Lagrangian for this nonconvex system is

$$L(u, p, \tau) = \int_I \left(pu' - \tau(\frac{1}{2}u^2 - \lambda) - \frac{1}{2m_0}p^2 + \frac{1}{2\mu}\tau^2 \right) \mathrm{d}t - \int_I fu\mathrm{d}t. \quad (3.167)$$

The criticality condition $D_u L(\bar{u}, p, \tau) = 0$ leads to the equilibrium equation

$$-p' - \bar{u}\tau = f \ \ \forall t \in I.$$

For any given $(p, \tau) \in \mathcal{T}_\varnothing$, this equation has solution

$$\bar{u} = -\frac{p' + f}{\tau}$$

for $\bar{u}$, $t \in I$. Thus, the pure complementary gap functional is easily obtained as

$$\begin{aligned} -G^c(p, \tau) &= \mathrm{sta}_u \int_I \left[pu' - \tau(\frac{1}{2}u^2 - \lambda) - fu \right] \mathrm{d}t \\ &= p(T)u(T) - m_0 u_0^2 + \int_I \left[\lambda\tau + \frac{(p' + f)^2}{2\tau} \right] \mathrm{d}t. \end{aligned}$$

On $\mathcal{T}_\varnothing$, the dual action is well defined as

$$\Pi^d(p, \tau) = p(T)u(T) - m_0 v_0 u_0 + \int_I [\frac{1}{2a}\tau^2 + \lambda\tau + \frac{(p' + f)^2}{2\tau} - \frac{1}{2m_0}p^2]\mathrm{d}t.$$

The criticality condition for Π^d leads to the *dual Duffing system* in the time domain $I \subset \mathbb{R}$

$$\left(\frac{1}{\tau}(p' + f) \right)' + \frac{1}{m_0}p = 0, \quad (3.168)$$

$$2\tau^2(\frac{1}{a}\tau + \lambda) = (f + p')^2. \quad (3.169)$$

This system consists of the so-called *differential-algebraic equations* (DAE's), which arise naturally in many applications (cf., e.g., Brenan *et al*, 1996; Beardmore and Song, 1998). Although the numerical solution of these types of systems has been the subject of intense research activity in the past few years, the solvability of each problem depends mainly on the so-called *index* of the system (see Chapter 5).

Remark 3.6.1 (Dynamical Bifurcation Criterion)

By Theorem 3.4.4 we know that for the given external force f and initial condition v_0, if at a certain time period $t \in I_c$,

$$\lambda > \frac{3}{2}\left(\frac{f + p'}{a}\right)^{2/3}, \tag{3.170}$$

then, the cubic algebraic equation (3.169) possesses three different real solutions for $\tau(t)$ and the system is in dynamical bifurcation state during the period I_c. The two negative solutions $\tau(t)$ are the sourse that leads to the chaotic motion of the system. Thus, the inequality (3.170) provides a *bifurcation (or chaotic) criterion* for the Duffing system. ■

Theoretically speaking, for the same initial conditions, the Duffing equation (3.166) and its dual system (3.169) should have the same solution set. Numerically, the primal and dual Duffing problems give complementary bounding approaches to the real solution. For the given data $a = 1, f = 0, u_0 = 0, v_0 = 0.2$, Figures 3.22a, 3.22b and 3.22c show that the differences between the primal (solid line) and dual (dashed line) numerical solutions vary with the parameter λ. Both primal and dual solutions are very sensitive to the initial data.

But for mathematicians it is also very important to have a profound knowledge of the physics background, *because it is possible that a mathematical result lies in a parameter space which is physically meaningless.*

—Eberhard Zeidler

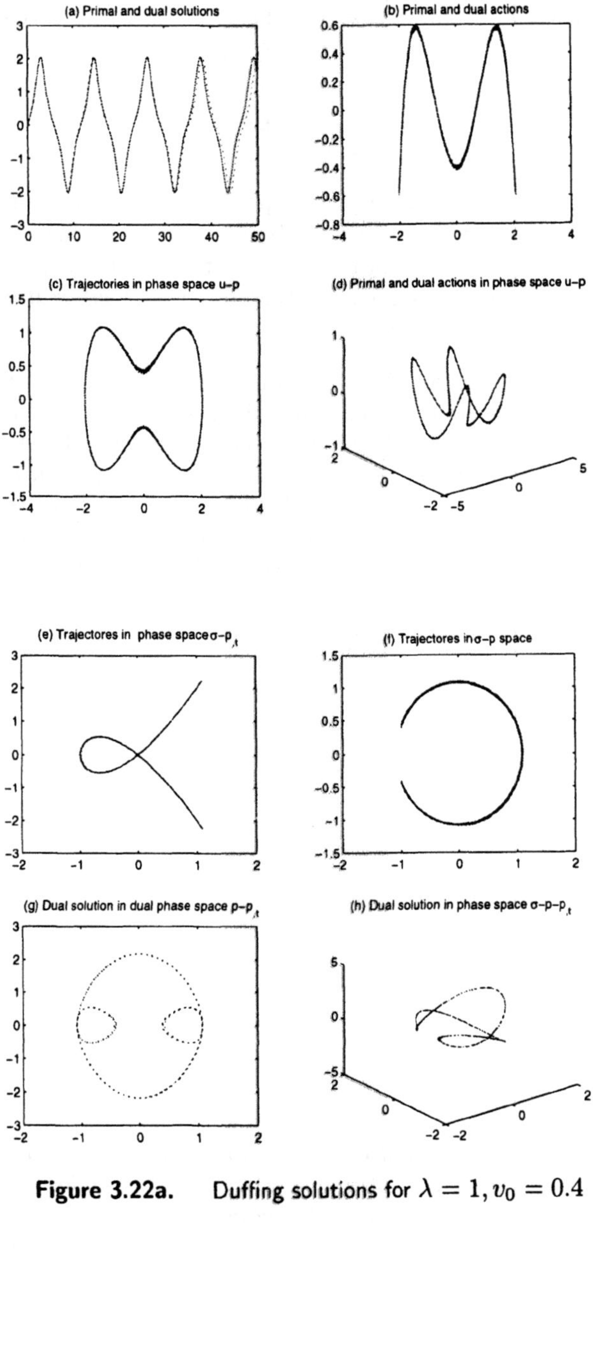

Figure 3.22a. Duffing solutions for $\lambda = 1, v_0 = 0.4$

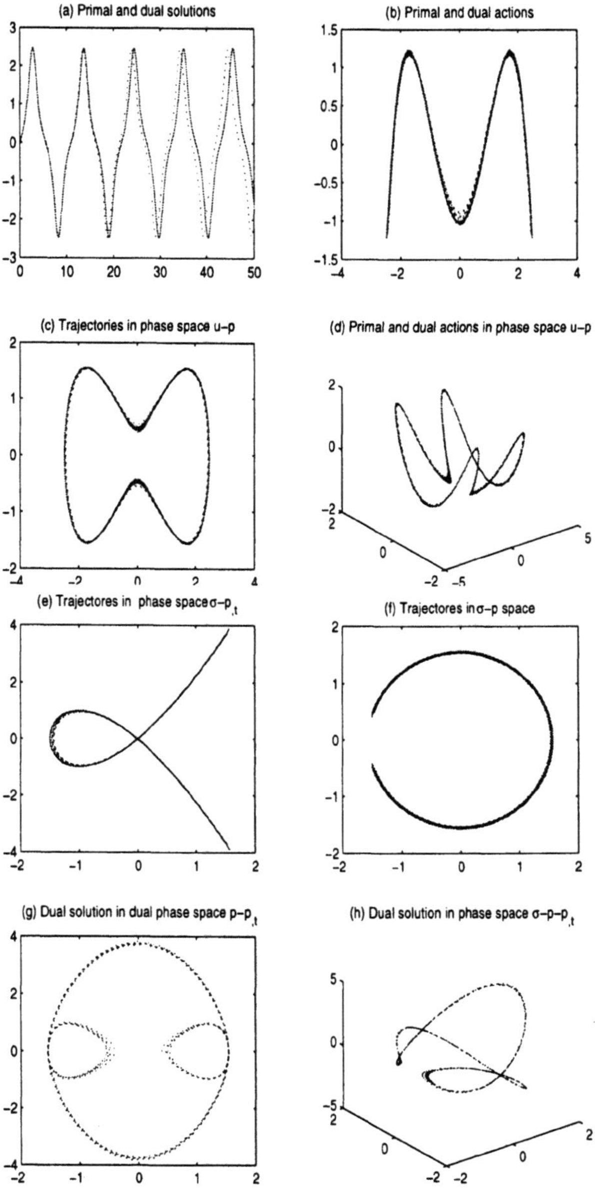

Figure 3.22b. Duffing solutions for $\lambda = 1.5, v_0 = 0.4$

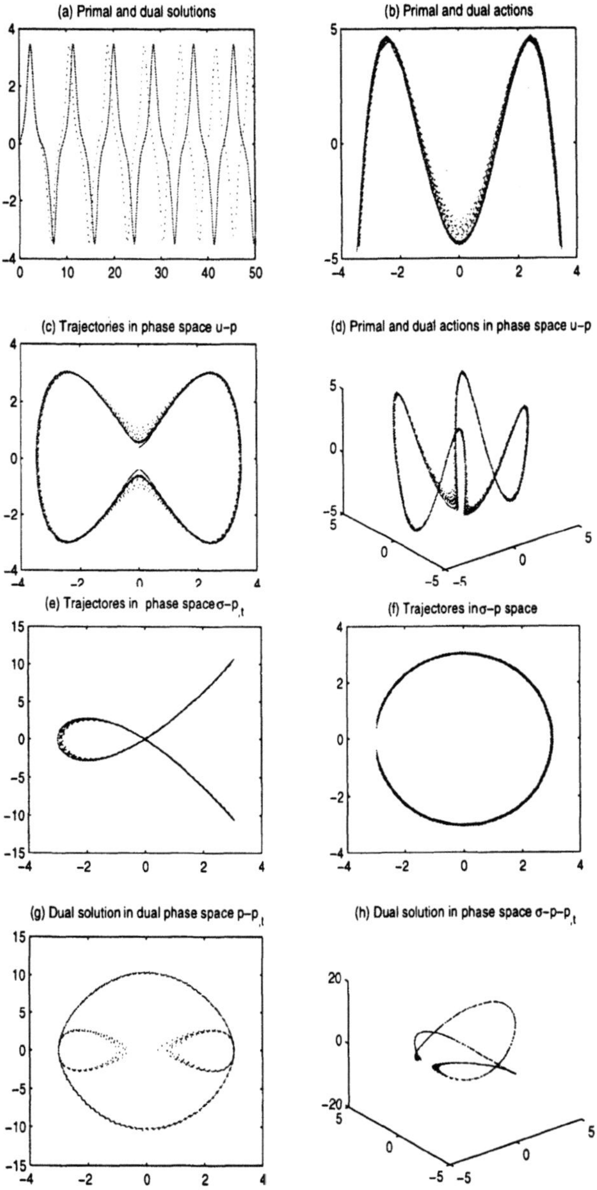

Figure 3.22c. Duffing solutions for $\lambda = 3, v_0 = 0.4$

3.7 Applications and Commentary

Following the tradition, we now present some applications of the theory developed in this chapter. By two examples from dissipative dynamics and nonsmooth bifurcation mechanics, we will show that the canonical dual transformation method plays potentially important role in nonconvex and nonsmooth analysis.

3.7.1 Dissipative Systems and Nonsmooth Bifurcation Problems

I. Dual Formulations of Forced Dissipative Duffing Systems.

The forced Duffing equation with dissipation,

$$q_{,tt} + \mu q_{,t} - \lambda q + \frac{1}{2} a q^3 = f(t), \tag{3.171}$$

arises for a number of electronic circuits. It describes the response of a nonlinear oscillator to a periodic driving force, for instance $f(t) = R\sin(\omega_0 t)$ with $R > 0, \quad \omega_0 \in \mathbb{R}$. Even this still rather simple equation describes a variety of phenomena including period doubling and tripling. The total action associated with this dissipative system is the nonconvex functional

$$\Pi(q) = \int_I \frac{1}{2} e^{\mu t} \left(q_{,t}^2 - (\frac{1}{2} a q^2 - \lambda)^2 \right) \mathrm{d}t - \int_I e^{\mu t} f(t) q \mathrm{d}t. \tag{3.172}$$

Let $\mathcal{U} = \mathcal{E} = \mathcal{L}^2(I)$ and define weighted bilinear forms

$$\langle u, u^* \rangle = \int_I e^{\mu t} u(t) u^*(t) \mathrm{d}t, \quad \langle v; p \rangle = \int_I e^{\mu t} v(t) p(t) \mathrm{d}t.$$

By introducing the nonlinear geometrical operator $\Lambda : \mathcal{U} \to \mathcal{E}$, defined by

$$\boldsymbol{\xi} = \begin{pmatrix} v \\ \xi \end{pmatrix} = \Lambda(q) = \begin{pmatrix} \frac{\mathrm{d}}{\mathrm{d}t} q \\ \frac{1}{2} a q^2 - \lambda \end{pmatrix}, \tag{3.173}$$

we find that the action $\bar{W}(v, \xi)$ is the saddle functional

$$\bar{W}(\boldsymbol{\xi}) = \int_I e^{\mu t} \frac{1}{2} \left(v^2 - \xi^2 \right) \mathrm{d}t. \tag{3.174}$$

For the initial-value problem, the admissible spaces $\mathcal{U}_a$ and $\mathcal{E}_a$ can be defined as

$$\mathcal{U}_a = \{ q \in \mathcal{L}^2(I) | \ q(0) = q_0 \},$$

$$\mathcal{E}_a = \{ (v, \xi) \in \mathcal{L}^2(I, \mathbb{R}^2) | \ v(0) = v_0, \ \xi(0) = \frac{1}{2} a q_0 - \lambda \}.$$

The range of the duality mapping $\varsigma = (p, \varsigma) = D\bar{W}(\xi)$ is

$$\mathcal{T}_a = \{(p,\varsigma) \in \mathcal{L}^2(I,\mathbb{R}^2) |\; p(0) = v_0,\;\; \varsigma(0) = \lambda - \frac{1}{2}aq_0\}.$$

On $\mathcal{T}_a$, the complementary action $\bar{W}^c$ is defined by

$$\bar{W}^c(\varsigma) = \langle \xi ; \varsigma \rangle - \bar{W}(\xi) = \int_I e^{\mu t}\frac{1}{2}\left(p^2 - \varsigma^2\right) \mathrm{d}t. \tag{3.175}$$

The nonlinear Lagrangian $L : \mathcal{U}_a \times \mathcal{T}_a \to \mathbb{R}$ reads

$$L(q,p,\varsigma) = \int_I e^{\mu t}\left(q_{,t}p + (\frac{1}{2}aq^2 - \lambda)\varsigma - \frac{1}{2}p^2 + \frac{1}{2}\varsigma^2 - fq\right) \mathrm{d}t. \tag{3.176}$$

The partial criticality condition $D_q L(\bar{q},\varsigma) = 0$ leads to the nonlinear equilibrium equation

$$\Lambda_t^*(\bar{q})\varsigma = f \;\;\Rightarrow\;\; -p_{,t} - \mu p + a\bar{q}\varsigma = f.$$

Solving this for $\bar{q}$ yields

$$\bar{q} = \frac{p_{,t} + \mu p + f}{a\varsigma}. \tag{3.177}$$

Thus, on the admissible space $\mathcal{T}_\emptyset$, defined by

$$\mathcal{T}_\emptyset = \{(p,\varsigma) \in \mathcal{T}_a |\;\; p_{,t} \in \mathcal{L}^2(I),\;\; \varsigma(t) \neq 0 \;\; \forall t \in I\},$$

the dual action is obtained by

$$\begin{aligned}\Pi^d(\varsigma) &= \mathrm{sta}_{q\in\mathcal{U}_a} L(q,\varsigma) = L(\bar{q},p,\varsigma) \\ &= J_q(p) + \int_I \frac{1}{2}e^{\mu t}\left(\varsigma^2 - 2\lambda\varsigma - p^2 - \frac{(p_{,t} + \mu p + f)^2}{a\varsigma}\right) \mathrm{d}t,\end{aligned}$$

where the bilinear concomitant $J_q(p)$ is given by

$$J_q(p) = e^{\mu T} q(T)p(T) - u_0 v_0.$$

The dual Euler-Lagrange equations for this forced dissipative Duffing system are the coupled differential-algebraic equations

$$\left(\frac{p_{,t} + \mu p + f}{a\varsigma}\right)_{,t} - p = 0, \tag{3.178a}$$

$$2a\varsigma^2(\varsigma - \lambda) + (p_{,t} + \mu p + f)^2 = 0. \tag{3.178b}$$

II. Nonsmooth Bifurcation Problem.

Let us consider the equilibrium problem of a one-dimensional elastoplastic bar, which is fixed at $x = 0$, free at $x = \ell$, and subjected to the distributed axial load $f(x)$ $\forall x \in I = (0, \ell)$. The extension $u(x)$ is a real-valued function. The canonical strain measure ξ is defined by the nonlinear geometrical equation

$$\xi = \Lambda(u) = \frac{1}{2}(u'^2 - 1), \tag{3.179}$$

where Λ is a special quadratic operator with $a = 1,\ b = 0, c = -1/2 = -\lambda$ corresponding to the general one defined by (3.67). Let $\xi_e > 0$ be the elastic strain limit. If $\xi \leq \xi_e$, the bar is in a state of elastic deformation, while if $\xi > \xi_e$ the bar is in a plastic state. Suppose that the constitutive relation is piecewise linear (see Fig. 3.23), of the form

$$\varsigma = \begin{cases} k_e\xi & \text{if } \xi \leq \xi_e, \\ k_p(\xi - \xi_e) + \varsigma_p & \text{if } \xi > \xi_e, \end{cases} \tag{3.180}$$

where k_e, k_p and ς_p are material constants. Then, the stored-energy $\bar{V}$ is a nonsmooth function (see Fig. 3.24(a)):

$$\bar{V}(\xi) = \begin{cases} \bar{V}_e(\xi) = \frac{1}{2}k_e\xi^2 & \text{if } \xi \leq \xi_e, \\ \bar{V}_p(\xi) = \frac{1}{2}k_e\xi_e^2 + \frac{1}{2}k_p(\xi - \xi_e)^2 + \varsigma_p(\xi - \xi_e) & \text{if } \xi > \xi_e. \end{cases} \tag{3.181}$$

Since $\bar{V}(\xi)$ is convex, its Fenchel-conjugate is easily obtained as

$$\bar{V}^c(\varsigma) = \begin{cases} \bar{V}_1^c(\varsigma) = \frac{1}{2k_e}\varsigma^2 & \text{if } \varsigma \leq \varsigma_e, \\ \bar{V}_2^c(\varsigma) = \frac{1}{2k_e}\varsigma_e^2 + \xi_e(\varsigma - \varsigma_e) & \text{if } \varsigma_e < \varsigma \leq \varsigma_p, \\ \bar{V}_3^c(\varsigma) = \frac{1}{2k_e}\varsigma_e^2 + \xi_e(\varsigma - \varsigma_e) + \frac{1}{2k_p}(\varsigma - \varsigma_p)^2 & \text{if } \varsigma > \varsigma_p, \end{cases} \tag{3.182}$$

which is a smooth function of ς (see Fig. 3.24(b)).

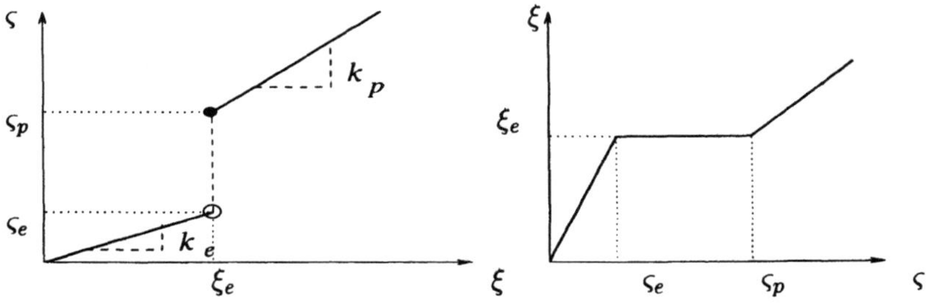

Figure 3.23. Discontinuous constitutive relation.

The inverse constitutive relation $\xi = D\bar{V}^c$ is then a continuous function, defined by

$$\xi = \begin{cases} \frac{1}{k_e}\varsigma & \text{if } \varsigma \leq \varsigma_e, \\ \xi_e & \text{if } \varsigma_e < \varsigma \leq \varsigma_p, \\ \xi_e + \frac{1}{k_p}(\varsigma - \varsigma_p) & \text{if } \varsigma > \varsigma_p. \end{cases} \tag{3.183}$$

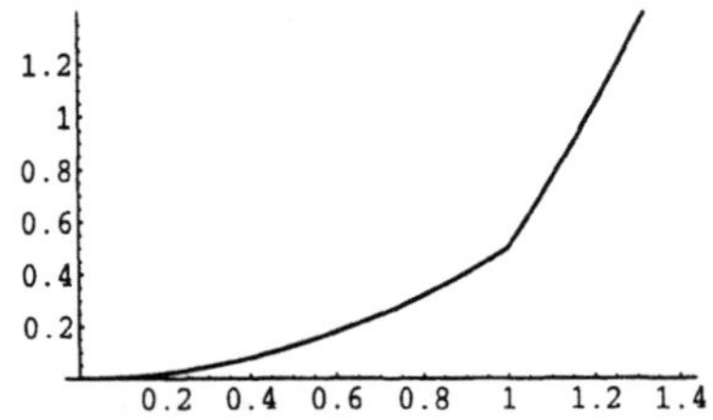

(a) Graph of $\bar{V}(\xi)$.

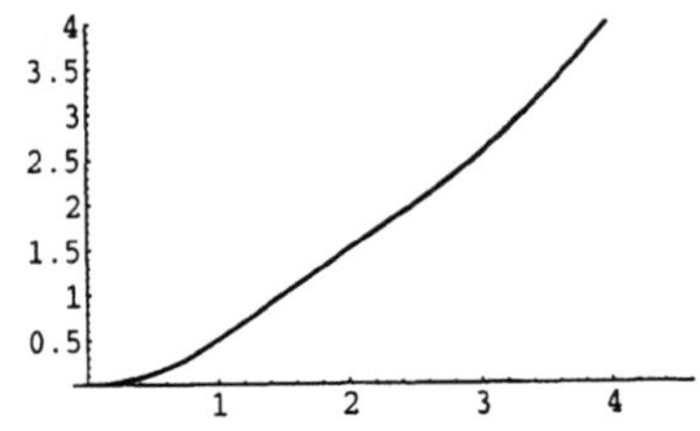

(b) Graph of $\bar{V}^c(\varsigma)$.

Figure 3.24. Nonsmooth $\bar{V}$ and smooth $\bar{V}^c$.

In terms of the infinitesimal strain $\epsilon = u'$, the stored energy $\bar{V}(\frac{1}{2}(\epsilon^2 - 1))$ is a double-well function of ϵ. The kinematically admissible space $\mathcal{U}_k$ in this problem is the convex set

$$\mathcal{U}_k = \{u \in \mathcal{L}^2(0,1)|\ \Lambda(u) \in \mathcal{L}^2(I),\ \ u(0) = 0\}.$$

Let $I_e, I_p \subset I$ be two subsets, defined by

$$I_e = \{x \in I|\ \xi(x) \le \xi_e\},$$

$$I_p = \{x \in I|\ \xi(x) > \xi_e\}.$$

Clearly $I_e \cup I_p = I$. The total potential energy $\Pi : \mathcal{U}_k \to \mathbb{R}$ is given as

$$\begin{aligned}\Pi(u) &= \bar{W}(\Lambda(u) - F(u) \\ &= \int_{I_e} \bar{V}_e(\frac{1}{2}u'^2 - \frac{1}{2})\ dx + \int_{I_p} \bar{V}_p(\frac{1}{2}u'^2 - \frac{1}{2})\ dx - \int_I \bar{f}u\ dx.\end{aligned} \quad (3.184)$$

For a given distributed external load $\bar{f}(x) : I \to \mathbb{R}$, the primal problem for this elastoplastic bar can be proposed as

$$(\Pi_{\inf}) : \quad \Pi(u) \to \min\ \ \forall u \in \mathcal{U}_k. \quad (3.185)$$

This is a nonsmooth, nonconvex variational problem. The interface I_{ep} between the elastic region I_e and the plastic region I_p is defined by

$$\partial I_{ep} = \{x_0 \in I|\ \xi(x_0) = \xi_e\},$$

which is unknown before the problem is solved. From the viewpoint of classical direct analysis, to solve this problem analytically seems very difficult. However, by the theorems proved above, the analytic solution of this nonsmooth problem can be determined.

Since $\xi = \frac{1}{2}(u'^2 - 1) \ge -\frac{1}{2}\ \ \forall u \in \mathcal{U}_k$, the ranges of the constitutive mapping $\varsigma = D\bar{V}$ are

$$\mathcal{R}_e = [-\frac{1}{2}k_e, \varsigma_e],\ \ \forall \xi \le \xi_e,$$

$$\mathcal{R}_p = (\varsigma_p, +\infty), \quad \forall \xi > \xi_e,$$

where $\varsigma_e = k_e \xi_e$. Then, the dual feasible spaces can be given as

$$\begin{aligned}
\mathcal{T}_a{}^1 &= \{\varsigma \in \mathcal{H}(I) | \ -\frac{1}{2}k_e \le \varsigma(x) \le \varsigma_e, \ \ \varsigma(x) \neq 0 \ \forall x \in I\}, \\
\mathcal{T}_a{}^2 &= \{\varsigma \in \mathcal{H}(I) | \ \varsigma_e < \varsigma(x) \le \varsigma_p, \ \ \varsigma(x) \neq 0 \ \forall x \in I\}, \\
\mathcal{T}_a{}^3 &= \{\varsigma \in \mathcal{H}(I) | \ \varsigma_p < \varsigma(x) < +\infty, \ \ \varsigma(x) \neq 0 \ \forall x \in I\}.
\end{aligned}$$

The complementary gap functional is then quadratic

$$G(u, \varsigma) = \int_I \frac{1}{2}\varsigma(1 + u'^2) \, dx.$$

By the relation $u'\varsigma = \sigma$, for any $\varsigma \in \mathcal{T}_a{}^i$, $u' = \sigma/\varsigma$. For the homogeneous mixed boundary conditions $u_0 = 0$, $\sigma_\ell = 0$, and the coefficients $a = 1, b = 0, c = -\frac{1}{2}$, the dual functional Π^d can be written simply as

$$\Pi^d(\varsigma) = -\sum_i^3 \int_{I_i} [\bar{V}_i^c(\varsigma) + \frac{1}{2}\varsigma + \frac{\sigma_i^2(x)}{2\varsigma}] \, dx, \tag{3.186}$$

where $\sigma_i(x)$ is a continuous function defined by

$$\sigma_i(x) = -\int \bar{f}(x) \, dx + d_i, \quad x \in I_i, \quad i = 1, 2, 3. \tag{3.187}$$

The domains I_i and the integral constants d_i are determined by the dual Euler-Lagrange equations

$$2\varsigma^2(\frac{1}{k_e}\varsigma + \frac{1}{2}) = \sigma_1^2 \quad \forall \varsigma \in \mathcal{T}_a{}^1, \tag{3.188}$$

$$2\varsigma^2(\xi_e + \frac{1}{2}) = \sigma_2^2 \ \forall \varsigma \in \mathcal{T}_a{}^2, \tag{3.189}$$

$$2\varsigma^2(\frac{1}{k_p}(\varsigma - \varsigma_p) + \xi_e + \frac{1}{2}) = \sigma_3^2 \quad \forall \varsigma \in \mathcal{T}_a{}^3 \tag{3.190}$$

as follows. For $\varsigma = \varsigma_e$ and $\varsigma = \varsigma_p$, equation (3.189) gives

$$\sigma_e = \varsigma_e\sqrt{2\xi_e + 1}, \quad \sigma_p = \varsigma_p\sqrt{2\xi_e + 1}. \tag{3.191}$$

For the given external load $\bar{f}(x)$ and boundary conditions, the domains I_i ($i = 1, 2, 3$) are then determined through the function $\sigma(x)$, i.e.

$$I_1 = \{x \in I | \ |\sigma(x)| \le \sigma_e\},$$

$$I_2 = \{x \in I | \ \sigma_e < |\sigma(x)| \le \sigma_p\},$$

$$I_3 = \{x \in I | \ |\sigma(x)| > \sigma_p\}.$$

Since the bar is free at $x = \ell$, we have $d_1 = \sigma_\ell = 0$, and

$$\sigma_1(x) = \int_x^\ell f(s)ds, \quad x \in [x_e, \ell], \quad \sigma_1(\ell) = 0.$$

The interface x_e is well-determined by

$$\sigma_1(x_e) = \int_{x_e}^\ell f(x)\, dx = \sigma_e \quad \Rightarrow x_e.$$

Similarly, the conditions $\sigma_2(x_e) = \sigma_e, \ \ \sigma_3(x_p) = \sigma_p$ lead to the integral constants $d_2 = \sigma_e$ and $d_3 = \sigma_p$, and

$$\begin{aligned} \sigma_2(x) &= \int_x^{x_e} f(s)ds + \sigma_e, \quad x \in [x_p, x_e], \\ \sigma_3(x) &= \int_x^{x_p} f(s)ds + \sigma_p, \quad x \in [0, x_p]. \end{aligned}$$

The condition $\sigma_2(x_p) = \sigma_p$ leads to the interface x_p

$$\int_{x_p}^{x_e} \bar{f}(x)\, dx + \sigma_e - \sigma_p = 0 \quad \Rightarrow x_p.$$

In the subdomain I_1, the system is controlled by equation (3.188). The graph of ς against $\sigma_1(x)$ is shown in Fig. 3.25(a). Since $a = 1$, $\sigma_{1\,\max} = \sigma_e$, by Theorem 3.4.4, if

$$\lambda = \frac{1}{2} \geq 1.5(\sigma_e/k_e)^{2/3} = 1.5(2\xi_e^3 + \xi_e^2)^{1/3} \ \Rightarrow \ \xi_e \leq \frac{1}{6},$$

then in I_1 the equation (3.188) may have three solutions (i.e., elastic buckling), depending on $\sigma(x)$. By Theorem 1, the analytic solution in I_1 is

$$u_1(x) = \int_{x_e}^x \frac{\sigma_1(s)}{\varsigma(s)}\, dx + u_1(x_e), \quad x \in I_1, \tag{3.192}$$

where $u_1(x_e)$ is determined by the interface condition.

In the domain I_2, the dual Euler-Lagrange equation (3.189) is linear (see Fig. 3.25(b)). Hence, in $I_2 = (x_p, x_e)$, the solution is unique:

$$u_2(x) = \text{sgn}[\sigma_2]\sqrt{2\xi_e + 1}(x - x_e) + u_1(x_e), \quad x \in I_2. \tag{3.193}$$

In the domain I_3, $\sigma_{3\,\min} = \sigma_p$. If $\lambda = \xi_e + \frac{1}{2} - \varsigma_p/k_p \leq 1.5(\sigma_p/k_p)^{2/3}$, then the algebraic equation (3.190) has a unique root ς (see Fig. 3.25(c)), which leads to the analytic solution

$$u_3(x) = \int_0^x \frac{\sigma_3(s)}{\varsigma(s)} ds, \quad x \in [0, x_p]. \tag{3.194}$$

However, if $\xi_e + \frac{1}{2} - \varsigma_p/k_p > 1.5(\sigma_p/k_p)^{2/3}$, the solution ς may not be unique. In this case, if a given $\bar{f}(x)$ is such that the dual Euler-Lagrange equation (3.190) has three real roots $\varsigma(x)$, then the system is in a plastic buckling state. The interface condition $u_3(x_p) = u_2(x_p)$ gives $u_1(x_e)$.

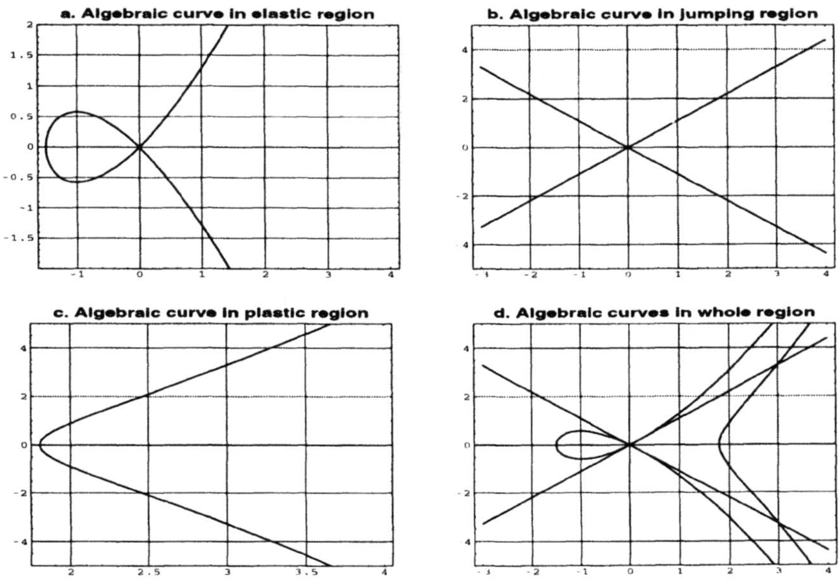

Figure 3.25. Graphs of the dual Euler-Lagrange equations (3.188)-(3.190)

3.7.2 Historical Notes and Commentary

It seems that the nonconvex variational problem with double-well energy was first studied by van der Waals in 1893 for a compressible fluid whose free energy at constant temperature depends not only on the density but also on the density gradient (cf., e.g., Rowlinson, 1979). Such problems also appear frequently in association with cosmology (Kibble, 1997), phase transitions and shape memory alloys (cf., e.g., Ericksen, 1975, 1986; Carr *et al*, 1984; Ball and James, 1987; Fonseca, 1988; Lurie and Cherkave, 1988; James and Kinderlehrer, 1989; Fried and Gurtin, 1996; Lazopoulos, 1995; Brokate and Sprekels, 1996; Lazopoulos and Ogden, 1998; Allaire and Lods, 1999), and post-buckling of structural mechanics (see, for example, Pipkin, 1993; Wu, 1994; Gao, 1996; Steigmann and Ogden, 1997). Direct approaches and relaxation methods for solving nonconvex variational problems have been discussed extensively during the last thirty years (see, for exmaple, Chang, 1981; Chow and Hale, 1982; Dacorogna, 1989 and many others). But even for very simple one-dimensional problems like (3.31), to obtain all analytic solutions is usually very difficult with traditional methods, and therefore, some numerical discretization approaches have been examined (cf., e.g., Rogers and Truskinovsky, 1997). Relaxation methods, however, can be used mainly for obtaining global minimizers of the relaxed problems. In phase transitions or post-bifurcation analysis, local maximizers usually play an important role in understanding the physical mechanism of systems. As was indicated by

Kohn (1991), the relaxation method for solving nonconvex variational problems with three or more phases (potential wells) is fundamentally more difficult.

As it was pointed by Rabinowitz (1986) that there are at least two sets of methods that have been developed to find critical points in nonconvex functionals: (1) Morse theory and its generalizations and (2) minimax theory of type

$$\varrho = \inf_{A \in \mathcal{S}} \max_{u \in A} P(u) \tag{3.195}$$

over a suitable class of sets $\mathcal{S}$. Morse's work (cf., e.g., see Milnor, 1963) reveals a deep relation between the topology of a space and the number and types of critical points of any functional defined on it. Today, as it pointed by Struwe (1996) that inspired by the deep work of Conley (1978), Morse theory seems to be turning back to its origins. Since the pioneering paper of Paul Rabinowitz in 1978, the minimax problems of the type (3.195) has been studied extensively in critical point theory during the passed twenty years (see the recent books by Chang, 1986, 1993; Mawhin and Willem, 1989; Ekeland, 1990; Struwe, 1996). The so-called *Mountain Pass Theorem* is the main minimax result proposed by Rabinowitz (1978), which involves a technical assumption: Palais-Smale condition. A right-saddle type maxmin duality theory was studied by Ghoussoub (1993).

The canonical dual transformation method for solving nonconvex variational problems was originally studied by Gao and Strang (1989) for large deformation variational/boundary-value problems with convex stored energy $\bar{W}$. In order to recover the lost symmetry between the primal and dual variational problems, they introduced the so-called *complementary gap function*, which leads to a nonlinear Lagrange duality theory in fully nonlinear variational problems. They proved that if this gap function is positive on the equilibrium admissible field, the system has a unique dual problem. Applications of this general duality theory have been given in a series of publications on finite elastoplasticity (see Gao *et al* 1990-99). Some open problems in large deformation plastic limit analysis were partially solved in Gao and Strang (1989b) and Gao (1994). In the study of geometrically nonlinear variational inequality problems, it was realized that by using this nonlinear Lagrange duality theory, the nonlinear von Karman equations can be transformed into a coupled quadratic variational problem (Yau and Gao, 1992). Recently, in the study of post-buckling analysis of large deformation beam theory, it was discovered (see Gao,1996b, 1997) that for convex stored energy, if the gap function is negative, then the generalized complementary energy $L(u, \epsilon^*)$ is a super-critical point functional, and a pure complementary energy principle (in terms of the conjugate stress only) was established for an extended beam theory. But, in general fully nonlinear systems, it remains unclear how to construct a truly dual variational principle. Actually, in finite deformation theory, the question concerning the existence of a pure complementary energy variational principle has been the subject of argument for more than forty years.

In this chapter we have developed systematically a tri-duality theory for one-dimensional nonconvex systems. The interesting triality theory reveals the intrinsic symmetry in nonconvex systems. By using the canonical dual transformation method, many nonconvex, nonsmooth variational problems and associated nonlinear differential equations are able to converted into certain algebraic systems. The geometrical properties of the algebraic curves play important role in study post-bifurcation phenomena of nonconvex systems. Applications in Duffing equation, the duality theory for nonconvex dynamical systems is first time proposed. The dynamical bifurcation criterion discovered via dual Duffing system may serve as an important role in understanding the chaotic phenomena in nonconvex dynamics.

Appendix 3.A: Mapping Properties of the Canonical Composition Operator $\bar{V}(\xi)$ between $\mathcal{L}^\beta(I)$ Spaces

Suppose I is a bounded domain in $\mathbb{R}$, and $\mathcal{U} = \mathcal{L}^\alpha(I)$ and $\mathcal{E} = \mathcal{L}^\beta(I)$ denote two Banach spaces of functions defined over I. Let $V(x,u)$ be a real-valued function defined on $I \times \mathbb{R}$, and satisfying the Carathéodory continuity conditions. For a given nonlinear mapping $\xi : \mathcal{U} \to \mathcal{E}$, consider the nonlinear composition operator $\bar{V}(\xi)$ defined by setting

$$V(x,u(x)) = \left(\bar{V} \circ \xi\right)(x,u(x)) = \bar{V}(\xi(x)) \quad \text{for } \xi(x) \text{ Lebesgue measurable.} \tag{3.A.1}$$

Clearly, $\bar{V}(\xi(x))$ is Lebesgue measurable and is continuous with respect to convergence in measure. Moreover, under the growth condition

$$|V(x,u)| \le a_1 + b_1|u|^\alpha, \quad |\xi(u)| \le a_2 + b_2|u|^{\alpha/\beta}, \quad 1 \le \alpha, \beta < +\infty \tag{3.A.2}$$

for some absolute constants $a_i, b_i \ge 0 \quad (i = 1,2)$, the following statements concerning the operator $\bar{V}(\xi(x))$ are equivalent.

(1) *$\bar{V}$ maps $\mathcal{E} = \mathcal{L}^\beta(I)$ into $\mathcal{L}(I)$.*
(2) *$\bar{V}$ is a continuous mapping of $\mathcal{E}$ into $\mathcal{L}(I)$.*
(3) *$\bar{V}$ is a bounded mapping of $\mathcal{E}$ into $\mathcal{L}(I)$.*

The proof of these results follows from straightforward measure theoretic considerations (see, for example, Krasnoselskii, 1963).

An easy induction argument shows that under suitable Carathéodory continuity conditions, the multi-variable composition operator

$$\bar{V}(\xi_1,\cdots,\xi_k) = V(x,\xi_1,\cdots,\xi_k) : \Pi_{i=1}^k \mathcal{L}^{\alpha_i}(\Omega) \to \mathcal{L}^\alpha(\Omega)$$

is continuous and bounded if and only if the function $V(x,\xi_1,\cdots,\xi_k)$ satisfies the growth condition

$$|V(x,\xi_1,\cdots,\xi_k)| \le \left(c_0 + \sum_{i=1}^k c_i|\xi|^{\alpha_i/\alpha}\right),$$

for constant coefficients $c_i \in \mathbb{R}$.

What cannot be seen is called evanescent;
What cannot be heard is called rarefied;
What cannot be touched is called minute.
These three cannot be fathomed
And so they are confused and looked upon as one.
Its upper part is not dazzling;
Its lower part is not obscure.
Dimly visible, it cannot be named
And returns to that which is without substance.
This is called the shape that has no shape,
The image that is without substance.
This is called indistinct and shadowy.
Go up to it and you will not see its head;
Follow behind it and you will not see its rear.
Hold fast to the way of antiquity
In order to keep in control the realm of today.
The ability to know the beginning of antiquity
Is called the thread running through the Tao.

—Lao Chi, *Tao De Chi* XIV

4 MULTI-DUALITY AND CLASSIFICATIONS OF GENERAL SYSTEMS

And the rain was upon the earth forty days and forty nights.

—Genesis, 7.12

The more one learns, the more one realizes that one know less; and one does less and less until one does nothing at all. When one does nothing at all there is nothing that is undone.

—Lao Chi

Everyone knows what a curve is, until he has studied enough mathematics to become confused by the countless exceptions.

—Felix Klein

If you weren't confused by quantum mechanics you didn't really understand it.

—Niels Bohr

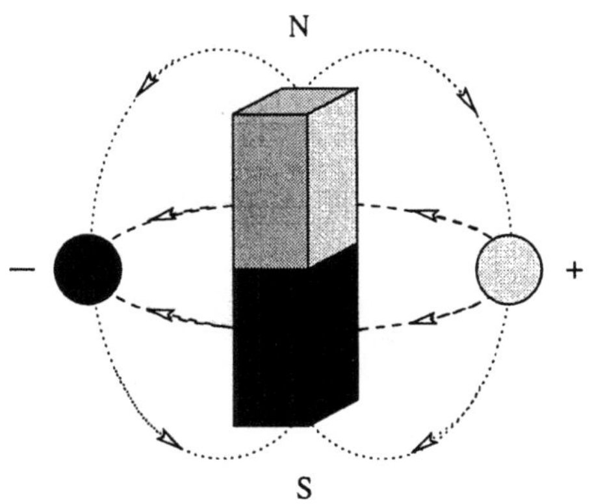

Figure 4.1. Duality, polarity and complementarity in electromagnetism

The quadratic canonical dual transformation method, presented in Chapter 3, can be generalized in different ways for solving general nonconvex systems. We begin with a special nonconvex variational problem with multi-well energy. Then, we present two powerful sequential dual transformations for general nonconvex systems with multi-well energy. These two methods produce two different duality theories. The classifications are discussed for general systems and a generalized tri-duality principle is presented in the central part of this section, which plays a core role in this book. Examples from discrete systems and continuum fields are listed. We end up by giving an application for nonconvex systems.

4.1 The First Type of Sequential Canonical Dual Transformation

4.1.1 Nonconvex Lagrangian and Discrete Dual Problems

We first consider the special nonconvex variational problem

$$(\mathcal{P}_{\text{sta}}): \quad P(u) = \int_0^1 V(x, u')\, \mathrm{d}x - \int_0^1 fu\, \mathrm{d}x \quad \to \text{sta} \;\; \forall u \in \mathcal{U}_k, \tag{4.1}$$

where the energy density $V(x, \epsilon)$ is a nonconvex function of the linear deformation $\epsilon = u'$, given by

$$V(x, \epsilon) = \frac{1}{2}a(x)(\frac{1}{8}\epsilon^4 - \frac{1}{2}b(x)\epsilon^2 + c(x))^2, \quad a(x) > 0 \quad \forall x \in I = (0, 1).$$

For the given distributed parameters $b(x)$ and $c(x) \in \mathbb{R}$, $V(x, \epsilon)$ has at most four wells (see Fig. 4.2). Nonconvex variational problems with multi-well energies appear frequently in phase transitions and hysteresis, shape memory alloys, post-bifurcation and buckling problems, and elsewhere. From the point view of traditional direct methods, nonlinear problems with three or more potential wells are fundamentally more difficult, as was explained in Kohn (1991). But, by using the canonical dual transformation method, we can solve this problem in simple ways and obtain all the possible solutions.

Let $\mathcal{U} = \mathcal{L}^2(I)$, $\mathcal{E} = \mathcal{L}^2(I) = \mathcal{E}^*$, and

$$\xi = \Lambda(u) = \frac{1}{8}u'^4 - \frac{1}{2}bu'^2 + c. \tag{4.2}$$

Clearly, $\Lambda : \mathcal{U} \to \mathcal{E}$ is a nonlinear, Gâteaux differentiable operator, and its Gâteaux derivative at $\bar{u}$ is

$$\Lambda_t(\bar{u}) = \left(\frac{1}{2}\bar{u}'^3 - b\bar{u}'\right)\frac{\mathrm{d}}{\mathrm{d}x}. \tag{4.3}$$

The complementary operator Λ_c is then, given by

$$\Lambda_c(u) = \Lambda(u) - \Lambda_t(u)u = c - \frac{3}{8}u'^4 + \frac{1}{2}bu'^2. \tag{4.4}$$

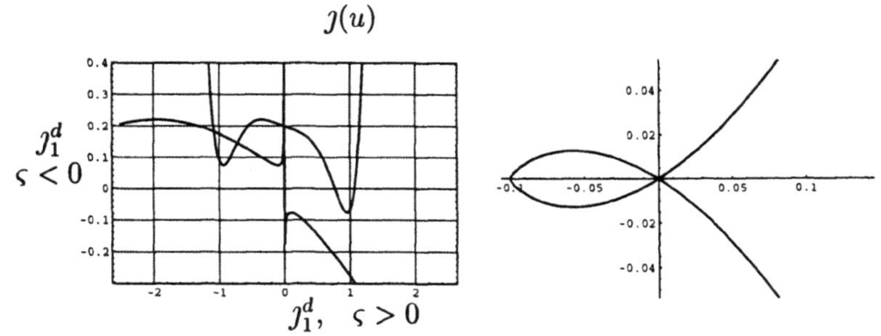

(a) Double-well energy and cubic algebraic curve ($c = -0.2,\ b = 0,\ \sigma = 0.11$)

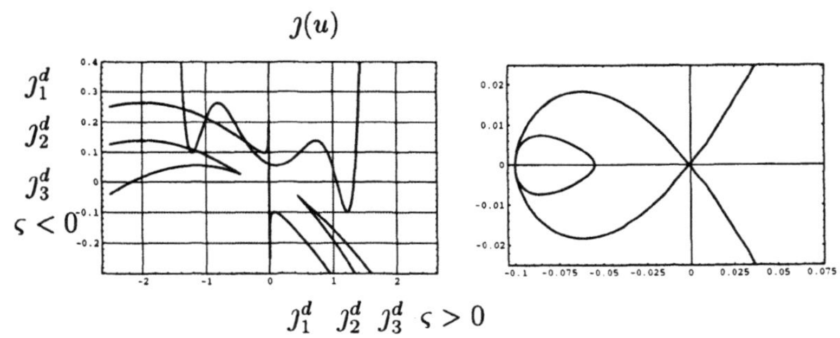

(b) Triple-well energy ($b = 0.3,\ c = -0.055,\ \sigma = 0.08$)

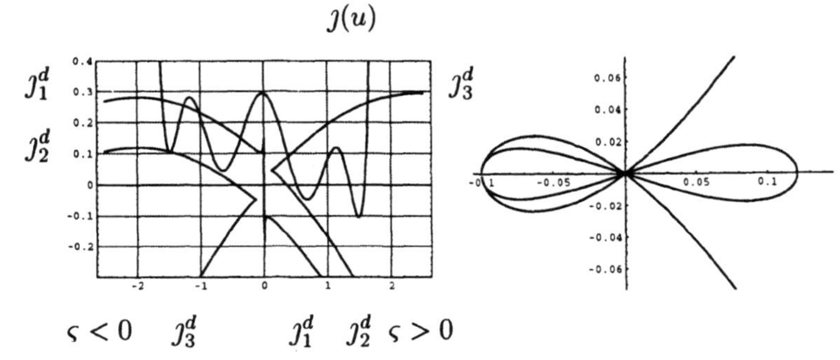

(c) Quadruple-well energy ($b = 0.65,\ c = 0.12,\ \sigma = 0.07$)

Figure 4.2. Multi-well energies and algebraic curves of equation (4.9)

In terms of $\xi = \Lambda(u)$, the stored energy density $\bar{V} : \mathcal{E} \to \mathcal{L}(I)$ and its Legendre conjugate are quadratic functions, given by

$$\bar{V}(\xi) = \frac{1}{2}a\xi^2, \quad \bar{V}^c(\varsigma) = \frac{1}{2a}\varsigma^2. \tag{4.5}$$

The complementary gap functional in this case is

$$G(u,\varsigma) = -\langle \Lambda_c(u); \varsigma \rangle = \int_0^1 \varsigma(\frac{3}{8}u'^4 - \frac{1}{2}bu'^2 - c)\, dx.$$

The nonlinear Lagrangian associated with $\Lambda(u)$ is

$$L(u,\varsigma) = \int_0^1 \left((\frac{1}{8}u'^4 - \frac{1}{2}bu'^2 + c)\varsigma - \frac{1}{2a}\varsigma^2 \right) dx - \int_0^1 fu\, dx. \tag{4.6}$$

The criticality condition $D_u L(\bar{u}, \varsigma) = 0$ leads to the Euler-Lagrange equation

$$((u'(x)^3 - 2bu'(x))\varsigma)' + f(x) = 0, \quad \forall x \in I.$$

Let $\epsilon = u'$ and $(\epsilon^3 - 2b\epsilon)\varsigma = \sigma$. Then, for the given external force $f(x)$ and boundary condition such that

$$\sigma(x) = \int_x^1 f(t)dt + \sigma_\ell,$$

where σ_ℓ is a constant depending on the boundary condition, the cubic algebraic equation

$$\epsilon^3 - 2b\epsilon = 2\sigma/\varsigma \tag{4.7}$$

has at most three roots $\epsilon_i(\varsigma)$, $(i = 1, 2, 3)$. For each $\epsilon_i(\varsigma)$, the "discrete dual functional" Π_i^d

$$\Pi_i^d(\varsigma) = -\int_0^1 \left[\frac{1}{2a}\varsigma^2 + \varsigma\left(\frac{3}{8}\epsilon_i^4(\varsigma) - \frac{1}{2}b\epsilon_i^2(\varsigma) - c \right) \right] dx, \quad i = 1, 2, 3 \tag{4.8}$$

is well-defined on the dual feasible space

$$\mathcal{T}_\emptyset = \{\varsigma \in \mathcal{L}^2(I) |\ \varsigma(x) \neq 0 \ \ \forall x \in I\}.$$

The dual Euler-Lagrange equation for this example is the nonlinear algebraic equation

$$\varsigma^4 \left(\frac{1}{a}\varsigma + \lambda \right)^2 \left(\frac{1}{a}\varsigma - c \right) + \frac{1}{4}b\sigma^2\varsigma^2 \left(\frac{1}{a}\varsigma + \lambda \right) = \frac{1}{32}\sigma^4, \tag{4.9}$$

where $\lambda = \frac{1}{2}b^2 - c$. For the given load f and parameters b, c, this algebraic equation has at most seven real roots ς_j $(j = 1, 2, \cdots, 7)$ (see Fig. 4.2).

To see the extremality properties of this nonconvex variational problem, let us consider the optimization problem

$$\jmath(u) = \frac{1}{2}a(\frac{1}{8}u^4 - \frac{1}{2}bu^2 + c)^2 - \sigma u \ \to \text{sta} \ \forall u \in \mathbb{R}. \tag{4.10}$$

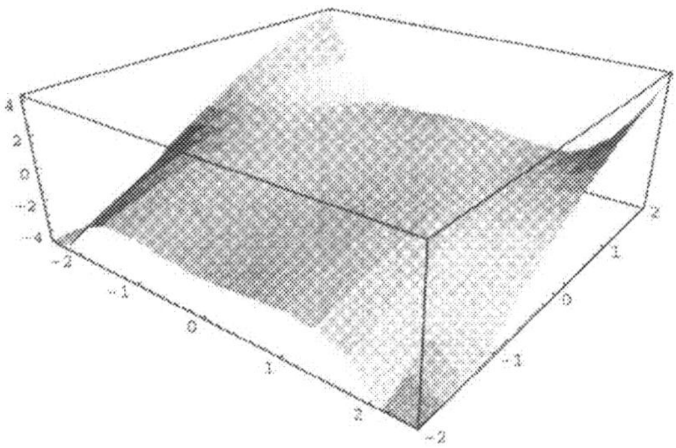

Figure 4.3. Lagrangian L with two wells ($\varsigma > 0$) and two hills ($\varsigma < 0$)

The dual function is

$$\jmath_i^d(\varsigma) = -\frac{1}{2a}\varsigma^2 - \varsigma[\frac{3}{8}\epsilon_i^4(\varsigma) - \frac{1}{2}b\xi_i^2(\varsigma) - c], \quad i = 1,2,3. \tag{4.11}$$

Fig. 4.2 shows that for each ϵ_j, $\jmath(u_j) = \jmath^d(\varsigma_j)$. The associated Lagrangian form is defined by

$$L_\jmath(u,\varsigma) = \varsigma(\frac{1}{8}u^4 - \frac{1}{2}bu^2 + c) - \frac{1}{2a}\varsigma^2 - \sigma u. \tag{4.12}$$

The Lagrangian L has at most seven critical points (u_j, ς_j). For the fixed values $a = 1, b = 0.65, c = 0.12$, Fig. 4.3 shows that L is concave in ς. However, for any given $\varsigma \in \mathbb{R}$, $L(u,\varsigma)$ is nonconvex in u with two wells (for $\varsigma > 0$) and two hills (for $\varsigma < 0$).

4.1.2 Second-Order Lagrangian and Coupled Dual Problem

Let $\lambda_1 = b$, $\lambda_2 = \frac{1}{2}b^2 - c$, and $\varsigma_2 = \varsigma$. The Lagrangian (4.6) can then be rewritten as

$$L(u,\varsigma_2) = \int_I \left(\frac{1}{2}\varsigma_2(\frac{1}{2}u'^2 - \lambda_1)^2 - \frac{1}{2a}\varsigma_2^2 - \lambda_2\varsigma_2^2\right)\, dx - \int_I fu\, dx. \tag{4.13}$$

For fixed $\varsigma_2 \in \mathcal{E}^*$, $L : \mathcal{U} \to \mathbb{R}$ is nonconvex. We can continue using the canonical dual transformation and let

$$\xi_1 = \Lambda(u) = \frac{1}{2}u'^2 - \lambda_1, \quad \bar{V}_1(\xi_1) = \varsigma_2\xi_1^2. \tag{4.14}$$

Thus, the conjugate variable ς_1

$$\varsigma_1 = D\bar{V}_1(\xi_1) = \varsigma_2\xi_1$$

dual to ξ_1 depends on both ς_2 and ξ_1. By the Legendre transformation, we then have

$$\bar{V}_1^c(\varsigma_1) = \varsigma_1\xi_1 - \bar{V}_1(\xi_1) = \frac{1}{2\varsigma_2}\varsigma_1^2.$$

Then, the *second-order Lagrangian* L_2 can be defined by

$$L_2(u,\varsigma_1,\varsigma_2) = \int_I \left(\varsigma_1(\frac{1}{2}u'^2 - \lambda_1) - \frac{1}{2\varsigma_2}\varsigma_1^2 - \frac{1}{2a}\varsigma_2^2 - \lambda_2\varsigma_2\right) \, dx - \int_I fu \, dx. \tag{4.15}$$

For fixed $\varsigma_1, \ \varsigma_2 \in \mathcal{E}^*$, the criticality condition $D_u L_2(u,\varsigma_1,\varsigma_2) = 0$ leads to

$$u'\varsigma_1 = \sigma.$$

If u' is replaced by σ/ς_1, the coupled dual functional Π_2^d can be putting in the form

$$\Pi_2^d(\varsigma_1,\varsigma_2) = -\int_I \left(\lambda_1\varsigma_1 + \lambda_2\varsigma_2 + \frac{1}{2\varsigma_2}\varsigma_1^2 + \frac{1}{2a}\varsigma_2^2 + \frac{1}{2\varsigma_1}\sigma^2\right) \, dx. \tag{4.16}$$

The criticality condition $D\Pi_2^d(\varsigma_1,\varsigma_2) = 0$ leads to a coupled cubic algebraic system

$$\varsigma_1^2(\frac{1}{\varsigma_2}\varsigma_1 + \lambda_1) = \frac{1}{2}\sigma^2, \tag{4.17a}$$

$$\varsigma_2^2(\frac{1}{a}\varsigma_2 + \lambda_2) = \frac{1}{2}\varsigma_1^2. \tag{4.17b}$$

This algebraic system is equivalent to (4.9). For given σ and each $\varsigma_2 = \varsigma$, equation (4.17a) has at most three real roots $\varsigma_1^i(\sigma,\varsigma)$, $i = 1,2,3$. Substituting ς_1^i into to Π_2^d leads to back $\Pi_i^d(\varsigma)$. For the given parameters $\lambda_1, \ \lambda_2$, this algebraic system has at most seven real roots for $\varsigma_2 = \varsigma^j$, $j = 1,\cdots,7$. For each ς^j, the analytic solution of this problem is

$$u^j(x) = \int_0^x \frac{\sigma(t)}{\varsigma_1(\varsigma^j)} dt + u_0. \tag{4.18}$$

4.1.3 Sequential Nonlinear Lagrangian Dual Transformations

The method can easily be generalized for nonconvex stored-energy density with n parameters λ_k $(k = 1,\cdots,n)$, by writing

$$V_n(x,\epsilon) = \frac{1}{2}a_n \left(\frac{1}{2}a_{n-1}\left(\cdots\frac{1}{2}a_1\left(\frac{1}{2}a_0\epsilon^2 - \lambda_1\right)^2 \cdots - \lambda_{n-1}\right)^2 - \lambda_n\right)^2. \tag{4.19}$$

Let

$$\xi_k = \Lambda^{(k)}(u) = \frac{1}{2}a_{k-1}\left(\cdots\frac{1}{2}a_1\left(\frac{1}{2}a_0 u'^2 - \lambda_1\right)^2 - \cdots\right)^2 - \lambda_k, \quad k = 1,2,\cdots,n \tag{4.20}$$

Thus, for $k = n$, the canonical stored energy density and its complementary energy are quadratic functions, namely

$$\bar{V}_n(\xi_n) = \frac{1}{2}a_n\xi_n^2, \quad \bar{V}_n^c(\varsigma_n) = \frac{1}{2a_n}\varsigma_n^2, \tag{4.21}$$

and

$$\varsigma_n = D\bar{V}_n(\xi_n) = a_n \xi_n.$$

However, for $k \in (1, n-1)$, $\bar{V}_k$ depends on ς_{k+1}. Thus

$$\bar{V}_k(\xi_k) = \frac{1}{2} a_k \varsigma_{k+1} \xi_k^2, \quad \bar{V}_k^c(\varsigma_k, \varsigma_{k+1}) = \frac{1}{2a_k \varsigma_{k+1}} \varsigma_k^2. \tag{4.22}$$

and

$$\varsigma_k = a_k \xi_k \varsigma_{k+1}.$$

The Lagrangian L_n of order n is then

$$L_n(u, \varsigma_1, \cdots, \varsigma_n) = \int_I \left(\frac{1}{2} a_0 u'^2 \varsigma_1 - \sum_{k=1}^{n} \left(\lambda_k \varsigma_k + \bar{V}_k^c \right) \right) \, dx - \int_I f u \, dx. \tag{4.23}$$

The criticality condition $D_u L_n(u, \varsigma_1, \cdots, \varsigma_n) = 0$ leads to the equilibrium condition

$$u' = \frac{\sigma}{a_0 \varsigma_1}, \quad \text{in } I. \tag{4.24}$$

The dual functional

$$\Pi^d(\varsigma_1, \cdots, \varsigma_n) = -\int_I \sum_{k=1}^{n} \left(\lambda_k \varsigma_k + \bar{V}_k^c \right) \, dx - \int_I \frac{\sigma^2}{2a_0 \varsigma_1} \, dx \tag{4.25}$$

for this nonconvex problem has n variational arguments. The criticality condition leads to the algebraic system

$$\begin{aligned}
&2a_0 \varsigma_1^2 \left(\tfrac{1}{a_1 \varsigma_2} \varsigma_1 + \lambda_1 \right) = \sigma^2, \\
&\vdots \\
&2a_{n-2} \varsigma_{n-1}^2 \left(\tfrac{1}{a_{n-1} \varsigma_n} \varsigma_{n-1} + \lambda_{n-1} \right) = \varsigma_{n-2}^2, \\
&2a_{n-1} \varsigma_n^2 \left(\tfrac{1}{a_n} \varsigma_n + \lambda_n \right) = \varsigma_{n-1}^2.
\end{aligned} \tag{4.26}$$

For $n = 3$, this system has at most 8-wells. The algebraic curves are shown in Fig. 4.4. Since, for each given ς_k, ς_{k+1} is a multi-valued function of ς_k, it is difficult to solve this coupled algebra system when n is large. Also, the dual functionals Π_i^d are very sensitive to the parameters. To overcome these difficulties, we need to develop a more powerful dual transformation method.

4.2 The Second Type of Sequential Canonical Dual Transformation

The canonical dual transformation method can be generalized in different ways to solve nonconvex problems with different types of geometrical operator Λ and stored energy V. In this section, we present another method for solving nonconvex problems with multi-well energy.

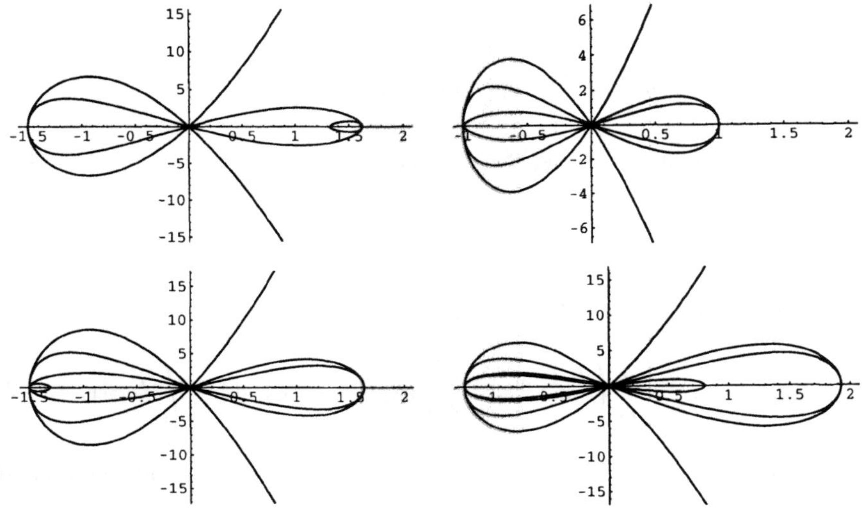

Figure 4.4. Dual algebraic curves for nonconvex problem with three parameters

4.2.1 General Nonconvexity and Composite Transformation

If the geometrical operator $\Lambda : \mathcal{U} \to \mathcal{E}$ is a general nonlinear, nonconvex mapping, we can continue to use the canonical dual transformation method discussed in Section 3.3, such that the general stored energy $V(x, u')$ can be written as

$$V(x, u') = \bar{V}(\Lambda(u)) = \bar{V}_n(\xi_n(\xi_{n-1}(\dots(\xi_1(u))\dots))), \tag{4.27}$$

where $\xi_k(\xi_{k-1})$ is either a convex or a concave function of ξ_{k-1}, and we write

$$\bar{V}_k(\xi_k) = \xi_{k+1}(\xi_k), \quad k = 1, \dots, n-1.$$

Thus, the geometrical operator $\Lambda : \mathcal{U} \to \mathcal{E}$ in this problem is a sequential composition of nonlinear mappings $\Lambda^{(k)} : \mathcal{E}_{k-1} \to \mathcal{E}_k, (k = 1, \cdots, n)$, $\mathcal{E}_0 = \mathcal{U}$, $\mathcal{E}_n = \mathcal{E}$, i.e.

$$\xi_n(u) = \Lambda(u) = \left[\Lambda^{(n)} \circ \Lambda^{(n-1)} \circ \dots \circ \Lambda^{(1)}\right](u).$$

Without loss of generality, we assume that $\xi_1(u)$ is quadratic function of $\xi_0 = \epsilon = u'$ and we write

$$\xi_1(u') = \Lambda^{(1)}(u) = \frac{1}{2} a_0 u'^2 + b_0 u' + c_0, \quad a_0(x) > 0.$$

The sequential constitutive mapping $D\bar{V}_k(\xi_k)$ is one-to-one and such that

$$\varsigma_k = D\bar{V}_k(\xi_k) \quad \Leftrightarrow \quad \xi_k = D\bar{V}_k^c(\varsigma_k) \Leftrightarrow \quad \bar{V}_k(\xi_k) + \bar{V}_k^c(\varsigma_k) = \xi_k \varsigma_k, \quad k = 1, \cdots, n-1.$$

Then, by using the chain rule, we have

$$\begin{aligned}\sigma &= \varsigma_0 = D\bar{V}_n(\xi_n)D\bar{V}_{n-1}(\xi_{n-1})\dots D\bar{V}_1(\xi_1)\frac{\partial \xi_1}{\partial \xi_0} \\ &= \varsigma_n\varsigma_{n-1}\dots\varsigma_1(a_0\xi_0 + b_0) = \varsigma_n!(a_0\xi_0 + b_0). \end{aligned} \tag{4.28}$$

Since $\varsigma_k = D\xi_{k+1}(\xi_k)$, the Gâteaux derivative of $\Lambda^{(n)}$ is

$$\delta\xi^{(n)}(u;v) = \xi_t^{(n)}(u)v = \varsigma_{n-1}!\Lambda_t^{(1)}(u)v, \tag{4.29}$$

and the tri-canonical equations in this nonconvex problem can be written as

$$\begin{cases} \xi_n = \left[\Lambda^{(n)}\circ\Lambda^{(n-1)}\circ\dots\circ\Lambda^{(1)}\right](u), \\ \varsigma_k = D\bar{V}_k(\xi_k), \quad k = 1,\cdots,n, \\ f = [\xi_t^{(n)}(u)]^*\varsigma_n = -\left[\varsigma_n\dots\varsigma_1(a_0u' + b_0)\right]'. \end{cases} \tag{4.30}$$

4.2.2 Canonical Lagrangian and Analytic Solution

Let $\varsigma = \{\varsigma_1,\cdots,\varsigma_n\} : \prod_{k=1}^n \mathcal{E}_k^* \to \mathbb{R}^n$. Since

$$\xi_{k+1}(\xi_k) = \bar{V}_k(\xi_k) = \varsigma_k\xi_k - \bar{V}_k^c(\varsigma_k),$$

the Lagrangian associated with the primal problem (Π_{sta}) can be written as

$$\begin{aligned} L(u,\varsigma) &= \int_I [\varsigma_n\xi_n(\xi_{n-1}) - \bar{V}_n^c(\varsigma_n)]\, dx - \int_I fu\, dx \\ &= \int_I \left[\xi_1(u)\varsigma_n! - \bar{V}_n^c(\varsigma_n) - \varsigma_n\bar{V}_{n-1}^c(\varsigma_{n-1}) - \dots - \frac{\varsigma_n!}{\varsigma_1}\bar{V}_1^c(\varsigma_1)\right] dx \\ &\quad - \int_I fu\, dx. \end{aligned} \tag{4.31}$$

For fixed ς, the criticality condition $D_uL(u,\varsigma) = 0$ leads to the equilibrium equation

$$\left(\Lambda_t^{(1)}(u)\right)^* \varsigma_n! = -\left((a_0u' + b_0)\varsigma_n!\right)' = f \quad \text{in } I. \tag{4.32}$$

Solving for u' and substituting

$$u'(x) = \frac{\sigma}{a_0\varsigma_n!} - \frac{b_0}{a_0} \tag{4.33}$$

into L, we obtain the dual functional

$$\Pi^d(\varsigma) = -\int_I \left[\bar{V}_n^c(\varsigma_n) + \varsigma_n\bar{V}_{n-1}^c(\varsigma_{n-1}) + \dots + \frac{\varsigma_n!}{\varsigma_1}\bar{V}_1^c(\varsigma_1)\right] dx - G^c(\varsigma), \tag{4.34}$$

where $G^c(\varsigma) = G(u(\varsigma),\varsigma)$ is the pure complementary gap functional associated with $\Lambda^{(n)}$, which is obtained by solving the stationary problem

$$\begin{aligned} G^c(\varsigma) &= -\text{sta}\left\{\int_I [\varsigma_n!\Lambda^{(1)}(u) - fu]\, dx \mid u \in \mathcal{U}_a\right\} \\ &= \int_I \left(\frac{\sigma^2}{2a_0\varsigma_n!} + \lambda_0\varsigma_n! - \frac{b_0\sigma}{a_0}\right) dx \end{aligned} \tag{4.35}$$

with $\lambda_0 = b_0^2/a_0 - c_0$.

The criticality condition $D\Pi^d(\varsigma) = 0$ leads to the dual Euler-Lagrange equations

$$2a_0\varsigma_n^{2}! \left[D\bar{V}_1^c(\varsigma_1) + \lambda_0\right] = \sigma^2, \tag{4.36}$$

$$D\bar{V}_{k+1}^c(\varsigma_{k+1}) = \varsigma_k D\bar{V}_k^c(\varsigma_k) - \bar{V}_k^c(\varsigma_k), \quad k = 1, \cdots, n-1. \tag{4.37}$$

Since $D\bar{V}_k^c(\varsigma_k)$ is one-to-one, for each ς_k, ς_{k+1} can be determined uniquely in terms of ς_k via equation (4.37). By Theorem 3.4.3, for each solution $\varsigma = \{\varsigma_1, \cdots, \varsigma_n\}$ of this algebraic system, the analytic solution is

$$u(x) = \int_0^x \left[\frac{g}{a_0\varsigma_n!} - \frac{b_0}{a_0}\right] ds + u_0. \tag{4.38}$$

4.2.3 Multi-Well Energy and Algebraic Curves

If $V(x, u')$ is the nonconvex function

$$V(x, u') = \frac{1}{2}a_n(\frac{1}{2}a_{n-1}(\cdots(\frac{1}{2}a_1(\frac{1}{2}a_0u'^2 - \lambda_1)^2 \cdots)^2 - \lambda_{n-1})^2 - \lambda_n)^2 \tag{4.39}$$

with n parameters $\lambda_k \in \mathbb{R}$, $k = 1, \cdots, n$, we can always choose the coefficients $b_0 = c_0 = \lambda_0 = 0$,

$$b_k = -a_k\lambda_k, \quad c_k = \frac{1}{2}a_k\lambda_k^2 \quad (k = 1, \ldots, n), \tag{4.40}$$

and let $\xi_0 = u'$, so that the geometrical operator $\Lambda^{(k)}$ is the quadratic operator

$$\begin{aligned} \xi_k &= \Lambda^{(k)}(\xi_{k-1}) = \frac{1}{2}a_{k-1}\xi_{k-1}^2 + b_{k-1}\xi_{k-1} + c_{k-1} \\ &= \frac{1}{2}a_{k-1}(\xi_{k-1} - \lambda_k)^2. \end{aligned} \tag{4.41}$$

Then, for each k, the stored energy $\bar{V}_k$ is a purely quadratic function of $\xi_k - \lambda_k$, i.e.

$$\bar{V}_k(\xi_k) = \frac{1}{2}a_k(\xi_k - \lambda_k)^2 = \xi_{k+1}(\xi_k), \quad k = 0, \ldots, n-1,$$

and

$$\varsigma_k = D\bar{V}_k(\xi_k) = a_k(\xi_k - \lambda_k),$$

$$\xi_k = D\bar{V}_k^c(\varsigma_k) = \frac{1}{a_k}\varsigma_k + \lambda_k,$$

$$\bar{V}_k^c(\varsigma_k) = \frac{1}{2a_k}\varsigma_k^2 + \lambda_k\varsigma_k.$$

Thus, the dual algebraic system can be written simply as

$$\begin{aligned} &2a_0\varsigma_n^{2}! \left[\tfrac{1}{a_1}\varsigma_1^2 + \lambda_1\right] = \sigma^2, \\ &\varsigma_{k+1} = \tfrac{a_{k+1}}{2a_k}\varsigma_k^2 - \lambda_{k+1}a_{k+1}, \quad k = 1, \cdots, n-1. \end{aligned} \tag{4.42}$$

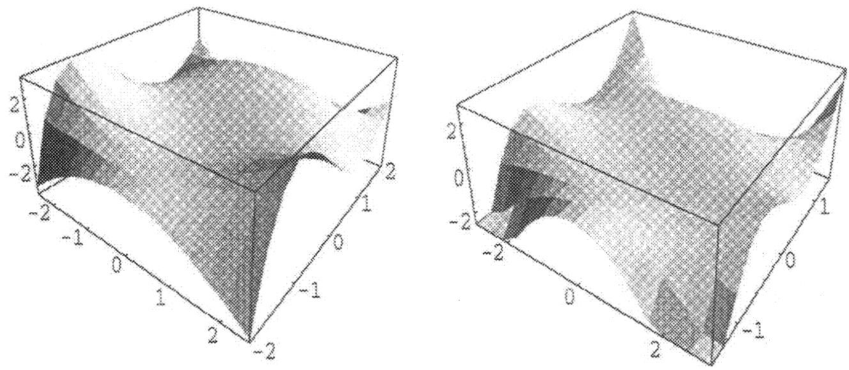

(a) $\lambda_1 = \lambda_2 = 1, a_2 = 1, a_1 = 1.5.$ (b) $\lambda_1 = 1.55, \lambda_2 = 0.45, a_1 = 1, a_2 = 1.6$

Figure 4.5. Lagrangians L_J for (a) triple and (b) quadruple well energies ($a_0 = 1$)

For $n = 2$, we have

$$\Pi_2^d(\varsigma_1, \varsigma_2) = -\int_I \left(\frac{1}{a_2}\varsigma_2^2 + \lambda_2\varsigma_2 + \varsigma_2(\frac{1}{2a_1}\varsigma_1^2 + \lambda_1\varsigma_1) \right) dx - \int_I \frac{\sigma^2}{2a_0\varsigma_1\varsigma_2} dx, \tag{4.43}$$

and the dual Euler-Lagrange equations are

$$2a_0\varsigma_1^2\varsigma_2^2(\frac{1}{a_1}\varsigma_1 + \lambda_1) = \sigma^2, \tag{4.44a}$$

$$\varsigma_2 = \frac{a_2}{2a_1}\varsigma_1^2 - \lambda_2 a_2. \tag{4.44b}$$

In this dual transformation, ς_2 is defined uniquely by ς_1. For the given parameters, the algebraic equation

$$2a_0\varsigma_1^2 \left(\frac{a_2}{2a_1}\varsigma_1^2 - \lambda_2 a_2 \right)^2 (\frac{1}{a_1}\varsigma_1 + \lambda_1) = \sigma^2 \tag{4.45}$$

has at most seven real roots for ς_1. For each root $\bar{\varsigma}_j$, the analytic solution is

$$u_j(x) = \int \frac{\sigma(x)}{a_0\bar{\varsigma}_j(a_2\bar{\varsigma}_j^2/2a_1 - \lambda_2 a_2)} dx + u_0, \quad j = 1, \cdots, 7. \tag{4.46}$$

Let

$$J_2(u) = \frac{1}{2}a_2(\frac{1}{2}a_1(\frac{1}{2}a_0u^2 - \lambda_1)^2 - \lambda_2)^2 - \sigma u.$$

The nonlinear Lagrangian and dual function are

$$L_J(u, \varsigma) = \left(\frac{a_2}{2a_1}\varsigma^2 - \lambda_2 a_2 \right) \left(\frac{1}{2}a_0u^2\varsigma - a\varsigma^2 - \lambda_1\varsigma + b \right) - \sigma u,$$

$$\mathcal{J}_2^d(\varsigma) = \left(\frac{a_2}{2a_1}\varsigma^2 - \lambda_2 a_2\right)\left(b - a\varsigma^2 - \lambda_1\varsigma\right) - \frac{a_1 g^2}{a_0 a_2 \varsigma(\varsigma^2 - 2a_1 a_2 \lambda_2)},$$

where $a = (a_2/2a_0 + 1)/2a_1$, $b = \lambda_2(a_2/2a_0 - 1)$, and $\varsigma_2 = a_2\varsigma_1^2/(2a_1) - \lambda_2 a_2$. For the given parameters, the graph of $L_{\mathcal{J}}(u, \varsigma)$ is shown in Fig. 4.5. Graphs of $\mathcal{J}_2$, $\mathcal{J}_2^d$ and algebraic curves of equation (4.44a) are shown in Fig. 4.6.

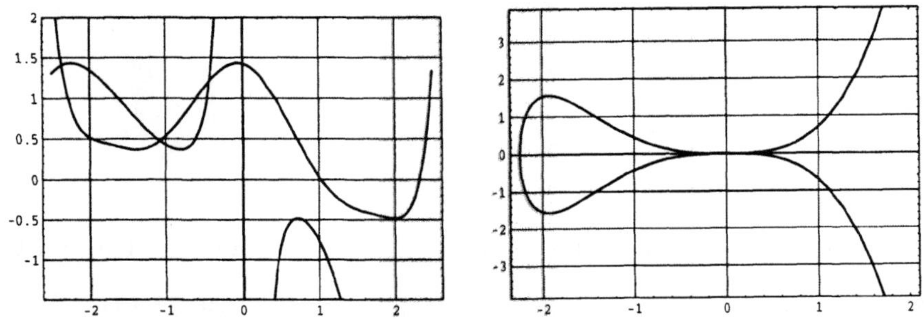

(a) Double-well energy: $\lambda_1 = 1.5, \lambda_2 = 0, a_0 = a_1 = a_2 = 1$

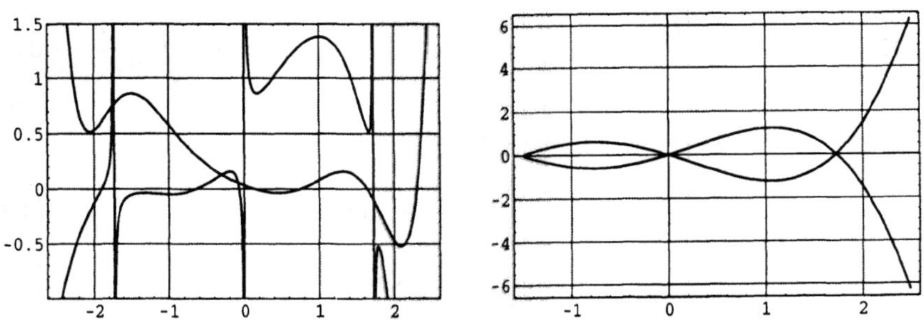

(b) Triple-well energy: $\lambda_1 = \lambda_2 = 1, a_0 = a_2 = 1, a_1 = 1.5$

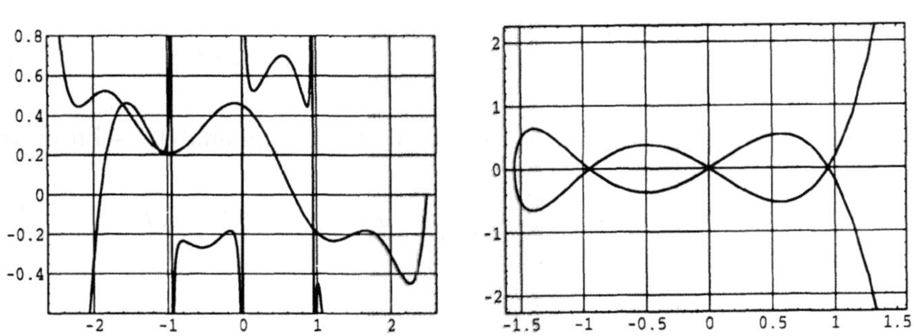

(c) Quadruple-well energy : $\lambda_1 = 1.55, \lambda_2 = .45, a_0 = a_1 = 1, a_2 = 1.6$

Figure 4.6. Multi-well energies and their algebraic curves

4.3 Canonical Systems: The Classifications

In order to study the mathematical theory of duality in natural phenomena, we need to make some mathematical definitions for general nonlinear systems. However, it is very difficult, and the author hopes that this section will not prove to be confusing.

4.3.1 Elementary System and Fundamental Equation

Generally speaking, for any physical system, at least two types of dependent variables can be identified. They are

(a) *source variables,* which represent the source (or the input) of the system. We use $\mathcal{F}$ to denote the source space. Elements $f \in \mathcal{F}$ might be, for example, a given external force vector $\bar{\mathbf{f}}$ in mechanics, or an electric charge $\bar{\rho}$ in electromagnetics.

(b) *configuration variables,* which represent the response (or the output) of the system. We denote, by $\mathcal{U}$, the space of configuration variables. Its elements $u \in \mathcal{U}$ might be the Lagrange coordinates $\mathbf{x}$ in analytical dynamics, or the displacements $\mathbf{u}$ in continuum mechanics, for example. These variables describe the state of the system.

Since physical variables in most systems usually appear in pairs, then for each given configuration variable $u \in \mathcal{U}$, there exists a variable u^* which is dual to u. All these dual variables form a *dual configuration* space, denoted by $\mathcal{U}^*$. The source space $\mathcal{F}$ is usually a subspace of $\mathcal{U}^*$. For each pair consisting of the configuration variable $u \in \mathcal{U}$ and its dual, $u^* \in \mathcal{U}^*$, we introduce the first type bilinear form $\langle *, * \rangle$, which is a scalar-valued function defined on the paired space $\mathcal{U} \times \mathcal{U}^*$. We say that paired spaces $\mathcal{U}$ and $\mathcal{U}^*$ are placed in duality by the bilinear form $\langle *, * \rangle : \mathcal{U} \times \mathcal{U}^* \to \mathbb{R}$. In static systems, this bilinear form is defined on a space domain $\Omega \subset \mathbb{R}^n$. However, in dynamical systems, the domain $\Omega_t = \Omega \times (0, T)$ is a subset of space-time $\mathbb{R}^{n+1}$. A space $\mathcal{U}$ is *reflexive* if $\mathcal{U}^{**} = \mathcal{U}$.

For given paired spaces $\mathcal{U}$ and $\mathcal{U}^*$, an *operator* $A : \mathcal{U} \to \mathcal{U}^*$ is a mapping from $\mathcal{U}$ into $\mathcal{U}^*$ which sends each $u \in \mathcal{U}$ to an element $u^* = A(u) \in \mathcal{U}^*$. We admit the possibility that an operator may be defined only on a subset $\mathcal{D}_A \subset \mathcal{U}$, called the *domain* of A. The set of elements given by $A(u)$ is called the *range* of A and denoted by $\mathcal{R}_A$, which is a subset of $\mathcal{U}^*$. The mapping $A : \mathcal{D}_A \subseteq \mathcal{U} \to \mathcal{R}_A \subseteq \mathcal{U}^*$ is invertible on $\mathcal{D}_A$ if there exists an inverse operator $A^{-1} : \mathcal{R}_A \to \mathcal{D}_A$ that associates a unique $u = A^{-1}(u^*) \in \mathcal{D}_A$ with each $u^* \in \mathcal{R}_A$.

Definition 4.3.1 (Elementary System and Source Spaces)

(D1) An *elementary system* is a collection of

(a) the objects: paired spaces $\mathcal{U}$ and $\mathcal{U}^*$, and
(b) the relations: the operator $A : \mathcal{U} \to \mathcal{U}^*$ and the nondegenerate bilinear form $\langle *, * \rangle : \mathcal{U} \times \mathcal{U}^* \to \mathbb{R}$ which puts $\mathcal{U}$ and $\mathcal{U}^*$ in duality.

Symbolically, an elementary system $\mathbb{S}^e$ is denoted by

$$\mathbb{S}^e := \{(\mathcal{U}, \mathcal{U}^*);\ \langle *, * \rangle, A\}, \quad \text{or simply} \quad \mathbb{S}^e := \{\langle \mathcal{U}, \mathcal{U}^* \rangle;\ A\}. \tag{4.47}$$

(D2) The *source space* $\mathcal{F}$ of an elementary system $\mathbb{S}^e$ is said to be

(a) a *dead source space* if $\mathcal{F} \subset \mathcal{U}^*$ and every element in $\mathcal{F}$ is time-independent;
(b) an *active source space* if $\mathcal{F} \subset \mathcal{U}^*$ and its elements are time-dependent;
(c) a *follower source space* if the elements $f \in \mathcal{F}$ depend on the configuration variables $u \in \mathcal{U}$. ◊

An element in dead source space is also called the *dead load.* For a given elementary system $\mathbb{S}^e$, the operator $A : \mathcal{U} \to \mathcal{U}^*$ is called the *fundamental operator*, which describes the basic property of the system. For a given source $\bar{u}^* \in \mathcal{F} \subset \mathcal{U}^*$, the equation

$$A(u) = \bar{u}^*. \tag{4.48}$$

is called the *fundamental equation.*

Recall that an operator $A : \mathcal{D}_A \to \mathcal{R}_A$ is said to be a *potential operator* if there exists a Gâteaux differentiable functional $\Xi : \mathcal{U}_k \subset \mathcal{U} \to \mathbb{R}$ such that $A(u) = D\Xi(u)$.

Definition 4.3.2 (Equilibrium, Potential and Reflexive Systems) Let $\mathbb{S}^e = \{\langle \mathcal{U}, \mathcal{U}^* \rangle;\ A\}$ be an elementary system, and let $\mathbb{S}^e_a = \{\langle \mathcal{U}_a, \mathcal{U}^*_a \rangle; A\} \subset \mathbb{S}^e$ be a subsystem.

(D1) $\mathbb{S}^e_a$ is said to be an *equilibrium system* if for each given source $\bar{u}^* \in \mathcal{F}$, there exists at least one configuration $u \in \mathcal{U}_a \subset \mathcal{U}$ such that the equilibrium equation

$$A(u) = \bar{u}^* \tag{4.49}$$

holds on $\mathcal{U}_a$.

(D2) $\mathbb{S}^e_a$ is said to be a *potential system* if the operator $A : \mathcal{D}_A \subset \mathcal{U} \to \mathcal{R}_A \subset \mathcal{U}^*$ is a potential operator. The associated functional $\Xi : \mathcal{U}_k \to \mathbb{R}$ is called the *stored action.*

(D3) The sub-system $\mathbb{S}^e_a$ is said to be a *reflexive system* if $\mathcal{U}_a$ is reflexive; $\mathbb{S}^e_a$ is said to be a *strictly reflexive system* if the operator $A : \mathcal{U}_a \to \mathcal{U}^*_a$ is invertible. ◊

A potential system with either dead source space or active source space is called a generalized conservative system. However, the system with a follower source space is not in general a conservative system. Complementary variational principles in rate plasticity with follower source were discussed by Gao and Onat (1990).

Physically speaking, a "well-defined" system should be an equilibrium system, i.e. for each given input $\bar{u}^* \in \mathcal{U}^*$, the system should have at least one response $\bar{u}$ which is a solution of the equilibrium equation (4.49). If $\mathbb{S}^e$ is strictly reflexive, the solution is unique. Hence, in this philosophy, if the solution of a mathematical system cannot be found or does not exist, then, this system may not be well-defined.

4.3.2 Canonical Systems and Classifications

Besides the configuration and source variables, in virtually every elementary system there exist some *intermediate variables* such as strain and stress in continuum mechanics, velocity and momentum in dynamics. A characteristic of these intermediate variables is that they always appear in pairs. The duality relation between each pair of intermediate variables describes certain interior properties of the system, typically contains characteristic parameters of the medium and physical constants. By $\mathcal{E}$ and $\mathcal{E}^*$ we denote the paired intermediate spaces. They are placed in duality by the second type of bilinear form $\langle *; * \rangle : \mathcal{E} \times \mathcal{E}^* \to \mathbb{R}$.

To link the two sets of paired spaces $(\mathcal{U}, \mathcal{U}^*)$ and $(\mathcal{E}, \mathcal{E}^*)$, three operators are introduced.

(A) *The geometrical operator* $\Lambda : \mathcal{U} \to \mathcal{E}$. This describes the "topology" of the system, i.e. the geometrical transformation of the system, which does not concern physical property of the system. The associated *geometrical equation*

$$\epsilon = \Lambda(u) \tag{4.50}$$

is also called the *definition equation* (see Tonti, 1972), or *kinematic equation* (see Oden and Reddy, 1974).

(B) *The constitutive operator* $C : \mathcal{E} \to \mathcal{E}^*$. This describes the "intrinsic" physical property of the problem. The *constitutive equation*

$$\epsilon^* = C(\epsilon), \tag{4.51}$$

provides the duality relation between the paired spaces $\mathcal{E}$ and $\mathcal{E}^*$.

(C) *The balance operator* $\Upsilon : \mathcal{E}^* \to \mathcal{U}^*$. This describes the local state of the system. For a given $u^* \in \mathcal{U}^*$, the balance equation

$$u^* = \Upsilon(\epsilon^*) \tag{4.52}$$

is also called the *equilibrium equation.* Like the geometrical equation, the balance equation involves only the "dual topology" of the system.

The sequence of mappings $\Lambda : \mathcal{U} \to \mathcal{E}$, $C : \mathcal{E} \to \mathcal{E}^*$, $\Upsilon : \mathcal{E}^* \to \mathcal{U}^*$ induces the *fundamental operator*: $A = \Upsilon \circ C \circ \Lambda : \mathcal{U} \to \mathcal{U}^*$, and the fundamental equation can be decomposed into three equations:

$$A(u) = u^* \quad \Rightarrow \quad \begin{cases} \epsilon = \Lambda(u), \\ \epsilon^* = C(\epsilon), \\ u^* = \Upsilon(\epsilon^*). \end{cases} \tag{4.53}$$

The definition of an elementary system can be extended into general systems that contain more objects and relations. For example, the system $\mathbb{S}^1$, defined by

$$\mathbb{S}^1 := \{\langle \mathcal{U}, \mathcal{U}^* \rangle, \langle \mathcal{E}; \mathcal{E}^* \rangle; \Lambda, C, \Upsilon\} \tag{4.54}$$

may be called the *first-order system* generated from $\mathbb{S}^e$, which is simply written as $\mathbb{S}$. It is an equilibrium system if for each given source $\bar{u}^* \in \mathcal{F} \subset \mathcal{U}^*$,

$$A(u) = (\Upsilon \circ C \circ \Lambda)(u) = \bar{u}^*.$$

Higher-order systems $\mathbb{S}^n$ $(n \geq 2)$ can be constructed in the same way. The structure of the n-th order system is shown in Fig. 4.7.

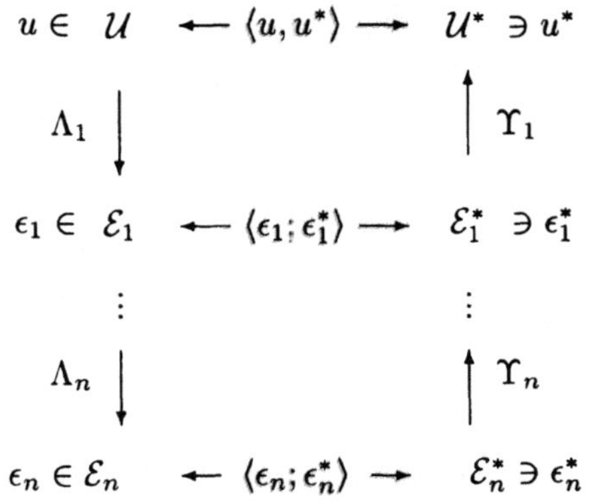

Figure 4.7. Structure of the n-th order system

Recall that a geometric operator $\Lambda : \mathcal{U} \to \mathcal{E}$ is said to be Gâteaux differentiable on $\mathcal{U}_a \subset \mathcal{U}$ if there exists an operator $\Lambda_t := D\Lambda : \mathcal{U}_a \to \mathcal{E}$ such that the directional derivative of Λ at $\bar{u}$ in the direction u can be written as

$$\delta\Lambda(\bar{u}; u) = \Lambda_t(\bar{u})u \quad \forall \bar{u}, u \in \mathcal{U}_a.$$

Thus, if $\mathbb{S}^e$ is a potential system then the relations among the operators are given by the following theorem.

Theorem 4.3.1 (Generalized Virtual Work Principles)
Let $\mathbb{S}^e = \{\langle \mathcal{U}, \mathcal{U}^* \rangle; \; A\}$ be a potential system such that there exists a stored action $\Xi : \mathcal{U}_a \to \mathbb{R}$, $A(u) = D\Xi(u) \;\; \forall u \in \mathcal{U}_a$. Let $\mathbb{S} = \{\langle \mathcal{U}, \mathcal{U}^* \rangle, \langle \mathcal{E}; \mathcal{E}^* \rangle; \Lambda, C, \Upsilon\}$ be the first-order system generated from $\mathbb{S}^e$.

(T1) If $\mathbb{S}^e_a = \{\langle \mathcal{U}_a, \mathcal{U}^*_a \rangle; \; A\} \subset \mathbb{S}^e$ is an equilibrium system then, for each given $\bar{u}^* \in \mathcal{F}$, there exists a $\bar{u} \in \mathcal{K} \subset \mathcal{U}_a$ such that

$$\delta\Xi(\bar{u}; u) = \langle u, \; A(\bar{u}) \rangle = \langle u, \bar{u}^* \rangle \quad \forall u \in \mathcal{K}. \tag{4.55}$$

(T2) If $\Lambda : \mathcal{U}_a \to \mathcal{E}_a \subset \mathcal{E}$ is Gâteaux differentiable and there exists a functional $\bar{\Xi} : \mathcal{E}_a \to \mathbb{R}$ such that $\Xi(u) = \bar{\Xi}(\Lambda(u)) \;\; \forall u \in \mathcal{U}_k = \{u \in \mathcal{U}_a | \;\; \Lambda(u) \in \mathcal{E}_a\}$, then, the constitutive operator $C : \mathcal{E}_a \to \mathcal{E}^*_a \subset \mathcal{E}^*$ is a potential operator defined by

$$C(\epsilon) = D\bar{\Xi}(\epsilon) \quad \forall \epsilon \in \mathcal{E}_a, \tag{4.56}$$

and the balance operator Υ is a linear operator in ϵ^* depending on $\bar{u} \in \mathcal{U}_a$, i.e.

$$\Upsilon = \Lambda^*_t(\bar{u}) : \mathcal{E}^*_a \times \mathcal{U}_a \to \mathcal{U}^*_a, \tag{4.57}$$

where Λ_t^* is the adjoint operator of the Gâteaux derivative $\Lambda_t(\bar{u})$ defined by

$$\langle \Lambda_t(\bar{u})\delta u; \epsilon^* \rangle = \langle \delta u, \Lambda_t^*(\bar{u})\epsilon^* \rangle \quad \forall \delta u \in \mathcal{U}_k. \tag{4.58}$$

Proof. By definition, the potential system $\mathbb{S}^e$ is an equilibrium system and hence that there exists a subset $\mathcal{K} \subset \mathcal{U}_a$ such that, for each given $\bar{u}^* \in \mathcal{F}$, there exists a $\bar{u} \in \mathcal{K}$ satisfying

$$A(\bar{u}) = D\Xi(\bar{u}) = \bar{u}^*.$$

By the fundamental lemma in the calculus of variations this equilibrium equation can be written in the *weak form*

$$\langle u, A(\bar{u}) \rangle = \langle u, D\Xi(\bar{u}) \rangle = \delta\Xi(\bar{u}; u) = \langle u, \bar{u}^* \rangle \quad \forall u \in \mathcal{K}. \tag{4.59}$$

If $\Lambda : \mathcal{U}_a \to \mathcal{E}_a \subset \mathcal{E}$ is Gâteaux differentiable and there exists a potential $\bar{\Xi} : \mathcal{E}_a \to \mathbb{R}$ such that $\Xi(u) = \bar{\Xi}(\Lambda(u))$ then, by the chain rule,

$$\delta\Xi(\bar{u}; u) = \delta\bar{\Xi}(\Lambda(\bar{u}); \delta\Lambda(\bar{u}; u)) = \langle \Lambda_t(\bar{u})u; D\bar{\Xi}(\bar{\epsilon}) \rangle = \langle u, A(\bar{u}) \rangle \quad \forall u \in \mathcal{U}_k,$$

with $\bar{\epsilon} = \Lambda(\bar{u})$ and

$$A(\bar{u}) = \Lambda_t^*(\bar{u}) D_\epsilon \bar{\Xi}(\Lambda(\bar{u})). \tag{4.60}$$

Thus, $A = \Upsilon \circ C \circ \Lambda$ leads to $C(\bar{\epsilon}) = D\bar{\Xi}(\bar{\epsilon})$ and $\Upsilon = \Lambda_t^*(\bar{u})$. □

Definition 4.3.3 (Canonical Systems and Canonical Constitutive Law)

Let $\mathbb{S}^e = \{\langle \mathcal{U}, \mathcal{U}^* \rangle; A\}$ be a potential elementary system. If there exists a Gâteaux differentiable geometrical operator $\Lambda : \mathcal{U}_a \to \mathcal{E}_a$ such that the constitutive operator $C : \mathcal{E}_a \to \mathcal{E}_a^*$ is invertible then the system defined by

$$\mathbb{S}_a := \{\langle \mathcal{U}_a, \mathcal{U}_a^* \rangle, \langle \mathcal{E}_a; \mathcal{E}_a^* \rangle; \Lambda, C\} \subset \mathbb{S}^2 \tag{4.61}$$

is called a *canonical system* of $\mathbb{S}^e$ generated by Λ, or simply a *canonical system*, and the invertible equation $\epsilon^* = C(\epsilon)$ is called the *canonical constitutive law*. $\mathbb{S}_a$ is called a *strictly canonical system* if the operator Λ is linear and the mappings $\Lambda : \mathcal{U}_a \to \mathcal{E}_a$ and $\Lambda^* : \mathcal{E}_a^* \to \mathcal{U}_a^*$ are bijective. ◇

Clearly, if $\mathbb{S} = \{\langle \mathcal{U}, \mathcal{U}^* \rangle, \langle \mathcal{E}; \mathcal{E}^* \rangle; \Lambda, C\}$ is a canonical system then $C : \mathcal{E} \to \mathcal{E}^*$ is invertible and $\mathcal{E}$ is reflexive.

Remark 4.3.1 For a given elementary system $\mathbb{S}^e$, its canonical system may not be unique, i.e. by choosing different Λ, we may have different canonical systems. The constitutive operator C depends on the choice of Λ. But, for strictly canonical systems, these choices must be equivalent to each other. Since the equilibrium equation $\Lambda_t^*(\bar{u})\epsilon^* = \bar{u}^*$ is a linear equation of ϵ^* for each given $\bar{u}$, the dual problem might be easier than the primal problem. This is the key idea of the so-called canonical dual transformation method discussed in previous chapters. ■

Definition 4.3.4 (Nonlinear Canonical Systems)
Let $\mathbb{S} = \{\langle \mathcal{U}, \mathcal{U}^* \rangle, \langle \mathcal{E}; \mathcal{E}^* \rangle; \Lambda, C\}$ be a canonical system.

(D1) $\mathbb{S}$ is said to be a *geometrically nonlinear (resp. linear) system* if $\Lambda : \mathcal{U} \to \mathcal{E}$ is a nonlinear (resp. linear) operator.

(D2) $\mathbb{S}$ is said to be a *constitutively (or physically) nonlinear (resp. linear) system* if the constitutive operator $C : \mathcal{E} \to \mathcal{E}^*$ is nonlinear (resp. linear).

(D3) $\mathbb{S}$ is said to be a *fully nonlinear (resp. linear) system* if it is both geometrically and physically nonlinear (resp. linear). ◇

In fully nonlinear canonical systems, the virtual work principle gives $\Upsilon = \Lambda_t^*$, and the *weak form* of the equilibrium equation is

$$\langle \Lambda_t(\bar{u})u; C(\Lambda\bar{u}) \rangle = \langle u, \bar{u}^* \rangle \quad \forall u \in \mathcal{U}_a. \tag{4.62}$$

By the fundamental lemma in the calculus of variations, it leads to the fully nonlinear *fundamental equation of equilibrium*

$$A(\bar{u}) = \Lambda_t^*(\bar{u})C(\Lambda\bar{u}) = \bar{u}^* \tag{4.63}$$

under suitable assumptions.

In geometrically linear canonical systems, the operator $\Lambda : \mathcal{U} \to \mathcal{E}$ is linear, and $\Upsilon = \Lambda^* : \mathcal{E}^* \to \mathcal{U}^*$. The relation

$$\langle \Lambda u; \epsilon^* \rangle = \langle u, \Lambda^* \epsilon^* \rangle, \tag{4.64}$$

is the infinite-dimensional analogue of transposing a matrix Λ in finite dimensional systems. When the geometric mapping Λ is a gradient-like operator, the equilibrium mapping Λ^* is a divergence-like operator and equation (4.64) reduces to the *Gauss-Green formula* for integration by parts. For brevity we refer to the more general abstract form (4.64) as Gauss' law. In a geometrically linear system $\mathbb{S} = \{\langle \mathcal{U}, \mathcal{U}^* \rangle, \langle \mathcal{E}; \mathcal{E}^* \rangle; \Lambda, C\}$ the fundamental equation (4.63) takes the symmetrical form

$$A(u) = \Lambda^* C(\Lambda\bar{u}) = \bar{u}^*, \tag{4.65}$$

which can be decomposed into the three equations

$$\epsilon = \Lambda u, \quad \epsilon^* = C(\epsilon), \quad \bar{u}^* = \Lambda^* \epsilon^*. \tag{4.66}$$

We refer to (4.66) as the *primal canonical equations* associated with the linear operator Λ, or simply the *primal canonical forms*.

For a linear system $\mathbb{S}$, we have a more symmetrical form, namely

$$Au = \Lambda^* C \Lambda \bar{u} = \bar{u}^*, \tag{4.67}$$

and $A = \Lambda^* C \Lambda : \mathcal{U} \to \mathcal{U}^*$ is a linear operator.

The relationship between spaces and operators in a fully nonlinear system is shown symbolically in Fig. 4.8

Example 4.3.1 Let $\Omega \subset \mathbb{R}^n$ be a simply connected open domain with boundary $\Gamma = \partial\Omega$, which is a $(n-1)$-dimensional manifold. Let the geometrically linear

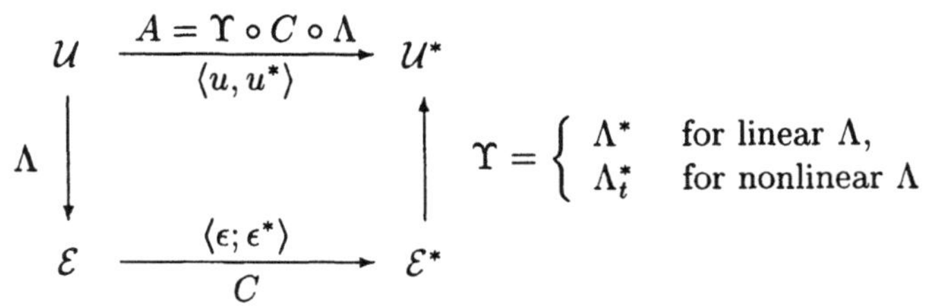

Figure 4.8. Diagrammatic representation for fully nonlinear systems.

operator Λ be a gradient on Ω, i.e. $\Lambda = \nabla = \{\partial_{x_i}\}$. Then, for each scalar-valued function $u \in \mathcal{U} = \mathcal{L}^\alpha(\Omega; \mathbb{R})$,

$$\epsilon = \Lambda u = \left\{ \frac{\partial u}{\partial x_i} \right\} = \{\epsilon_i(x)\}, \quad x \in \Omega, \quad i = 1, 2, \cdots, n$$

is a vector-valued function on $\mathcal{E} = \mathcal{L}^\beta(\Omega; \mathbb{R}^n)$. The dual spaces $\mathcal{U}^* = \mathcal{L}^{\alpha^*}(\Omega; \mathbb{R})$ and $\mathcal{E}^* = \mathcal{L}^{\beta^*}(\Omega; \mathbb{R}^n)$ are defined respectively by the bilinear forms

$$\langle u, u^* \rangle = \int_\Omega u(x) u^*(x) \mathrm{d}\Omega, \quad \langle \epsilon; \epsilon^* \rangle = \int_\Omega \epsilon \cdot \epsilon^* \mathrm{d}\Omega. \tag{4.68}$$

Let $\mathcal{U}_a = \{u \in \mathcal{U} | \ u(x) = 0 \ \ \forall x \in \Gamma\}$ and $\mathcal{U}_k = \{u \in \mathcal{U}_a | \ \Lambda u \in \mathcal{E}\}$. By using the identity

$$\mathrm{div}(\epsilon^* u) = u \mathrm{div} \epsilon^* + \epsilon^* \mathrm{grad} u$$

and the Gauss-Green formula

$$\int_\Omega \mathrm{div} \epsilon^* \mathrm{d}\Omega = \int_\Gamma \mathbf{n} \cdot \epsilon^* \mathrm{d}\Gamma,$$

where $\mathbf{n} : \Gamma \to \mathbb{R}^n$ is a unit vector normal outward to the boundary Γ, we obtain

$$\langle \Lambda u; \epsilon^* \rangle = -\int_\Omega u \ \mathrm{div} \epsilon^* \mathrm{d}\Omega + \int_\Gamma u(\mathbf{n} \cdot \epsilon) \mathrm{d}\Gamma = \langle u, \Lambda^* \epsilon^* \rangle \ \ \forall u \in \mathcal{U}_k. \tag{4.69}$$

The bilinear concomitant $J(u, \epsilon^*) = \int_\Gamma u(\mathbf{n} \cdot \epsilon^*) \mathrm{d}\Gamma$ vanishes on $\mathcal{U}_k$. Thus, the adjoint of Λ = grad is $\Lambda^* = -$div. Let $C : \mathcal{E} \to \mathcal{E}^*$ be a symmetric linear operator, i.e. $C = C^T$ is a symmetric matrix function in $\mathbb{R}^n \times \mathbb{R}^n$. Then, $\mathbb{S} = \{\langle \mathcal{U}, \mathcal{U}^* \rangle, \langle \mathcal{E}; \mathcal{E}^* \rangle; \ \ \mathrm{grad}, C\}$ is a linear canonical system over the space domain $\Omega \subset \mathbb{R}^n$. The action $\Xi : \mathcal{U}_k \to \mathbb{R}$ is the quadratic functional

$$\Xi(u) = \frac{1}{2} \langle \mathrm{grad} u; C \mathrm{grad} u \rangle = \int_\Omega \frac{1}{2} (\mathrm{grad} u) \cdot C \cdot (\mathrm{grad} u) \mathrm{d}\Omega,$$

and we have the results

$\Xi : \mathcal{U}_k \to \mathbb{R}$ is (strictly) convex $\Leftrightarrow$ C is (strictly) positive-definite;
$\Xi : \mathcal{U}_k \to \mathbb{R}$ is (strictly) concave $\Leftrightarrow$ C is (strictly) negative-definite.

For any given $\bar{u}^* \in \mathcal{U}^*$, the virtual work principle leads to the second-order partial differential equation

$$Au(x) = -\mathrm{div}(C\mathrm{grad}\ u(x)) = \bar{u}^*(x) \quad x \in \Omega.$$

The fundamental operator $A : \mathcal{U} \to \mathcal{U}^*$ is a self-adjoint linear operator, and

$$\begin{aligned} &A : \mathcal{U}_k \to \mathcal{U}_a^* \text{ is elliptic if } C \text{ is strictly (either positive or negative) definite;} \\ &A : \mathcal{U}_k \to \mathcal{U}_a^* \text{ is hyperbolic if } C \text{ is strictly indefinite;} \\ &A : \mathcal{U}_k \to \mathcal{U}_a^* \text{ is parabolic if } C \text{ is singular.} \end{aligned} \tag{4.70}$$

In particular, if $C = \mathbf{I}$, the identity matrix in $\mathbb{R}^n \times \mathbb{R}^n$, then

$$A = -\mathrm{div\ grad} = -\Delta = -(\partial^2_{x_1^2} + \cdots + \partial^2_{x_n^2})$$

is the well-known Lapalacian in $\Omega \subset \mathbb{R}^n$. ♡

4.3.3 Polar Systems

Let $\mathbb{S} = \{\langle \mathcal{U}, \mathcal{U}^* \rangle, \langle \mathcal{E}; \mathcal{E}^* \rangle; \Lambda, C\}$ be a geometrically linear canonical system. By the duality of nature, there might exists a "dual system", controlled by the same intrinsic constitutive law as that in $\mathbb{S}$ but with the "polar" input and output. Let $\mathcal{U}^o$ be a linear space, in duality with $\mathcal{U}^{o*}$ through the third type of bilinear form

$$\langle u^o \ ; \ u^{o*} \rangle : \mathcal{U}^o \times \mathcal{U}^{o*} \to \mathbb{R}. \tag{4.71}$$

For a given linear mapping $\Lambda^o : \mathcal{U}^o \to \mathcal{E}^*$, its adjoint $\Lambda^{o*} : \mathcal{E} \to \mathcal{U}^{o*}$ is defined by

$$\langle \epsilon; \Lambda^o u^o \rangle = \langle \Lambda^{o*} \epsilon \ ; \ u^o \rangle = \langle u^o \ ; \ \Lambda^{o*} \epsilon \rangle. \tag{4.72}$$

We have the following definition.

Definition 4.3.5 (Polar System)
Let $\mathbb{S} = \{\langle \mathcal{U}, \mathcal{U}^* \rangle, \langle \mathcal{E}; \mathcal{E}^* \rangle; \Lambda, C\}$ be a given geometrically linear canonical system with source space $\mathcal{F} \subset \mathcal{U}^*$.

(D1) A system $\mathbb{S}^o$ defined by

$$\mathbb{S}^o = \{\langle \mathcal{U}^o \ ; \ \mathcal{U}^{o*} \rangle, \langle \mathcal{E}^*; \mathcal{E} \rangle; \Lambda^o, C^o\} \tag{4.73}$$

is said to be a *polar system* of $\mathbb{S}$ if and only if $C^o = C^{-1}$. The space $\mathcal{U}^o$ is called the *polar configuration space*, the operator $\Lambda^o : \mathcal{U}^o \to \mathcal{E}^*$ the *polar geometrical operator*, the adjoint $\Lambda^{o*} : \mathcal{E} \to \mathcal{U}^{o*}$ the *polar equilibrium operator*, and the composition $A^o := \Lambda^{o*} \circ C^o \circ \Lambda^o : \mathcal{U}^o \to \mathcal{U}^{o*}$ the *polar fundamental operator* .

(D2) $\mathbb{S}^o$ is said to be an *iso-source polar system* of $\mathbb{S}$ if, for each given $f \in \mathcal{F} \subset \mathcal{U}^*$, the weak form

$$\langle \Lambda u; \Lambda^o u^o \rangle = \langle u, f \rangle \quad \forall u \in \mathcal{U} \tag{4.74}$$

holds for all $u^o \in \mathcal{U}^o$. ◇

By the fundamental lemma in the calculus of variations, under certain conditions on the spaces $\mathcal{U}$ and $\mathcal{U}^o$, the equality (4.74) leads to the so-called *polar equilibrium equation*

$$\Lambda^* \Lambda^o u^o = f. \tag{4.75}$$

Thus, the iso-source polar configuration space provides a set of solutions of the equation

$$\mathcal{U}^o = \{u^o | \ \Lambda^* \Lambda^o u^o = f \ \forall f \in \mathcal{F}\}. \tag{4.76}$$

Dually, if $\mathbb{S}^o$ defined by (4.73) is a geometrically linear canonical system with the *polar source space* $\mathcal{F}^o \subset \mathcal{U}^{o*}$, then, $\mathbb{S} = \{\langle \mathcal{U}, \mathcal{U}^* \rangle, \langle \mathcal{E}; \mathcal{E}^* \rangle; \Lambda, C\}$ is an iso-source polar system of $\mathbb{S}^o$ if $C = C^{o-1}$, and, for each given polar source $\bar{u}^{o*} \in \mathcal{F}^o$ and $u \in \mathcal{U}$,

$$\langle \Lambda u; \Lambda^o u^o \rangle = \langle u^o \ ; \ \bar{u}^{o*} \rangle \ \forall u^o \in \mathcal{U}^o. \tag{4.77}$$

The strong form of this equation is

$$\Lambda^{o*} \Lambda u = \bar{u}^{o*}, \tag{4.78}$$

which is called the *compatibility equation.* The polar source $\bar{u}^{o*} \in \mathcal{F}^o$ is also called the *incompatibility source.*

For a given polar source $\bar{u}^{o*} \in \mathcal{F}^o \subset \mathcal{U}^{o*}$, the canonical equations for the polar system $\mathbb{S}^o$ are

$$\epsilon^* = \Lambda^o u^o, \quad \epsilon = C^o(\epsilon^*), \quad \bar{u}^{o*} = \Lambda^{o*} \epsilon. \tag{4.79}$$

We refer to (4.79) as the *polar canonical equations* associated with the linear operator Λ^o. Similarly, we refer to the abstract problem of finding $u^o \in \mathcal{U}^o$ such that

$$A^o(u^o) = (\Lambda^{o*} \circ C^o \circ \Lambda^o)(u^o) = \bar{u}^{o*} \tag{4.80}$$

as the *polar canonical problem* associated with (4.65), and equation (4.80) as the *polar fundamental equation* .

The relationship between the spaces and operators for the canonical system $\mathbb{S}$ and its polar system $\mathbb{S}^o$ are presented symbolically in Fig. 4.9.

Example 4.3.2 (Fluid Mechanics) Let $\Omega \subset \mathbb{R}^3$ be an open bounded domain with boundary $\partial\Omega = \Gamma = \Gamma_u \cup \Gamma_f$. The equation for steady barotropic irrotational flow of a perfect fluid is

$$\text{div}(\rho_o \text{grad}\phi) = \sigma, \tag{4.81}$$

where ϕ is the velocity potential, σ is the mass production and ρ_o is the mass density. In this system, the primal canonical equations are given by

$$\left.\begin{array}{lll} \mathbf{v} = \text{grad}\phi & \text{in } \Omega, & \phi = 0 \text{ on } \Gamma_u \\ \mathbf{p} = \rho_o \mathbf{v} & \text{in } \Omega, & \\ \sigma = -\text{div}\mathbf{p} & \text{in } \Omega, & \mathbf{n} \cdot \mathbf{p} = 0 \text{ on } \Gamma_f, \end{array}\right\} \tag{4.82}$$

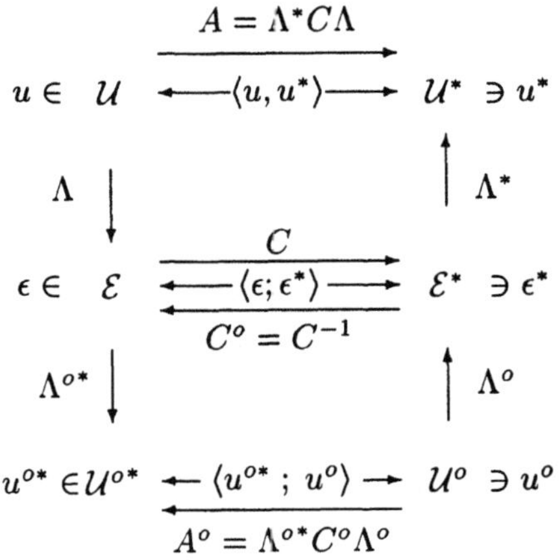

Figure 4.9. Structure of the canonical system and its polar

where $\mathbf{v}$ is the velocity vector, and $\mathbf{p}$ is the momentum density vector. The geometrical operator for this system is $\Lambda = \text{grad}$, and the balance operator associated with the standard bilinear forms (4.68) is $\Lambda^* = -\text{div}$. Let

$$\begin{aligned}
\mathcal{U}_a &= \{\phi \in \mathcal{L}^2(\Omega) |\ \phi(x) = 0 \ \forall x \in \Gamma_u\}, \\
\mathcal{U}_a^* &= \{\phi^* \in \mathcal{L}^2(\Omega) |\ \phi^*(x) = \sigma \ \forall x \in \Omega, \ \phi^*(x) = 0 \ \forall x \in \Gamma_f\}.
\end{aligned}$$

With $\mathcal{E}_a = \mathcal{E} = \mathcal{L}^2(\Omega; \mathbb{R}^3) = \mathcal{E}_a^*$, the system

$$\mathbb{S}_a = \{\langle \mathcal{U}_a, \mathcal{U}_a^* \rangle, \langle \mathcal{E}_a; \mathcal{E}_a^* \rangle; \text{grad}, \rho_o\}$$

is a geometrically linear reflexive canonical system.

Let us consider another canonical problem, namely

$$\left.\begin{array}{lll}
\mathbf{p} = \text{curl}\, \boldsymbol{\psi} & \text{in } \Omega, & \mathbf{n} \times \boldsymbol{\psi} = 0 \text{ on } \Gamma_u^o, \\
\mathbf{v} = \frac{1}{\rho_o} \mathbf{p} & \text{in } \Omega, & \\
\eta = \text{curl}\, \mathbf{v} & \text{in } \Omega, & \mathbf{n} \times \mathbf{v} = 0 \text{ on } \Gamma_f^o,
\end{array}\right\} \tag{4.83}$$

where $\boldsymbol{\psi} \in \mathcal{U}^o = \mathcal{L}^2(\Omega; \mathbb{R}^3)$ is the vector potential, its dual variable $u^{o*} = \eta \in \mathcal{U}^{o*} = \mathcal{L}^2(\Omega; \mathbb{R}^3)$ is called the *abnormality vector*, and Γ_u^o and Γ_f^o are complementary boundaries, satisfying $\Gamma_u^o \cup \Gamma_f^o = \Gamma$, $\Gamma_u^o \cap \Gamma_f^o = \emptyset$. Let

$$\begin{aligned}
\mathcal{U}^o{}_a &= \{\boldsymbol{\psi} \in \mathcal{L}^2(\Omega; \mathbb{R}^3) |\ B^o \boldsymbol{\psi} = \mathbf{n} \times \boldsymbol{\psi}(x) = 0 \ \forall x \in \Gamma_u^o\}, \\
\mathcal{U}^{o*}_a &= \{\boldsymbol{\psi}^* \in \mathcal{L}^2(\Omega; \mathbb{R}^3) |\ \boldsymbol{\psi}^*(x) = \eta \ \forall x \in \Omega, \ \boldsymbol{\psi}^*(x) = 0 \ \forall x \in \Gamma_f^o\}.
\end{aligned}$$

Then, the polar system $\mathbb{S}_a^o = \{\langle \mathcal{U}^o{}_a, \mathcal{U}^{o*}_a \rangle, \langle \mathcal{E}_a^*; \mathcal{E}_a \rangle; \text{curl}, \ 1/\rho_o\}$ of $\mathbb{S}$ is also a geometrically linear reflexive canonical system. The kinematically admissible space $\mathcal{U}_k^o$ and statically admissible space $\mathcal{S}_s^o$ in the polar canonical problem can be defined as

$$\begin{aligned}\mathcal{U}_k^o &= \{\psi \in \mathcal{U}^o{}_a |\ \Lambda^o\psi = \text{curl}\psi \in \mathcal{L}^2(\Omega;\mathbb{R}^3)\},\\ \mathcal{S}_s^o &= \{\mathbf{v} \in \mathcal{L}^2(\Omega;\mathbb{R})|\ \Lambda^{o*}\mathbf{v} = \text{curl}\mathbf{v} \in \mathcal{U}^{o*}_a\}.\end{aligned}$$

Using the identity $(\nabla\times\psi)\cdot\mathbf{v} = \nabla\cdot(\psi\times\mathbf{v}) + \psi\cdot(\nabla\times\mathbf{v})$, and the Gauss-Green theorem, we obtain

$$\int_\Omega (\nabla\times\psi)\cdot\mathbf{v}\mathrm{d}\Omega = \int_{\partial\Omega} \mathbf{n}\cdot(\psi\times\mathbf{v})\mathrm{d}\Gamma + \int_\Omega \psi\cdot(\nabla\times\mathbf{v})\mathrm{d}\Omega.$$

Thus, for any given $\psi \in \mathcal{K}_a^o$ and $\mathbf{v} \in \mathcal{S}_a^o$,

$$\langle \text{curl}\ \psi; \mathbf{v}\rangle = \int_\Omega (\nabla\times\psi)\cdot\mathbf{v}\mathrm{d}\Omega = \int_\Omega \psi\cdot(\nabla\times\mathbf{v})\mathrm{d}\Omega = \langle \psi\ ;\ \text{curl}\ \mathbf{v}\rangle.$$

In this polar system, the geometrical operator is therefore the self-adjoint operator $\Lambda^o = \Lambda^{o*} = \text{curl}$. The fundamental equation reads

$$\text{curl}\left(\frac{1}{\rho_o}\text{curl}\psi(x)\right) = \eta\ \ \forall x \in \Omega. \tag{4.84}$$

By definition, for any given mass production $\sigma(x)$ and $\psi \in \mathcal{U}^o{}_a$, if

$$\langle \text{grad}\ \phi; \text{curl}\ \ \psi\rangle = \langle \phi, \sigma\rangle\ \ \forall \phi \in \mathcal{U}_a,$$

then, $\mathbb{S}^o$ is an iso-source polar system of $\mathbb{S}$. This requires that

$$-\text{div curl}\psi(x) = \sigma(x)\ \ \forall x \in \Omega.$$

Dually, for any given $\eta(x)$ and $\psi(x) \in \mathcal{U}$, if

$$\langle \text{grad}\ \phi; \text{curl}\ \ \psi\rangle = \langle \eta\ ;\ \psi\rangle\ \ \forall \psi \in \mathcal{U}^o{}_a,$$

then, $\mathbb{S}$ is an iso-source polar system of $\mathbb{S}^o$. This requires that the configuration variable satisfies

$$\text{curl div}\phi(x) = \eta(x)\ \ \forall x \in \Omega.$$

Clearly,

$$\langle \text{grad}\ \phi; \text{curl}\ \ \psi\rangle = 0\ \ \forall \phi \in \mathcal{C}^2(\Omega;\mathbb{R}),\quad \psi \in \mathcal{C}^2(\Omega,\mathbb{R}^3). \tag{4.85}$$

This leads to the well-known formula

$$\text{curl grad}\ \ \phi = 0\ \ \forall \phi \in \mathcal{C}^2(\Omega;\mathbb{R}), \tag{4.86}$$

$$\text{div curl}\ \psi = 0\ \ \forall \psi \in \mathcal{C}^2(\Omega,\mathbb{R}^3) \tag{4.87}$$

in vector analysis. Thus, if $\sigma(x) = 0$ and $\eta = 0 \in \mathbb{R}^3$, then, the systems $\mathbb{S}$ and $\mathbb{S}^o$ are mutually polar systems. It is interesting to note that in classical field theory with the standard bilinear forms, the triple operators (grad, div, curl) satisfy the formal adjoint relations

$$\mathrm{grad}^* = -\mathrm{div}, \quad \mathrm{curl}^* = \mathrm{curl}. \tag{4.88}$$

Also, by the classical *Helmholtz decomposition*, any vector field $\mathbf{v}$ may be written in the form

$$\mathbf{v} = \nabla\phi + \mathrm{curl}\psi \tag{4.89}$$

and specifies the choice of ϕ and ψ. This decomposition plays an important role in fluid mechanics and elasticity (cf., e.g., Auchmuty, 1994). ♡

$$\begin{array}{ccc} \phi \in \mathcal{U} & \xrightarrow{A = -\mathrm{div}\rho_o\mathrm{grad}} & \mathcal{U}^* \ni \sigma \\ \Lambda = \mathrm{grad} \downarrow & & \uparrow \Lambda^* = -\mathrm{div} \\ \mathbf{v} \in \mathcal{E} & \underset{C^o = 1/\rho_o}{\overset{C=\rho_o}{\rightleftarrows}} & \mathcal{E}^* \ni \mathbf{p} \\ \Lambda^{o*} = \mathrm{curl} \downarrow & & \uparrow \Lambda^o = \mathrm{curl} \\ \eta \in \mathcal{U}^{o*} & \xleftarrow[A^o = \mathrm{curl}\frac{1}{\rho_o}\mathrm{curl}]{} & \mathcal{U}^o \ni \psi \end{array}$$

Figure 4.10. Canonical system of fluid mechanics and its polar

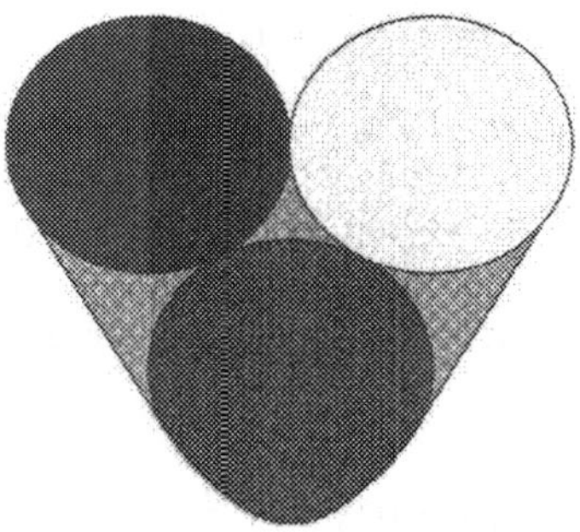

Figure 4.11. Topology III: a way to prove the four color problem?

So teach us to number our days, that we may apply our hearts unto wisdom.
—Psalms, 90.12

4.4 Generalized Tri-Duality Principles

The Emperor of Center was called Hundun (Chaos)

—Chuang Chi

I want to know how God created this world. I am not interested in this or that phenomenon, in the spectrum of this or that element. I want to know His thoughts; the reset are details.

—Albert Einstein

To summarize the results obtained so far, we shall present an informal generalized duality principle for canonical systems. We need first to classify the transformations.

Definition 4.4.1 (Conjugate, Dual and Polar Transformations) Let $\mathbb{S} = \{\langle \mathcal{U}, \mathcal{U}^* \rangle, \langle \mathcal{E}; \mathcal{E}^* \rangle; \Lambda, C\}$ be a geometrically linear canonical system and $\mathbb{S}^o$ its polar system. Let $F(u)$ and $W(\epsilon)$ be given functions, defined on $\mathcal{U}$ and $\mathcal{E}$, respectively.

(D1) If there exists a certain operation "*" such that

$$F^*(u^*) = (F(u))^* \;\; \forall u^* \in \mathcal{U}^*, \quad W^*(\epsilon^*) = (W(\epsilon))^* \;\; \forall \epsilon^* \in \mathcal{E}^* \tag{4.90}$$

are well-defined, then the operator "*" is called the *conjugate transformation*, and the functions F^*, W^* are called the *conjugates* of F and W, respectively.

(D2) If there exists a certain operation "*d*" such that

$$F^d(\epsilon^*) = (F(u))^d, \quad F^d(u^*) = (F(\epsilon))^d \;\; \forall (u^*, \epsilon^*) \in \mathcal{U}^* \times \mathcal{E}^* \tag{4.91}$$

are well-defined on $\mathcal{E}^*$, then, the operator "*d*" is called the *dual transformation*, and the function F^d is called the *dual* of F.

(D3) If there exists a certain operation "*o*" such that for a given $u^o \in \mathcal{U}^o$

$$F^o(u^o) = (F(u))^o \tag{4.92}$$

is a well-defined function on $\mathcal{U}^o$, then, the operator "*o*" is called the *polar transformation* and the function F^o is called the *polar* of F. ◊

According to this definition, the conjugate of a function $\Phi(u, \epsilon) : \mathcal{U} \times \mathcal{E} \to \mathbb{R}$ is a function $\Phi^*(u^*, \epsilon^*) = (\Phi(u, \epsilon))^*$ on $\mathcal{U}^* \times \mathcal{E}^*$. However, the dual of $\mathcal{J}(u, \epsilon)$ is $\Phi^d(\epsilon^*, u^*)$ on $\mathcal{E}^* \times \mathcal{U}^*$.

The operations "$*$", "*d*" and "*o*" can also be used for systems. For example, if $\mathbb{S}_p = \{\mathcal{U}, \mathcal{E}; \Lambda\}$ is a primal subsystem of a geometrically linear canonical system $\mathbb{S}$, then, its conjugate system is

$$\mathbb{S}_p^* = \{\mathcal{U}, \mathcal{E}; \Lambda\}^* = \{\mathcal{U}^*, \mathcal{E}^*; \Lambda^*\},$$

where $\Lambda^* : \mathcal{U}^* \to \mathcal{E}^*$ is the inverse of $\Lambda^d : \mathcal{E}^* \to \mathcal{U}^*$, i.e. $\Lambda^* = (\Lambda^d)^{-1}$. Since, in almost all textbooks, the notation for adjoint operator is Λ^*, instead of Λ^d, in this book, we follow this tradition. However, whenever we use this operator, we should keep in mind that Λ^* is an adjoint operator, not a conjugate operator.

With this definition, the dual transformation of the primal system $\mathbb{S}_p$ is

$$(\mathbb{S}_p)^d = \{\mathcal{U}, \mathcal{E}; \Lambda\}^d = \{\mathcal{E}^d, \mathcal{U}^d; \Lambda^d\}.$$

Since the dual space of a vector space is its conjugate space, we have $\mathcal{U}^d = \mathcal{U}^*$, $\mathcal{E}^d = \mathcal{E}^*$. The dual operator $\Lambda^d : \mathcal{E}^d \to \mathcal{U}^d$ is the adjoint operator of Λ, i.e.

$$\Lambda^d = \Lambda^*.$$

Thus, the dual system $(\mathbb{S}_p)^d$ can be written as

$$\mathbb{S}_d = \{\mathcal{E}^*, \mathcal{U}^*; \Lambda^*\}. \tag{4.93}$$

The polar transformation of the canonical system $\mathbb{S}$ is the polar system

$$\begin{aligned} \mathbb{S}^o &= \{\langle \mathcal{U}, \mathcal{U}^* \rangle, \langle \mathcal{E}; \mathcal{E}^* \rangle; \Lambda, C\}^o \\ &= \{\langle \mathcal{U}^o; \mathcal{U}^{*o} \rangle, \langle \mathcal{E}^o; \mathcal{E}^{*o} \rangle; \Lambda^o, C^o\}. \end{aligned}$$

Thus, $\mathcal{E}^o = \mathcal{E}^*$, $\mathcal{E}^{*o} = \mathcal{E}^{o*} = \mathcal{E}$. Since

$$\langle \Lambda u; \epsilon^* \rangle = \langle \epsilon; \Lambda^o w \rangle,$$

we have

$$A^o = (\Lambda^* C \Lambda)^o = \Lambda^{*o} C^o \Lambda^o = \Lambda^{o*} C^{-1} \Lambda^o.$$

Note that A is a potential operator if and only if $C = C^T$. If A is symmetric ($A = A^T$), then, $\Lambda^{*o} = \Lambda^{o*} = \Lambda^o$.

Definition 4.4.2 (Complementarity Conditions)

(D1) Let $P(u)$ be a given function on a subset $\mathcal{C} \subset \mathcal{U}$, and let $P^c(u^*)$ be a function defined on a subset $\mathcal{C}^c \subset \mathcal{U}^*$. For a given constant c, the condition

$$P(u) + P^c(u^*) = c, \quad \forall u \in \mathcal{C}, \quad \forall u^* \in \mathcal{C}^c \tag{4.94}$$

is called the *(addition) complementarity condition.* For a given function P and a constant c, the function P^c defined by (4.94) is called the *complementary* of the function P associated with c.

(D2) Let $\{(\mathcal{U}, \mathcal{U}^*); \langle *, * \rangle\}$ and $\{(\mathcal{E}, \mathcal{E}^*); \langle *; * \rangle\}$ be two duality pairings, and let $Y : \mathcal{C} \subset \mathcal{U} \to \mathcal{E}$ and $Y^c : \mathcal{C}^c \subset \mathcal{U}^* \to \mathcal{E}^*$ be two mappings defined on the subsets $\mathcal{C}$ and $\mathcal{C}^c$, respectively. For a given constant $c \in \mathbb{R}$, the condition

$$\langle Y(u); Y^c(u^*) \rangle = c \quad \forall u \in \mathcal{C}, \quad \forall u^* \in \mathcal{C}^c \tag{4.95}$$

is called the *(polar) complementarity condition.* For a given $Y(u)$ and a constant c, the mapping $Y^c : \mathcal{C}^c \to \mathcal{E}^*$ defined by (4.95) is called the *(polar) complementary* of the mapping $Y : \mathcal{C} \to \mathcal{E}$ associated with c. ◇

We now generalize the duality results obtained so far, formulating them as an imprecise (but suggestive) "principle".

Generalized Tri-Duality Principle

Let $\mathbb{S}^e = \{\langle \mathcal{U}, \mathcal{U}^* \rangle; A\}$ *be a strictly reflexive system,* $\mathbb{S} = \{\langle \mathcal{U}^*, \mathcal{U} \rangle, \langle \mathcal{E}; \mathcal{E}^* \rangle; \Lambda, C\}$ *a strictly canonical system, and let* $\mathbb{S}^o$ *be its strictly polar system. Then, the following principles hold under certain conditions.*

(T1) **Conjugate-Duality Principle**

If the exists a statement defined on the space $\mathcal{U} \subset \mathbb{S}^e$*, then, its conjugate statement, obtained simply by replacing the concepts by their conjugate is also valid on the conjugate space* $\mathcal{U}^* \subset \mathbb{S}^e$*.*

(T2) **Complementary-Duality Principle**

(i) If there exists a statement in the primal system $\mathbb{S}_p \subset \mathbb{S}$*, then, the dual statement, obtained by replacing the concepts in this statement by their dual concepts, is also valid for the dual system* $\mathbb{S}_d$*.*

(ii) If there exists a statement in the system $\mathbb{S}$*, then, the dual statement, obtained by exchanging the concepts in this statement by their dual concepts, is also valid for the system* $\mathbb{S}$*.*

(T3) **Polar-Duality Principle**

If there exists a statement in $\mathbb{S}$*, then, the polar statement in* $\mathbb{S}^o$*, obtained simply by replacing the concepts in* $\mathbb{S}$ *by their polar concepts is also valid for the polar system.*

Though the generalized tri-duality principle is informal, its applications will make it clear. The precise formulation of this generalized duality principle would require an application of formal logic. The interested reader can find this in monographs on category theory (cf., e.g., MacLane (1971) for mathematicians; Geroch (1985) for physicists; Lawvere and Schanuel (1997) for general readers).

Philosophy is the product of wonder.... And, at the end, when philosophic thought has done its best, the wonder remains.

—Philip Morrison, *Nothing is too Wonderful to be True*, 1995

4.5 Framework for Geometrically Linear Canonical Systems

It is interesting to note that the framework constructed for one-dimensional systems also generally exists for n-dimensional discrete and mathematical physical problems. In this section, we will pass from the electrical networks, through discrete structural mechanics, continuum mechanics to complicated electronic-meganetic system.

4.5.1 Discrete Systems: Electrical Networks and Structural Mechanics

4.5.1.1 Framework for Electrical Networks.

The laws governing an electric network are well-known. They are

(a) Kirchhoff's voltage law: the voltage drops in each branch is equal to the difference in the potentials at two ends of the branch;

(b) Ohm's law: the voltage drop in each branch is proportional to the current in that branch;

(c) Kirchhoff's current law: the net flow of current at each node must be zero.

The associated equations are analogues of the tri-canonical equations we have seen for continuous systems, i.e. the deformation equation, the physical equation and the balance (or equilibrium) equation. This similarity does not arise by an accident. To see this, let us consider a simple electrical network with five branches and four nodes, plus a voltage source at one branch and some current sources at the nodes (see Fig. 4.12).

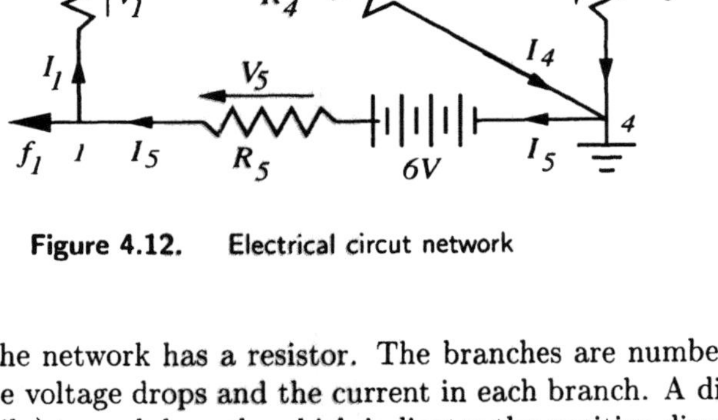

Figure 4.12. Electrical circut network

Each branch of the network has a resistor. The branches are numbered 1-5 and V_i, I_i denote the voltage drops and the current in each branch. A direction is assigned (arbitrarily) to each branch, which indicates the positive direction of the current in it. In the fifth branch, a battery with voltage source (6 volt) is applied. One of the nodes, say the number 4 (chosen arbitrarily), is grounded, i.e. at zero potential. This corresponds to the displacement boundary condition in boundary-value problems of continuum mechanics. Suppose that the other nodes are numbered 1,2,3 and that the potentials and input current sources at these nodes are denoted by ϕ_i, f_i, $(i = 1,2,3)$ respectively. By the three laws for electrical networks, the relations among the potentials $\boldsymbol{\phi} = \{\phi_i\}$, voltage drops $\mathbf{V} = \{V_i\}$, and the current $\mathbf{I} = \{I_i\}$ can be written as the follows (with $\phi_4 = 0 = f_4$).

(a) Kirchhoff's voltage law:

$$\mathbf{V} = \Lambda_t \boldsymbol{\phi} + \mathbf{b} = \begin{pmatrix} 1 & -1 & 0 & 0 \\ 0 & 1 & -1 & 0 \\ 0 & 0 & 1 & -1 \\ 0 & 1 & 0 & -1 \\ -1 & 0 & 0 & 1 \end{pmatrix} \begin{pmatrix} \phi_1 \\ \phi_2 \\ \phi_3 \\ \phi_4 \end{pmatrix} + \begin{pmatrix} 0 \\ 0 \\ 0 \\ 0 \\ 6 \end{pmatrix}. \tag{4.96}$$

(b) Ohm's law:

$$\mathbf{I} = C\mathbf{V} = \begin{pmatrix} 1/R_1 & 0 & 0 & 0 & 0 \\ 0 & 1/R_2 & 0 & 0 & 0 \\ 0 & 0 & 1/R_3 & 0 & 0 \\ 0 & 0 & 0 & 1/R_4 & 0 \\ 0 & 0 & 0 & 0 & 1/R_5 \end{pmatrix} \begin{pmatrix} V_1 \\ V_2 \\ V_3 \\ V_4 \\ V_5 \end{pmatrix}. \tag{4.97}$$

(c) Kirchhoff's current law:

$$\mathbf{f} = \Lambda_t^T \mathbf{I} = \begin{pmatrix} 1 & 0 & 0 & 0 & -1 \\ -1 & 1 & 0 & 1 & 0 \\ 0 & -1 & 1 & 0 & 0 \\ 0 & 0 & -1 & -1 & 1 \end{pmatrix} \begin{pmatrix} I_1 \\ I_2 \\ I_3 \\ I_4 \\ I_5 \end{pmatrix} = \begin{pmatrix} -f_1 \\ f_2 \\ f_3 \\ 0 \end{pmatrix}. \tag{4.98}$$

Thus, the mapping Λ in this problem is the affine operator $\Lambda\boldsymbol{\phi} = \Lambda_t\boldsymbol{\phi} + \mathbf{b}$, where Λ_t is the (Gâteaux) derivative of Λ with respective to $\boldsymbol{\phi}$. The balance operator is Λ_t^T. Combining these three equations, the governing equation for the electrical network is

$$A\boldsymbol{\phi} = \Lambda_t^T C \Lambda \boldsymbol{\phi} = \mathbf{f}, \tag{4.99}$$

where $A = \Lambda_t^T C \Lambda \in \mathbb{R}^{4\times 4}$, which is not symmetric since Λ is not linear. Let

$$\mathbf{K} = \Lambda_t^T C \Lambda_t = \mathbf{K}^T, \quad \bar{\mathbf{f}} = \mathbf{f} - \Lambda_t^T C \mathbf{b}. \tag{4.100}$$

The equilibrium equation (4.99) can be simply written in our favorite symmetric form

$$\mathbf{K}\boldsymbol{\phi} = \bar{\mathbf{f}}. \tag{4.101}$$

For a general electrical network with n nodes and m branches, plus the voltage source $\mathbf{b} \in \mathbb{R}^m$ and current source $\mathbf{f} \in \mathbb{R}^n$, the solution $\bar{\boldsymbol{\phi}} \in \mathbb{R}^n$ of this equation gives the potentials at the nodes and $\bar{\mathbf{I}} = C\Lambda\bar{\boldsymbol{\phi}}$ gives the currents.

$$\begin{array}{ccc}
\phi \text{ or } \mathbf{u} = \mathbf{x} & \xrightarrow{\Lambda_t^T C \Lambda} & \mathbf{x}^* = \mathbf{f} \\
\Big\downarrow \; \Lambda\phi = \Lambda_t\phi + \mathbf{b} & & \Lambda_t^* = \Lambda_t^T \; \Big\uparrow \\
\mathbf{V} \text{ or } \epsilon = \mathbf{y} & \xrightarrow{\quad C \quad} & \mathbf{y}^* = \mathbf{I} \text{ or } \sigma
\end{array}$$

Figure 4.13. Framework for general discrete networks

The two bilinear forms in this discrete system are defined by

$$\langle \phi, \mathbf{f} \rangle := \sum_{i=1}^{n} \phi_i f_i = \phi \cdot \mathbf{f} \;\; \forall \phi, \mathbf{f} \in \mathbb{R}^n, \tag{4.102}$$

$$\langle \mathbf{V}; \mathbf{I} \rangle := \sum_{i=1}^{m} V_i I_i = \mathbf{V} \cdot \mathbf{I} \;\; \forall \mathbf{V}, \mathbf{I} \in \mathbb{R}^m. \tag{4.103}$$

In these notations, the internal and external potential energies are

$$W(\mathbf{V}) = \frac{1}{2}\langle C\mathbf{V}; \mathbf{V} \rangle,$$

$$F(\phi) = \langle \mathbf{f}, \phi \rangle.$$

Let $\mathcal{U}_a \subset \mathbb{R}^n$ be an admissible potential space such that any element $\phi \in \mathcal{U}_a$ satisfies the ground conditions at the given nodes. Then, on $\mathcal{U}_a$, the total potential energy of this network system is

$$P(\phi) = W(\Lambda\phi) - F(\phi) = \frac{1}{2}\langle \phi, \mathbf{K}\phi \rangle - \langle \bar{\mathbf{f}}, \phi \rangle + \frac{1}{2}\langle \mathbf{b}, C\mathbf{b} \rangle. \tag{4.104}$$

Since $\mathbf{K} = \mathbf{K}^T$, the directional derivative of P at $\bar{\phi}$ in the direction $\phi \in \mathcal{U}_a$ is

$$\delta P(\bar{\phi}; \phi) = \langle C\Lambda\bar{\phi}; \Lambda_t\phi \rangle - \langle \mathbf{f}, \phi \rangle = \langle \Lambda_t^T C(\Lambda_t\bar{\phi} + \mathbf{b}) - \mathbf{f}, \phi \rangle \;\; \forall \phi \in \mathcal{U}_a.$$

Hence, the criticality condition

$$\delta P(\bar{\phi}; \phi) = 0 \;\; \forall \phi$$

gives the Euler-Lagrange equation (4.99). Thus, the system

$$\mathbb{S} = \{ \langle \mathbb{R}^n, \mathbb{R}^n \rangle, \langle \mathbb{R}^m; \mathbb{R}^m \rangle; \mathbf{K}, \Lambda_t \}$$

is a linear canonical system. Fig. 4.13 shows the framework structure for network systems.

4.5.1.2 Framework for Structural Mechanics.

In the electrical network given in Fig. 4.12, if we replace the resistors and the current sources by elastic bars and external nodal forces, respectively, a two-dimensional elastic structure is obtained (see Fig. 4.14). Each bar resists a change in its length. The nodal connections are by hinges, which can transmit forces only along the bars. The problem of statical structural mechanics is to find the internal stresses in m bars and the displacement vector of each node. To make the structure stable (no rigid motion), some nodes need to be grounded. The fixed supports will fix two (i.e. horizontal and vertical) displacements, say node 5 in Fig. 4.14; however, rollers can only resist the vertical displacements (say, node 1 in Fig.4.14). At each ungrounded node an external force $\mathbf{f}_i = \{f_i^h, f_i^v\}$ is applied, and a displacement vector $\mathbf{u}^i = \{u_i, v_i\}$ is to be determined.

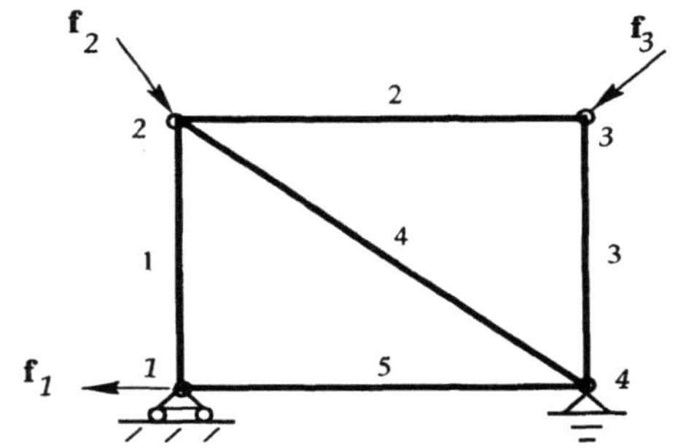

Figure 4.14. Frame structure

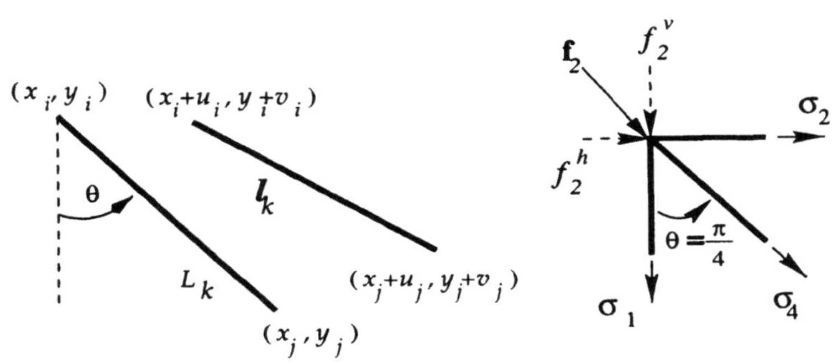

(a) Geometrical deformation. (b) Equilibrium state.

Figure 4.15. Geometrical and equilibrium analysis of the frame

In the structure shown in Fig. 4.14, the mechanical system has five bars ($m = 5$) and four nodes ($N = 4$). Two supports provide $r = 3$ known displacements (boundary conditions). Thus, the system has $n = 2N - r = 5$ degrees of freedom in the displacements. We refer to this displacement vector of n unknowns as $\mathbf{u}$, and the vector of external forces as $\mathbf{f}$. Let $\boldsymbol{\epsilon} = \{\epsilon_i\}$ and $\boldsymbol{\sigma} = \{\sigma_i\}, (i = 1, ..., m)$, denote the strain and stress vectors, respectively. Each component of $\boldsymbol{\epsilon}$ represent the elongation in each bar. If the stress component $\sigma_i > 0$, then, the bar i is in tension; otherwise, the bar i is in compression.

The analogies so far are clear. The displacement and force vectors, $\mathbf{u}, \mathbf{f} \in \mathbb{R}^n$, correspond respectively to potential and current source vectors in an electrical circut, while the strain and stress vectors, $\boldsymbol{\epsilon}, \boldsymbol{\sigma} \in \mathbb{R}^m$, correspond to voltage and current vectors, respectively. Suppose that some bars have *residual strain* (or *pre-elongations*) before the deformation; *the internal voltage source vector* $\mathbf{b}$ now becomes the residual strain vector. The tri-canonical equations for this mechanical network are as follows.

(a) compatibility equation:	$\boldsymbol{\epsilon} = \Lambda_t \mathbf{u} + \mathbf{b}$,
(b) Hooke's law:	$\boldsymbol{\sigma} = C\boldsymbol{\epsilon}$,
(c) equilibrium equation:	$\mathbf{f} = \Lambda_t^* \boldsymbol{\sigma}$.

The affine operator $\Lambda \in \mathbb{R}^{m \times n}$ in this system is called the *deformation matrix*, which can be determined by pure geometrical analysis. On the other hand, the *equilibrium matrix* $\Lambda_t^* \in \mathbb{R}^{n \times m}$ is determined completely by statical analysis.

Consider the k-th bar with original length L_k. The coordinates of its two ends are (x_i, y_i) and (x_j, y_j). After deformation, the two ends moved to (x_i+u_i, y_i+v_i) and (x_j+u_j, y_j+v_j), respectively (Fig.4.15(a)). The length of the stretched bar is given by the Pythagoras formula

$$\ell_k = \sqrt{(L_k \cos\theta + u_i - u_j)^2 + (L_k \sin\theta + v_i - v_j)^2},$$

where

$$L_k = \sqrt{(x_j - x_i)^2 + (y_j - y_i)^2}, \quad \theta = \arctan\frac{|y_j - y_i|}{|x_j - x_i|}.$$

For small deformations, the second order terms $u_i^2/L_k, ...$ can be ignored. Then, the first-order Taylor expansion gives

$$\ell_k = L_k + (u_i - u_j)\cos\theta + (v_i - v_j)\sin\theta.$$

The difference $\epsilon_k = \ell_k - L_k$ is the elongation of the k-th bar. If this bar has pre-elongation ϵ_k^o, then, the total elongation is $\epsilon_k = (u_i - u_j)\cos\theta + (v_i - v_j)\sin\theta + \epsilon_k^o$.

The equilibrium matrix Λ_t^* must be the transpose of Λ_t in the deformation matrix. However, it can be obtained directly by the method of statical equilibrium. Since each node is in equilibrium, the net force on it in each direction must be zero. We assume that all bars are in tension. In Fig. 4.15(b), for the applied external forces f_h^2 and f_v^2, the equilibrium equations are

$$-\sigma_2 - \frac{\sqrt{2}}{2}\sigma_4 = f_h^2,$$

$$\sigma_1 + \frac{\sqrt{2}}{2}\sigma_4 = -f_v^2.$$

Now, for the structure shown in Fig. 4.14, the three equations can be written as follows.

(a) Compatibility equation:

$$\boldsymbol{\epsilon} = \Lambda_t \mathbf{u} + \mathbf{b} = \begin{pmatrix} 0 & 0 & 1 & 0 & 0 \\ 0 & -1 & 0 & 1 & 0 \\ 0 & 0 & 0 & 0 & 1 \\ 0 & -\frac{\sqrt{2}}{2} & \frac{\sqrt{2}}{2} & 0 & 0 \\ -1 & 0 & 0 & 0 & 0 \end{pmatrix} \begin{pmatrix} u_1 \\ u_2 \\ v_2 \\ u_3 \\ v_3 \end{pmatrix} + \begin{pmatrix} b_1 \\ b_2 \\ b_3 \\ b_4 \\ b_5 \end{pmatrix}. \tag{4.105}$$

(b) Hooke's law:

$$\boldsymbol{\sigma} = C\boldsymbol{\epsilon} = \begin{pmatrix} E_1 & 0 & 0 & 0 & 0 \\ 0 & E_2 & 0 & 0 & 0 \\ 0 & 0 & E_3 & 0 & 0 \\ 0 & 0 & 0 & E_4 & 0 \\ 0 & 0 & 0 & 0 & E_5 \end{pmatrix} \begin{pmatrix} \epsilon_1 \\ \epsilon_2 \\ \epsilon_3 \\ \epsilon_4 \\ \epsilon_5 \end{pmatrix}. \tag{4.106}$$

(c) Equilibrium equation:

$$\mathbf{f} = \Lambda_t^T \boldsymbol{\sigma} = \begin{pmatrix} 0 & 0 & 0 & 0 & -1 \\ 0 & -1 & 0 & -\frac{\sqrt{2}}{2} & 0 \\ 1 & 0 & 0 & \frac{\sqrt{2}}{2} & 0 \\ 0 & 1 & 0 & 0 & 0 \\ 0 & 0 & 1 & 0 & 0 \end{pmatrix} \begin{pmatrix} \sigma_1 \\ \sigma_2 \\ \sigma_3 \\ \sigma_4 \\ \sigma_5 \end{pmatrix} = \begin{pmatrix} -f_1^h \\ f_2^h \\ -f_2^v \\ -f_3^h \\ -f_3^v \end{pmatrix}. \tag{4.107}$$

Together, the global equilibrium equation for this structural mechanics system has the same formula as for the electric network, i.e.

$$A\mathbf{u} = \Lambda_t^T C \Lambda \mathbf{u} = \mathbf{f}, \quad \text{or} \quad \mathbf{K}\mathbf{u} = \bar{\mathbf{f}} = \mathbf{f} - \Lambda_t^T C \mathbf{b}, \tag{4.108}$$

where $\mathbf{K} = \Lambda_t^T C \Lambda_t = \mathbf{K}^T$ is called the stiffness matrix in structural mechanics.

In this example, the deformation operator Λ_t is a square matrix ($m = n = 5$), and the equilibrium equation $\Lambda_t^* \boldsymbol{\sigma} = \mathbf{f}$ alone is enough to determine the stresses in the bars. Hooke's law then gives the elongation vector $\boldsymbol{\epsilon} = C^{-1}\boldsymbol{\sigma}$, and the displacements come from $\mathbf{u} = \Lambda_t^{-1}(\boldsymbol{\epsilon} - \mathbf{b})$.

A structure with the property of nonsingular square equilibrium matrix is called *statically determinate*. This requires that $m = n = 2N - r$. The structure is called *statically indeterminate* if $m > n$ and rank $\Lambda_t = n$. In this case, the equation $\Lambda_t^* C \Lambda \mathbf{u} = \mathbf{f}$ has a unique solution. If $m < n$, the system has infinite number of solutions and the structure becomes a mechanism (see Fig. 4.16).

4.5.2 Continuum Mechanics and Stress Functions

Let us now move from discrete systems in finite dimensional spaces to continuum systems in infinite dimensional spaces to see how the same intrinsic framework arises.

Let $\Omega \subset \mathbb{R}^3$ be a simply-connected domain with boundary $\Gamma = \partial\Omega = \Gamma_u \cup \Gamma_f$. For a given body force field $\mathbf{f} = \{f_i\} : \Omega \to \mathbb{R}^3$, and surface traction $\mathbf{t} = \{t_i(\mathbf{x})\}$:

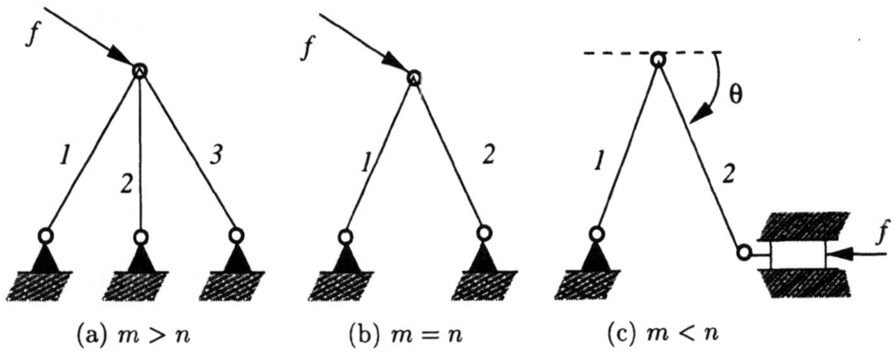

Figure 4.16. Stability of truss: (a) indeterminate, (b) determinate and (c) mechanism

$\Gamma_f \to \mathbb{R}^3$, the displacement $\mathbf{u} = \{u_i(\mathbf{x})\} \in \mathbb{R}^3$ is a vector field from Ω into $\mathbb{R}^3$. The fundamental field equation of the classical elastostatics of homogeneous isotropic bodies is the well-known *Navier equation*

$$\mu\nabla^2\mathbf{u} + (\lambda + \mu)\nabla(\nabla \cdot \mathbf{u}) + \mathbf{f} = 0 \text{ in } \Omega, \tag{4.109}$$

where λ and μ are Lamé constants. Introduction of the strain tensor $\boldsymbol{\epsilon} \in \mathcal{E} = \mathcal{L}^\beta(\Omega; \mathbb{R}^{3\times3})$ and the stress tensor $\boldsymbol{\sigma} \in \mathcal{E}^* = \mathcal{L}^{\beta^*}(\Omega; \mathbb{R}^{3\times3})$, enables this fundamental equation to be written in the canonical forms

$$\begin{array}{ll} \text{(a) strain-displacement relation:} & \boldsymbol{\epsilon} = \frac{1}{2}[\nabla\mathbf{u} + (\nabla\mathbf{u})^t] \quad \text{in } \Omega, \\ \text{(b) the generalized Hooke's law:} & \boldsymbol{\sigma} = \mathbf{H}\boldsymbol{\epsilon} \text{ in } \Omega, \\ \text{(c) equilibrium equation:} & \mathbf{f} = -\nabla \cdot \boldsymbol{\sigma} \quad \text{in } \Omega. \end{array} \tag{4.110}$$

In (4.110) $\mathbf{H}$ is a fourth-order tensor. For homogeneous isotropic materials,

$$\mathbf{H} = \{H_{ijkl}\}, \;\; H_{ijkl} = \lambda\delta_{ij}\delta_{kl} + \mu(\delta_{ik}\delta_{jl} + \delta_{il}\delta_{jk}),$$

and

$$\delta_{ij} = \begin{cases} 1 & \text{if } i = j \\ 0 & \text{if } i \neq j \end{cases}$$

is the Kronecker delta. Let $\mathcal{U} = \mathcal{L}^2(\Omega; \mathbb{R}^3) = \mathcal{U}^*$ and

$$\mathcal{E}_a = \{\boldsymbol{\epsilon} \in \mathcal{E} |\;\; \boldsymbol{\epsilon} = \{\epsilon_{ij}\} = \boldsymbol{\epsilon}^T\}, \;\; \mathcal{E}_a^* = \{\boldsymbol{\sigma} \in \mathcal{E}^* |\;\; \boldsymbol{\sigma} = \{\sigma_{ij}\} = \boldsymbol{\sigma}^T\}.$$

The two standard bilinear forms can be defined by

$$\begin{aligned} \langle \mathbf{u}, \mathbf{u}^* \rangle &:= \int_\Omega \mathbf{u} \cdot \mathbf{f} d\Omega + \int_\Gamma \mathbf{u} \cdot \mathbf{t} d\Gamma = \int_\Omega u_i f_i d\Omega + \int_\Gamma u_i t_i d\Gamma, \\ \langle \boldsymbol{\epsilon}; \boldsymbol{\sigma} \rangle &:= \int_\Omega \boldsymbol{\epsilon} : \boldsymbol{\sigma} d\Omega = \int_\Omega \epsilon_{ij}\sigma_{ij} d\Omega, \end{aligned}$$

where the double dot product is defined by

$$\boldsymbol{\epsilon} : \boldsymbol{\sigma} = \mathrm{tr}(\boldsymbol{\epsilon\sigma}) = \sum_{i=1}^{3}\sum_{j=1}^{3} \epsilon_{ij}\sigma_{ij} = \epsilon_{ij}\sigma_{ij},$$

which denotes the trace of the product of two second-order tensors. Here, Einstein's convention of summation over repeated sub-index is used. The geometrical mapping $\Lambda = \mathrm{grad}_s = \frac{1}{2}[\nabla + \nabla^t] : \mathcal{U} \to \mathcal{E}$ is a symmetrical linear partial differential operator. For the system with a given dead load field $\bar{\mathbf{u}}^* = \{\bar{\mathbf{f}}(\mathbf{x})\ \forall \mathbf{x} \in \Omega,\ \bar{\mathbf{t}}(\mathbf{x})\ \forall \mathbf{x} \in \Gamma_f\}$, and homogeneous displacement boundary condition on Γ_u, the admissible spaces can be defined by

$$\begin{aligned} \mathcal{U}_a &= \{\mathbf{u} \in \mathcal{U} |\ \mathbf{u}(\mathbf{x}) = 0\ \ \forall \mathbf{x} \in \Gamma_u\}, \\ \mathcal{U}_a^* &= \{\mathbf{u}^* \in \mathcal{U}^* |\ \mathbf{u}^*(\mathbf{x}) = \bar{\mathbf{f}}\ \ \forall \mathbf{x} \in \Omega,\ \ \mathbf{u}^*(\mathbf{x}) = \bar{\mathbf{t}}\ \ \forall \mathbf{x} \in \Gamma_f\}. \end{aligned}$$

Let $\mathcal{U}_k \subset \mathcal{U}_a$ be the kinematically admissible space

$$\mathcal{U}_k = \{\mathbf{u} \in \mathcal{U}_a |\ \mathrm{grad}_s \mathbf{u} \in \mathcal{E}\}.$$

Using the identity

$$(\nabla \mathbf{u}) : \boldsymbol{\sigma} = \mathbf{u} \cdot (-\mathrm{div}\boldsymbol{\sigma}) + \mathrm{div}(\mathbf{u} \cdot \boldsymbol{\sigma})$$

and the *Gauss-Green formula*

$$\int_\Omega \mathrm{div}(\mathbf{u} \cdot \boldsymbol{\sigma}) d\Omega = \int_\Gamma \mathbf{n} \cdot (\mathbf{u} \cdot \boldsymbol{\sigma}) d\Gamma,$$

where $\mathbf{n} : \Gamma \to \mathbb{R}^3$ is a unit vector normal outward to the boundary Γ, we have

$$\begin{aligned} \langle \Lambda \mathbf{u}; \boldsymbol{\sigma} \rangle &= \int_\Omega (\mathrm{grad}_s\ \mathbf{u}) : \boldsymbol{\sigma} d\Omega = \int_\Omega \mathbf{u} \cdot (-\mathrm{div}\boldsymbol{\sigma}) d\Omega + \int_{\Gamma_f} \mathbf{u} \cdot (\mathbf{n} \cdot \boldsymbol{\sigma}) d\Gamma \\ &= \langle \mathbf{u}, \bar{\Lambda}^* \boldsymbol{\sigma} \rangle \quad \forall \mathbf{u} \in \mathcal{U}_k, \end{aligned} \tag{4.111}$$

where the adjoint $\bar{\Lambda}^* : \mathcal{E}^* \to \mathcal{U}^*$ of the linear operator $\Lambda = \mathrm{grad}_s$ is defined by

$$\bar{\Lambda}^* = \begin{cases} \Lambda^* = -\mathrm{div} & \text{in } \Omega, \\ \mathbf{n} \cdot & \text{on } \Gamma, \end{cases}$$

and $\Lambda^* = -\mathrm{div}$ is the formal adjoint of Λ. For nonlinear elasticity, the constitutive mapping $\mathbf{H} : \mathcal{E}_a \to \mathcal{E}_a^*$ is one-to-one and the system defined by

$$\mathbb{S}_a = \{\langle \mathcal{U}_a, \mathcal{U}_a^* \rangle, \langle \mathcal{E}_a; \mathcal{E}_a^* \rangle; \mathrm{grad}_s, \mathbf{H}\}$$

is a geometrically linear canonical system. The fundamental equation for this system can be written in the symmetrical abstract form

$$\bar{\Lambda}^* \mathbf{H}(\Lambda \mathbf{u}) = \bar{\mathbf{u}}^* \quad \Rightarrow \quad \begin{cases} -\mathrm{div}\ \mathbf{H}(\mathrm{grad}_s\ \mathbf{u}) = \bar{\mathbf{f}} & \text{in } \Omega, \\ \mathbf{n} \cdot \mathbf{H}\ (\mathrm{grad}_s\ \mathbf{u}) = \bar{\mathbf{t}} & \text{on } \Gamma_f. \end{cases} \tag{4.112}$$

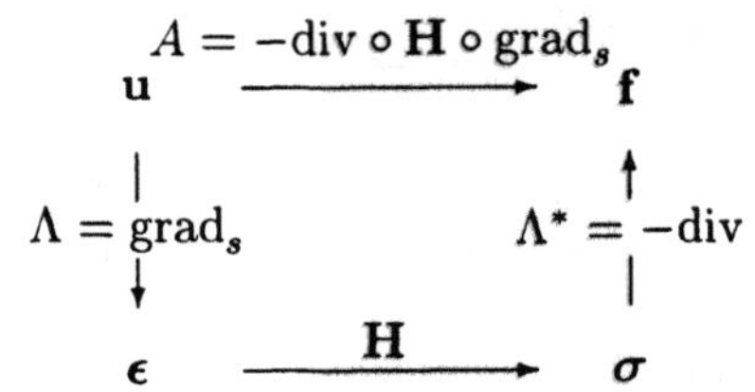

Figure 4.17. Framework structure for classical linear elasticity

For homogeneous and isotropic Hooke's materials, this is Navier's equation (4.109). If $\mathbf{H} = \mathbf{I}$, the identity tensor, then, this abstract equilibrium equation is just Poisson's equation

$$\Lambda^*\Lambda\mathbf{u} = -\Delta\mathbf{u} = \bar{\mathbf{f}} \quad \text{in } \Omega.$$

The canonical forms for the mixed boundary value problem can be written as

$$\begin{aligned} &\epsilon = \Lambda\mathbf{u} = \text{grad}_s\mathbf{u} \quad \text{in } \Omega, \qquad \mathbf{u}(\mathbf{x}) = 0 \quad \text{on } \Gamma_u, \\ &\boldsymbol{\sigma} = \mathbf{H}\epsilon \quad \text{in } \Omega, \\ &\bar{\mathbf{f}} = \Lambda^*\boldsymbol{\sigma} = -\text{div}\,\boldsymbol{\sigma} \quad \text{in } \Omega, \quad \mathbf{n}\cdot\boldsymbol{\sigma} = \bar{\mathbf{t}} \text{ on } \Gamma_f. \end{aligned} \tag{4.113}$$

Fig. 4.17 shows the relations among the variables and operators in this generalized Hookean system.

On introduction of the so-called *tensor stress function* $\psi \in \mathcal{U}^o = \mathcal{L}^2(\Omega; \mathbb{R}^{3\times 3})$, the polar canonical equations associated with (4.113) can be proposed as

$$\begin{aligned} &\sigma = \text{curl}\,\psi \quad \text{in } \Omega, \quad B^o\psi = \psi_0 \quad \text{on } \Gamma_\psi, \\ &\epsilon = \mathbf{H}^{-1}\sigma \quad \text{in } \Omega, \\ &\bar{\psi}^* = \text{curl}\,\epsilon \quad \text{in } \Omega, \quad B^{o*}\epsilon = \epsilon_0 \quad \text{on } \Gamma_\epsilon, \end{aligned} \tag{4.114}$$

where $B^o : \Omega \to \Gamma$ is a boundary operator and B^{o*} is its adjoint, $\Gamma_\psi, \Gamma_\epsilon \subset \Gamma$ such that $\Gamma_\psi \cup \Gamma_\epsilon = \Gamma$, and $\Gamma_\psi \cap \Gamma_\epsilon = \emptyset$, $\bar{\psi}^* \in \mathcal{U}^{o*}$ is a given polar source field. The polar fundamental equation is

$$\text{curl}(\mathbf{H}^{-1}\text{curl}\,\psi) = \bar{\psi}^*. \tag{4.115}$$

In the case that $\bar{\psi}^* = 0 \in \mathcal{U}^{o*}$, the polar equilibrium equation $\Lambda^{o*}\epsilon = 0$ leads to the well-known *Beltrami-Mitchell compatibility condition*

$$\text{curl grad}_s\mathbf{u} = 0.$$

Let

$$\mathcal{U}^o{}_a = \{\psi \in \mathcal{U}^o |\ B^o\psi = \psi_0 \quad \text{on } \Gamma_\psi\},$$

$$\mathcal{U}_a^{o*} = \{\psi^* \in \mathcal{U}^{o*} |\ \psi^*(\mathbf{x}) = \bar{\psi}^* \ \forall \mathbf{x} \in \Omega, \ \psi^* = \epsilon_0 \ \forall \mathbf{x} \in \Gamma_\epsilon\}.$$

The system $\mathbb{S}_a^o = \{\langle \mathcal{U}^o{}_a\ ;\ \mathcal{U}^{o*}_a\rangle, \langle \mathcal{E}_a^*; \mathcal{E}_a\rangle; \text{curl}, \mathbf{H}^{-1}\}$ is a polar system of $\mathbb{S}_a$. By definition, if

$$\langle \text{grad}_s \mathbf{u};\ \boldsymbol{\psi}\rangle = \langle \mathbf{u}, \bar{\mathbf{u}}^*\rangle \quad \forall \mathbf{u} \in \mathcal{U}_k,$$

i.e. if

$$\bar{\Lambda}^* \bar{\Lambda^o} \boldsymbol{\psi} = \begin{cases} -\text{div curl}\ \boldsymbol{\psi} = \bar{\mathbf{f}} & \text{in } \Omega, \\ \mathbf{n} \cdot \text{curl}\ \boldsymbol{\psi} = \bar{\mathbf{t}} & \text{on } \Gamma_f = \Gamma_\psi, \end{cases}$$

then, $\mathbb{S}_a^o$ is an iso-source polar system of $\mathbb{S}$.

Introduction of the *Eddington or permutation tensor* $\mathbf{e} = \{e^{ijk}\}$

$$e^{ijk} = \begin{cases} 1 & i,j,k \text{ are even permutation} \\ -1 & i,j,k \text{ are odd permutation} \\ 0 & \text{otherwise} \end{cases}$$

the relation between the symmetrical stress $\boldsymbol{\sigma} = \boldsymbol{\sigma}^T$ and the tensor stress function $\boldsymbol{\psi}$ can be written as

$$\sigma^{ij} = e^{ims} e^{jkr} \psi_{sr,mk}.$$

We list three well-known tensor stress functions.

I. *Maxwell Tensor Stress Function*

$$\boldsymbol{\psi} = \begin{pmatrix} \psi^1 & 0 & 0 \\ & \psi^2 & 0 \\ Sym & & \psi^3 \end{pmatrix}. \tag{4.116}$$

In this case, the polar configuration equation reads

$$\boldsymbol{\sigma} = \Lambda^o \boldsymbol{\psi} = \begin{pmatrix} \psi^3_{,22} + \psi^2_{,33} & -\psi^3_{,12} & -\psi^2_{,31} \\ & \psi^1_{,33} + \psi^3_{,11} & -\psi^1_{,23} \\ Sym & & -\psi^2_{,11} + \psi^1_{,22} \end{pmatrix}. \tag{4.117}$$

It is easy to check that $\Lambda^{o*}\boldsymbol{\sigma} = \text{curl}\boldsymbol{\sigma} = 0$.

II. *Morera Tensor Stress Function*

$$\boldsymbol{\psi} = \begin{pmatrix} 0 & -\psi^3 & -\psi^2 \\ & 0 & -\psi^1 \\ Sym & & 0 \end{pmatrix}. \tag{4.118}$$

In this case, the tensor stress can be represented in terms of $\boldsymbol{\psi}$ as

$$\boldsymbol{\sigma} = \Lambda^o \boldsymbol{\psi} = \begin{pmatrix} \psi^1_{,23} & -\frac{1}{2}(\psi^1_{,1} + \psi^2_{,2} - \psi^3_{,3})_{,3} & -\frac{1}{2}(\psi^1_{,1} - \psi^2_{,2} + \psi^3_{,3})_{,2} \\ & \psi^2_{,31} & -\frac{1}{2}(-\psi^1_{,1} + \psi^2_{,2} + \psi^3_{,3})_{,1} \\ Sym & & \psi^3_{,12} \end{pmatrix}.$$

III. *Blokh-Krulkov Polar Function.*

Let $\psi \in C^5(\Omega;\mathbb{R})$ be any given real valued function, and $a,b,c \in \mathbb{R}$ are constants. The polar geometrical equation can be defined by

$$\boldsymbol{\sigma} = \Lambda^o \psi = \begin{pmatrix} (b+c)\psi_{,2233} & -c\psi_{,1233} & -b\psi_{,3122} \\ & (c+a)\psi_{,3311} & -a\psi_{,2311} \\ Sym & & (a+b)\psi_{,1122} \end{pmatrix}.$$

For general materials, the constitutive relation $\boldsymbol{\sigma} = \mathbf{H}(\boldsymbol{\epsilon}) = DV(\boldsymbol{\epsilon})$ could be nonlinear, and $V(\boldsymbol{\epsilon})$ is the stored-energy density.

Let $\bar{\mathbf{f}} \in \mathcal{U}_a^*$ be a given external force field and $(\bar{\mathbf{u}}, \bar{\boldsymbol{\sigma}}) \in \mathcal{U}_a \times \mathcal{E}_a^*$ be the corresponding equilibrium state. Then, the virtual work principle leads to the identity

$$\langle \Lambda \bar{\mathbf{u}}; \bar{\boldsymbol{\sigma}} \rangle = \langle \bar{\mathbf{u}}, \bar{\mathbf{f}} \rangle.$$

By the general duality principle, this identity can be separated into two dual principles as follows.

(I1) *Virtual displacement principle*:

$$\langle \Lambda \delta \mathbf{u}; \bar{\boldsymbol{\sigma}} \rangle = \langle \delta \mathbf{u}, \bar{\mathbf{f}} \rangle \ \ \forall \delta \mathbf{u} \in \mathcal{U}_k \Rightarrow \Lambda^* \bar{\boldsymbol{\sigma}} = \bar{\mathbf{f}}.$$

(I2) *Virtual stress principle*:

$$\langle \Lambda \bar{\mathbf{u}}; \delta \boldsymbol{\sigma} \rangle = \langle \bar{\boldsymbol{\epsilon}}; \delta \boldsymbol{\sigma} \rangle \ \ \forall \delta \boldsymbol{\sigma} \in \mathcal{S}_s \Rightarrow \Lambda \bar{\mathbf{u}} = \bar{\boldsymbol{\epsilon}},$$

where

$$\mathcal{S}_s = \{\boldsymbol{\sigma} \in \mathcal{E}^* | -\operatorname{div}\boldsymbol{\sigma}(\mathbf{x}) = \bar{\mathbf{f}} \ \ \forall \mathbf{x} \in \Omega, \ \ \mathbf{n} \cdot \boldsymbol{\sigma}(\mathbf{x}) = \bar{\mathbf{t}} \ \ \forall \mathbf{x} \in \Gamma_f\}$$

is the statically admissible space.

4.5.3 Coupled Electromagnetic Field: Maxwell's equations

Maxwell's equations, which describe electric and magnetic fields in a homogeneous and isotropic medium, are

$$\nabla \cdot [\epsilon(\nabla \phi + \frac{\partial \mathbf{A}}{\partial t})] + \rho = 0, \tag{4.119}$$

$$\frac{\partial}{\partial t}[\epsilon(\nabla \phi + \frac{\partial \mathbf{A}}{\partial t})] + \nabla \times [\frac{1}{\mu} \nabla \times \mathbf{A}] = \mathbf{j}. \tag{4.120}$$

Here ϕ is the scalar potential, $\mathbf{A}$ is the vector magnetic potential, ρ is the charge density, $\mathbf{j}$ is the current density, ϵ is the inductive capacity of the medium, and μ is the permeability of free space. Introduction of two pairs of variables $\{\mathbf{E}, \mathbf{D}\}$ (i.e. the electric field intensity and electric flux density) and $\{\mathbf{B}, \mathbf{H}\}$ (i.e. the magnetic flux density and magnetic field intensity), the complicated Maxwell equations can also be written as the tri-canonical systems of equations

$$\begin{aligned}
&\text{(a) definition equation:} && \begin{pmatrix} \mathbf{E} \\ \mathbf{B} \end{pmatrix} = \begin{pmatrix} \text{grad} & \frac{\partial}{\partial t} \\ 0 & \text{curl} \end{pmatrix} \begin{pmatrix} \phi \\ \mathbf{A} \end{pmatrix}; \\
&\text{(b) constitutive equation:} && \begin{pmatrix} \mathbf{D} \\ \mathbf{H} \end{pmatrix} = \begin{pmatrix} \epsilon & 0 \\ 0 & -\frac{1}{\mu} \end{pmatrix} \begin{pmatrix} \mathbf{E} \\ \mathbf{B} \end{pmatrix}; \\
&\text{(c) equilibrium equation:} && \begin{pmatrix} \rho \\ -\mathbf{j} \end{pmatrix} = \begin{pmatrix} -\text{div} & 0 \\ -\frac{\partial}{\partial t} & \text{curl} \end{pmatrix} \begin{pmatrix} \mathbf{D} \\ \mathbf{H} \end{pmatrix}.
\end{aligned} \tag{4.121}$$

For this four-dimensional space-time dynamical system, we set

$$\mathbf{u} = \begin{pmatrix} \phi \\ \mathbf{A} \end{pmatrix}, \ \ \mathbf{u}^* = \begin{pmatrix} \rho \\ -\mathbf{j} \end{pmatrix},$$

$$\mathbf{e} = \begin{pmatrix} \mathbf{E} \\ \mathbf{B} \end{pmatrix}, \quad \mathbf{e}^* = \begin{pmatrix} \mathbf{D} \\ \mathbf{H} \end{pmatrix}.$$

The two standard bilinear forms between the two parings $(\mathbf{u}, \mathbf{u}^*)$ and $(\mathbf{e}, \mathbf{e}^*)$ are defined respectively as

$$\begin{aligned} \langle \mathbf{u}, \mathbf{u}^* \rangle &= \int_\Omega [\phi\rho - \mathbf{A} \cdot \mathbf{j}] d\Omega, \\ \langle \mathbf{e}; \mathbf{e}^* \rangle &= \int_\Omega [\mathbf{E} \cdot \mathbf{D} + \mathbf{B} \cdot \mathbf{H}] d\Omega. \end{aligned}$$

Using the identity

$$(\nabla \times \mathbf{A}) \cdot \mathbf{H} = \nabla \cdot (\mathbf{A} \times \mathbf{H}) + \mathbf{A} \cdot (\nabla \times \mathbf{H}),$$

and the Gauss-Green theorem, we obtain

$$\int_\Omega (\nabla \times \mathbf{A}) \cdot \mathbf{H} d\Omega = \int_{\partial\Omega} \mathbf{n} \cdot (\mathbf{A} \times \mathbf{H}) d\partial\Omega + \int_\Omega \mathbf{A} \cdot (\nabla \times \mathbf{H}) d\Omega.$$

If $\mathbf{n} \times \mathbf{A}(\mathbf{x}) = \mathbf{0} \ \forall \mathbf{x} \in \partial\Omega$, then

$$\langle \text{curl } \mathbf{A}; \mathbf{H} \rangle = \int_\Omega (\nabla \times \mathbf{A}) \cdot \mathbf{H} d\Omega = \int_\Omega \mathbf{A} \cdot (\nabla \times \mathbf{H}) d\Omega = \langle \mathbf{A}; \text{curl } \mathbf{H} \rangle. \quad (4.122)$$

Let

$$\mathcal{U}_a = \{\mathbf{u} = (\phi, \ \mathbf{A})^t \in C(\Omega; \mathbb{R}^4) | \ \phi(\mathbf{x}) = 0, \ \ \mathbf{A}(\mathbf{x}) = \mathbf{0} \ \ \forall \mathbf{x} \in \partial\Omega\},$$

and

$$\Lambda = \begin{pmatrix} \text{grad} & \frac{\partial}{\partial t} \\ 0 & \text{curl} \end{pmatrix}.$$

Then, for any given $\mathbf{u} \in \mathcal{U}_a$,

$$\begin{aligned} \langle \Lambda\mathbf{u}; \mathbf{e}^* \rangle &= \langle \text{grad}\phi + \frac{d}{dt}\mathbf{A}, \mathbf{D} \rangle + \langle \nabla \times \mathbf{A}, \mathbf{H} \rangle \\ &= \langle \phi, -\text{div}\mathbf{D} \rangle + \langle \mathbf{A}, -\frac{d}{dt}\mathbf{D} + \nabla \times \mathbf{H} \rangle \\ &=: \langle \mathbf{u}, \Lambda^* \mathbf{e}^* \rangle \ \ \forall \mathbf{u} \in \mathcal{U}_a, \end{aligned}$$

where the formal adjoint of Λ is defined by

$$\Lambda^* = \begin{pmatrix} -\text{div} & 0 \\ -\frac{\partial}{\partial t} & \text{curl} \end{pmatrix}.$$

The constitutive operator in this problem is the indefinite matrix

$$C = \begin{pmatrix} \epsilon & 0 \\ 0 & -\frac{1}{\mu} \end{pmatrix}.$$

With these abstract notations, the Maxwell's equations can be written in the symmetric form

$$\Lambda^* C \Lambda \mathbf{u} = \mathbf{u}^*.$$

General speaking, in dynamical systems, the constitutive matrix C is usually indefinite, while in statical systems, C is usually positive definite. More examples can be found in the celebrated textbooks by Strang (1986) and by Oden and Reddy (1976). There we can see that many physical theories show formal similarities due to the existence of a common mathematical structure.

4.6 Framework for Boundary-Value Problems

4.6.1 Sobolev Spaces

Now let us consider partial differential operators in n-dimensional domain. Suppose that Ω is an open, bounded region in $\mathbb{R}^n$ with boundary $\Gamma = \partial\Omega$. We say that Ω belongs to the class C^r if its boundary Γ is an r-times continuously differential manifold of dimension $n-1$ and Ω is locally situated on one side only of Γ. The variable $x = \{x_1, \cdots, x_n\}$ denotes a point in $\Omega \subset \mathbb{R}^n$. If $\alpha = (\alpha_1, \cdots, \alpha_n)$ $(\alpha_i \geq 0)$ is a multi-integer, and $|\alpha| = \sum_{i=1}^n \alpha_i$, we write

$$D^\alpha u = D_1^{\alpha_1} \cdots D_n^{\alpha_n} = \frac{\partial^{|\alpha|} u}{\partial x_1^{\alpha_1} \cdots \partial x_n^{\alpha_n}}. \tag{4.123}$$

We denote by $\mathcal{L}^p(\Omega)$ $(1 \leq p < +\infty)$ the space of real functions $u : \Omega \to \mathbb{R}$. It is a Banach space with norm

$$\|u\|_{\mathcal{L}^p(\Omega)} = \left(\int_\Omega |u(x)|^p \, dx \right)^{1/p}.$$

For an integer $m > 0$, and $1 \leq p \leq +\infty$, the *Sobolev space* of order (m, p), denoted $\mathcal{W}^{m,p}(\Omega)$, is defined as the space of functions on Ω whose derivatives of order $\leq m$ are in $\mathcal{L}^p(\Omega)$, i.e.

$$\mathcal{W}^{m,p}(\Omega) = \{u \in \mathcal{L}^p(\Omega) | \; D^\alpha u \in \mathcal{L}^p(\Omega) \;\; \forall \, |\alpha| \leq m\}. \tag{4.124}$$

It is a Banach space for the Sobolev norm

$$\|u\|_{\mathcal{W}^{m,p}(\Omega)} = \left(\sum_{|\alpha| \leq m} \|D^\alpha u\|^p_{\mathcal{L}^p(\Omega)} \right)^{1/p}. \tag{4.125}$$

It is clear that

$$\mathcal{W}^{m,p}(\Omega) \subset \mathcal{W}^{k,r}(\Omega), \;\; m \geq k, \;\; p \geq r. \tag{4.126}$$

For $p = 2$, we write $\mathcal{H}^m(\Omega) = \mathcal{W}^{m,2}(\Omega)$. The Sobolev spaces then become Hilbert spaces, endowed with the Hilbert norm

$$\langle u, v \rangle_{\mathcal{H}^m(\Omega)} = \sum_{|\alpha| \leq m} \langle D^\alpha u, D^\alpha v \rangle. \tag{4.127}$$

If the open set Ω is regular (e.g., $\Omega \in \mathcal{C}^{m+2}$), then for a given $u \in \mathcal{L}^p(\bar{\Omega})$, the *trace operator* $\gamma : \mathcal{W}^{m,p}(\Omega) \to (\mathcal{L}^p(\Gamma))^m$ is defined by

$$\gamma = (\gamma_0, \cdots, \gamma_{m-1}), \quad \gamma_j u = \frac{\partial^j u}{\partial \nu^j} |_\Gamma, \quad 0 \leq j \leq m-1, \tag{4.128}$$

where ν is the unit vector outward normal to Γ.

We let

$$\mathcal{W}_0^{m,p}(\Omega) = \text{Ker } \gamma. \tag{4.129}$$

Thus,

$$u \in \mathcal{W}_0^{m,p}(\Omega) \quad \Rightarrow \quad \frac{\partial^j u}{\partial \nu^j} |_\Gamma = 0, \quad \forall\, 0 \leq j \leq m-1.$$

The dual spaces of $\mathcal{W}_0^{m,p}(\Omega)$ are the "negative" spaces

$$\mathcal{W}^{-m,q}(\Omega) = (\mathcal{W}_0^{m,p}(\Omega))^* \quad \frac{1}{p} + \frac{1}{q} = 1, \quad \mathcal{H}^{-m}(\Omega) = (\mathcal{H}_0^m(\Omega))^*, \tag{4.130}$$

and

$$\|u\|_{\mathcal{W}^{-m,q}(\Omega)} = \sup_{v \in \mathcal{W}_0^{m,p}(\Omega)} \frac{|\langle u, v \rangle|}{\|v\|_{\mathcal{W}^{m,p}(\Omega)}}, \tag{4.131}$$

where $\langle *, * \rangle$ denotes a bilinear form on $\mathcal{W}^{-m,q}(\Omega) \times \mathcal{W}_0^{m,p}(\Omega)$.

4.6.2 Boundary Value Problems

We now consider boundary-value problems for canonical systems. For the sake of convenience, we let $\mathbb{S} = \{\langle \mathcal{U}, \mathcal{U} \rangle, \langle \mathcal{E}; \mathcal{E}^* \rangle; \Lambda, C\}$ be a geometrically linear system, the configuration space $\mathcal{U} = \mathcal{H} = \mathcal{U}^*$ be a Hilbert space with $m = 1$, and Λ a continuous linear operator from $\mathcal{U}$ into a Hilbert space $\mathcal{E}$ of functions on Ω. For a given source function $f \in \mathcal{F} \subset \mathcal{U}^*$, the fundamental equation is

$$\Lambda^* C \Lambda u(x) = f(x) \quad x \in \Omega. \tag{4.132}$$

In order to consider the boundary conditions, we need to introduce some boundary operators. Let $\partial\mathcal{U}$ denote a space of functions $u(s), s \in \Gamma$, and $\partial\mathcal{E}^*$ a space of functions $\sigma(s)$ defined on the boundary Γ. We assume that any $u \in \mathcal{U}$ and $\sigma \in \mathcal{E}^*$ can be continuously extended so that they agree with the functions $u_o \in \partial\mathcal{U}$ and $\sigma_o \in \partial\mathcal{E}^*$ on Γ. That is, there are trace operators γ_o and δ_o such that

$$\begin{aligned} \gamma_o u = u_o, &\quad \gamma_o : \mathcal{U} \to \partial\mathcal{U}, \\ \delta_o \sigma = \sigma_o, &\quad \delta_o : \mathcal{E}^* \to \partial\mathcal{E}^*. \end{aligned} \tag{4.133}$$

On the product spaces $\partial\mathcal{U}^* \times \partial\mathcal{U}$ and $\partial\mathcal{E}^* \times \partial\mathcal{E}$, we introduce the bilinear pairings

$$\langle *, * \rangle_{\partial\mathcal{U}} : \partial\mathcal{U}^* \times \partial\mathcal{U} \to \mathbb{R}, \quad \langle *; * \rangle_{\partial\mathcal{E}} : \partial\mathcal{E}^* \times \partial\mathcal{E} \to \mathbb{R}.$$

Let B_o be a continuous linear boundary operator mapping $\partial\mathcal{U}$ into $\partial\mathcal{E}$. Its adjoint B_o^* is defined by

$$\langle \sigma_o; B_o u_o \rangle_{\partial\mathcal{E}} = \langle B_o^* \sigma_o, u_o \rangle_{\partial\mathcal{U}} \quad \forall u_o \in \partial\mathcal{U}, \quad \sigma_o \in \partial\mathcal{E}^*. \tag{4.134}$$

Then,

$$B_o : \partial\mathcal{U} \to \partial\mathcal{E}, \quad B_o^* : \partial\mathcal{E}^* \to \partial\mathcal{U}^*, \tag{4.135}$$

and the compositions

$$B = -B_o\gamma_o : \mathcal{U} \to \partial\mathcal{E}, \quad B^* = B_o^*\delta_o : \mathcal{E}^* \to \partial\mathcal{U}^* \tag{4.136}$$

are boundary operators (see Fig. 4.18).

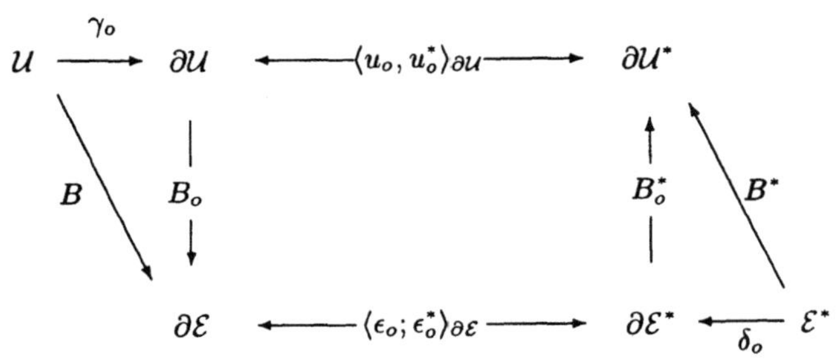

Figure 4.18. Boundary operators and spaces.

In these abstract notations, two types of boundary conditions can be proposed, namely

$$Bu = u_o, \quad u_o \in \partial\mathcal{U}; \tag{4.137}$$

$$B^*\sigma = t_o, \quad t_o \in \partial\mathcal{U}^*. \tag{4.138}$$

The first one is the generalized Dirichlet boundary condition for the operator $A = \Lambda^* C \Lambda$. The second is the generalized Neumann boundary condition for A. These two types of boundary conditions cannot be arbitrarily and simultaneously imposed on all of Γ: if u_o is given on Γ, then, t_o cannot be given; it is $t_o = B_o^*\delta_o\sigma$ on Γ. Conversely, if t_o is given on Γ, then, u_o is $B_o\gamma_o u$ on Γ. However, if the boundary can be split into two parts Γ_u and Γ_f, satisfying

$$\Gamma = \Gamma_u \cup \Gamma_f, \qquad \Gamma_u \cap \Gamma_f = \emptyset,$$

then, the specification of a configuration variable $\bar{u}$ on Γ_u, and a source variable $\bar{t}$ on Γ_f yields the *mixed boundary conditions*

$$Bu = u_o = \begin{cases} \bar{u} & \text{on } \Gamma_u, \\ -B_o\gamma_o u & \text{on } \Gamma_f, \end{cases} \tag{4.139}$$

$$B^*\sigma = t_o = \begin{cases} B_o^*\delta_o\sigma & \text{on } \Gamma_u, \\ \bar{t} & \text{on } \Gamma_f. \end{cases} \tag{4.140}$$

Then, the general statement of such a problem is as follows.

Problem 4.6.1 Given the input data $\bar{f}$ in Ω, $\bar{u}$ on Γ_u, and $\bar{t}$ on Γ_f, find the response $u \in \mathcal{U}$ and associated $\epsilon \in \mathcal{E}$, $\sigma \in \mathcal{E}^*$ such that

$$\begin{array}{lll} \epsilon = \Lambda u & \text{in } \Omega, & Bu = \bar{u} \quad \text{on } \Gamma_u, \\ \sigma = C(\epsilon) & \text{in } \Omega, & \\ \bar{f} = \Lambda^* \sigma & \text{in } \Omega, & B^*\sigma = \bar{t} \quad \text{on } \Gamma_f. \end{array} \tag{4.141}$$

♠

4.6.3 Formal Adjoints and Bilinear Concomitants

We now introduce the product spaces

$$\bar{\mathcal{U}} = \mathcal{U} \times \partial\mathcal{U},; \quad \bar{\mathcal{E}} = \mathcal{E} \times \partial\mathcal{E}$$

and the operators,

$$\bar{\Lambda}_o = \begin{pmatrix} \Lambda & 0 \\ 0 & B_o \end{pmatrix}, \quad \bar{\Lambda}_o^* = \begin{pmatrix} \Lambda^* & 0 \\ 0 & B_o^* \end{pmatrix}. \tag{4.142}$$

Then, if we let $\mathbf{u} = (u, \gamma_o u) \in \bar{\mathcal{U}}$ and $\mathbf{s} = (\sigma, \delta_o \sigma) \in \bar{\mathcal{E}}^*$, and

$$\begin{array}{ll} \langle \bar{\Lambda}_o \mathbf{u}; \mathbf{s} \rangle_{\bar{\mathcal{E}}} & = \langle \Lambda u; \sigma \rangle + \langle B_o \gamma_o u; \delta_o \sigma \rangle_{\partial\mathcal{E}}, \\ \langle \mathbf{u}, \bar{\Lambda}_o^* \mathbf{s} \rangle_{\bar{\mathcal{U}}} & = \langle u, \Lambda^* \sigma \rangle + \langle \gamma_o u, B_o^* \delta_o \sigma \rangle_{\partial\mathcal{U}} \end{array} \tag{4.143}$$

denote duality pairings on $\bar{\mathcal{U}}^* \times \bar{\mathcal{U}}$ and $\bar{\mathcal{E}}^* \times \bar{\mathcal{E}}$, respectively, we have

$$\langle \bar{\Lambda}_o \mathbf{u}; \mathbf{s} \rangle_{\bar{\mathcal{E}}} = \langle \mathbf{u}, \bar{\Lambda}_o^* \mathbf{s} \rangle_{\bar{\mathcal{U}}}. \tag{4.144}$$

This gives the *generalized Gauss-Green formula*

$$\langle \Lambda u; \sigma \rangle = \langle u, \Lambda^* \sigma \rangle + J_\Gamma(u, \sigma) \tag{4.145}$$

for the operator Λ, where Λ^* is the formal adjoint of Λ associated with the duality pairings $\langle *; * \rangle$ and $\langle *, * \rangle$. The *bilinear concomitant* $J_\Gamma(u, \sigma)$ in this general setting reads

$$J_\Gamma(u, \sigma) = \langle \gamma_o u, B^* \sigma \rangle_{\partial\mathcal{U}} + \langle Bu; \delta_o \sigma \rangle_{\partial\mathcal{E}}. \tag{4.146}$$

Similarly to the boundary potentials discussed in Chapter 1, we set

$$J_\sigma(u) = \langle \gamma_o u, \bar{t} \rangle_{\partial\mathcal{U}} =: \langle u, \bar{t} \rangle_{\Gamma_t}, \tag{4.147}$$

$$J_u(\sigma) = \langle \bar{u}; \delta_o \sigma \rangle_{\partial\mathcal{E}} =: \langle \bar{u}; \sigma \rangle_{\Gamma_u}. \tag{4.148}$$

The boundary conditions in Problem 4.6.1 can then be written in the complementary forms

$$Bu = DJ_u(\sigma) = \bar{u} \quad \text{on } \Gamma_u, \tag{4.149}$$

$$B^*\sigma = DJ_\sigma(u) = \bar{t} \quad \text{on } \Gamma_t. \tag{4.150}$$

These are equivalent to the Fenchel-Young equality

$$J_\sigma(u) + J_u(\sigma) = J_\Gamma(u, \sigma) \tag{4.151}$$

on the boundary. Since the constitutive mapping $C : \mathcal{E}_a \subset \mathcal{E} \to \mathcal{E}_a^* \subset \mathcal{E}^*$ is one-to-one in a canonical systems, the stored energy $W : \mathcal{E}_a \to \mathbb{R}$ can be defined generally by

$$W(\epsilon) = \int_\Omega \int C(\epsilon) \mathrm{d}\epsilon \mathrm{d}\Omega. \tag{4.152}$$

It is Gâteaux differentiable on $\mathcal{E}_a$, and the complementary energy is well defined by the Legendre transformation

$$W^c(\sigma) = \langle \epsilon(\sigma); \sigma \rangle - W(\epsilon(\sigma)) \;\; \forall \sigma \in \mathcal{E}_a^*. \tag{4.153}$$

The admissible spaces for the mixed boundary-value problem 4.6.1 are

$$\begin{aligned} \mathcal{U}_a &= \{u \in \mathcal{U} |\; Bu = \bar{u} \;\; \text{on } \Gamma_u\}, \\ \mathcal{U}_a^* &= \{u^* \in \mathcal{U}^* |\; u^* = \bar{f} \;\; \text{in } \Omega, \;\; u^* = \bar{t} \;\; \text{on } \Gamma_f\}, \\ \mathcal{U}_k &= \{u \in \mathcal{U}_a |\; \Lambda u \in \mathcal{E}_a\}, \\ \mathcal{S}_s &= \{\sigma \in \mathcal{E}_a^* |\; \Lambda^* \sigma \in \mathcal{U}_a^*\}. \end{aligned} \tag{4.154}$$

In linear systems, $A = \Lambda^* C \Lambda$ is a linear operator, and the generalized Gauss-Green formula can be written as

$$\langle v, Au \rangle = \langle A^* v, u \rangle + J_\Gamma(u, v), \tag{4.155}$$

where

$$J_\Gamma(u, v) = \langle \gamma_o u, B^* C \Lambda v \rangle_{\partial \mathcal{U}} + \langle Bu; \delta_o C \Lambda v \rangle_{\partial \mathcal{E}}.$$

Clearly, A is formally self-adjoint if and only if C is symmetric.

Example 4.6.1 In $\Omega \subset \mathbb{R}^3$, if we let $\Lambda = \text{grad}$, $\mathcal{U} = \mathcal{H}(\Omega)$ and $\mathcal{E} = \mathcal{H}(\Omega; \mathbb{R}^3)$, then, by using the identity

$$\text{div}(\boldsymbol{\sigma} u) = u \text{div} \boldsymbol{\sigma} + \boldsymbol{\sigma} \text{grad} u$$

and the Gauss-Green formula

$$\int_\Omega \text{div} \boldsymbol{\sigma} \mathrm{d}\Omega = \int_\Gamma \mathbf{n} \cdot \boldsymbol{\sigma} \mathrm{d}\Gamma,$$

where $\mathbf{n} : \Gamma \to \mathbb{R}^3$ is a unit vector normal outward to the boundary Γ, we obtain

$$\langle \Lambda u; \boldsymbol{\sigma} \rangle = -\int_\Omega u \; \text{div} \boldsymbol{\sigma} \mathrm{d}\Omega + \int_\Gamma u(\mathbf{n} \cdot \boldsymbol{\sigma}) \mathrm{d}\Gamma = \langle u, \Lambda^* \boldsymbol{\sigma} \rangle + J_\Gamma(u, \boldsymbol{\sigma}), \tag{4.156}$$

with

$$J_\Gamma(u, \boldsymbol{\sigma}) = \int_\Gamma u(\mathbf{n} \cdot \boldsymbol{\sigma}) \mathrm{d}\Gamma.$$

The formal adjoint of Λ = grad is therefore $\Lambda^* = -$div, and the boundary operator $B_o = \mathbf{n}$ has adjoint $B_o^* = \mathbf{n}\cdot$.

Let $C = 1$, so that $A = \Lambda^*\Lambda = -\Delta = A^*$ is formally self-adjoint. In this case, the generalized Gauss-Green formula is

$$\int_\Omega v(\Delta u)\mathrm{d}\Omega = \int_\Omega (\Delta v)u\mathrm{d}\Omega + \int_\Gamma n \cdot (v\nabla u - u\nabla v)\mathrm{d}\Gamma, \tag{4.157}$$

which is, of course, the classical Gauss-Green formula. The Lagrange identity in this case is

$$v(\Delta u) - (\Delta v)u = \mathrm{div}(v\nabla u - u\nabla v). \tag{4.158}$$

♡

In the notation defined in (4.123), an arbitrary linear operator of order m in n variables can be written as

$$A = \sum_{|\alpha|\le m} a_\alpha(x)D^\alpha, \quad |\alpha| = \sum_{i=1}^n \alpha_i, \tag{4.159}$$

where the coefficients $a_\alpha(x) = a_{|\alpha|}(x_1, \cdots, x_n)$ are arbitrary functions. If $m = n = 2$, then

$$\begin{aligned} A &= \sum_{|\alpha|\le 2} a_\alpha(x)D^\alpha = \sum_{|\alpha|=0} a_\alpha(x)D^\alpha + \sum_{|\alpha|=1} a_\alpha(x)D^\alpha + \sum_{|\alpha|=2} a_\alpha(x)D^\alpha \\ &= a_{00} + a_{10}\frac{\partial}{\partial x_1} + a_{01}\frac{\partial}{\partial x_2} + a_{20}\frac{\partial^2}{\partial x_1^2} + a_{11}\frac{\partial^2}{\partial x_1 \partial x_2} + a_{02}\frac{\partial^2}{\partial x_2^2}. \end{aligned}$$

The corresponding forms of the Gauss-Green formula and the Lagrange identity are, respectively

$$\int_\Omega (vAu - uA^*v)\mathrm{d}\Omega = \int_\Gamma n \cdot J(u,v)\mathrm{d}\Gamma$$

and

$$vAu - uA^*v = \mathrm{div}J(u,v).$$

The formal adjoint A^*, defined by

$$A^*v = \sum_{|\alpha|\le m} (-1)^{|\alpha|}D^\alpha(a_\alpha v), \tag{4.160}$$

is obtained by integration by parts. If $A = A^*$, A is said to be formally self-adjoint. In general, the expression for the vector bilinear conjunct J is somewhat complicated, so we present some specific examples.

Example 4.6.2 For $m = 4$, $A = \Delta\Delta$, A is formally self-adjoint and the Lagrange identity

$$v\Delta\Delta u - u\Delta\Delta v = \mathrm{div}[v\mathrm{grad}\Delta u - u\mathrm{grad}\Delta v + (\Delta v)\mathrm{grad}u - (\Delta u)\mathrm{grad}v].$$

holds for all $u, v \in C^4(\Omega)$. ♡

Example 4.6.3 In the space-time space $(t_1, t_2) \times \Omega \subset \mathbb{R}^{n+1}$, $x = (t, x_1, \cdots, x_n)$, the diffusion operator is

$$A = \frac{\partial}{\partial t} - \Delta, \quad \Delta = \left(\frac{\partial^2}{\partial x_1^2} + \cdots + \frac{\partial^2}{\partial x_n^2} \right). \tag{4.161}$$

Its formal adjoint is

$$A^* = -\frac{\partial}{\partial t} - \Delta. \tag{4.162}$$

We have

$$v(u_t - \Delta u) - u(-v_t - \Delta v) = \text{div} J(u, v), \tag{4.163}$$

with

$$J(u, v) = n_t uv + (u \,\text{grad} v - v \,\text{grad} u), \quad \text{grad} u = (u_{x_1}, \cdots, u_{x_n}),$$

where n_t is a unit vector in the t direction. The Gauss-Green formula becomes

$$\int_{t_1}^{t_2} \int_\Omega (vAu - uA^*v) \, dxdt = \int_\Omega (uv)|_{t=t_1}^{t_2} \, dx + \int_{t_1}^{t_2} \int_\Omega \left(u \frac{\partial v}{\partial n_x} - v \frac{\partial u}{\partial n_x} \right) \, dxdt. \quad ♡$$

Example 4.6.4 The wave operator

$$\square^2 := \frac{\partial^2}{\partial t^2} - \Delta \tag{4.164}$$

is formally self-adjoint under the standard bilinear forms, and the Lagrange identity becomes

$$v\square^2 u - u\square^2 v = \text{div}[n_t(vu_t - uv_t) + u \,\text{grad} v - v \,\text{grad} u]. \tag{4.165}$$

The Gauss-Green formula for this operator is then

$$\begin{aligned} \int_{t_1}^{t_2} \int_\Omega (v\square^2 u - u\square^2 v) \, dxdt \;\; &= \;\; \int_{t_1}^{t_2} \int_\Omega \left(u \frac{\partial v}{\partial n_x} - v \frac{\partial u}{\partial n_x} \right) \, dxdt \\ &\quad + \int_\Omega (vu_t - uv_t)|_{t=t_1}^{t_2} \, dx. \quad ♡ \end{aligned}$$

4.7 Nonlinear Systems and Commentary

4.7.1 Superconductivity: Ginzburg-Landau Equation

Let $\Omega \subset \mathbb{R}^2$ be a smooth, bounded simply-connected domain with boundary $\partial\Omega$. The configuration $u : \Omega \to \mathbb{R}$ is a real-valued function. On the boundary, u is prescribed. Then, the Ginzburg-Landau equation in superconductivity can be written as

$$\begin{aligned} &\Delta u(x) + \tfrac{1}{\mu^2} u(x)(1 - u(x)^2) = 0 \;\; \forall x \in \Omega, \\ &u(x) = \bar{u}(x) \;\; \forall x \in \partial\Omega. \end{aligned} \tag{4.166}$$

The total potential energy associated with this nonlinear partial differential equation is the nonconvex energy

$$P(u) = \int_\Omega \frac{k_0}{2}|\nabla u|^2 d\Omega + \int_\Omega \frac{1}{4\mu^2}(1-u^2)^2 d\Omega \tag{4.167}$$

where

$$F(u) = \int_\Omega \frac{1}{4\mu^2}(1-u^2)^2 d\Omega,$$

is a double-well energy. In this problem, if we define a generalized finite deformation strain vector $\boldsymbol{\xi}$ by

$$\boldsymbol{\xi} = \Lambda(u) = (\text{grad}\, u\ ,\ \frac{1}{2}u^2)^T = (\boldsymbol{\epsilon},\ \xi)^T, \tag{4.168}$$

then the stored-energy density is the quadratic function

$$\bar{V}(\boldsymbol{\xi}) = \frac{k_0}{2}\boldsymbol{\epsilon}\cdot\boldsymbol{\epsilon} + \frac{1}{2\mu^2}\xi^2 = \frac{1}{2}\boldsymbol{\xi}^t C \boldsymbol{\xi}$$

of $\boldsymbol{\xi} = (\boldsymbol{\epsilon},\ \xi)^T$, where

$$C = \begin{pmatrix} k_0 & 0 \\ 0 & \frac{1}{\mu^2} \end{pmatrix}.$$

Hence, a Kirchhoff-type stress $\boldsymbol{\varsigma}$ can be defined by

$$\boldsymbol{\varsigma} = (\boldsymbol{\sigma},\ \varsigma)^T = DV(\boldsymbol{\xi}) = \left(k_0\boldsymbol{\epsilon}, \frac{1}{\mu^2}\xi\right)^T.$$

Since Λ is a quadratic operator, $\boldsymbol{\xi}$ can be considered as a Green-type strain vector, and its dual variable $\boldsymbol{\varsigma}$ as a Kirchhoff-type stress. The directional derivative of $\Lambda(u)$ at $\bar{u}$ in the direction u is given simply as

$$\delta\boldsymbol{\xi}(\bar{u};u) = (\text{grad}\, u\ , \bar{u}u)^T = \Lambda_t(\bar{u})u. \tag{4.169}$$

Let

$$\mathcal{U}_k = \{u \in \mathcal{L}^4(\Omega)|\ \Lambda(u) \in \mathcal{L}^2(\Omega;\mathbb{R}^3),\ \ u(x) = 0\ \ \forall x \in \partial\Omega\}$$

be the admissible space. From the Gauss-Green formula,

$$\begin{aligned} \langle \Lambda_t(\bar{u})u; \boldsymbol{\varsigma}\rangle &= \int_\Omega [(\boldsymbol{\sigma})\cdot(\text{grad} u) + \varsigma\bar{u}u] d\Omega \\ &= \int_\Omega (-\text{div}\boldsymbol{\sigma} + \varsigma\bar{u})u d\Omega + \int_{\partial\Omega} \mathbf{n}\cdot\boldsymbol{\sigma} u d\Gamma \\ &= \langle u, \Lambda_t^*(\bar{u})\boldsymbol{\varsigma}\rangle\ \ \forall u \in \mathcal{U}_k, \end{aligned}$$

it can be seen that the adjoint Λ_t^* of Λ_t is defined as

$$\Lambda_t^*(\bar{u})\boldsymbol{\varsigma} = -\text{div}\boldsymbol{\sigma} + \bar{u}\varsigma.$$

The fundamental equation in this nonlinear problem can be written in the same form as for other in nonconvex problems discussed in this section, namely

$$\Lambda_t^*(\bar{u})C\Lambda(\bar{u}) = 0. \tag{4.170}$$

Thus, the Ginzburg-Landau equation belongs to a geometrically nonlinear canonical system.

4.7.2 Commentary and Bibliographic Notes

The development of $\Lambda^*\Lambda$-operator theory was apparently initiated by von Neumann in 1932, and was subsequently extended and put into a more general setting in the studies of complementary variational principles by Rall (1969), Arthurs (1970, 1980), Noble and Sewell (1972), Tonti (1972), Oden and Reddy (1974, 1983) and Sewell (1987). In the excellent textbook by Strang (1986), the tri-canonical form $\Lambda^* C \Lambda$ can be seen through continuous theories to discrete systems. In biology systems, similar framework also exists (see Novak and Tyson, 1993). Duality theory for semigroups were developed as early as 1955 by Philipps (see Hille and Philipps, 1965). Recent monograph by Neerven (1992) gives a good account of modern developments on the adjoint of a semigroup of linear operators. Generalization is made to Volterra integral equations by Jung (1997).

In this chapter, we presented a further generalization of the $\Lambda^* C \Lambda$-operator theory in classical convex systems. By identifying the intrinsic constitutive relations of systems and by using the Legendre duality theory, a fully nonlinear canonical system theory has been proposed and clarified. The relationship between the canonical system and its various generalizations are depicted symbolically by means of diagrams. The potentially powerful sequential dual canonical transformation method can be used to solve many nonconvex, nonsmooth problems in differential fields.

The imprecise but suggestive generalized tri-duality principle proposed in the central part of this book possesses certain significance in methodology and scientific philosophy. The original version of this principle was proposed in the author's Ph.D. thesis as a "philosophical by-product" of the duality theory for nonsmooth elastoplastic systems and pan-penalty finite element methods (Gao, 1986a). Applications were made to general systems (Gao, 1986b, 1996). This principle can be used to check, to construct and to predict conjugate, complementary and polar statements of the primal ones in general systems. For example, one of the important current mathematical progress in enumerative geometry and theoretical physics is the so-called *mirror principle, or mirror symmetry* (cf., e.g., Candelas *et al*, 1991; Lian *et al*, 1997), which asserts that a certain physical theory defined on a three-dimensional *Calabi-Yau manifold* is equivalent to another theory defined on its "mirror manifold". This principle is actually a kind of polar-duality principle. Also as we know that the Ghoussoub's duality theorem (Ghoussoub, 1993) in critical point theory is only a the right-saddle type minimax duality theorem. By the complementary-duality principle, it can be predicted that there surely exists a certain tri-duality theorem associated with the Mountain Pass Theorem and its dual statement. Interested readers can certainly find substantial examples through the related references and their own research subjects. Here we list only a few works on duality theory in different areas.

Abstract algebra and analysis: Weyl (1946), Hofmann (1970), Nakagami and Takesaki (1979), Constantinescu (1980), Huang *et al* (1989), Huang (1990), Olver (1995), and Elliott (1997).

Applied mathematics: Arthurs (1980), Strang (1986), Sewell (1987).

Approximation-transforming theory and pansystem methodology: Wu (1984, 1987).

Category theory: Kaijser and Pelletier (1986), Lawvere and Schanuel (1997).

Continuum mechanics: Oden and Reddy (1983), Ogden (1984), Temam (1985).

Economics: Blackorby *et al* (1978), Cornes (1992), Aubin (1993), and Yuan *et al* (1998).

Dynamical systems: (1) for mathematicians, see Mawhin and Willem (1989), Ekeland (1990), Marsden and Ratiu (1994); (2) for engineers, see Bui (1992), Tabarrok and Rimrott (1994), Yoshimura (1995).

General symmetry and complementarity: Weyl (1952), Gruber *et al* (1980-93), Kuyk (1977), Hargittai (1986, 1989), Bluman and Kumei (1989), Bunch (1989), Rosen (1995), Olver (1995).

Mathematical programming: Walk (1989), Hiriart-Urruty and Lemaréchal (1993), Du and Pardalos (1995), Wright (1997), Kaplan (1999).

Numerical algorithms and computations: Strang and Fix (1973), Glowinski, Lions and Trémolières (1981), Wright (1997).

Physics: Yang (1984), Geroch (1985).

Stochastic linear and dynamic programming: Klein Haneveld (1986).

Variational analysis and methods: Ekeland and Temam (1976), Logan (1977), Cottle *et al* (1980), Ghoussoub (1993), Rockafellar and Wets (1998).

The whole world recognizes the beautiful as the beautiful, yet this is only the ugly;
The whole world recognizes the good as the good, yet this is only the bad.
Thus, Something and Nothing produce each other;
The difficult and the easy complementary each other;
The long and the short off-set each other;
The high and the low incline towards each other;
Note and sound harmonize with each other;
Before and after follow each other.

—Lao Chi, II

Nature is the realization of the simplest conceivable mathematical ideas. I am convinced we can discover by means of purely mathematical constructions the concepts and laws connecting them with each other, which furnish the key to understanding of natural phenomena

—Albert Einstein, 1933

III Duality in Canonical Systems

5 DUALITY IN GEOMETRICALLY LINEAR SYSTEMS

I have hardly every known a mathematician who was able to reason.
—Plato, 370 BC

I love mathematics not only because it is applicable to technology but also because it is beautiful.
— Rósza Péter

A great science is mathematics, but mathematicians are often only blockheads.
—George Christoph Lichtenberg, 1799

It is true that a mathematician, who is not somewhat of a poet, will never be a perfect mathematician.
—Karl Weierstrass, 1815-1897

Mathematical laws governing nature are sources of the symmetry in nature.
—H. Weyl, 1952

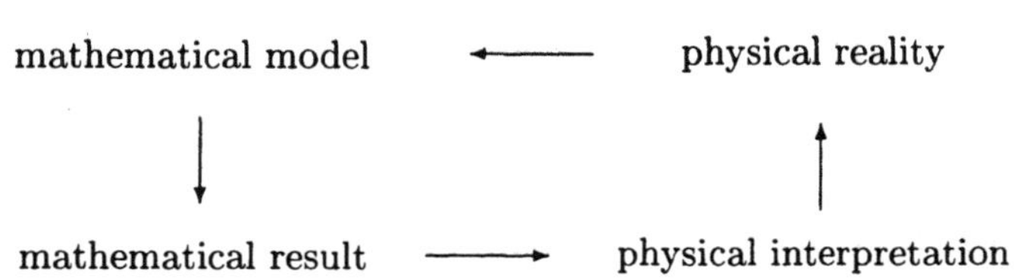

Figure 5.1. Duality in feedback mathematical physics system: E. Zeidler's scheme (1988)

This chapter presents some important mathematical duality theories in geometrically linear, nonconvex and nonsmooth variational problems. By introduction of a so-called indicator functional of the feasible sets, many different problems subject to equality or inequality constraints can be put in a unified framework. The intrinsic relations among these well-known duality theories are revealed and the duality theory in geometrically linear systems is presented in a systematical way. The goal of this chapter is construction not proof. For readers who are seriously interested in the mathematical details of these topics, we refer to the some well-known books by Ekeland and Temam (1976), Zeidler (1988), Ekeland (1990), Aubin (1993) as well as the recent monograph by Motreanu and Panagiotopoulos (1999) for details. However, readers who is not interested in these mathematical details can certainly omit this chapter and continue to the next chapter without difficulty.

5.1 Extended Variational Problems and Fenchel Duality Theory

Variational problems and duality theory can be considerably enriched by allowing functionals defined on a space $\mathcal{U}$ to take on the values $\pm\infty$. Thus, we will frequently consider minimization problems in which $P : \mathcal{U} \to \bar{\mathbb{R}}$, where $\bar{\mathbb{R}} = \mathbb{R} \cup \{-\infty\} \cup \{+\infty\}$ is the extended real line. We also use the notations $\overrightarrow{\mathbb{R}} = \mathbb{R} \cup \{+\infty\}$, $\overleftarrow{\mathbb{R}} = \mathbb{R} \cup \{-\infty\}$. The domain of a real-valued functional $P_0 : \mathcal{K} \subset \mathcal{U} \to \mathbb{R}$ can be extended by defining a new functional P according to

$$P(u) = \begin{cases} P_0(u) & u \in \mathcal{K} \\ +\infty & u \notin \mathcal{K}, \end{cases} \tag{5.1}$$

which is defined on all of $\mathcal{U}$. By this definition, a constrained optimization problem

$$P_0(u) \to \min \quad s.t. \ u \in \mathcal{K},$$

can be written as the unconstrained optimization problem

$$P(u) \to \min \quad \forall u \in \mathcal{U}.$$

In this section, we shall discuss briefly the main results for constrained variational problems, Fenchel duality theory and calculus on Banach spaces. For further references we cite Zeidler (1984), Struwe (1996) and Rockafellar and Wets (1998).

5.1.1 Extended Functionals and Semicontinuities

In the following, we let $P : \mathcal{U} \to \bar{\mathbb{R}}$ be a real-valued functional defined on a linear space $\mathcal{U}$. We list some definitions and notations in convex analysis.

Definition 5.1.1 (Effective domain, Graph and Epigraph)

(D1) The *effective domain* of P , written as dom P, is a subset of $\mathcal{U}$ defined by

$$\text{dom} P := \{u \in \mathcal{U} | \quad -\infty < P(u) < \ +\infty \ \}. \tag{5.2}$$

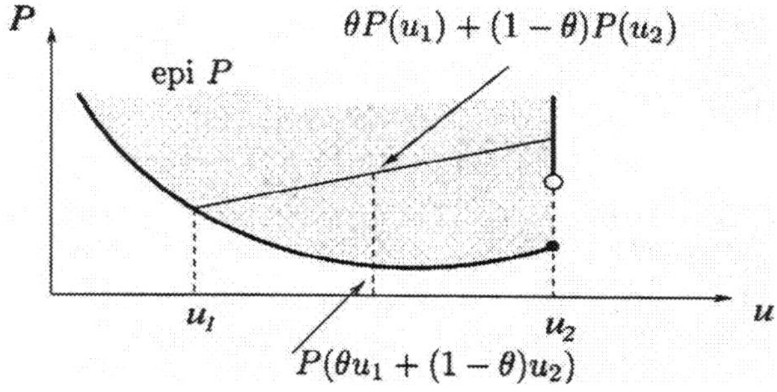

Figure 5.2. Convex functional and its epigraph

(D2) The *graph* of P is a subset of $\mathcal{U} \times \mathbb{R}$ defined by

$$\text{graph}P := \{(v, \alpha) \in \mathcal{U} \times \mathbb{R} | \ \alpha = P(v)\}. \tag{5.3}$$

(D3) The *epigraph* of P is a subset of points above or on the graph of P, defined by

$$\text{epi}P := \{(v, \alpha) \in \mathcal{U} \times \mathbb{R} | \ \alpha \geq P(v)\}. \tag{5.4}$$

◊

The projection of epi P on $\mathcal{U}$ is none other than dom P. For the case in which $\mathcal{U} = \mathbb{R}$, the epigraph is the area above the curve representing the graph of the functional (see Fig. 5.2).

The notion of epigraph can facilitate conversion of many statements on convex functionals into statements on convex sets.

Theorem 5.1.1 *A functional $P : \mathcal{U} \to \overrightarrow{\mathbb{R}}$ is convex if and only if its epigraph is a convex set.*

Proof: Assuming that P is convex, we shall prove that epi$P$ is a convex set. Let (u, a) and (v, b) be two points in epi$P$. Then, by the convexity of P, we have

$$P(\theta u + (1 - \theta)v) \leq \theta P(u) + (1 - \theta)P(v) \leq \theta a + (1 - \theta)b,$$

so that the point $((\theta u + (1 - \theta)v), \theta a + (1 - \theta)b) \in$ epi P for all $\theta \in [0, 1]$ and the set epi$P$ is convex.

Conversely, let $(u, P(u))$ and $(v, P(v)) \in$ epi P; then, $((\theta u + (1 - \theta)v), (\theta P(u) + (1 - \theta)P(v)) \in$ epi P, $\forall \theta \in [0, 1]$, and this shows that P is convex. □

Definition 5.1.2 (Indicator functional) If $\mathcal{K}$ is a subset of $\mathcal{U}$, the functional from $\mathcal{U}$ to $\overrightarrow{\mathbb{R}}$ defined by

$$\mathcal{I}_{\mathcal{K}}(u) := \begin{cases} 0 & \text{if } u \in \mathcal{K}, \\ +\infty & \text{if } u \notin \mathcal{K}, \end{cases} \tag{5.5}$$

is called the *indicator functional* $\mathcal{I}_{\mathcal{K}}$ of $\mathcal{K}$. ◇

Clearly, we have

$$\text{dom } \mathcal{I}_{\mathcal{K}} = \mathcal{K}, \quad \text{epi } \mathcal{I}_{\mathcal{K}} = \mathcal{K} \times \overrightarrow{\mathbb{R}},$$

and the indicator functional $\mathcal{I}_{\mathcal{K}}$ is convex if and only if $\mathcal{K}$ is a convex set. Thus, the study of convex sets is equivalent to the study of convex functionals. If $\{P_i\}(i \in \mathbb{I} := \{1, 2, \cdots\})$ is any family of convex functionals of $\mathcal{U}$ into $\overrightarrow{\mathbb{R}}$, then their pointwise supremum $P = \sup_{i \in \mathbb{I}} P_i$ is convex.

Remark 5.1.1 If P and G are convex functionals from $\mathcal{U}$ to $\overrightarrow{\mathbb{R}}$, then $\alpha P + \beta G$ is convex for any $\alpha, \beta \geq 0$. If $P(u) = -G(u) = \pm\infty$, we stipulate that $(P+G)(u) = +\infty$.

Dually, if P and G are concave functionals from $\mathcal{U}$ to $\overleftarrow{\mathbb{R}}$, then $\alpha P + \beta G$ is concave for any $\alpha, \beta \geq 0$. If $P(u) = -G(u) = \pm\infty$, we stipulate that $(P + G)(u) = -\infty$. ■

In particular, if $P : \mathcal{U} \to \mathbb{R}$ is a linear functional, graph P and epi P have some important geometrical properties.

Definition 5.1.3 (Affine Hyperplane, Half-Spaces and Separation)
Let $l : \mathcal{U} \to \mathbb{R}$ be a nonzero linear functional and let $\alpha \in \mathbb{R}$.
(D1) An *affine hyperplane* $\mathcal{H}_\alpha$ is a set of the form

$$\mathcal{H}_\alpha := \{u \in \mathcal{U} \mid l(u) = \alpha\}$$

(D2) The *open half spaces* bounded by $\mathcal{H}_\alpha$ are

$$\check{\mathcal{H}}_\alpha := \{u \in \mathcal{U} | \ l(u) > \alpha\}, \ \ \hat{\mathcal{H}}_\alpha := \{u \in \mathcal{U} | \ l(u) < \alpha\}.$$

The *closed half spaces* bounded by $\mathcal{H}_\alpha$ are

$$\text{cl}\check{\mathcal{H}}_\alpha := \{u \in \mathcal{U} | \ l(u) \geq \alpha\}, \ \ \text{cl}\hat{\mathcal{H}}_\alpha := \{u \in \mathcal{U} | \ l(u) \leq \alpha\}.$$

(D3) An affine hyperplane H_α is said to *separate the sets* $\mathcal{K}, \mathcal{M} \subset \mathcal{U}$ if

$$l(u) \leq \alpha \ \ \forall u \in \mathcal{K}, \quad l(u) \geq \alpha \ \ \forall u \in \mathcal{M}.$$

If the strict forms of the above inequalities hold, then the sets $\mathcal{K}$ and $\mathcal{M}$ are *strictly separated* by the affine hyperplane $\mathcal{H}_\alpha$. ◇

It is obvious that both affine hyperplanes and half-spaces are convex subsets, and that the affine hyperplane $\mathcal{H}_\alpha$ is closed if and only if the linear functional l is continuous. If $\mathcal{U}$ is a normed space, and $\mathcal{U}^*$ is its dual, then the linear functional $l(u)$ can be written as $l(u) = \langle u, u^* \rangle$ for some $u^* \in \mathcal{U}^*$. A geometrical version of the *Hahn-Banach* Theorem A.3.2 can now be given as:

Theorem 5.1.2 (Hahn-Banach) *Let $\mathcal{U}$ be a Banach space, and suppose $\mathcal{K}$ and $\mathcal{M}$ are nonempty, disjoint, and convex subsets of $\mathcal{U}$.*

(T1) If $\mathcal{K}$ is open, then there exists a closed affine hyperplane that separates $\mathcal{K}$ and $\mathcal{M}$.

(T2) If $\mathcal{K}$ is closed and $\mathcal{M}$ is compact, then there exists a closed affine hyperplane that strictly separates $\mathcal{K}$ and $\mathcal{M}$.

(T3) Every closed convex set is the intersection of the closed half spaces that contain it.

An important geometrical concept is that of a cone in a linear space.

Definition 5.1.4 (Cone, Proper Cone and Conjugate Cones)

(D1) A subset $\mathcal{C}$ of a linear space $\mathcal{U}$ is a *cone* if and only if $\lambda\mathcal{C} \subset \mathcal{C}\ \forall\lambda > 0$.

(D2) $\mathcal{C}$ is a cone with vertex at the origin if and only if $v \in \mathcal{C}$ implies $\lambda v \in \mathcal{C}\ \forall\lambda \geq 0$. A cone $\mathcal{C}$ is said to be *proper* if $\mathcal{C}$ is convex with vertex at the origin.

(D3) Let $\mathcal{K}$ be a set in a normed space $\mathcal{U}$. The sets

$$\mathcal{K}^{\oplus} = \{u^* \in \mathcal{U}^* |\ \langle u, u^*\rangle \geq 0\ \forall u \in \mathcal{K}\}, \tag{5.6}$$

$$\mathcal{K}^{\ominus} = \{u^* \in \mathcal{U}^* |\ \langle u, u^*\rangle \leq 0\ \forall u \in \mathcal{K}\} \tag{5.7}$$

are called the *positive and negative conjugate cones* of $\mathcal{K}$, respectively. ◇

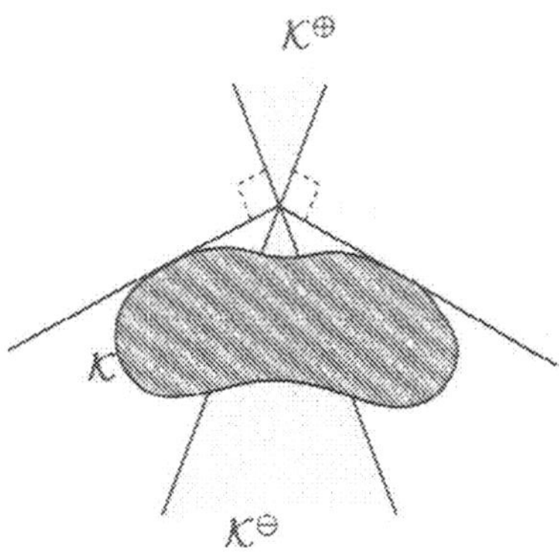

Figure 5.3. Set $\mathcal{K}$ and its conjugate cones.

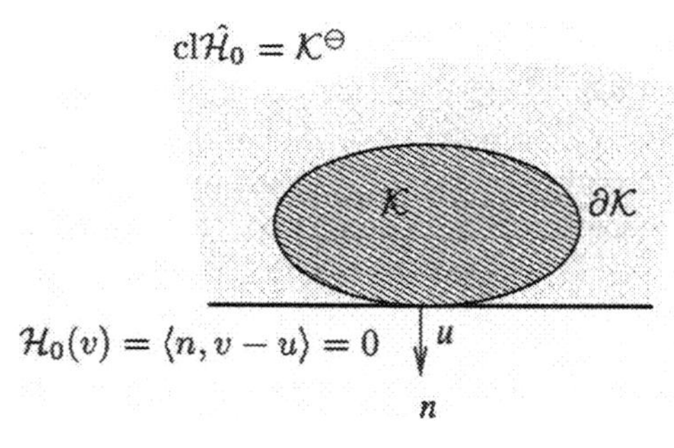

Figure 5.4. Support hyperplane and exterior normal vector at u

From this definition, it follows that any linear subspace of $\mathcal{U}$ is a cone. It is a simple matter to verify that both $\mathcal{K}^{\oplus}$ and $\mathcal{K}^{\ominus}$ are proper cones. They are always closed even if $\mathcal{K}$ is not. If $\mathcal{K}$ is a subspace of Hilbert space $\mathcal{H}$, then,

$$\mathcal{K}^{\oplus} = \mathcal{K}^{\ominus} = \mathcal{K}^{\perp} \subset \mathcal{H}.$$

The conjugate cones can thus be regarded as generalizations of the *orthogonal complement* (see Appendix A.3) of a set. In this situation, the definition is illustrated in Fig. 5.3

If $\mathcal{K}$ and $\mathcal{T}$ are two subsets of a normed space $\mathcal{U}$ and $\mathcal{K} \subset \mathcal{T}$, then $\mathcal{T}^{\ominus} \subset \mathcal{K}^{\ominus}$. In the general case the conjugate cone can be interpreted as a collection of half-spaces. If $\mathcal{K} = \mathcal{H}_0$, then $\mathcal{K}^{\oplus} = \text{cl}\check{\mathcal{H}}_0$ and $\mathcal{K}^{\ominus} = \text{cl}\hat{\mathcal{H}}_0$.

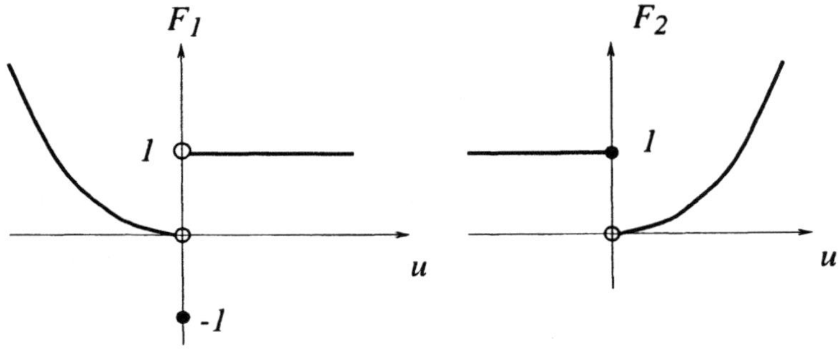

(a) Lower semicontinuous at $u = 0$. (b) Upper semicontinuous at $u = 0$

Figure 5.5. Semicontinuous functions

Definition 5.1.5 (Exterior Normal and Support Hyperplane)
Let $\mathcal{K}$ be a subset of $\mathcal{U}$ with boundary $\partial\mathcal{K}$. A unit vector $n \in \mathcal{U}$ is said to be an *exterior normal* to $\mathcal{K}$ at the point $u \in \mathcal{K}$ if $\langle n, v - u\rangle \leq 0$ for every $v \in \mathcal{K}$, i.e. if $\mathcal{K}$ is contained in the closed half-space $\mathrm{cl}\hat{\mathcal{H}}_0 = \{v \in \mathcal{U}|\ \ \langle n, v - u\rangle \leq 0\}$. The hyperplane $\mathcal{H}_0 = \{v \in \mathcal{U}|\ \ \langle n, v - u\rangle = 0\} = \partial\hat{\mathcal{H}}_0$ is called a *support hyperplane* for $\mathcal{K}$ at u. ◇

We recall from calculus that, a real-valued function $f : I \subset \mathbb{R} \to \mathbb{R}$ is continuous at a point $\bar{x} \in \mathbb{R}$ if and only if $\lim_{n\to\infty} f(x_n) = f(\bar{x})$ for every sequence of real numbers $\{x_n\}$ converging to $\bar{x}$. Moreover, if $f(\bar{x}) \leq f(x)\ \ \forall x \in I$, then f assumes its minimum at $\bar{x}$. We may extend these elementary ideas to extended real-valued functionals defined on complete normed spaces and relax the continuity condition by introducing the following concepts.

Definition 5.1.6 (Lower and Upper Semicontinuity)
Let $\mathcal{U}$ be a Banach space.

(D1) A functional $P : \mathcal{U} \to \bar{\mathbb{R}}$ is said to be *lower semicontinuous (l.s.c)* at a point u_0 in its domain if and only if, for any sequence $\{u_n\}$ converging to u_0,

$$\lim_{n\to\infty} \inf P(u_n) \geq P(u_0). \tag{5.8}$$

(D2) Similarly, P is *upper semicontinuous (u.s.c)* at u_0 if and only if

$$\lim_{n\to\infty} \sup P(u_n) \leq P(u_0). \tag{5.9}$$

(D3) If P is lower (or upper) semicontinuous at all points in its domain, we simply say that P is lower (upper) semicontinuous (on $\mathcal{U}$ or on dom P). ◇

Example 5.1.1 The discontinuous function $F_1 : \mathbb{R} \to \vec{\mathbb{R}}$ shown in Fig. 5.5(a) is defined by

$$F_1(u) = \begin{cases} u^2 & u < 0 \\ -1 & u = 0 \\ 1 & u > 0 \end{cases}$$

When $u_n \to 0$, the cluster points of $F_1(u_n)$ are 1 and 0, $F_1(0) = -1$ is below all cluster points of $F_i(u_n)$, so F is lower semicontinuous at $u = 0$.

The function defined by (see Fig. 5.5(b))

$$F_2(u) = \begin{cases} 1 & u \le 0 \\ u^2 & u > 0 \end{cases}$$

is upper semicontinuous at $u = 0$. The cluster points of $F_2(u_n)$ lie below or on $u = 0$. ♡

Another convenient way of recognizing lower semicontinuous functions is provided by the following fundamental theorem.

Theorem 5.1.3 $P : \mathcal{U} \to \overline{\mathbb{R}}$ *is lower semicontinuous if and only if* epi P *is closed.*

Proof: Assume P is l.s.c. and pick $(u_n, \alpha_n) \in$ epi P. Suppose $u_n \to u$ and $\alpha_n \to \alpha$. We must show that $(u, \alpha) \in$ epi$P$. Since $\alpha_n \ge P(u_n)$ $\forall n$, $\alpha = \liminf \alpha_n \ge \liminf P(u_n) \ge P(u)$. Thus, $(u, \alpha) \in$ epi P.

Next, assume epi$P$ is closed. Let $u_n \to u$. Since $\liminf P(u_n)$ is defined as the lowest cluster point of the sequence $\{P(u_n)\}$, there exists a subsequence $\{u_{n_k}\}$ such that $P(u_{n_k}) \to \liminf P(u_n)$ and clearly $(u_{n_k}, P(u_{n_k})) \in$ epi$P$. Then, $(u_{n_k}, P(u_{n_k})) \to (u, \alpha)$ where $\alpha = \liminf P(u_n)$. But epi$P$ is closed. Hence, $\liminf P(u_n) \ge P(u)$; i.e. P is l.s.c. □

Example 5.1.2 The function $P : \mathbb{R} \to \overline{\mathbb{R}}$, defined by

$$P(u) = \begin{cases} 0 & u \le 0, \\ +\infty & u > 0, \end{cases}$$

is lower semicontinuous on its domain since its epigraph is closed. But the function

$$G(u) = \begin{cases} 0 & u < 0 \\ +\infty & u \ge 0 \end{cases}$$

is not lower semicontinuous on its domain because epi G is not closed. ♡

Example 5.1.3 Let $\mathcal{K}$ be a nonempty subset of $\mathcal{U}$. Its indicator functional $\mathcal{I}_{\mathcal{K}}$ is lower semicontinuous if and only if $\mathcal{K}$ is closed. It is convex if and only if $\mathcal{K}$ is convex. It is continuous at v, where if v is in the interior of $\mathcal{K}$. ♡

Definition 5.1.7 (Canonical Functional Spaces)

(D1) The set of functionals $P : \mathcal{U} \to \bar{\mathbb{R}}$ which are either convex or concave is denoted by $\Gamma(\mathcal{U})$. In particular, let $\check{\Gamma}(\mathcal{U})$ denote the subset of $P \in \Gamma(\mathcal{U})$ which are convex and $\hat{\Gamma}(\mathcal{U})$ the subset of $P \in \Gamma(\mathcal{U})$ which are concave.

(D2) The *canonical functional space* $\Gamma_G(\mathcal{U})$ is a subset of functionals $P \in \Gamma(\mathcal{U})$ which are Gâteaux differentiable on $\mathcal{U}$.

(D3) The *extended canonical functional space* $\Gamma_0(\mathcal{U})$ is a subset of functionals $P \in \Gamma(\mathcal{U})$ which are either convex, lower semicontinuous or concave, upper semicontinuous, and if P takes the values $\pm\infty$, then P is identically equal to $\pm\infty$. ◊

According to this definition, a functional $\Phi \in \check{\Gamma}(\mathcal{U}\times\mathcal{E})$ is such that $\Phi : \mathcal{U}\times\mathcal{E} \to \bar{\mathbb{R}}$ is convex on the product space, while $\Phi \in \check{\Gamma}_0(\mathcal{U}) \times \check{\Gamma}(\mathcal{E})$ implies that for any fixed $\epsilon \in \mathcal{E}$, $\Phi : \mathcal{U} \to \bar{\mathbb{R}}$ is convex, proper and lower semicontinuous, and for any fixed $u \in \mathcal{U}$, $\Phi : \mathcal{E} \to \bar{\mathbb{R}}$ is concave. However, $\Phi \in \check{\Gamma}(\mathcal{U}) \times \check{\Gamma}(\mathcal{E})$ does not imply that $\Phi \in \check{\Gamma}(\mathcal{U} \times \mathcal{E})$.

In variational problems we are particularly interested in properties of functionals that are *weakly* semicontinuous on subsets of normed spaces.

Definition 5.1.8 (Weak and Weak-∗ Convergence)

Let $\mathcal{U}$ be a Banach space, $\mathcal{U}^*$ its dual, and $\langle *, * \rangle$ the inner product over $\mathcal{U}\times\mathcal{U}^*$.

(D1) We define *weak convergence* of a sequence $\{v_n\} \in \mathcal{U}$ as $n \to \infty$by

$$v_n \rightharpoonup v \quad \text{iff} \quad \langle v_n, v^* \rangle \to \langle v, v^* \rangle \ \ \forall v^* \in \mathcal{U}^*.$$

(D2) We define *weak-* convergence* of a sequence $\{v_n^*\} \in \mathcal{U}^*$ as $n \to \infty$ by

$$v_n^* \overset{*}{\rightharpoonup} v^* \quad \text{iff} \quad \langle v, v_n^* \rangle \to \langle v, v^* \rangle \ \ \forall v \in \mathcal{U}. \quad ◊$$

Norm convergence (see Appendix) in $\mathcal{U}$ (respectively, in $\mathcal{U}^*$) is also called *strong convergence*, and is denoted by $v_n \to v$ (respectively, $v_n^* \to v^*$).

Definition 5.1.9 (Weak Semicontinuity)

Let $\mathcal{K}$ be a subset of a normed linear space $\mathcal{U}$.

(D1) A functional $P : \mathcal{K} \to \bar{\mathbb{R}}$ is said to be *weakly l.s.c.* at a point $v_0 \in \mathcal{K}$ if, for every sequence $\{v_n\}$ converging weakly to a point $v_0 \in \mathcal{K}$,

$$\lim_{n\to\infty} \inf P(v_n) \geq P(v_0).$$

(D2) Similarly, if

$$\lim_{n\to\infty} \sup P(v_n) \leq P(v_0),$$

then P is *weakly u.s.c.* ◊

Let $\mathcal{U}$ be a Banach space with norm $\| * \|_{\mathcal{U}}$. If $v_n \rightharpoonup v \in \mathcal{U}$, then $\|v_n\|_{\mathcal{U}}$ is bounded and

$$\lim_{n\to\infty} \inf \|v_n\|_{\mathcal{U}} \geq \|v\|_{\mathcal{U}}.$$

Hence, the functional $P(v) = \|v\|_{\mathcal{U}}$ is weakly l.s.c.

Also, if $v_n^* \overset{*}{\rightharpoonup} v^* \in \mathcal{U}^*$, then $\|v_n^*\|_{\mathcal{U}^*}$ is bounded and

$$\lim_{n\to\infty} \inf \|v_n^*\|_{\mathcal{U}^*} \geq \|v^*\|_{\mathcal{U}^*}.$$

Theorem 5.1.4 *For a convex functional $P : \mathcal{K} \subset \mathcal{U} \to \vec{\mathbb{R}}$ on the Banach space $\mathcal{U}$, if $\mathcal{K}$ is closed and convex, then*

$$P \text{ is l.s.c. on } \mathcal{K} \ \Leftrightarrow \ P \text{ is weakly l.s.c. on } \mathcal{K}.$$

Moreover, if $u \in \text{dom } P$, then P is l.s.c. at u if and only if, for each $\omega > 0$ there exists a $\delta(\omega) > 0$ such that

$$\|v - u\| < \delta(\omega) \Rightarrow P(u) < P(v) + \omega \ \ \forall v \in \mathcal{K}.$$

Example 5.1.4 Let $\Omega \subset \mathbb{R}^n$ be an open bounded domain and $\mathcal{U} = \mathcal{H}^1(\Omega)$ a Sobolev space. The functional defined by

$$P(u) = \int_\Omega |\nabla u| \mathrm{d}\Omega, \ \ |\nabla u| \equiv \left(\sum_{i=1}^n \left| \frac{\partial u}{\partial x_i} \right|^2 \right)^{1/2}$$

is convex, weakly l.s.c. on $\mathcal{H}^1(\Omega)$. ♡

Weak convergence in normed spaces is very important in direct methods, and we refer to Yosida (1966) and Zeidler (1985) for details. However, in duality theory and methods, weak star convergence plays an important role. We list some standard results in the following proposition.

Proposition 5.1.1 *Let $\mathcal{U}$ be a Banach space. The following statements hold.*

(S1) If $\mathcal{U}$ is reflexive, then every bounded sequence in $\mathcal{U}$ has a weakly convergent subsequence.

(S2) If $\mathcal{U}$ is separable, then every bounded sequence in $\mathcal{U}^*$ has a weak-* convergent subsequence.

(S3) If $\mathcal{U}$ is reflexive, then on $\mathcal{U}^*$, weak-* convergence and weak convergence coincide. In particular, all Hilbert spaces and all finite-dimensional Banach spaces are reflexive.

(S4) If $\dim \mathcal{U} < \infty$, then strong convergence, weak convergence, and weak-* convergence coincide.

(S5) If $\mathcal{K}$ is a closed convex subset of $\mathcal{U}$, then for all $\{u_n\} \in \mathcal{K}$, $u_n \rightharpoonup u$ as $n \to \infty$ implies $u \in \mathcal{K}$.

(S6) As $n \to \infty$, we have the dual relations

$$\text{(a) if } u_n^* \to u^* \text{ in } \mathcal{U}^*, \text{ and } \ u_n \rightharpoonup u \text{ in } \mathcal{U}, \text{ then } \langle u_n, u_n^* \rangle \to \langle u, u^* \rangle,$$

$$\text{(b) if } u_n^* \overset{*}{\rightharpoonup} u^* \text{ in } \mathcal{U}^*, \text{ and } \ u_n \to u \text{ in } \mathcal{U}, \text{ then } \langle u_n, u_n^* \rangle \to \langle u, u^* \rangle.$$

The proofs of these propositions can be found in Yosida (1966).

Remark 5.1.2 According to the statement (S3), weak-* convergence plays a special role only in the dual spaces $\mathcal{U}^*$ of nonreflexive Banach spaces. For example, if $\mathcal{U} = \mathcal{L}^1(\Omega)$, $\Omega \subset \mathbb{R}^N$, then $\mathcal{U}^* = \mathcal{L}^\infty(\Omega)$ and weak-* convergence $u_n^* \overset{*}{\rightharpoonup} u^*$ in $\mathcal{U}^*$ means that

$$\int_\Omega u_n^* u \mathrm{d}\Omega \to \int_\Omega u^* u \mathrm{d}\Omega \ \ \forall u \in \mathcal{L}^1(\Omega) \text{ as } n \to \infty.$$

The space $\mathcal{L}^1(\Omega)$ is separable; therefore, by the statement (S2), every bounded sequence in $\mathcal{L}^\infty(\Omega)$ has a weak-* convergent subsequence. ■

5.1.2 Primal Variational Problems and General Theorems

Let $P : \mathcal{U} \to \overrightarrow{\mathbb{R}}$ be a given functional. We consider the general primal variational problem

$$(\mathcal{P}_{\text{inf}}) : \quad P(u) \to \inf \quad \forall u \in \mathcal{U}. \tag{5.10}$$

In this section, we wish to characterize the solution of the unconstrained primal problem $(\mathcal{P}_{\text{inf}})$ when the functional $P : \mathcal{U} \to \overrightarrow{\mathbb{R}}$ is Gâteaux differentiable on $\mathcal{K} = \text{dom } P \subset \mathcal{U}$.

We begin with a fundamental result.

Theorem 5.1.5 (Necessary Conditions)
Let $P : \mathcal{K} \subset \mathcal{U} \to \mathbb{R}$ have a Gâteaux variation on a normed linear space $\mathcal{K}$.
(T1) If P is convex, then it is weakly lower semicontinuous on $\mathcal{K}$.
(T2) If either $(\mathcal{P}_{\text{inf}})$ or $(\mathcal{P}_{\text{sta}})$ has a solution $\bar{u}$ on an open subset $\mathcal{K}$, then

$$\delta P(\bar{u}; u) = 0 \ \ \forall u \in \mathcal{K}, \tag{5.11}$$

i.e. the extrema of P on an open set $\mathcal{K}$ must be critical points of P.
(T3) If $(\mathcal{P}_{\text{inf}})$ has a solution $\bar{u}$ on a convex set $\mathcal{K}$, then

$$\delta P(\bar{u}; u - \bar{u}) \geq 0 \ \ \forall u \in \mathcal{K}.$$

Proof. (T1) If $\{u_n\} \in \mathcal{K}$ converges weakly to $\bar{u} \in \mathcal{K}$, then

$$\lim_{n \to \infty} \inf \langle DP(\bar{u}), u_n - \bar{u} \rangle = 0.$$

Hence, $P(\bar{u}) \leq \lim_{n\to\infty} \inf P(u_n)$, which means that P is weakly lower semicontinuous.

(T2) If $\bar{u}$ is a local extremum of P on the open subset $\mathcal{K} \subset \mathcal{U}$, then the functional $P(\bar{u} + \theta u)$ of the real-variable θ must achieve a local extremum at $\theta = 0$ for all $u \in \mathcal{K}$. Thus, by ordinary calculus,

$$\frac{d}{d\theta} P(\bar{u} + \theta u)|_{\theta=0} = 0,$$

and therefore $\bar{u}$ must be a critical point of P.

(T3) If $\mathcal{K}$ is convex, $\bar{u} + \theta(u - \bar{u}) \in \mathcal{K} \ \ \forall \theta \in [0, 1]$ and hence

$$\frac{d}{d\theta} P(\bar{u} + \theta(u - \bar{u}))|_{\theta=0} \geq 0$$

for a minimum at $\bar{u}$. □

Remark 5.1.3 The statement (T2) in this theorem shows that the Euler-Lagrange equation $\delta P(\bar{u}; u) = 0 \ \ \forall u \in \mathcal{K}$ is only a necessary condition for a local extrema of a Gâteaux differentiable functional P on an open set, which is not true if the set $\mathcal{K}$ is closed. A sufficient condition for a general functional is given by the following *generalized Weierstrass theorem.* ■

Theorem 5.1.6 (Sufficient Conditions)

Let $\mathcal{U}$ be a reflexive Banach space, $\mathcal{K}$ a nonempty bounded, closed and convex subset of $\mathcal{U}$. Let $P : \mathcal{K} \subset \mathcal{U} \to \overrightarrow{\mathbb{R}}$.

(T1) If P is weakly l.s.c. on $\mathcal{K}$, then P is bounded below and attains its minimum on $\mathcal{K}$, i.e. $(\mathcal{P}_{\inf})$ has at least one solution.

(T2) If P is convex, l.s.c. and proper, then the set of solutions of $(\mathcal{P}_{\inf})$ is a closed convex set (possibly empty) of $\mathcal{K}$.

(T3) If P is strictly convex over $\mathcal{K}$, then $(\mathcal{P}_{\inf})$ has at most one solution.

Proof. (T1) Let us first show by contradiction that P is bounded below. Suppose that P is not bounded below on $\mathcal{K}$. Then, there exists a sequence $\{u_n\}$ such that $P(u_n) < -n$. Since $\mathcal{K}$ is bounded, closed and convex, there exists a subsequence $\{u_{n_k}\}$ of $\{u_n\}$ which converges weakly to some point $u \in \mathcal{K}$. But, because P is weakly l.s.c., $P(u) \le \lim_{n_k} \inf P(u_{n_k})$, which contradicts the assumed property $\lim_{n_k \to \infty} P(u_{n_k}) \to -\infty$. Hence P is bounded below on $\mathcal{K}$.

(T2) Let $\alpha = \inf\{P(u)|\ \forall u \in \mathcal{K}\}$ and let $\{u_n\}$ be a sequence taken from $\mathcal{K}$ such that $\alpha = \lim_{n\to\infty} P(u_n)$. Since $\mathcal{K}$ is convex and closed, there exists a subsequence $\{u_{n_j}\}$ of $\{u_n\}$ which converges weakly to an element $\bar{u} \in \mathcal{K}$. But, because P is weakly l.s.c., we must have $P(\bar{u}) \le \lim_{n_j \to\infty} \inf P(u_{n_j}) = \alpha$. Since α is the greatest lower bound of P on $\mathcal{K}$, i.e. $\alpha \le P(\bar{u})$, we must have $\alpha = P(\bar{u})$, i.e. $\bar{u} \in \mathcal{K}$ is a solution of $(\mathcal{P}_{\inf})$.

(T3) Let $\alpha = \inf P(u)\ \ \forall u \in \mathcal{K}$. Then, the solution set of $(\mathcal{P}_{\inf})$ is

$$\{\bar{u} \in \mathcal{K}|\ P(\bar{u}) \le \alpha\},$$

which is the epigraph of P for fixed α. Since P is convex, lower semicontinuous, this solution set is convex and closed.

Suppose that P is strictly convex and that $(\mathcal{P}_{\inf})$ has two solutions, $\bar{u}_1$ and $\bar{u}_2$. By the strict convexity of P,

$$P(\theta\bar{u}_1 + (1-\theta)\bar{u}_2) < \theta P(\bar{u}_1) + (1-\theta)P(\bar{u}_2)\ \ \forall \bar{u}_1 \neq \bar{u}_2,\ \ \theta \in [0,1],$$

we arrive at a contradiction for $P(\bar{u}_1) = P(\bar{u}_2) = \inf\{P(u)|\ \forall u \in \mathcal{K}\}$ and $\bar{u}_1 \neq \bar{u}_2$ when $\theta = \frac{1}{2}$. □

Remark 5.1.4 Conditions on $\mathcal{K}$ in the theorem can be replaced by the requirement that $\mathcal{K}$ be weakly sequentially compact. However, the requirement that $\mathcal{K}$ be bounded can be relaxed if P is coercive on $\mathcal{K}$, i.e.

$$\lim_{\|u\|\to\infty} P(u) = +\infty\ \ \forall u \in \mathcal{K}.$$

■

Corollary 5.1.1 *If $P : \mathcal{U} \to \bar{\mathbb{R}}$ is a coercive, weakly l.s.c. functional, its effective domain $\mathcal{K} = \mathrm{dom}\, P$ is a nonempty weakly sequentially closed subset of a reflexive Banach space, and P is bounded below and attains its minimum on $\mathcal{K}$.*

Let

$$\mathcal{K}_r := \{u \in \mathcal{K}|\ \ P(u) \le r\}. \tag{5.12}$$

Definition 5.1.10 (Quasiconvexity and Lower Semi-Compactness)

(D1) A functional $P : \mathcal{K} \subset \mathcal{U} \to \mathbb{R}$ is said to be *quasiconvex* if and only if $\mathcal{K}_r$ is convex for all $r \in \mathbb{R}$.

(D2) $P : \mathcal{K} \subset \mathcal{U} \to \mathbb{R}$ is said to be *lower semicompact* if and only if $\mathcal{K}_r$ is compact for all $r \in \mathbb{R}$. ♢

Theorem 5.1.7 (Main Theorem) *Let $\mathcal{K} \neq \emptyset$. For the functional $P : \mathcal{K} \subset \mathcal{U} \to \overrightarrow{\mathbb{R}}$, the problem $(\mathcal{P}_{\text{inf}})$ has a solution if either (a) P is lower semicompact or (b) P is l.s.c. on the compact set $\mathcal{K}$.*

Proof. (a) If P is lower semicompact, the set $\mathcal{K}_r = \{u \in \mathcal{K} | \ P(u) \leq r\}$ is compact for all $r \in \mathbb{R}$. Let $\alpha = \inf_{u \in \mathcal{K}} P(u)$. If $\alpha = +\infty$, the assertion is trivial because $P \equiv +\infty$. Therefore, we let $\alpha < r_o < +\infty$ for a fixed r_o. The set $\mathcal{K}_{r_o}$ is compact. Since the intersection of a finite number of sets $\mathcal{K}_r$ with $\alpha < r \leq r_o$ is always nonempty, it follows that there is a point $\bar{u}$ such that

$$\bar{u} \in \bigcap_{\alpha < r \leq r_o} \mathcal{K}_r.$$

Hence, $P(\bar{u}) = \alpha$, i.e. $\bar{u}$ is a solution of $(\mathcal{P}_{\text{inf}})$.

(b) This is a special case of (a). □

Corollary 5.1.2 *If the functional $P : \mathcal{K} \to \mathbb{R}$ is continuous on the nonempty compact set $\mathcal{K}$, then both problems $(\mathcal{P}_{\text{inf}})$ and $(\mathcal{P}_{\text{sup}})$ have a solution on $\mathcal{K}$.*

Proof. Since $P : \mathcal{K} \to \mathbb{R}$ is continuous if and only if P is both l.s.c. and u.s.c. on $\mathcal{K}$, then this corollary follows from the statement (a) upon applying it to $-P$. □

Definition 5.1.11 Let $\mathcal{U}$ be a Banach space. A subset $\mathcal{K} \subset \mathcal{U}$ is said to be *weak sequentially compact* if each sequence in $\mathcal{K}$ possesses a weakly convergent subsequence with limit value in $\mathcal{K}$. ♢

The main theorem has various alternative versions.

Theorem 5.1.8 *The problem $(\mathcal{P}_{\text{inf}})$ has a solution on the nonempty set $\mathcal{K}$ if one of the following three conditions is fulfilled.*

(T1) $\mathcal{U} = \mathbb{R}^n$, $n \geq 1$, $\mathcal{K}$ is closed and bounded, and P is lower semicontinuous.

(T2) $\mathcal{K}$ is a reflexive Banach space, $\mathcal{K}$ is closed, bounded and convex, P is lower semicontinuous and convex or, more generally, lower semicontinuous and quasiconvex.

(T3) $\mathcal{U}$ is a Banach space, P is weak sequentially lower semicontinuous on $\mathcal{K}$, and $\mathcal{K}$ is weak sequentially compact.

Definition 5.1.12 (Palais-Smale (PS) Condition)

Let the functional $P : \mathcal{K} \subset \mathcal{U} \to \mathbb{R}$ be Gâteaux differentiable on the subset $\mathcal{K}$ of a Banach space $\mathcal{U}$. If, for any given sequence $\{u_n\} \in \mathcal{K}$,

$$\left. \begin{array}{l} P(u_n) \text{ is bounded} \\ \lim_{n \to \infty} \|P(u_n)\| = 0 \end{array} \right\} \Rightarrow \{u_n\} \text{ has a convergent subsequence,} \qquad (5.13)$$

then P is said to satisfy the *Palais-Smale condition*, or simply, (PS) condition.◊

Theorem 5.1.9 *Let $P : \mathcal{K} \subset \mathcal{U} \to \mathbb{R}$ be a l.s.c. and Gâteaux differentiable functional on the subset $\mathcal{K}$ of a Banach space $\mathcal{U}$. If P is bounded below and satisfies (PS), then, the problem $(\mathcal{P}_{\inf})$ has a solution $\bar{u}$ which is also a solution of $(\mathcal{P}_{\rm sta})$, i.e.*

$$P(\bar{u}) = \inf_{u \in \mathcal{U}} P(u), \quad DP(\bar{u}) = 0.$$

Example 5.1.5 For a scalar-valued function $u : \Omega \subset \mathbb{R}^n \to \mathbb{R}$, we use the notation $|\nabla u|^\alpha = (\nabla u \cdot \nabla u)^{\alpha/2}$ for $2 \leq \alpha \leq \infty$. We consider the problem

$$P(u) = \int_\Omega \frac{1}{\alpha} |\nabla u|^\alpha \mathrm{d}\Omega - \int_\Omega f u \mathrm{d}\Omega \to \min \quad \forall u \in \mathcal{W}_0^{1,\alpha}. \tag{5.14}$$

Since $u \in \mathcal{W}_0^{1,\alpha}(\Omega) \Rightarrow u = 0$ on $\partial\Omega$, $|u|_{1,\alpha,\Omega} = 0$ if and only if $u = 0$, so the seminorm

$$|u|_{1,\alpha} = \left[\int_\Omega |\nabla u|^\alpha \, \mathrm{d}x \right]^{1/\alpha} = \|\nabla u\|_{0,\alpha,\Omega}$$

defines a norm on $\mathcal{W}_0^{1,\alpha}(\Omega)$. Hence, with this norm, $\mathcal{W}_0^{1,\alpha}(\Omega)$ is a reflexive Banach space. We can prove that on $\mathcal{W}_0^{1,\alpha}(\Omega)$, P is Gâteaux differentiable , strictly convex, and coercive. If $\bar{u}$ is a critical point of P, then $\bar{u}$ must be a unique global minimizer of the problem (5.14) and satisfies the Euler-Lagrange equation

$$A(u) = -\nabla \cdot \left(|\nabla \bar{u}|^{\alpha-2} \nabla \bar{u} \right) = f \quad \text{in } \Omega. \tag{5.15}$$

For $\alpha = 2$, P is the quadratic functional

$$P(u) = \frac{1}{2} \int_\Omega |\nabla u|^2 \, \mathrm{d}x - \int_\Omega f u \, \mathrm{d}x,$$

and then the Euler-Lagrange equation $Au = -\Delta u = f$ is the well-known Poisson equation. ♡

In the case that $Au = \Lambda^* DW(\Lambda u)$ is a linear operator and $DF(u) = f \in \mathcal{U}^*$, we have the following result in partial differential equations.

Proposition 5.1.2 *Let $\mathcal{U}$ be a Hilbert space, and let a linear operator $A : \mathcal{U} \to \mathcal{U}$ satisfy*

$$\langle Au, u \rangle \geq \theta \|u\|^2 \quad \forall u \in \mathcal{U}, \quad \theta > 0. \tag{5.16}$$

Then, for each $f \in \mathcal{U}^* = \mathcal{U}$ there exists a unique $u \in \mathcal{U}$ satisfying

$$Au = f. \tag{5.17}$$

5.1.3 Fenchel Transformation

Fenchel duality plays a control role in modern analysis and optimization. Our goal in this section is to present an introduction to convex analysis.

Let $\{(\mathcal{U},\mathcal{U}^*), \langle *,*\rangle\}$ be a duality pairing, that is two real vector spaces $\mathcal{U}$ and $\mathcal{U}^*$ placed in duality with the bilinear form $\langle *,*\rangle : \mathcal{U} \times \mathcal{U}^* \to \mathbb{R}$. On $\mathcal{U}$ any continuous affine functional can be written as

$$l(u) = \langle u, u^* \rangle - \alpha, \quad u^* \in \mathcal{U}^*, \quad \alpha \in \mathbb{R}.$$

Let P be a functional from $\mathcal{U}$ into $\bar{\mathbb{R}}$, not necessarily convex. For a fixed $u^* \in \mathcal{U}^*$, l belongs to a family of continuous affine functionals everywhere less than or equal to P if and only if

$$\alpha \geq \langle u, u^* \rangle - P(u) \;\; \forall u \in \mathcal{U}.$$

The lowest value of α for a fixed u^* has a special significance.

Definition 5.1.13 (Fenchel Transformations and Conjugate functionals)

(D1) The functional $P^\sharp : \mathcal{U}^* \to \overrightarrow{\mathbb{R}}$ defined by

$$P^\sharp(u^*) = \sup_{u \in \mathcal{U}} \{\langle u, u^* \rangle - P(u)\}, \tag{5.18}$$

associated with every $u^* \in \mathcal{U}^*$ is called the *super-conjugate functional* of P.

(D2) The functional $P^\flat : \mathcal{U}^* \to \overleftarrow{\mathbb{R}}$ defined by

$$P^\flat(u^*) = \inf_{u \in \mathcal{U}} \{\langle u, u^* \rangle - P(u)\}, \tag{5.19}$$

associated with every $u^* \in \mathcal{U}^*$ is called the *sub-conjugate functional* of P.

(D3) The operation $P \to P^* = \{P^\sharp, \; P^\flat\}$ defined by either (5.18) or (5.19) is called the *Fenchel transformation*, written as

$$P^*(u^*) = \underset{u \in \mathcal{U}}{\text{ext}} \{\langle u, u^* \rangle - P(u)\}. \tag{5.20}$$

Both the super- and the sub-conjugates are called *Fenchel conjugate functionals* of P. ◊

In finite dimensional space $\mathcal{U} = \mathbb{R}^n$, let P be finite, differentiable, either convex or concave functional, and such that for each $u^* \in \mathbb{R}^n$ the equation $DP(u) = u^*$ has a unique solution $u \in \mathbb{R}^n$. If we denote this solution by $s(u^*)$, then

$$P^*(u^*) = s(u^*)u^* - P(s(u^*)). \tag{5.21}$$

This is known as the *Legendre transformation*, and thus $P^*(u^*) = P^c(u^*)$ is the Legendre conjugate of P. A geometrical interpretation of $P^*(u^*)$ is shown in Fig. 5.6.

By the definition of the super-conjugate functional, we always have the *Fenchel-Young inequality*

$$\langle u, u^* \rangle \leq P(u) + P^\sharp(u^*) \quad \forall (u, u^*) \in \mathcal{U} \times \mathcal{U}^*. \tag{5.22}$$

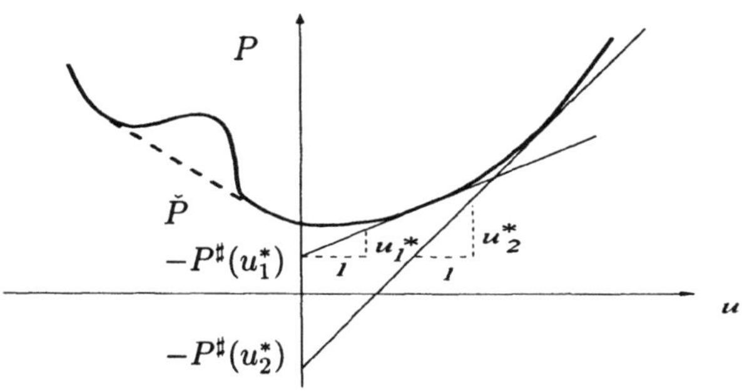

Figure 5.6. Nonconvex function and its super-conjugate and convex hull.

Dually, from the definition of the sub-conjugate functional,

$$\langle u, u^* \rangle \geq P(u) + P^\flat(u^*) \quad \forall (u, u^*) \in \mathcal{U} \times \mathcal{U}^*. \tag{5.23}$$

Definition 5.1.14 (Biconjugate and Convex Hull)
(D1) The functional $P^{**} : \mathcal{U} \to \bar{\mathbb{R}}$ defined by

$$P^{**}(u) = \operatorname*{ext}_{u^* \in \mathcal{U}^*} \{\langle u, u^* \rangle - P^*(u^*)\} \tag{5.24}$$

is called the *biconjugate functional* of P.

(D2) The functional $\check{P} : \mathcal{U} \to \overrightarrow{\mathbb{R}}$ defined by a set of convex functionals $G : \mathcal{U} \to \overrightarrow{\mathbb{R}}$ such that

$$\check{P}(u) = \sup\{G(u) | \; G(u) \leq P(u) \;\; \forall u \in \mathcal{U}\} \tag{5.25}$$

is called the *(lower) convex envelope (or hull)* of P.

(D3) The functional $\hat{P} : \mathcal{U} \to \overleftarrow{\mathbb{R}}$ defined by a set of concave functional $G : \mathcal{U} \to \overleftarrow{\mathbb{R}}$ such that

$$\hat{P}(u) = \inf\{G(u) | \; G(u) \geq P(u) \;\; \forall u \in \mathcal{U}\} \tag{5.26}$$

is called the *(upper) concave envelope* of P. ◊

The following important properties were established, essentially, by Fenchel and Moreau.

Theorem 5.1.10 *Let $P : \mathcal{U} \to \overrightarrow{\mathbb{R}}$. Then*
(T1) $P^\sharp(0) = -\inf_{u \in \mathcal{U}} P(u)$.
(T2) $P^\sharp \in \check{\Gamma}(\mathcal{U}^*)$. *If* $P \equiv +\infty$, *then dom* $P = \emptyset$, $P^\sharp \equiv +\infty$.
(T3) If $P \leq G$, *then* $P^\sharp \geq G^\sharp$.
(T4) In general, $P^{\sharp\sharp\sharp} = P^\sharp$.
(T5) $P^\sharp = (\check{P})^\sharp$.
(T6) In general $P^{\sharp\sharp} \leq \check{P} \leq P$. *If* $P \in \check{\Gamma}(\mathcal{U})$, *then* $P^{\sharp\sharp} = \check{P} = P$. *In particular, if P takes only finite values, then* $P^{\sharp\sharp} = \check{P}$.

(T7) If $\{P_i\}$ $(i \in I)$ is an arbitrary family of functionals on $\mathcal{U}$, then

$$\left(\inf_{i\in I} P_i\right)^\sharp \geq \sup_{i\in I} P_i^\sharp, \quad \left(\sup_{i\in I} P_i\right)^\sharp \leq \inf_{i\in I} P_i^\sharp.$$

Geometrically, the epigraph of $\check{P}$ is obtained from the convex hull of epi P. Moreover, the equalities

$$(\alpha P)^*(u^*) = \alpha P^*(u^*/\alpha), \quad \alpha > 0, \tag{5.27}$$

$$(P+\alpha)^* = P^* + \alpha, \quad \alpha \in \mathbb{R} \tag{5.28}$$

are easy to verify.

Definition 5.1.15 (Fenchel Conjugate Pair)
Let $P \in \Gamma(\mathcal{U})$ and $G \in \Gamma(\mathcal{U}^*)$ be two canonical functionals. If they correspond in the bijection:

$$P = G^*, \qquad G = P^*, \tag{5.29}$$

then P and G are said to be a *Fenchel conjugate pair* . ◊

The Fenchel conjugation establishes a bijection between $\Gamma(\mathcal{U})$ and $\Gamma(\mathcal{U}^*)$ and a one-to-one correspondence between $\Gamma_0(\mathcal{U})$ and $\Gamma_0(\mathcal{U}^*)$.

Definition 5.1.16 (Support Functional)
Let $\mathcal{K}$ be a nonempty subset of $\mathcal{U}$. The conjugate functional of the indicator $\mathcal{I}_{\mathcal{K}}$, defined by

$$\mathcal{I}_{\mathcal{K}}^\sharp(u^*) := \sup_{u\in\mathcal{K}} \langle u, u^*\rangle = \sup_{u\in\mathcal{U}} [\langle u, u^*\rangle - \mathcal{I}_{\mathcal{K}}(u)] = (\mathcal{I}_{\mathcal{K}})^\sharp(u^*) \tag{5.30}$$

is called a *support functional* of $\mathcal{K}$. ◊

Applying the duality again, we have

$$\mathcal{I}_{\mathcal{K}}^{\sharp\sharp}(v) = \mathcal{I}_{\mathrm{cl}\check{\mathcal{K}}}(v),$$

where cl$\check{\mathcal{K}}$ denotes the closed convex hull of $\mathcal{K}$.

Proposition 5.1.3 *If $\mathcal{K}$ is closed and convex, then $\mathcal{I}_{\mathcal{K}}^{\sharp\sharp}(v) = \mathcal{I}_{\mathcal{K}}(v)$.*

In general, $\mathcal{I}_{\mathcal{K}}^\sharp$ may be infinite. The geometrical interpretation of the support functional can be understood as follows.

Given an element $u^* \in \mathcal{U}^*$, we consider the family of half-spaces $\hat{\mathcal{H}}_\alpha = \{u \in \mathcal{K} |\ \langle u, u^*\rangle \leq \alpha\}$. As the parameter α increases, these half-spaces get larger and $\mathcal{I}_{\mathcal{K}}^\sharp$ is defined as the infimum of those constants α such that $\mathcal{K}$ is contained within the half-space (see Fig. 5.4)

Example 5.1.6 Let a discontinuous function $P : \mathbb{R} \to \overrightarrow{\mathbb{R}}$ be given by

$$P(u) = \begin{cases} |u| & \text{if } u \neq 0 \\ 1 & \text{if } u = 0. \end{cases}$$

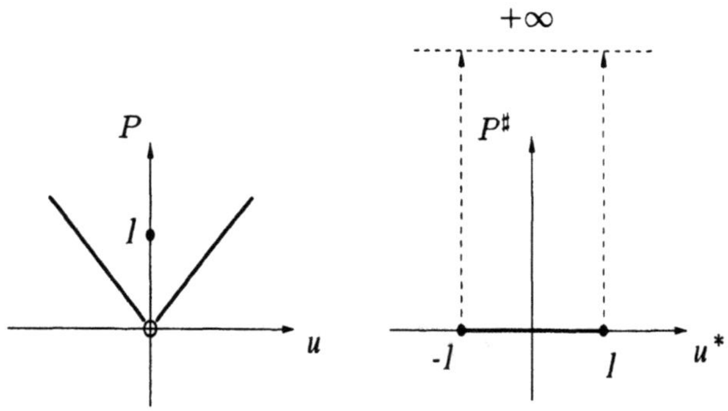

Figure 5.7. Example of a discontinuous function and its conjugate

By definition, $P^\sharp : \mathbb{R} \to \overline{\mathbb{R}}$ is defined by

$$P^\sharp(u^*) = \sup_{u \in \mathbb{R}}\{u^* u - P(u)\} = \begin{cases} \sup_{u \neq 0}\{u^* u - |u|\}, \\ -1 \end{cases}$$

Clearly,

$$P^\sharp(0) = -\inf_{u \in \mathbb{R}}\{|u|, \ 1\} = 0.$$

If $u^* \neq 0$, we have

$$P^\sharp(u^*) = \sup_{u \neq 0}\{|u|(|u^*| - 1)\} = \begin{cases} 0 & \text{if } |u^*| \leq 1 \\ +\infty & \text{if } |u^*| > 1. \end{cases}$$

Hence, the super-conjugate of P is

$$P^\sharp(u^*) = \begin{cases} 0 & \text{if } |u^*| \leq 1 \\ +\infty & \text{if } |u^*| > 1, \end{cases}$$

which is convex, l.s.c. as shown in Fig. 5.7.

Example 5.1.7 Let $\mathcal{U}$ be a Banach space with norm $\|*\|_{\mathcal{U}}$, and $P(u) = \frac{1}{\alpha}\|u\|_{\mathcal{U}}^{\alpha} : \mathcal{U} \to \mathbb{R}$, with $\alpha \geq 1$. Then,

$$\begin{aligned} P^\sharp(u^*) &= \sup_u\{\langle u, u^*\rangle - \frac{1}{\alpha}\|u\|^{\alpha}\} \\ &= \sup_{\theta \geq 0}\sup_{\|u\| = \theta}\{\langle u, u^*\rangle - \frac{1}{\alpha}\|u\|^{\alpha}\} = \sup_{\theta \geq 0}\{\|u^*\|_{\mathcal{U}^*}.\theta - \frac{1}{\alpha}\theta^{\alpha}\} \\ &= \begin{cases} \frac{1}{\alpha^*}\|u^*\|_{\mathcal{U}^*}^{\alpha^*} & \text{if } \alpha > 1, \ \alpha^* = \frac{\alpha}{\alpha - 1}, \\ \mathcal{I}_I(\|u^*\|_{\mathcal{U}^*}) & \text{if } \alpha = 1, \ I = [0, 1]. \end{cases} \end{aligned}$$

Hence,

$$P^\sharp(u^*) = \frac{1}{\alpha^*}\|u^*\|_{\mathcal{U}^*}^{\alpha^*}, \quad \frac{1}{\alpha} + \frac{1}{\alpha^*} = 1.$$

Example 5.1.8 Let $A \in \mathbb{R}^{n\times n}$ (the set of $n \times n$ matrices) and $P(A) = \det A$. Then,

$$P^\sharp(A^*) = \sup_{A\in\mathbb{R}^{n\times n}} \{A : A^* - \det(A)\} \equiv +\infty, \quad P^{\sharp\sharp}(A) \equiv -\infty,$$

where the $A : A^* = \sum_{i,j=1}^n A_{ij}A^*_{ij} = \langle A, A^*\rangle$ is a standard bilinear form in $\mathbb{R}^{n\times n}$. ♡

5.1.4 Generalized Differentials and Fenchel Duality

Sub-gradients and sub-differentials represent generalizations of the classical differentiation notions. Let $P : \mathcal{U} \to \mathbb{R}$ be a continuous and convex functional. By the Fenchel-Young inequality we know that for a given $u \in \mathcal{U}$, there exists a $u^* \in \mathcal{U}^*$ such that

$$P^\sharp(u^*) \geq \langle v, u^*\rangle - P(v) \;\; \forall v \in \mathcal{U}.$$

From the definition of a conjugate functional, we know that if the supremum of the right-hand side is actually attained at u then

$$P^\sharp(u) = \langle u, u^*\rangle - P(u).$$

Substituting this equality into the Fenchel-Young inequality, we obtain

$$P(v) \geq \langle v - u, u^*\rangle - P(u) \;\; \forall v \in \mathcal{U}.$$

If P is differentiable, the slope u^* is the derivative of P at u. For extended functionals, we have the following definition.

Definition 5.1.17 (Generalized Differentials and Gradients)
Let $P : \mathcal{U} \to \bar{\mathbb{R}}$ be a functional with nonempty domain, $u \in \operatorname{dom} P$.

(D1) The *sub-differential* $\partial^- P(u)$ of P at u is the subset (which may be empty) of $\mathcal{U}^*$ defined by

$$\partial^- P(u) := \{\, u^* \in \mathcal{U}^* \mid P(v) - P(u) \geq \langle v - u, u^*\rangle \;\; \forall v \in \mathcal{U} \,\}. \tag{5.31}$$

The elements $u^* \in \partial^- P(u)$ are also called *sub-gradients*. If $\partial^- P(u) \neq \emptyset$, then P is said to be *sub-differentiable* at u.

(D2) The *super-differential* $\partial^+ P(u)$ of P at u is the subset (which may be empty) of $\mathcal{U}^*$ defined by

$$\partial^+ P(u) := \{\, u^* \in \mathcal{U}^* \mid P(v) - P(u) \leq \langle v - u, u^*\rangle \;\; \forall v \in \mathcal{U} \,\}. \tag{5.32}$$

The elements $u^* \in \partial^+ P(u)$ are also called *super-gradients*. If $\partial^+ P(u) \neq \emptyset$, then P is said to be *super-differentiable* at u. ◇

Clearly, $\partial^+ P(u) = -\partial^-(-P(u))$. In convex analysis, it is convention that ∂^- is simply written as ∂. In this book, ∂ stands for either ∂^- or ∂^+, i.e.

$$\partial = \{\partial^-, \partial^+\}.$$

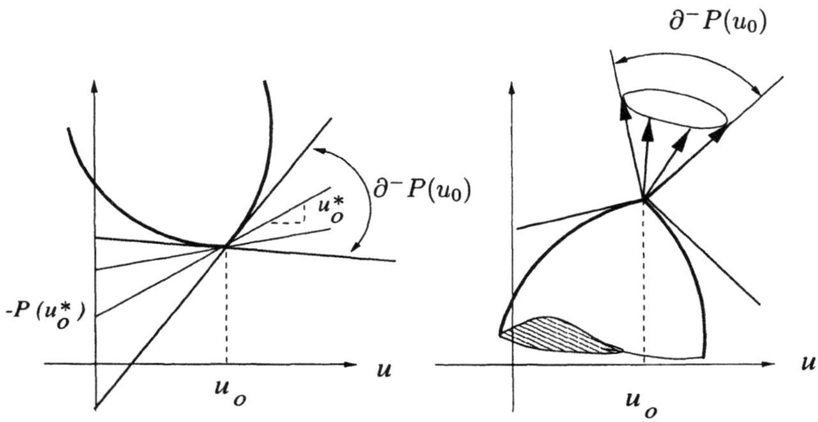

(a) The graph of $P(u)$. (b) The level surface $P(u) = \alpha$.

Figure 5.8. Geometrical interpretation of the sub-differentials for graph and level surface.

The super-differential ∂^+ is also called the *over-differential*, written as $\bar{\partial}$ (see Aubin and Ekeland, 1984).

Remark 5.1.5 Although there is no convexity requirement for P in the definition, the sub-differential is actually works only for convex functional. If P is not convex at u_0, the supremum in

$$\sup_v \{\langle v, u^* \rangle - P(v)\}$$

cannot achieved at u_0; hence, $\partial^- P(u_0) = \emptyset$. Of course, the supremum is achived at $P^{\sharp\sharp}(u_0)$. Thus, for (locally) concave functionals, the dual definition ∂^+ should be used. ■

An equivalent definition of the sub-differential of P is

$$u^* \in \partial^- P(u) \quad \Leftrightarrow \quad u \text{ maximizes } u \to \langle u^*, u \rangle - P(u) \text{ on } \mathcal{U}. \tag{5.33}$$

To see the geometrical interpretation of $\partial^- P(u)$, let us consider a convex functional $P(u)$, which is not differentiable at u_0 (see Fig. 5.8). The sub-gradient of the graph of P at u_0 is shown in (a). We can see that the tangents to the graph of P on either side of $(u, P(u))$ form cones in which lie all continuous affine functionals $\ell(u)$ such that $\ell(u_0) = P(u_0)$. The slopes of these affine functionals are easily seen to be sub-gradients of P at u_0. However, the sub-gradient of P for the level surface $P(u) = c$ forms a normal cone at u_0 (see Fig. 5.8(b)). For the level surface of $P(u)$, $\partial^- P(u)$ denotes a normal cone at $P(u_0) = c$.

The definition of sub-differential plays an important role in optimization problems. It is obvious from the definition that if

$$P(v) - P(u) \geq 0 = \langle 0, v - u \rangle \;\; \forall v \in \mathcal{U}.$$

then

$$P(u) = \inf_{v \in \mathcal{U}} P(v) \text{ if and only if } 0 \in \partial^- P(u). \tag{5.34}$$

Similarly,

$$P(u) = \sup_{v \in \mathcal{U}} P(v) \text{ if and only if } 0 \in \partial^+ P(u). \tag{5.35}$$

Recalling the definition of the conjugate functional we obtain the following theorem.

Theorem 5.1.11 (Fenchel Duality Theorem)
Let $P : \mathcal{U} \to \bar{\mathbb{R}}$ and let P^* be its Fenchel-conjugate. Then,
(T1) $u^* \in \partial P(u) \;\Rightarrow\; u \in \partial P^*(u^*)$;
(T2) $u^* \in \partial P(u) \;\Leftrightarrow\; P^*(u^*) = \langle u, u^* \rangle - P(u)$;
(T3) if $P \in \Gamma(\mathcal{U})$, then $u \in \partial P^*(u^*) \;\Leftrightarrow\; u^* \in \partial P(u)$.

If $\ell(u) = \langle u^*, u \rangle + c$ is a continuous affine functional everywhere below P, then $\ell(u) \le P^{\sharp\sharp}(u)$ $\forall u \in \mathcal{U}$. For some u, if $\ell(u) = P(u)$, then $\ell(u) = P^{\sharp\sharp}(u)$. Therefore, we have

$$\partial P(u) \neq \emptyset \;\Rightarrow\; P(u) = P^{**}(u),$$

$$P(u) = P^{**}(u) \;\Rightarrow\; \partial P(u) = \partial P^{**}(u).$$

The geometrical interpretation of these properties is shown in Fig. 5.6.

Some further properties of sub-gradients and their relation to Gâteaux differentiability are listed in the following theorem.

Theorem 5.1.12 *Let $P \in \Gamma_0(\mathcal{U})$. Then,*
(T1) $\partial P(u)$ (possible empty) is a convex, weak-closed subset of $\mathcal{U}^*$;*
(T2) if P is continuous at u, then $\partial P \neq \emptyset$;
(T3) if P is Gâteaux differentiable at u, then it is sub-differentiable at u and

$$\partial P = \{DP(u)\}.$$

Conversely, if at the point $u \in \mathcal{U}$, P is continuous and has only one sub-gradient, then P is Gâteaux differentiable at u and $\partial P(u) = DP(u)$.

Example 5.1.9 Let $P(u) = |u|$, $u \in \mathbb{R}$. Since P is Gâteaux differentiable for $u \neq 0$, we have $\partial^- P(u) = DP(u) = \operatorname{sgn}(u)$. At $u = 0$, by definition, if $u^* \in \partial^- P(0)$, we have $|v| \ge u^* v$ $\forall v \in \mathbb{R}$. Thus, $u^* \in [-1, 1]$ and

$$\partial^- P(u) = \begin{cases} 1 & u > 0 \\ [-1, 1] & u = 0 \\ -1 & u < 0, \end{cases}$$

which forms a convex subset of $\mathbb{R}$. ♡

Example 5.1.10 Let $P : \mathbb{R} \to \overrightarrow{\mathbb{R}}$ be given by

$$P(u) = \begin{cases} 0 & u \le 0, \\ +\infty & u > 0. \end{cases}$$

For $u < 0$, $\partial^- P(u) = 0$. At $u = 0$, if $u^* \in \partial^- P(0)$, then $P(v) \geq u^* v \;\; \forall v \in \mathbb{R}$. Thus, $u^* \in [0, +\infty)$. If $u^* \in \partial^- P(u)$, $u > 0$, then $P(v) - \infty \geq u^*(v - u) \;\; \forall v \in \mathbb{R} \Rightarrow \partial^- P(u) = \emptyset$, and hence

$$\partial^- P(u) = \begin{cases} 0 & u < 0 \\ [0, +\infty) & u = 0 \\ \emptyset & u > 0. \end{cases}$$

Again, this forms a convex subset of $\mathbb{R}$. ♡

Definition 5.1.18 (Normal and Tangent Cones)
(D1) If $\mathcal{K}$ is a closed convex subset of $\mathcal{U}$, then the *normal cone* to $\mathcal{K}$ at u is a subset of $\mathcal{U}^*$ defined by

$$N_{\mathcal{K}}(u) = \{u^* \in \mathcal{U}^* | \;\; \langle u^*, u - v\rangle \geq 0 \;\; \forall v \in \mathcal{K}\}. \tag{5.36}$$

(D2) The closure of the cone generated by $\mathcal{K} - \{u\}$, defined by

$$T_{\mathcal{K}}(u) = \mathrm{cl}\left\{ \cup_{\theta > 0} \frac{\mathcal{K} - u}{\theta} \right\}, \tag{5.37}$$

is said to be the *tangent cone* to $\mathcal{K}$ at u. ◇

Example 5.1.11 Let $\mathcal{U}$ be a Banach space and $\mathcal{I}_{\mathcal{K}}$ the indicator functional of a nonempty, closed, convex set, i.e.

$$\mathcal{I}_{\mathcal{K}}(u) = \begin{cases} 0 & \text{if } u \in \mathcal{K}, \\ +\infty & \text{if } u \notin \mathcal{K}. \end{cases}$$

Then, $u^* \in \partial^- \mathcal{I}_{\mathcal{K}}(u)$ implies that $\mathcal{I}_{\mathcal{K}}(v) - \mathcal{I}_{\mathcal{K}}(u) \geq \langle u^*, v - u\rangle \;\; \forall v \in \mathcal{U}$. Thus, if $u \in \mathcal{K}$, $\langle u^*, u - v\rangle \geq 0 \;\; \forall v \in \mathcal{K}$ and hence

$$\partial^- \mathcal{I}_{\mathcal{K}}(u) = \begin{cases} \{0\} & \text{if } u \in \mathcal{K} \\ N_{\mathcal{K}}(u) & \text{if } u \in \partial\mathcal{K} \\ \emptyset & \text{if } u \notin \mathcal{K}, \end{cases}$$

where $N_{\mathcal{K}}(u)$ stands for a cone of outward normals to $\mathcal{K}$ at u. ♡

Proposition 5.1.4 *If $\mathcal{K}$ is a closed convex subset of $\mathcal{U}$, then*

$$N_{\mathcal{K}}(u) = \partial^- \mathcal{I}_{\mathcal{K}}(u), \tag{5.38}$$

and the tangent cone is the negative conjugate of the normal cone, i.e.

$$T_{\mathcal{K}}(u) = N_{\mathcal{K}}^{\ominus}(u) \;\; \forall u \in \mathcal{K}. \tag{5.39}$$

5.1.5 Calculus on Banach Spaces

We next investigate how the operators ∂ and $*$ operate on convex functionals. First, it is easily seen that

$$\partial(\lambda P)(x) = \lambda \partial P(x) \ \ \forall \lambda > 0. \tag{5.40}$$

Also, the following result follows directly from standard differential calculus.

Theorem 5.1.13 *Let $P_1, P_2 \in \check{\Gamma}_0(\mathcal{U})$. Then at every $u \in \mathcal{U}$,*

$$\partial^-(P_1 + P_2)(u) \supset \partial^- P_1(u) + \partial^- P_2(u). \tag{5.41}$$

However, if there exists a point $u \in$ dom $P_1 \cap$ dom P_2, where P_1 is continuous, then,

$$\partial^-(P_1 + P_2)(v) = \partial^- P_1(v) + \partial^- P_2(v) \ \ \forall v \in \mathcal{U}. \tag{5.42}$$

We now go back to discuss the Fenchel duality for integral functionals on $\mathcal{L}^\alpha$, $(\alpha \in [1, +\infty])$. Let $\Omega \subset \mathbb{R}^n$ be endowed with the Lebesgue measure, and let $f : \Omega \times \mathbb{R}^n \to \overrightarrow{\mathbb{R}}$ be some non-negative function, i.e.

$$f(x, u) \geq 0 \ \ \forall (x, u) \in \Omega \times \mathbb{R}^n. \tag{5.43}$$

For every $u \in \mathcal{L}^\alpha(\Omega; \mathbb{R}^n)$, henceforth abbreviated as to $\mathcal{L}^\alpha$, define $P : \mathcal{L}^\alpha \to \overrightarrow{\mathbb{R}}$ by

$$P(u) = \int_\Omega f(x, u(x)) d\Omega. \tag{5.44}$$

We shall collect some of its main properties to serve for future reference. The proofs of these properties can be found in Ekeland (1990 §2.2 - §2.3).

Proposition 5.1.5 *Assume that f is a non-negative function on Ω, If $f(x, \cdot)$ is l.s.c. on $\mathbb{R}^n$, then P is l.s.c. on $\mathcal{L}^\alpha$. If f is convex in u, then P is convex on $\mathcal{L}^\alpha$.*

Proof. The convexity is clear. To prove the lower semicontinuity, we need Fatou's lemma. If u_n is a sequence converging to u in $\mathcal{L}^\alpha(\Omega; \mathbb{R}^n)$, then we can choose a subsequence $u_{n'}$ such that

$$\lim_{n \to \infty} \inf P(u_n) = \lim_{n' \to \infty} P(u_{n'}).$$

Since f is non-negative function, Fatou's lemma yields

$$\lim_{n' \to \infty} \int_\Omega f(x, u_{n'}) d\Omega \geq \int_\Omega \lim_{n' \to \infty} \inf f(x, u_{n'}(x)) d\Omega.$$

Using the pointwise convergence $u_{n'}(x) \to u(x)$ and the lower semi-continuity of $f(x, u)$, we have

$$\lim_{n \to \infty} \inf P(u_n) \geq \int_\Omega \lim_{n \to \infty} \inf f(x, u_n(x)) \, dx \geq \int_\Omega f(x, u(x)) d\Omega = P(u),$$

i.e., P is l.s.c. □

We note that $(\mathcal{L}^\alpha)^* = \mathcal{L}^{\alpha^*}$ with $1/\alpha + 1/\alpha^* = 1$. By using the obvious notations

$$f^*(x; u^*) = [f(x,\cdot)]^*(u^*), \quad \partial f(x;u) = \partial f(x,\cdot)(u),$$

we have the following result.

Theorem 5.1.14 *Assume that Ω has finite measure, that the non-negative function f is convex, l.s.c. on $\mathbb{R}^n$, and that there exists $\bar{u} \in \mathcal{L}^\infty$ such that*

$$\int_\Omega f(x, \bar{u}(x))d\Omega < +\infty. \tag{5.45}$$

Then,

$$P^*(u^*) = \int_\Omega f^*(x; u^*(x))d\Omega \;\; \forall u^* \in \mathcal{L}^{\alpha^*}. \tag{5.46}$$

Corollary 5.1.3 *Assume that Ω has finite measure, that f is convex, l.s.c. on $\mathbb{R}^n$, and that there exists $\bar{u}$, $\bar{u}^*$ such that*

$$\int_\Omega |f(x, \bar{u}(x))|d\Omega < \infty, \quad \int_\Omega |f^*(x; \bar{u}^*(x))|d\Omega < \infty. \tag{5.47}$$

Then,

$$\partial P(u) = \{u^* \in \mathcal{L}^{\alpha^*} |\ u^*(x) \in \partial f(x; u(x)) \quad in\ \Omega\}. \tag{5.48}$$

Proof. From the convexity of P, we know that $u^* \in \partial P(u) \;\Leftrightarrow\; u \in \partial P^*(u^*)$ and

$$P(u) + P^*(u^*) = \langle u, u^* \rangle.$$

Explicitly, this

$$\int_\Omega [f(x, u(x)) + f^*(x; u^*(x)) - u(x)u^*(x)]d\Omega = 0.$$

By Fenchel-Young's inequality, the integrand is non-negative almost everywhere (a.e.). So we must have

$$f(x, u(x)) + f^*(x; u^*(x)) - \langle u(x), u^*(x) \rangle = 0 \quad \text{in } \Omega.$$

This means that $u^*(x) \in \partial^- f(x; u(x))$ a.e. in Ω. □

The following continuity and differentiability results are due to Krasnoselskii (1963).

Theorem 5.1.15 *Assume that Ω has finite measure, and $f : \Omega \times \mathbb{R}^n \to \mathbb{R}$ is continuous on $\mathbb{R}^n$ for a.e. $x \in \Omega$. Suppose that there are constants a, c and $r > 0$ such that*

$$|f(x,u)| \le \max\{a, c\|u\|^r\}.$$

Let $\alpha > r$ and define a map $G : \mathcal{L}^\alpha \to \mathcal{L}^{\alpha/r}$ by

$$[G(u)](x) = f(x, u(x)).$$

Then, G is continuous in the norm topology.

Corollary 5.1.4 *Assume that Ω has finite measure, that the non-negative convex function $f(x,u)$ is C^1 with respect to u for a.e. $x \in \Omega$, and that the conditions in (5.47) are satisfied. Let $\alpha \in (1,+\infty)$ and $\alpha^* = \alpha/(\alpha-1)$. If, for some constants a and c, the condition*

$$\|f'(x;u(x))\| \leq \max\{a, c\|u\|^{\alpha-1}\} \tag{5.49}$$

holds, then $DP : \mathcal{L}^\alpha \to \mathcal{L}^{\alpha^*}$ is C^1 and

$$[DP(u)](x) = f'(x;u(x)). \tag{5.50}$$

Let $\mathcal{U}$ and $\mathcal{E}$ be two Banach spaces with topological duals $\mathcal{U}^*$ and $\mathcal{E}^*$, and $\Lambda : \mathcal{D}(\Lambda) \subset \mathcal{U} \to \mathcal{E}$ a closed linear operator with adjoint $\Lambda^* : \mathcal{E}^* \to \mathcal{U}^*$. Let $W \in \check{\Gamma}(\mathcal{E})$, and the composite functional $W \circ \Lambda : \mathcal{U} \to \mathbb{R}$ belong to $\check{\Gamma}(\mathcal{U})$.

Theorem 5.1.16 *If there exists a point $u_0 \in \mathcal{D}(\Lambda) \subset \mathcal{U}$ at which $W(\Lambda u_0)$ is continuous and finite, then,*

$$(W \circ \Lambda)^*(u^*) = \min\{W^*(\epsilon^*)|\ \Lambda^*\epsilon^* = u^*\}. \tag{5.51}$$

The minimum is $+\infty$ if $(\Lambda^*)^{-1}(u^*) = \emptyset$, and is achieved at some point $e_0^* \in (\Lambda^*)^{-1}(u^*)$ otherwise.

Corollary 5.1.5 *With the same assumption, we have*

$$\partial(W \circ \Lambda)(u) = \Lambda^*\partial W(\Lambda u) \ \ \forall u \in \mathcal{U}. \tag{5.52}$$

Proof. Let $u^* \in \partial(W \circ \Lambda)(u)$. By definition,

$$W(\Lambda u) + (W \circ \Lambda)^*(u^*) = \langle u, u^*\rangle.$$

from the proposition that there is some ϵ^* such that $\Lambda^*\epsilon^* = u^*$ and

$$(W \circ \Lambda)^*(u^*) = W^*(\epsilon^*),$$

we then obtain

$$W(\Lambda u) + W^*(\epsilon^*) = \langle u, \Lambda^*\epsilon^*\rangle = \langle \Lambda u, \epsilon^*\rangle.$$

This means precisely that $\epsilon^* \in \partial W(\Lambda u)$. Equation $\Lambda^*\epsilon^* = u^*$ gives $u^* \in \Lambda^*\partial W(\Lambda u)$, and the result follows. □

Now we let

$$P(u) = W(\Lambda u) - F(u). \tag{5.53}$$

Theorem 5.1.17 *Let $F \in \hat{\Gamma}_0(\mathcal{U})$, $W \in \check{\Gamma}(\mathcal{E})$ and $\Lambda : \mathcal{U} \to \mathcal{E}$. If there exists a point $u_0 \in \mathcal{D}(\Lambda) \subset \mathcal{U}$ at which $W(\Lambda u_0)$ is continuous and finite and $F(u_0)$ is finite, then $0 \in \partial^- P(\bar{u})$ if and only if*

$$0 \in \Lambda^*\partial^- W(\Lambda\bar{u}) - \partial^+ F(\bar{u}). \tag{5.54}$$

Proof. By Theorem 5.1.13

$$\Lambda^* \partial^- W(\Lambda \bar{u}) - \partial^+ F(\bar{u}) \subset \partial^-(W \circ \Lambda - F)(\bar{u}).$$

We need to prove that

$$u^* \in \partial^-(W \circ \Lambda - F)(u) \Rightarrow \quad u^* \in \Lambda^* \partial^- W(\Lambda u) - \partial^+ F(u).$$

Under assumption, $\Lambda(\text{dom } F) \cup \text{dom } W \neq \emptyset$, we take $u^* \in \partial^-(W \circ \Lambda - F)(u)$. Then, there exists $\bar{\epsilon}^* \in \mathcal{E}^*$ such that

$$(W \circ \Lambda)^*(u^*) = F^*(u^* - \Lambda^* \bar{\epsilon}^*) + W(\bar{\epsilon}^*).$$

Thus, from the Fenchel-Young equality, we have

$$\begin{aligned} \langle u, u^* \rangle &= W(\Lambda u) - F(u) + (W \circ \Lambda - F)^*(u^*) \\ &= (W(\Lambda u) + W^*(\bar{\epsilon}^*)) - (F(u) + F^*(u^* - \Lambda^* \bar{\epsilon}^*)). \end{aligned}$$

Consequently,

$$0 = [\langle \Lambda u; \bar{\epsilon}^* \rangle - W(\Lambda u) - W^*(\bar{\epsilon}^*)] - [\langle u, u^* - \Lambda^* \bar{\epsilon}^* \rangle - F(u) - F^*(u^*)].$$

Since the first bracket is negative or zero, and the second one is positive or zero, it follows that they are both zero, and hence

$$\bar{\epsilon}^* \in \partial^- W(\Lambda u), \quad u^* - \Lambda^* \bar{\epsilon}^* \in \partial^+ F(u).$$

Thus, we have shown that $u^* = (u^* - \Lambda^* \bar{\epsilon}^*) + \Lambda^* \bar{\epsilon}^* \in \Lambda^* \partial^- W(\Lambda u) - \partial^+ F(u)$. □

5.2 Perturbation and Rockafellar Duality Theory

Rockafellar duality theory is based on an elegant perturbation method which was proposed by Rockafellar (1969). Although this duality is a special case of the generalized Lagrange duality in geometrically linear convex systems, which will be discussed in the next section, the perturbation method has many important applications in constrained nonlinear optimization, parametric optimal control and variational problems. Our discussion in this section is mainly for convex primal problems.

5.2.1 Perturbation, Normality and Stability

Given the paired reflexive Banach spaces $\{(\mathcal{U}, \mathcal{U}^*), \langle *, * \rangle\}$ and $\{(\mathcal{E}, \mathcal{E}^*), \langle *; * \rangle\}$, we shall assume the existence of a bounded linear operator $\Lambda : \mathcal{U} \to \mathcal{E}$ with adjoint $\Lambda^* : \mathcal{E}^* \to \mathcal{U}^*$. We shall also assume that the functional P can be written as

$$P(u) = \Phi(u, \Lambda u)$$

where $\Phi : \mathcal{U} \times \mathcal{E} \to \bar{\mathbb{R}}$ is a canonical functional. Then, the problem $(\mathcal{P}_{\text{inf}})$ takes the form

$$(\mathcal{P}_{\text{inf}}) : \quad P(\bar{u}) = \inf_{u \in \mathcal{U}} \Phi(u, \Lambda u). \tag{5.55}$$

For every $\epsilon \in \mathcal{E}$, we consider a functional $\Psi(u, \epsilon)$ defined on $\mathcal{U} \times \mathcal{E}$ such that

$$\Psi(u, 0) = P(u) = \Phi(u, \Lambda u).$$

Thus, the perturbation of problem $(\mathcal{P}_{\text{inf}})$ can be proposed as a family of the following type of problem.

Definition 5.2.1 (Perturbed Problem)
For every given $\epsilon \in \mathcal{E}$, the problem of finding $h(\epsilon)$ such that

$$(\mathcal{P}_{\text{inf}})_\epsilon : \quad h(\epsilon) = \inf_{u \in \mathcal{U}} \Psi(u, \epsilon) \tag{5.56}$$

is called the *perturbed problem* of $(\mathcal{P}_{\text{inf}})$. ◇

Clearly, $(\mathcal{P}_{\text{inf}}) = (\mathcal{P}_{\text{inf}})_0$. Two main types of useful perturbations will be described later.

Proposition 5.2.1 *If $\Psi \in \check{\Gamma}_0(\mathcal{U} \times \mathcal{E})$, then the functional $h : \mathcal{E} \to \overrightarrow{\mathbb{R}}$ is convex.*

Proof. We need to show that for any given $\theta \in [0, 1]$,

$$h(\theta\epsilon + (1 - \theta)e) \leq \theta h(\epsilon) + (1 - \theta)h(e), \quad \forall \epsilon, e \in \mathcal{E}. \tag{5.57}$$

The inequality is obvious if either $h(\epsilon)$ or $h(e) = +\infty$, so we assume that $h(\epsilon) < +\infty$ and $h(e) < +\infty$. Then, for every $a > h(\epsilon)$ and every $b > h(e)$, there are $u, v \in \mathcal{U}$ such that

$$h(\epsilon) \leq \Psi(u, \epsilon) \leq a, \quad h(e) \leq \Psi(v, e) \leq b.$$

Since $\Psi \in \check{\Gamma}_0(\mathcal{U} \times \mathcal{E})$,

$$\begin{aligned} h(\theta\epsilon + (1 - \theta)e) &= \inf_{w \in \mathcal{U}} \Psi(w, \theta\epsilon + (1 - \theta)e) \leq \Psi(\theta u + (1 - \theta)v, \theta\epsilon + (1 - \theta)e) \\ &\leq \theta\Psi(u, \epsilon) + (1 - \theta)\Psi(v, e) \leq \theta a + (1 - \theta)b. \end{aligned}$$

Let a and b decrease towards $h(\epsilon)$ and $h(e)$, respectively. The inequality

$$h(\theta\epsilon + (1 - \theta)e) \leq \theta a + (1 - \theta)b$$

then gives (5.57) as the limit. □

Definition 5.2.2 (Normal and Stable Problems)
(D1) Problem $(\mathcal{P}_{\text{inf}})$ is said to be *normal* if $h(0)$ is finite and h is l.s.c. at 0.
(D2) Problem $(\mathcal{P}_{\text{inf}})$ is said to be *stable* if $h(0)$ is finite and $h(\epsilon) - h(0) \geq 0 \ \forall \epsilon \in \mathcal{E}$. ◇

Theorem 5.2.1 (Stability Criterion) *We assume that (A1) Ψ is convex, (A2) $(\mathcal{P}_{\text{inf}})$ is normal, and (A3) there exists $u_0 \in \mathcal{U}$ such that $\epsilon \mapsto \Psi(u_0, \epsilon)$ is finite and continuous at $0 \in \mathcal{E}$. Then, problem $(\mathcal{P}_{\text{inf}})$ is stable.*

Proof. By Proposition 5.2.1, h is convex. Since $(\mathcal{P}_{\text{inf}})$ is normal, $h(0)$ is finite.

Since the function $\epsilon \mapsto \Psi(u_0, \epsilon)$ is convex and continuous at $0 \in \mathcal{E}$, there exists a neighborhood $\mathcal{E}_b$ of 0 in $\mathcal{E}$ on which this functional is bounded above, i.e.

$$\Psi(u_0, \epsilon) \le M < +\infty \ \ \forall \epsilon \in \mathcal{E}_b.$$

But $h(\epsilon) = \inf_{u \in \mathcal{U}} \Psi(u, \epsilon) \le \Psi(u_0, \epsilon) \le M$, $\forall \epsilon \in \mathcal{E}_b$, and, therefore, h is continuous at 0. □

Here we describe two important particular cases of the perturbation problem.

I. The first one is based on the general framework of Fenchel (1949) and Rockafellar (1967). The primal functional is assumed to be $P(u) = \Phi(u, \Lambda u)$ and the perturbation is

$$\Psi(u, \epsilon) = \Phi(u, \Lambda u + \epsilon), \tag{5.58}$$

where $\epsilon \in \mathcal{E}$ can be consider as a perturbation parameter, or an *internal source*. The relationships between the properties of Ψ and those of Φ are

$$\Phi \text{ is convex} \quad \Rightarrow \quad \Psi \text{ is convex}, \tag{5.59}$$

$$\Phi \in \check{\Gamma}_0(\mathcal{U} \times \mathcal{E}) \quad \Rightarrow \quad \Psi \in \check{\Gamma}_0(\mathcal{U} \times \mathcal{E}). \tag{5.60}$$

If $P(u) = \Phi(u, \Lambda u) = W(\Lambda u) - F(u)$, and $W \in \check{\Gamma}(\mathcal{E})$, $F \in \hat{\Gamma}(\mathcal{U})$, then $\Psi(u, \epsilon) = W(\Lambda u + \epsilon) - F(u)$. In this case, the stability criterion theorem shows that if there exists at least one $u \in \mathcal{U}$ such that $F(u)$ is finite and $W(\Lambda u)$ is finite and continuous at Λu, then $(\mathcal{P}_{\inf})$ is stable.

II. The second one, due to Arrow-Hurwitz (cf., e.g., Ekeland and Temam, 1967) in constrained programming, is

$$\begin{aligned} &\text{Minimize} \quad F_0(u) \ \ \forall u \in \mathcal{U}_a, \\ &s.t. \ \Lambda u \le 0, \end{aligned} \tag{5.61}$$

where $F_0 \in \Gamma_0(\mathcal{U}_a)$ is a canonical functional, $\mathcal{U}_a$ is a non-empty closed convex subset of $\mathcal{U}$, and $\Lambda : \mathcal{U} \to \mathcal{E}$ is a linear operator. Let $\mathcal{I}_{\mathcal{U}_a}(u)$ be the indicator functional of $\mathcal{U}_a$ and $\mathcal{I}_{\mathcal{E}_a}(\epsilon)$ the indicator functional of the set

$$\mathcal{E}_a = \{\epsilon \in \mathcal{E} \mid \epsilon \le 0\}. \tag{5.62}$$

We write $F = -F_0(u) - \mathcal{I}_{\mathcal{U}_a}(u)$, $W(\Lambda u) = \mathcal{I}_{\mathcal{E}_a}(\Lambda u)$, and let

$$P(u) = \Phi(u, \Lambda u) = W(\Lambda u) - F(u) = \mathcal{I}_{\mathcal{E}_a}(\Lambda u) + F_0(u) + \mathcal{I}_{\mathcal{U}_a}(u).$$

Thus, the constrained problem (5.61) can be converted into the extended unconstrained problem $(\mathcal{P}_{\inf})$.

Lemma 5.2.1 *For each given $\epsilon \in \mathcal{E}$, let $\mathcal{K}_\epsilon = \{u \in \mathcal{U} \mid \ u \in \mathcal{U}_a, \ \ \Lambda u \le \epsilon\}$. Then,*

(L1) $\mathcal{K}_\epsilon$ is closed and convex in $\mathcal{U}$ for all $\epsilon \in \mathcal{E}$.

(L2) The set $\mathcal{U}_\epsilon = \{(u,\epsilon) \in \mathcal{U} \times \mathcal{E} |\ u \in \mathcal{U}_a,\ \ \Lambda u \le \epsilon\}$ *is closed and convex in* $\mathcal{U} \times \mathcal{E}$.

The perturbation functional Ψ in this case is chosen to be

$$\Psi(u,\epsilon) = F_0(u) + \mathcal{I}_{\mathcal{K}_\epsilon}(u) = \begin{cases} F_0(u) & \text{if } u \in \mathcal{U}_a,\ \ \Lambda u \le \epsilon \\ +\infty & \text{otherwise.} \end{cases} \tag{5.63}$$

Lemma 5.2.2 *If* $F_0 \in \check{\Gamma}(\mathcal{U}_a)$*, then the perturbation functional* Ψ *defined by* (5.63) *is convex, l.s.c. on* $\mathcal{U} \times \mathcal{E}$*, i.e.* $\Psi \in \check{\Gamma}_0(\mathcal{U} \times \mathcal{E})$.

Proof. Since $\mathcal{U}_\epsilon$ is a closed convex subset of $\mathcal{U} \times \mathcal{E}$, $\mathcal{I}_{\mathcal{U}_\epsilon}(u,\epsilon)$ is convex, l.s.c. on $\mathcal{U} \times \mathcal{E}$. If $F_0 \in \check{\Gamma}_0(\mathcal{U}_a)$, $F_0(u) + \mathcal{I}_{\mathcal{U}_a}(u) \in \check{\Gamma}_0(\mathcal{U})$, then, since $\mathcal{U}_a$ is closed and convex, $\Phi(u,\epsilon) = F_0(u) + \mathcal{I}_{\mathcal{U}_a}(u) + \mathcal{I}_{\mathcal{E}_a}(\epsilon)$ is convex, l.s.c. on $\mathcal{U} \times \mathcal{E}$. This leads to $\Psi(u,\epsilon) = \Phi(u, \Lambda u - \epsilon) = F_0(u) + \mathcal{I}_{\mathcal{U}_a}(u) + \mathcal{I}_{\mathcal{E}_a}(\Lambda u - \epsilon) \in \check{\Gamma}_0(\mathcal{U} \times \mathcal{E})$. □

Applications of this perturbed problem will be illustrated in Section 5.5.

5.2.2 Rockafellar Transformation and Duality Theory

By using the perturbation functional $\Psi : \mathcal{U} \times \mathcal{E} \to \overrightarrow{\mathbb{R}}$ associated with the primal problem $(\mathcal{P}_{\inf})$, the dual problem $(\mathcal{P}_{\sup})$ can be obtained via the Fenchel transformation. Our discussions here are subject to the assumption $\Psi \in \check{\Gamma}_0(\mathcal{U} \times \mathcal{E})$.

Definition 5.2.3 (Rockafellar Dual Transformation)
For a given $\epsilon \in \mathcal{E}$, suppose that $h(\epsilon) = \inf_{u \in \mathcal{U}} \Psi(u,\epsilon)$ is a solution of the perturbed problem $(\mathcal{P}_{\inf})_\epsilon$. Then, the operation $P(u) = \Psi(u,0) \to P^d(\epsilon^*)$, defined by

$$P^d(\epsilon^*) = \inf_{\epsilon \in \mathcal{E}} \{h(\epsilon) - \langle \epsilon ; \epsilon^* \rangle\} = -h^\sharp(\epsilon^*) \ \ \epsilon^* \in \mathcal{E}^*, \tag{5.64}$$

is called the *Rockafellar dual transformation*, and $P^d : \mathcal{E}^* \to \overleftarrow{\mathbb{R}}$ is called the dual functional of P in the sense of Rockafellar. ◇

Since $h^* : \mathcal{E}^* \to \overrightarrow{\mathbb{R}}$ is convex, l.s.c., $P^d : \mathcal{E}^* \to \overleftarrow{\mathbb{R}}$ is always concave, u.s.c. The problem $\inf P^d$ is usually trivial, i.e. is $-\infty$. Thus, only one set of primal-dual variational problems can be proposed, as follows.

$$(\mathcal{P}_{\inf}) : \quad P(u) \to \inf \ \ \forall u \in \mathcal{U}, \tag{5.65}$$

$$(\mathcal{P}^d_{\sup}) : \quad P^d(\epsilon^*) \to \sup \ \ \forall \epsilon^* \in \mathcal{E}^*. \tag{5.66}$$

First of all, we have the following result.

Theorem 5.2.2 (Rockafellar Weak Duality Theorem)
Suppose that $\Phi \in \check{\Gamma}(\mathcal{U}) \times \check{\Gamma}(\mathcal{E})$.
(T1) For any given $(u,\epsilon^*) \in \mathcal{U} \times \mathcal{E}^*$,

$$-\infty \le \sup P^d(\epsilon^*) \le \inf P(u) \le +\infty. \tag{5.67}$$

(T2) If the problem $(\mathcal{P}_{\inf})$ is non-trivial (i.e., $P(u) \not\equiv \infty$), then

$$\sup P^d(\epsilon^*) \leq \inf P(u) < +\infty \quad \forall u \in \mathcal{U}, \quad \epsilon^* \in \mathcal{E}^*. \tag{5.68}$$

(T3) If the problem $(\mathcal{P}^d_{\sup})$ is non-trivial (i.e., $P^d(\epsilon^*) \not\equiv -\infty$), then

$$-\infty < \sup P^d(\epsilon^*) \leq \inf P(u) \quad \forall u \in \mathcal{U}, \quad \epsilon^* \in \mathcal{E}^*. \tag{5.69}$$

Proof. Let $\epsilon^* \in \mathcal{E}^*$. By definition,

$$h^\sharp(\epsilon^*) = \sup_{u \in \mathcal{U}} \sup_{\epsilon \in \mathcal{E}} \{\langle \epsilon; \epsilon^* \rangle - \Psi(u, \epsilon)\}.$$

In particular, for every $u \in \mathcal{U}$,

$$h^\sharp(\epsilon^*) = \Psi^\sharp(0, \epsilon^*) \geq \langle 0, \epsilon^* \rangle - \Psi(u, 0) \quad \forall \epsilon^* \in \mathcal{E}^*,$$

i.e.

$$P^d(\epsilon^*) = -\Psi^\sharp(0, \epsilon^*) \leq \Psi(u, 0) \quad \forall u \in \mathcal{U}, \quad \forall \epsilon^* \in \mathcal{E}^*. \tag{5.70}$$

This implies $\sup P^d(\epsilon^*) \leq \inf P(u)$.

If $(\mathcal{P}_{\inf})$ is non-trivial, there exists a $u_0 \in \mathcal{U}$ such that $P(u_0) < +\infty$ and thus, with (5.67), we have (5.68). The proof for the other statements are analogous.□

Lemma 5.2.3 *The solution set of $(\mathcal{P}^d_{\sup})$ is identical to $\partial^- h^{\sharp\sharp}(0)$.*

Proof. If $\bar{\epsilon}^* \in \mathcal{E}^*$ is a solution of $(\mathcal{P}^d_{\sup})$, then

$$\Psi^\sharp(0, \bar{\epsilon}^*) \leq \Psi^\sharp(0, \epsilon^*) \quad \forall \epsilon^*.$$

Thus, $-h^\sharp(\bar{\epsilon}^*) \geq -h^\sharp(\epsilon^*) \quad \forall \epsilon^* \in \mathcal{E}^*$,

$$P^d(\bar{\epsilon}^*) = \sup\{-h^\sharp(\epsilon^*)\} = \sup_{\epsilon^* \in \mathcal{E}^*} \{-h^\sharp(\epsilon^*)\} = h^{\sharp\sharp}(0), \tag{5.71}$$

and this is equivalent to $\bar{\epsilon}^* \in \partial^- h^{\sharp\sharp}(0)$. □

It is clear from the proof of this Lemma that the inequality (5.67) is equivalent to the well-known inequality $\sup_{\epsilon^*} P^d(\epsilon^*) = h^{\sharp\sharp}(0) \leq h(0) = P(u)$. The difference

$$g = \inf P - \sup P^d \neq 0$$

is called the ***duality gap***. If $\Psi \in \check{\Gamma}_0(\mathcal{U} \times \mathcal{E})$, the extremality condition $(0, \bar{\epsilon}^*) \in \partial^- \Psi(\bar{u}, 0)$ implies that

$$\Psi(\bar{u}, 0) + \Psi^\sharp(0, \bar{\epsilon}^*) = \langle (\bar{u}, 0), (0, \bar{\epsilon}^*) \rangle = 0. \tag{5.72}$$

In this case we have $P(\bar{u}) = P^d(\bar{\epsilon}^*)$. Hence, the duality gap vanishes if the functional P is convex.

Theorem 5.2.3 (Rockafellar Strong Duality Theorem)
(T1) $(\mathcal{P}_{\inf})$ *is stable if and only if* $(\mathcal{P}^d_{\sup})$ *has at least one solution and*

$$\inf P = \max P^d. \tag{5.73}$$

(T2) Dually, $(\mathcal{P}^d_{\sup})$ *is stable if and only if* $(\mathcal{P}_{\inf})$ *has at least one solution and*

$$\min P = \sup P^d. \tag{5.74}$$

(T3) If $(\mathcal{P}_{\inf})$ *and* $(\mathcal{P}^d_{\sup})$ *are both stable, then both have solutions and*

$$+\infty > \min P = \max P^d > -\infty. \tag{5.75}$$

Proof. We only prove the statements (T1) and (T3) since the statement (T2) follows by duality.

If $(\mathcal{P}_{\inf})$ is stable, $h(0)$ is finite and $\partial^- h(0)$ is non-empty, we have in this case the value $h(0) = h^{\sharp\sharp}(0)$ is finite, which implies that $\partial^- h^{\sharp\sharp}(0) = \partial^- h(0) \neq \emptyset$ and, in addition, the problem $(\mathcal{P}_{\inf})$ is normal. Thus, by means of Lemma 5.2.3, the solution set of the dual problem is not empty and $\inf P = \max P^d$.

Conversely, if $h(0) = h^{\sharp\sharp}(0) \in \mathbb{R}$ and if $(\mathcal{P}^d_{\sup})$ has some solutions, then $\partial^- h^{\sharp\sharp}(0) = \partial^- h(0) \neq \emptyset$, which implies that $(\mathcal{P}_{\inf})$ is stable.

If both $(\mathcal{P}_{\inf})$ and $(\mathcal{P}^d_{\sup})$ are stable, then $h(0)$ and $h^{\sharp\sharp}(0)$ are finite and $\min P = \max P^d$. Since $\partial^- h(0) \neq \emptyset$, there exists some solutions such that $\inf P = \min P > -\infty$. Dually, $\max P^d < +\infty$. □

This theorem shows that as long as the primal problem is stable, the dual problem is sure to have at least one solution. However, the existence conditions for the primal solution are stronger. If the primal solution does not exist, the dual problem can provide a *generalized solution* of the problem.

Theorem 5.2.4 (Dual Equivalence Theorem)
The following statements are equivalent to each other.

(S1) $(\mathcal{P}_{\inf})$ *is stable and has a solution* $\bar{u}$.
(S2) $(\mathcal{P}^d_{\sup})$ *is stable and has a solution* $\bar{\epsilon}^*$.
(S3) The extremality relation $P(\bar{u}) = P^d(\bar{\epsilon}^*)$ *is satisfied.*

Proof. By the strong duality theorem, the condition that $(\mathcal{P}_{\inf})$ be stable and have a solution is equivalent to having $\min P = \max P^d$. Dually, this is equivalent to the condition that $(\mathcal{P}^d_{\sup})$ be stable and has a solution. □

Thus, we have the following results.

Theorem 5.2.5 (Complementarity and Extremality Relations)
The following three conditions are equivalent to each other.
(T1) $\bar{u}$ *is a solution of* $(\mathcal{P}_{\inf})$, $\bar{\epsilon}^*$ *is a solution of* $(\mathcal{P}^d_{\sup})$ *and*

$$P(\bar{u}) = \min P(u) = \max P^d(\epsilon^*) = P^d(\bar{\epsilon}^*).$$

(T2) $\Psi(\bar{u},0)$ *and* $\Psi^\sharp(0,\bar{\epsilon}^*)$ *satisfy the complementarity condition*

$$\Psi(\bar{u},0) + \Psi^\sharp(0,\bar{\epsilon}^*) = 0. \tag{5.76}$$

(T3) $\bar{u} \in \mathcal{U}$ *and* $\bar{\epsilon}^* \in \mathcal{E}^*$ *satisfy the extremality condition*

$$(0,\bar{\epsilon}^*) \in \partial^-\Psi(\bar{u},0) \quad \text{or} \quad (\bar{u},0) \in \partial^-\Psi^\sharp(0,\bar{\epsilon}^*). \tag{5.77}$$

Proof. In fact, if $P(\bar{u}) = \inf P(u)$, $P^d(\bar{\epsilon}^*) = \sup P^d(\epsilon^*)$ and $P(\bar{u}) = P^d(\bar{\epsilon}^*)$, then

$$\inf P(u) = \Psi(\bar{u},0) = \sup P^d(\epsilon^*) = -\Psi^\sharp(0,\bar{\epsilon}^*).$$

This gives the complementarity condition. On the product space $(\mathcal{U},\mathcal{E})\times(\mathcal{U}^*,\mathcal{E}^*)$, the complementarity condition can also be written as

$$\Psi(\bar{u},0) + \Psi^\sharp(0,\bar{\epsilon}^*) = \langle(\bar{u},0),(0,\bar{\epsilon}^*)\rangle,$$

which, by the Fenchel duality theorem, is equivalent to the extremality conditions.

Conversely, if the complementarity condition is true, it follows from (5.70) that

$$\Psi(u,0) + \Psi^\sharp(0,\epsilon^*) \geq \Psi(\bar{u},0) + \Psi^\sharp(0,\bar{\epsilon}^*) = 0 \;\; \forall u \in \mathcal{U}, \;\; \forall \epsilon^* \in \mathcal{E}^*,$$

which leads to $P(\bar{u}) = \Psi(\bar{u},0) = \inf_u \Psi(u,0)$ and $P^d(\bar{\epsilon}^*) = -\Psi^\sharp(0,\bar{\epsilon}^*) = \sup_{\epsilon^*}\{-\Psi^\sharp(0,\epsilon^*)\}$. □

In the case of $\Psi(u,\epsilon) = \Phi(u,\Lambda u + \epsilon)$, we have the following corollary.

Corollary 5.2.1 *The following three statements are equivalent to each other.*

(S1) $\bar{u}$ *is a solution of* $(\mathcal{P}_{\inf})$, $\bar{\epsilon}^*$ *is a solution of* $(\mathcal{P}^d_{\sup})$ *and*

$$P(\bar{u}) = \min P(u) = \max P^d(\epsilon^*) = P^d(\bar{\epsilon}^*).$$

(S2) $\Phi(\bar{u},\Lambda\bar{u})$ *and* $\Phi^\sharp(-\Lambda^*\bar{\epsilon}^*,\bar{\epsilon}^*)$ *satisfy the complementary condition*

$$\Phi(\bar{u},\Lambda\bar{u}) + \Phi^\sharp(-\Lambda^*\bar{\epsilon}^*,\bar{\epsilon}^*) = 0. \tag{5.78}$$

(S3) $\bar{u} \in \mathcal{U}$ *and* $\bar{\epsilon}^* \in \mathcal{E}^*$ *satisfy the extremality condition*

$$(-\Lambda^*\bar{\epsilon}^*,\bar{\epsilon}^*) \in \partial^-\Phi(\bar{u},\Lambda\bar{u}). \tag{5.79}$$

Proof. By definition, the pair $(-\Lambda^*\bar{\epsilon}^*,\bar{\epsilon}^*) \in \partial^-\Phi(\bar{u},\Lambda\bar{u})$ if and only if

$$0 = \langle\bar{u},-\Lambda^*\bar{\epsilon}^*\rangle + \langle\Lambda\bar{u};\bar{\epsilon}^*\rangle = \Phi(\bar{u},\Lambda\bar{u}) + \Phi^\sharp(-\Lambda^*\bar{\epsilon}^*,\bar{\epsilon}^*). \tag{5.80}$$

From the inequality (5.67), we have

$$0 = \langle u,-\Lambda^*\epsilon^*\rangle + \langle\Lambda u;\epsilon^*\rangle \leq \Phi(u,\Lambda u) + \Phi^\sharp(-\Lambda^*\epsilon^*,\epsilon^*) \;\; \forall (u,\epsilon^*) \in \mathcal{U}\times\mathcal{E}^*,$$

and it follows that

$$\Phi(\bar{u}, \Lambda\bar{u}) = \inf_u \Phi(u, \Lambda u), \quad \Phi^\sharp(-\Lambda^*\bar{\epsilon}^*, \bar{\epsilon}^*) = \inf_{\epsilon^*} \Phi^\sharp(-\Lambda^*\epsilon^*, \epsilon^*),$$

i.e. $\bar{u}$ minimizes $P(u)$ and $\bar{\epsilon}^*$ is a solution of $(\mathcal{P}^d_{\sup})$.

Conversely, if $P(\bar{u}) = \inf P(u)$, $P^d(\bar{\epsilon}^*) = \sup P^d(\epsilon^*)$ and $P(\bar{u}) = P^d(\bar{\epsilon}^*)$, then $P \in \Gamma_0(\mathcal{U})$ and equality (5.80) is clearly true. □

Theorem 5.2.6 *Let $\mathcal{U}$ be a reflexive Banach space. We assume that*
(A1) $\Phi \in \check{\Gamma}_0(\mathcal{U}, \mathcal{E})$,
(A2) there exists $u_0 \in \mathcal{U}$ such that $\epsilon \mapsto \Phi(u_0, \epsilon)$ is finite at $0 \in \mathcal{E}$,
(A3) $\lim P(u) = +\infty$ if $u \in \mathcal{U}$, $\|u\| \to \infty$.
Under these conditions, $(\mathcal{P}_{\inf})$ and $(\mathcal{P}^d_{\sup})$ each have at least one solution,

$$\inf P(v) = \sup P^d(\epsilon^*), \tag{5.81}$$

and the extremality condition (5.77) is satisfied.

In the case of $\Phi(u, \epsilon) = W(\epsilon) - F(u)$ with $W \in \check{\Gamma}(\mathcal{E})$ and $F \in \hat{\Gamma}(\mathcal{U})$, the dual functional takes the form

$$P^d(\epsilon^*) = -\Phi^\sharp(-\Lambda^*\epsilon^*, \epsilon^*) = F^\flat(\Lambda^*\epsilon^*) - W^\sharp(\epsilon^*). \tag{5.82}$$

The primal and dual problems have the familiar forms

$$(\mathcal{P}_{\inf}): \qquad P(u) = W(\Lambda u) - F(u) \quad \to \inf \quad \forall u \in \mathcal{U}, \tag{5.83}$$

$$(\mathcal{P}^d_{\sup}): \qquad P^d(\epsilon^*) = F^\flat(\Lambda^*\epsilon^*) - W^\sharp(\epsilon^*) \quad \to \sup \quad \forall \epsilon^* \in \mathcal{E}^*. \tag{5.84}$$

The effective domains of P and P^d are respectively

$$\begin{aligned} \mathcal{K} &= \{u \in \mathcal{U} |\ u \in \text{dom } F,\ \Lambda u \in \text{dom } W \subset \mathcal{E}\}, \\ \mathcal{S} &= \{\epsilon^* \in \mathcal{E}^* |\ \epsilon^* \in \text{dom } W^\sharp,\ \Lambda^*\epsilon^* \in \text{dom } F^\flat \subset \mathcal{U}^*\}. \end{aligned}$$

If $\mathcal{K}_a$ is a open subset of $\mathcal{K}$ such that P is Gâteaux differentiable on $\mathcal{K}_a$, then the critical condition $DP(\bar{u}) = 0$ leads to the *Euler-Lagrange equation*

$$\Lambda^* DW(\Lambda\bar{u}) - DF(\bar{u}) = 0 \tag{5.85}$$

for $(\mathcal{P}_{\inf})$. Since $P \in \check{\Gamma}(\mathcal{K}_a)$, any critical point $\bar{u}$ is a global minimizer on $\mathcal{K}_a$.

Dually, on an open subset $\mathcal{S}_a \subset \mathcal{S}$, the critical condition for the Gâteaux differentiable functional P^d leads to the *dual Euler-Lagrange equation*

$$\Lambda DF^\flat(\Lambda^*\bar{\epsilon}^*) - DW^\sharp(\bar{\epsilon}^*) = 0 \tag{5.86}$$

for $(\mathcal{P}^d_{\sup})$. On $\mathcal{K}_a$ and $\mathcal{S}_a$, the extremality condition $P(\bar{u}) = P^d(\bar{\epsilon}^*)$ holds and the Euler-Lagrange equations (5.85), (5.86) are equivalent to each other. The extremality condition (5.77) now leads to the following result.

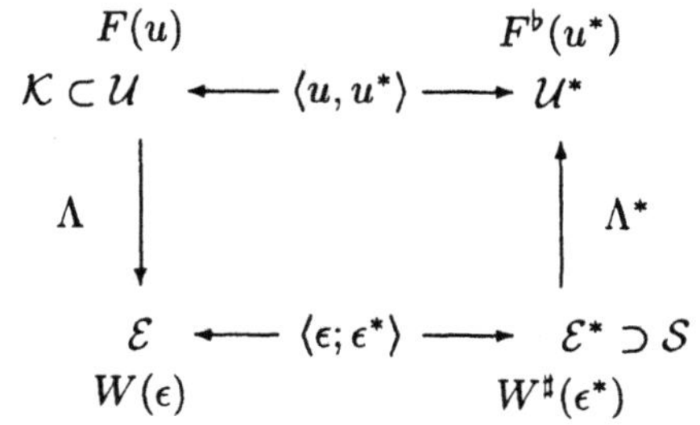

Figure 5.9. Variational structure in a geometrically linear convex system

Proposition 5.2.2 *Suppose that $F \in \hat{\Gamma}_0(\mathcal{U})$, $W \in \check{\Gamma}_0(\mathcal{E})$ and there exists a $u_0 \in \mathcal{K}_a$ such that $W(\Lambda u_0)$ is continuous. Then, $\bar{u}$ minimizes P and $\bar{\epsilon}^*$ maximizes P^d if and only if the extremality conditions*

$$\Lambda \bar{u} \in \partial^- W^\sharp(\bar{\epsilon}^*), \qquad \Lambda^* \bar{\epsilon}^* \in \partial^+ F(\bar{u}), \tag{5.87}$$

hold.

Proof. Since $F \in \hat{\Gamma}_0(\mathcal{U})$, $W \in \check{\Gamma}_0(\mathcal{E})$, we know that $P : \mathcal{U} \to \overrightarrow{\mathbb{R}}$ is l.s.c. and convex, and $P^d : \mathcal{E}^* \to \underleftarrow{\mathbb{R}}$ is concave, u.s.c. From Theorem 5.1.17, $\bar{u}$ minimizes P if and only if

$$0 \in \partial^- P(\bar{u}) = \Lambda^* \partial W(\Lambda \bar{u}) - \partial^+ F(\bar{u}),$$

i.e. there exists $\bar{u}^* \in \Lambda^* \partial^- W(\Lambda \bar{u}) \cap \partial^+ F(\bar{u})$. Hence, there exists $\bar{\epsilon}^* \in \partial^- W(\Lambda \bar{u})$ such that $\Lambda^* \bar{\epsilon}^* \in \partial^+ F(\bar{u})$.

Dually, $\bar{\epsilon}^*$ maximizes P^d, i.e. $P^d(\bar{\epsilon}^*) = \sup P^d(\epsilon^*)$, if and only if

$$0 \in \partial^+ P^d(\bar{\epsilon}^*) = \Lambda \partial^+ F^\flat(\Lambda^* \bar{\epsilon}^*) - \partial^- W^\sharp(\bar{\epsilon}^*).$$

Then, there exists $\bar{\epsilon} \in \partial^- W^\sharp(\bar{\epsilon}^*) \cap \Lambda \partial^+ F^\flat(\Lambda^* \bar{\epsilon}^*)$. Hence, there exists $\bar{u}$ such that $\bar{\epsilon} = \Lambda \bar{u} \in \partial^- W^\sharp(\bar{\epsilon}^*)$, and the extremality conditions are satisfied.

Conversely, if $\bar{u}$ and $\bar{\epsilon}^*$ satisfy the extremality conditions (5.87), then by the Fenchel duality theorem, we have the inverse extremality conditions

$$\bar{\epsilon}^* \in \partial^- W(\Lambda \bar{u}), \quad \bar{u} \in \partial^+ F^\flat(\Lambda^* \bar{\epsilon}^*).$$

Combining the extremality conditions and their inverses, we have $0 \in \partial^- P(\bar{u})$ and $0 \in \partial^+ P^d(\bar{\epsilon}^*)$. Hence, $\bar{u}$ minimizes P and $\bar{\epsilon}^*$ maximizes P^d. □

The diagrammatic framework for this special primal and dual variational structure is shown in Fig. 5.9.

Example 5.2.1 Let $\Omega \subset \mathbb{R}^n$ be a bounded open set with Lipschitz boundary $\partial\Omega = \Gamma$, and let $\Lambda = \text{grad}$. If $u : \Omega \subset \mathbb{R}^n \to \mathbb{R}^m$ is a vector-valued function, then

$$\mathbf{A} = \nabla u = \left(\frac{\partial u^i}{\partial x_\alpha} \right)_{1 \le i \le m,\ 1 \le \alpha \le n}$$

is an $m \times n$ matrix. Let $f : \Omega \times \mathbb{R}^m \times \mathbb{R}^{m\times n} \to \mathbb{R}$ be a $\mathcal{C}^2$ function and let

$$\mathcal{K}_a = \{u \in \mathcal{L}^p(\Omega;\mathbb{R}^m)|\ \nabla u \in \mathcal{L}^p(\Omega;\mathbb{R}^{m\times n}),\ u(x) = 0\ \forall x \in \Gamma\} = \mathcal{W}_0^{1,p}.$$

If $\bar{u} \in \mathcal{U}$ is a solution of

$$P(u) = \Phi(u, \Lambda u) = \int_\Omega f(x, u(x), \nabla u(x)) d\Omega \quad \to \min \quad \forall u \in \mathcal{K}_a,$$

then we must have

$$P(\bar{u}) \le P(\bar{u} + \theta u)\ \ \forall u \in \mathcal{C}_0^2(\Omega;\mathbb{R}^m),\ \ \forall \theta \in \mathbb{R}.$$

The $\mathcal{C}^2$ hypothesis on f ensures that the functional $P(\bar{u} + \theta w)$ is continuous and differentiable in θ. Since $\bar{u}$ minimizes P, we have

$$\frac{d}{d\theta} P(\bar{u} + \theta u)|_{\theta=0} = \langle DP(\bar{u}), u\rangle = 0\ \ \forall u \in \mathcal{C}_0^2(\Omega;\mathbb{R}^m).$$

Since f and $\bar{u}$ are $\mathcal{C}^2$, the fundamental lemma of the calculus of variations leads to $DP(\bar{u}) = 0$, i.e.

$$DP(\bar{u}) = -\text{div}\left(\frac{\partial f}{\partial \epsilon}(x, \bar{u}, \nabla \bar{u})\right) + \frac{\partial}{\partial u} f(x, \bar{u}, \nabla \bar{u}) = 0 \quad \forall x \in \Omega. \tag{5.88}$$

This shows that the solution $\bar{u}$ of the Euler-Lagrange equation is a global minimizer of P on $\mathcal{K}_a$. ♡

5.3 Extended Lagrangian Duality Theory

Lagrange duality theory plays a central role in general variational problems and constrained optimization theory. In geometrically linear systems, Lagrangian formulations and theorems take particular symmetrical forms. We let $\mathbb{S} = \{\langle \mathcal{U}, \mathcal{U}^*\rangle, \langle \mathcal{E}; \mathcal{E}^*\rangle; \Lambda, C\}$ be a canonical system, $\Phi(u, \epsilon)$ and $\Psi(u, \epsilon)$ given functionals on $\mathcal{U} \times \mathcal{E}$ such that $\Psi(u, 0) = P(u) = \Phi(u, \Lambda u)$. In this section, we will study the Lagrangian duality theory for the extended extremum problem

$$(\mathcal{P}_{\text{ext}}): \quad P(u) \ \to \text{ext}\ \ \forall u \in \mathcal{U}, \tag{5.89}$$

where $P : \mathcal{U} \to \bar{\mathbb{R}}$ is a given extended functional in general geometrically linear systems, i.e. there exists a linear operator $\Lambda : \mathcal{U} \to \mathcal{E}$ such that $P(u) = \Phi(u, \Lambda u)$ and for any given $u \in \mathcal{U}$, $\Phi : \mathcal{E} \to \bar{\mathbb{R}}$ is a canonical functional. Our approach is the combination of the elegant Rockafellar concept of perturbation for convex systems and Auchmuty's nice duality in general geometrically linear, nonconvex systems. We first consider the general Lagrangian forms and extremality conditions in infinite-dimensional spaces, and then give applications in finite-dimensional problems. The general Lagrange duality theory in fully nonlinear systems will be discussed in the next chapter.

5.3.1 Lagrangian Forms and Critical Points

Let $\{(\mathcal{U},\mathcal{U}^*),\langle *,*\rangle\}$ and $\{(\mathcal{E},\mathcal{E}^*),\langle *;*\rangle\}$ be two duality pairings, and $\Psi:\mathcal{U}\times\mathcal{E}\to\bar{\mathbb{R}}$ a canonical functional. The partial conjugate of $\Psi(u,\epsilon)$ with respect to ϵ is defined by

$$\Psi^*_\epsilon(u,\epsilon^*)=\operatorname*{ext}_{\epsilon\in\mathcal{E}}\{\langle\epsilon;\epsilon^*\rangle-\Psi(u,\epsilon)\}.$$

Recall that if

$$\Psi(u,0)=P(u)\ \ \forall u\in\mathcal{U},$$

then, for each given $\epsilon\in\mathcal{E}$,

$$(\mathcal{P}_{\rm ext})_\epsilon:\quad h(\epsilon)=\operatorname{ext}\{\Psi(u,\epsilon)|\ \ \forall u\in\mathcal{U}\}$$

is the pertubated extremum problem associated with the primal extremum problem $(\mathcal{P}_{\rm ext})$.

Definition 5.3.1 (Lagrangian Form)
For a given problem $(\mathcal{P}_{\rm ext})$ such that $\Psi:\mathcal{U}\times\mathcal{E}\to\bar{\mathbb{R}}$ is the associated perturbation functional, the functional $L:\mathcal{U}\times\mathcal{E}^*\to\bar{\mathbb{R}}$ defined by

$$L(u,\epsilon^*)=\operatorname{ext}\{\Psi(u,\epsilon)-\langle\epsilon;\epsilon^*\rangle|\ \ \forall\epsilon\in\mathcal{E}\}=-\Psi^\sharp_\epsilon(u,\epsilon^*)\tag{5.90}$$

is called the *Lagrangian form* of the problem $(\mathcal{P}_{\rm ext})$ relative to the given perturbation. ◇

To see the relations between the primal functional $P(u)$ and its associated Lagrangian $L(u,\epsilon^*)$, we assume that $\Psi\in\Gamma_0(\mathcal{U}\times\mathcal{E})$. Then for any $u\in\mathcal{U}$, $\Psi_u:\epsilon\mapsto\Psi(u,\epsilon)$ belongs to $\Gamma(\mathcal{E})$. Thus $\Psi^{**}_\epsilon(u,\epsilon)=\Psi(u,\epsilon)$, whence

$$\begin{aligned}\Psi(u,\epsilon)&=\Psi^{**}_\epsilon(u,\epsilon)=\operatorname*{ext}_{\epsilon^*\in\mathcal{E}^*}\{\langle\epsilon;\epsilon^*\rangle-\Psi^*_\epsilon(u,\epsilon^*)\}\\&=\operatorname*{ext}_{\epsilon^*\in\mathcal{E}^*}\{\langle\epsilon;\epsilon^*\rangle+L(u,\epsilon^*)\}.\end{aligned}$$

Hence,

$$P(u)=\Psi(u,0)=\operatorname*{ext}_{\epsilon^*\in\mathcal{E}^*}L(u,\epsilon^*),$$

and the primal problem $(\mathcal{P}_{\rm ext})$ is equivalent to

$$(L_{\rm ext}):\ \ L(u,\epsilon^*)\ \ \to\operatorname{ext}\ \forall(u,\epsilon^*)\in\mathcal{U}\times\mathcal{E}^*.\tag{5.91}$$

On the other hand, without assuming anything about Ψ, we have

$$\begin{aligned}\Psi^*(u^*,\epsilon^*)&=\operatorname*{ext}_{u\in\mathcal{U}}\operatorname*{ext}_{\epsilon\in\mathcal{E}}\{\langle u,u^*\rangle+\langle\epsilon;\epsilon^*\rangle-\Psi(u,\epsilon)\}\\&=\operatorname*{ext}_{u\in\mathcal{U}}\{\langle u,u^*\rangle+\operatorname*{ext}_{\epsilon\in\mathcal{E}}[\langle\epsilon;\epsilon^*\rangle-\Psi(u,\epsilon)]\}\\&=\operatorname*{ext}_{u\in\mathcal{U}}\{\langle u,u^*\rangle-L(u,\epsilon^*)\}.\end{aligned}$$

The functional dual to P is then obtained as

$$P^d(\epsilon^*)=\operatorname*{ext}_{u\in\mathcal{U}}L(u,\epsilon^*)=-\Psi^*(0,\epsilon^*)=-h^*(\epsilon^*).$$

The problem dual to $(\mathcal{P}_{\text{ext}})$ can be posed as

$$(\mathcal{P}^d_{\text{ext}}): \quad P^d(\epsilon^*) \ \to \text{ext} \ \forall \epsilon^* \in \mathcal{E}^*, \tag{5.92}$$

which is equivalent to (L_{ext}) for an arbitrary functional Ψ.

For the important particular case $\Psi(u,\epsilon) = \Phi(u, \Lambda u + \epsilon)$, we have

$$L(u,\epsilon^*) = -\underset{\epsilon\in\mathcal{E}}{\text{ext}}\{\langle\epsilon;\epsilon^*\rangle - \Phi(u, \Lambda u + \epsilon)\}.$$

For a fixed $u \in \mathcal{U}$, setting $e = \Lambda u + \epsilon$, we find that

$$\begin{aligned} L(u,\epsilon^*) &= -\underset{e\in\mathcal{E}}{\text{ext}}\{\langle e;\epsilon^*\rangle + \langle -\Lambda u;\epsilon^*\rangle - \Phi(u,e)\} \\ &= \langle \Lambda u;\epsilon^*\rangle - \Phi^*_\epsilon(u,\epsilon^*) \end{aligned} \tag{5.93}$$

Hence, the dual functional now reads

$$P^d(\epsilon^*) = \underset{u\in\mathcal{U}}{\text{ext}}\, L(u,\epsilon^*) = -h^\sharp(\epsilon^*) = -\Phi^*(-\Lambda^*\epsilon^*, \epsilon^*). \tag{5.94}$$

Clearly, if $\Phi \in \check{\Gamma}_0(\mathcal{U}\times\mathcal{E})$ and $\Phi^*_u(u^*,\epsilon)$ and $\Phi^*_\epsilon(u,\epsilon^*)$ are its partial super-conjugates, then the extremality conditions $\Lambda^*\epsilon^* \in \partial^-_u\Phi(u,\epsilon)$ and $\epsilon^* \in \partial^-_\epsilon\Phi(u,\epsilon)$ lead to

$$\Phi(u,\Lambda u) + \Phi^*_u(\Lambda^*\epsilon^*, \Lambda u) = \langle u, \Lambda^*\epsilon^*\rangle, \quad \Phi(u,\Lambda u) + \Phi^*_\epsilon(u,\epsilon^*) = \langle \Lambda u;\epsilon^*\rangle.$$

From these, we obtain

$$\langle \Lambda u;\epsilon^*\rangle - \Phi^*_\epsilon(u,\epsilon^*) = \langle u, \Lambda^*\epsilon^*\rangle - \Phi^*_u(\Lambda^*\epsilon^*, \Lambda u).$$

Thus, the so-called *complementary Lagrangian* $L^c : \mathcal{U}\times\mathcal{E}^* \to \bar{\mathbb{R}}$ can be defined by

$$L^c(u,\epsilon^*) = \Phi^*_u(\Lambda^*\epsilon^*, \Lambda u) - \langle u, \Lambda^*\epsilon^*\rangle. \tag{5.95}$$

Clearly, the complementarity condition $L(u,\epsilon^*) + L^c(u,\epsilon^*) = 0$ holds for all critical points (u,ϵ^*) of L.

Introducing a so-called *pseudo-Lagrangian* $L_p : \mathcal{U}\times\mathcal{E}\times\mathcal{E}^* \to \bar{\mathbb{R}}$, defined by

$$L_p(u,\epsilon,\epsilon^*) = \langle \Lambda u - \epsilon;\epsilon^*\rangle + \Phi(u,\epsilon), \tag{5.96}$$

we see that, for any $(u,\epsilon^*) \in \mathcal{U}\times\mathcal{E}^*$, eqn (5.93) gives

$$L(u,\epsilon^*) = \underset{\epsilon\in\mathcal{E}}{\text{ext}} L_p(u,\epsilon,\epsilon^*).$$

A vector $\bar{u} \in \mathcal{U}$ is said to be a ***Lagrange multiplier*** for $(\mathcal{P}^d_{\text{ext}})$ if $\bar{u}$ is a solution of $(\mathcal{P}_{\text{ext}})$.

Dually, a vector $\bar{\epsilon}^* \in \mathcal{E}^*$ is said to be a Lagrange multiplier for $(\mathcal{P}_{\text{ext}})$ if $\bar{\epsilon}^*$ is a solution to $(\mathcal{P}^d_{\text{ext}})$.

This can be seen clearly in the special case when $P(u) = \Phi(u,\Lambda u) = W(\Lambda u) - F(u)$, where $F(u) : \mathcal{U} \to \bar{\mathbb{R}}$, and $W(\epsilon) : \mathcal{E} \to \bar{\mathbb{R}}$ are two given arbitrary functionals. The dual functional in this special case takes the form (5.82). Using the admissible Lagrange multiplier $\epsilon^* \in \mathcal{E}^*$ for relaxing the "geometrical constraint" $\epsilon = \Lambda u$ in the feasible set $\mathcal{K}$ leads to the pseudo-Lagrangian

$$L_p(u, \epsilon^*, \epsilon) = \langle \Lambda u - \epsilon; \epsilon^* \rangle + W(\epsilon) - F(u).$$

While the admissible Lagrange multiplier for the equilibrium constraint $\Lambda^* \epsilon^* = \bar{u}^*$ in the dual feasible set $\mathcal{S}$ is just $u \in \mathcal{U}$. Thus, the Lagrangian form associated with $(\mathcal{P}^d_{\text{ext}})$ is

$$L(u, \epsilon^*) = \langle \Lambda u; \epsilon^* \rangle - W^*(\epsilon^*) - F(u).$$

Its complementary form in this special case reads

$$L^c(u, \epsilon^*) = -\langle u, \Lambda^* \epsilon^* \rangle - W(\Lambda u) - F^*(-\Lambda^* \epsilon^*).$$

The convexity relations between Ψ and L are listed in the following table.

Table 5.1. Convexity relation between Ψ and L

$\Psi \in \check{\Gamma}(\mathcal{U} \times \mathcal{E})$	$\Psi \in \check{\Gamma}(\mathcal{U}) \times \hat{\Gamma}(\mathcal{E})$	$\Psi \in \hat{\Gamma}(\mathcal{U}) \times \check{\Gamma}(\mathcal{E})$	$\Psi \in \hat{\Gamma}(\mathcal{U}) \times \hat{\Gamma}(\mathcal{E})$
$L \in \check{\Gamma}(\mathcal{U}) \times \hat{\Gamma}(\mathcal{E}^*)$	$L \in \check{\Gamma}(\mathcal{U}) \times \check{\Gamma}(\mathcal{E}^*)$	$L \in \hat{\Gamma}(\mathcal{U}) \times \hat{\Gamma}(\mathcal{E}^*)$	$L \in \hat{\Gamma}(\mathcal{U}) \times \check{\Gamma}(\mathcal{E}^*)$

By the definition of the generalized differentials, the following statements are obviously true for an arbitrarily given functional $L : \mathcal{U} \times \mathcal{E}^* \to \bar{\mathbb{R}}$.

A point $(\bar{u}, \bar{\epsilon}^*)$ is a right-saddle point of L if and only if

$$0 \in \partial_u^- L(\bar{u}, \bar{\epsilon}^*), \quad 0 \in \partial_{\epsilon^*}^+ L(\bar{u}, \bar{\epsilon}^*). \tag{5.97}$$

A point $(\bar{u}, \bar{\epsilon}^*)$ is a left-saddle point of L if and only if

$$0 \in \partial_u^+ L(\bar{u}, \bar{\epsilon}^*), \quad 0 \in \partial_{\epsilon^*}^- L(\bar{u}, \bar{\epsilon}^*). \tag{5.98}$$

A point $(\bar{u}, \bar{\epsilon}^*)$ is a super-critical point of L if and only if

$$0 \in \partial_u^+ L(\bar{u}, \bar{\epsilon}^*), \quad 0 \in \partial_{\epsilon^*}^+ L(\bar{u}, \bar{\epsilon}^*). \tag{5.99}$$

A point $(\bar{u}, \bar{\epsilon}^*)$ is a sub-critical point of L if and only if

$$0 \in \partial_u^- L(\bar{u}, \bar{\epsilon}^*), \quad 0 \in \partial_{\epsilon^*}^- L(\bar{u}, \bar{\epsilon}^*). \tag{5.100}$$

Theorem 5.3.1 (Critical Points Theorem)

Let $L \in \Gamma(\mathcal{U}) \times \Gamma(\mathcal{E}^*)$ be a given arbitrary functional and $(\bar{u}, \bar{\epsilon}^*)$ be either a saddle, super- or sub-critical point of L.

(T1) If L is partially Gâteaux differentiable at $(\bar{u}, \bar{\epsilon}^*)$, then $(\bar{u}, \bar{\epsilon}^*)$ is a critical point of L.

(T2) If P and P^d are Gâteaux differentiable at $\bar{u}$ and $\bar{\epsilon}^*$, respectively, then $DP(\bar{u}) = 0$, $DP^d(\bar{\epsilon}^*) = 0$ and

$$P(\bar{u}) = L(\bar{u}, \bar{\epsilon}^*) = P^d(\bar{\epsilon}^*). \tag{5.101}$$

As we have seen in one-dimensional geometrically linear systems, if Ψ is convex then the system has only one set of primal and dual problems. However, if Ψ is nonconvex, the system has two sets of primal and dual extremal problems. We shall discuss the duality theory in general geometrically linear canonical systems. If $\Lambda : \mathcal{U} \to \mathcal{E}$ and $\Lambda^* : \mathcal{E}^* \to \mathcal{U}^*$ are surjective, then the following results also hold dually for L^c.

5.3.2 Extended Saddle Lagrangian Duality Theory

We now turn our attentions to the extremality relations between primal and dual functionals that are linked by the extended Lagrangian form. We first consider a convex functional $\Psi \in \check{\Gamma}(\mathcal{U} \times \mathcal{E})$.

Lemma 5.3.1 *If $\Psi \in \check{\Gamma}_0(\mathcal{U} \times \mathcal{E})$, then the associated Lagrangian form $L : \mathcal{U} \times \mathcal{E}^* \to \bar{\mathbb{R}}$ is a right saddle-Lagrangian, i.e. $L \in \check{\Gamma}(\mathcal{U}) \times \hat{\Gamma}_0(\mathcal{E}^*)$.*

Proof. Since $\epsilon^* \mapsto L(u, \epsilon^*)$ is precisely the conjugate functional of $\epsilon \mapsto \Psi(u, \epsilon)$, then if Ψ is convex on $\mathcal{E}$,

$$\epsilon^* \mapsto L(u, \epsilon^*) \text{ is concave, u.s.c.} \quad \forall u \in \mathcal{U}.$$

From the definition of L, we obtain

$$L(u, \epsilon^*) = \inf_{\epsilon \in \mathcal{E}} \{\Psi(u, \epsilon) - \langle \epsilon; \epsilon^* \rangle\} \leq \Psi(u, \epsilon) - \langle \epsilon; \epsilon^* \rangle \ \ \forall \epsilon \in \mathcal{E}.$$

For any given $\theta \in [0, 1]$, the convexity of $\Psi(u, \epsilon)$ gives the inequality

$$\begin{aligned} L(\theta u_1 + (1-\theta)u_2, \epsilon^*) &= \Psi(\theta u_1 + (1-\theta)u_2, \epsilon) - \langle \epsilon; \epsilon^* \rangle \\ &\leq \theta[\Psi(u_1, \epsilon) - \langle \epsilon, \epsilon^* \rangle] + (1-\theta)[\Psi(u_2, \epsilon) - \langle \epsilon, \epsilon^* \rangle], \end{aligned}$$

which holds for all $\epsilon \in \mathcal{E}$. Letting $\Psi(u_1, \epsilon) - \langle \epsilon, \epsilon^* \rangle$ decrease towards to $L(u_1, \epsilon^*)$ and $\Psi(u_2, \epsilon) - \langle \epsilon, \epsilon^* \rangle$ decrease towards $L(u_2, \epsilon^*)$, we obtain

$$L(\theta u_1 + (1-\theta)u_2, \epsilon^*) \leq \theta L(u_1, \epsilon^*) + (1-\theta)L(u_2, \epsilon^*).$$

Thus, $u \mapsto L(u, \epsilon^*)$ is convex. □

Remark 5.3.1 We cannot assert that $u \mapsto L(u, \epsilon^*)$ $\forall \epsilon^* \in \mathcal{E}^*$ is l.s.c. (or u.s.c.) on $\mathcal{U}$ even if $u \mapsto \Psi(u, \epsilon)$ is l.s.c. (or u.s.c.) on $\mathcal{U}$. ■

The following result shows the relationship between the Rockafellar dual functional P^d and L.

Proposition 5.3.1 *Let $\Psi \in \check{\Gamma}(\mathcal{U} \times \mathcal{E})$ be a perturbation of P, and $L : \mathcal{U} \times \mathcal{E}^* \to \bar{\mathbb{R}}$ the associated Lagrangian form. Then, for any given $\epsilon^* \in \mathcal{E}^*$,*

$$P^d(\epsilon^*) = \sup_{u \in \mathcal{U}} L(u, \epsilon^*) \in \hat{\Gamma}_0(\mathcal{E}^*).$$

Theorem 5.3.2 *Let $\Psi \in \check{\Gamma}_0(\mathcal{U} \times \mathcal{E})$ such that $L : \mathcal{U} \times \mathcal{E}^* \to \bar{\mathbb{R}}$ is the associated Lagrangian form.*

(S1) $(\bar{u}, \bar{\epsilon}^*)$ *is a saddle point of* L *if and only if* $\bar{u}$ *is a solution of* $(\mathcal{P}_{\inf})$, $\bar{\epsilon}^*$ *is a solution of* $(\mathcal{P}^d_{\sup})$ *and* $P(\bar{u}) = P^*(\bar{\epsilon}^*)$, *i.e.*

$$L(\bar{u}, \bar{\epsilon}^*) = \inf_{u \in \mathcal{U}} \sup_{\epsilon^* \in \mathcal{E}^*} L(u, \epsilon^*) = \sup_{\epsilon^* \in \mathcal{E}^*} \inf_{u \in \mathcal{U}} L(u, \epsilon^*) \quad \Leftrightarrow$$
$$P(\bar{u}) = \inf_{u \in \mathcal{U}} P(u) = \sup_{\epsilon^* \in \mathcal{E}^*} P^d(\epsilon^*) = P^d(\bar{\epsilon}^*).$$

(S2) Dually, if $(\mathcal{P}_{\inf})$ *is stable, then* $\bar{u} \in \mathcal{U}$ *is a solution of* $(\mathcal{P}_{\inf})$ *if and only if there exists* $\bar{\epsilon}^* \in \mathcal{E}^*$ *such that* $(\bar{u}, \bar{\epsilon}^*)$ *is a saddle point of* L.

The following theorem gives the conditions of existence of the saddle points of L.

Theorem 5.3.3 *If* $\mathcal{U}_a$ *and* $\mathcal{E}^*_a$ *are bounded, nonempty, closed convex subsets of* $\mathcal{U}$ *and* $\mathcal{E}^*$, *respectively, then* $L(u, \epsilon^*)$ *has at least one saddle point* $(\bar{u}, \bar{\epsilon}^*) \in \mathcal{U}_a \times \mathcal{E}^*_a$ *such that*

$$L(\bar{u}, \bar{\epsilon}^*) = \min_{u \in \mathcal{U}_a} \max_{\epsilon^* \in \mathcal{E}^*_a} L(u, \epsilon^*). \tag{5.102}$$

If $\mathcal{U}_s \times \mathcal{E}^*_s$ *is the saddle point set of* L, *then it must be a convex set.*

If $\{\epsilon^* \to L(u, \epsilon^*) \ \ \forall u \in \mathcal{U}\}$ *is strictly concave, then* $\mathcal{E}^*_s$ *contains at most one point.*

If $\{u \to L(u, \epsilon^*) \ \ \forall \epsilon^* \in \mathcal{E}^*\}$ *is strictly convex, then* $\mathcal{U}_s$ *contains at most one point.*

In the special case when $\Phi(u, \epsilon) = W(\epsilon) - F(u)$, the *pseudo-Lagrangian* L_p, given by

$$L_p(u, \epsilon^*, \epsilon) = \langle \Lambda u - \epsilon; \epsilon^* \rangle + W(\epsilon) - F(u),$$

is the well-known *Hu-Washizu potential energy* in solid mechanics. For convex $W : \mathcal{E} \to \overline{\mathbb{R}}$, the Fenchel-Young inequality leads to

$$L(u, \epsilon^*) \leq L_p(u, \epsilon^*, \epsilon) \ \ \forall \epsilon \in \mathcal{E},$$

and the equality holds if and only if $\epsilon^* \in \partial^- W(\epsilon)$. Furthermore, we have the following result.

Proposition 5.3.2 *If* $\Phi \in \check{\Gamma}(\mathcal{U} \times \mathcal{E})$, *then, on the product space* $\mathcal{U} \times \mathcal{E}^* \times \mathcal{E}$,

$$\sup_{\epsilon^*} \inf_{u} \inf_{\epsilon} L_p(u, \epsilon^*, \epsilon) = \inf_{u} \sup_{\epsilon^*} \inf_{\epsilon} L_p(u, \epsilon^*, \epsilon), \tag{5.103}$$

and

$$\sup_{\epsilon^*} L_p(u, \epsilon^*, \epsilon) = \begin{cases} P(u) & \text{if } u \in \mathcal{K}_a, \\ +\infty & \text{otherwise.} \end{cases} \tag{5.104}$$

Proof. Since $L(u, \epsilon^*) = \sup_\epsilon L_p(u, \epsilon^*, \epsilon)$, the statement (5.103) follows directly from the saddle Lagrange theory.

In fact, if $u \in \mathcal{K}_a$ and with $\epsilon = \Lambda u$, we have $L_p(u, \epsilon^*, \epsilon) = P(u)$. Otherwise, if $u \notin \mathcal{K}_a$ then there exists at least one $e^* \in \mathcal{E}^*$ such that $\langle e^*, \Lambda u - \epsilon \rangle = \alpha > 0$. Since $\mathcal{E}^*$ is a cone, i.e. $\{\lambda e^*\} \in \mathcal{E}^* \ \forall \lambda > 0$, let $\epsilon^* = \lambda e^*$. Then we have

$$\begin{aligned} \sup_{\epsilon^* \in \mathcal{E}^*} L_p(u, \epsilon^*, \epsilon) &\geq \sup_{\lambda > 0} L_p(u, \epsilon^*, \epsilon) = \Phi(u, \epsilon) + \sup_{\lambda > 0} \lambda \{\langle e^*, \Lambda u - \epsilon \rangle\} \\ &= \Phi(u, \epsilon) + \alpha \sup_{\lambda > 0} \lambda = +\infty, \end{aligned}$$

and the statement (5.104) is proved. □

5.3.3 Super- and Sub-Lagrangian Duality

For the nonconvex perturbed functional $\Psi \in \Gamma(\mathcal{U} \times \mathcal{E})$, two cases can be identified.

I. $L \in \hat{\Gamma}(\mathcal{U}) \times \hat{\Gamma}(\mathcal{E}^*)$ is a super-Lagrangian. In this case,

$$P(u) = \sup_{\epsilon^* \in \mathcal{E}^*} L(u, \epsilon^*) \quad \forall u \in \mathcal{U}, \tag{5.105}$$

$$P^d(\epsilon^*) = \sup_{u \in \mathcal{U}} L(u, \epsilon^*) \quad \forall \epsilon^* \in \mathcal{E}^*. \tag{5.106}$$

II. $L \in \check{\Gamma}(\mathcal{U}) \times \check{\Gamma}(\mathcal{E}^*)$ is a sub-Lagrangian. In this case,

$$P(u) = \inf_{\epsilon^* \in \mathcal{E}^*} L(u, \epsilon^*) \quad \forall u \in \mathcal{U}, \tag{5.107}$$

$$P^d(\epsilon^*) = \inf_{u \in \mathcal{U}} L(u, \epsilon^*) \quad \forall \epsilon^* \in \mathcal{E}^*. \tag{5.108}$$

In either case, both primal and dual problems are generally nonconvex. The following lemma is due to Auchmuty (1983).

Lemma 5.3.2 *If $L \in \hat{\Gamma}(\mathcal{U}) \times \hat{\Gamma}(\mathcal{E}^*)$ is a super-Lagrangian, then we have the following results.*

(L1) $0 \in \partial^+_{\epsilon^*} L(u, \bar{\epsilon}^*) \Rightarrow P(u) = L(u, \bar{\epsilon}^*) \leq P^d(\bar{\epsilon}^*) \ \forall u \in \mathcal{U}.$

(L2) $0 \in \partial^+_u L(\bar{u}, \epsilon^*) \Rightarrow P(\bar{u}) \geq L(\bar{u}, \epsilon^*) = P^d(\epsilon^*) \ \forall \epsilon^* \in \mathcal{E}^*.$

Theorem 5.3.4 (Super-Critical Point Theorem)

Let $L \in \hat{\Gamma}(\mathcal{U}) \times \hat{\Gamma}(\mathcal{E}^*)$ be a super-Lagrangian, and $(\bar{u}, \bar{\epsilon}^*)$ be a super-critical point of L.

(T1) $(\bar{u}, \bar{\epsilon}^*)$ maximizes L on $\mathcal{U} \times \mathcal{E}^*$ if and only if $\bar{u}$ is a solution of $(\mathcal{P}_{\sup})$, $\bar{\epsilon}^*$ is a solution of $(\mathcal{P}^d_{\sup})$ and $P(\bar{u}) = P^d(\bar{\epsilon}^*) = L(\bar{u}, \bar{\epsilon}^*)$, i.e.

$$\begin{aligned} L(\bar{u}, \bar{\epsilon}^*) = \sup_{u \in \mathcal{U}} \sup_{\epsilon^* \in \mathcal{E}^*} L(u, \epsilon^*) \Leftrightarrow \\ 0 \in \partial^+ P(\bar{u}) \quad \Leftrightarrow \quad 0 \in \partial^+ P^d(\bar{\epsilon}^*). \end{aligned}$$

(T2) $(\bar{u}, \bar{\epsilon}^*)$ is a super-minimax point of L if and only if $\bar{u} \in \mathcal{U}$ is a solution of $(\mathcal{P}_{\inf})$, $\bar{\epsilon}^*$ is a solution of $(\mathcal{P}^d_{\inf})$ and $P(\bar{u}) = P^d(\bar{\epsilon}^*) = L(\bar{u}, \bar{\epsilon}^*)$, i.e.

$$\begin{aligned} & L(\bar{u}, \bar{\epsilon}^*) = \inf_{u \in \mathcal{U}} \sup_{\epsilon^* \in \mathcal{E}^*} L(u, \epsilon^*) = \inf_{\epsilon^* \in \mathcal{E}^*} \sup_{u \in \mathcal{U}} L(u, \epsilon^*) \quad \Leftrightarrow \\ & 0 \in \partial^- P(\bar{u}) \quad \Leftrightarrow \quad 0 \in \partial^- P^d(\bar{\epsilon}^*). \end{aligned}$$

By the general duality principle, the following dual results hold for sub-Lagrangians.

Theorem 5.3.5 (Sub-Critical Point Theorem)
Let $L \in \check{\Gamma}(\mathcal{U}) \times \check{\Gamma}(\mathcal{E})$ and $(\bar{u}, \bar{\epsilon}^*)$ be a sub-critical point of L.

(T1) $(\bar{u}, \bar{\epsilon}^*)$ minimizes L on $\mathcal{U} \times \mathcal{E}^*$ if and only if $\bar{u}$ is a solution of $(\mathcal{P}_{\inf})$, $\bar{\epsilon}^*$ is a solution of $(\mathcal{P}^d_{\inf})$ and $P(\bar{u}) = P^d(\bar{\epsilon}^*) = L(\bar{u}, \bar{\epsilon}^*)$, i.e.

$$L(\bar{u}, \bar{\epsilon}^*) = \inf_{u \in \mathcal{U}} \inf_{\epsilon^* \in \mathcal{E}^*} L(u, \epsilon^*) \quad \Leftrightarrow$$
$$0 \in \partial^- P(\bar{u}) \quad \Leftrightarrow \quad 0 \in \partial^- P^d(\bar{\epsilon}^*).$$

(T2) $(\bar{u}, \bar{\epsilon}^*)$ is a sub-maxmin point of L if and only if $\bar{u} \in \mathcal{U}$ is a solution of $(\mathcal{P}_{\sup})$, $\bar{\epsilon}^*$ is a solution of $(\mathcal{P}^d_{\sup})$ and $P(\bar{u}) = P^d(\bar{\epsilon}^*) = L(\bar{u}, \bar{\epsilon}^*)$, i.e.

$$L(\bar{u}, \bar{\epsilon}^*) = \sup_{u \in \mathcal{U}} \inf_{\epsilon^* \in \mathcal{E}^*} L(u, \epsilon^*) = \sup_{\epsilon^* \in \mathcal{E}^*} \inf_{u \in \mathcal{U}} L(u, \epsilon^*) \quad \Leftrightarrow$$
$$0 \in \partial^+ P(\bar{u}) \quad \Leftrightarrow \quad 0 \in \partial^+ P^d(\bar{\epsilon}^*).$$

5.3.4 Duality in Finite-dimensional Systems and Index Theory

Duality theory takes particularly symmetrical forms when $\mathcal{U} = \mathbb{R}^n = \mathcal{U}^*$ and $\mathcal{E} = \mathbb{R}^m = \mathcal{E}^*$ are finite dimensional.

Let $L : \mathbb{R}^n \times \mathbb{R}^m \to \bar{\mathbb{R}}$ be a given arbitrary functional, such that

$$P(u) = L^*_\epsilon(u, \epsilon^*) = \operatorname*{ext}_{\epsilon^* \in \mathcal{E}^*} L(u, \epsilon^*), \tag{5.109}$$
$$P^d(\epsilon^*) = L^*_u(u, \epsilon^*) = \operatorname*{ext}_{u \in \mathcal{U}} L(u, \epsilon^*). \tag{5.110}$$

We need the following assumption.

Assumption 5.3.1 Let $\mathcal{K}_o \subset \mathbb{R}^n$ and $\mathcal{S}_o \subset \mathbb{R}^m$ be two subsets on which P and P^d are finite.

(A1) For each $u \in \mathcal{K}_o$, there exists a unique $\sigma(u) \in \mathbb{R}^m$ such that

$$P(u) = L(u, \sigma(u)). \tag{5.111}$$

(A2) For each $\epsilon^* \in \mathcal{S}_o$, there exists a unique $\xi(\epsilon^*) \in \mathbb{R}^n$ such that

$$P^d(\epsilon^*) = L(\xi(\epsilon^*), \epsilon^*). \tag{5.112}$$

♣

By this assumption, if $(\bar{u}, \bar{\epsilon}^*)$ is a critical point of L and P is Gâteaux differentiable at $\bar{u}$, then by applying the chain rule to (5.111) we obtain

$$DP(\bar{u}) = D_u L(\bar{u}, \sigma(\bar{u})) + D_\sigma L(\bar{u}, \sigma(\bar{u})) D\sigma(\bar{u}) = 0$$

since $\sigma(\bar{u}) = \bar{\epsilon}^*$. This shows that $\bar{u}$ is also a critical point of P. Dually, $\bar{\epsilon}^*$ is a critical point of P^d.

Recall from Section 1.4 that if $P \in C^2$, then a point $\bar{u} \in \mathbb{R}^n$ is called *regular* if

$$\det D^2 P(\bar{u}) \neq 0.$$

Definition 5.3.2 (Morse Index) Let $P : \mathcal{K}_a \subset \mathbb{R}^n \to \bar{\mathbb{R}}$ be C^2 and $\bar{u} \in \mathcal{K}_a$ a regular critical point of P. The *Morse index* $i_d(\bar{u}) \in \mathbb{R}$ of $\bar{u}$ is defined to be the number of negative eigenvalues of $D^2 P(\bar{u})$. ◊

From this definition, it is follows that a critical point $\bar{u}$ of P is a local minimum if and only if $i_d(\bar{u}) = 0$; it is a local maximum if and only if $i_d(\bar{u}) = n$.

Morse index plays an important role in nonconvex analysis. The surprising fact is that if $(\bar{u}, \bar{\epsilon}^*)$ is a critical point of L, and $\bar{u}$ and $\bar{\epsilon}^*$ are regular critical point of P and P^d, respectively, then their indices are related. The following results are due to Auchmuty (1983).

Lemma 5.3.3 *Suppose, in addition to Assumption 5.3.1, that $(\bar{u}, \bar{\epsilon}^*) \in \mathcal{K}_o \times \mathcal{S}_o$ is a critical point of L, that L is C^k $(k \geq 2)$ on a neighborhood of $(\bar{u}, \bar{\epsilon}^*)$, that $\bar{u} = \xi(\bar{\epsilon}^*) \in \text{int}\mathcal{K}_o$, $\bar{\epsilon}^* = \sigma(\bar{u}) \in \text{int}\mathcal{S}_o$ and that ξ and σ are C^k functions on neighborhoods of $\bar{u}$ and $\bar{\epsilon}^*$, respectively. Let*

$$D^2 L(\bar{u}, \bar{\epsilon}^*) = \begin{pmatrix} K & \Lambda^* \\ \Lambda & M \end{pmatrix}, \tag{5.113}$$

with $K \in \mathbb{R}^{n\times n}$ and $M \in \mathbb{R}^{m\times m}$ non-singular. Then,

$$D^2 P(\bar{u}) = K - \Lambda M^{-1} \Lambda^*, \tag{5.114}$$
$$D^2 P^d(\bar{\epsilon}^*) = M - \Lambda^* K^{-1} \Lambda. \tag{5.115}$$

Theorem 5.3.6 (Iso-Index Theorem)
Let the assumptions of Lemma 5.3.3 hold with $n = m$ and assume that $\Lambda : \mathbb{R}^n \to \mathbb{R}^n$ is non-singular and that either K or M is negative-definite or positive-definite. Then, $\bar{u}$ is a regular critical point of P if and only if $\bar{\epsilon}^*$ is a regular critical point of P^d. When this holds, $i_d(\bar{u}) = i_d(\bar{\epsilon}^*)$.

Corollary 5.3.1 *Suppose $(\bar{u}, \bar{\epsilon}^*)$ is a critical point of L and the assumptions of Theorem 5.3.6 hold. Then, a regular critical point $\bar{u}$ of P is a local minimum (or maximum) of P if and only if $\bar{\epsilon}^*$ is regular and a local minimum (or maximum) of P^d.*

Proof. From Theorem 5.3.6, $\bar{u}$ is regular if and only if $\bar{\epsilon}^*$ is regular and $i_d(\bar{u}) = i_d(\bar{\epsilon}^*)$. Thus, if $i_d(\bar{u}) = 0$ if and only if $i_d(\bar{\epsilon}^*) = 0$ then the local minima correspond. Similarly, if $i_d(\bar{u}) = i_d(\bar{\epsilon}^*) = n$. □

This corollary shows that if $i_d(\bar{u}) = i_d(\bar{\epsilon}^*) = \{0, n\}$, then $(\bar{u}, \bar{\epsilon}^*)$ is either a super- or sub-critical point of L.

Usually for saddle-point problems, $(\bar{u}, \bar{\epsilon}^*)$ being a right-saddle point of L implies that K is positive-definite and M is negative-definite. Thus, $\bar{u}$ is a local minimizer of P and $\bar{\epsilon}^*$ must be a local maximizer of P^d.

Example 5.3.1 Let us now demonstrate how the above scheme fits in with the classical Lagrangian multiplier methods in finite dimensional linear programming. Let $\mathcal{U} = \mathcal{U}^* = \mathbb{R}^n$, $\mathcal{E} = \mathcal{E}^* = \mathbb{R}^m$, with the standard inner products $(\mathbf{x}, \mathbf{x}^*) = \mathbf{x} \cdot \mathbf{x}^*$ in $\mathbb{R}^n$, and $\langle \mathbf{y}; \mathbf{y}^* \rangle = \mathbf{y} \cdot \mathbf{y}^*$ in $\mathbb{R}^m$. For fixed $\bar{u}^* = \mathbf{f} \in \mathbb{R}^n$, we consider the optimization problem

$$(\mathcal{P}_{\min}): \quad P(\mathbf{x}) = \frac{1}{2}\langle \Lambda\mathbf{x}; M_0\Lambda\mathbf{x}\rangle - \frac{1}{2}\langle \mathbf{x}, K_0\mathbf{x}\rangle - \langle \mathbf{x}, \mathbf{f}\rangle \quad \to \min \;\; \forall \mathbf{x} \in \mathbb{R}^n,$$

where $\Lambda : \mathbb{R}^n \to \mathbb{R}^m$ is a linear mapping and $K_0 \in \mathbb{R}^{n\times n}$ and $M_0 \in \mathbb{R}^{m\times m}$ are two non-singular symmetric matrices. Let $\Phi : \mathbb{R}^n \times \mathbb{R}^m \to \mathbb{R}$ be defined by

$$\Phi(\mathbf{x}, \mathbf{y}) = \frac{1}{2}\langle \mathbf{y}; M_0\mathbf{y}\rangle - \frac{1}{2}\langle \mathbf{x}, K_0\mathbf{x}\rangle - \langle \mathbf{x}, \mathbf{f}\rangle.$$

Then, $P(\mathbf{x}) = \Phi(\mathbf{x}, \Lambda\mathbf{x})$ and the Lagrangian form associated with this problem is

$$L(\mathbf{x}, \mathbf{y}^*) = \langle \Lambda\mathbf{x}; \mathbf{y}^*\rangle - \frac{1}{2}\langle \mathbf{y}^*; M_0^{-1}\mathbf{y}^*\rangle - \frac{1}{2}\langle \mathbf{x}, K_0\mathbf{x}\rangle - \langle \mathbf{x}, \mathbf{f}\rangle.$$

The criticality condition $DL(\bar{\mathbf{x}}, \bar{\mathbf{y}}^*) = 0$ leads to the Lagrange equations

$$\Lambda\mathbf{x} = M_0^{-1}\mathbf{y}^*, \quad \Lambda^*\mathbf{y}^* = K_0\mathbf{x} + \mathbf{f}.$$

For a fixed $\mathbf{y}^* \in \mathbb{R}^m$, the dual function $P^d : \mathbb{R}^m \to \mathbb{R}$ is given by

$$P^d(\mathbf{y}^*) = \operatorname*{ext}_{\mathbf{x}\in\mathbb{R}^n} L(\mathbf{x}, \mathbf{y}^*) = \frac{1}{2}\langle \Lambda^*\mathbf{y}^* - \mathbf{f}, K_0^{-1}(\Lambda^*\mathbf{y}^* - \mathbf{f})\rangle - \frac{1}{2}\langle \mathbf{y}^*; M_0^{-1}\mathbf{y}^*\rangle.$$

Then, $L : \mathbb{R}^n \times \mathbb{R}^m \to \mathbb{R}$ is a right saddle function if M_0 is positive-definite and K_0 is negative-definite. In this case, $P : \mathbb{R}^n \to \mathbb{R}$ is convex and $P^d : \mathbb{R}^m \to \mathbb{R}$ is concave.

If both M_0 and K_0 are positive-definite (negative-definite), then L is a super-Lagrangian (sub-Lagrangian). In this case, both P and P^d are nonconvex. By Theorem 5.3.6, if $n = m$ and $\Lambda : \mathbb{R}^n \to \mathbb{R}^n$ is non-singular, we have $i_d(\mathbf{x}) = i_d(\mathbf{y}^*)$, and $\bar{\mathbf{x}}$ minimizes (maximizes) P if and only if $\bar{\mathbf{y}}^*$ minimizes (maximizes) P^d. ♡

5.4 Hamilton and Clarke Duality Theories

This section shall discuss the connection between the Lagrangian and Hamiltonian in geometrically linear systems, as well as the Clarke duality theory in dynamical systems.

5.4.1 Hamiltonian and Canonical Inclusions

Let $\{(\mathcal{U}, \mathcal{U}^*), \langle *, * \rangle\}$ and $\{(\mathcal{E}, \mathcal{E}^*), \langle *; * \rangle\}$ be two duality pairings, and $\Phi \in \Gamma(\mathcal{U} \times \mathcal{E})$ a real-valued canonical functional. The partial conjugates of $\Phi(u, \epsilon)$ are defined by

$$\Phi^*_\epsilon(u, \epsilon^*) = \operatorname*{ext}_{\epsilon\in\mathcal{E}}\{\langle \epsilon; \epsilon^*\rangle - \Phi(u, \epsilon)\}, \tag{5.116}$$

$$\Phi^*_u(u^*, \epsilon) = \operatorname*{ext}_{u\in\mathcal{U}}\{\langle u, u^*\rangle - \Phi(u, \epsilon)\}. \tag{5.117}$$

Definition 5.4.1 (Hamiltonian and Complementary Hamiltonian)
(D1) The functional $H : \mathcal{U} \times \mathcal{E}^* \to \bar{\mathbb{R}}$ defined by

$$H(u,\epsilon^*) = \Phi^*_\epsilon(u,\epsilon^*) \tag{5.118}$$

is called the *Hamiltonian* associated with Φ .

(D2) Let $\Lambda : \mathcal{U} \to \mathcal{E}$ be a linear operator and $\Lambda^* : \mathcal{E}^* \to \mathcal{U}^*$ be its adjoint. The functional $H^c : \mathcal{U} \times \mathcal{E}^* \to \bar{\mathbb{R}}$ defined by

$$H^c(u,\epsilon^*) = -\Phi^*_u(\Lambda^*\epsilon^*, \Lambda u) \tag{5.119}$$

is called the *complementary Hamiltonian* associated with Φ . ◊

Remark 5.4.1 Since $\epsilon^* \mapsto H(u,\epsilon^*)$ is precisely the conjugate functional of $\epsilon \mapsto \Phi(u,\epsilon)$, then, if $\Phi \in \check{\Gamma}(\mathcal{U} \times \mathcal{E})$, $H : \mathcal{U} \times \mathcal{E}^* \to \bar{\mathbb{R}}$ is a left saddle functional and

$$\epsilon^* \mapsto H(u,\epsilon^*) \text{ is l.s.c.} \quad \forall u \in \mathcal{U}.$$

However, we cannot assert that for any given $\epsilon^* \in \mathcal{E}^*$, $u \mapsto H(u,\epsilon^*)$ is u.s.c. (or l.s.c.) on $\mathcal{U}$ even if $u \mapsto \Phi(u,\epsilon)$ is l.s.c. (or u.s.c.) on $\mathcal{U}$. ■

To see the relationship between H and $P(u) = \Phi(u, \Lambda u)$, we assume $\Phi \in \Gamma_0(\mathcal{U} \times \mathcal{E})$. Then, for any $u \in \mathcal{U}$, $\Phi_u : \epsilon \mapsto \Phi(u,\epsilon)$ belongs to $\Gamma(\mathcal{E})$. Thus, $\Phi^{**}_\epsilon(u,\epsilon) = \Phi(u,\epsilon)$, whence

$$\Phi(u,\epsilon) = \Phi^{**}_\epsilon(u,\epsilon) = \operatorname*{ext}_{\epsilon^* \in \mathcal{E}^*} \{\langle \epsilon;\epsilon^*\rangle - \Phi^*_\epsilon(u,\epsilon^*)\},$$

and

$$P(u) = \Phi^{**}(u,\Lambda u) = \operatorname*{ext}_{\epsilon^* \in \mathcal{E}^*} \{\langle \Lambda u;\epsilon^*\rangle - H(u,\epsilon^*)\} = H^*_{\epsilon^*}(u,\Lambda u).$$

On the other hand, without assuming anything about Φ, we have

$$\begin{aligned}
\Phi^*(u^*,\epsilon^*) &= \operatorname*{ext}_{u \in \mathcal{U}} \operatorname*{ext}_{\epsilon \in \mathcal{E}} \{\langle u,u^*\rangle + \langle \epsilon;\epsilon^*\rangle - \Phi(u,\epsilon)\} \\
&= \operatorname*{ext}_{u \in \mathcal{U}} \{\langle u,u^*\rangle + \operatorname*{ext}_{\epsilon \in \mathcal{E}}[\langle \epsilon;\epsilon^*\rangle - \Phi(u,\epsilon)]\} \\
&= \operatorname*{ext}_{u \in \mathcal{U}} \{\langle u,u^*\rangle - H(u,\epsilon^*)\},
\end{aligned}$$

and the functional dual to P is obtained as

$$P^d(\epsilon^*) = \Phi^*(\Lambda^*\epsilon^*,\epsilon^*) = \operatorname*{ext}_{u \in \mathcal{U}} \langle u,\Lambda^*\epsilon^*\rangle - H(u,\epsilon^*)\} = H^*_u(\Lambda^*\epsilon^*,\epsilon^*).$$

The relationship between the Lagrangian and Hamiltonian is given by the following proposition.

Proposition 5.4.1 *If* $\Psi(u,\epsilon^*) = \Phi(u,\Lambda u + \epsilon^*)$, *then*

$$L(u,\epsilon^*) = \langle \Lambda u;\epsilon^*\rangle - H(u,\epsilon^*),$$

$$L^c(u,\epsilon^*) = -\langle u,\Lambda^*\epsilon^*\rangle - H^c(u,\epsilon^*).$$

Proof. By definition of the Lagrangian, with $e = \Lambda u + \epsilon$, we have

$$\begin{aligned} L(u,\epsilon^*) &= -\operatorname*{ext}_{\epsilon\in\mathcal{E}}\{\langle\epsilon;\epsilon^*\rangle - \Phi(u,\Lambda u+\epsilon)\} \\ &= -\operatorname*{ext}_{e\in\mathcal{E}}\{\langle e-\Lambda u;\epsilon^*\rangle - \Phi(u,e)\} \\ &= \langle\Lambda u;\epsilon^*\rangle - H(u,\epsilon^*), \end{aligned}$$

as required. Similar for L^c. □

It is clearly that if $(\bar{u},\bar{\epsilon}^*)$ is a critical point of L, then $(\bar{u},\bar{\epsilon}^*)$ satisfies the Hamiltonian canonical forms

$$D_u L(\bar{u},\bar{\epsilon}^*) = 0 \to \quad \Lambda^*\bar{\epsilon}^* = D_u H(\bar{u},\bar{\epsilon}^*), \tag{5.120}$$

$$D_{\epsilon^*} L(\bar{u},\bar{\epsilon}^*) = 0 \to \quad \Lambda\bar{u} = D_{\epsilon^*} H(\bar{u},\bar{\epsilon}^*). \tag{5.121}$$

In this case,

$$H(\bar{u},\bar{\epsilon}^*) + H^c(\bar{u},\bar{\epsilon}^*) = 0 = -L(\bar{u},\bar{\epsilon}^*) - L^c(\bar{u},\bar{\epsilon}^*).$$

If $H(u,\epsilon^*)$ is Gâteaux differentiable on $\mathcal{U}\times\mathcal{E}^*$, then the canonical inclusions can be written as the canonical forms

$$\begin{pmatrix} \Lambda u \\ \Lambda^*\epsilon^* \end{pmatrix} = \begin{pmatrix} D_{\epsilon^*}H(u,\epsilon^*) \\ D_u H(u,\epsilon^*) \end{pmatrix} = \mathbf{J}^{\oplus} DH(u,\epsilon^*), \tag{5.122}$$

where

$$\mathbf{J}^{\oplus} = \begin{pmatrix} 0 & 1 \\ 1 & 0 \end{pmatrix}.$$

If $\Phi \in \Gamma(\mathcal{U}\times\mathcal{E})$ can be split into $\Phi(u,\epsilon) = W(\epsilon) - F(u)$, then the Hamiltonian $H : \mathcal{U}\times\mathcal{E}^* \to \overrightarrow{\mathbb{R}}$ and its complementary form take the forms

$$H(u,\epsilon^*) = W^*(\epsilon^*) + F(u), \tag{5.123}$$

$$H^c(u,\epsilon^*) = W(\Lambda u) + F^*(-\Lambda^*\epsilon^*). \tag{5.124}$$

If $W \in \Gamma_0(\mathcal{E})$ and $F \in \Gamma_0(\mathcal{U})$ are Gâteaux differentiable, then the canonical forms (5.122) can be written as

$$\Lambda u = D_{\epsilon^*}H(u,\epsilon^*) = DW^*(\epsilon^*), \tag{5.125}$$

$$\Lambda^*\epsilon^* = D_u H(u,\epsilon^*) = DF(u). \tag{5.126}$$

It is clear that the convexities of $H(u,\epsilon^*)$ and $H^c(u,\epsilon^*)$ are the same and depend on the functional $\Phi : \mathcal{U}\times\mathcal{E} \to \bar{\mathbb{R}}$ according to Table 5.2.

We discuss only the first two cases.

I. $\Phi : \mathcal{U}\times\mathcal{E} \to \overrightarrow{\mathbb{R}}$ is convex, i.e. $\Phi \in \check{\Gamma}(\mathcal{U}\times\mathcal{E})$.

In this case, $H \in \hat{\Gamma}(\mathcal{U})\times\check{\Gamma}(\mathcal{E}^*)$ is a left-saddle functional. The partial conjugates of H are defined as

$$H^*_{\epsilon^*}(u,\epsilon) = \sup_{\epsilon^*\in\mathcal{E}^*}\{\langle\epsilon;\epsilon^*\rangle - H(u,\epsilon^*)\}, \tag{5.127}$$

$$H^*_u(u^*,\epsilon^*) = \inf_{u^*\in\mathcal{U}^*}\{\langle u,u^*\rangle - H(u,\epsilon^*)\}. \tag{5.128}$$

Table 5.2. The convexity relations among Φ, H and H^c

$\Phi \in \check{\Gamma}(\mathcal{U} \times \mathcal{E})$	$\Phi \in \hat{\Gamma}(\mathcal{U}) \times \check{\Gamma}(\mathcal{E})$	$\Phi \in \check{\Gamma}(\mathcal{U}) \times \hat{\Gamma}(\mathcal{E})$	$\Phi \in \hat{\Gamma}(\mathcal{U}) \times \hat{\Gamma}(\mathcal{E})$
$H \in \hat{\Gamma}(\mathcal{U}) \times \check{\Gamma}(\mathcal{E}^*)$ $H^c \in \check{\Gamma}(\mathcal{U}) \times \hat{\Gamma}(\mathcal{E}^*)$	$H \in \check{\Gamma}(\mathcal{U}) \times \check{\Gamma}(\mathcal{E}^*)$ $H^c \in \check{\Gamma}(\mathcal{U}) \times \check{\Gamma}(\mathcal{E}^*)$	$H \in \hat{\Gamma}(\mathcal{U}) \times \hat{\Gamma}(\mathcal{E}^*)$ $H^c \in \hat{\Gamma}(\mathcal{U}) \times \hat{\Gamma}(\mathcal{E}^*)$	$H \in \check{\Gamma}(\mathcal{U}) \times \hat{\Gamma}(\mathcal{E}^*)$ $H^c \in \hat{\Gamma}(\mathcal{U}) \times \check{\Gamma}(\mathcal{E}^*)$

Since $\epsilon^* \mapsto H(u, \epsilon^*)$ is precisely the conjugate functional of $\epsilon \mapsto \Phi(u, \epsilon)$, it is clear that the Hamiltonian H satisfies $H_\epsilon^*(u, \epsilon) = \Phi(u, \epsilon)$ and the functionals

$$P(u) = H_\epsilon^*(u, \Lambda u), \quad P^d(\epsilon^*) = H_u^*(\Lambda^* \epsilon^*, \epsilon^*) \tag{5.129}$$

are well-defined.

We denote by $\partial_{\epsilon^*}^- H(\bar{u}, \bar{\epsilon}^*)$ the sub-differential of $\epsilon^* \mapsto H(\bar{u}, \epsilon^*)$ at $\bar{\epsilon}^*$ and by $\partial_u^+ H(\bar{u}, \bar{\epsilon}^*)$ the super-differential of $u \mapsto H(u, \bar{\epsilon}^*)$ at $\bar{u}$. We have the following result.

Proposition 5.4.2 *If $\Phi \in \check{\Gamma}(\mathcal{U} \times \mathcal{E})$, then the extremality relation*

$$(-\Lambda^* \bar{\epsilon}^*, \bar{\epsilon}^*) \in \partial^- \Phi(\bar{u}, \Lambda \bar{u}) \tag{5.130}$$

is equivalent to the canonical inclusions

$$\Lambda \bar{u} \in \partial_{\epsilon^*}^- H(\bar{u}, \bar{\epsilon}^*), \qquad \Lambda^* \bar{\epsilon}^* \in \partial_u^+ H(\bar{u}, \bar{\epsilon}^*). \tag{5.131}$$

Proof. From the definition of the sub-differential, the extremality relation (5.130) can be written as

$$\Phi(u, \Lambda u) - \Phi(\bar{u}, \Lambda \bar{u}) \geq \langle u - \bar{u}, -\Lambda^* \bar{\epsilon}^* \rangle + \langle \epsilon - \bar{\epsilon}; \bar{\epsilon}^* \rangle \;\; \forall (u, \epsilon) \in \mathcal{U} \times \mathcal{E}. \tag{5.132}$$

Taking $\bar{u} = u$, we then obtain

$$H(\bar{u}, \bar{\epsilon}^*) = \sup_{\epsilon \in \mathcal{E}} \{ \langle \epsilon; \bar{\epsilon}^* \rangle - \Phi(\bar{u}, \epsilon) \} = \langle \Lambda \bar{u}; \bar{\epsilon}^* \rangle - \Phi(\bar{u}, \Lambda \bar{u}). \tag{5.133}$$

The definition of H implies that

$$H(\bar{u}, \epsilon^*) - H(\bar{u}, \bar{\epsilon}^*) \geq \langle \Lambda \bar{u}; \epsilon^* - \bar{\epsilon}^* \rangle \;\; \forall \epsilon^* \in \mathcal{E}^*,$$

i.e. $\Lambda \bar{u} \in \partial_{\epsilon^*}^- H(\bar{u}, \bar{\epsilon}^*)$. Moreover, on use of (5.133), (5.132) can be written as

$$\langle \epsilon; \bar{\epsilon}^* \rangle - \Phi(u, \epsilon) \leq \langle u, \Lambda^* \bar{\epsilon}^* \rangle - \Phi(\bar{u}, \Lambda \bar{u}) = \langle u - \bar{u}, \Lambda^* \bar{\epsilon}^* \rangle + H(\bar{u}, \bar{\epsilon}^*). \tag{5.134}$$

Taking the supremum on $\mathcal{U}$, we obtain

$$H(u, \bar{\epsilon}^*) - H(\bar{u}, \bar{\epsilon}^*) \leq \langle u - \bar{u}, \Lambda^* \bar{\epsilon}^* \rangle \;\; \forall u \in \mathcal{U}, \tag{5.135}$$

i.e. $\Lambda^* \bar{\epsilon}^* \in \partial_u^+ H(\bar{u}, \bar{\epsilon}^*)$.

Conversely, suppose that $\Lambda\bar{u} \in \partial^-_{\epsilon^*} H(\bar{u}, \bar{\epsilon}^*)$ and that $\Lambda^*\bar{\epsilon}^* \in \partial^+_u H(\bar{u}, \bar{\epsilon}^*)$. Since $\Phi \in \Gamma(\mathcal{U}, \mathcal{E})$, $\Phi(u, \epsilon) = \sup_{\epsilon^* \in \mathcal{E}^*}\{\langle \epsilon; \epsilon^* \rangle - H(u, \epsilon^*)\}$, and, since

$$\langle \Lambda\bar{u}; \bar{\epsilon}^* \rangle = \Phi(\bar{u}, \Lambda\bar{u}) + H(\bar{u}, \bar{\epsilon}^*),$$

it follows that

$$\Phi(\bar{u}, \Lambda\bar{u}) - \Phi(u, \epsilon) + \langle \epsilon - \Lambda\bar{u}; \bar{\epsilon}^* \rangle \leq H(u, \bar{\epsilon}^*) - H(\bar{u}, \bar{\epsilon}^*). \tag{5.136}$$

Moreover, since $\Lambda^*\bar{\epsilon}^* \in \partial^+_u H(\bar{u}, \bar{\epsilon}^*)$, we have

$$\Phi(u, \epsilon) - \Phi(\bar{u}, \Lambda\bar{u}) \geq \langle \epsilon - \Lambda\bar{u}; \bar{\epsilon}^* \rangle + \langle u - \bar{u}, -\Lambda^*\bar{\epsilon}^* \rangle. \tag{5.137}$$

Hence, $(-\Lambda^*\bar{\epsilon}^*, \bar{\epsilon}^*) \in \partial\Phi(\bar{u}, \Lambda\bar{u})$. □

II. $\Phi : \mathcal{U} \times \mathcal{E} \to \bar{\mathbb{R}}$ is a left-saddle functional, i.e. $\Phi \in \hat{\Gamma}(\mathcal{U}) \times \check{\Gamma}(\mathcal{E})$.

In this case, $H \in \check{\Gamma}(\mathcal{U}) \times \check{\Gamma}(\mathcal{E}^*)$ and the extremality condition

$$(\Lambda^*\bar{\epsilon}^*, \Lambda\bar{u}) \in \partial^- H(\bar{u}, \bar{\epsilon}^*) \tag{5.138}$$

implies the *Hamiltonian canonical sub-inclusions*

$$\begin{aligned} \Lambda\bar{u} &\in \partial^-_{\epsilon^*} H(\bar{u}, \bar{\epsilon}^*), \\ \Lambda^*\bar{\epsilon}^* &\in \partial^-_u H(\bar{u}, \bar{\epsilon}^*). \end{aligned} \tag{5.139}$$

The following definition is due to Auchmuty (1983).

Definition 5.4.2 (Anomalous Critical Point)
Let $H : \mathcal{U} \times \mathcal{E}^* \to \bar{\mathbb{R}}$ be a given Hamiltonian and $L : \mathcal{U} \times \mathcal{E}^* \to \bar{\mathbb{R}}$ the associated Lagrangian. A point $(\bar{u}, \bar{\epsilon}^*)$ is said to be an *anomalous critical point* of L if the equations in (5.139) hold. ◇

Proposition 5.4.3 *A point $(\bar{u}, \bar{\epsilon}^*)$ is an anomalous critical point of L if and only if it is a super-critical point of L.*

Lemma 5.4.1 *Let $H \in \check{\Gamma}(\mathcal{U}) \times \check{\Gamma}(\mathcal{E}^*)$ be a Hamiltonian associated with $\Phi \in \hat{\Gamma}(\mathcal{U}) \times \check{\Gamma}(\mathcal{E}^*)$ such that $P(u) = H^*_\epsilon(u, \Lambda u)$, $P^d(\epsilon^*) = H^*_u(\Lambda^*\epsilon^*, \epsilon^*)$. Then,*
(L1) $\Lambda u \in \partial^-_{\epsilon^*} H(u, \bar{\epsilon}^*) \;\Rightarrow\; P(u) \leq P^d(\bar{\epsilon}^*)$,
(L2) $\Lambda^*\epsilon^* \in \partial^-_u H(\bar{u}, \epsilon^*) \;\Rightarrow\; P(\bar{u}) \geq P^d(\epsilon^*)$.

Proof. (L1) When $\Lambda u \in \partial^-_{\epsilon^*} H(u, \bar{\epsilon}^*)$, one has

$$H(u, \epsilon^*) - H(u, \bar{\epsilon}^*) \geq \langle \Lambda u; \epsilon^* - \bar{\epsilon}^* \rangle \;\; \forall \epsilon^* \in \mathcal{E}^*,$$

which is equivalent to

$$\langle \Lambda u; \epsilon^* \rangle - H(u, \epsilon^*) \leq \langle \Lambda u; \bar{\epsilon}^* \rangle - H(u, \bar{\epsilon}^*) \;\; \forall \epsilon^* \in \mathcal{E}^*.$$

Taking the supremum over $\mathcal{E}^*$, using the definition of P^d and the Fenchel-Young inequality

$$P^d(\bar{\epsilon}^*) = \sup_{u\in\mathcal{U}}\{\langle u, \Lambda^*\bar{\epsilon}^*\rangle - H(u,\bar{\epsilon}^*)\} \geq \langle u, \Lambda^*\bar{\epsilon}^*\rangle - H(u,\bar{\epsilon}^*),$$

we obtain

$$P(u) = \sup_{\epsilon^*\in\mathcal{E}^*}\{\langle \Lambda u; \epsilon^*\rangle - H(u,\epsilon^*)\} \leq P^*(\bar{\epsilon}^*) \ \ \forall u \in \mathcal{U},$$

as required.

Similarly for (L2). □

Theorem 5.4.1 *If the point $(\bar{u}, \bar{\epsilon}^*) \in \mathcal{U}\times\mathcal{E}^*$ satisfies the canonical sub-inclusion forms*

$$\Lambda\bar{u} \in \partial^-_{\epsilon^*} H(\bar{u},\bar{\epsilon}^*), \ \ \Lambda^*\bar{\epsilon}^* \in \partial^-_u H(\bar{u},\bar{\epsilon}^*), \tag{5.140}$$

then $P(\bar{u}) = P^d(\bar{\epsilon}^*)$.

5.4.2 Clarke Duality Theorem

Let $\mathcal{U}$ be a Banach space and $F : \mathcal{U} \to \overline{\mathbb{R}}$ be a convex, l.s.c. functional on $\mathcal{U}$. Let $\Lambda : \mathcal{U} \to \mathcal{E}$ be a linear operator, and $C : \mathcal{E} \to \mathcal{E}^*$ be linear and symmetric such that the linear operator $A = \Lambda^* C\Lambda : \mathcal{U} \to \mathcal{U}^*$ is closed, i.e. the graph $A \subset \mathcal{U}\times\mathcal{U}^*$ is closed. We assume further that A is self-adjoint, i.e.

$$\langle u, Av\rangle = \langle Au, v\rangle \ \ \forall u, v \in \text{dom } A. \tag{5.141}$$

We take $W : \mathcal{E} \to \bar{\mathbb{R}}$ to be a quadratic form $W(\epsilon) = \frac{1}{2}\langle\epsilon; C\epsilon\rangle$ such that

$$P(u) = \Phi(u, \Lambda u) = \frac{1}{2}\langle\Lambda u; C\Lambda u\rangle + F(u) = \frac{1}{2}\langle Au, u\rangle + F(u). \tag{5.142}$$

Let $\mathcal{K} = \text{dom } A \cap \text{dom } F$ be an open subset of $\mathcal{U}$. If P is Gâteaux differentiable at $\bar{u} \in \mathcal{K}$, and

$$A\bar{u} + DF(\bar{u}) = 0, \tag{5.143}$$

then $\bar{u}$ is a critical point of P. This is the standard definition.

If P is nonsmooth, a critical point of P is defined to be a point $\bar{u} \in \text{dom } A$ where

$$0 \in A\bar{u} + \partial^- F(\bar{u}). \tag{5.144}$$

Proposition 5.4.4 *If, along any straight line through $\bar{u}$, P has a local minimum or a local maximum at $\bar{u}$, then $\bar{u}$ is a critical point of P.*

Let $W(\epsilon) = \frac{1}{2}\langle\epsilon; C\epsilon\rangle$ and $\Phi(u,\epsilon) = W(\epsilon) + F(u)$. The Hamiltonian associated with $P(u) = \Phi(u, \Lambda u)$ is

$$H(u,\epsilon^*) = \Phi^*_\epsilon(u,\epsilon^*) = W^*(\epsilon^*) - F(u), \tag{5.145}$$

where $W^*(\epsilon^*) = \text{sta}\{\langle \epsilon; \epsilon^* \rangle - W(\epsilon) | \ \forall \epsilon \in \mathcal{E}\} = \frac{1}{2}\langle C^{-1}\epsilon^*; \epsilon^* \rangle$. By definition, the dual functional of P is

$$\begin{aligned} P^d(\epsilon^*) &= H_u^*(\Lambda^*\epsilon^*, \epsilon^*) = \inf_{u \in \mathcal{U}} \{\langle u, \Lambda^*\epsilon^* \rangle - H(u, \epsilon^*)\} \\ &= -\sup_{u \in \mathcal{U}} \{\langle u, -\Lambda^*\epsilon^* \rangle - F(u)\} - W^*(\epsilon^*) \\ &= \begin{cases} -F^*(-\Lambda^*\epsilon^*) - W^*(\epsilon^*) & \text{if } -\Lambda^*\epsilon^* \in \text{dom } F^* \\ +\infty & \text{if } -\Lambda^*\epsilon^* \notin \text{dom } F^*. \end{cases} \end{aligned}$$

On the dual feasible space $\mathcal{S} = \{\epsilon^* \in \mathcal{E}^* | \ -\Lambda^*\epsilon^* \in \text{dom } F^*\}$,

$$P^d(\epsilon^*) = -\frac{1}{2}\langle C^{-1}\epsilon^*; \epsilon^* \rangle - F^*(-\Lambda^*\epsilon^*).$$

Replacing ϵ^* by $C\Lambda u$, and setting $P^c = -P^d(C\Lambda u)$, we obtain

$$P^c(u) := \frac{1}{2}\langle Au, u \rangle + F^*(-Au). \tag{5.146}$$

This is the so-called *Clarke dual action.* Since $F^* \in \Gamma(\mathcal{U}^*)$, the critical points of P^c are the solutions of

$$0 \in \partial^-[F^* \circ (A)](u) - Au, \quad u \in \text{dom } A. \tag{5.147}$$

Let $\mathcal{K}^c = \{u \in \mathcal{U} | \ u \in \text{dom } A, \ \ Au \in \text{dom } F^*\}$ be the complementary feasible space. We have the following general result.

Theorem 5.4.2 (Generalized Clarke Duality Theorem)

(T1) If $\bar{u}$ is a critical point of P, then all $u^c \in Ker\ A + \bar{u}$ are critical points of P^c.

(T2) Conversely, if $\mathcal{K}^c \neq \emptyset$ and if u^c is a critical point of P^c, then there is some $\bar{u} \in Ker\ A + u^c$ that is a critical point of P.

(T3) If u^c is a critical point of P^c, then the complementary condition

$$P(\bar{u}) + P^c(u^c) = 0 \quad \forall \bar{u} \in u^c + Ker\ A \tag{5.148}$$

holds.

Proof. (T1) If $\bar{u}$ is a critical point of P, then $-A\bar{u} \in \partial^- F(\bar{u})$. By the Legendre duality formula, this is equivalent to

$$\bar{u} \in \partial^- F^*(-A\bar{u}).$$

From the formula (5.52), we know that $\Lambda^*\partial^- F^*(\Lambda u) = \partial^-(F^* \circ \Lambda)(u)$. For a self-adjoint operator $A = A^*$, and

$$A\bar{u} \in A\partial^- F^*(-A\bar{u}) = -\partial^- \left(F^* \circ (-A)\right)(\bar{u}).$$

This means that $\bar{u}$ is a critical point of P^c. Since $P^c(u)$ is invariant under translation in Ker A, i.e. for any given $v \in$ Ker A,

$$P^c(u + v) = P^c(u) \ \ \forall u \in \mathcal{K}^c,$$

then all the points in $\bar{u} + \text{Ker } A$ are also critical points of P^c.

(T2) Conversely, let u^c be a critical point of P^c. Since $\mathcal{K}^c$ is not empty, we have

$$-Au^c \in \partial^- (F^* \circ (-A))(u^c) = -A\partial^- F^*(-Au^c), \tag{5.149}$$

which can be rewritten as the system

$$Av = 0, \qquad v + u^c = \partial^- F^*(-Au^c)$$

of two equations. By the Legendre duality formula, we obtain

$$-Au^c \in \partial^- F(u^c + v),$$

and hence

$$-A(u^c + v) = -Au^c \in \partial^- F(u^c + v) \;\; \forall v \in \text{Ker } A.$$

This shows that $u^c + v$ is a critical point of P.

(T3) By the Fenchel duality theory, u^c is a critical point of P^c if and only if

$$F(\bar{u}) + F^*(-Au^c) = \langle \bar{u}, -Au^c \rangle.$$

Since $A\bar{u} = Au^c$ for all $\bar{u} \in u^c + \text{Ker } A$, we can rewrite this as

$$\left(\frac{1}{2}\langle A\bar{u}, \bar{u}\rangle + F(\bar{u})\right) + \left(\frac{1}{2}\langle Au^c, u^c\rangle + F^*(-Au^c)\right) = 0.$$

This is the complementary condition in the statement (T3). □

5.5 Duality in Variational Inequality and Complementarity Problems

In this section we shall give some simple results and applications related to primal-dual variational problems with inequality constraints. We shall see that many different theories and problems involving variational inequalities, complementarity and nonlinear programming can be put in the same unified framework.

5.5.1 Primal Variational Inequality Problems

Let $\mathcal{U}$ be a reflexive Banach space and $\mathcal{K}$ a non-empty closed convex subset of $\mathcal{U}$. For any given real-valued functional $P : \mathcal{K} \subset \mathcal{U} \to \mathbb{R}$, we consider the constrained variational problem

$$(\mathcal{P}_{\text{inf}}): \;\; P(\bar{u}) = \inf_{u \in \mathcal{K}} P(u). \tag{5.150}$$

Since $\mathcal{K}$ is convex, then for $u_1, u_2 \in \mathcal{K}$, $\theta u_1 + (1-\theta)u_2 \in \mathcal{K} \;\; \forall \theta \in [0,1]$. Hence, if $\bar{u}$ is a solution of $(\mathcal{P}_{\text{inf}})$, then

$$P(\bar{u} + \theta(u - \bar{u})) - P(\bar{u}) \geq 0 \;\; \forall u \in \mathcal{K}.$$

Thus, the minimizer $\bar{u}$ of the minimization problem $(\mathcal{P}_{\text{inf}})$ can be characterized by the *variational inequality*

$$(\mathcal{P}_{\text{vi}}): \;\; P(\bar{u}) \leq P(u) \;\; \forall u \in \mathcal{K}. \tag{5.151}$$

If P is Gâteaux differentiable, then

$$\lim_{\theta\to 0^+}\frac{1}{\theta}\{P(\bar u+\theta(u-\bar u))-P(\bar u)\}=\langle DP(\bar u),u-\bar u\rangle\geq 0\ \ \forall u\in\mathcal{K},$$

and the variational inequality $(\mathcal{P}_{\rm vi})$ reduces to the standard form

$$(\mathcal{P}_{\rm vi})_1:\quad \langle DP(\bar u),u-\bar u\rangle\geq 0\ \ \forall u\in\mathcal{K}. \tag{5.152}$$

By symmetry, an alternative form is

$$(\mathcal{P}_{\rm vi})_2:\quad \langle DP(u),u-\bar u\rangle\geq 0\ \ \forall u\in\mathcal{K}. \tag{5.153}$$

For convex P, we have the following result.

Proposition 5.5.1 *Assume that $P\in\check{\Gamma}_G(\mathcal{K})$ is convex, Gâteaux differentiable with continuous derivative $DP\in\mathcal{U}^*$. Then, for a given $\bar u\in\mathcal{K}$,*

$$(\mathcal{P}_{\rm vi})\ \Leftrightarrow\ (\mathcal{P}_{\rm vi})_1\ \Leftrightarrow\ (\mathcal{P}_{\rm vi})_2. \tag{5.154}$$

Proof. We first prove that $(\mathcal{P}_{\rm vi})\ \Leftrightarrow\ (\mathcal{P}_{\rm vi})_1$. If $\bar u\in\mathcal{K}$ minimizes P on $\mathcal{K}$, then, for any $u\in\mathcal{K}$ and $\theta\in(0,1)$,

$$P(\bar u+\theta(u-\bar u))-P(\bar u)\geq 0\ \ \forall u\in\mathcal{K}.$$

Since P is Gâteaux differentiable,

$$\lim_{\theta\to 0^+}\frac{1}{\theta}\{P(\bar u+\theta(u-\bar u))-P(\bar u)\}=\langle DP(\bar u),u-\bar u\rangle\geq 0\ \ \forall u\in\mathcal{K}.$$

Conversely, for any $u_1,u_2\in\mathcal{K}$ and $\theta\in(0,1)$, the convexity of P gives

$$P(u_1)-P(u_2)\geq\frac{1}{\theta}[P(\theta u_1+(1-\theta)u_2)-P(u_2)].$$

Taking the limit as $\theta\to 0^+$, we find that if $\bar u$ satisfies the variational inequality $(P2)$, then

$$P(u)-P(\bar u)\geq\langle DP(\bar u),u-\bar u\rangle\geq 0$$

which shows that $\bar u$ minimizes P on $\mathcal{K}$.

Now we shall prove that $(\mathcal{P}_{\rm vi})_1\ \Leftrightarrow\ (\mathcal{P}_{\rm vi})_2$. Since P is convex, $DP:\mathcal{U}\to\mathcal{U}^*$ is a monotone operator, i.e.

$$\langle DP(u)-DP(\bar u),u-\bar u\rangle\geq 0\ \ \forall\bar u,u\in\mathcal{U}, \tag{5.155}$$

and hence

$$\langle DP(u),u-\bar u\rangle\geq\langle DP(\bar u),u-\bar u\rangle\geq 0\ \ \forall\bar u,u\in\mathcal{U}.$$

If $\bar u$ is a solution of $(\mathcal{P}_{\rm vi})_1$, then $\bar u$ satisfies $(\mathcal{P}_{\rm vi})_2$.

Now, if $\bar u$ satisfies $(\mathcal{P}_{\rm vi})_2$, then, by taking $u=(1-\theta)\bar u+\theta w$, $w\in\mathcal{K}$, $\theta\in(0,1)$, we find that

$$\theta\langle DP((1-\theta)\bar u+\theta w),w-\bar u\rangle=\theta\langle DP(\bar u+\theta(w-\bar u)),w-\bar u\rangle\geq 0.$$

The function $\theta \mapsto \langle DP(\bar{u} + \theta(w - \bar{u})), w - \bar{u}\rangle$ is the derivative of the scalar function $\theta \mapsto P(\bar{u} + \theta(w - \bar{u}))$; it is continuous and hence, when $\theta \to 0^+$, we have

$$\langle DP(\bar{u}), w - \bar{u}\rangle \geq 0 \ \forall w \in \mathcal{K}.$$

This shows that $(\mathcal{P}_{\rm vi})_1 \Leftrightarrow (\mathcal{P}_{\rm vi})_2$. □

For a given linear operator $\Lambda : \mathcal{U} \to \mathcal{E}$, if $P(u) = W(\Lambda u) - F(u)$ and $\mathcal{K} = \{u \in \mathcal{U} | \ u \in \text{dom } F, \ \ \Lambda u \in \text{dom } W\}$ is a feasible space, then the problems $(\mathcal{P}_{\rm vi})_1$ and $(\mathcal{P}_{\rm vi})_2$ can be written respectively as

$$(\mathcal{P}_{\rm vi})'_1 : \quad \langle \Lambda^* DW(\Lambda\bar{u}), u - \bar{u}\rangle \geq F(u) - F(\bar{u}) \ \ \forall u \in \mathcal{K}, \tag{5.156}$$

$$(\mathcal{P}_{\rm vi})'_2 : \quad \langle \Lambda^* DW(\Lambda u), u - \bar{u}\rangle \geq F(u) - F(\bar{u}) \ \ \forall u \in \mathcal{K}. \tag{5.157}$$

Corollary 5.5.1 *Assume that* $W \in \check{\Gamma}(\mathcal{E})$, $F \in \hat{\Gamma}_0(\mathcal{U})$, *and that* W *is Gâteaux differentiable on the range* $\mathcal{R}_\Lambda(\mathcal{K})$. *Then, for a given* $\bar{u} \in \mathcal{K}$,

$$(\mathcal{P}_{\rm vi}) \ \Leftrightarrow \ (\mathcal{P}_{\rm vi})'_1 \ \Leftrightarrow \ (\mathcal{P}_{\rm vi})'_2.$$

Replacing $F(u)$ by $F(u) - \mathcal{I}_{\mathcal{K}}(u)$, the constraint in the convex set $\mathcal{K}$ can be relaxed. Let $A(u) = \Lambda^* DW(\Lambda u)$. Then, a general existence theory for solutions of abstract variational inequalities on Banach spaces $\mathcal{U}$ can be proposed by the well-known theorem (cf., e.g., Oden, 1986)

Theorem 5.5.1 *Let* $\mathcal{U}$ *be a reflexive Banach space and* $A : \mathcal{U} \to \mathcal{U}^*$ *be bounded, pseudomonotone, and coercive in the sense that there exists a* $u_0 \in$ *dom* F *such that*

$$\lim_{\|u\| \to \infty} \frac{\langle A(u), u - u_0\rangle - F(u) + F(u_0)}{\|u\|} = +\infty. \tag{5.158}$$

Then, there exists at least one solution $\bar{u} \in$ *dom* F *of the variational inequality*

$$\langle A(\bar{u}), u - \bar{u}\rangle \geq F(u) - F(\bar{u}) \ \ \forall u \in \mathcal{U}. \tag{5.159}$$

The solution is unique if A *is strictly monotone.*

If $\mathcal{K}$ is bounded, then the coercivity condition on A can be relaxed. In particular, for a given $f \in \mathcal{U}^*$ such that

$$F(u) = \langle f, u\rangle - \mathcal{I}_{\mathcal{K}}(u), \tag{5.160}$$

Theorem 5.5.1 has the following special case.

Proposition 5.5.2 *Let* $A : \mathcal{K} \to \mathcal{U}^*$ *be bounded, pseudomonotone, and coercive. Then, the following statements hold.*

(S1) For all $f \in \mathcal{U}^*$, *there exists a* $\bar{u} \in \mathcal{U}$ *such that*

$$A\bar{u} = f. \tag{5.161}$$

(S2) If $\mathcal{K}$ *is a closed convex set of* $\mathcal{U}$, *then, for all* $f \in \mathcal{U}^*$, *there exists a* $\bar{u} \in \mathcal{K}$ *such that*

$$\langle A\bar{u} - f, u - \bar{u}\rangle \geq 0 \ \ \forall u \in \mathcal{K}. \tag{5.162}$$

Proof. Since $\mathcal{K} = \mathcal{U}$, we set $u = \bar{u} + w$ and $u = \bar{u} - w$ in turn. This yields

$$\langle A\bar{u} - f, w \rangle = 0 \ \ \forall w \in \mathcal{U}.$$

It follows that $A\bar{u} = f$. □

5.5.2 Complementarity and Karush-Kuhn-Tucker Conditions

Recall that a cone $\mathcal{C} \in \mathcal{U}$ is said to be proper if it is convex with vertex at the origin. The importance of a proper cone in $\mathcal{U}$ is that we can introduce a partial ordering relation $\geq$ on $\mathcal{C}$ defined by

$$u_1 \geq u_2 \Leftrightarrow u_2 - u_1 \in \mathcal{C}.$$

The partial ordering relation $\geq$ is reflexive ($u - u = 0 \in \mathcal{C}$), antisymmetric ($u_1 \geq u_2$ and $u_2 \geq u_1 \Rightarrow u_1 = u_2$) and transitive ($u_1 \geq u_2,\ u_2 \geq u_3 \Rightarrow u_1 \geq u_3$). These relations imply that the partial ordering $\geq$ is compatible with the structure of a linear space in the sense that

$$u \geq 0 \Rightarrow \theta u \geq 0 \ \ \forall \theta > 0 \in \mathbb{R},$$

$$u_1 \geq u_2 \Rightarrow \ u_1 + w \geq u_2 + w \ \ \forall w \in \mathcal{U}.$$

Thus, the *positive cone* $\mathcal{C} = \mathcal{C}^+ \subset \mathcal{U}$ is the set of positive elements for the ordering relation $\geq$, i.e.

$$\mathcal{C}^+ = \{u \in \mathcal{U} |\ u \geq 0\}.$$

Likewise, the *negative cone* $\mathcal{C} = \mathcal{C}^- = \{-\mathcal{C}^+\}$ is the set of negative elements for the ordering relation $\leq$

$$\mathcal{C}^- = \{u \in \mathcal{U} |\ u \leq 0\}.$$

Recall the definitions

$$\mathcal{C}^\oplus = \{u^* \in \mathcal{U}^* |\ \langle u, u^* \rangle \geq 0 \ \ \forall u \in \mathcal{C}\}, \quad \mathcal{C}^\ominus = \{u^* \in \mathcal{U}^* |\ \langle u, u^* \rangle \leq 0 \ \ \forall u \in \mathcal{C}\}$$

of the positive and negative conjugate cones. It is easy to see that the positive (resp. negative) conjugate of a positive (resp. negative) cone is also a positive cone in $\mathcal{U}^*$, denoted by $\mathcal{C}^{*+}$, while the negative (resp. positive) conjugate of a positive (resp. negative) cone in $\mathcal{U}$ is a negative cone in $\mathcal{U}^*$, denoted by $\mathcal{C}^{*-}$, i.e.

$$\mathcal{C}^{*+} = (\mathcal{C}^+)^\oplus = (\mathcal{C}^-)^\ominus, \quad \mathcal{C}^{*-} = (\mathcal{C}^-)^\oplus = (\mathcal{C}^+)^\ominus. \tag{5.163}$$

Lemma 5.5.1 *Let $\mathcal{C}$ be a closed proper (either positive or negative) cone in $\mathcal{U}$ and let $\mathcal{C}^\ominus$ be its negative conjugate cone in $\mathcal{U}^*$. Then,*

(L1) the support functional of $\mathcal{C}$ is the indicator of its negative conjugate cone, i.e.

$$\mathcal{I}^\sharp_{\mathcal{C}}(u^*) = \mathcal{I}_{\mathcal{C}^\ominus}(u^*),$$

(L2) the equivalent relations

$$u^* \in \partial^- \mathcal{I}_{\mathcal{C}}(u) \ \ \Leftrightarrow \ \ u \in \partial^- \mathcal{I}_{\mathcal{C}^\ominus}(u^*) \ \Leftrightarrow$$

$$u \in \mathcal{C}, \quad \langle u, u^* \rangle = 0, \quad u^* \in \mathcal{C}^\ominus \tag{5.164}$$

hold.

Proof. Since $\mathcal{C}$ is a closed convex subset of $\mathcal{U}$, its indicator $\mathcal{I}_\mathcal{C} : \mathcal{U} \to \overline{\mathbb{R}}$ is convex, l.s.c. The Fenchel super-conjugate can be computed at once:

$$(\mathcal{I}_\mathcal{C})^\sharp (u^*) = \sup_{u \in \mathcal{U}} \{\langle u, u^* \rangle - \mathcal{I}_\mathcal{C}(u)\} = \sup_{u \in \mathcal{C}} \langle u, u^* \rangle = \begin{cases} 0 & \text{if } u^* \in \mathcal{C}^\ominus \\ +\infty & \text{otherwise.} \end{cases}$$

Recall that the Fenchel super-conjugate of $\mathcal{I}_\mathcal{C}$ is the support functional, so that $\mathcal{I}_\mathcal{C}^\sharp(u^*) = \mathcal{I}_{\mathcal{C}^\ominus}(u^*)$. Since $\mathcal{I}_\mathcal{C}^\sharp(u^*)$ is convex, l.s.c. on $\mathcal{U}^*$, $(L2)$ follows from the Fenchel duality theorem. □

Remark 5.5.1 This lemma shows a very important fact that if $u^* \in \partial^- \mathcal{I}_\mathcal{C}(u)$, then u and its dual u^* are orthogonal. The complementarity condition $\langle u, u^* \rangle = 0$ can also be written as $u \perp u^*$. In finite dimensional space $\mathcal{U} = \mathcal{U}^* = \mathbb{R}^n$, equation and inequalities in (5.164) are the well-known *KKT conditions*. In the case that $\mathcal{C}$ is a positive cone, this equation reads

$$0 \leq u, \quad u \perp u^*, \quad u^* \leq 0.$$

In mathematical programming, this is also referred as the *complementarity condition.* ■

Now let us consider the extremality conditions for the constrained variational problem $(\mathcal{P}_{\inf})$. On introduction the indicator functional of the feasible set $\mathcal{K}$ in (5.150), and the extended functional $\bar{P}(u) = P(u) + \mathcal{I}_\mathcal{K}(u) : \mathcal{U} \to \overline{\mathbb{R}}$, the primal problem $(\mathcal{P}_{\inf})$ given in (5.150) is seen to be equivalent to the Euler-Lagrange inclusion

$$0 \in \partial^- \bar{P} = \partial^- \left(P(\bar{u}) + \mathcal{I}_\mathcal{K}(\bar{u}) \right). \tag{5.165}$$

First, we consider $\mathcal{K}$ to be a non-empty proper cone in $\mathcal{U}$ and $P(u)$ to be Gâteaux differentiable on $\mathcal{K}$. Thus the extremality condition (5.165) then leads to

$$DP(\bar{u}) = \bar{u}^*, \qquad \bar{u} \in \mathcal{K}, \ \langle \bar{u}, \bar{u}^* \rangle = 0, \ \bar{u}^* \in \mathcal{K}^\ominus. \tag{5.166}$$

This is the so-called *complementarity problem,* which is a strong form of a special case of the extended variational problem $(\mathcal{P}_{\inf})$.

Next, we let the feasible set $\mathcal{K} = \{u \in \mathcal{U}_a | \ \Lambda u \leq 0 \in \mathcal{E}\}$, in which $\mathcal{U}_a$ is a non-empty closed convex subset of $\mathcal{U}$ and $\Lambda : \mathcal{U} \to \mathcal{E}$, be a linear operator. In this case $(\mathcal{P}_{\inf})$ is the constrained problem proposed in (5.61), i.e.

$$(\mathcal{P}_{\inf}) : \quad \begin{cases} P(u) \to \inf \ \forall u \in \mathcal{U}_a, \\ s.t. \ \Lambda u \leq 0. \end{cases}$$

Let $\mathcal{E}_a = \{\epsilon \in \mathcal{E} | \ \epsilon \leq 0\}$ be a negative cone. Thus on $\mathcal{U}$, the extended problem is

$$\bar{P}(u) = \mathcal{I}_{\mathcal{E}_a}(\Lambda u) + P(u) + \mathcal{I}_{\mathcal{K}_a}(u) \quad \to \inf \ \forall u \in \mathcal{U}.$$

In this case, since $P \in \check{\Gamma}_G(\mathcal{U}_a)$ is Gâteaux differentiable, then the extremality condition $0 \in \partial^- \bar{P}(\bar{u})$ leads to

$$\begin{cases} \Lambda^* \bar{\epsilon}^* = -DP(\bar{u}), \\ \Lambda\bar{u} \leq 0 \in \mathcal{E}, \ \langle \Lambda\bar{u}; \bar{\epsilon}^* \rangle = 0, \ \ \bar{\epsilon}^* \geq 0 \in \mathcal{E}^*. \end{cases} \tag{5.167}$$

This is the so-called *mixed complementarity problem* in mathematical programming. Actually, if we set $W(\epsilon) = \mathcal{I}_{\mathcal{E}_a}(\epsilon)$ and $F(u) = -P(u)$, this mixed complementarity problem is then expressible as the standard tri-canonical forms

$$\begin{array}{ll} \epsilon = \Lambda u, & u \in \mathcal{U}_a, \\ \epsilon^* \in \partial^- W(\epsilon), & \\ u^* = \Lambda^* \epsilon^*, & u^* = DF(u). \end{array} \tag{5.168}$$

Since $\mathcal{E}_a$ is a negative cone, its negative conjugate cone is a positive cone in $\mathcal{E}^*$, i.e.

$$\mathcal{E}_a^\ominus = \{\epsilon^* \in \mathcal{E}^* | \ \langle \epsilon; \epsilon^* \rangle \leq 0 \ \ \forall \epsilon \leq 0 \in \mathcal{E}\} = \{\epsilon^* \in \mathcal{E}^* | \ \epsilon^* \geq 0\}.$$

Thus, $W^*(\epsilon^*) = \mathcal{I}_{\mathcal{E}_a^\ominus}(\epsilon^*)$ and the Lagrangian $L : \mathcal{U} \times \mathcal{E}^* \to \bar{\mathbb{R}}$ associated with this constrained problem reads

$$L(u, \epsilon^*) = \langle \Lambda u; \epsilon^* \rangle - \mathcal{I}_{\mathcal{E}_a^\ominus}(\epsilon^*) - F(u) + \mathcal{I}_{\mathcal{U}_a}(u).$$

On the effective domain $\mathcal{Z}_a = \mathcal{U}_a \times \mathcal{E}_a^\ominus$, this Lagrangian form reduces to

$$L(u, \epsilon^*) = \langle \Lambda u; \epsilon^* \rangle - F(u).$$

For any given $F \in \Gamma_0(\mathcal{U}_a)$, the functional dual to $\bar{P}$ is simply the Fenchel conjugate

$$\bar{P}^d(\epsilon^*) = \operatorname*{ext}_{u \in \mathcal{U}_a} L(u, \epsilon^*) = F^*(\Lambda^* \epsilon^*)$$

of $F(u)$.

Theorem 5.5.2 *Suppose that $P \in \Gamma_0(\mathcal{U}_a)$, and there exists $u_0 \in \mathcal{U}_a$ such that $\Lambda u_0 < 0$.*

(T1) $(\bar{u}, \bar{\epsilon}^*)$ is a critical point of L if and only if it is a solution of the mixed complementarity problem (5.167).

(T2) If $P = -F \in \check{\Gamma}_0(\mathcal{U}_a)$, then $\bar{u} \in \mathcal{U}_a$ is a minimizer of $\bar{P}$ if and only if there exists a maximizer $\bar{\epsilon}^* \in \mathcal{E}_a^\ominus$ of $\bar{P}^d$ such that $(\bar{u}, \bar{\epsilon}^*)$ is a saddle point of L, i.e.

$$\langle \Lambda u; \bar{\epsilon}^* \rangle - F(u) \geq \langle \Lambda \bar{u}; \bar{\epsilon}^* \rangle - F(\bar{u}) \geq \langle \Lambda \bar{u}; \epsilon^* \rangle - F(\bar{u}) \ \ \forall (u, \epsilon^*) \in \mathcal{Z}_a. \tag{5.169}$$

(T3) If $P = -F \in \hat{\Gamma}_0(\mathcal{U}_a)$, then $\bar{u} \in \mathcal{U}_a$ is a minimizer of $\bar{P}$ if and only if there exists a minimizer $\bar{\epsilon}^* \in \mathcal{E}_a^\ominus$ of $\bar{P}^d$ such that $(\bar{u}, \bar{\epsilon}^*)$ is a super-critical point of L, i.e.

$$\langle \Lambda u; \bar{\epsilon}^* \rangle - F(u) \geq \langle \Lambda \bar{u}; \bar{\epsilon}^* \rangle - F(\bar{u}) \leq \langle \Lambda \bar{u}; \epsilon^* \rangle - F(\bar{u}) \ \ \forall (u, \epsilon^*) \in \mathcal{U}_a \times \mathcal{E}_a^\ominus. \tag{5.170}$$

Proof. Clearly, if $P = -F \in \check{\Gamma}(\mathcal{U}_a)$ then $L \in \check{\Gamma}(\mathcal{U}_a) \times \hat{\Gamma}(\mathcal{E}_a^\ominus)$ is a right-saddle functional. However, if $P \in \hat{\Gamma}(\mathcal{U}_a)$ then $L \in \hat{\Gamma}(\mathcal{U}_a) \times \hat{\Gamma}(\mathcal{E}_a^\ominus)$ is a super-Lagrange form. Thus, this theorem follows by the Lagrange duality theory. □

5.5.3 Bi-Complementarity and Duality

Now let us consider the more general canonical variational problem

$$P(u) = W(\Lambda u) - F(u) \to \inf \quad \forall u \in \mathcal{K}_a, \tag{5.171}$$

where, as usual, $\Lambda : \mathcal{U} \to \mathcal{E}$ is a linear operator and $W \in \Gamma(\mathcal{E}_a)$ and $F \in \Gamma(\mathcal{U}_a)$ are two canonical functionals. But there is something new in the feasible space $\mathcal{K}_a = \{u \in \mathcal{U}_a | \ \Lambda u \in \mathcal{E}_a\}$. We introduce two more duality pairs of locally convex topological vector spaces $\{(\mathcal{V}, \mathcal{V}^*), \langle *, * \rangle\}$, $\{(\mathcal{Y}, \mathcal{Y}^*), \langle *; * \rangle\}$ and the two mappings

$B : \mathcal{U} \to \mathcal{V}$ which is linear, and

$Y : \mathcal{E} \to \mathcal{Y}$ which is Gâteaux differentiable and convex with respect to the relation $\leq$, i.e.

$$Y(\theta\epsilon_1 + (1-\theta)\epsilon_2) \leq \theta Y(\epsilon_1) + (1-\theta)Y(\epsilon_2) \quad \forall \epsilon_1, \epsilon_2 \in \mathcal{E}_a, \quad \forall \theta \in (0,1).$$

The admissible spaces $\mathcal{U}_a \subset \mathcal{U}$ and $\mathcal{E}_a \subset \mathcal{E}$ are defined respectively by

$$\begin{aligned} \mathcal{U}_a &= \{u \in \mathcal{U} | \ Bu \geq 0 \in \mathcal{V}\}, \\ \mathcal{E}_a &= \{\epsilon \in \mathcal{E} | \ Y(\epsilon) \geq 0 \in \mathcal{Y}\}. \end{aligned}$$

Clearly, $\mathcal{U}_a$ is a positive cone in $\mathcal{V}$ and $\mathcal{E}_a$ is a positive cone in $\mathcal{Y}$. Let

$$\bar{W}(\epsilon) = W(\epsilon) + \mathcal{I}_{\mathcal{E}_a}(\epsilon), \quad \bar{F}(u) = F(u) - \mathcal{I}_{\mathcal{U}_a}(u).$$

The constraints in (5.171) can be relaxed, and the extended problem then reads

$$\bar{P}(u) = \Phi(u, \Lambda u) = \bar{W}(\Lambda u) - \bar{F}(u) \quad \to \inf \quad \forall u \in \mathcal{U}.$$

If the effective domain $\mathcal{K} = \operatorname{dom} \bar{P}$ is not empty and there exists at least one $u_0 \in \mathcal{K}$ such that $Y(\Lambda u_0) < 0 \in \mathcal{Y}$, $Bu_0 < 0 \in \mathcal{V}$, then the extremality condition $0 \in \partial^- \bar{P}(\bar{u})$ can be written as

$$0 \in \Lambda^* \partial^- \bar{W}(\Lambda \bar{u}) - \partial^+ \bar{F}(\bar{u}).$$

If $F \in \hat{\Gamma}_0(\mathcal{U}_a)$ is Gâteaux differentiable, the condition $u^* \in \partial^+ \bar{F}(u)$ leads to

$$u^* = DF(u) - B^* v^*,$$

$$Bu \geq 0 \in \mathcal{V}, \quad \langle Bu, v^* \rangle = 0, \quad v^* \leq 0 \in \mathcal{V}^*,$$

where $B^* : \mathcal{V}^* \to \mathcal{U}^*$ is the adjoint of $B : \mathcal{U} \to \mathcal{V}$, and v^* is a Lagrange multiplier for the constraint $Bu \geq 0$.

If $W \in \check{\Gamma}_0(\mathcal{E}_a)$ is Gâteaux differentiable, then the condition $\epsilon^* \in \partial^- \bar{W}(\epsilon)$ gives

$$\epsilon^* = DW(\epsilon) - Y_t^*(\epsilon)\mu^*,$$

$$Y(\epsilon) \geq 0 \in \mathcal{Y}, \quad \langle Y(\epsilon); \mu^* \rangle = 0, \quad \mu^* \leq 0 \in \mathcal{Y}^*,$$

$Y_t(\epsilon) = DY(\epsilon)$ being the Gâteaux derivative of Y, $Y_t^*(\epsilon)$ the adjoint of Y_t at ϵ and $\mu^* \in \mathcal{Y}^*$ is a Lagrange multiplier for $Y(\epsilon) \geq 0$.

Hence, the tri-canonical forms associated with the constrained variational problem (5.171) are

$$\begin{cases} \epsilon = \Lambda u, & Bu \in \mathcal{U}_a, \\ \epsilon^* \in \partial \bar{W}(\epsilon), & \\ u^* = \Lambda^* \epsilon^*, & u^* \in \partial \bar{F}(u). \end{cases} \tag{5.172}$$

The fundamental form for these tri-canonical forms is the so-called *bi-complementarity problem*

$$\Lambda^* \left(DW(\Lambda u) - Y_t^*(\Lambda u)\mu^* \right) = DF(u) - B^* v^*, \tag{5.173}$$

$$Bu \geq 0 \in \mathcal{V}, \ \langle Bu, v^* \rangle = 0, \ v^* \leq 0 \in \mathcal{V}^*, \tag{5.174}$$

$$Y(\Lambda u) \geq 0 \in \mathcal{Y}, \ \langle Y(\Lambda u); \mu^* \rangle = 0, \ \mu^* \leq 0 \in \mathcal{Y}^*. \tag{5.175}$$

This problem was originally proposed as an elastoplastic contact problem for an extended beam model (see Gao, 1995, 1998), where (5.174) is called the *external complementarity condition*, while (5.175) is called the *internal complementarity condition*. The diagrammatic representation for this bi-complementarity system is shown in Fig. 5.10, in which, $B^c : \mathcal{U}^* \to \mathcal{V}^*$ is called the *complementary operator* of B. Similarly for Y^c.

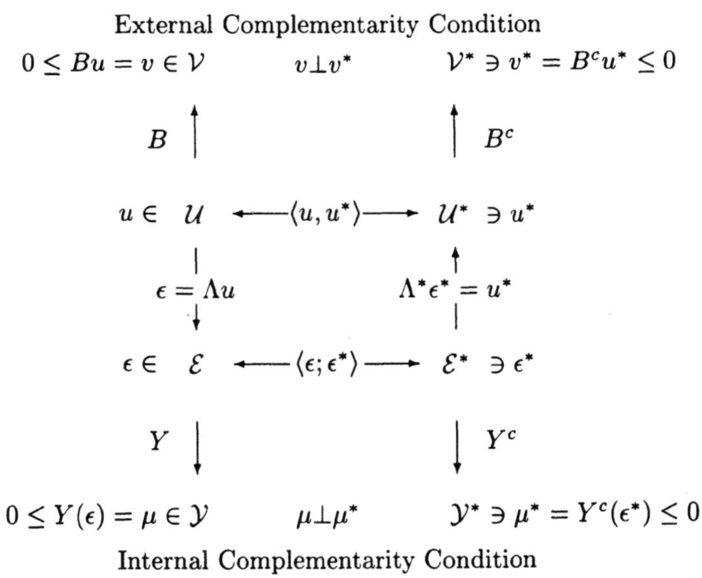

Figure 5.10. Framework for bi-complementarity problems in geometrically linear systems

The extended Lagrangian $L : \mathcal{U} \times \mathcal{E}^* \to \bar{\mathbb{R}}$ associated with this bi-complementarity problem has the standard form

$$L(u, \epsilon^*) = \langle \Lambda u; \epsilon^* \rangle - \bar{W}^*(\epsilon^*) - \bar{F}(u).$$

When particular forms of $W(\epsilon)$ and $Y(\epsilon)$ are given, the Fenchel conjugate $\bar{W}^*$ can be computed at once, and the duality theory follows.

5.5.4 Mathematical Programming and Primal-Dual Methods

We now turn our attention to nonlinear optimization problems in finite dimensional space. Extensive applications will be given in the last chapter. Here we only provide some simple examples for illustration.

Let $\mathcal{U} = \mathcal{U}^* = \mathbb{R}^n$, $\mathcal{E} = \mathcal{E}^* = \mathbb{R}^m$, with the standard coordinatewise partial ordering, and let $\Lambda \in \mathbb{R}^{m\times n}$ be a matrix. First, let us consider a global optimization problem.

Problem 5.5.1 For a given $\mathbf{b} \in \mathbb{R}^n$, find $\bar{\mathbf{x}} \in \mathbb{R}^n$ such that

$$(\mathcal{P}_{\min}): \quad F_0(\bar{\mathbf{x}}) = \min_{\mathbf{x}\in\mathbb{R}^n} F_0(\mathbf{x}), \quad s.t. \ \Lambda\mathbf{x} = \mathbf{b}, \quad \mathbf{x} \geq 0, \tag{5.176}$$

where $F_0(\mathbf{x})$ is a given canonical function, and $\Lambda : \mathcal{U} = \mathbb{R}^n \to \mathcal{E} = \mathbb{R}^m$ is a linear mapping. ♠

This is a standard nonlinear programming problem with both equality and inequality constraints. The canonical function F_0 is known as a *cost function* in mathematical economics systems, while in optimal design and control it is also referred as the *target function*. If $F_0(\mathbf{x})$ is convex and differentiable, both direct approaches and duality theory have been well developed for this convex and smooth problem. However, for concave, piecewise differentiable F_0, the primal problem is nonconvex and nonsmooth. The bi-duality theory developed in this book plays an important role in the solution of this very difficult problem.

To relax the inequality constraint, we introduce an admissible feasible set

$$\mathcal{U}_a = \{\mathbf{x} \in \mathbb{R}^n | \ \mathbf{x} \geq 0\}, \tag{5.177}$$

which is a convex cone in $\mathbb{R}^n$. On $\mathcal{U}_a$, the external energy $F : \mathbb{R}^n \to \overleftarrow{\mathbb{R}}$ is defined by

$$F(\mathbf{x}) = -F_0(\mathbf{x}) - \mathcal{I}_{\mathcal{U}_a}(\mathbf{x}).$$

To relax the equality constraint $\mathbf{y} = \Lambda\mathbf{x} = \mathbf{b}$, we introduce an "internal energy" $W : \mathbb{R}^m \to \overrightarrow{\mathbb{R}}$, defined by

$$W(\mathbf{y}) = \mathcal{I}_{\mathcal{E}_a}(\mathbf{y}) = \begin{cases} 0 & \text{if } \mathbf{y} = \mathbf{b}, \\ +\infty & \text{if } \mathbf{y} \neq \mathbf{b}, \end{cases} \tag{5.178}$$

where the admissible space $\mathcal{E}_a = \{\mathbf{y} \in \mathbb{R}^m | \ \mathbf{y} = \mathbf{b}\}$ has only one vector in $\mathbb{R}^m$. Thus, the extended problem reads

$$P(\mathbf{x}) = \Phi(\mathbf{x}, \Lambda\mathbf{x}) = W(\Lambda\mathbf{x}) - F(\mathbf{x}) \quad \to \min \quad \forall \mathbf{x} \in \mathbb{R}^n. \tag{5.179}$$

The implicit constraint for this extended problem is

$$\mathbf{x} \in \mathcal{K}_a = \{\mathbf{x} \in \mathcal{U}_a | \ \Lambda\mathbf{x} \in \mathcal{E}_a\} = \{\mathbf{x} \in \mathbb{R}^n | \ \mathbf{x} \geq 0, \ \Lambda\mathbf{x} = \mathbf{b}\}.$$

We know that $\mathcal{K}_a$ is a convex, closed subset of $\mathbb{R}^n$.

Since $W : \mathcal{E}_a \to \overrightarrow{\mathbb{R}}$ is convex, l.s.c. its super-Fenchel conjugate can be computed as

$$W^*(\mathbf{y}^*) = \max_{\mathbf{y}\in\mathbb{R}^m}\{\langle \mathbf{y};\mathbf{y}^*\rangle - W(\mathbf{y})\} = \langle \mathbf{b};\mathbf{y}^*\rangle, \tag{5.180}$$

which is well-defined on the whole domain $\mathcal{E}_a^* = \mathbb{R}^m$. Then, the extended Lagrangian $L : \mathbb{R}^n \times \mathbb{R}^m \to \bar{\mathbb{R}}$ associated with $(\mathcal{P}_{\min})$ has the simple form

$$L(\mathbf{x},\mathbf{y}^*) = \langle \Lambda\mathbf{x};\mathbf{y}^*\rangle - W^*(\mathbf{y}^*) - F(\mathbf{x}) = \langle \Lambda\mathbf{x} - \mathbf{b};\mathbf{y}^*\rangle + F_0(\mathbf{x}) + \mathcal{I}_{\mathcal{U}_a}(\mathbf{x}).$$

If F_0 is (piecewise) Gâteaux differentiable and $\mathcal{K}_a \neq \emptyset$, the criticality condition $DL(\mathbf{x},\mathbf{y}^*) = 0$ leads to the canonical Lagrange equations

$$\Lambda\mathbf{x} = \mathbf{b}, \quad \Lambda^*\mathbf{y}^* + DF_0(\mathbf{x}) = \boldsymbol{\mu}$$

subject to the KKT condition

$$\mathbf{x} \geq 0, \quad \langle \mathbf{x},\boldsymbol{\mu}\rangle = 0, \quad \boldsymbol{\mu} \leq 0.$$

In this finite dimensional problem, the adjoint operator of Λ is simply $\Lambda^* = \Lambda^T$.

Clearly, the convexity of L depends only on that of F_0. For the canonical function F_0, two cases should be discussed.

I. $F_0 \in \check{\Gamma}_0(\mathcal{U}_a)$ is convex. In this case, L is a saddle Lagrangian. For $F(\mathbf{x}) = -F_0(\mathbf{x}) - \mathcal{I}_{\mathcal{U}_a}(\mathbf{x})$, the Fenchel conjugate of the extended function $F : \mathbb{R}^n \to \overleftarrow{\mathbb{R}}$ becomes

$$F^\flat(\mathbf{x}^*) = \inf_{\mathbf{x}\in\mathbb{R}^n}\{\langle \mathbf{x},\mathbf{x}^*\rangle - F(\mathbf{x})\} = \inf_{\mathbf{x}\in\mathcal{U}_a}\{\langle \mathbf{x},\mathbf{x}^*\rangle - F_0(\mathbf{x})\}.$$

Thus, for any fixed $\mathbf{y}^* \in \mathbb{R}^m$, the dual function $P^d : \mathbb{R}^m \to \overleftarrow{\mathbb{R}}$, given by

$$P^d(\mathbf{y}^*) = \inf\{L(\mathbf{x},\mathbf{y}^*)|\ \mathbf{x} \in \mathbb{R}^n\} = F^\flat(\Lambda^*\mathbf{y}^*) - \langle \mathbf{b};\mathbf{y}^*\rangle,$$

is concave, u.s.c., the problem dual to $(\mathcal{P}_{\min})$ is

$$(\mathcal{P}^d_{\max}): \ P^d(\mathbf{y}^*) \ \to \max \ \forall \mathbf{y}^* \in \mathbb{R}^m,$$

and the saddle Lagrange duality theory leads to

$$\inf_{\mathbf{x}\in\mathbb{R}^n} P(\mathbf{x}) = \sup_{\mathbf{y}^*\in\mathbb{R}^m} P^d(\mathbf{y}^*).$$

As we have shown in Chapter 3, if F_0 is nonsmooth, then its Legendre conjugate F^* is usually smooth. Thus, this classical dual approach can be used effectively to solve some nonsmooth convex problems.

II. $F_0 \in \hat{\Gamma}(\mathcal{U}_a)$ is concave. In this case, L is a super-Lagrangian and by letting $F(\mathbf{x}) = -F_0(\mathbf{x}) + \mathcal{I}_{\mathcal{U}_a}(\mathbf{x})$, we obtain the Fenchel super-conjugate of the extended function $F : \mathbb{R}^n \to \overrightarrow{\mathbb{R}}$ in the form

$$F^\sharp(\mathbf{x}^*) = \sup_{\mathbf{x}\in\mathbb{R}^n}\{\langle \mathbf{x},\mathbf{x}^*\rangle - F(\mathbf{x})\} = \sup_{\mathbf{x}\in\mathcal{U}_a}\{\langle \mathbf{x},\mathbf{x}^*\rangle - F_0(\mathbf{x})\}.$$

The dual function $P^d : \mathbb{R}^m \to \overrightarrow{\mathbb{R}}$ defined by

$$P^d(\mathbf{y}^*) = \sup_{\mathbf{x}\in\mathbb{R}^n} L(\mathbf{x},\mathbf{y}^*) = F^\sharp(\Lambda^*\mathbf{y}^*) - \langle \mathbf{b};\mathbf{y}^*\rangle$$

is convex, l.s.c., the problem dual to $(\mathcal{P}_{\min})$ reads

$$(\mathcal{P}^d_{\min}): \quad P^d(\mathbf{y}^*) \quad \to \min \ \ \forall \mathbf{y}^* \in \mathbb{R}^m.$$

In this case, the primal problem is nonconvex and the minimizer can be found only on the boundary $\partial\mathcal{K}_a$. However, the dual problem is convex! The super-Lagrange duality theory leads to

$$\inf_{\mathbf{x}\in\mathbb{R}^n} P(\mathbf{x}) = \inf_{\mathbf{y}^*\in\mathbb{R}^m} P^d(\mathbf{y}^*).$$

In the very special case that the cost function $F_0(\mathbf{x}) = \langle \mathbf{x}, \mathbf{f}\rangle$ is linear for a given vector $\mathbf{f} \in \mathbb{R}^n$, the primal problem is the linear programming problem

$$(\mathcal{P}_{\min})_l: \quad \min_{\mathbf{x}\in\mathbb{R}^n} \langle \mathbf{x}, \mathbf{f}\rangle, \quad s.t. \ \ \Lambda\mathbf{x} = \mathbf{b}, \ \ \mathbf{x} \geq 0.$$

The associated Lagrangian $L : \mathcal{U}_a \times \mathbb{R}^m \to \mathbb{R}$ is simply a linear function, given by

$$L(\mathbf{x},\mathbf{y}^*) = \langle \Lambda\mathbf{x} - \mathbf{b};\mathbf{y}^*\rangle - \langle \mathbf{x},\mathbf{f}\rangle.$$

Since F_0 is linear, it can be considered as either convex or concave. In the convex case, the Lagrangian should be considered as a right-saddle function, and the Fenchel inf-conjugate of $F(\mathbf{x})$ reads

$$F^\flat(\mathbf{x}^*) = \inf_{\mathbf{x}\geq 0} \langle \mathbf{x}^* + \mathbf{f}, \mathbf{x}\rangle = \begin{cases} 0 & \text{if } \mathbf{x}^* + \mathbf{f} \geq 0, \\ -\infty & \text{if } \mathbf{x}^* + \mathbf{f} < 0. \end{cases}$$

Let $\mathbf{x}^o = -\mathbf{y}^*$ and define

$$\mathcal{X}^o_- = \{\mathbf{x}^o \in \mathbb{R}^m | \ \Lambda^*\mathbf{x}^o \leq \mathbf{f}\}.$$

Then, the problem dual to $(\mathcal{P}_{\min})_l$ is

$$(\mathcal{P}^d_{\max})_l: \quad P^d(\mathbf{x}^o) = \langle \mathbf{b};\mathbf{x}^o\rangle \quad \to \max \ \ \forall \mathbf{x}^o \in \mathcal{X}^o_-.$$

The weak duality theorem yields to

$$\langle \mathbf{x}, \mathbf{f}\rangle \geq \langle \mathbf{b};\mathbf{x}^o\rangle \ \forall \mathbf{x} \in \mathcal{K}_a, \ \ \forall \mathbf{x}^o \in \mathcal{X}^o_-.$$

The application of this theorem to economics reveals the unfavorable fact:

the minimal cost $F(\bar{\mathbf{x}}) = \langle \bar{\mathbf{x}}, \mathbf{f}\rangle$ *you have to pay is great than or equal to the maximal benefit* $F^o(\bar{\mathbf{x}}^o) = \langle \mathbf{b};\bar{\mathbf{x}}^o\rangle$ *you can gain.*

On the other hand, if we consider the linear function F_0 to be concave then the associated L is a super-Lagrangian. In this case,

$$F^\sharp(\mathbf{x}^*) = \sup_{\mathbf{x}\geq 0}\langle \mathbf{x}^* + \mathbf{f}, \mathbf{x}\rangle = \begin{cases} 0 & \text{if } \mathbf{x}^* + \mathbf{f} \leq 0, \\ -\infty & \text{if } \mathbf{x}^* + \mathbf{f} > 0. \end{cases}$$

The dual feasible space is defined by

$$\mathcal{X}_+^o = \{\mathbf{x}^o \in \mathbb{R}^m |\ \Lambda^*\mathbf{x}^o \geq \mathbf{f}\},$$

and the dual problem is the constrained minimization

$$(\mathcal{P}_{\min}^d)_l : \quad P^d(\mathbf{x}^o) = \langle \mathbf{b}; \mathbf{x}^o\rangle \quad \to \min \ \ \forall \mathbf{x}^o \in \mathcal{X}_+^o.$$

By the strong duality theorem, we have

$$P(\bar{x}) = \min_{\mathbf{x}\in\mathcal{K}_a}\langle \mathbf{x}, \mathbf{f}\rangle = \max_{\mathbf{x}^o\in\mathcal{X}_-^o}\langle \mathbf{b}; \mathbf{x}^o\rangle = \min_{\mathbf{x}^o\in\mathcal{X}_+^o}\langle \mathbf{b}; \mathbf{x}^o\rangle.$$

This strong duality theorem highlights an important fact in linear programming, i.e.

the dual solution $\bar{\mathbf{x}}^o$ has to be found on the boundary of the dual feasible set, i.e.

$$\bar{\mathbf{x}}^o \in \partial\mathcal{X}_+^o \cap \mathcal{X}_-^o.$$

This fact plays an important role in the development of primal-dual algorithms for large-scale computations.

Finally, let us consider the nonhomogeneous linear programming problem

$$(\mathcal{P}_{\min}) : \qquad \min\langle \mathbf{x}, \mathbf{f}\rangle \quad s.t. \ \mathbf{x} - \mathbf{x}_0 \geq 0, \ \ \Lambda\mathbf{x} - \mathbf{y}_0 \geq 0,$$

where $\mathbf{x}_0 \in \mathbb{R}^n$ and $\mathbf{y}_0 \in \mathbb{R}^m$ are two given vectors. In this problem, the mappings B and Y are affine operators. The admissible space $\mathcal{U}_a = \{\mathbf{x} \in \mathbb{R}^n | \mathbf{x} \geq \mathbf{x}_0\}$ is a convex cone with vertex at $\mathbf{x}_0$, while $\mathcal{E}_a = \{\mathbf{y} \in \mathbb{R}^m |\ \mathbf{y} - \mathbf{y}_0 \geq 0\}$ is a convex cone in $\mathbb{R}^m$ with vertex at $\mathbf{y}_0$.

Let $W(\mathbf{y}) = \mathcal{I}_{\mathcal{E}_a}(\mathbf{y})$. Its super-conjugate can be calculated at once as

$$W^\sharp(\mathbf{y}^*) = \sup_{\mathbf{y}\geq\mathbf{y}_0}\langle \mathbf{y}; \mathbf{y}^*\rangle = \langle \mathbf{y}_0; \mathbf{y}^*\rangle + \mathcal{I}_{\mathcal{E}_a^*}(\mathbf{y}^*),$$

where the dual admissible space $\mathcal{E}_a^*$ is a negative cone in $\mathbb{R}^m$, i.e.

$$\mathcal{E}_a^* = \{\mathbf{y}^* \in \mathbb{R}^m |\ \mathbf{y}^* \leq 0\}.$$

The Lagrangian $L : \mathcal{U}_a \times \mathcal{E}_a^* \to \mathbb{R}$ associated with this linear programming is the linear function

$$L(\mathbf{x}, \mathbf{y}^*) = \langle \Lambda\mathbf{x} - \mathbf{y}_0; \mathbf{y}^*\rangle + \langle \mathbf{x}, \mathbf{f}\rangle.$$

The criticality condition for L leads to

$$\mathbf{x} \geq \mathbf{x}_0, \ \ \Lambda^*\mathbf{y}^* + \mathbf{f} \leq 0, \ \ \langle \mathbf{x}, \Lambda^*\mathbf{y}^* + \mathbf{f}\rangle = 0,$$
$$\mathbf{y}^* \leq 0, \ \ \Lambda\mathbf{x} - \mathbf{y}_0 \geq 0, \ \ \langle \Lambda\mathbf{x} - \mathbf{y}_0; \mathbf{y}^*\rangle = 0.$$

This is a bi-complementarity problem. In this case, the complementary operator $B^c : \mathcal{U}^* = \mathbb{R}^n \to \mathcal{V}^* = \mathbb{R}^n$ of $Bu = \mathbf{x} - \mathbf{x}_0$ is also an affine mapping, and $B^c\mathbf{x}^* = \mathbf{x}^* + \mathbf{f}$. For the affine operator $Y(\mathbf{y}) = \mathbf{y} - \mathbf{y}_0$, the complementary Y^c is simply an identity in $\mathcal{E}^* = \mathbb{R}^m$: $Y^c(\mathbf{y}^*) = \mathbf{y}^*$.

If we let $F(\mathbf{x}) = -\langle \mathbf{x}, \mathbf{f} \rangle - \mathcal{I}_{\mathcal{U}_a}(\mathbf{x})$, then its inf-conjugate function is

$$F^\flat(\mathbf{x}^*) = \inf_{\mathbf{x} \geq \mathbf{x}_0} \langle \mathbf{x}, \mathbf{x}^* \rangle = \langle \mathbf{x}_0, \mathbf{x}^* + \mathbf{f} \rangle - \mathcal{I}_{\mathcal{U}_a^{*-}}(\mathbf{x}^*)$$

with

$$\mathcal{U}_a^{*-} = \{\mathbf{x}^* \in \mathbb{R}^n \mid \mathbf{x}^* + \mathbf{f} \leq 0\}.$$

In this case, on the dual feasible space

$$\mathcal{S}_a^- = \{\mathbf{y}^* \in \mathbb{R}^m \mid \mathbf{y}^* \leq 0, \ \Lambda^*\mathbf{y}^* + \mathbf{f} \leq 0\},$$

the classical saddle Lagrange theory leads to the max-dual problem

$$(\mathcal{P}_{\max}): \quad P^d(\mathbf{y}^*) = \langle \mathbf{x}_0, \Lambda^*\mathbf{y}^* + \mathbf{f} \rangle - \langle \mathbf{y}_0; \mathbf{y}^* \rangle \quad \to \max \quad \forall \mathbf{y}^* \in \mathcal{S}_a^-.$$

However, if we write $F(\mathbf{x}) = -\langle \mathbf{x}, \mathbf{f} \rangle + \mathcal{I}_{\mathcal{U}_a}(\mathbf{x})$, which is a convex, l.s.c., its sup-conjugate is

$$F^\sharp(\mathbf{x}^*) = \sup_{\mathbf{x} \geq \mathbf{x}_0} \langle \mathbf{x}, \mathbf{x}^* \rangle = \langle \mathbf{x}_0, \mathbf{x}^* + \mathbf{f} \rangle + \mathcal{I}_{\mathcal{U}_a^{*+}}(\mathbf{x}^*)$$

with

$$\mathcal{U}_a^{*+} = \{\mathbf{x}^* \in \mathbb{R}^n \mid \mathbf{x}^* + \mathbf{f} \geq 0\}.$$

In this case, the dual feasible space is

$$\mathcal{S}_a^+ = \{\mathbf{y}^* \in \mathbb{R}^m \mid \mathbf{y}^* \leq 0, \ \Lambda^*\mathbf{y}^* + \mathbf{f} \geq 0\},$$

and the super-Lagrange duality theory yields the min-dual problem

$$(\mathcal{P}_{\min}): \quad P^d(\mathbf{y}^*) = \langle \mathbf{x}_0, \Lambda^*\mathbf{y}^* + \mathbf{f} \rangle - \langle \mathbf{y}_0; \mathbf{y}^* \rangle \quad \to \min \quad \forall \mathbf{y}^* \in \mathcal{S}_a^+.$$

From the point view of numerical computations, direct methods for solving primal problems can only provide upper bounds on the total cost. The max-dual problem leads to a lower bound approach to the cost function within the feasible set $\mathcal{X}_-^o$, while the min-dual problem yields an upper bound approach within $\mathcal{X}_+^o$. Based upon the classical max-dual problem, the so-called *primal-dual interior point method* has been studied extensively in recent years. Since the min-dual problem plays an equally important role in linear programming, it can be expected that by using both the max-dual and min-dual problems some powerful algorithms can be developed for large-scale constrained linear optimization.

5.5.5 Commentary

Mathematical duality theory for geometrically linear problems has been studied extensively during the last forty years. Many elegant theorems and methods have been well developed. Associated algorithms and computer software are widely used in nonlinear optimization, dynamical systems, mathematical economics, nonconvex-nonsmooth variational problems, convex control with time delay and many other fields (cf., e.g., Du and Pardalos, 1995, Horst and Pardalos, 1995, Tsoutsinos and Vinter, 1995; Lee and Yung, 1997). However, most of the well-known duality theorems and methods belong to the classical saddle Lagrange duality catalogue. Double-min duality theory for the difference of two convex functions was first studied by Toland (1979). The generalizations of extended Lagrange duality for solving nonconvex problems are due to Auchmuty (1983). Many important applications have been made to operator equation and initial value problems (Auchmuty, 1988), evolution equations (Auchmuty, 1993), nonsymmetric linear equations (Auchmuty, 1995) and non-potential operator equations (Auchmuty, 1997). However, as many other authors, only double-min type duality were studied for these geometrically linear systems. The so-called nonconvex conjugate and related topics in global optimizations were studied, among many others, by Ekeland (1977), Flores-Bazan (1995), and by Oettli and Schlager (1998).

Based on the Rockafellar perturbation for convex problems and Auchmuty's extended Lagrange duality theory, this chapter has presented a unified duality theory for geometrically linear systems. In applications to any given nonlinear and nonconvex problems, the key step is to choose a suitable geometrical operator Λ. As long as the problem belongs to a geometrically linear system, the canonical dual transformation method and the extended Lagrange duality theory presented in this chapter can be used to develop many important dual formulations and algorithms for solving both convex and nonconvex problems.

As a special case of a variational inequality, the complementarity problem (NCP) was introduced in Cottle's Ph.D. dissertation (Cottle, 1964). The subject of complementarity theory has become a rich source of inspiration in both mathematical and engineering sciences. Many different kinds of complementarity problems have been proposed and the theory has been extended and generalized in various directions to study a wide class of problems arising in optimization and control, mechanics, operation research, fluid dynamics, economics and transportation. Several monographs have documented the historical details of the origin of these problems and their evolution (e.g. Cottle *et al*, 1992, Isac, 1992, Luo *et al*, 1996). Comprehensive up-to-date treatments of complementarity problems in finite dimensions and extensive documentation of applications in engineering and equilibrium modeling are given in the recent survey articles by Ferris and Pang (1997).

Motivated by engineering mechanics problems, the bi-complementarity problem was first proposed by Gao (1998). Since the tri-canonical forms exist in many systems and theories, this symmetrical structure can be considered as a unified framework, and with a suitable (sequential) extension this framework covers almost all the complementarity problems proposed so far. The associated duality theory can be used to clarify the global extremality of the problems.

The five colors make man's eyes blind;
The five tones make his ears deaf;
The five tastes injure his palate.

—Lao Chi, *Tao De Chin*, XII

The reciprocal relationship of epistemology and science is of noteworthy kind. They are dependent upon each other. Epistemology without contact with science becomes an empty scheme. Science without epistemology is–insofar as it is thinkable at all–primitive and muddled.

—Albert Einstein, 1946

Mathematics is about theorems: how to find them; how to prove them; how to generalize them; how to use them; how to understand them.

—John L. Casti, *Five Golden Rules*, 1996

6 DUALITY IN FINITE DEFORMATION SYSTEMS

"I am the Alpha and the Omega", says the Sovereign God, who is and who was and who is to come, the Almighty.

—Revelation, 1.8

He who understands geometry may understand anything in this world.

—Galileo Galilei, 1564-1642

The general laws of nature are to be expressed in equations which are valid for all coordinate systems

—Albert Einstein, 1916

As any human activity needs goals, mathematical research needs problems

—David Hilbert, 1932

Elasticity led to a vast range of mathematical problems involving linear algebra, differential geometry, ordinary and partial differential equations (mostly nonlinear), elliptic functions and the calculus of variations.

—Clifford Ambrose Truesdell, 1983

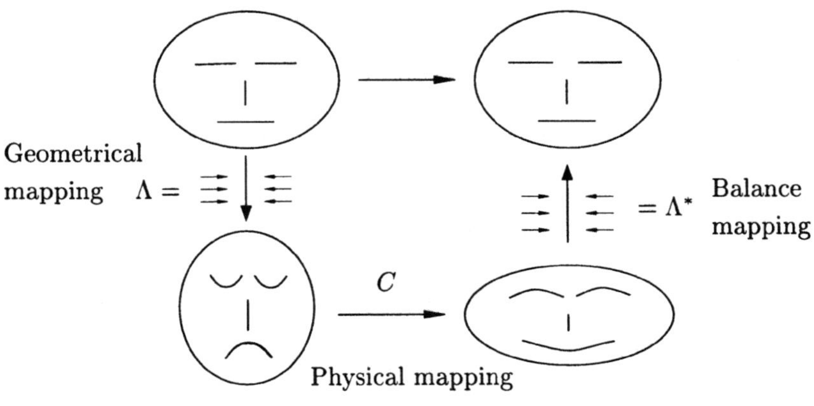

Figure 6.1. Duality in finite deformation system.

In this chapter we shall select topics from finite deformation continuum mechanics and minimum surface type problems in differential geometry, and use them to illustrate a general duality theory for n-dimensional nonconvex finite deformation systems in which the geometrical mapping Λ is a nonlinear partial differential operator. The methods and ideas can certainly be generalized to many other problems.

6.1 Finite deformation theory

6.1.1 Deformation Geometry and Equilibrium Principles

Consider a deformable body in an undeformed state, or *reference configuration*, occupying an open, bounded, connected region $\Omega \subset \mathbb{R}^n$ with boundary $\Gamma = \partial\Omega$.

A *deformation* of the body is a smooth, vector-valued mapping χ from the reference configuration Ω to a configuration ω in $\mathbb{R}^m$ with boundary $\partial\omega$. We write

$$\psi = \chi(\mathbf{X}) = \{\chi^i(X_\alpha)\}, \quad i = 1, 2, ..., m, \quad \alpha = 1, 2, ..., n. \tag{6.1}$$

A mapping with domain Ω in $\mathbb{R}^n$ is a *Lagrangian field*, or *material field*; a mapping with domain $\omega = \chi(\Omega)$ in $\mathbb{R}^m$ is an *Eulerian field*, or *spatial field*.

The deformation is said to be one-to-one if, for any two distinct points $\mathbf{X}_1$, $\mathbf{X}_2 \in \bar{\Omega} := \Omega \cup \partial\Omega$, $\psi_1 = \chi(\mathbf{X}_1)$ and $\psi_2 = \chi(\mathbf{X}_2)$ are distinct points of $\bar{\omega} := \omega \cup \partial\omega$. The deformation is said to be *onto* if, for each point ψ of $\bar{\omega}$, there is a point $\mathbf{X}$ of $\bar{\Omega}$ such that $\psi = \chi(\mathbf{X})$. The deformation is said to be *admissible* if it is one-to-one and onto. This requirement ensures that the deformation is invertible, and we then have

$$\mathbf{X} = \chi^{-1}(\psi).$$

The *admissible configuration space*, denoted $\mathcal{X}$, is the collection of all smooth, one-to-one configuration fields (6.1), defined by

$$\mathcal{X} := \{\chi \in C^1(\Omega; \mathbb{R}^m) | \text{ rank } (\nabla\chi(\mathbf{X})) = \min\{m, n\} \ \forall \mathbf{X} \in \Omega\}. \tag{6.2}$$

On $\mathcal{X}$, the *deformation gradient* is defined by

$$\mathbf{A}(\mathbf{X}) = \nabla\chi(\mathbf{X}) = \frac{\partial\chi(\mathbf{X})}{\partial\mathbf{X}} = \{A^i_\alpha\} \in \mathbb{R}^{m\times n}, \quad A^i_\alpha = \frac{\partial\chi^i}{\partial X_\alpha}, = \chi^i_{,\alpha}, \tag{6.3}$$

which is a *two-point tensor field*, i.e. a second-order tensor with both Lagrangian and Eulerian coordinates.

Consider two neighboring material points in Ω having position vectors $\mathbf{X}$ and $\mathbf{X} + \mathrm{d}\mathbf{X}$. The square of the distance between them is

$$|\mathrm{d}\mathbf{X}|^2 = \mathrm{d}\mathbf{X}^T\mathrm{d}\mathbf{X} = \mathrm{d}\mathbf{X} \cdot \mathrm{d}\mathbf{X}.$$

After deformation, the same two material points are separated by the vector

$$\mathrm{d}\psi = \frac{\partial\psi}{\partial\mathbf{X}}\mathrm{d}\mathbf{X} = \mathbf{A}\mathrm{d}\mathbf{X}, \tag{6.4}$$

so that the square of the distance separating them in ω is

$$|d\boldsymbol{\psi}|^2 = d\boldsymbol{\psi}^T d\boldsymbol{\psi} = d\mathbf{X}^T \mathbf{A}(\mathbf{X})^T \mathbf{A}(\mathbf{X}) d\mathbf{X}. \tag{6.5}$$

The *metric tensor* is a symmetric Lagrangian tensor field

$$\mathbf{C} = \mathbf{A}^T \mathbf{A} \in \mathbb{R}^{n\times n}, \quad C_{\alpha\beta} = A^i_\alpha A^i_\beta \tag{6.6}$$

that measures the length of an elementary vector defined in the deformed configuration ω relative to that in the reference configuration Ω. For any given admissible deformation $\chi \in \mathcal{X}$, we have

$$\mathbf{C} = \mathbf{C}^T, \quad \text{rank } \mathbf{C} = \min\{m, n\}.$$

In differential geometry, $\mathbf{C}$ is called the *Riemannian metric tensor.* Very often $m \geq n$. If $m = n + 1$, the admissible deformation is a *Riemannian hypersurface.*

In continuum mechanics, $\mathbf{C}$ is called the *right Cauchy-Green strain tensor.* Clearly, if at a point $\mathbf{X} \in \Omega$, $\mathbf{C}(\mathbf{X}) = \mathbf{I}$, the identity tensor in $\mathbb{R}^{n\times n}$, then $|d\chi(\mathbf{X})|^2 = |d\mathbf{X}|^2$, the material is said to be *unstrained* at $\mathbf{X}$. Otherwise, the material is said to be *strained* at $\mathbf{X}$ if

$$|d\chi(\mathbf{X})|^2 - |d\mathbf{X}|^2 = d\mathbf{X}^T(\mathbf{C} - \mathbf{I})d\mathbf{X} \neq 0, \quad \mathbf{X} \in \Omega,$$

and the tensor $\mathbf{C} - \mathbf{I}$ can be regarded as a *strain tensor.*

The so-called *left Cauchy-Green strain tensor* is defined by

$$\mathbf{B} = \mathbf{A}\mathbf{A}^T \in \mathbb{R}^{m\times m}, \quad B^{ij} = A^i_\alpha A^j_\alpha,$$

which is a symmetric Eulerian tensor field.

In the case $m = n$, the *deformation Jacobian* $J(\mathbf{A})$. defined by

$$J = \det \mathbf{A}(\chi(\mathbf{X})) > 0 \ \ \forall \chi \in \mathcal{X},$$

is a positive scalar for an admissible deformation. The *displacement field* is defined as

$$\mathbf{u}(\mathbf{X}) = \chi(\mathbf{X}) - \mathbf{X},$$

which is a material vector field. An area element dA of the reference configuration, with (unit) normal vector $\mathbf{N} \in \mathbb{R}^n$, is transformed by the deformation into an area element

$$da = J|\mathbf{A}^{-T}\mathbf{N}|dA, \tag{6.7}$$

with (unit) normal vector

$$\nu = \frac{\mathbf{A}^{-T}\mathbf{N}}{|\mathbf{A}^{-T}\mathbf{N}|} \in \mathbb{R}^m.$$

Similarly, a volume element $d\Omega$ is transformed by the deformation into the volume element

$$d\omega = J(\mathbf{A})d\Omega.$$

The material is said to be *incompressible* if $J(\mathbf{A}) = 1$.

A *motion* of the body is a time-dependent mapping

$$\chi : \Omega \times \mathbb{R} \to \mathbb{R}^m, \quad \psi = \chi(\mathbf{X}, t) \tag{6.8}$$

where $\chi(*, t)$ is the deformation undergone by the body at time t.

The *velocity* of a material point $\mathbf{X}$ at time t is defined by

$$\mathbf{v}(\mathbf{X}, t) := \frac{\partial \chi}{\partial t}(\mathbf{X}, t) = \dot{\chi}(\mathbf{X}, t) = \{\chi^i_{,t}(X_\alpha, t)\}. \tag{6.9}$$

The *acceleration* of a particle is the time rate of change of its velocity and is given by

$$\mathbf{a}(\mathbf{X}, t) = \frac{\partial \mathbf{v}}{\partial t}(\mathbf{X}, t) = \{\chi^i_{,tt}(X_\alpha, t)\}. \tag{6.10}$$

The mass of a body is a measure of its inertia and is invariant during deformation. If ρ_o is the mass density in Ω and ρ that in ω, then we have the *law of conservation of mass*

$$\rho(\chi)\mathrm{d}\omega = \rho_o(\mathbf{X})\mathrm{d}\Omega \;\Rightarrow\; J\rho(\chi) = \rho_o(\mathbf{X}) \in \mathbb{R}. \tag{6.11}$$

Let $\Omega_t = \Omega \times (t_0, t_1) \in \mathbb{R}^{n+1}$ be space-time, $\Gamma_t = \partial\Omega \times (t_0, t_1)$, and let $\bar{\Omega} = \Omega \cup \Gamma$ denote the closure of Ω. The *external force* is a vector field $\chi^* : \bar{\Omega} \times (t_0, t_1) \to \mathbb{R}^m$, which is the source of the deformation. A *system of external forces* acting at a given time t on a given body consists of

(a) the *body force:* $\mathbf{b} : \Omega_t \to \mathbb{R}^m$, and

(b) the *surface traction:* $\mathbf{t} : \Gamma_t \to \mathbb{R}^m$.

Theorem 6.1.1 (Cauchy's Stress Principle) *For a given external force field* $\bar{\chi}^* = \{\bar{\mathbf{b}}$ (*in* Ω_t), $\bar{\mathbf{t}}$ (*on* $\Gamma \times (t_0, t_1))\}$, *there exists an equilibrium state* χ *and a second-order tensor field* $\mathbf{S} : \bar{\Omega} \times (t_0, t_1) \to \mathbb{R}^{n\times m}$, *called the nominal stress tensor, such that the following three equilibrium conditions hold.*

(T1) *External equilibrium condition:*

$$\mathbf{N} \cdot \mathbf{S} = \bar{\mathbf{t}} \quad \text{on } \Gamma_t. \tag{6.12}$$

(T2) *Internal equilibrium condition:*

$$-\nabla \cdot \mathbf{S} = \bar{\mathbf{b}} - \rho_o \chi_{,tt} \quad \text{in } \Omega_t. \tag{6.13}$$

(T3) *Equilibrium of angular momentum:*

$$(\nabla\chi)\mathbf{S} = ((\nabla\chi)\mathbf{S})^T \quad \text{in } \Omega_t. \tag{6.14}$$

The *nominal stress* $\mathbf{S} = \{S^\alpha_i\} \in \mathbb{R}^{n\times m}$ is a two-point tensor, which describes the equilibrium state of the deformed body in $\mathbb{R}^m$ in terms of the Lagrangian coordinates $\mathbf{X} = \{X_\alpha\} \in \mathbb{R}^n$. Its transpose $\mathbf{S}^T \in \mathbb{R}^{m\times n}$ is very often called the *first Piola-Kirchhoff stress tensor*. The equilibrium conditions can be transformed into the Eulerian coordinate system $\psi = \{\psi^i\} \in \mathbb{R}^n$ via the so-called *Cauchy stress tensor*

$$\boldsymbol{\sigma} = J^{-1}\mathbf{A}\mathbf{S} \in \mathbb{R}^{m\times m}. \tag{6.15}$$

Clearly, $\boldsymbol{\sigma} = \boldsymbol{\sigma}^T$ is a symmetrical Eulerian tensor field.

The two-point tensor fields $\mathbf{A} = \{A^i_\alpha\}$ and $\mathbf{S} = \{S^\alpha_i\}$ are introduced by pure geometrical analysis and equilibrium principle, respectively. Since they are not symmetric tensors, the additional equilibrium condition of angular momentum (6.14) is required. In order to describe the physical properties of materials we need to introduce some symmetrical strain and stress measures.

6.1.2 Hyperelasticity and Constitutive Laws

We use the notation $\mathcal{M}(\Omega; \mathbb{R}^{m\times n})$, or simply $\mathcal{M}$, to denote the space of all second-order tensor functions with domain Ω in $\mathbb{R}^n$ and range in $\mathbb{R}^{m\times n}$. Let $\mathcal{A} \subset \mathcal{M}$ be an admissible deformation gradient space, defined by

$$\mathcal{A} = \{\mathbf{A} \in \mathcal{M} |\ \text{rank}\, \mathbf{A}(\mathbf{X}) = \min\{m, n\}\ \forall \mathbf{X} \in \Omega\}. \tag{6.16}$$

In particular, we need the important tensor spaces in $\mathbb{R}^{n\times n}$, defined by

$$\begin{aligned}
\mathcal{M}_+ &= \{\mathbf{A} \in \mathcal{A} \mid\ \det \mathbf{A}(\mathbf{X}) > 0\ \forall \mathbf{X} \in \Omega\},\\
\mathcal{M}_{sym} &= \{\mathbf{A} \in \mathcal{M} \mid\ \mathbf{A}(\mathbf{X}) = \mathbf{A}(\mathbf{X})^T\ \forall \mathbf{X} \in \Omega\},\\
\mathcal{M}_{ort} &= \{\mathbf{A} \in \mathcal{A} \mid\ \mathbf{A}(\mathbf{X})^T = \mathbf{A}(\mathbf{X})^{-1}\ \forall \mathbf{X} \in \Omega\},\\
\mathcal{M}^+_{ort} &= \{\mathbf{A} \in \mathcal{M}_{ort} \mid\ \det \mathbf{A}(\chi) = 1\}.
\end{aligned} \tag{6.17}$$

Thus, for a given admissible deformation $\chi \in \mathcal{X}$ such that $\mathbf{A} \in \mathcal{M}_+$, we have the (unique) polar decompositions

$$\mathbf{A} = \mathbf{R} \cdot \mathbf{U} = \mathbf{V} \cdot \mathbf{R}, \tag{6.18}$$

where $\mathbf{R} \in \mathcal{M}^+_{ort}$ is called the *rotation tensor*, the positive definite and symmetric tensors $\mathbf{U}, \mathbf{V} \in \mathcal{M}_{sym}$ are called the *right stretch tensor* and *left stretch tensor*, respectively.

Definition 6.1.1 (Hyperelastic materials) An elastic material is said to be a *hyperelastic material* if there exists a real-valued function $U(\mathbf{X}, \mathbf{A})$ corresponding to the internal elastic energy stored in the deformed body, such that the constitutive equation is characterized by

$$\mathbf{S}(\mathbf{X}) = \frac{\partial U}{\partial \mathbf{A}}(\mathbf{X}, \mathbf{A}(\mathbf{X})). \tag{6.19}$$

This real-valued function is called the *stored-energy density* per unit initial volume. ◇

Not every function U is suitable for defining the internal energy. For hyperelastic materials, the stored-energy density U must satisfy certain requirements.

Definition 6.1.2 (Objectivity and Isotropy)

(D1) *Objective Set and Objective Function:* A subset $\mathcal{A}_a \subset \mathcal{M}$ is said to be *objective* if for every $\mathbf{A} \in \mathcal{A}_a$ and every $\mathbf{Q} \in \mathcal{M}^+_{ort}$, $\mathbf{QA} \in \mathcal{A}_a$. A scalar-valued function $U : \Omega \times \mathcal{A}_a \to \mathbb{R}$ is said to be *objective* if its domain is objective and

$$U(\mathbf{X}, \mathbf{QA}) = U(\mathbf{X}, \mathbf{A}) \ \ \forall \mathbf{A} \in \mathcal{A}_a, \ \ \forall \mathbf{Q} \in \mathcal{M}^+_{ort}. \tag{6.20}$$

(D2) *Isotropic Set and Isotropic Function* . A subset $\mathcal{A}_a \subset \mathcal{M}$ is said to be *isotropic* if for every $\mathbf{A} \in \mathcal{A}_a$ and every $\mathbf{Q} \in \mathcal{M}^+_{ort}$ also $\mathbf{AQ}^T \in \mathcal{A}_a$. A scalar-valued function $U : \Omega \times \mathcal{A}_a \to \mathbb{R}$ is said to be *isotropic* if its domain is isotropic and

$$U(\mathbf{X}, \mathbf{AQ}^T) = U(\mathbf{X}, \mathbf{A}) \ \ \forall \mathbf{A} \in \mathcal{A}_a, \ \ \forall \mathbf{Q} \in \mathcal{M}^+_{ort}. \tag{6.21}$$

◇

Objectivity is also referred as *material frame indifference.* For a hyperelastic material, the stored-energy density has to be an objective function. Objectivity implies that the equilibrium condition of angular momentum (symmetry of the Cauchy stress tensor) holds. When $m = n$, if we write $\mathbf{A} = \mathbf{RU} \in \mathcal{A}$ and choose $\mathbf{Q} = \mathbf{R}^T$, we see that $U(\mathbf{X}, \mathbf{A}) = U(\mathbf{X}, \mathbf{U})$. Clearly, for any $\mathbf{A} \in \mathcal{M}_+$, the right Cauchy-Green tensor is an objective tensor-valued function

$$\mathbf{C}(\mathbf{QA}) = (\mathbf{QA})^T(\mathbf{QA}) = \mathbf{A}^T\mathbf{A} = \mathbf{C}(\mathbf{A}) \ \ \forall \mathbf{Q} \in \mathcal{M}_{ort}.$$

Isotropy means that the material possesses a certain symmetry in Ω. Many explicit models of hyperelastic materials are not only objective, but also isotropic. Thus, objective, isotropic hyperelasticity is equivalent to the assertion that U depends on $\mathbf{A}$ only through the left stretch tensor $\mathbf{V}$ and that

$$U(\mathbf{X}, \mathbf{QVQ}^T) = U(\mathbf{X}, \mathbf{V}) \ \ \forall \mathbf{Q} \in \mathcal{M}_{ort}.$$

In the case $n = 3$, the three invariants of the right Cauchy-Green strain tensor $\mathbf{C} = \mathbf{A}^T\mathbf{A} \in \mathcal{M}_+ \cap \mathcal{M}_{sym}$ are defined by

$$\begin{aligned}
I_1 &= \mathrm{tr}(\mathbf{C}) = C_1 + C_2 + C_3, \\
I_2 &= \mathrm{tr}(\mathrm{cof}(\mathbf{C})) = (\det \mathbf{C})\mathrm{tr}(\mathbf{C}^{-1}) = \tfrac{1}{2}[(\mathrm{tr}\mathbf{C})^2 - \mathrm{tr}(\mathbf{C}^2)] \\
&= C_1C_2 + C_2C_3 + C_3C_1, \\
I_3 &= \det \mathbf{C} = \tfrac{1}{6}[(\mathrm{tr}\mathbf{C})^3 - 3\mathrm{tr}(\mathbf{C})\mathrm{tr}(\mathbf{C}^2) + 2\mathrm{tr}(\mathbf{C}^3)] = C_1C_2C_3,
\end{aligned} \tag{6.22}$$

where $\mathrm{cof}(\mathbf{C})$ denotes the matrix of *cofactors* of $\mathbf{C}$, and C_1, C_2, C_3 are eigenvalues of $\mathbf{C}$. It is easy to prove that these three invariants are also isotropic functions of $\mathbf{A}$.

Theorem 6.1.2 (Isotropic Hyperelasticity Law)

For an isotropic hyperelastic material, the stored-energy function $U(\mathbf{X}, \mathbf{A})$ is an isotropic function of the right Cauchy-Green strain tensor $\mathbf{C} = \mathbf{A}^T\mathbf{A}$, i.e. there exist real-valued non-negative functions $\bar{U}$ and $\tilde{U}$ such that

$$U(\mathbf{X}, \mathbf{A}) = \bar{U}(\mathbf{X}, \mathbf{C}) = \tilde{U}(\mathbf{X}, I_1, I_2, I_3). \tag{6.23}$$

Using this theorem, a tensor-valued function for hyperelastic materials can be defined by

$$\mathbf{C}^* := \frac{\partial \bar{U}(\mathbf{X}, \mathbf{C})}{\partial \mathbf{C}} \in \mathbb{R}^{n \times n}. \tag{6.24}$$

Clearly, for any given $\mathbf{C} \in \mathcal{M}_{sym}$, $\mathbf{C}^* = \{C^{*\alpha\beta}\} \in \mathcal{M}_{sym}$ is a symmetrical Lagrangian tensor, and $2\mathbf{C}^*$ is called the *second Piola-Kirchhoff stress tensor*. Actually, this definition does not require isotropy.

The material is said to be *homogeneous* if $U(\mathbf{X}, \mathbf{A}(\mathbf{X})) = U(\mathbf{A}(\mathbf{X}))$, i.e. U depends on $\mathbf{X}$ only through $\mathbf{A}(\mathbf{X})$. It is clear that the relation between $\mathbf{C}^*$ and $\mathbf{S}$ is

$$\mathbf{S} = DU(\mathbf{A}) = \frac{\partial \bar{U}(\mathbf{A}^T \mathbf{A})}{\partial \mathbf{A}} = 2\mathbf{C}^* \mathbf{A}^T. \tag{6.25}$$

Theorem 6.1.3 (Rivlin-Ericksen, 1955) *The most general constitutive law for a homogeneous isotropic body is given by*

$$\mathbf{C}^* = \alpha_1 \mathbf{I} + \alpha_2 \mathbf{C} + \alpha_3 \mathbf{C}^2, \tag{6.26}$$

where

$$\alpha_i = \alpha_i(I_1, I_2, I_3), \quad i = 1, 2, 3,$$

are scalar functions of the three invariants of $\mathbf{C}$.

The hyperelastic constitutive law (6.19) is a useful model of the elastic behavior of many rubber-like materials. We list some isotropic hyperelastic materials for $n = m = 3$, characterized by the associated stored-energy density $U(\mathbf{A}) = \bar{U}(\mathbf{C}) = \tilde{U}(I_1, I_2, I_3)$.

Example 6.1.1 (St. Venant-Kirchhoff Material) (see Ciarlet, 1987)

$$\bar{U}(\mathbf{C}) = \frac{1}{2}\lambda \left(\mathrm{tr} \frac{1}{2}(\mathbf{C} - \mathbf{I}) \right)^2 + \mu \mathrm{tr} \left(\frac{1}{2}(\mathbf{C} - \mathbf{I}) \right)^2, \tag{6.27}$$

where λ, μ are the classical Lamé constants. Clearly, the stored energy $\bar{U}$ for St. Venant-Kirchhoff materials is a strictly convex (quadratic) function of the right Cauchy-Green strain tensor $\mathbf{C}$. It has particularly important applications in the large deformation of structures with small strains, such as thin-walled plates and shells. ♡

Example 6.1.2 (Blatz-Ko Material)

The generalized *Blatz-Ko material* is defined by

$$\begin{aligned} \tilde{U} &= \frac{1}{2}\mu\alpha \left(I_1 - 1 - \frac{1}{\nu} + \frac{(1-2\nu)}{\nu} I_3^{-\nu/(1-2\nu)} \right) \\ &+ \frac{1}{2}\mu(1-\alpha) \left(I_2 I_3^{-1} - 1 - \frac{1}{\mu} + \frac{(1-2\nu)}{\nu} I_3^{\nu/(1-2\nu)} \right), \end{aligned} \tag{6.28}$$

where $\mu > 0$, $\nu \in (0, 0.5]$ are the shear modulus and Poisson's ratio for infinitesimal deformations, while $\alpha \in [0, 1]$ is related to the volume fraction of voids in the foam-rubber material modeled by (6.28). This stored-energy function was first proposed by Blatz and Ko (1962) as a model for foam rubber compressible materials undergoing large deformations. ♡

Example 6.1.3 (Ogden Material) Ogden's general (unconstrained) material (see Ogden, 1984) is often used to model rubber-like materials. In terms of the principal stretches $\lambda_1, \lambda_2, \lambda_3$ of $\mathbf{A}$, i.e. the eigenvalues of $\mathbf{C}^{1/2}$ the Ogden general model takes the particularly nice mathematical form

$$\begin{aligned} U &= \sum_{i=1}^{M} a_i(\lambda_1^{\alpha_i} + \lambda_2^{\alpha_i} + \lambda_3^{\alpha_i} - 3) + \sum_{j=1}^{N} b_j\left((\lambda_1\lambda_2)^{\beta_j} + (\lambda_2\lambda_3)^{\beta_j} + (\lambda_1\lambda_3)^{\beta_j} - 3\right) \\ &\quad + g(\lambda_1\lambda_2\lambda_3), \end{aligned} \tag{6.29}$$

where $\alpha_i, \beta_j \geq 1$, $a_i, b_j > 0$ are material constants, and $g : (0, \infty) \to \mathbb{R}$ is convex with

$$\lim_{d \to 0+} g(d) = +\infty.$$

It is easy to check that U is a symmetric function in λ_i, i.e.

$$U(\lambda_1, \lambda_2, \lambda_3) = U(\lambda_2, \lambda_1, \lambda_3) = U(\lambda_3, \lambda_2, \lambda_1) = \cdots.$$

In terms of $\mathbf{C} = \mathbf{A}^T\mathbf{A}$, the Ogden material reads (Ogden, 1976)

$$\begin{aligned} \bar{U}(\mathbf{C}) &= \sum_{i=1}^{3} \mu_i \mathrm{tr}(\mathbf{C}^{\alpha_i/2}) + \mu_4\alpha_4^{-2}(\det \mathbf{C})^{-\alpha_4/2} \\ &+ \frac{1}{2}\left(\mu_4\alpha_4^{-1}\sum_{i=1}^{3}\alpha_i\mu_i\right)\ln(\det \mathbf{C}) + \mu_5(\det \mathbf{C})^{1/3}((\det \mathbf{C})^{1/2} - 1)\mathrm{tr}\mathbf{C}^{-2}. \end{aligned}$$

In certain cases, the stored-energy density of Ogden materials is a convex function of $\mathbf{C}$.

If $M = N = 1$, $\alpha_1 = p$, $\beta_1 = q$, then in terms of $\mathbf{U} = \mathbf{C}^{\frac{1}{2}}$, equation (6.29) yields

$$\breve{U}(\mathbf{U}) = a_1\mathrm{tr}(\mathbf{U}^p - \mathbf{I}) + b_1\mathrm{tr}(\mathrm{cof}(\mathbf{U}^q) - \mathbf{I}) + g(\det \mathbf{U}), \tag{6.30}$$

which is usually a convex function of $\mathbf{U}$ for certain p and q. ♡

In terms of the invariants I_i $(i = 1, 2, 3)$, an alternative to the expansion (6.29), but not in general equivalent to it, can be written as

$$\tilde{U}(I_1, I_2, I_3) = \sum_{p,q=0}^{\infty} c_{pq}(I_1 - 3)^p(I_2 - 3)^q + \sum_{r=1}^{\infty} \mu_r(I_3 - 1)^r, \tag{6.31}$$

where $p, q, r = 0, 1, 2, \cdots$, and the coefficients c_{pq} and μ_r are independent of the deformation, on any closed region of the space of I_i $(i = 1, 2, 3)$. The terms "3" and "1" are normalization constants such that the stored energy $\tilde{U}$ vanishes in the reference (undeformed) configuration where $I_1 = I_2 = 3$, and $I_3 = 1$.

For an incompressible material $I_3 = 1$ and $\tilde{U}$ depends on only two independent deformation invariants. In particular, the expansion (6.31) is replaced by

$$\tilde{U}(I_1, I_2) = \sum_{p,q=0}^{M} c_{pq}(I_1 - 3)^p (I_2 - 3)^q, \tag{6.32}$$

with $c_{00} = 0$. For $I_3 = 1$, it is easy to show that

$$I_1^2 \geq 3I_2, \quad I_2^2 \geq 3I_1.$$

There are many important special cases of (6.32).

I. *Knowles-Sternberg material* (1980)

$$\tilde{U}(I_1) = \frac{\mu}{2\alpha}\left\{\left(1 + \frac{\alpha}{n}(I_1 - 3)\right)^n - 1\right\},$$

subject to the incompressibility condition $I_3 = 1$.

II. *Neo-Hookean material* (see Knowles and Sternberg (1980))

$$\tilde{U}(I_1) = c_{10}(I_1 - 3), \quad s.t. \ I_3 = 1.$$

A typical numerical value of c_{10} is $0.183MPa$.

III. *Mooney-Rivlin material* ($M = 1$)

$$\tilde{U}(I_1, I_2) = c_{10}(I_1 - 3) + c_{01}(I_2 - 3), \quad s.t. \ I_3 = 1,$$

with typical numerical values $c_{10} = 0.183MPa$ and $c_{01} = 0.0034MPa$ (see Chadwick and Haddon, 1972). We note that in the limit as $I_3 \to 1$, the Blatz-Ko material (6.28) also approaches the Mooney-Rivlin material with $c_{10} = 0.5\mu\alpha$, $c_{01} = 0.5\mu(1 - \alpha)$.

IV. *Haines-Wilson (1979) material* ($M = 3$)

$$\begin{aligned}\tilde{U}(I_1, I_2) &= c_{10}(I_1 - 3) + c_{01}(I_2 - 3) + c_{20}(I_1 - 3)^2 + c_{02}(I_2 - 3)^2 \\ &\quad + c_{11}(I_1 - 3)(I_2 - 3) + c_{30}(I_1 - 3)^3.\end{aligned}$$

V. *Ciarlet-Geymonat material* (1982)

$$\tilde{U}(I_1, I_2, I_3) = c_{10}(I_1 - 3) + c_{01}(I_2 - 3) + \alpha(I_3 - 1) - (c_{10} + 2c_{01} + \alpha)\ln I_3.$$

In terms of $\mathbf{C}$, a particular case of the Ogden incompressible material reads

$$\bar{U}(\mathbf{C}) = \sum_{i=1}^{N} \mu_i \alpha_i^{-1}\left(\mathrm{tr}(\mathbf{C}^{\alpha_i/2}) - 3\right), \tag{6.33}$$

subject to $I_3 = 1$; typical numerical values are

$$\begin{cases} N = 3, \ \alpha_1 = 1.3, \ \alpha_2 = 5, \ \alpha_3 = -2, \\ \mu_1 = 0.63MPa, \ \mu_2 = 0.0012MPa, \ \mu_3 = -0.01MPa. \end{cases}$$

The Blatz-Ko and Ogden materials are two of the most widely used constitutive models for compressible isotropic nonlinearly elastic solids. One of the interesting features of these models is that the system of the second-order partial differential equations governed by these models can lose ellipticity at sufficiently severe deformations. Such phenomena have been noticed in the context of crack problems in finite elastoplasticity (Knowles and Sternberg, 1978, 1980) as well as in phase transitions in nonsmooth elastic solids (see, e.g., Abeyaratne and Knowles, 1989; Rosakis and Jiang, 1993). An extensive discussion of hyperelasticity of rubber, elastomers and biological tissues can be found in a survey paper by Beatty (1987).

6.1.3 Boundary-Value Problems, Regularity and Uniqueness

In *mixed boundary-value problems*, the boundary $\Gamma = \partial\Omega$ is usually split into two parts, $\Gamma = \Gamma_\chi \cup \Gamma_f$, such that $\Gamma_\chi \cap \Gamma_f = \emptyset$. On Γ_f, the surface traction $\bar{\mathbf{t}}$ is applied, while, on Γ_χ, the deformation is prescribed. Thus, for a given function $\bar{\chi}$ on Γ_χ, the admissible configuration space $\mathcal{X}_a \subset \mathcal{X}$ reads

$$\mathcal{X}_a = \{\chi \in \mathcal{X} |\ \chi(\mathbf{X}) = \bar{\chi}(\mathbf{X})\ \ \forall \mathbf{X} \in \Gamma_\chi\}.$$

Let $\mathcal{X}^* \subset \mathbb{R}^m$ be a vector space in duality with $\mathcal{X}$ by the bilinear form $\langle *, * \rangle$: $\mathcal{X} \times \mathcal{X}^* \to \mathbb{R}$ defined by

$$\langle \chi, \chi^* \rangle = \int_\Omega \chi \cdot \chi^* \mathrm{d}\Omega + \int_\Gamma \chi \cdot \chi^* \mathrm{d}\Gamma \quad \forall (\chi, \chi^*) \in \mathcal{X} \times \mathcal{X}^*.$$

Let $\mathcal{S} = \mathcal{A}^* \subset \mathbb{R}^{n \times m}$ be the dual space of $\mathcal{A}$ with the standard bilinear form $\langle *; * \rangle : \mathcal{A} \times \mathcal{S} \to \mathbb{R}$ defined by

$$\langle \mathbf{A}; \mathbf{S} \rangle = \int_\Omega \mathbf{A} : \mathbf{S} \mathrm{d}\Omega = \int_\Omega \mathrm{tr}(\mathbf{AS}) \mathrm{d}\Omega = \int_\Omega A^i_\alpha S^\alpha_i \mathrm{d}\Omega \ \ \forall (\mathbf{A}, \mathbf{S}) \in \mathcal{A} \times \mathcal{S}.$$

Thus, for a given linear operator $\Lambda = \mathrm{grad} : \mathcal{X} \to \mathcal{A}$, the Gauss-Green formula reads

$$\langle \mathrm{grad}\chi; \mathbf{S} \rangle = \int_\Omega \chi \cdot (-\nabla \cdot \mathbf{S}) \mathrm{d}\Omega + \int_\Gamma \chi \cdot (\mathbf{N} \cdot \mathbf{S}) \mathrm{d}\Gamma = \langle \chi, \mathrm{grad}^* \mathbf{S} \rangle,$$

where the adjoint operator $\mathrm{grad}^* : \mathcal{S} \to \mathcal{X}^*$ associated with these two standard bilinear forms is defined by

$$\mathrm{grad}^* \mathbf{S} = \begin{cases} -\mathrm{div}\mathbf{S} & \text{in } \Omega, \\ \mathbf{N} \cdot \mathbf{S} & \text{on } \Gamma. \end{cases}$$

Let $\mathcal{X}^*_a \subset \mathcal{X}^* \to \mathbb{R}$ be the source space

$$\mathcal{X}^*_a = \{\chi^* \in \mathcal{X}^* |\ \chi^*(\mathbf{X}) = \bar{\mathbf{b}}(\mathbf{X})\ \ \forall \mathbf{X} \in \Omega,\ \ \chi^*(\mathbf{X}) = \bar{\mathbf{t}}\ \ \forall \mathbf{X} \in \Gamma_f\}.$$

For hyperelastic materials, we let $\mathcal{A}_a \subset \mathcal{A}$ be an admissible space of deformation gradients such that, for any given $\mathbf{A} \in \mathcal{A}_a$, the stored-energy density $U(\mathbf{X}, \mathbf{A}) \in$

$\mathcal{L}(\Omega \times \mathbb{R}^{m\times n})$ satisfies the generalized Carathéodory continuity conditions. Let $\mathcal{S}_a \subset \mathcal{S}$ be the range of the constitutive mapping $DU : \mathcal{A}_a \to \mathcal{S}$. Suppose that for a given force field $\chi^* \in \mathcal{X}_a^*$ the deformation $\bar{\chi}$ is an equilibrium state and $\bar{\mathbf{S}} = DU(\mathbf{X}, \mathbf{A}(\bar{\chi}))$ is the associated stress field. Then, by Cauchy's stress principle, the external and internal equilibrium equations (6.12) and (6.13) can be written together in the weak form

$$\langle \chi, \text{grad}^* \bar{\mathbf{S}} - \chi^* \rangle = \int_\Omega \chi \cdot (-\nabla \cdot \bar{\mathbf{S}} - \bar{\mathbf{b}}) d\Omega + \int_{\Gamma_f} (\mathbf{N} \cdot \bar{\mathbf{S}} - \bar{\mathbf{t}}) \cdot \chi d\Gamma = 0 \ \ \forall \chi \in \mathcal{X}_a,$$

or, equivalently,

$$\int_\Omega \mathbf{A}(\chi) : \bar{\mathbf{S}}(\bar{\chi}) d\Omega = \int_\Omega \bar{\mathbf{b}} \cdot \chi d\Omega + \int_{\Gamma_f} \bar{\mathbf{t}} \cdot \chi d\Gamma \ \ \forall \chi \in \mathcal{X}_a. \tag{6.34}$$

This is the *virtual work principle*. By the fact that

$$\mathbf{S} = DU(\mathbf{X}, \mathbf{A}) = \partial \bar{U}(\mathbf{X}, \mathbf{C}) / \partial \mathbf{A} = 2\mathbf{C}^* \mathbf{A}^T,$$

the momentum equilibrium equation

$$\mathbf{AS} = 2\mathbf{AC}^* \mathbf{A}^T = \mathbf{S}^T \mathbf{A}^T$$

is automatically satisfied for hyperelasticity. Thus, the mixed boundary-value problem for hyperelasticity can be proposed as follows.

Problem 6.1.1 (Mixed Boundary-Value Problem)
Suppose that the stored-energy density $U(\mathbf{X}, \mathbf{A})$ of a hyperelastic material is piecewise differentiable on $\mathcal{A}_a$ for any given $\mathbf{X} \in \Omega$. For a given external force field $\chi^* = \{\bar{\mathbf{b}} \text{ in } \Omega, \ \bar{\mathbf{t}} \text{ on } \Gamma_f\}$, find the equilibrium state $\chi \in \mathcal{X}_a$ and the associated $(\mathbf{A}, \mathbf{S}) \in \mathcal{A} \times \mathcal{S}$ such that

$$(\mathcal{P}_{\text{bv}})_\chi : \quad \begin{cases} \mathbf{A} = \nabla \chi \text{ in } \Omega, & \chi = \bar{\chi} \ \text{ on } \Gamma_\chi, \\ \mathbf{S} = DU(\mathbf{X}, \mathbf{A}) \ \text{ in } \Omega, & \\ \bar{\mathbf{b}} = -\nabla \cdot \mathbf{S} \ \text{ in } \Omega, & \mathbf{N} \cdot \mathbf{S} = \bar{\mathbf{t}} \ \text{ on } \Gamma_f. \end{cases} \tag{6.35}$$

In continuum mechanics, where $m = n$, we let $\mathcal{U} \subset \mathcal{L}^\alpha(\Omega; \mathbb{R}^n)$ be an displacement space such that for any given $\mathbf{u} \in \mathcal{U}$, $\chi = \mathbf{u} + \mathbf{X} \in \mathcal{X}$. Thus, in terms of $\mathbf{u}(\mathbf{X})$, the deformation gradient tensor reads

$$\mathbf{A} = \Lambda(\mathbf{u}) = \nabla \mathbf{u} + \mathbf{I} \in \mathbb{R}^{n\times n}.$$

In this case, $\Lambda : \mathcal{U} \to \mathcal{A}$ is an affine operator and $\Lambda_t = \text{grad}$, $\Lambda_c = -\mathbf{I}$. Thus, the above mixed boundary-value problem can also be expressed as follows.

Problem 6.1.2 (Mixed Boundary-Value Problem)
For a given external force field $\bar{\mathbf{u}}^* = \{\bar{\mathbf{b}} \text{ in } \Omega, \ \bar{\mathbf{t}} \text{ on } \Gamma_f\}$, find the equilibrium state $\mathbf{u} \in \mathcal{U}_a$ and the associated $(\mathbf{A}, \mathbf{S}) \in \mathcal{A} \times \mathcal{S}$ such that

$$(\mathcal{P}_{\text{bv}})_u : \quad \begin{cases} \mathbf{A} = \nabla \mathbf{u} + \mathbf{I} \text{ in } \Omega, & \mathbf{u} = \bar{\mathbf{u}} \ \text{ on } \Gamma_u, \\ \mathbf{S} = DU(\mathbf{X}, \mathbf{A}) \ \text{ in } \Omega, & \\ \bar{\mathbf{b}} = -\nabla \cdot \mathbf{S} \ \text{ in } \Omega, & \mathbf{N} \cdot \mathbf{S} = \bar{\mathbf{t}} \ \text{ on } \Gamma_f. \end{cases} \tag{6.36}$$

Since the deformation gradient $\mathbf{A}$ is not a strain measure, the stored-energy U is not a convex function of $\mathbf{A}$ unless some restriction is imposed. As pointed out by Hill (1957), such convexity would imply a unique large deformation χ in problems where bifurcations are known to intervene. Thus, in nonconvex finite deformation problems, the constitutive mapping $\mathbf{S} = DU(\mathbf{X}, \mathbf{A}) : \mathcal{A} \to \mathcal{S}$ is not one-to-one. A review of restrictions which have been proposed for hyperelastic strain-energy functions was given by Truesdell and Noll (1965, §52). Ogden (1977) has concluded that for all deformations of practical interest in rubberlike materials, it is reasonable to suppose that there exists a strictly convex function $\breve{U}(\mathbf{X}, \mathbf{U})$ defined on the domain $\mathcal{D}_{\mathbf{U}}$ in positive definite symmetric set such that

$$U(\mathbf{X}, \mathbf{A}) = \breve{U}(\mathbf{X}, \mathbf{U}).$$

In this case, the conjugate stress associated with $\mathbf{U}$ is defined uniquely by

$$\mathbf{U}^*(\mathbf{X}) = \frac{\partial \breve{U}(\mathbf{X}, \mathbf{U})}{\partial \mathbf{U}} \in \mathcal{M}_{sym}.$$

This symmetric Lagrangian tensor field is the well-known *Biot-Jaumann stress.* By the polar decomposition $\mathbf{F} = \mathbf{R}\mathbf{U}$ and the chain rule, it is easy to show that

$$\mathbf{U}^* = \frac{1}{2}(\mathbf{S}\mathbf{R} + \mathbf{R}^T\mathbf{S}^T). \tag{6.37}$$

Since $\mathbf{S}$ is undetermined, this equation shows that $\mathbf{U}^*$ is in general neither positive nor negative definite although its dual $\mathbf{U}$ is strictly positive definite. Let $\mathcal{R}_{\mathbf{U}} \subset \mathcal{M}_{sym}$ be the range of the constitutive mapping $D\breve{U} : \mathcal{D}_{\mathbf{U}} \to \mathcal{M}_{sym}$. Then, by the convexity of $\breve{U}(\mathbf{U})$, the Legendre duality relations

$$\breve{U}^*(\mathbf{U}^*) = \mathbf{U}^* : \mathbf{U} - \breve{U}(\mathbf{U}) \;\Leftrightarrow\; \mathbf{U} = D\breve{U}^*(\mathbf{U}^*) \;\Leftrightarrow\; \mathbf{U}^* = D\breve{U}(\mathbf{U}) \tag{6.38}$$

hold on $\mathcal{D}_{\mathbf{U}} \times \mathcal{R}_{\mathbf{U}}$.

Definition 6.1.3 (Ogden Manifolds)

For a given homogeneous isotropic hyperelastic material governed by the constitutive law $\mathbf{S} = DU(\mathbf{A}) : \mathcal{A}_a \subset \mathcal{A} \to \mathcal{S}_a \subset \mathcal{S}$, let $\mathcal{D}_{\mathbf{U}} \times \mathcal{R}_{\mathbf{U}}$ be a Legendre canonical duality domain on which the Legendre duality relations (6.38) hold. For each pair $(\mathbf{U}, \mathbf{U}^*) \in \mathcal{D}_{\mathbf{U}} \times \mathcal{R}_{\mathbf{U}}$, let (λ_i, λ_i^*), $i = 1, 2, 3$, be their eigenvalues.

(D1) The subspace $\mathcal{O}_s \subset \mathcal{A}_a$ defined by

$$\mathcal{O}_s = \{\mathbf{A} \in \mathcal{A}_a | \;\; \mathbf{U} = \sqrt{\mathbf{A}^T\mathbf{A}} \in \mathcal{D}_{\mathbf{U}}, \;\; \mathbf{U}^* = D\breve{U}(\mathbf{U}), \;\; \lambda_i^* + \lambda_j^* = 0 \;\; \forall i \neq j\} \tag{6.39}$$

is called the *Ogden singular manifold.*

(D2) The complementary space of $\mathcal{O}_s$, defined by

$$\mathcal{O}_r = \{\mathbf{A} \in \mathcal{A}_a | \;\; \mathbf{A} \notin \mathcal{O}_s\} = \mathcal{A}_a \setminus \mathcal{O}_s, \tag{6.40}$$

is called the *Ogden regular manifold.* ◇

The following result, proposed originally by Ogden (1977), is of fundamental importance in that it clarifies the extent to which non-uniqueness of solution of certain boundary-value problems can be expected.

Theorem 6.1.4 (Ogden Regularity Theorem)

Suppose that, for a given homogeneous isotropic hyperelastic material, $\mathcal{D}_{\mathbf{U}} \subset \mathcal{M}_{sym} \cap \mathcal{M}_{+}$ *is a convex, regular domain of a* C^2 *function* $\breve{U}(\mathbf{U})$ *such that, on* $\mathcal{D}_{\mathbf{U}} \times \mathcal{R}_{\mathbf{U}}$*, the Legendre duality relations* (6.38) *hold.*

(T1) In the local sense, each given $\mathbf{A} \in \mathcal{O}_s$ *is a singular point of* $U(\mathbf{A})$*. Dually, the constitutive relation* $\mathbf{S} = DU(\mathbf{A})$ *is locally invertible for all* $\mathbf{A} \in \mathcal{O}_r$.

(T2) In the global sense within $\mathcal{O}_r$*, there are four distinct branches of the inversion of* $\mathbf{S} = DU(\mathbf{A})$.

(T3) Within $\mathcal{O}_s$*, for each given* $\mathbf{S} = DU(\mathbf{A})$ *such that* $\lambda_i^* + \lambda_j^* = 0$, $i \neq j$, $\mathbf{U}^*$ *is undetermined by (6.37) to the extent that the orientation of its principal axes (and hence those of* $\mathbf{U}$*) is determined only to within an arbitrary rotation about the* λ_k^**-direction,* $\{ijk\}$ *being a cyclic permutation of* $\{123\}$.

Remark 6.1.1 This theorem has a very nice geometrical version, i.e. when the Biot-Jaumann stress vector moves on the planes $\lambda_i^* + \lambda_j^* = 0$ $(i \neq j)$ in λ^*-space, the corresponding polar envelops a singular surface in λ_i-space, on which, $U(\mathbf{A}) = \breve{U}(\mathbf{U})$ is singular in $\mathbf{A}$.

For an *isotropic material*, equation (6.37) simplifies to $\mathbf{U}^* = \mathbf{SR}$. ∎

This theorem also plays an important role in the formulation of complementary variational principle.

6.2 Primal, Dual and Polar Variational Problems

In this section we shall study the classical duality theory for nonconvex finite deformation problems. We first discuss briefly the generalized convexities in traditional direct methods. We then show that due to the nonconvexity of the stored-energy density, there exists a duality gap between the total potential and the Fenchel-Rockafellar dual energy. Finally, the so-called polar variational principles are introduced. Two important cases that play important roles in continuum mechanics and differential geometry are discussed. We shall assume for simplicity that the stored-energy density $U : \Omega \times \mathcal{A}_a \to \mathbb{R}$ is homogeneous. Thus, $U = U(\mathbf{A})$.

6.2.1 Generalized Convexities

We first let $\Lambda = \text{grad} : \mathcal{X} \to \mathcal{A}$ be a linear geometrical mapping, $W(\mathbf{A}) = \int_\Omega U(\mathbf{A}) \mathrm{d}\Omega$ the internal stored energy and

$$F(\chi) = \langle \chi, \bar{\chi}^* \rangle = \int_\Omega \bar{\mathbf{b}} \cdot \chi \mathrm{d}\Omega + \int_{\Gamma_f} \bar{\mathbf{t}} \cdot \chi \mathrm{d}\Gamma$$

the external energy. On the kinematically admissible space

$$\mathcal{X}_k = \{\chi \in \mathcal{X}_a | \ \nabla\chi \in \mathcal{A}_a\},$$

the total potential energy $P : \mathcal{X}_k \to \mathbb{R}$ associated with the boundary-value problem $(\mathcal{P}_{\text{bv}})_\chi$ reads

$$P(\chi) = W(\Lambda\chi) - F(\chi) = \int_\Omega U(\nabla\chi)\mathrm{d}\Omega - \int_\Omega \chi \cdot \bar{\mathbf{b}}\mathrm{d}\Omega - \int_{\Gamma_f} \chi \cdot \bar{\mathbf{t}}\mathrm{d}\Gamma.$$

The minimum potential principle leads to the primal variational problem

$$(\mathcal{P}_{\text{inf}})_\chi : \quad P(\chi) \quad \to \min \;\; \forall \chi \in \mathcal{X}_k,$$

and the criticality condition $DP(\bar{\chi}) = 0$ leads to the nonlinear system of partial differential equations

$$\begin{aligned} -\nabla \cdot (DU(\nabla\bar{\chi})) &= \bar{\mathbf{b}} \quad \text{in } \Omega, \\ \mathbf{N} \cdot (DU(\nabla\bar{\chi})) &= \bar{\mathbf{t}} \quad \text{on } \Gamma_f. \end{aligned}$$

Since the stored-energy density $U(\mathbf{A})$ is usually nonconvex, the mixed boundary-value problem $(\mathcal{P}_{\text{bv}})_\chi$ and the variational problem $(\mathcal{P}_{\text{inf}})_\chi$ are not equivalent. Engineers and scientists usually try to understand the physical meaning of the very difficult mathematical problems and to develop some new physical variables to make problems easier. Therefore, many important strain and stress measures for the convex stored-energy density have been discovered and some well-known generalized complementary variational principles were proposed during this century. However, it seems that mathematicians prefer to solve the challenging problems directly, and therefore some generalizations of the notation of convexity of the stored-energy density have been proposed.

Definition 6.2.1 (Rank-One, Quasi- and Poly-Convexity)

(D1) A function $U : \mathbb{R}^{m\times n} \to \mathbb{R}$ is said to be *rank-one convex* if

$$U(\theta\mathbf{A}_1 + (1-\theta)\mathbf{A}_2) \le \theta U(\mathbf{A}_1) + (1-\theta)U(\mathbf{A}_2) \;\; \forall \theta \in [0,1], \tag{6.41}$$

for every $\mathbf{A}_1, \mathbf{A}_2 \in \mathbb{R}^{m\times n}$ with rank $\{\mathbf{A}_1 - \mathbf{A}_2\} \le 1$.

(D2) A continuous function $U : \mathcal{A} \to \mathbb{R}$ defined on an open set $\mathcal{A} \subset \mathbb{R}^{m\times n}$ is said to be *quasiconvex* if

$$\int_\Omega U(\mathbf{A}_0 + \nabla\phi(\mathbf{X}))\mathrm{d}\Omega \ge U(\mathbf{A}_0)\int_\Omega \mathrm{d}\Omega = U(\mathbf{A}_0)\mathrm{meas}(\Omega) \tag{6.42}$$

for every open bounded domain $\Omega \subset \mathbb{R}^n$, for every constant matrix $\mathbf{A}_0 \in \mathcal{A}$ and for every vector $\phi \in \mathcal{W}_0^{1,\infty}(\Omega; \mathbb{R}^m)$ with $\mathbf{A}_0 + \nabla\phi \in \mathcal{A}$.

(D3) A function $U : \mathcal{E} \to \mathbb{R}$ is said to be *polyconvex* if there exists a real-valued convex function U_{cof} of $\mathbf{V}(\mathbf{A})$ such that

$$U(\mathbf{A}) = U_{cof}(\mathbf{V}(\mathbf{A})), \tag{6.43}$$

where $\mathbf{V}(\mathbf{A})$ denotes the vector consisting of all minors (i.e. subdeterminants) of $\mathbf{A}$, namely

$$\mathbf{V}(\mathbf{A}) = (\mathbf{A}, \mathrm{cof}_2\mathbf{A}, \cdots, \mathrm{cof}_{n\wedge m}\mathbf{A}),$$

and $\mathrm{cof}_s\mathbf{A}$ stands for the matrix of all $s \times s$ cofactors of $\mathbf{A}$, $2 \le s \le n \wedge m := \min\{n, m\}$. ◇

For example, if $m = n = 3$, $U(\mathbf{A})$ is polyconvex if and only if there exists a convex function $U_{cof} : \mathbb{R}^{3\times 3} \times \mathbb{R}^{3\times 3} \times \mathbb{R} \to \mathbb{R}$ with

$$U(\mathbf{A}) = U_{cof}(\mathbf{A}, \mathrm{cof}\mathbf{A}, \det \mathbf{A}) \ \ \forall \mathbf{A} \in \mathbb{R}^{3\times 3}.$$

In two dimensions $n = m = 2$, $U(\mathbf{A})$ is polyconvex if and only if there exists a convex function $U_{cof} : \mathbb{R}^{2\times 2} \times \mathbb{R} \to \mathbb{R}$ with

$$U(\mathbf{A}) = V_{cof}(\mathbf{A}, \det \mathbf{A}) \ \ \forall \mathbf{A} \in \mathbb{R}^{2\times 2}.$$

Example 6.2.1 In terms of $\mathbf{A}$, the stored-energy function for the Ogden material (6.30) reads

$$U(\mathbf{A}) = c_1 \mathrm{tr}\left((\mathbf{A}^T\mathbf{A})^{p/2} - \mathbf{I}\right) + c_2 \mathrm{tr}\left(\mathrm{cof}(\mathbf{A}^T\mathbf{A})^{q/2} - \mathbf{I}\right) + g(\det \mathbf{A}). \quad (6.44)$$

Clearly, for $c_i > 0$, $p, q \ge 1$, $U(\mathbf{A})$ is polyconvex. In the special case of the Mooney-Rivlin material, the polyconvexity of $U(\mathbf{A})$ follows easily from

$$U(\mathbf{A}) = a_1 \mathrm{tr}\left(\mathbf{A}^T\mathbf{A} - \mathbf{I}\right) + a_2 \mathrm{tr}\left(\mathrm{cof}(\mathbf{A}^T\mathbf{A}) - \mathbf{I}\right) + g(\det \mathbf{A}), \quad (6.45)$$

which corresponds to $p = q = 2$ in (6.44). ♡

Quasiconvexity was proposed by Morrey (1952), while rank-one convexity and polyconvexity were introduced by Ball (1977). There are important relations among these generalized convexities.

Theorem 6.2.1 (Ball, 1977) *Let $U : \mathbb{R}^{m\times n} \to \mathbb{R}$ be a given real-valued function.*

(T1) U convex $\Rightarrow$ U polyconvex $\Rightarrow$ U quasiconvex $\Rightarrow$ U rank-one convex.

(T2) If $m = 1$ or $n = 1$, then all these notions are equivalent.

(T3) If $U \in C^2(\mathbb{R}^{m\times n})$, then rank-one convexity is equivalent to the Legendre-Hadamard condition

$$\sum_{i,j=1}^{m} \sum_{\alpha,\beta=1}^{n} \frac{\partial^2 U(\mathbf{A})}{\partial A^i_\alpha \partial A^j_\beta} \lambda^i \lambda^j \mu_\alpha \mu_\beta \ge 0 \ \ \forall \boldsymbol{\lambda} \in \mathbb{R}^m, \ \boldsymbol{\mu} \in \mathbb{R}^n. \quad (6.46)$$

The strong form of the *Legendre-Hadamard condition* for $\boldsymbol{\lambda}$, $\boldsymbol{\mu} \ne 0 \in \mathbb{R}^n$ is referred to as the ***strong ellipticity condition***. If (6.46) holds, then the Euler-Lagrange equations for the problem $(\mathcal{P}_{\inf})_\chi$ constitute an elliptic system of equations.

An important case arises when the stored-energy $U : \mathbb{R}^{m\times n} \to \mathbb{R}$ is the quadratic function

$$U(\mathbf{A}) = \frac{1}{2}\mathbf{A} : \mathbf{H} : \mathbf{A}, \quad (6.47)$$

where $\mathbf{H} \in \mathbb{R}^{n\times m} \times \mathbb{R}^{n\times m}$ is a given symmetric fourth-order tensor. This special case is of particular interest since the Euler equations for the primal variational problem are then linear.

Theorem 6.2.2 (Convexity for Quadratic Forms)
Let $U \in \mathbb{R}^{m\times n} \to \mathbb{R}$ be a quadratic function.
(T1) U is convex if and only if

$$U(\mathbf{A}) \geq 0 \ \ \forall \mathbf{A} \in \mathbb{R}^{m\times n}.$$

(T2) U is rank-one convex if and only if U is quasiconvex.
(T3) If $n = 2$ or $m = 2$, then

$$U \textit{ polyconvex} \Leftrightarrow U \textit{ quasiconvex} \Leftrightarrow U \textit{ rank-one convex}. \tag{6.48}$$

The proof of the statement (T1) is trivial. The first proof of (T2) was given by van Hove (1947). The proof of (T3) relies on an important algebraic lemma (cf., e.g., Uhlig (1979)). The examples given in Šilhavý (1997, §17.7, 17.8 and §21.6) show that for fluids and for crystals with the nonorthogonal symmetry group of a Bravais lattice, the equivalences (6.48) hold. However, for $n = m \geq 3$, the examples due to Terpstra (1938) and Serre (1983) show that there are quasiconvex quadratic functions that are not polyconvex. The following example, due to Alibert and Dacorogna (1992), is particularly impressive.

Example 6.2.2 Let $n = m = 2$, $\gamma \in \mathbb{R}$ and let $U_\gamma : \mathcal{M} \to \mathbb{R}$ be defined as

$$U_\gamma(\mathbf{A}) = |\mathbf{A}|^2 \left(|\mathbf{A}|^2 - \gamma \det \mathbf{A}\right), \quad \mathbf{A} \in \mathcal{M},$$

where $|\mathbf{A}| := \sqrt{\mathbf{A} : \mathbf{A}}$ denotes the length of $\mathbf{A} \in \mathcal{M}$. Then

$$\begin{array}{ll} U_\gamma \text{ is convex} & \Leftrightarrow |\gamma| \leq \frac{2}{3}\sqrt{2}, \\ U_\gamma \text{ is polyconvex} & \Leftrightarrow |\gamma| \leq 1. \end{array}$$

There exists a $\theta > 0$ such that

$$\begin{array}{ll} U_\gamma \text{ is quasiconvex} & \Leftrightarrow |\gamma| \leq 1 + \theta, \\ U_\gamma \text{ is rank-one convex} & \Leftrightarrow |\gamma| \leq 2/\sqrt{3}. \end{array}$$

Since U_γ is a nonquadratic function, these equivalences are not trivial. Other examples, given by Aubert (1987) and Šverák (1991, 1992), show that the differences between polyconvexity, quasiconvexity and rank-one convexity must be nonquadratic functions. All these examples, including U_γ, are in the class of $\mathcal{M}^+_{ort}$ invariant functions, i.e. functions satisfying

$$U(\mathbf{QAR}) = U(\mathbf{A}) \ \ \forall \mathbf{Q}, \mathbf{R} \in \mathcal{M}^+_{ort},$$

for every point $\mathbf{A}$ in the domain Ω. ♡

Theorem 6.2.1 shows that if either $m = 1$ or $n = 1$ the generalized convexities of U are the same as classical convexity. However, this is definitely not the case for $m > 1$ and $n > 1$, and this makes the multi-dimensional calculus of variations an altogether more difficult subject. Hence, the duality approaches for solving nonconvex finite deformation problems are of fundamental importance.

6.2.2 Fenchel-Rockafellar Dual Variational Principle and Duality Gap

From the Fenchel transformation, the conjugate function of the stored-energy density W can be obtained as

$$W^*(\mathbf{S}) = \sup_{\mathbf{A}\in\mathcal{M}} \{\langle \mathbf{A}; \mathbf{S}\rangle - W(\mathbf{A})\}. \tag{6.49}$$

The Fenchel-conjugate function $W^*(\mathbf{S})$ is always convex and lower semicontinuous. Let $\mathcal{S}_s \subset \mathcal{M}$ be the so-called statically admissible space

$$\mathcal{S}_s := \{\mathbf{S} \in \mathcal{M} |\ -\nabla\cdot\mathbf{S} = \bar{\mathbf{b}} \ \text{ in } \Omega;\ \mathbf{N}\cdot\mathbf{S} = \bar{\mathbf{t}} \ \text{ on } \Gamma_f\}. \tag{6.50}$$

By the Fenchel-Rockafellar duality theory, the functional $P^d : \mathcal{S}_s \to \mathbb{R}$ dual to the total potential P is the concave functional

$$P^d(\mathbf{S}) = \int_{\Gamma_\chi} \bar{\chi}\cdot\mathbf{S}\cdot\mathbf{N}\mathrm{d}\Gamma - \int_\Omega U^*(\mathbf{S})\mathrm{d}\Omega. \tag{6.51}$$

If $W : \mathcal{A}_a \to \mathbb{R}$ is convex and Gâteaux differentiable, then $W^* = W^c$ is the classical complementary energy density, which can be obtained by the Legendre transformation. In this case, P^d can be written as

$$P^c(\mathbf{S}) = W^c(\mathbf{S}) - \int_{\Gamma_\chi} \bar{\chi}\cdot\mathbf{S}\cdot\mathbf{N}\mathrm{d}\Gamma.$$

This is the well-known Levinson total complementary energy in continuum mechanics (Levinson, 1965; Zubov, 1970). For nonconvex W, $-P^d$ can be considered as the generalized complementary energy. Thus, the dual problem is to maximize P^d over $\mathcal{S}_s$,

$$(\mathcal{P}^d_{\text{sup}}): \qquad P^d(\mathbf{S}) \ \to \sup \ \ \forall \mathbf{S} \in \mathcal{S}_s. \tag{6.52}$$

Lemma 6.2.1 *Suppose that for a given external load $\bar{\chi}^* = \{\bar{\mathbf{b}} \text{ in } \Omega,\ \bar{\mathbf{t}} \text{ on } \Gamma_f\}$, the statically admissible set $\mathcal{T}_a$ is nonempty. Then, the critical point $\bar{\mathbf{S}}$ of P^d is a global maximizer on $\mathcal{S}_s$ and*

$$\inf_{\chi\in\mathcal{X}_k} P(\chi) \geq \sup_{\mathbf{S}\in\mathcal{S}_s} P^d(\mathbf{S}). \tag{6.53}$$

The equality in (6.53) holds if and only if $W(\mathbf{A})$ is convex in $\mathbf{A}$. If this is the case, and there exists a function $\bar{U}^* : \mathcal{M}_{sym} \to \mathbb{R}$ such that $U^*(\mathbf{S}) = \bar{U}^*(\mathbf{S}\mathbf{S}^T)$, then the critical point $\bar{\mathbf{S}}$ of P^d and the associated $\bar{\chi}$ defined by $\mathbf{A}(\bar{\chi}) = DW^*(\bar{\mathbf{S}})$ solves $(\mathcal{P}_{\text{bv}})_\chi$. If $W(\mathbf{A})$ is strictly convex, the solution (if it exists) is unique.

Proof Since the Fenchel-conjugate $W^* : \mathcal{M} \to \overrightarrow{\mathbb{R}}$ is always convex and lower semicontinuous, $P^d : \mathcal{S}_s \to \underleftarrow{\mathbb{R}}$ is concave and upper semicontinuous, and the critical point (if it exists) of P^d is a global maximizer of P^d over the nonempty, convex set $\mathcal{S}_s$. The inequality (6.53) follows directly from the Fenchel-Young inequality

$$W^*(\mathbf{S}) \geq \langle \mathbf{A}; \mathbf{S} \rangle - W(\mathbf{A}), \ \ \forall \mathbf{A}, \mathbf{S} \in \mathcal{M}.$$

The equality $\inf P = \sup P^d$ holds if and only if W is convex in $\mathbf{A}$. In this case, the constitutive equation $\mathbf{S} = DW(\mathbf{A})$ is invertible and we have the equivalences

$$\mathbf{A} = DW^*(\mathbf{S}) \ \ \Leftrightarrow \mathbf{S} = DW(\mathbf{A}) \ \ \Leftrightarrow W(\mathbf{A}) = \langle \mathbf{A}; \mathbf{S} \rangle - W^*(\mathbf{S}).$$

If $\bar{\mathbf{S}} \in \mathcal{S}_s$ is a critical point of P^d, and $\mathbf{A}(\bar{\chi}) = DW^*(\bar{\mathbf{S}})$ is the associated deformation gradient, then

$$\begin{aligned} P^d(\bar{\mathbf{S}}) &= W(\mathbf{A}(\bar{\chi})) - \langle \nabla \bar{\chi}; \bar{\mathbf{S}} \rangle + \int_{\Gamma_\chi} \bar{\chi} \cdot \bar{\mathbf{S}} \cdot \mathbf{N} d\Gamma \\ &= W(\mathbf{A}(\bar{\chi})) - \langle \bar{\chi}, \bar{\chi}^* \rangle = P(\bar{\chi}). \end{aligned}$$

Moreover, if there exists a function $\bar{U}^*$ such that $U^*(\mathbf{S}) = \bar{U}^*(\mathbf{S}\mathbf{S}^T)$, then

$$\mathbf{A} = DU^*(\mathbf{S}) = 2\mathbf{S}^T \frac{\partial \bar{U}^*(\mathbf{S}\mathbf{S}^T)}{\partial (\mathbf{S}^T\mathbf{S})} = 2\mathbf{S}^T D\bar{U}^*(\mathbf{S}\mathbf{S}^T),$$

where $D\bar{U}^*(\mathbf{S}\mathbf{S}^T) \in \mathcal{M}_{sym}$ is a symmetric second-order tensor. The critical point $\bar{\mathbf{S}}$ and the associated $\bar{\chi}$ therefore satisfy the equilibrium equation of angular momentum

$$\mathbf{A}(\bar{\chi})\bar{\mathbf{S}} = 2\bar{\mathbf{S}}^T D\bar{U}^*(\bar{\mathbf{S}}\bar{\mathbf{S}}^T)\bar{\mathbf{S}} = (\mathbf{A}(\bar{\chi})\bar{\mathbf{S}})^T.$$

This shows that if $W(\mathbf{A})$ is convex, then the critical point $\bar{\mathbf{S}}$ of P^d and the associated $\bar{\chi}$ satisfy all equations in $(\mathcal{P}_{\text{bv}})_\chi$, and $\bar{\mathbf{S}}$ is therefore a solution. If W is strictly convex then $P(\chi)$ is strictly convex on $\mathcal{X}_k$ and P^d is strictly concave on $\mathcal{T}_k$. In this case, P^d has at most one maximizer. □

Remark 6.2.1 By the Ogden regularity theorem, we know that the invertibility of the constitutive law $\mathbf{S} = DU(\mathbf{A})$ depends on the signs of the eigenvalues of $\mathbf{U}^*(\mathbf{X})$. If $\bar{\chi}$ is a critical point of P such that $\nabla \bar{\chi} \in \mathcal{O}_r$, then the stored-energy density $U(\mathbf{A})$ is convex. In this case, if $\bar{\mathbf{S}}$ is a critical point of P^d, then

$$P(\bar{\chi}) = \min_{\chi \in \mathcal{X}_k} P(\chi) = \max_{\mathbf{S} \in \mathcal{S}_s} P^d(\mathbf{S}) = P^d(\bar{\mathbf{S}}).$$

Moreover, the axiom of objectivity ensures that there exists a non-negative function $\bar{U}^*$ such that $\nabla \bar{\chi} = 2\bar{\mathbf{S}}^T \cdot D\bar{U}^*(\bar{\mathbf{S}}\bar{\mathbf{S}}^T)$ (Ogden, 1977). ■

Lemma 6.2.1 shows that if the total potential P is nonconvex, then there exists a duality gap between the primal and dual problems. In this case, the maximizer of P^d does not solve the boundary-value problem $(\mathcal{P}_{\text{bv}})_\chi$. Since P^d is always concave, then if $\mathcal{S}_s$ is a non-empty, bounded subspace of a reflexive Banach space the theory of convex analysis ensures that the dual problem $(\mathcal{P}^d_{\text{sup}})$ has at least one maximizer $\bar{\mathbf{S}}$ on $\mathcal{S}_s$. However, to obtain the deformation field $\bar{\chi}$ by integrating $\nabla\bar{\chi} = \bar{\mathbf{A}} = DU(\bar{\mathbf{S}})$ along any path in Ω, the tensor field $\bar{\mathbf{A}}$ has to satisfy the *deformation compatibility condition*

$$\nabla \times DU^*(\bar{\mathbf{S}}) = 0 \in \mathbb{R}^{m\times n}.$$

In order to overcome this difficulty, the so-called polar variational problem is proposed.

6.2.3 Polar Variational Problems

In the case that the constitutive relation $\mathbf{S} = DU(\mathbf{A})$ is pointwise invertible in Ω, the polar system of $\mathbb{S}$ can be defined by $\mathbb{S}^o = \{\langle\mathcal{X}^o\,;\,\mathcal{X}^{o*}\rangle, \langle\mathcal{S};\mathcal{A}\rangle; \Lambda^o, DU^*\}$. If, for a given source $\bar{\chi}^* \in \mathcal{X}_a^*$,

$$\langle \text{grad}\chi; \Lambda^o\chi^o\rangle = \langle\chi, \bar{\chi}^*\rangle \;\; \forall\chi \in \mathcal{X}_k, \tag{6.54}$$

then $\mathbb{S}^o$ is an iso-source polar system of $\mathbb{S}$. Let $\mathcal{X}_k^o$ be the subset of $\mathcal{X}^o$ defined by

$$\mathcal{X}_k^o = \{\chi^o \in \mathcal{X}^o |\;\, \text{grad}^*\Lambda^o\chi^o = \bar{\chi}^*\}.$$

The polar canonical boundary-value problem associated with $(\mathcal{P}_{\text{bv}})_\chi$ can now be proposed.

Problem 6.2.1 (Polar Mixed Boundary-Value Problem) For a given source $\bar{\chi}^* = \{\bar{\mathbf{b}} \text{ in } \Omega,\; \bar{\mathbf{t}} \text{ on } \Gamma_f\}$ and $\bar{\chi}$ on Γ_χ, find χ^o such that

$$(\mathcal{P}_{\text{bv}}^o): \quad \left\{ \begin{array}{lll} \mathbf{S} = \Lambda^o\chi^o \;\; \text{in } \Omega, & B^o\chi^o = \bar{\mathbf{t}} & \text{on } \Gamma_f, \\ \mathbf{A} = DW^*(\mathbf{S}) \;\; \text{in } \Omega, & & \\ 0 = \Lambda^{o*}\mathbf{A} \;\; \text{in } \Omega, & B^{o*}\mathbf{A} = \bar{\chi} & \text{on } \Gamma_\chi, \end{array} \right. \tag{6.55}$$

where $B^o = \mathbf{N}\cdot\Lambda^o : \mathcal{X}^o \to \partial\mathcal{X}^o$ is a polar boundary operator and $B^{o*} : \mathcal{A} \to \partial\mathcal{X}$ is a properly defined adjoint operator. ♠

Replacing $\mathbf{S}$ by $\Lambda^o\chi^o$ and letting

$$F^o(\chi^o) = \int_{\Gamma_\chi} \mathbf{N}\cdot\Lambda^o\chi^o\cdot\bar{\chi}d\Gamma,$$

we define the polar functional $P^o : \mathcal{X}_k^o \to \mathbb{R}$ by

$$P^o(\chi^o) = W^*(\Lambda^o\chi^o) - F^o(\chi^o). \tag{6.56}$$

Thus, the polar variational problem associated with $(\mathcal{P}_{\text{bv}}^o)$ can be proposed as

$$(\mathcal{P}_{\text{inf}}^o): \quad P^o(\chi^o) \;\to \inf \;\; \forall\chi^o \in \mathcal{X}_k^o. \tag{6.57}$$

Theorem 6.2.3 *Suppose that $U : \mathcal{A}_a \to \mathbb{R}$ is convex and both $\mathcal{X}_k$ and $\mathcal{X}_k^o$ are not empty. Then,*

$$\inf_{\chi\in\mathcal{X}_k} P(\chi) + \inf_{\chi^o\in\mathcal{X}_k^o} P^o(\chi^o) = 0. \tag{6.58}$$

This theorem can be proved easily by using Lemma 6.2.1 and the strong duality theorem. The choice of the polar operator depends on the external source field $\bar{\chi}^*$. Here, we consider only two cases.

I. $\Lambda^o = \text{grad} = \{\frac{\partial}{\partial X_\alpha}\} : \mathbb{R}^m \to \mathbb{R}^{n\times m}$.

In this case, the polar variable $\chi^o = \{\chi^{oi}(X_\alpha)\} \in \mathbb{R}^m$ is a vector-valued function from $\Omega \subset \mathbb{R}^n$ into $\mathbb{R}^m$, and $\Lambda^o\chi^o = \nabla\chi^o$ should be understood as a two point tensor in $\mathcal{S} \subset \mathbb{R}^{n\times m}$. The equilibrium conditions are

$$\text{grad}^*\Lambda^o\chi^o = \begin{cases} -\nabla\cdot(\nabla\chi^o) = -\Delta\chi^o = \bar{\mathbf{b}} & \text{in } \Omega, \\ \mathbf{N}\cdot(\nabla\chi^o) = \bar{\mathbf{t}} & \text{on } \Gamma_f. \end{cases}$$

The polar kinematically admissible space $\mathcal{X}_k^o$ can be written as

$$\mathcal{X}_k^o = \{\chi^o \in \mathcal{X}^o(\Omega;\mathbb{R}^m)|\ \Delta\chi^o + \bar{\mathbf{b}} = 0 \ \text{ in } \Omega,\ \mathbf{N}\cdot(\nabla\chi^o) = \bar{\mathbf{t}} \ \text{ on } \Gamma_f\}.$$

To relax the equilibrium constraints in $\mathcal{X}_k^o$, we introduce an admissible Lagrange multiplier $\chi \in \mathcal{X}_a$ such that the Lagrangian $L^o : \mathcal{X}^o \times \mathcal{X}_a \to \mathbb{R}$ associated with the polar variational problem $(\mathcal{P}^o_{\text{inf}})$ is obtained as

$$\begin{aligned} L^o(\chi^o,\chi) &= P^o(\chi^o) - \langle \chi, \text{grad}^*\nabla\chi^o - \bar{\chi}^*\rangle \\ &= \int_\Omega [U^*(\nabla\chi^o) - (\nabla\chi) : (\nabla\chi^o) + \chi\cdot\bar{\mathbf{b}}]d\Omega + \int_{\Gamma_f} \mathbf{N}\cdot(\nabla\chi^o)\cdot\chi d\Gamma. \end{aligned}$$

It is not difficult to show that the criticality condition $\delta P^o(\bar{\chi}^o;\chi^o) = 0\ \forall\chi^o \in \mathcal{X}_k^o$ is equivalent to the Lagrange equations

$$\begin{aligned} &\Lambda\bar{\chi} = \nabla\bar{\chi} = DU^*(\nabla\bar{\chi}^o) \ \text{ in } \Omega, \\ &-\nabla\cdot(\nabla\bar{\chi}^o) = \bar{\mathbf{b}} \ \text{ in } \Omega, \\ &\mathbf{N}\cdot(\nabla\bar{\chi}^o) = 0 \ \text{ on } \Gamma_f. \end{aligned}$$

Since the constitutive equation $\mathbf{S} = DU(\mathbf{A})$ is invertible in Ω, the first equation is equivalent to $\Lambda^o\bar{\chi}^o = DU(\Lambda\bar{\chi})$. Thus, the critical point $\bar{\chi}^o$ of P^o is a solution of the boundary-value problem $(\mathcal{P}_{\text{bv}})_\chi$.

Recall that the Lagrangian associated with the primal problem $(\mathcal{P}_{\text{inf}})_\chi$ is

$$\begin{aligned} L(\chi,\mathbf{S}) &= \langle\Lambda\chi;\mathbf{S}\rangle - W^*(\mathbf{S}) - F(\chi) \\ &= \int_\Omega [(\nabla\chi) : \mathbf{S} - U^*(\mathbf{S}) - \chi\cdot\bar{\mathbf{b}}]d\Omega - \int_{\Gamma_f} \mathbf{N}\cdot\mathbf{S}\cdot\chi d\Gamma. \end{aligned}$$

Clearly,

$$L(\chi,\nabla\chi^o) + L^o(\chi^o,\chi) = 0.$$

This equation shows that L^o is actually a complementary functional of L.

II. $\Lambda^o = \text{curl} : \mathbb{R}^{n\times m} \to \mathcal{S}$.

In this case, the polar variable $\chi^o = \{\chi_i^{o\alpha}\} \in \mathbb{R}^{n\times m}$ is a two-point tensor field. If $m = n$, χ^o can be regarded as the *stress tensor function*, as discussed in Chapter 4. If $\bar{\mathbf{b}} = 0$ in Ω, then the internal equilibrium equation

$$\text{grad}^* \Lambda^o \chi^o = -\text{div curl}\chi^o = -\nabla \cdot (\nabla \times \chi^o) = 0 \in \mathbb{R}^m \quad \text{in } \Omega$$

is automatically satisfied, and the polar kinematically admissible space $\mathcal{X}_k^o$ simply takes the form

$$\mathcal{X}_k^o = \{\chi^o \in \mathcal{M}(\Omega; \mathbb{R}^{n\times m}) |\ \mathbf{N} \cdot (\text{curl}\chi^o) = \bar{\mathbf{t}} \quad \text{on } \Gamma_f\}.$$

The Gâteaux variation of P^o at $\bar{\chi}^o \in \mathcal{X}_k^o$ in the direction $\chi^o \in \mathcal{X}_k^o$ is

$$\delta P^o(\bar{\chi}^o; \chi^o) = \langle DU^*(\bar{\mathbf{S}}); \text{curl}\chi^o \rangle - \int_{\Gamma_\chi} \mathbf{N} \cdot (\text{curl}\chi^o) \cdot \bar{\chi} d\Gamma,$$

where $\bar{\mathbf{S}} = \text{curl}\bar{\chi}^o$. Let $\bar{\mathbf{A}} = DU^*(\bar{\mathbf{S}}) = \text{grad}\chi(\bar{\chi}^o)$. Using the identities

$$\mathbf{A} : (\text{curl}\chi^o) = (\text{curl}\mathbf{A}^T) : \chi^o + \text{tr}[\text{curl}(\chi^o \cdot \mathbf{A})], \tag{6.59}$$

$$(\text{grad}\chi) : (\chi^{oT} \times \mathbf{N}) = \mathbf{N} \cdot (\text{curl}\chi^o) \cdot \chi - \mathbf{N} \cdot [\text{curl}(\chi^o \cdot \chi)]$$

and Stokes theorem

$$\int_\Omega \text{tr}[\text{curl}(\chi^o \cdot \mathbf{A})] d\Omega = \oint_\Gamma \mathbf{A} : (\chi^{oT} \times \mathbf{N}) d\Gamma, \tag{6.60}$$

we obtain

$$\delta P^o(\bar{\chi}^o; \delta\chi^o) = \int_\Omega (\text{curl}\bar{\mathbf{A}}^T(\bar{\mathbf{S}})) : \delta\chi^o d\Omega + \int_{\Gamma_\chi} \mathbf{N} \cdot (\text{curl}\delta\chi^o) \cdot (\chi(\bar{\chi}^o) - \bar{\chi}) d\Gamma \tag{6.61}$$

for all $\delta\chi^o \in \mathcal{X}_k^o$ such that $\text{curl}\delta\chi^o$ vanishes on Γ_f. Thus, the criticality condition $\delta P^o(\bar{\chi}^o; \delta\chi^o) = 0 \ \forall \delta\chi^o \in \mathcal{X}_k^o$ leads to the *polar Euler-Lagrange equation*

$$\text{curl}[DU^*(\text{curl}\bar{\chi}^o)]^T = 0 \quad \text{in } \Omega, \tag{6.62}$$

$$\chi(\bar{\chi}^o) = \bar{\chi} \quad \text{on } \Gamma_\chi. \tag{6.63}$$

This shows that the critical point $\bar{\chi}^o \in \mathcal{X}_k^o$ of P^o satisfies all the governing equations and boundary conditions in $(\mathcal{P}_{\text{bv}})_\chi$, and is therefore a solution of $(\mathcal{P}_{\text{bv}})_\chi$.

Besides these two polar operators, there are many other choices. One which plays an important role in differential geometry will be discussed later.

6.3 Canonical Strain Measures and Complementary Gap Functional

By the duality of nature, the laws that govern natural phenomena should be simple, possess certain symmetries and perhaps even have elegant formulations. It is reasonable to assume that for any given real hyperelastic material, there exists a certain strain measure such that the stored energy density is a strictly convex function of this measure. Since the deformation gradient **A** is not a strain measure, the stored-energy density in finite deformation theory is usually not convex in **A** and there exists a duality gap between the total potential and the Fenchel-Rockafellar dual energy. In order to recover this duality gap, the canonical transformation and Hill's conjugate strain-stress functions will be discussed in this section. The complementary gap function associated with these nonlinear geometrical measures plays an important role in fully nonlinear systems. Finally, we present a unified complementary energy principle, with the help of which the well-known generalized variational principles of continuum mechanics are embraced by a unified framework.

6.3.1 Canonical Strain Measures and Canonical Boundary-Value Problems

Assume that for a given finitely deformed system, there exists a Gâteaux differentiable, objective function $\mathbf{E} = \mathbf{E}(\nabla\chi)$ defined on an objective subset $\mathcal{E}_a$ and that there exists a canonical function $\bar{U} : \mathcal{E}_a \to \mathbb{R}$ satisfying the isotropy condition

$$\bar{U}(\mathbf{E}(\mathbf{AQ})) = \bar{U}(\mathbf{E}) \quad \forall \mathbf{A} \in \mathcal{A}, \quad \forall \mathbf{Q} \in \mathcal{M}^+_{ort},$$

such that

$$U(\mathbf{X}, \nabla\chi) = \bar{U}(\mathbf{E}(\nabla\chi)), \quad \forall\chi \in \mathcal{U}.$$

We call such a function $\mathbf{E}(\mathbf{A})$ the *canonical strain measure*, and the canonical function $\bar{U}(\mathbf{E})$ the *canonical strain-energy density.* The canonical strain measure $\mathbf{E}(\mathbf{A})$ could be a tensor-valued, vector-valued, or scalar-valued function of the deformation gradient $\mathbf{A} \in \mathcal{A}$.

We use $\mathcal{E}$ to denote the function space of all possible canonical strain measures. Then, the geometrical mapping $\Lambda : \mathcal{X} \to \mathcal{E}$ is a Gâteaux differentiable operator such that the geometrical equation reads

$$\Lambda(\chi) = \mathbf{E}(\nabla\chi).$$

The directional derivative of Λ at $\bar{\chi}$ in the direction $\chi \in \mathcal{X}$ can be written as

$$\begin{aligned}\delta\Lambda(\bar{\chi};\chi) &= \lim_{\theta\to 0^+} \frac{\mathbf{E}(\nabla(\bar{\chi}+\theta\chi)) - \mathbf{E}(\nabla\bar{\chi})}{\theta} \\ &= (\mathbf{E}_t(\nabla\bar{\chi}))\nabla\chi = \Lambda_t(\bar{\chi})\chi,\end{aligned}$$

where $\Lambda_t(\bar{\chi})$ denotes the Gâteaux derivative of Λ at $\bar{\chi}$, and $\mathbf{E}_t(\mathbf{A}) = D\mathbf{E}(\mathbf{A})$ is the Gâteaux derivative of $\mathbf{E}$ with respect to its argument $\mathbf{A}$. Let $\mathcal{E}_a$ be a subset of $\mathcal{E}$ on which $\bar{U} : \mathcal{E}_a \to \mathbb{R}$ is differentiable. The associated *canonical stress measure* is then defined by

$$\mathbf{T} = D\bar{U}(\mathbf{E}),$$

which is also an isotropic function. By the chain rule, the relationship between the nominal stress $\mathbf{S}$ and the canonical stress $\mathbf{T}$ is obtained in the form

$$\mathbf{S} = \frac{\partial U(\mathbf{X}, \mathbf{A})}{\partial \mathbf{A}} = D\bar{U}(\mathbf{E})\frac{\partial \mathbf{E}}{\partial \mathbf{A}} = \mathbf{T}\mathbf{E}_t(\mathbf{A}). \tag{6.64}$$

As usual, we use $\mathcal{T} = \mathcal{E}^*$ to denote the canonical stress space, in duality with $\mathcal{E}$ via the bilinear form $\langle *; * \rangle : \mathcal{E} \times \mathcal{T} \to \mathbb{R}$. Furthermore, we let $\mathcal{T}_a \subset \mathcal{T}$ be the range of the canonical constitutive mapping. For a given source $\bar{\chi}^* \in \mathcal{X}_a^*$, such that $\bar{\chi} \in \mathcal{X}_k$ and the associated $\bar{\mathbf{T}} \in \mathcal{T}_a$ define the equilibrium state, the virtual work principle

$$\langle \delta\Lambda(\bar{\chi};\chi); \bar{\mathbf{T}} \rangle = \langle \Lambda_t(\bar{\chi})\chi; \bar{\mathbf{T}} \rangle = \langle \chi, \bar{\Lambda}_t^*(\bar{\chi})\bar{\mathbf{T}} \rangle = \langle \chi, \bar{\chi}^* \rangle \ \ \forall \chi \in \mathcal{X}_k$$

leads to the canonical equilibrium equations

$$\bar{\Lambda}_t^*(\bar{\chi})\bar{\mathbf{T}} = \begin{cases} -\mathrm{div}[\bar{\mathbf{T}}\mathbf{E}_t(\nabla\bar{\chi})] = \bar{\mathbf{b}} & \text{in } \Omega, \\ \mathbf{N} \cdot [\bar{\mathbf{T}}\mathbf{E}_t(\nabla\bar{\chi})] = \bar{\mathbf{t}} & \text{on } \Gamma_f. \end{cases}$$

Thus, the canonical boundary-value problem, which is equivalent to $(\mathcal{P}_{\mathrm{bv}})_\chi$, may be proposed as follows.

Problem 6.3.1 (Canonical Boundary-Value Problem) For a given external force field $\bar{\chi}^* = \{\bar{\mathbf{b}} \text{ in } \Omega, \ \bar{\mathbf{t}} \text{ on } \Gamma_f\}$, find the equilibrium state $\bar{\chi} \in \mathcal{X}_k$ and the associated $(\bar{\mathbf{E}}, \bar{\mathbf{T}}) \in \mathcal{E}_a \times \mathcal{T}_a$ such that

$$(\mathcal{P}_{\mathrm{cbv}})_\chi : \begin{cases} \bar{\mathbf{E}} = \Lambda(\bar{\chi}) \text{ in } \Omega, & \chi = \bar{\chi} \ \text{ on } \Gamma_\chi, \\ \bar{\mathbf{T}} = D\bar{U}(\bar{\mathbf{E}}) \ \text{ in } \Omega, & \\ \bar{\mathbf{b}} = -\nabla \cdot [\bar{\mathbf{T}}\mathbf{E}_t(\nabla\bar{\chi})] \ \text{ in } \Omega, & \mathbf{N} \cdot [\bar{\mathbf{T}}\mathbf{E}_t(\nabla\bar{\chi})] = \bar{\mathbf{t}} \ \text{ on } \Gamma_f. \end{cases}$$

♠

Suppose that the mapping $\Lambda : \mathcal{X}_a \to \mathcal{E}$ is Gâteaux differentiable. By defining $\Lambda_c(\chi) = \Lambda(\chi) - \Lambda_t(\chi)\chi)$, we obtain the so-called *Λ-canonical decomposition*

$$\Lambda(\chi) = \Lambda_t(\chi)\chi + \Lambda_c(\chi) \ \ \forall \chi \in \mathcal{X}_a. \tag{6.65}$$

Then, the *complementary gap functional*, or simply the gap functional, associated with the canonical operator Λ is defined naturally by

$$G(\chi, \mathbf{T}) = \langle -\Lambda_c(\chi); \mathbf{T} \rangle.$$

Thus, for any given external force $\bar{\chi}^* \in \mathcal{X}_a^*$ such that $(\bar{\chi}, \bar{\mathbf{T}})$ is an equilibrium state, the relationship between the two bilinear forms is given by

$$\langle \Lambda(\bar{\chi}); \bar{\mathbf{T}} \rangle = \langle \bar{\chi}, \bar{\chi}^* \rangle - G(\bar{\chi}, \bar{\mathbf{T}}).$$

Actually, this is the way that the complementary gap functional was originally introduced in Gao and Strang (1989). In terms of the deformation gradient $\mathbf{A}$, the canonical decomposition (6.65) can be written as

$$\mathbf{E}(\mathbf{A}) = \mathbf{E}_t(\mathbf{A})\mathbf{A} + \mathbf{E}_c(\mathbf{A}), \ \ \forall \mathbf{A} \in \mathcal{A} \tag{6.66}$$

with $\mathbf{E}_c(\nabla\chi) = \Lambda_c(\chi)$.

Example 6.3.1 In the case that the stored-energy density $\bar{U}(\mathbf{C})$ is a canonical function of the right Cauchy-Green strain tensor, the canonical strain measure

$\mathbf{E} = \mathbf{C} = (\nabla\chi)^T(\nabla\chi)$ is a quadratic function of $\mathbf{A}$ and $\Lambda : \mathcal{X}_a \to \mathcal{E}$ is a quadratic function of the deformation $\chi \in \mathcal{X}_a$. Its directional derivative at $\bar{\chi}$ in the direction $\chi \in \mathcal{X}$ reads

$$\delta\Lambda(\bar{\chi};\chi) = \Lambda_t(\bar{\chi})\chi = (\nabla\bar{\chi})^T(\nabla\chi) + (\nabla\chi)^T(\nabla\bar{\chi}).$$

By the virtual work principle

$$\langle \Lambda_t(\bar{\chi})\chi; \bar{\mathbf{C}}^* \rangle = \langle \chi, \bar{\Lambda}_t^*(\bar{\chi})\bar{\mathbf{C}}^* \rangle = \langle \chi, \bar{\chi}^* \rangle \quad \forall \chi \in \mathcal{X}_a,$$

the equilibrium equation can be written in terms of $\mathbf{C}^*$ as

$$\bar{\Lambda}_t^*(\bar{\chi})\bar{\mathbf{C}}^* = \bar{\chi}^* \quad \Rightarrow \quad \begin{cases} -\nabla \cdot (2\bar{\mathbf{C}}^*(\nabla\bar{\chi})^T) = \bar{\mathbf{b}} & \text{in } \Omega, \\ \mathbf{N} \cdot (2\bar{\mathbf{C}}^*(\nabla\bar{\chi})^T) = \bar{\mathbf{t}} & \text{on } \Gamma. \end{cases}$$

By the canonical decomposition,

$$\Lambda_c(\chi) = \Lambda(\chi) - \Lambda_t(\chi)\chi = -(\nabla\chi)^T(\nabla\chi),$$

and the gap functional associated with Λ is therefore also a quadratic functional of χ, namely

$$G(\chi, \mathbf{C}^*) = \langle -\Lambda_c(\chi); \mathbf{C}^* \rangle = \int_\Omega \mathrm{tr}[(\nabla\chi)\mathbf{C}^*(\nabla\chi)^T] \mathrm{d}\Omega.$$

In terms of the right Cauchy-Green strain tensor and its conjugate stress, the tri-canonical equations for hyperelasticity can written as

$$\mathbf{C} = \Lambda(\chi), \quad \mathbf{C}^* = D\breve{U}(\mathbf{C}), \quad \bar{\chi}^* = \bar{\Lambda}_t^*(\chi)\mathbf{C}^*. \tag{6.67}$$

This canonical problem is certainly equivalent to the mixed boundary-value problem $(\mathcal{P}_{\mathrm{bv}})_\chi$. ♡

In the case that $U(\mathbf{A}) = \breve{U}(\mathbf{C})$ with $\breve{U}(\mathbf{C})$ convex, we have the following result.

Lemma 6.3.1 *Suppose that $\breve{U}(\mathbf{C})$ is strictly convex on $\mathcal{C} \subset \mathcal{E}$ and for a given deformation $\chi \in \mathcal{X}$ such that $\mathbf{A} = \nabla\chi$, $\mathbf{S} = DU(\mathbf{A}) = 2D\breve{U}(\mathbf{C})\mathbf{A}^T = 2\mathbf{C}^*\mathbf{A}^T$. If $\mathbf{C}^*(\mathbf{X})$ is strictly positive-definite at $\mathbf{X} \in \Omega$, then the constitutive relation $\mathbf{S} = DU(\mathbf{A})$ is locally invertible at $\mathbf{X}$ and*

$$U^*(\mathbf{S}) = \breve{U}^*(\mathbf{C}^*) + \mathrm{tr}[(\nabla\chi)\mathbf{C}^*(\nabla\chi)]. \tag{6.68}$$

Proof Since $DU(\mathbf{A}) = \mathbf{S} = 2\mathbf{C}^*\mathbf{A}^T$, the second-order directional derivative of U at $\mathbf{A}$ in the direction $\delta\mathbf{A}$ can be written as

$$\delta^2 U(\mathbf{A}; \delta\mathbf{A}^2) = \delta\mathbf{A} : \mathbf{H}(\mathbf{A}) : \delta\mathbf{A}, \tag{6.69}$$

where $\mathbf{H}$ is the elastic tensor

$$\mathbf{H}(\mathbf{A}) = \frac{\partial^2 U(\mathbf{A})}{\partial \mathbf{A}^2} = 2(\mathbf{I} \otimes \mathbf{T}^{(1)} + \mathbf{A}\mathbf{D}\mathbf{A}^T),$$

or in terms of the components

$$H_{ij}^{\alpha\beta} = \delta_{ij} C^{*\alpha\beta} + A_i^{\gamma} D_{\gamma\theta}^{\alpha\beta} A_j^{\theta},$$

in which

$$\mathbf{D} = \frac{\partial^2 \breve{U}(\mathbf{C})}{\partial \mathbf{C}^2}$$

is the Hessian tensor of $\breve{U}$ at the strain tensor $\mathbf{C} \in \mathcal{C}$. The strict convexity of $\breve{U}(\mathbf{C})$ ensures that $\mathbf{D}$ is strictly positive-definite on $\mathcal{C}$. If $\mathbf{C}^*$ is positive-definite at $\mathbf{X}$, then, for any non-zero $\delta\mathbf{A}$,

$$\delta^2 U(\mathbf{A}; \delta\mathbf{A}^2) = \delta\mathbf{A} : \mathbf{H}(\mathbf{A}) : \delta\mathbf{A} > 0 \ \ \forall \delta\mathbf{A} \neq 0.$$

This shows that if $\mathbf{C}^*$ is positive definite at $\mathbf{X}$, then the stored-energy density $U(\mathbf{A})$ is strictly convex at $\mathbf{X} \in \Omega$, and the constitutive relation is therefore invertible at $\mathbf{X}$. Moreover, by the Legendre duality theorem, we have

$$\begin{aligned} U^*(\mathbf{S}) &= \mathbf{S} : \mathbf{A} - U(\mathbf{A}) = 2(\mathbf{A}\mathbf{C}^*) : \mathbf{A} - \bar{U}(\mathbf{C}(\mathbf{A})) \\ &= \mathbf{C}^* : [2\mathbf{A}^T\mathbf{A} - \mathbf{C}(\mathbf{A})] + \breve{U}^*(\mathbf{C}^*) \\ &= \breve{U}^*(\mathbf{C}^*) + \mathbf{A}\mathbf{C}^*\mathbf{A}^T, \end{aligned}$$

which proves the lemma. □

This lemma was originally proposed in Gao (1992) and will be used frequently in the rest of this chapter.

6.3.2 Hill-Seth Strain Family and Hill's Conjugate Pairing

We now turn to the discussion of a class of important stress and strain measures, which play a very important role in continuum mechanics. In the following, we assume that $m = n$.

Definition 6.3.1 (Hill-Seth Strain Family)
For any given parameter $\eta \in \mathbb{R}$, the function defined by

$$\mathbf{E}^{(\eta)} := \frac{1}{2\eta}[\mathbf{C}^{\eta} - \mathbf{I}] = \frac{1}{2\eta}[(\mathbf{A}^T\mathbf{A})^{\eta} - \mathbf{I}], \tag{6.70}$$

is called the *Hill-Seth strain measure*, and the function space

$$\mathcal{E}^{(\eta)} = \{\mathbf{E}^{(\eta)} \ \ \forall \eta \in \mathbb{R}\}$$

is called the *Hill-Seth strain space (or family)*. ◇

Clearly, any Hill-Seth strain is an objective function of $\mathbf{A}$, and the stored energy $\bar{U}$ is objective, i.e.

$$U(\mathbf{Q}\mathbf{A}) = \bar{U}(\mathbf{E}^{(\eta)}(\mathbf{Q}\mathbf{A})) = \bar{U}(\mathbf{E}^{(\eta)}(\mathbf{A})) = U(\mathbf{A}) \ \ \forall \mathbf{A} \in \mathcal{M}_+, \ \ \mathbf{Q} \in \mathcal{M}_{ort}^+.$$

Thus, if $\bar{U}(\mathbf{E}^{(\eta)})$ is Gâteaux differentiable on $\mathcal{E}^{(\eta)}$, then we have

$$\delta\bar{U}(\mathbf{E}^{(\eta)};\delta\mathbf{E}^{(\eta)}) = D\bar{U}(\mathbf{E}^{(\eta)}) : \delta\mathbf{E}^{(\eta)} \;\; \forall\mathbf{E}^{(\eta)}, \;\; \delta\mathbf{E}^{(\eta)} \in \mathcal{E}^{(\eta)},$$

where the Gâteaux derivative $D\bar{U}(\mathbf{E}^{(\eta)})$ is a stress measure, which maps the Hill-Seth strain space into its dual space, denoted by $\mathcal{T}^{(\eta)} = (\mathcal{E}^{(\eta)})^*$.

Definition 6.3.2 (Hill's Conjugate Strain-Stress Measures)
Let $\{(\mathcal{E}^{(\eta)}, \mathcal{T}^{(\eta)}; \langle *; * \rangle\}$ be a duality paring. If there exists a Gâteaux differentiable function $\bar{U} : \mathcal{E}^{(\eta)} \to \mathbb{R}$ satisfying $U(\mathbf{A}) = \bar{U}(\mathbf{E}^{(\eta)})$, then the Gâteaux derivative

$$\mathbf{T}^{(\eta)} = D\bar{U}(\mathbf{E}^{(\eta)}) \tag{6.71}$$

of $\bar{U}$ at $\mathbf{E}^{(\eta)}$ is called *Hill's conjugate stress*, and for all $\eta \in \mathbb{R}$, the pairs $(\mathbf{E}^{(\eta)}, \mathbf{T}^{(\eta)})$ defined in this way are said to be *Hill's conjugate strain and stress measures*.

Moreover, if, for a given $\eta \in \mathbb{R}$, (6.71) is globally invertible and the Legendre duality relation holds on the product space $\mathcal{E}^{(\eta)} \times \mathcal{T}^{(\eta)}$, then we have

$$\mathbf{T}^{(\eta)} = D\bar{U}(\mathbf{E}^{(\eta)}) \Leftrightarrow \mathbf{E}^{(\eta)} = D\bar{U}^*(\mathbf{T}^{(\eta)}) \;\; \Leftrightarrow$$
$$\bar{U}(\mathbf{E}^{(\eta)}) + \bar{U}^*(\mathbf{T}^{(\eta)}) = \mathbf{E}^{(\eta)} : \mathbf{T}^{(\eta)},$$

and $\mathcal{E}^{(\eta)} \times \mathcal{T}^{(\eta)}$ is then said to be a *Hill canonical conjugate space.* Moreover, any conjugate pair $(\mathbf{E}^{(\eta)}, \mathbf{T}^{(\eta)}) \in \mathcal{E}^{(\eta)} \times \mathcal{T}^{(\eta)}$ is said to be a *Hill canonical conjugate pair*, and $\bar{U} : \mathcal{E}^{(\eta)} \to \mathbb{R}$ is said to be a *Hill canonical strain-energy density.* $\diamond$

The notion of conjugacy in this context was introduced by Hill (1968). Hill's conjugate strains and stresses play an important role in finite deformation variational principles. Actually, according to Hill (1968, 1978) and Ogden (1977), the strain energy function $\bar{U}$ of hyperelasticity is a strictly convex function on certain Hill-Seth strain spaces $\mathcal{E}^{(\eta)}$.

We list some of important conjugate parings.

I. If $\eta = 1$, the Hill-Seth strain is

$$\mathbf{E}^{(1)} = \frac{1}{2}[\mathbf{C} - \mathbf{I}] = \frac{1}{2}[\mathbf{A}^T\mathbf{A} - \mathbf{I}]. \tag{6.72}$$

This is the well-known *Green-St. Venant strain tensor.* The associated stress

$$\mathbf{T}^{(1)} = D\bar{U}(\mathbf{E}^{(1)}) = 2\frac{\partial\bar{U}(\frac{1}{2}(\mathbf{C} - \mathbf{I}))}{\partial\mathbf{C}} = 2\mathbf{C}^* \tag{6.73}$$

is called the *second Piola-Kirchhoff stress tensor.* The connections between $\mathbf{T}^{(1)}$ and certain other stress tensors are given by

$$\mathbf{T}^{(1)} = J\mathbf{A}^{-1}\boldsymbol{\sigma}\mathbf{A}^{-1T} = \mathbf{S}\mathbf{A}^{-1T} = 2\mathbf{C}^*.$$

Since the Green-St. Venant strain $\mathbf{E}^{(1)}$ is a quadratic tensor function of $\mathbf{A}$, it plays a very important role in finite deformation theory. According to Hill

(1968, 1978), the stored-energy density $\bar{U}$ is a strictly convex function for many materials. For the St. Venant-Kirchhoff material (6.27), for example, $\bar{U}$ is a quadratic function of $\mathbf{E}^{(1)}$, i.e.

$$\bar{U}(\mathbf{E}^{(1)}) = \frac{1}{2}\lambda\left(\mathrm{tr}\mathbf{E}^{(1)}\right)^2 + \mu\mathrm{tr}\left(\mathbf{E}^{(1)}\right)^2.$$

Substituting $\mathbf{A} = \nabla\mathbf{u} + \mathbf{I}$ into $\mathbf{E}^{(1)}$, we obtain

$$\mathbf{E}^{(1)}(\nabla\nu) = \Lambda(\mathbf{u}) = \frac{1}{2}[\nabla\mathbf{u} + (\nabla\mathbf{u})^T + (\nabla\mathbf{u})^T(\nabla\mathbf{u})],$$

which is a quadratic function of $\nabla\mathbf{u}$. The directional derivative of $\mathbf{E}^{(1)}$ at $\bar{\mathbf{u}}$ in the direction $\mathbf{u} \in \mathcal{U}$ is

$$\delta\mathbf{E}^{(1)}(\nabla\bar{\mathbf{u}}; \nabla\mathbf{u}) = \mathbf{E}_t^{(1)}(\mathbf{A}(\bar{\mathbf{u}}))(\nabla\mathbf{u}) = \Lambda_t(\bar{\mathbf{u}})\mathbf{u},$$

where $\mathbf{E}_t^{(1)}(\mathbf{A})$ is a tensor operator such that

$$\mathbf{E}_t^{(1)}(\bar{\mathbf{A}})\mathbf{A} = \frac{1}{2}[\bar{\mathbf{A}}^T\mathbf{A} + \mathbf{A}^T\bar{\mathbf{A}} + \mathbf{A} + \mathbf{A}^T],$$

and $\Lambda_t(\bar{\mathbf{u}}) : \mathcal{U} \to \mathcal{M}_{sym}$ is the Gâteaux derivative of Λ at $\bar{\mathbf{u}}$. Thus,

$$\Lambda_t(\bar{\mathbf{u}})\mathbf{u} = \frac{1}{2}[\nabla\mathbf{u} + (\nabla\mathbf{u})^T + (\nabla\bar{\mathbf{u}})^T(\nabla\mathbf{u}) + (\nabla\mathbf{u})^T(\nabla\bar{\mathbf{u}})]. \tag{6.74}$$

From the canonical decomposition $\Lambda(\mathbf{u}) = \Lambda_t(\mathbf{u})\mathbf{u} + \Lambda_c(\mathbf{u})$, it is easy to see that

$$\Lambda_c(\mathbf{u}) = -\frac{1}{2}(\nabla\mathbf{u})^T(\nabla\mathbf{u}) = \mathbf{E}_c^{(1)}(\nabla\mathbf{u}). \tag{6.75}$$

The adjoint operator of Λ_t is defined by

$$\Lambda_t^*(\mathbf{u})\mathbf{T}^{(1)} = \begin{cases} -\nabla\cdot[\mathbf{T}^{(1)}(\mathbf{I} + \nabla\mathbf{u})^T] & \text{in } \Omega \\ \mathbf{N}\cdot[\mathbf{T}^{(1)}(\mathbf{I} + \nabla\mathbf{u})^T] & \text{on } \Gamma. \end{cases} \tag{6.76}$$

Using Λ_c, we may express the complementary gap functional associated with Λ as

$$G(\mathbf{u}, \mathbf{T}^{(1)}) = \langle -\Lambda_c(\mathbf{u}); \mathbf{T}^{(1)}\rangle = \int_\Omega \frac{1}{2}\mathrm{tr}[(\nabla\mathbf{u})\mathbf{T}^{(1)}(\nabla\mathbf{u})^T]d\Omega. \tag{6.77}$$

II. When $\eta = \frac{1}{2}$, the polar decomposition $\mathbf{A} = \mathbf{R}\cdot\mathbf{U}$ enable us to write

$$\mathbf{E}^{(\frac{1}{2})} = \mathbf{U} - \mathbf{I} = \{U_{\alpha\beta} - \delta_{\alpha\beta}\} \in \mathcal{M}_{sym}, \tag{6.78}$$

which is the so-called *Biot strain tensor*, or *right extensional strain tensor* in continuum mechanics. The associated conjugate stress

$$\mathbf{T}^{(\frac{1}{2})}(\mathbf{X}) = D\bar{U}(\mathbf{E}^{(\frac{1}{2})}) = \frac{\partial\bar{U}(\mathbf{U} - \mathbf{I})}{\partial\mathbf{U}} = \mathbf{U}^* \in \mathcal{T}^{(\frac{1}{2})}$$

is just the Biot-Jaumann stress, and we have

$$\mathbf{T}^{(\frac{1}{2})} = \frac{1}{2}[\mathbf{U}\mathbf{T}^{(1)} + \mathbf{T}^{(1)}\mathbf{U}] = \frac{1}{2}[\mathbf{S}\mathbf{R} + \mathbf{R}^T\mathbf{S}^T]. \tag{6.79}$$

III. If $\eta = 0$, the Hill-Seth strain measure

$$\mathbf{E}^{(0)} = \lim_{\eta \to 0+} \frac{1}{2\eta}[(\mathbf{U}^T\mathbf{U})^\eta - \mathbf{I}) = \ln \mathbf{U}, \tag{6.80}$$

which is called the *Hencky strain* or the *logarithmic strain tensor*. The conjugate stress $\mathbf{T}^{(0)}$ was discussed by Hoger (1987). In the case $\mathbf{T}^{(\eta)}\mathbf{E}^{(\eta)} = \mathbf{E}^{(\eta)}\mathbf{T}^{(\eta)}$ (which is true if and only if the material is *isotropic*), we have (see Ogden, 1984)

$$\mathbf{T}^{(\eta)} = \mathbf{T}^{(\frac{1}{2})}\mathbf{U}^{1-\eta}. \tag{6.81}$$

This special case leads to

$$\mathbf{T}^{(0)} = \mathbf{T}^{(\frac{1}{2})}\mathbf{U}, \quad \mathbf{T}^{(\frac{1}{2})} = \mathbf{S}\mathbf{R}. \tag{6.82}$$

IV. When $\eta = -1$,

$$\mathbf{E}^{(-1)} = \frac{1}{2}[\mathbf{I} - (\mathbf{A}^T \cdot \mathbf{A})^{-1}] = \frac{1}{2}(\mathbf{I} - \mathbf{U}^{-2}) \in \mathcal{M}_{sym} \tag{6.83}$$

is called the *Almansi strain tensor.* The associated conjugate stress

$$\mathbf{T}^{(-1)} = D\bar{U}(\mathbf{E}^{(-1)}) \in \mathbb{R}^{n \times n}$$

is called the *weighted convected stress tensor*, and

$$\mathbf{T}^{(-1)} = J\mathbf{A}^T\sigma\mathbf{A} = \mathbf{U}^2\mathbf{T}^{(1)}\mathbf{U}^2.$$

From the canonical decomposition (6.66), the complementary gap functional associated with the Hill-Seth strain family is obtained in the form

$$G(\chi, \mathbf{T}^{(\eta)}) = \langle -\mathbf{E}_c^{(\eta)}(\nabla\chi); \mathbf{T}^{(\eta)} \rangle, \tag{6.84}$$

where $\mathbf{E}_c^{(\eta)}(\nabla\chi) = \mathbf{E}^{(\eta)}(\nabla\chi) - \mathbf{E}_t^{(\eta)}(\nabla\chi)(\nabla\chi)$. For $\eta = 1$ and $\frac{1}{2}$, the complementary gap functional can be written in terms of the deformation $\chi \in \mathcal{X}_a$ as follows.

$$\eta = 1 \;\Rightarrow\; G(\chi, \mathbf{T}^{(1)}) = \int_\Omega \frac{1}{2}\mathrm{tr}[(\nabla\chi)\mathbf{T}^{(1)}(\nabla\chi)^T]\mathrm{d}\Omega. \tag{6.85}$$

$$\eta = \frac{1}{2} \;\Rightarrow\; G(\chi, \mathbf{T}^{(\frac{1}{2})}) = \int_\Omega \mathrm{tr}\left(\mathbf{T}^{(\frac{1}{2})}\right)\mathrm{d}\Omega. \tag{6.86}$$

In terms of the displacement $\mathbf{u} \in \mathcal{U}_a$, the gap functional for the quadratic operator $\mathbf{E}^{(1)}$ is the quadratic functional defined by (6.77). However, for $\eta = \frac{1}{2}$, the gap functional is

$$G(\mathbf{u}, \mathbf{T}^{(\frac{1}{2})}) = \int_\Omega [\mathrm{tr}(\mathbf{T}^{(\frac{1}{2})}) - \mathrm{tr}\left(\mathbf{S}(\mathbf{T}^{(\frac{1}{2})}, \mathbf{R}(\mathbf{u}))\right)]\mathrm{d}\Omega, \tag{6.87}$$

where the nominal stress $\mathbf{S}$ depends on the Biot stress and the rotation tensor through equation (6.79). For other $\eta \in \mathbb{R}$, the complementary gap functional may have more complicated forms.

Clearly, $\bar{U}$ is convex in $\mathbf{U}$ if and only if it is convex in $\mathbf{E}^{(\frac{1}{2})} = \mathbf{U} - \mathbf{I}$. Thus, the Biot strain measure and its conjugate stress $\mathbf{T}^{(\frac{1}{2})} = \mathbf{U}^*$ play an important role in complementary-duality variational problems. On the Hill canonical conjugate space $\mathcal{E}^{(\eta)} \times \mathcal{T}^{(\eta)}$ we have, for convex $\bar{U}(\mathbf{U})$,

$$\bar{U}^*(\mathbf{T}^{(\frac{1}{2})}) = \bar{U}^*(\mathbf{T}^{(\eta)}) + \mathbf{T}^{(\frac{1}{2})} : (\mathbf{U} - \mathbf{I}) - \mathbf{T}^{(\eta)} : \mathbf{E}^{(\eta)}. \tag{6.88}$$

This equation can be used to find the connections between the complementary energies $\bar{U}^*(\mathbf{T}^{(\eta)})$ on Hill's conjugate spaces. For example, since

$$\mathbf{T}^{(\frac{1}{2})} : \mathbf{E}^{(\frac{1}{2})} = \mathbf{T}^{(\frac{1}{2})} : (\mathbf{U} - \mathbf{I}) = 2\mathbf{T}^{(1)} : \mathbf{E}^{(1)} + \mathrm{tr}\mathbf{T}^{(1)} - \mathbf{T}^{(1)} : \mathbf{U},$$

it is easy to prove that

$$\bar{U}^*(\mathbf{T}^{(1)}) = \bar{U}^*(\mathbf{T}^{(\frac{1}{2})}) - \frac{1}{2}\mathbf{T}^{(1)} : (\mathbf{U} - \mathbf{I})^2 = \bar{U}^*(\mathbf{T}^{(-1)}). \tag{6.89}$$

6.3.3 Generalized Variational Principles in Continuum Mechanics

We are now able to present generalized complementary energy variational principles in continuum mechanics with $m = n$. For the sake of convenience, we assume that for a given admissible displacement $\mathbf{u} \in \mathcal{U}_a$, there exists a strain measure $\mathbf{E} = \Lambda(\mathbf{u}) \in \mathcal{E}_a$ such that the stored-energy density $\bar{U} : \mathcal{E}_a \to \mathbb{R}$ is a piecewise differentiable canonical function. Let $\mathcal{T}_a$ be the range of $D\bar{U}(\mathbf{E})$ $\forall \mathbf{E} \in \mathcal{E}_a$. Then, the Legendre duality relation

$$\mathbf{T} = D\bar{U}(\mathbf{E}) \;\Leftrightarrow\; \mathbf{E} = D\bar{U}^*(\mathbf{E}) \;\Leftrightarrow\; \mathbf{E} : \mathbf{T} = \bar{U}(\mathbf{E}) + \bar{U}^*(\mathbf{T}) \tag{6.90}$$

holds on $\mathcal{E}_a \times \mathcal{T}_a$. Thus, on $\mathcal{U}_a \times \mathcal{E}_a$, the total potential

$$\Phi(\mathbf{u}, \mathbf{E}) = \bar{W}(\mathbf{E}) - F(\mathbf{u}) = \int_\Omega \bar{U}(\mathbf{E}) d\Omega - \int_\Omega \mathbf{u}^T \bar{\mathbf{b}} d\Omega - \int_{\Gamma_t} \mathbf{u}^T \bar{\mathbf{t}} d\Gamma \tag{6.91}$$

is well defined. For homogeneous boundary conditions, the kinematically admissible space $\mathcal{U}_k$ can be written as

$$\mathcal{U}_k = \{\mathbf{u} \in \mathcal{U} |\; \Lambda(\mathbf{u}) \in \mathcal{E}_a, \quad \mathbf{u} = 0 \text{ on } \Gamma_u \;\}. \tag{6.92}$$

The total canonical potential energy $\Pi : \mathcal{U}_k \to \mathbb{R}$, defined by

$$\Pi(\mathbf{u}) = \Phi(\mathbf{u}, \Lambda(\mathbf{u})) = \int_\Omega \bar{U}(\Lambda(\mathbf{u})) d\Omega - \int_\Omega \mathbf{u}^T \bar{\mathbf{b}} d\Omega - \int_{\Gamma_t} \mathbf{u}^T \bar{\mathbf{t}} d\Gamma, \tag{6.93}$$

is Gâteaux differentiable. The primal problem associated with the canonical boundary-value problem $(\mathcal{P}_{\mathrm{cbv}})_u$ is then the nonconvex stationary problem

$$(\Pi_{\mathrm{sta}})_u : \quad \Pi(\mathbf{u}) \to \mathrm{sta} \;\; \forall \mathbf{u} \in \mathcal{U}_k. \tag{6.94}$$

If $\bar{W}(\mathbf{E})$ is replaced by $\langle \mathbf{E}; \mathbf{T}\rangle - \bar{W}^*(\mathbf{T})$, the canonical Lagrangian $L : \mathcal{U}_a \times \mathcal{T}_a \to \mathbb{R}$ associated with this problem takes the standard form

$$L(\mathbf{u}, \mathbf{T}) = \langle \Lambda(\mathbf{u}); \mathbf{T}\rangle - \bar{W}^*(\mathbf{T}) - F(\mathbf{u}). \tag{6.95}$$

The criticality condition $\delta L(\bar{\mathbf{u}}, \bar{\mathbf{T}}; \mathbf{u}, \mathbf{T}) = 0 \quad \forall(\mathbf{u}, \mathbf{T}) \in \mathcal{U}_a \times \mathcal{T}_a$ leads to the canonical Lagrangian equations

$$\begin{aligned} \Lambda(\bar{\mathbf{u}}) &= D\bar{W}^*(\bar{\mathbf{T}}), \\ \bar{\Lambda}_t^*(\bar{\mathbf{u}})\bar{\mathbf{T}} &= DF(\bar{\mathbf{u}}), \end{aligned}$$

which are equivalent to the governing equations in $(\mathcal{P}_{\rm bv})_u$. Thus, the critical points $(\bar{\mathbf{u}}, \bar{\mathbf{T}})$ of L are solutions of $(\mathcal{P}_{\rm bv})_u$.

Let $\mathcal{Z}_e \subset \mathcal{U}_a \times \mathcal{T}_a$ be the *equilibrium space*

$$\mathcal{Z}_e = \{(\mathbf{u}, \mathbf{T}) \in \mathcal{U}_a \times \mathcal{T}_a |\ \bar{\Lambda}_t^*(\mathbf{u})\mathbf{T} = DF(\mathbf{u})\}. \tag{6.96}$$

On $\mathcal{Z}_e$, the canonical Lagrangian L reduces to the generalized complementary energy $\Pi^c : \mathcal{Z}_e \to \mathbb{R}$, given by

$$\Pi^c(\mathbf{u}, \mathbf{T}) = \bar{W}^*(\mathbf{T}) + G(\mathbf{u}, \mathbf{T}).$$

Theorem 6.3.1 (Unified Complementary Variational Principle)
Suppose that for the given strain measure $\mathbf{E} = \Lambda(\mathbf{u})$, $U(\mathbf{A}) = \bar{U}(\mathbf{E})$ *and the Legendre duality relations* (6.90) *hold. Then, among all admissible* $(\mathbf{u}, \mathbf{T}) \in \mathcal{Z}_e$, *the critical point* $(\bar{\mathbf{u}}, \bar{\mathbf{T}})$ *of* Π^c *solves the boundary-value problem* $(\mathcal{P}_{\rm bv})_u$.

This theorem can easily be proved by using the canonical Lagrangian L. The generalized complementary energy Π^c was proposed by Gao and Strang (1989). Using the complementary gap functional, the duality gap, which exists in the Fenchel-Rockafellar duality theory can be recovered. For certain choices of Λ, this theorem leads to some well-known complementary energy principles in continuum mechanics.

I. $\Lambda(\mathbf{u}) = \mathbf{E}^{(1)}$ is the Green strain tensor. In this case, the conjugate stress is the second Piola-Kirchhoff stress $\mathbf{T}^{(1)}$, the gap functional (6.77) is a quadratic functional of $\nabla\mathbf{u}$, and the generalized complementary energy Π^c can be written as

$$\Pi^c_{hr}(\mathbf{u}, \mathbf{T}^{(1)}) = \int_\Omega \bar{U}^*(\mathbf{T}) d\Omega + \int_\Omega \frac{1}{2} \mathrm{tr}[(\nabla\mathbf{u})\mathbf{T}^{(1)}(\nabla\mathbf{u})^T] d\Omega. \tag{6.97}$$

This is the well-known Hellinger-Reissner total complementary energy.

Corollary 6.3.1 (Hellinger-Reissner Complementary Energy Principle)

Among all admissible $(\mathbf{u}, \mathbf{T}) \in \mathcal{Z}_e$, *the critical point* $(\bar{\mathbf{u}}, \bar{\mathbf{T}})$ *of* Π^c_{hr} *solves the boundary-value problem* $(\mathcal{P}_{\rm bv})_u$.

Since the Green strain tensor $\mathbf{E}^{(1)} = \Lambda(\mathbf{u})$ is a quadratic operator, the stored-energy density $\bar{U}(\mathbf{E}^{(1)})$ is a strictly convex function for many materials. Thus, the Hellinger-Reissner principle has many applications in the large deformation of structures. The Lagrangian L associated with this quadratic operator is called the *generalized Hellinger-Reissner complementary energy.* In this case, if we replace $\bar{W}^*(\mathbf{T}^{(1)})$ by the Legendre transformation form $\langle \mathbf{E}^{(1)}; \mathbf{T}^{(1)} \rangle - \bar{W}(\mathbf{E}^{(1)})$, then the functional $L_{hw} : \mathcal{U}_a \times \mathcal{T}_a^{(1)} \times \mathcal{E}^{(1)} \to \mathbb{R}$, defined by

$$L_{hw}(\mathbf{u}, \mathbf{T}^{(1)}, \mathbf{E}^{(1)}) = \langle \Lambda(\mathbf{u}) - \mathbf{E}^{(1)}; \mathbf{T}^{(1)} \rangle + \bar{W}(\mathbf{E}^{(1)}) - F(\mathbf{u}), \tag{6.98}$$

is the well-known *Hu-Washizu potential energy,* proposed by Hu (1955) and Washizu (1955), independently.

II. For $\Lambda(\mathbf{u}) = \mathbf{E}^{(\frac{1}{2})}$, the conjugate stress is $\mathbf{T}^{(\frac{1}{2})}$. With the gap functional given by (6.87), the generalized complementary energy Π^c becomes

$$\Pi^c(\mathbf{u}, \mathbf{T}^{(\frac{1}{2})}) = \int_\Omega \bar{U}^*(\mathbf{T}^{(\frac{1}{2})}) \mathrm{d}\Omega + \int_\Omega \mathrm{tr}[\mathbf{T}^{(\frac{1}{2})} - \mathbf{S}(\mathbf{T}^{(\frac{1}{2})}, \mathbf{R}(\mathbf{u}))] \mathrm{d}\Omega. \tag{6.99}$$

By use of the relation $\mathbf{T}^{(\frac{1}{2})} = \mathbf{T}^{(\frac{1}{2})}(\mathbf{S}, \mathbf{R})$ (6.79), and with the nominal stress tensor $\mathbf{S} \in \mathcal{A}_a^*$ and the rotation tensor $\mathbf{R} \in \mathcal{M}_{ort}$ as the independent variables, the total complementary energy Π^c reduces to the well-known *de Veubeke complementary energy*

$$\Pi^c_{dv}(\mathbf{S}, \mathbf{R}) = \int_\Omega \bar{U}^*(\mathbf{T}^{(\frac{1}{2})}(\mathbf{S}, \mathbf{R})) \mathrm{d}\Omega + \int_\Omega \mathrm{tr}[\mathbf{T}^{(\frac{1}{2})}(\mathbf{S}, \mathbf{R}) - \mathbf{S})] \mathrm{d}\Omega. \tag{6.100}$$

Corollary 6.3.2 (de Veubeke Complementary Energy Principle)
Among all the statically admissible fields $\mathbf{S} \in \mathcal{S}_s$ *and independent rotation fields* $\mathbf{R} \in \mathcal{M}^+_{ort}$, *any stationary point* $(\bar{\mathbf{S}}, \bar{\mathbf{R}})$ *of* Π^c_{dv} *solves the boundary-value problem* $(\mathcal{P}_{\mathrm{bv}})_u$.

III. In the case that $\Lambda(\mathbf{u}) = \nabla \mathbf{u} + \mathbf{I} = \mathbf{A}$ is an affine operator and the constitutive relation $\mathbf{S} = DU(\mathbf{A})$ is pointwise invertible, $\Lambda_t = \mathrm{grad}$, and $\Lambda_c = -\mathbf{I}$. The gap functional then depends only on the nominal stress $\mathbf{S}$ and is given by

$$G(\mathbf{S}) = \langle -\Lambda_c; \mathbf{S} \rangle = \int_\Omega \mathrm{tr}(\mathbf{S}) \mathrm{d}\Omega. \tag{6.101}$$

For the nonhomogeneous boundary-value problem $(\mathcal{P}_{\mathrm{bv}})_u$, the generalized complementary energy $\Pi^c : \mathcal{S}_s \to \mathbb{R}$ is

$$\Pi^c_L(\mathbf{S}) = \int_\Omega [U^c(\mathbf{S}) + \mathrm{tr}(\mathbf{S})] \mathrm{d}\Omega - \int_{\Gamma_u} \mathbf{N} \cdot \mathbf{S} \cdot \bar{\mathbf{u}} \mathrm{d}\Gamma. \tag{6.102}$$

This is the well-known *Levinson complementary energy.* If $U(\mathbf{A})$ is strictly convex on $\mathcal{A}_a$, then $\Pi^c_L(\mathbf{S}) = -P^d(\mathbf{S})$. By Lemma 6.2.1, any critical point of Π^c_L on $\mathcal{S}_s$ solves the boundary-value problem $(\mathcal{P}_{\mathrm{bv}})_u$ and minimizes Π^c_L on $\mathcal{S}_s$. However, in nonconvex finite deformation systems, the Legendre complementary energy function $U^c(\mathbf{S})$ does not usually have a simple algebraic expression. In this case, Levinson's complementary energy principle is impractical to use.

As we have seen thus far, introduction of the complementary gap functional enables the critical points of the generalized complementary energy Π^c to solve $(\mathcal{P}_{bv})$. However, unlike the complementary energy principles in linear deformation systems, the total complementary energy Π^c depends on both the canonical stress field $\mathbf{T}$ and either the displacement $\mathbf{u}$ or the rotation field $\mathbf{R}$. The questions of existence of a pure complementary energy and the extremality conditions of the well-known complementary energy principles in finite deformation theory were open problems for many years and generated much argument.

6.4 Tri-Duality Theory in Finite Deformation Problems

The main goal of this section is to solve the open problems mentioned above. We first present an extremality theorem for the nonlinear Lagrangian in finite deformation theory. Based on the canonical Lagrangian duality theory, the pure complementary energy can actually be constructed, and a general analytic solution form is obtained. For simplicity, we let $\mathcal{E} = \mathcal{L}^{\beta}(\Omega; \mathcal{M}_{sym}), \beta \in (1, +\infty)$, with $\mathcal{E}_a \subset \mathcal{E}$ a subspace, and let $\Lambda : \mathcal{U} \to \mathcal{E}_a$ be the quadratic tensor operator such that the canonical strain measure $\mathbf{E} : \mathcal{A} \to \mathcal{E}_a$ can be defined by

$$\mathbf{E}(\nabla\chi) = \Lambda(\chi) = \frac{1}{2}a(\mathbf{X})(\nabla\chi)^T(\nabla\chi) + \beta(\mathbf{X}), \tag{6.103}$$

where $a(\mathbf{X}) > 0 \ \forall \mathbf{X} \in \Omega$ is a strictly positive real-valued function in $\mathcal{C}^1(\Omega)$ and $\beta(\mathbf{X}) \in \mathcal{M}_{sym}$ is a symmetric tensor-valued function. The results presented in this section will be based on the following assumption.

Assumption 6.4.1 For a given hyperelastic material governed by the stored-energy function $U : \Omega \times \mathcal{A} \to \mathbb{R}$, assume that the functions $a \in \mathcal{C}^1(\Omega)$ and $\beta \in \mathcal{C}^1(\Omega; \mathcal{M}_{sym})$ can be chosen in such a way that there exists a non-negative function $\bar{U} : \mathcal{E}_a \to \mathcal{L}(\Omega)$ satisfying:

(A1) $U(\mathbf{X}, \mathbf{A}) = \bar{U}(\mathbf{E}(\mathbf{A})) \ \forall (\mathbf{X}, \mathbf{A}) \in \Omega \times \mathcal{A}$;

(A2) $\bar{U} : \mathcal{E}_a \to \mathcal{L}(\Omega)$ is piecewisely differentiable such that $D\bar{U} : \mathcal{E}_a \to \mathcal{T}_a \subset \mathcal{T} = \mathcal{E}^*$ is one-to-one and onto, i.e. the Legendre duality relation

$$\mathbf{T} = D\bar{U}(\mathbf{E}) \ \Leftrightarrow \ \mathbf{E} = D\bar{U}^*(\mathbf{T}) \ \Leftrightarrow \ \bar{U}(\mathbf{E}) + \bar{U}^*(\mathbf{T}) = \mathbf{E} : \mathbf{T}. \tag{6.104}$$

holds on $\mathcal{E}_a \times \mathcal{T}_a \subset \mathcal{M}_{sym} \times \mathcal{M}_{sym}$. ♣

6.4.1 Triality Theorem

For the quadratic operator Λ defined in (6.103), its directional derivative at $\bar{\chi}$ in the direction χ is

$$\delta\Lambda(\bar{\chi}; \chi) = \frac{1}{2}a(\mathbf{X})[(\nabla\bar{\chi})^T(\nabla\chi) + (\nabla\chi)^T(\nabla\bar{\chi})] = \mathbf{E}_t(\nabla\bar{\chi})\nabla\chi,$$

where $\mathbf{E}_t(\bar{\mathbf{A}}) = \left(a(\mathbf{X})\bar{\mathbf{A}}\right)_{Sym}$ is a symmetric fourth-order tensor operator such that

$$\mathbf{E}_t(\bar{\mathbf{A}})\mathbf{A} = \frac{1}{2}a(\mathbf{X})[\bar{\mathbf{A}}^T\mathbf{A} + \mathbf{A}^T\bar{\mathbf{A}}].$$

Thus, the Gâteaux derivative of Λ at $\bar{\chi}$ depends linearly on $\bar{\chi}$: $\Lambda_t(\bar{\chi}) = \mathbf{E}_t(\nabla\bar{\chi})\nabla$. By the canonical decomposition $\Lambda = \Lambda_t + \Lambda_c$, the complementary operator Λ_c is also a symmetric quadratic operator, namely

$$\Lambda_c(\chi) = \Lambda(\chi) - \Lambda_t(\chi)\chi = \Lambda_n(\chi) + \beta(\mathbf{X}), \tag{6.105}$$

where Λ_n is the pure quadratic operator

$$\Lambda_n(\chi) = -\frac{1}{2}a(\mathbf{X})(\nabla\chi)^T(\nabla\chi).$$

For $\chi = \bar{\chi} + \delta\chi$, Taylor's formula for the quadratic operator Λ reads (see Gao and Strang (1989))

$$\begin{aligned}\Lambda(\bar{\chi} + \delta\chi) &= \Lambda(\bar{\chi}) + \delta\Lambda(\bar{\chi};\delta\chi) + \delta^2\Lambda(\bar{\chi};\delta\chi,\delta\chi) \\ &= \Lambda(\bar{\chi}) + \Lambda_t(\bar{\chi})\delta\chi - \Lambda_n(\delta\bar{\chi}).\end{aligned} \tag{6.106}$$

The complementary gap functional associated with Λ is a quadratic function of χ, i.e.

$$G(\chi,\mathbf{T}) = \langle -\Lambda_c(\chi);\mathbf{T}\rangle = \int_\Omega \frac{1}{2}\mathrm{tr}[a(\mathbf{X})(\nabla\chi)\mathbf{T}(\nabla\chi)^T - \mathbf{T}\beta]\mathrm{d}\Omega, \tag{6.107}$$

and we let

$$G_n(\chi,\mathbf{T}) = \langle -\Lambda_n(\chi);\mathbf{T}\rangle = \int_\Omega \frac{1}{2}a(\mathbf{X})\mathrm{tr}[(\nabla\chi)\mathbf{T}(\nabla\chi)^T]\mathrm{d}\Omega, \tag{6.108}$$

which can be called the *normalized gap functional.*

The canonical Lagrangian $L : \mathcal{Z}_a = \mathcal{X}_a \times \mathcal{T}_a \to \mathbb{R}$ associated with the canonical strain (6.103) is defined by

$$\begin{aligned}L(\chi,\mathbf{T}) &= \langle\Lambda(\chi);\mathbf{T}\rangle - \bar{W}^*(\mathbf{T}) - F(\chi) \\ &= \int_\Omega [\Lambda(\chi) : \mathbf{T} - \bar{U}^*(\mathbf{T}) - \bar{\mathbf{b}}\cdot\chi]\mathrm{d}\Omega - \int_{\Gamma_t} \bar{\mathbf{t}}\cdot\chi\mathrm{d}\Gamma.\end{aligned} \tag{6.109}$$

Theorem 6.4.1 (Quadruple-Critical Point Theorem)

Suppose that the assumption 6.4.1 holds. Then, any critical point $(\bar{\chi},\bar{\mathbf{T}})$ *of* L *solves the boundary-value problem* $(\mathcal{P}_{\mathrm{cbv}})_\chi$.

Moreover, if $\bar{W} : \mathcal{E}_a \to \mathbb{R}$ *is convex and Gâteaux differentiable, then*
(T1) $(\bar{\chi},\bar{\mathbf{T}})$ *is a right saddle point of* L *if and only if* $G_n(\chi,\bar{\mathbf{T}}) \geq 0 \;\; \forall\chi \in \mathcal{X}_a$;
(T2) $(\bar{\chi},\bar{\mathbf{T}})$ *is a super-critical point of* L *if and only if* $G_n(\chi,\bar{\mathbf{T}}) < 0 \;\; \forall\chi \in \mathcal{X}_a$.

If $\bar{W} : \mathcal{E}_a \to \mathbb{R}$ *is concave and Gâteaux differentiable, then*
(T1)* $(\bar{\chi},\bar{\mathbf{T}})$ *is a left saddle point of* L *if and only if* $G_n(\chi,\bar{\mathbf{T}}) \leq 0 \;\; \forall\chi \in \mathcal{X}_a$;
(T2)* $(\bar{\chi},\bar{\mathbf{T}})$ *is a sub-critical point of* L *if and only if* $G_n(\chi,\bar{\mathbf{T}}) > 0 \;\; \forall\chi \in \mathcal{X}_a$.

Proof The criticality condition $\delta L(\bar{\chi},\bar{\mathbf{T}};\chi,\mathbf{T}) = 0 \;\; \forall(\chi,\mathbf{T}) \in \mathcal{Z}_a$ leads to the canonical Lagrange equations

$$\Lambda(\bar{\chi}) = D\bar{W}^*(\bar{\mathbf{T}}) \quad \text{in } \Omega, \tag{6.110}$$

$$\Lambda_t^*(\bar{\chi})\bar{\mathbf{T}} = \begin{cases} -\nabla\cdot[a(\mathbf{X})(\nabla\bar{\chi})\bar{\mathbf{T}}]^T = \bar{\mathbf{b}} & \text{in } \Omega, \\ \mathbf{N}\cdot[a(\mathbf{X})(\nabla\bar{\chi})\bar{\mathbf{T}}]^T = \bar{\mathbf{t}} & \text{on } \Gamma_f. \end{cases} \tag{6.111}$$

By the Legendre duality relation (6.104), equation (6.110) is equivalent to $\bar{\mathbf{T}} = D\bar{W}(\Lambda(\bar{\chi}))$. Thus, the critical point $(\bar{\chi}, \bar{\mathbf{T}})$ of L satisfies all the governing equations in $(\mathcal{P}_{\text{cbv}})_{\chi}$, and is therefore a solution.

Since $F : \mathcal{X}_a \to \mathbb{R}$ is a linear functional, $F(\bar{\chi} + \delta\chi) = F(\bar{\chi}) + \langle DF(\bar{\chi}), \delta\chi \rangle$. Let $\chi = \bar{\chi} + \delta\chi$. Then using (6.106), we obtain

$$\begin{aligned} L(\chi, \bar{\mathbf{T}}) - L(\bar{\chi}, \bar{\mathbf{T}}) &= \langle \Lambda_t^*(\bar{\chi})\bar{\mathbf{T}} - DF(\bar{\chi}), \delta\chi \rangle + \langle -\Lambda_n(\delta\chi); \bar{\mathbf{T}} \rangle \\ &= \langle D_{\chi} L(\bar{\chi}, \bar{\mathbf{T}}), \delta\chi \rangle + \langle -\Lambda_n(\delta\chi); \bar{\mathbf{T}} \rangle. \end{aligned}$$

Noting that $(\bar{\chi}, \bar{\mathbf{T}})$ is a critical point of L, so that $D_{\chi} L(\bar{\chi}, \bar{\mathbf{T}}) = 0$, we then have

$$L(\chi, \bar{\mathbf{T}}) - L(\bar{\chi}, \bar{\mathbf{T}}) = \langle -\Lambda_n(\chi - \bar{\chi}); \bar{\mathbf{T}} \rangle \quad \forall \chi \in \mathcal{X}_a.$$

This leads to

$$\begin{aligned} L(\chi, \bar{\mathbf{T}}) \geq L(\bar{\chi}, \bar{\mathbf{T}}) &\quad \text{iff } G_n(\chi, \bar{\mathbf{T}}) \geq 0 \ \ \forall \chi \in \mathcal{U}, \\ L(\chi, \bar{\mathbf{T}}) \leq L(\bar{\chi}, \bar{\mathbf{T}}) &\quad \text{iff } G_n(\chi, \bar{\mathbf{T}}) \leq 0 \ \ \forall \chi \in \mathcal{U}. \end{aligned} \tag{6.112}$$

If $\bar{W} : \mathcal{E}_a \to \mathbb{R}$ is convex and Gâteaux differentiable, then $L : \mathcal{T}_a \to \mathbb{R}$ is concave for any given $\chi \in \mathcal{X}_a$. If $(\bar{\chi}, \bar{\mathbf{T}})$ is a critical point of L, we have

$$L(\bar{\chi}, \mathbf{T}) \leq L(\bar{\chi}, \bar{\mathbf{T}}) \quad \forall \mathbf{T} \in \mathcal{T}. \tag{6.113}$$

The statements (T1) and (T2) are proved by combining this with (6.112).

However, if $\bar{W} : \mathcal{E}_a \to \mathbb{R}$ is concave and Gâteaux differentiable, then $L : \mathcal{T}_a \to \mathbb{R}$ is convex for any given $\chi \in \mathcal{X}_a$. If $(\bar{\chi}, \bar{\mathbf{T}})$ is a critical point of L, we have

$$L(\bar{\chi}, \mathbf{T}) \geq L(\bar{\chi}, \bar{\mathbf{T}}) \quad \forall \mathbf{T} \in \mathcal{T}_a. \tag{6.114}$$

The statements (T1*) and (T2*) can then be proved by combining this with (6.112). □

Let $\mathcal{Z}_c \subset \mathcal{X}_a \times \mathcal{T}_a$ be the critical point set of L defined by

$$\mathcal{Z}_c = \{(\bar{\chi}, \bar{\mathbf{T}}) \in \mathcal{X}_a \times \mathcal{T}_a | \ \delta L(\bar{\chi}, \bar{\mathbf{T}}; \chi, \mathbf{T}) = 0 \ \ \forall \chi \in \mathcal{X}_a, \ \ \mathbf{T} \in \mathcal{T}_a\}. \tag{6.115}$$

Theorem 6.4.2 (Triality Theorem in Finite Deformation Systems)
Let $(\bar{\chi}, \bar{\mathbf{T}}) \in \mathcal{Z}_c$ be a critical point of L, and $\mathcal{Z}_r = \mathcal{X}_r \times \mathcal{T}_r$ be a neighborhood such that on $\mathcal{Z}_r$, $(\bar{\chi}, \bar{\mathbf{T}})$ is the only critical point of L.

(T1) For convex $\bar{W} : \mathcal{E}_a \to \mathbb{R}$, if $G_n(\chi, \bar{\mathbf{T}}) \geq 0 \ \forall \chi \in \mathcal{X}_r$, then

$$\min_{\chi \in \mathcal{X}_r} \max_{\mathbf{T} \in \mathcal{T}_r} L(\chi, \mathbf{T}) = L(\bar{\chi}, \bar{\mathbf{T}}) = \max_{\mathbf{T} \in \mathcal{T}_r} \min_{\chi \in \mathcal{X}_r} L(\chi, \mathbf{T}). \tag{6.116}$$

However, if $G_n(\chi, \bar{\mathbf{T}}) < 0 \ \forall \chi \in \mathcal{X}_r$, then either

$$\min_{\chi \in \mathcal{X}_r} \max_{\mathbf{T} \in \mathcal{T}_r} L(\chi, \mathbf{T}) = L(\bar{\chi}, \bar{\mathbf{T}}) = \min_{\mathbf{T} \in \mathcal{T}_r} \max_{\chi \in \mathcal{X}_r} L(\chi, \mathbf{T}). \tag{6.117}$$

or

$$\max_{\chi \in \mathcal{X}_r} \max_{\mathbf{T} \in \mathcal{T}_r} L(\chi, \mathbf{T}) = L(\bar{\chi}, \bar{\mathbf{T}}) = \max_{\mathbf{T} \in \mathcal{T}_r} \max_{\chi \in \mathcal{X}_r} L(\chi, \mathbf{T}). \tag{6.118}$$

(T2) For concave $\bar{W} : \mathcal{E}_a \to \mathbb{R}$, if $G_n(\chi, \bar{\mathbf{T}}) \leq 0 \;\; \forall \chi \in \mathcal{X}_r$, then

$$\max_{\chi \in \mathcal{X}_r} \min_{\mathbf{T} \in \mathcal{T}_r} L(\chi, \mathbf{T}) = L(\bar{\chi}, \bar{\mathbf{T}}) = \min_{\mathbf{T} \in \mathcal{T}_r} \max_{\chi \in \mathcal{X}_r} L(\chi, \mathbf{T}). \tag{6.119}$$

However, if $G_n(\chi, \bar{\mathbf{T}}) > 0 \;\; \forall \chi \in \mathcal{X}_r$, then either

$$\max_{\chi \in \mathcal{X}_r} \min_{\mathbf{T} \in \mathcal{T}_r} L(\chi, \mathbf{T}) = L(\bar{\chi}, \bar{\mathbf{T}}) = \max_{\mathbf{T} \in \mathcal{T}_r} \min_{\chi \in \mathcal{X}_r} L(\chi, \mathbf{T}) \tag{6.120}$$

or

$$\min_{\chi \in \mathcal{X}_r} \min_{\mathbf{T} \in \mathcal{T}_r} L(\chi, \mathbf{T}) = L(\bar{\chi}, \bar{\mathbf{T}}) = \min_{\mathbf{T} \in \mathcal{T}_r} \min_{\chi \in \mathcal{X}_r} L(\chi, \mathbf{T}). \tag{6.121}$$

Proof We only prove the statement (T1). Since $G_n(\chi, \mathbf{T})$ is a pure quadratic function of χ, then

$$G_n(\chi, \bar{\mathbf{T}}) \geq 0 \;\; \forall \chi \in \mathcal{X}_r \;\; \Leftrightarrow \;\; \bar{\mathbf{T}}(\mathbf{X}) \text{ is positive semi-definite in } \Omega,$$
$$G_n(\chi, \bar{\mathbf{T}}) \leq 0 \;\; \forall \chi \in \mathcal{X}_r \;\; \Leftrightarrow \;\; \bar{\mathbf{T}}(\mathbf{X}) \text{ is negative semi-definite in } \Omega.$$

Thus, if $\bar{W} : \mathcal{E}_a \to \mathbb{R}$ is convex, Theorem 6.4.1 shows that the Lagrangian L is a right saddle functional at $(\bar{\chi}, \bar{\mathbf{T}})$ if and only if the gap functional $G_n(\chi, \bar{\mathbf{T}}) \geq 0$ for all $\chi \in \mathcal{X}_r$. By the saddle Lagrangian theorem, the statement (6.116) holds on the (open) neighborhood $\mathcal{Z}_r$.

If $\bar{\mathbf{T}}(\mathbf{X})$ is non-positive definite in Ω, the critical point $(\bar{\chi}, \bar{\mathbf{T}})$ is a super-critical point of L on $\mathcal{Z}_r$. The statements (6.117) and (6.118) follow from the super-critical Lagrangian theorem proved in Chapter 3. □

Remark 6.4.1 This theorem solves the open problem in finite deformation theory left by Hellinger (1914) and Reissner (1953). Note that if $a(\mathbf{X}) = 1$ and $\beta = -\frac{1}{2}\mathbf{I}$, then $\mathbf{E} = \Lambda(\chi)$ is the Green strain tensor $\mathbf{E}^{(1)}$, and $L(\chi, \mathbf{T})$ is the generalized Hellinger-Reissner energy. This theorem shows that L may possess two types of critical points $(\bar{\chi}, \bar{\mathbf{T}})$. For convex $\bar{W}$, if the normalized gap functional $G_n(\bar{\chi}, \bar{\mathbf{T}}) \geq 0$, $(\bar{\chi}, \bar{\mathbf{T}})$ is a right saddle point. However, if $G_n(\bar{\chi}, \bar{\mathbf{T}}) < 0$, $(\bar{\chi}, \bar{\mathbf{T}})$ is a super-critical point. In this case, the total potential energy Π is nonconvex and the critical point $\bar{\chi}$ could be either a local minimizer or a local maximizer of Π. ■

6.4.2 Tri-Duality Theorems in Finite Deformation Theory

Similar to the Lagrangian duality developed in one-dimensional nonconvex systems, the canonical primal and dual problems can be defined through the canonical Lagrangian $L(\chi, \mathbf{T})$.

First, for a fixed $\chi \in \mathcal{X}_k = \{\chi \in \mathcal{X}_a | \;\; \Lambda(\chi) \in \mathcal{E}_a\}$, the total potential associated with $(\mathcal{P}_{\text{bv}})_\chi$ is obtained as

$$\Pi(\chi) = \text{sta}\{L(\chi, \mathbf{T}) | \;\; \forall \mathbf{T} \in \mathcal{T}_a\} = \bar{W}(\Lambda(\chi)) - F(\chi). \tag{6.122}$$

The primal canonical variational problem is to find $\bar{\chi} \in \mathcal{X}_k$ such that

$$(\Pi_{\text{sup}}): \quad \Pi(\bar{\chi}) = \text{sta}\Pi(\chi) \;\; \forall \chi \in \mathcal{X}_k. \tag{6.123}$$

Let $\mathcal{X}_c = \{\bar{\chi} \in \mathcal{X}_k | \ \delta\Pi(\bar{\chi}; \chi) = 0 \ \ \forall \chi \in \mathcal{X}_k\}$ be a critical point set of Π. The following theorem is due to Gao and Strang (1989).

Theorem 6.4.3 (Minimum Potential Energy Theorem)
Let us assume, in addition to 6.4.1, that $\bar{W} : \mathcal{E}_a \to \mathbb{R}$ is convex, $\mathcal{X}_k$ is an open set, $\bar{\chi} \in \mathcal{X}_k$ is a critical point of Π and that $\bar{\mathbf{T}} = D\bar{W}(\Lambda(\bar{\chi}))$. If $G_n(\chi, \bar{\mathbf{T}}) \geq 0 \ \forall \chi \in \mathcal{X}_k$, then $\bar{\chi}$ is a global minimizer of Π. If $\bar{\mathbf{T}}(\bar{\chi})$ is positive-definite for all $\bar{\chi} \in \mathcal{X}_c$, then all critical points $\bar{\chi} \in \mathcal{X}_c$ are global minimizers of Π, and $\mathcal{X}_c$ is a convex subset of $\mathcal{X}_k$. If $\bar{\mathbf{T}}(\bar{\chi})$ is strictly positive-definite for all $\bar{\chi} \in \mathcal{X}_c$, then $\mathcal{X}_c$ contains at most one critical point of Π.

Proof By the convexity of $\bar{W}(\mathbf{E})$, for any given $\chi \in \mathcal{X}_k$ such that $\mathbf{E} = \Lambda(\chi)$, we have

$$\bar{W}(\Lambda(\chi)) - \bar{W}(\Lambda(\bar{\chi})) \geq \langle D\bar{W}(\Lambda(\bar{\chi})); \Lambda(\chi) - \Lambda(\bar{\chi})\rangle \ \ \forall \Lambda(\chi) \in \mathcal{E}_a.$$

Substituting equation (6.106) for the quadratic operator Λ, we have

$$\bar{W}(\Lambda(\chi)) - \bar{W}(\Lambda(\bar{\chi})) \geq \langle \Lambda_t(\bar{\chi})\delta\chi; \bar{\mathbf{T}}\rangle + \langle -\Lambda_n(\delta\bar{\chi}); \bar{\mathbf{T}}\rangle \ \ \forall \chi \in \mathcal{X}_k,$$

where $\delta\chi = \chi - \bar{\chi}$. For a linear functional $F(\chi)$,

$$F(\chi) - F(\bar{\chi}) = \langle DF(\bar{\chi}), \chi - \bar{\chi}\rangle \ \ \forall \chi \in \mathcal{X}_a.$$

We then obtain

$$\begin{aligned} \Pi(\chi) - \Pi(\bar{\chi}) &\geq \langle \bar{\mathbf{T}}(\bar{\chi}); \Lambda_t(\bar{\chi})\delta\chi\rangle + G_n(\delta\chi, \bar{\mathbf{T}}(\bar{\chi})) - \langle DF(\bar{\chi}), \delta\chi\rangle \\ &= \langle \Lambda_t^*(\bar{\chi})\bar{\mathbf{T}}(\bar{\chi}) - DF(\bar{\chi}), \delta\chi\rangle + G_n(\delta\chi, \bar{\mathbf{T}}(\bar{\chi})) \\ &= G_n(\chi - \bar{\chi}, \bar{\mathbf{T}}) \ \ \forall \chi \in \mathcal{X}_k \end{aligned}$$

since $\bar{\chi} \in \mathcal{X}_c$ is a critical point of Π. If $G_n(\chi, \bar{\mathbf{T}}) \geq 0 \ \ \forall \chi \in \mathcal{X}_k$, then $\Pi(\chi) - \Pi(\bar{\chi}) \geq 0 \ \forall \chi \in \mathcal{X}_k$, and $\bar{\chi}$ is therefore a global minimizer of Π.

If $\bar{\mathbf{T}}(\bar{\chi})$ is positive-definite in Ω for all $\bar{\chi} \in \mathcal{X}_c$, similar to Lemma 6.3.1 we can prove that $\Pi : \mathcal{X}_k \to \mathbb{R}$ is convex. Thus, by Theorem 5.1.6 we know that the set $\mathcal{X}_c$ of critical points of Π is a closed convex subset of $\mathcal{X}$.

If $\bar{\mathbf{T}}(\mathbf{X})$ is strictly positive-definite in Ω, then

$$\Pi(\chi) - \Pi(\bar{\chi}) \geq G_n(\chi - \bar{\chi}, \bar{\mathbf{T}}) > 0 \ \ \forall \chi \neq \bar{\chi} \in \mathcal{X}_s,$$

in which case Π has at most one critical point. □

Dually, for a fixed $\mathbf{T} \in \mathcal{T}_a$, the total complementary energy dual to Π can be defined by

$$\Pi^d(\mathbf{T}) = \text{sta}\{L(\chi, \mathbf{T}) | \ \forall \chi \in \mathcal{X}_a\} = F_\Lambda^d(\mathbf{T}) - \bar{W}^*(\mathbf{T}), \tag{6.124}$$

where $F_\Lambda^d : \mathcal{T}_a \to \mathbb{R}$ is the dual functional of F associated with Λ and defined by

$$F_\Lambda^d(\mathbf{T}) = \text{sta}\{\langle \Lambda(\chi); \mathbf{T}\rangle - F(\chi) | \ \forall \chi \in \mathcal{X}_a\}. \tag{6.125}$$

For the quadratic operator $\Lambda(\chi)$, the relation between Π^d and L is well defined by the following proposition.

Proposition 6.4.1 *Suppose that $\Lambda : \mathcal{X} \to \mathcal{E}$ is defined by (6.103) and Assumption 6.4.1 holds. Then, for any given $\mathbf{T} \in \mathcal{T}_a$,*

$$\Pi^d(\mathbf{T}) = \begin{cases} \inf_{\chi \in \mathcal{X}_a} L(\chi, \mathbf{T}) & \text{if } \mathbf{T}(\mathbf{X}) \text{ is positive-definite in } \Omega, \\ \sup_{\chi \in \mathcal{X}_a} L(\chi, \mathbf{T}) & \text{if } \mathbf{T}(\mathbf{X}) \text{ is negative-definite in } \Omega. \end{cases} \tag{6.126}$$

Proof Since $\Lambda : \mathcal{X} \to \mathcal{E}$ is quadratic, then, if $\mathbf{T}(\mathbf{X})$ is positive-definite in Ω, $L : \mathcal{X}_a \to \mathbb{R}$ is convex. Thus, the solution $\bar{\chi}$ of the stationary problem (6.125) minimizes L. Similarly for negative-definite $\mathbf{T}$. □

Let $\mathcal{T}_s$ be a subset of $\mathcal{T}_a$ such that F_Λ^d is finite and well defined. The canonical dual variational problem can be proposed as

$$(\Pi^d_{\text{sta}}): \quad \Pi^d(\mathbf{T}) \to \text{sta } \forall \mathbf{T} \in \mathcal{T}_s. \tag{6.127}$$

The general critical point lemma for one-dimensional nonconvex systems (Lemma 3.5.1) also holds in finite deformation theory.

Lemma 6.4.1 *If $(\bar{\chi}, \bar{\mathbf{T}})$ is any one of the right or left saddle points, the super- or sub-critical points of L, and L is partially Gâteaux differentiable at $(\bar{\chi}, \bar{\mathbf{T}})$, then $(\bar{\chi}, \bar{\mathbf{T}})$ is a critical point of L. If Π and Π^d are Gâteaux differentiable at $\bar{\chi}$ and $\bar{\mathbf{T}}$, respectively, then $D\Pi(\bar{\chi}) = 0$, $D\Pi^d(\bar{\mathbf{T}}) = 0$ and*

$$\Pi(\bar{\chi}) = L(\bar{\chi}, \bar{\mathbf{T}}) = \Pi^d(\bar{\mathbf{T}}). \tag{6.128}$$

The proof of this lemma is an analogous to the proof of Lemma 3.5.1. Similarly to Theorem 3.5.3, the tri-duality theorem in n-dimensional finite deformation systems depends on the positivity of the canonical stress.

Theorem 6.4.4 (Tri-Duality Theorem in Finite Deformation Theory)

Suppose that $(\bar{\chi}, \bar{\mathbf{T}}) \in \mathcal{Z}_c$ is a critical point of L, and $\mathcal{Z}_r = \mathcal{X}_r \times \mathcal{T}_r$ is a neighborhood of $(\bar{\chi}, \bar{\mathbf{T}})$.

(T1) If $\bar{W} : \mathcal{E}_a \to \mathbb{R}$ is convex and $\bar{\mathbf{T}}(\chi)$ is positive-definite in Ω, then

$$\Pi(\bar{\chi}) = \min_{\chi \in \mathcal{X}_r} \Pi(\chi) = \max_{\mathbf{T} \in \mathcal{T}_r} \Pi^d(\mathbf{T}) = \Pi^d(\bar{\mathbf{T}}). \tag{6.129}$$

(T2) If $\bar{W} : \mathcal{E}_a \to \mathbb{R}$ is concave and $\bar{\mathbf{T}}(\chi)$ is negative-definite in Ω, then

$$\Pi(\bar{\chi}) = \max_{\chi \in \mathcal{X}_r} \Pi(\chi) = \min_{\mathbf{T} \in \mathcal{T}_r} \Pi^d(\mathbf{T}) = \Pi^d(\bar{\mathbf{T}}). \tag{6.130}$$

(T3) If $\bar{W} : \mathcal{E}_a \to \mathbb{R}$ is convex and $\bar{\mathbf{T}}(\chi)$ is negative-definite in Ω, or $\bar{W} : \mathcal{E}_a \to \mathbb{R}$ is concave and $\bar{\mathbf{T}}(\chi)$ is positive-definite in Ω, then either

$$\Pi(\bar{\chi}) = \max_{\chi \in \mathcal{X}_r} \Pi(\chi) = \max_{\mathbf{T} \in \mathcal{T}_r} \Pi^d(\mathbf{T}) = \Pi^d(\bar{\mathbf{T}}) \tag{6.131}$$

or

$$\Pi(\bar{\chi}) = \min_{\chi \in \mathcal{X}_r} \Pi(\chi) = \min_{\mathbf{T} \in \mathcal{T}_r} \Pi^d(\mathbf{T}) = \Pi^d(\bar{\mathbf{T}}). \tag{6.132}$$

Moreover, if $G_n(\bar{\chi}, \bar{\mathbf{T}}) \geq 0 \ \forall (\bar{\chi}, \bar{\mathbf{T}}) \in \mathcal{Z}_c$, then $\mathcal{Z}_c$ is a saddle point set, i.e. for any given $(\bar{\chi}, \bar{\mathbf{T}}) \in \mathcal{Z}_c$, $\bar{\chi}$ is a global minimizer of P and $\bar{\mathbf{T}}$ is a global maximizer of Π^d. If G_n is strictly positive on $\mathcal{Z}_c$, then $\mathcal{Z}_c$ has only one saddle point.

Proof The proof of this theorem follows by combining Theorem 6.4.2, Proposition 6.4.1 and Lemma 6.4.1. □

6.4.3 Pure Complementary Energies and Analytical Solutions

The consideration now is to formulate the pure complementary energy by solving the stationary problem (6.125). Recall that for any given source $\bar{\chi}^* \in \mathcal{X}_a^*$, if $\mathbf{S} \in \mathcal{S}_s$, then the nominal stress field is said to be statically admissible. This means that

$$\text{grad}^*\mathbf{S} = \bar{\chi}^* \Leftrightarrow \begin{cases} -\nabla \cdot \mathbf{S} = \bar{\mathbf{b}} & \text{in } \Omega, \\ \mathbf{N} \cdot \mathbf{S} = \bar{\mathbf{t}} & \text{on } \Gamma_f. \end{cases}$$

For any given $\mathbf{S} \in \mathcal{S}_s$, by replacing $\bar{\chi}^*$ by $\bar{\chi}^* = \text{grad}^*\mathbf{S}$, the canonical Lagrangian L can be written in the form

$$L(\chi, \mathbf{T}) = \langle \Lambda(\chi); \mathbf{T} \rangle - \int_\Omega \bar{U}^*(\mathbf{T}) d\Omega - \langle \chi, \text{grad}^*\mathbf{S} \rangle. \tag{6.133}$$

For a fixed $\mathbf{S} \in \mathcal{S}_s$, the criticality condition $\delta L(\bar{\chi}, \mathbf{T}; \chi) = 0 \ \forall \chi \in \mathcal{X}_a$ leads to

$$a(\mathbf{X}) \mathbf{T} (\nabla \bar{\chi})^T = \mathbf{S} \ \text{ in } \Omega.$$

Let $\mathcal{T}_s$ be a subset of $\mathcal{T}_a$ defined by

$$\mathcal{T}_s = \{ \mathbf{T} \in \mathcal{T}_a | \ \det \mathbf{T}(\mathbf{X}) \neq 0 \ \forall \mathbf{X} \in \Omega \}. \tag{6.134}$$

Then, on $\mathcal{T}_s \times \mathcal{S}_s$, the deformation gradient $\bar{\mathbf{A}} = \nabla \bar{\chi}$ defined by

$$\nabla \bar{\chi}(\mathbf{X}) = \frac{1}{a(\mathbf{X})} \mathbf{S}^T(\mathbf{X}) \mathbf{T}^{-1}(\mathbf{X}) \tag{6.135}$$

solves the stationary problem $\text{sta}\{L(\chi, \mathbf{T}) | \ \forall \chi \in \mathcal{X}_a\}$. When this is substituted into L, a construction of the dual canonical energy $P^d : \mathcal{T}_s \to \mathbb{R}$ is suggested as

$$\Pi^d(\mathbf{T}) = \int_\Omega \text{tr}(\beta \mathbf{T} - \frac{1}{2a} \mathbf{S}^T \mathbf{T}^{-1} \mathbf{S}) d\Omega - \int_\Omega \bar{U}^*(\mathbf{T}) d\Omega + \int_{\Gamma_\chi} \bar{\chi} \cdot \mathbf{S} \cdot \mathbf{N} d\Gamma, \tag{6.136}$$

in which the statically admissible nominal stress field $\mathbf{S} \in \mathcal{S}_s$ can be considered as a tensor-valued parameter. Since $\delta \mathbf{T}^{-1} = -\mathbf{T}^{-1}(\delta \mathbf{T}) \mathbf{T}^{-1}$, the Gâteaux variation of Π^d at $\mathbf{T}$ in the direction of $\delta \mathbf{T}$ is given by

$$\delta \Pi^d(\mathbf{T}; \delta \mathbf{T}) = \int_\Omega [\beta - D\bar{U}^*(\mathbf{T})] : \delta \mathbf{T} d\Omega + \int_\Omega \frac{1}{2a} \delta \mathbf{T} : (\mathbf{T}^{-1}(\mathbf{S}\mathbf{S}^T)\mathbf{T}^{-1}) d\Omega.$$

For any given $\mathbf{S} \in \mathcal{S}_s$, the solution of the stationary variational problem

$$(\Pi^d_{\text{sta}})_S : \quad \delta\Pi^d(\bar{\mathbf{T}}; \delta\mathbf{T}) = 0 \;\; \forall \delta\mathbf{T} \in \mathcal{T}_s \tag{6.137}$$

leads to the *dual canonical Euler-Lagrange equation*

$$2a\bar{\mathbf{T}}[D\bar{U}^*(\bar{\mathbf{T}}) - \beta]\bar{\mathbf{T}} = \mathbf{S}\mathbf{S}^T. \tag{6.138}$$

This is a symmetrical tensor equation. By the Legendre duality relation (6.104), for any fixed $\mathbf{S} \in \mathcal{S}_s$, the solution $\bar{\mathbf{T}} = \bar{\mathbf{T}}(\mathbf{S})$ of this tensor equation and the associated $\bar{\mathbf{A}}(\mathbf{S}) = \nabla\bar{\chi}(\mathbf{S})$ defined by (6.135) satisfies (1) the canonical constitutive law $\bar{\mathbf{T}} = D\bar{U}(\Lambda(\bar{\chi})$, (2) the equilibrium equation $\Lambda^*_t(\bar{\chi})\bar{\mathbf{T}} = \text{grad}^*\mathbf{S} = \bar{\chi}^*$ and (3) the equilibrium condition of angular momentum $\bar{\mathbf{A}}\mathbf{S} = (\bar{\mathbf{A}}(\mathbf{S})\mathbf{S})^T$. In addition, if $\bar{\chi}$ satisfies the boundary condition on $\mathcal{X}_a$, then $(\bar{\chi}, \bar{\mathbf{T}})$ is a solution of $(\mathcal{P}_{\text{cbv}})_\chi$.

Remark 6.4.2 In the dual canonical variational problem $(\Pi^d_{\text{sta}})_S$, there is also an inequality constraint $\det \mathbf{T}(\mathbf{X}) \neq 0 \;\; \forall \mathbf{X} \in \Omega$. But this constraint is much weaker than the inequality constraint $\det \mathbf{A}(\chi) > 0$ in the potential variational problems $(\mathcal{P}_{\text{sta}})$ and (Π_{sta}). The Lagrange multiplier for this inequality constraint is also identical to zero. Detailed discussions on this inequality constraint and general constructions of the pure complementary energy principle will be given in the paper by Ogden and Gao (2000). ■

The dual canonical Euler-Lagrange equation (6.138) reveals the algebraic relation between the nominal stress $\mathbf{S}$ and the canonical conjugate stress $\mathbf{T}$. Since the subset $\mathcal{S}_s$ is a hyperplane in $\mathbb{R}^{n\times m}$, the statically admissible field $\mathbf{S} \in \mathcal{S}_s$ can be written as

$$\mathbf{S} = \mathbf{S}_p + \mathbf{S}^o, \quad \text{with} \quad \text{grad}^*\mathbf{S}_p = \bar{\chi}^*, \quad \mathbf{S}^o \in \text{Ker}\,(\text{grad}^*), \tag{6.139}$$

where $\mathbf{S}_p \in \mathcal{S}_s$ is a particular solution of the linear equation $\text{grad}^*\mathbf{S} = \bar{\chi}^*$, while $\mathbf{S}^o$ is a homogeneous solution, i.e. $\text{grad}^*\mathbf{S}^o = 0 \in \mathbb{R}^m$. Let $\mathcal{X}^o$ be a polar configuration space, and $\Lambda^o : \mathcal{X}^o \to \mathcal{S}$ be the polar geometrical mapping such that $\mathcal{X}^o_0$ is the homogeneous polar configuration space

$$\mathcal{X}^o_0 = \{\chi^o \in \mathcal{X}^o | \;\langle \text{grad}\chi; \Lambda^o\chi^o\rangle = 0 \;\; \forall \chi \in \mathcal{X}_a\}. \tag{6.140}$$

Lemma 6.4.2 *For any given $\chi^o \in \mathcal{X}^o_0$ such that $\mathbf{S}(\chi^o) = \mathbf{S}_p + \Lambda^o\chi^o \in \mathcal{S}_s$, $\bar{\mathbf{T}} = \bar{\mathbf{T}}(\chi^o)$ is a solution of (6.138) and $\bar{\chi} = \bar{\chi}(\chi^o)$ is defined by*

$$\nabla\bar{\chi} = \frac{1}{a}\mathbf{S}^T(\chi^o)\bar{\mathbf{T}}^{-1}(\chi^o), \tag{6.141}$$

then, if $\bar{\chi} \in \mathcal{X}_a$, the pair $(\bar{\chi}, \bar{\mathbf{T}})$ is a critical point of L and the duality condition

$$\Pi(\bar{\chi}(\chi^o)) = \Pi^d(\bar{\mathbf{T}}(\chi^o)) \;\; \forall \chi^o \in \mathcal{X}^o_0 \tag{6.142}$$

holds.

Proof The critical condition $\delta L(\bar{\chi}, \bar{\mathbf{T}}; \chi, \mathbf{T}) = 0 \quad \forall(\chi, \mathbf{T}) \in \mathcal{Z}_a$ leads to the canonical Lagrange equations

$$\begin{aligned} \Lambda(\bar{\chi}) &= D\bar{W}^*(\bar{\mathbf{T}}), \\ \bar{\Lambda}_t^*(\bar{\chi})\bar{\mathbf{T}} &= \bar{\chi}^* = \text{grad}^*\mathbf{S}. \end{aligned} \tag{6.143}$$

For any given $\chi^o \in \mathcal{X}_0^o$ such that $\mathbf{S} = \mathbf{S}_p + \Lambda^o\chi^o \in \mathcal{S}_s$, the equations in (6.143) are equivalent to (6.138) and (6.141), respectively. This shows that the solution $(\bar{\chi}(\chi^o), \bar{\mathbf{T}}(\chi^o))$ of (6.138) and (6.141) satisfies (6.143). If such a $\bar{\chi}(\chi^o)$ satisfies the boundary condition in $\mathcal{X}_a$, then it must be a critical point of L.

Moreover, by Assumption 6.4.1, if $(\bar{\chi}, \bar{\mathbf{T}})$ is a critical point of the Lagrangian L, the canonical Euler-Lagrange equation $\mathbf{E}(\bar{\mathbf{A}}(\bar{\chi})) = \Lambda(\bar{\chi}) = D\bar{W}^*(\bar{\mathbf{T}})$ is equivalent to

$$\begin{aligned} \bar{U}(\mathbf{E}(\bar{\mathbf{A}})) &= \bar{\mathbf{T}} : \mathbf{E}(\bar{\mathbf{A}}) - \bar{U}^*(\bar{\mathbf{T}}) = \bar{\mathbf{T}} : (\frac{1}{2}a\bar{\mathbf{A}}^T\bar{\mathbf{A}} + \beta) - \bar{U}^*(\bar{\mathbf{T}}) \\ &= [\mathbf{E}_t(\bar{\mathbf{A}})\bar{\mathbf{T}}] : \bar{\mathbf{A}} + \mathbf{E}_c(\bar{\mathbf{A}}) : \bar{\mathbf{T}} - \bar{U}^*(\bar{\mathbf{T}}). \end{aligned}$$

Integrating both sides and replacing $\bar{\mathbf{A}} = \nabla\bar{\chi}$ by $\mathbf{S}^T\bar{\mathbf{T}}^{-1}/a$, we have

$$\int_\Omega [\bar{U}(\Lambda(\bar{\chi})) - \mathbf{S} : (\nabla\bar{\chi})]d\Omega = -\int_\Omega [\bar{U}^*(\bar{\mathbf{T}}) + \text{tr}[\frac{1}{2a}\mathbf{S}^T\bar{\mathbf{T}}^{-1}\mathbf{S} - \beta\bar{\mathbf{T}}]d\Omega.$$

For $\mathbf{S} = \mathbf{S}_p + \Lambda^o\chi^o \in \mathcal{S}_s$, the Gauss-Green formula leads to (6.142) when the condition $\bar{\chi} \in \mathcal{X}_a$ is fulfilled. □

Remark 6.4.3 Since $\mathcal{X}_0^o$ is a hyperplane in $\mathcal{X}^o$, then in order to use the criticality condition (6.141) to solve for $\bar{\chi}$, the *deformation compatibility condition*

$$\Lambda^{o*}\text{grad}\bar{\chi}(\chi^o) = \Lambda^{o*}\left(\frac{1}{a(\mathbf{X})}(\mathbf{S}_p + \Lambda^o\chi^o)^T\bar{\mathbf{T}}^{-1}(\chi^o)\right) = 0 \in \mathcal{X}^{o*} \tag{6.144}$$

is required. Generally speaking, if $\mathcal{S} \subset \mathbb{R}^{n\times m}$, the dimension of the *kernel* of the linear operator $\text{grad}^* : \mathcal{S} \to \mathbb{R}^m$ is

$$\dim \text{Ker}\ (\text{grad}^*) = (m-1) \times n.$$

Thus, for a given polar operator Λ^o, the polar configuration variable χ^o can be determined by the compatibility condition (6.144). ■

For example, if $m = n = 2$, then $\dim \text{Ker}\ (\text{grad}^*) = 2$, i.e. the statically admissible nominal stress field $\mathbf{S} \in \mathcal{S}_s$ has only two degrees of freedom. If $\mathcal{X}^o \subset \mathbb{R}^2$ and $\Lambda^o = \text{curl} : \mathcal{X}^o \to \mathbb{R}^2$, then $\chi^o \in \mathcal{X}_0^o$ can be determined by the vector equation (6.144).

Based on Lemma 6.4.2, the analytical solution of the problem $(\mathcal{P}_{\text{cbv}})_\chi$ can be proposed.

Theorem 6.4.5 *Suppose that for a given boundary deformation $\bar{\chi}(\mathbf{X}_0)$ $\forall\mathbf{X}_0 \in \Gamma_\chi$ and an external force field $\bar{\chi}^* = \{\bar{\mathbf{b}}$ in Ω, $\bar{\mathbf{t}}$ on $\Gamma_f\}$, there exists a $\mathbf{S}_p \in \mathcal{S}_s$.*

(T1) If there exists a $\bar{\chi}^o \in \mathcal{X}_0^o$ such that $\bar{\mathbf{T}} = \bar{\mathbf{T}}(\bar{\chi}^o)$ is a solution of the dual canonical Euler-Lagrange equation (6.138) associated with $\mathbf{S}(\bar{\chi}^o) = \mathbf{S}_p + \Lambda^o \bar{\chi}^o \in \mathcal{S}_s$, and $\bar{\chi}^o$ satisfies the compatibility condition

$$\nabla \times [\frac{1}{a}\bar{\mathbf{T}}^{-1}(\mathbf{S}_p + \Lambda^o \bar{\chi}^o)] = \mathbf{0},$$

then the path-independent integral

$$\bar{\chi}(\mathbf{X}) = \int_{\mathbf{X}_0}^{\mathbf{X}} \frac{1}{a}[(\mathbf{S}_p + \Lambda^o \bar{\chi}^o)^T \bar{\mathbf{T}}^{-1}] d\mathbf{X} + \bar{\chi}(\mathbf{X}_0) \quad \forall \mathbf{X}_0 \in \Gamma_\chi \tag{6.145}$$

is a solution of $(\mathcal{P}_{\rm cbv})_\chi$.

(T2) Conversely, each solution of $(\mathcal{P}_{\rm cbv})_\chi$ can be written in the form (6.145) for a certain solution $\bar{\mathbf{T}}$ of (6.138).

(T3) If $\bar{\mathbf{T}}(\mathbf{X})$ is positive-definite in Ω, then $\bar{\chi}(\mathbf{X})$ is a global minimizer of $\Pi(\chi)$. If $\bar{\mathbf{T}}(\mathbf{X})$ is negative-definite in Ω, then $\bar{\chi}(\mathbf{X})$ is either a local minimizer or a local maximizer of $\Pi(\chi)$. Moreover, if all the solutions of the dual Euler-Lagrange equation (6.138) are positive-definite in Ω, then $\bar{\chi}(\mathbf{X})$ is a unique global minimizer of Π.

Proof For the statement (T1) we need to prove that the deformation defined by (6.145) satisfies all the governing equations in $(\mathcal{P}_{\rm cbv})_\chi$. Taking the gradient on both sides of (6.145), we have $\nabla\bar{\chi} = \mathbf{A}(\bar{\chi}) = \mathbf{S}^T\bar{\mathbf{T}}^{-1}/a$. Since the canonical stress $\bar{\mathbf{T}}$ obtained by the dual canonical Euler-Lagrange equation (6.138) is symmetric, we have

$$\mathbf{A}(\bar{\chi})\mathbf{S} = \mathbf{S}^T\mathbf{T}^{-1}\mathbf{S} = (\mathbf{A}(\bar{\chi})\mathbf{S})^T \quad \forall \mathbf{S} \in \mathcal{S}_s.$$

The deformation $\bar{\chi}$ defined by (6.145) therefore satisfies the equilibrium law of angular momentum. For any given $\chi^o \in \mathcal{X}_0^o$, $\mathbf{S} = \mathbf{S}_p + \Lambda^o\chi^o \in \mathcal{S}_s$ is a statically admissible field, and hence $a\bar{\mathbf{T}}(\nabla\bar{\chi})^T = \mathbf{S}$ satisfies the equilibrium conditions. Substituting this into the dual canonical Euler-Lagrange equation (6.138), we obtain

$$D\bar{W}^*(\bar{\mathbf{T}}) = \frac{1}{2}[a(\nabla\bar{\chi})^T(\nabla\bar{\chi}) + \beta].$$

By the convexity of $\bar{W}(\mathbf{E})$, this inverse geometric-constitutive equation is equivalent to $\bar{\mathbf{T}} = D\bar{W}(\mathbf{E}(\bar{\chi}))$. Thus, $\nabla\bar{\chi} = \mathbf{S}^T\bar{\mathbf{T}}^{-1}/a$ satisfies all the governing equations. If $\nabla \times (\mathbf{S}^T\bar{\mathbf{T}}^{-1}/a)^T = \mathbf{0}$, then by vector calculus the displacement field $\bar{\chi}$ can be obtained by integrating $\mathbf{S}^T\bar{\mathbf{T}}^{-1}/a$ along any path in Ω. Hence, $\bar{\chi}$ determined by (6.145) is a general solution of $(\mathcal{P}_{\rm cbv})_\chi$.

The statement (T2) can be proved by the Legendre duality theorem, while the triality theorem leads to the statement (T3). □

Remark 6.4.4 Since the dual canonical Euler-Lagrange equation (6.138) is a symmetrical tensor equation, we can always, in practice, find an orthogonal transformation $\mathbf{R} \in \mathcal{M}_{ort}$ such that $\mathbf{R}^T\mathbf{S}\mathbf{S}^T\mathbf{R}$ is a diagonal tensor. Thus, the tensor equation (6.138) can be reduced to n decoupled algebraic equations, and the deformation stretches $\lambda_i, (i = 1, 2, 3)$, (i.e. the square roots of the eigenvalues of the right Cauchy-Green strain tensor $\mathbf{C} = \mathbf{A}^T\mathbf{A}$) can easily be obtained from the eigenvalues of $\mathbf{T}^{-1}\mathbf{S}\mathbf{S}^T\mathbf{T}^{-1}/a^2$. ■

6.5 Minimal Hyper-Surface Problems

We shall apply the results of the duality developed so far to the problems of calculus of variations arising from differential geometry. For a parametric minimal hypersurface, our results show that the solution of the polar variational problem leads to the so-called *polar (or adjoint, conjugate) minimal hypersurface*. For non-parametric hypersurfaces, the primal problem may or may not possess a solution and coerciveness is available only in a non-reflexive space ($\mathcal{L}^1(\Omega)$ or a space built on $\mathcal{L}^1(\Omega)$). The dual problem, on the other hand, possesses a unique solution.

6.5.1 Parametric Minimal Hypersurface and Primal-Dual Problems

Let $\Omega \subset \mathbb{R}^n$ be an open, bounded domain with boundary $\partial\Omega$. A finite deformation $\chi : \Omega \to \mathbb{R}^m$ is said to be a *parametric hypersurface* if $m = n + 1$, i.e.

$$\chi(\mathbf{X}) = \{\chi^i(X_\alpha)\}, \quad i = 1, 2, \cdots, n+1, \quad \alpha = 1, 2, \cdots, n. \tag{6.146}$$

The Lagrangian coordinates $\mathbf{X} = \{X_\alpha\}$ are called the *parameters*. The surface χ is said to be *regular* (i.e. admissible) if it is a C^2-mapping and

$$\nabla\chi \in \mathcal{A} = \{\mathbf{A} \in \mathcal{M}(\Omega; \mathbb{R}^{(n+1)\times n})| \text{ rank } \mathbf{A}(\mathbf{X}) = n \ \forall \mathbf{X} \in \Omega\}.$$

Clearly, for each given $\chi \in \mathcal{X}$ and the associated deformation gradient $\mathbf{A} = \nabla\chi \in \mathcal{A}$, the right Cauchy-Green strain tensor

$$\mathbf{C} = \mathbf{A}^T\mathbf{A} = \{C_{\alpha\beta}\}, \quad C_{\alpha\beta} = \chi_{,\alpha} \cdot \chi_{,\beta} = \sum_{i=1}^{n+1} \frac{\partial\chi^i}{\partial X_\alpha}\frac{\partial\chi^i}{\partial X_\beta}$$

is strictly positive-definite and invertible in Ω. In differential geometry, the right Cauchy-Green strain tensor $\mathbf{C} \in \mathcal{M}_{sym}$ is also referred as the *Riemannian metric tensor*, and we write

$$I_3(\mathbf{C}(\mathbf{X})) = \det \mathbf{C}(\mathbf{X}) > 0, \quad \mathbf{C}^{-1}(\mathbf{X}) = \{C^{\alpha\beta}(X_\gamma)\}.$$

For a given space contour $\bar{\chi} : \partial\Omega \to \mathbb{R}^m$, we let $\mathcal{X}_a \subset \mathcal{X}$ be defined as

$$\mathcal{X}_a = \{\chi \in \mathcal{X}| \ \chi(\mathbf{X}) = \bar{\chi}(\mathbf{X}) \ \forall \mathbf{X} \in \partial\Omega\},$$

and let $\mathcal{A}_a \subset \mathcal{A}$ be the admissible deformation gradient space

$$\mathcal{A}_a = \{\mathbf{A} \in \mathcal{A}| \ \mathbf{A} \in \mathcal{L}^1(\Omega; \mathbb{R}^{(n+1)\times n})\}.$$

On the kinematically admissible space $\mathcal{X}_k = \{\chi \in \mathcal{X}_a | \ \nabla\chi \in \mathcal{A}_a\}$, we wish to study the primal minimization problem

$$(\mathcal{P}_{\inf})_\chi : \ P(\chi) = \int_\Omega \sqrt{\det[(\nabla\chi)^T(\nabla\chi)]} d\Omega \to \inf \ \ \forall \chi \in \mathcal{X}_k. \tag{6.147}$$

For any given $\chi \in \mathcal{X}_k$, the total stored-energy $W(\nabla\chi)$ in this problem is actually the total area of the hypersurface lying on the contour $\bar{\chi}(\mathbf{X})$ on $\partial\Omega$. The solution $\bar{\chi}$ of this primal problem leads to a *minimal hypersurface*, which is controlled by the Dirichlet problem (i.e. the Euler-Lagrange equation for $(\mathcal{P}_{\inf})_\chi$)

$$\begin{aligned} \Delta_{\mathcal{M}}\chi(\mathbf{X}) &= 0 \quad \forall \mathbf{X} \in \Omega, \\ \chi(\mathbf{X}) &= \bar{\chi}(\mathbf{X}) \quad \forall \mathbf{X} \in \partial\Omega, \end{aligned} \tag{6.148}$$

where $\Delta_{\mathcal{M}}$ is the *Laplace-Beltrami operator*, defined by

$$\Delta_{\mathcal{M}}\chi = \frac{1}{\sqrt{I_3}} \frac{\partial}{\partial X_\beta} \left(\sqrt{I_3} C^{\alpha\beta} \frac{\partial}{\partial X_\alpha} \chi \right).$$

Minimal surface problems and related topics in differential geometry, theoretic physics and engineering have been studied extensively for centuries. Since the kinematically admissible space $\mathcal{X}_k \subset \mathcal{L}^1$ is not reflexive, existence theorem discussed in Chapter 5 cannot be applied to these kinds of problems. However, using the duality theory and methods already developed, some well-known theorems and formulations can be presented in a very nice way, and some interesting results can be obtained.

To set the problem in our framework, we first let $\Lambda = \text{grad}$. On $\mathcal{A}_a$, the total "energy" $W(\mathbf{A}) : \mathcal{A}_a \to \mathbb{R}$, given by

$$W(\mathbf{A}) = \int_\Omega \sqrt{\det(\mathbf{A}^T\mathbf{A})} d\Omega = \int_\Omega \sqrt{I_3(\mathbf{C}(\mathbf{A}))} d\Omega,$$

is at least twice Gâteaux differentiable. Thus, the "nominal stress tensor" can be defined by

$$\mathbf{S} = DW(\mathbf{A}) = \sqrt{\det(\mathbf{A}^T\mathbf{A})}(\mathbf{A}^T\mathbf{A})^{-1}\mathbf{A}^T = \sqrt{I_3}\mathbf{C}^{-1}\mathbf{A}^T \in \mathbb{R}^{n\times(n+1)}. \tag{6.149}$$

The Dirichlet problem is then written in the standard tri-forms

$$\left.\begin{aligned} \mathbf{A} &= \Lambda\chi = \text{grad}\chi, \\ \mathbf{S} &= DW(\mathbf{A}), \\ 0 &= \Lambda^*\mathbf{S} = -\text{div}\,\mathbf{S} \end{aligned}\right\} \ \text{in } \Omega, \tag{6.150}$$

on $\mathcal{X}_a$. Generally speaking, the functional $W : \mathcal{A}_a \to \mathbb{R}$ may not be convex. However, its Legendre conjugate can be defined by the following lemma.

Lemma 6.5.1 *For any given* $\mathbf{A} \in \mathcal{A}_a$ *such that* $\mathbf{S} = \dot{D}W(\mathbf{A})$,

$$\det(\mathbf{S}\mathbf{S}^T) = \left(\det(\mathbf{A}^T\mathbf{A})\right)^{n-1} \quad \forall n = 1, 2, \cdots, \tag{6.151}$$

and the Legendre conjugate of $W : \mathcal{A}_a \to \mathbb{R}$ *is*

$$W^c(\mathbf{S}) = \int_\Omega (n-1)\left(\det(\mathbf{S}\mathbf{S}^T)\right)^{\frac{1}{2(n-1)}} d\Omega. \tag{6.152}$$

Proof Since $\mathbf{S} = DW(\mathbf{A}) = I_3^{\frac{1}{2}}\mathbf{C}^{-1}\mathbf{A}^T$, we have

$$\mathbf{S}\mathbf{S}^T = I_3\mathbf{C}^{-1}\mathbf{A}^T\mathbf{A}\mathbf{C}^{-1},$$

and since $I_3 = \det(\mathbf{A}^T\mathbf{A})$, $\mathbf{C} = \mathbf{A}^T\mathbf{A}$, and $\det(\mathbf{C}^m) = (\det\mathbf{C})^m$, we obtain

$$\det(\mathbf{S}\mathbf{S}^T) = I_3^n(\det\mathbf{C})^{-2}\det(\mathbf{A}^T\mathbf{A}) = \left(\det(\mathbf{A}^T\mathbf{A})\right)^{n-1}.$$

Instead of solving the constitutive equation $\mathbf{S} = \mathbf{S}(\mathbf{A})$ for $\mathbf{A}$, we substitute $\mathbf{S} = \mathbf{S}(\mathbf{A})$ into the Legendre transformation $W^c(\mathbf{S}) = \langle\mathbf{A};\mathbf{S}\rangle - W(\mathbf{A})$ to obtain

$$W^c(\mathbf{S}) = \int_\Omega \sqrt{\det(\mathbf{A}^T\mathbf{A})}(\mathrm{tr}(\mathbf{C}^{-1}\mathbf{A}^T\mathbf{A}) - 1)\mathrm{d}\Omega = \int_\Omega (n-1)\sqrt{\det(\mathbf{A}^T\mathbf{A})}\mathrm{d}\Omega.$$

Replacing $\det(\mathbf{A}^T\mathbf{A})$ by $(\det(\mathbf{S}\mathbf{S}^T))^{1/(n-1)}$, we arrive at (6.152), as required. □

Let $\mathcal{S} = \mathcal{A}^* \subset \mathbb{R}^{n\times(n+1)}$ be a tensor space defined on Ω and $\mathcal{S}_a \subset \mathcal{S}$ the range of the constitutive mapping $DW : \mathcal{A}_a \to \mathcal{S}$, such that

$$\mathcal{S}_a = \{\mathbf{S} \in \mathcal{S} |\ \mathrm{rank}\ \mathbf{S}(\mathbf{X}) = n\ \ \forall\mathbf{X} \in \Omega\}. \tag{6.153}$$

In terms of the linear operator $\Lambda = \mathrm{grad} : \mathcal{X} \to \mathcal{A}$, the Lagrangian $L : \mathcal{X}_a \times \mathcal{S}_a \to \mathbb{R}$ associated with the primal problem $(\mathcal{P}_{\inf})_\chi$ is

$$\begin{aligned} L(\chi,\mathbf{S}) &= \langle\Lambda\chi;\mathbf{S}\rangle - W^*(\mathbf{S}) \\ &= \int_\Omega \left[(\nabla\chi) : \mathbf{S} - (n-1)\left(\det(\mathbf{S}\mathbf{S}^T)\right)^{\frac{1}{2(n-1)}}\right]\mathrm{d}\Omega. \end{aligned} \tag{6.154}$$

The criticality condition for L gives the Lagrange equations

$$\begin{aligned} \Lambda\chi &= DW^c(\mathbf{S}) = \left(\det(\mathbf{S}\mathbf{S}^T)\right)^{\frac{1}{2(n-1)}}\mathbf{S}^T(\mathbf{S}\mathbf{S}^T)^{-1}, \\ \Lambda^*\mathbf{S} &= -\nabla\cdot\mathbf{S} = 0\ \text{ in } \Omega. \end{aligned}$$

It is not difficult to prove that these Lagrange equations are equivalent to those in (6.150). Thus, the critical point $(\bar{\chi},\bar{\mathbf{S}})$ of L solves the Dirichlet problem. The statically admissible space $\mathcal{S}_s$ in this problem takes the simple form

$$\mathcal{S}_s = \{\mathbf{S} \in \mathcal{S}_a |\ \mathrm{div}\mathbf{S} = 0 \text{ in } \Omega\}. \tag{6.155}$$

On $\mathcal{S}_s$, the dual functional $P^d : \mathcal{S}_s \to \mathbb{R}$ is

$$\begin{aligned} P^d(\mathbf{S}) &= F^c(\bar{\Lambda}^*\mathbf{S}) - W^c(\mathbf{S}) \\ &= \int_{\Gamma_\chi} \mathbf{N}\cdot\mathbf{S}\cdot\bar{\chi}\mathrm{d}\Gamma - \int_\Omega (n-1)\left(\det(\mathbf{S}\mathbf{S}^T)\right)^{\frac{1}{2(n-1)}}\mathrm{d}\Omega. \end{aligned} \tag{6.156}$$

By the general critical point lemma, if $\mathcal{Z}_c \subset \mathcal{X}_a \times \mathcal{S}_a$ is the critical point set of L, the equality

$$P(\bar{\chi}) = L(\bar{\chi}, \bar{\mathbf{S}}) = P^d(\bar{\mathbf{S}}) \ \forall (\bar{\chi}, \bar{\mathbf{S}}) \in \mathcal{Z}_c \tag{6.157}$$

holds.

Now we let $\Lambda : \mathcal{X} \to \mathcal{E} = \mathcal{M}_{sym} \cap \mathcal{M}_+$ be a pure quadratic operator such that $\mathbf{E} = \mathbf{C} = \Lambda(\chi) = (\nabla\chi)^T(\nabla\chi)$. The canonical strain space $\mathcal{E}_a \subset \mathcal{M}_{sym}$ is defined by

$$\mathcal{E}_a = \{\mathbf{C} \in \mathcal{M}_{sym} | \ \mathbf{C}(\mathbf{X}) \text{ is strictly positive-definite } \sqrt{I_3} \in \mathcal{L}^1 \ \forall \mathbf{X} \in \Omega\}. \tag{6.158}$$

On $\mathcal{E}_a$, the stored energy

$$\bar{W}(\mathbf{C}) = \int_\Omega \bar{U}(\mathbf{C}) d\Omega, \ \ \bar{U}(\mathbf{C}) = \sqrt{\det \mathbf{C}},$$

is Gâteaux differentiable. Thus, the conjugate stress of $\mathbf{C}$ is a second Piola-Kirchhoff type stress,

$$\mathbf{T} = D\bar{W}(\mathbf{C}) = \frac{\partial \bar{U}(\mathbf{C})}{\partial \mathbf{C}} = \frac{1}{2}(I_3(\mathbf{C}))^{\frac{1}{2}} \mathbf{C}^{-1} \in \mathcal{M}_{sym} \tag{6.159}$$

and we have $\mathbf{S} = 2\mathbf{T}\mathbf{A}^T$.

Lemma 6.5.2 *Let $\mathcal{E}_a \subset \mathbb{R}^{2\times 2}$ be an admissible space defined by (6.158) and let $\mathcal{T}_a \subset \mathcal{T} = \mathcal{E}^*$ be the range of the constitutive mapping $D\bar{W} : \mathcal{E}_a \to \mathcal{T}$.*

(T1) The functional $\bar{W} : \mathcal{E}_a \to \mathbb{R}$ is concave.

(T2) For any given $\mathbf{T} \in \mathcal{T}_a$, the Fenchel conjugate of $\bar{W}(\mathbf{C})$ is

$$\bar{W}^*(\mathbf{T}) = \begin{cases} 0 & \text{if } \det \mathbf{T} \geq \frac{1}{4}, \\ -\infty & \text{otherwise.} \end{cases} \tag{6.160}$$

Proof (T1) For any given $\mathbf{C} \in \mathcal{E}_a$, there exists a rotation tensor $\mathbf{R} \in \mathcal{M}^+_{ort}$ such that

$$\mathbf{C} = \mathbf{R}^T\mathbf{D}\mathbf{R}, \ \ \mathbf{R} = \begin{pmatrix} \cos\theta & \sin\theta \\ -\sin\theta & \cos\theta \end{pmatrix}, \ \ \mathbf{D} = \begin{pmatrix} c_1 & 0 \\ 0 & c_2 \end{pmatrix},$$

where $\theta \in \mathbb{R}$, and $c_1, c_2 > 0$ are eigenvalues of $\mathbf{C} \in \mathcal{E}_a$. Thus, we have $\bar{U}(\mathbf{C}) = \sqrt{c_1 c_2}$. It is easy to show that the Hessian matrix of $\hat{U}(c_1, c_2) = \sqrt{c_1 c_2}$ is negative semi-definite for any given $c_1, c_2 > 0$. Thus, $\hat{U}(c_1, c_2)$ is a concave function of $c_1, c_2 > 0$. Since $\bar{U}(\mathbf{C}(c_1, c_2, \theta)) = \hat{U}(c_1, c_2)$ does not dependent on the rotation $\theta \in \mathbb{R}$, the Hessian matrix of $\bar{U}(\mathbf{C}(c_1, c_2, \theta))$ is also negative semi-definite for any $c_1, c_2 > 0$ and $\theta \in \mathbb{R}$. This shows that $\bar{U}(\mathbf{C})$ is concave on $\mathcal{E}_a$, and so is $\bar{W}(\mathbf{C})$.

(T2) From the concavity of $\bar{W}(\mathbf{C})$ on $\mathcal{E}_a$, we may write the Fenchel transformation as

$$\bar{W}^*(\mathbf{T}) = \inf_{\mathbf{C} \in \mathcal{E}_a} \{\langle \mathbf{C}; \mathbf{T} \rangle - \bar{W}(\mathbf{C})\}.$$

For a given $\mathbf{T} \in \mathcal{T}_a$, the extremality condition leads to $\mathbf{C} = \rho(\mathbf{C})\mathbf{T}^{-1}$ with $\rho(\mathbf{C}) > 0$. Substituting this into the above equation, we obtain

$$\begin{aligned} \bar{W}^*(\mathbf{T}) &= \inf_{\rho>0} \rho(2 - (\det \mathbf{T})^{-\frac{1}{2}}) \\ &= \begin{cases} 0 & \text{if } 2 - (\det \mathbf{T})^{-\frac{1}{2}} \geq 0, \\ -\infty & \text{otherwise,} \end{cases} \end{aligned}$$

as required. □

It follows that the admissible space $\mathcal{T}_a$ reads

$$\mathcal{T}_a = \{\mathbf{T} \in \mathcal{M}_{sym}(\Omega; \mathbb{R}^{2\times 2}) |\ \det \mathbf{T}(\mathbf{X}) \geq \frac{1}{4} \quad \text{in } \Omega\}.$$

For the pure quadratic operator $\Lambda(\chi) = (\nabla\chi)^T(\nabla\chi)$, the complementary gap functional is

$$G(\chi, \mathbf{T}) = \int_\Omega \text{tr}[(\nabla\chi)\mathbf{T}(\nabla\chi)^T]\text{d}\Omega. \tag{6.161}$$

Thus, for any given $\mathbf{S} \in \mathcal{S}_s$, the pure complementary energy in this parametric surface problem is

$$\Pi^d(\mathbf{T}) = \int_{\partial\Omega} \mathbf{N} \cdot \mathbf{S} \cdot \bar{\chi}\text{d}\Gamma - \int_\Omega \frac{1}{4}\text{tr}(\mathbf{S}^T\mathbf{T}^{-1}\mathbf{S})\text{d}\Omega \ \ \forall \mathbf{T} \in \mathcal{T}_a. \tag{6.162}$$

Use of the Lagrange multiplier $\mu \geq 0$ introduced for relaxing the inequality constraint in $\mathcal{T}_a$, enables the dual canonical Euler-Lagrange equation to be written

$$4\mu(\det \mathbf{T})\mathbf{T} = \mathbf{S}\mathbf{S}^T \tag{6.163}$$

subject to the KKT condition

$$\mu \geq 0, \quad \frac{1}{4} - \det \mathbf{T} \leq 0, \quad \mu(\frac{1}{4} - \det \mathbf{T}) = 0 \quad \text{in } \Omega.$$

The analytic solution for this inhomogeneous Dirichlet boundary-value problem has the form

$$\chi(\mathbf{X}) = \int_{\mathbf{X}_0}^{\mathbf{X}} \frac{1}{2}\mathbf{S}^T\mathbf{T}^{-1} \cdot \text{d}\mathbf{X} + \bar{\chi}(\mathbf{X}_0), \quad \forall \mathbf{X}_0 \in \partial\Omega. \tag{6.164}$$

Its extremality property is governed by the Triality Theorem.

6.5.2 Polar-Complementary Variational Problem and Polar Surfaces

If the domain Ω is a simply-connected subset in $\mathbb{R}^2$ such that $\mathbf{X} = (u, v) \in \Omega$ is a vector of parameters, then the finite deformation $\chi = (\chi^1, \chi^2, \chi^3) : \Omega \to \mathbb{R}^3$ is a parametric surface in $\mathbb{R}^3$. In this case

$$\mathbf{A}(\chi) = (\chi_{,u}, \chi_{,v}) = \begin{pmatrix} \chi^1_{,u} & \chi^1_{,v} \\ \chi^2_{,u} & \chi^2_{,v} \\ \chi^3_{,u} & \chi^3_{,v} \end{pmatrix},$$

$$\mathbf{C} = \begin{pmatrix} \chi_{,u} \cdot \chi_{,u} & \chi_{,u} \cdot \chi_{,v} \\ \chi_{,v} \cdot \chi_{,u} & \chi_{,v} \cdot \chi_{,v} \end{pmatrix}.$$

Clearly, if χ satisfies the so-called *conformality relations*

$$|\chi_{,u}|^2 = |\chi_{,v}|^2, \quad \chi_{,u} \cdot \chi_{,v} = 0 \quad \text{in } \Omega, \tag{6.165}$$

then the Laplace-Beltrami operator $\Delta_{\mathcal{M}}$ reduces to the Laplace operator Δ and (6.148) degenerates to the *harmonic equation*

$$\Delta\chi(u,v) = 0 \quad \text{in } \Omega. \tag{6.166}$$

Definition 6.5.1 (Minimal Surface and Polar Surface)

Let $\Omega \subset \mathbb{R}^2$ be a simply-connected parameter domain.

(D1) A surface $\chi : \Omega \to \mathbb{R}^3$ of class C^2 is said to be a *minimal surface* if it satisfies the conformality relations in (6.165) in addition to the harmonic equation (6.166).

(D2) A surface $\chi^o : \Omega \to \mathbb{R}^3$ of class C^2 is said to be a *polar surface* of the minimal surface χ on Ω if it is a solution of the *Cauchy-Riemann equations*

$$\chi_{,u} = \chi^o_{,v}, \quad \chi_{,v} = -\chi^o_{,u} \quad \text{in } \Omega. \tag{6.167}$$

In differential geometry, the polar surface is also referred as the *adjoint or conjugate surface.* ◊

Clearly, all polar surfaces of some given minimal surface χ differ only by a constant vector, and equations (6.165), (6.166) and (6.167) imply that

$$\Delta\chi^o = 0, \quad |\chi^o_{,u}|^2 = |\chi^o_{,v}|^2, \quad \chi^o_{,u} \cdot \chi^o_{,v} = 0 \quad \text{in } \Omega, \tag{6.168}$$

i.e. the polar surface to some minimal surface is a minimal surface.

Let $w = u + iv \in \Omega \subset \mathbb{R}^2 \cong \mathbb{C}$. Then, the mapping $\chi : \mathbb{C} \to \mathbb{R}$, $\chi(w) = (\chi^1(w), \chi^2(w), \chi^3(w))$, represents a *curve* in $\mathbb{C}^3$. Consider an arbitrary harmonic mapping $\chi : \Omega \to \mathbb{C}^3$ of a simply connected domain $\Omega \in \mathbb{C}$, and let χ^o be the polar harmonic mapping to χ, defined as a solution of (6.167). Then,

$$\Upsilon := \chi + i\chi^o : \Omega \to \mathbb{C}$$

is called a *holomorphic curve* in $\mathbb{C}^3$ with components: $\{\Upsilon^j(w)\}$, $j = 1,2,3$, $w \in \mathbb{C}$. Its complex derivative $\Upsilon' = \mathrm{d}\Upsilon/\mathrm{d}w$ is given by

$$\Upsilon'(w) = \chi_u(u,v) + i\chi^o_u(u,v) = \chi_u(u,v) - i\chi_v(u,v),$$

whence it follows that

$$\Upsilon' \cdot \Upsilon' = |\chi_u|^2 - |\chi_v|^2 - 2i\chi_u \cdot \chi_v.$$

Thus, the conformality relations (6.166) are satisfied if and only if the *isotropy relation*

$$\Upsilon' \cdot \Upsilon' = 0 \tag{6.169}$$

is fulfilled. A curve $\Upsilon : \Omega \to \mathbb{C}$ is said to be *isotropic* if the isotropy relation (6.169) is satisfied. Let $\mathcal{X}_n \subset \mathcal{X}_k$ be the admissible configuration space

$$\mathcal{X}_n = \{\chi \in \mathcal{X}_k |\ |\chi_{,u}|^2 = |\chi_{,v}|^2,\ \chi_{,u} \cdot \chi_{,v} = 0 \ \text{ in } \Omega\}. \tag{6.170}$$

Then, on $\mathcal{X}_n$, the primal functional takes the quadratic form

$$P(\chi) = \int_\Omega |\nabla \chi|^2 d\Omega. \tag{6.171}$$

Any critical point $\bar{\chi} \in \mathcal{X}_n$ of P is a minimal surface.

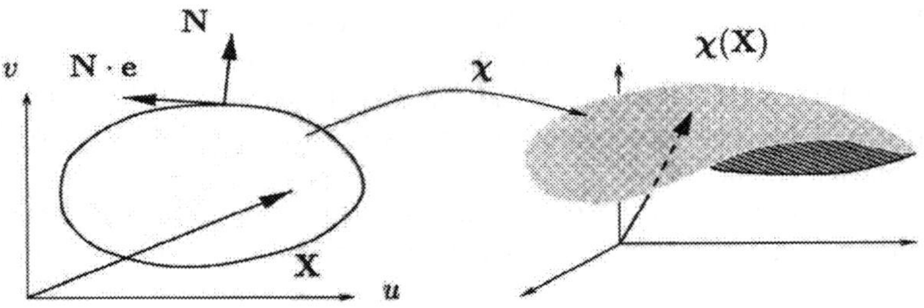

Figure 6.2. Finite deformation from $\mathbb{R}^2$ to $\mathbb{R}^3$.

Let the polar operator $\Lambda^o : \mathbb{R}^3 \to \mathbb{R}^{2\times 3}$ be

$$\Lambda^o = \nabla \mathbf{e}^T = \begin{pmatrix} 0 & -1 \\ 1 & 0 \end{pmatrix} \begin{pmatrix} \frac{\partial}{\partial u} \\ \frac{\partial}{\partial v} \end{pmatrix} = \begin{pmatrix} -\frac{\partial}{\partial v} \\ \frac{\partial}{\partial u} \end{pmatrix} = \{e^{\alpha\beta}\partial_\beta\}, \tag{6.172}$$

where $\mathbf{e}$ is the permutation tensor in $\mathbb{R}^{2\times 2}$,

$$\mathbf{e} = \begin{pmatrix} 0 & 1 \\ -1 & 0 \end{pmatrix} = \{e_{\alpha\beta}\}.$$

Clearly, $\mathbf{e} \in \mathcal{M}^+_{ort}$ is an antisymmetric rotation tensor, i.e.

$$\mathbf{e}^T = \mathbf{e}^{-1} = -\mathbf{e}, \quad \{e^{\alpha\beta}\} = \{e_{\beta\alpha}\} = -\{e_{\alpha\beta}\}, \quad \det \mathbf{e} = 1.$$

Let the polar configuration space be $\mathcal{X}^o = C^1(\Omega; \mathbb{R}^3)$. From the definition

$$\langle \mathbf{A}; \Lambda^o \chi^o \rangle = \langle \Lambda^{o*} \mathbf{A}, \chi^o \rangle \ \forall \mathbf{A} \in \mathcal{A},\ \chi^o \in \mathcal{X}^o,$$

the adjoint operator $\Lambda^{o*} : \mathcal{A} \to \mathcal{X}^{o*}$ is defined by

$$\Lambda^{o*} = \nabla^* \mathbf{e} = \begin{cases} -\nabla \cdot \mathbf{e} = \left(\frac{\partial}{\partial v},\ -\frac{\partial}{\partial u}\right) & \text{in } \Omega, \\ \mathbf{N} \cdot \mathbf{e} = (-N_v,\ N_u) & \text{on } \partial\Omega, \end{cases}$$

where the unit vector $\mathbf{N} \in \mathbb{R}^2$ normal to the boundary $\partial\Omega$ is defined by

$$\mathbf{N} = (N_u,\ N_v),\ N_u = \frac{dv}{\sqrt{(du)^2 + (dv)^2}},\quad N_v = \frac{-du}{\sqrt{(du)^2 + (dv)^2}},$$

and $\mathbf{N}\cdot\mathbf{e}\in\mathbb{R}^2$ is the unit vector tangent to the boundary, given by

$$\mathbf{N}\cdot\mathbf{e}=\left(\frac{du}{\sqrt{(du)^2+(dv)^2}},\ \frac{dv}{\sqrt{(du)^2+(dv)^2}}\right).$$

Thus, for any given $\chi\in\mathcal{X}_a$, the identity

$$\langle\Lambda\chi;\Lambda^o\chi^o\rangle=\int_{\partial\Omega}\mathbf{N}\cdot(\Lambda^o\chi^o)\cdot\bar{\chi}\mathrm{d}\Gamma\quad\forall\chi^o\in\mathcal{X}^o$$

leads to the polar equilibrium condition

$$\bar{\Lambda}^{o*}\Lambda\chi=\begin{cases}-\nabla\cdot(\mathbf{e}\nabla\chi)=-\chi_{,vu}+\chi_{,uv}\equiv 0 & \text{in }\Omega,\\ \mathbf{N}\cdot(\mathbf{e}\nabla\chi)=\chi_{,v}N_u-\chi_{,u}N_v=-\bar{\chi}_{,t} & \text{on }\partial\Omega,\end{cases}$$

where $\bar{\chi}_{,t}$ denotes the tangential derivative

$$\bar{\chi}_{,t}=\left(\bar{\chi}N_\alpha e^{\alpha\beta}\right)_{,\beta}=\nabla_t\bar{\chi}.$$

Let

$$\mathcal{X}^o_k=\{\chi^o\in\mathcal{X}^o|\ \operatorname{rank}(\mathbf{e}^T\nabla\chi^o(\mathbf{X}))=2\ \forall\mathbf{X}\in\Omega\}.\tag{6.173}$$

Replacing $\mathbf{S}$ in $P^d(\mathbf{S})$ by $\Lambda^o\chi^o$, the polar functional becomes

$$\begin{aligned}P^o(\chi^o)&=W^c(\Lambda^o\chi^o)-F^o(\chi^o)\\&=\int_\Omega\sqrt{\det((\nabla\chi^o)^T(\nabla\chi^o))}\mathrm{d}\Omega-\oint_{\partial\Omega}\mathbf{N}\cdot(\mathbf{e}^T\nabla\chi^o)\cdot\bar{\chi}\mathrm{d}\Gamma.\end{aligned}\tag{6.174}$$

Thus, the polar variational problem associated with $(\mathcal{P}_{\inf})$ can be proposed as

$$(\mathcal{P}^o_{\inf}):\quad P^o(\chi^o)\ \to\min\ \ \forall\chi^o\in\mathcal{X}^o_k.\tag{6.175}$$

The criticality condition for P^o leads to a *Neumann problem*

$$\Delta^o_{\mathcal{M}}\chi^o(\mathbf{X})=0\quad\text{in }\Omega,\tag{6.176}$$

$$\mathbf{N}\cdot\left(\sqrt{I_3(\mathbf{C}^o)}\mathbf{C}^{o-1}\nabla\chi^o\right)=-\bar{\chi}_{,t}\text{ on }\partial\Omega,\tag{6.177}$$

where $\Delta^o_{\mathcal{M}}$ is the *polar Laplace-Beltrami operator*

$$\Delta^o_{\mathcal{M}}=\frac{1}{(I_3(C^o_{\alpha\beta}))^{\frac{1}{2}}\partial X_\beta}\left((I_3(C^o_{\alpha\beta}))^{\frac{1}{2}}C^{o\alpha\beta}\frac{\partial}{\partial X_\alpha}\chi^o\right)$$

on the polar surface, and $I_3(C^o_{\alpha\beta})=\det\mathbf{C}^o$, $\mathbf{C}^o=(\nabla\chi^o)^T(\nabla\chi^o)=\{C^o_{\alpha\beta}\}$ is the induced Riemann metric tensor on the polar hypersurface $\chi^o\in\mathcal{X}^o_a$.

In terms of $(\chi,\chi^o)\in\mathcal{X}_a\times\mathcal{X}^o_a$, the Lagrangian $L(\chi,\mathbf{S})$ has the form

$$L(\chi,\Lambda^o\chi^o)=\langle\nabla\chi;\mathbf{e}^T\nabla\chi^o\rangle-W^*(\nabla\chi^o).\tag{6.178}$$

Theorem 6.5.1 *If* $(\bar{\chi}, \nabla\bar{\chi}^o) \in \mathcal{X}_a \times \mathcal{S}_a$ *is a critical point of* L *then the complementarity condition*

$$P(\bar{\chi}) + P^o(\bar{\chi}^o) = 0 \tag{6.179}$$

holds. Moreover, if $\bar{\chi}$ *is a minimal surface in* $\mathbb{R}^3$*, then* $\bar{\chi}^o$ *is its polar surface.*

Proof The criticality condition for L leads to

$$\mathbf{A}(\bar{\chi}) = \nabla\bar{\chi} = DW^c(\mathbf{S}(\bar{\chi}^o) = (I_3(\nabla\bar{\chi}^o))^{\frac{1}{2}}(C(\nabla\bar{\chi}^o))^{-1}\mathbf{e}^T(\nabla\chi^o) \quad \text{in } \Omega, \tag{6.180}$$

$$\nabla \cdot (\mathbf{e}^T\nabla\bar{\chi}^o) = 0 \quad \text{in } \Omega. \tag{6.181}$$

From the Legendre duality theorem we know that (6.180) is equivalent to $W(\nabla\bar{\chi}) + W^c(\nabla\bar{\chi}^o) = \langle \mathbf{A}(\bar{\chi}); \mathbf{S}(\bar{\chi}^o)\rangle$. Substituting this into L leads to

$$L(\bar{\chi}, \Lambda^o\bar{\chi}^o) = P(\bar{\chi}).$$

On the other hand, for any given $\bar{\chi} \in \mathcal{X}_a$, $L(\bar{\chi}, \Lambda^o\chi^o) = -P^o(\chi^o)$. Thus, (6.179) holds for any critical point of L.

If $\bar{\chi} \in \mathcal{X}_a$ is a minimal surface, then it satisfies the conformality relations (6.165) and the harmonic equation $\Delta\bar{\chi} = 0$. By the Legendre duality theorem, (6.180) is equivalent to

$$\nabla\bar{\chi}(\mathbf{X}) = \mathbf{e}^T\nabla\bar{\chi}^o(\mathbf{X}) \ \ \forall \mathbf{X} \in \Omega.$$

This leads to the Cauchy-Riemann equations in (6.167), and $\bar{\chi}^o$ is a polar surface of the minimal surface $\bar{\chi}$. □

This theorem shows that the minimal surface $\bar{\chi}$ is a solution of the Dirichlet problem, while its polar surface is a solution of the Neumann problem. In terms of W and W^c, the complementarity condition (6.179) can be written as

$$W(\nabla\chi) + W^c(\nabla\chi^o) = \oint_{\partial\Omega} \mathbf{N} \cdot (\mathbf{e}^T\nabla\chi^o) \cdot \bar{\chi} d\Gamma,$$

i.e. the value of the boundary integral is twice the area of the minimal surface bounded by the contour $\chi(\mathbf{X}_0)$.

Example 6.5.1 (Catenoid and Helicoid)

A *catenoid* of radius $d > 0$ is a doubly-connected minimal surface which can be parametrized by (cf., e.g., Dierkes *et al*, 1992)

$$\chi(u, v) = d(\cosh u \cos v, \quad -\cosh u \sin v, u)$$

on a simply connected domain $\Omega = (-\infty, +\infty) \times [0, 2\pi] \in \mathbb{R}^2$, if we choose $\chi^3(0,0) = 0$. By means of the formulas

$$\cosh(u + iv) = \cosh u \cos v + i \sinh u \sin v,$$

$$\sinh(u + iv) = \sinh u \cos v + i \cosh u \sin v,$$

it not difficult to verify that the mapping

$$\Upsilon(w) = d(\cosh w, i \sinh w, w) : \mathbb{C} \to \mathbb{C}^3$$

is a isotropic curve, and that

$$\chi(w) = \mathrm{Re}\Upsilon(w).$$

The polar surface $\chi^o : \Omega \to \mathbb{C}^3$ of χ is defined by

$$\chi^o(u, v) = \mathrm{Im}\Upsilon(w) = d(\sinh u \sin v, \sinh u \cos v, v). \tag{6.182}$$

It is easy to see that

$$P(\chi) = \int_\Omega |\chi_u|^2 d\Omega = \int_\Omega \cosh^2 u \, dudv = -P^o(\chi^o).$$

For the domain $\Omega = [-2, 2] \times [0, 2\pi]$ with $d = 1$, the graphs of χ and its polar χ^o are shown in Fig. 6.3. ♡

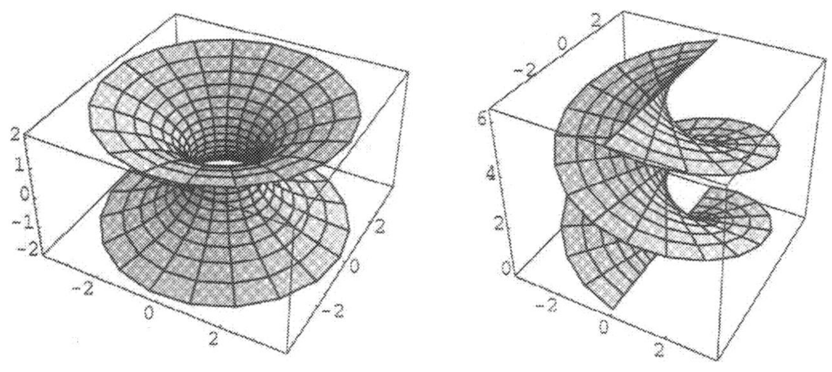

Figure 6.3. Catenoid and its polar surface, a helicoid.

Example 6.5.2 (Henneberg Surface and Its Polar Surface)
Consider an isotropic curve $\Upsilon : \mathbb{C} \to \mathbb{C}^3$ given by

$$\Upsilon(w) = \left(-1 + \cosh 2w, i \cosh w + \frac{1}{3} i \cosh 3w, -\sinh w + \frac{1}{3} \sinh 3w \right).$$

Its real part $\chi(u, v)$ =Re $\Upsilon(u + iv)$ is the well-known *Henneberg surface*:

$$\chi(u, v) = \begin{pmatrix} -1 + \cosh 2u \cos 2v, \\ -\sinh u \sin v - \frac{1}{3} \sinh 3u \sin 3v, \\ -\sinh u \cos v + \frac{1}{3} \sinh 3u \cos 3v \end{pmatrix}.$$

It is a minimal surface, and χ maps the whole v-axis onto a straight line segment of length 2 on the x-axis in $\mathbb{R}^3$ (see Fig. 6.4).

The polar surface χ^o to χ has the form

$$\chi^o(u,v) = \begin{pmatrix} \sinh 2u \sin 2v, \\ \cosh u \cos v + \frac{1}{3}\cosh 3u \cos 3v, \\ -\cosh u \sin v + \frac{1}{3}\cosh 3u \sin 3v \end{pmatrix}.$$

Both χ and χ^o are periodic in v with period 2π. ◊

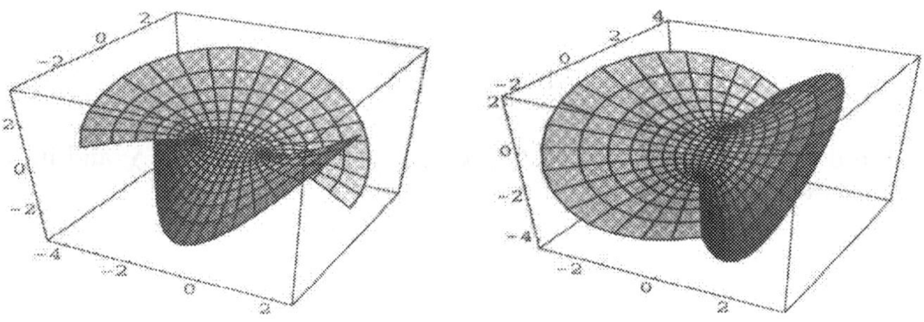

Figure 6.4. Henneberg's surface and its polar surface ($|u| \leq 3\pi/10,\ \ v \in [0, \frac{1}{2}\pi]$)

Example 6.5.3 (Catalan's Surface and Its Polar Surface)

Catalan's surface is periodic in the z-direction and defined by

$$\chi(u,v) = (1 - \cos u \cosh v, 4 \sin \frac{1}{2}u \sinh \frac{1}{2}v, u - \sin u \cosh v).$$

The translation in the parameter plane mapping $w = u + iv$ onto $u + 4\pi + iv$ corresponds to a 4π-shift of the surface along the z-axis. The associated isotropic curve is given by

$$\Upsilon(w) = \left(1 - \cosh(iw), 4i \cosh(\frac{1}{2}iw), w + i \sinh(iw)\right).$$

This implies that the polar surface $\chi^o = \mathrm{Im}\ \Upsilon$ has the representation

$$\chi^o(u,v) = \left(\sin u \sinh v, 4 \cos \frac{1}{2}u \cosh \frac{1}{2}v, v - \cos u \sinh v\right)$$

with the y-axis as line of symmetry and the y,z-plane as plane of symmetry (see Fig. 6.5). ◊

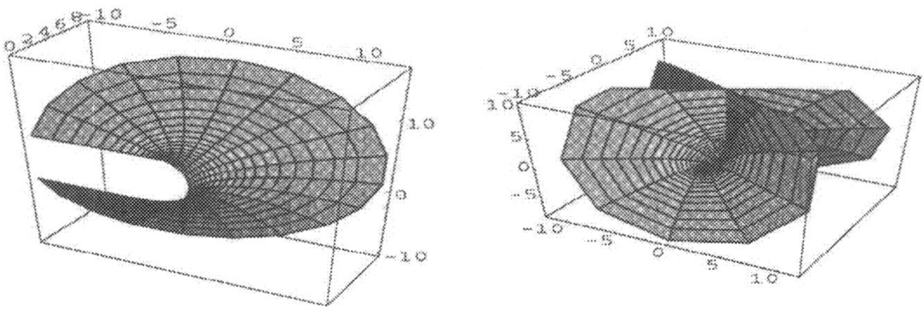

Figure 6.5. Catalan's surface and its polar surface ($u \in [0, 2\pi], \;\; v \in [0, \pi]$)

6.5.3 Non-Parametric Surfaces

For a nonparametric surface, the deformation $\chi(X_\alpha)$ takes the special form

$$\chi(x, y) = \{x, \; y, u(x, y)\}, \tag{6.183}$$

where $u : \Omega \to \mathbb{R}$ is the graph of the mapping $\chi : \Omega \to \mathbb{R}^3$. In this case, the deformation gradient $\mathbf{A}$ and the Riemannian metric tensor $\mathbf{C}$ have the very simple forms

$$\mathbf{A} = \nabla \chi = \begin{pmatrix} 1 & 0 \\ 0 & 1 \\ u_x & u_y \end{pmatrix} \in \mathbb{R}^{3\times 2}, \quad \mathbf{C} = \mathbf{I} + (\nabla u) \otimes (\nabla u) \in \mathbb{R}^{2\times 2}. \tag{6.184}$$

It is easy to verify that

$$\det \mathbf{C} = 1 + |\nabla u|^2 > 0. \tag{6.185}$$

To put the nonparametric minimal surface problem in our framework, we let $\mathcal{U} = \mathcal{L}^1(\Omega)$ and $\mathcal{E} = \mathcal{L}^1(\Omega; \mathbb{R}^2)$. In terms of the standard bilinear forms $\langle *, * \rangle$ and $\langle *; * \rangle$, their dual spaces are $\mathcal{U}^* = \mathcal{L}^\infty(\Omega)$, $\mathcal{E}^* = \mathcal{L}^\infty(\Omega; \mathbb{R}^2)$. For a given $\bar{u}$ on the boundary $\partial\Omega$, the admissible space is

$$\mathcal{U}_a = \{u \in \mathcal{L}^1(\Omega) | \;\; u(x, y) = \bar{u} \quad \text{on } \partial\Omega\}. \tag{6.186}$$

Using the linear geometrical mapping $\Lambda = \text{grad} : \mathcal{U} \to \mathcal{E}$, the primal problem for a minimal non-parametric surface can be written as

$$(\mathcal{P}_{\text{inf}})_u : \;\; P(u) = W(\Lambda u) = \int_\Omega \sqrt{1 + |\nabla u|^2} \mathrm{d}\Omega \to \inf \;\; \forall u \in \mathcal{U}_k. \tag{6.187}$$

The kinematically admissible space $\mathcal{U}_k$ defined by

$$\mathcal{U}_k = \{u \in \mathcal{U}_a | \;\; \nabla u \in \mathcal{E}_a = \mathcal{E}\}$$

is a subset of the Sobolev space $\mathcal{W}^{1,1}(\Omega; \mathbb{R}^2)$.

Lemma 6.5.3 *For any given $\epsilon \in \mathcal{E} = \mathcal{L}^1(\Omega; \mathbb{R}^2)$, the stored energy $W : \mathcal{E} \to \mathbb{R}$ defined by*

$$W(\epsilon) = \int_\Omega \sqrt{1 + |\epsilon|^2} d\Omega$$

is convex.

Proof The second-order Gâteaux variation of W at $\bar{\epsilon}$ in the direction ϵ is

$$\delta^2 W(\bar{\epsilon}; \epsilon^2) = \int_\Omega \epsilon_\alpha H^{\alpha\beta}(\bar{\epsilon}) \epsilon_\beta d\Omega,$$

where $H^{\alpha\beta}(\bar{\epsilon})$ is the Hessian of $U(\epsilon) := \sqrt{1 + |\epsilon|^2}$ defined by

$$H^{\alpha\beta}(\bar{\epsilon}) = \frac{\partial^2 U(\bar{\epsilon})}{\partial \bar{\epsilon}_\alpha \partial \bar{\epsilon}_\beta} = (U^2(\bar{\epsilon})\delta^{\alpha\beta} - \bar{\epsilon}^\alpha \bar{\epsilon}^\beta) U^{-\frac{3}{2}}(\bar{\epsilon}).$$

By the *Cauchy-Schwarz inequality*

$$\epsilon_\alpha \epsilon^\alpha \bar{\epsilon}_\beta \bar{\epsilon}^\beta - (\epsilon_\alpha \bar{\epsilon}^\alpha)^2 \geq 0 \quad \forall \epsilon \in \mathbb{R}^2,$$

it is easy to show that for any given $\bar{\epsilon} \in \mathcal{E}$,

$$\delta^2 U(\bar{\epsilon}; \epsilon^2) \geq U^{-\frac{3}{2}}(\bar{\epsilon}) |\epsilon|^2 \geq 0 \quad \forall \epsilon \in \mathcal{E}.$$

The inequality holds strictly for any non-zero $\epsilon \in \mathcal{E}$, and this shows that $W(\epsilon)$ is strictly convex on $\mathcal{E}$. □

By this lemma we know that $P : \mathcal{U}_k \to \mathbb{R}$ is strictly convex. The primal problem $(\mathcal{P}_{\text{inf}})_u$ has at most one solution and satisfies the Euler-Lagrange equation

$$A(u) = \Lambda^* DU(\Lambda u) = -\nabla \cdot \left(\frac{\nabla u}{\sqrt{1 + |\nabla u|^2}} \right) = 0 \quad \text{in } \Omega. \tag{6.188}$$

The fundamental operator $A : \mathcal{U} \to \mathcal{U}^*$ in this non-parametric surface problem is actually the mean curvature of the surface. If $A(u) = 0$ then the graph of $u : \Omega \to \mathbb{R}^3$ is a minimal surface. Since $\Lambda = \text{grad}$ is a linear operator, the problem $(\mathcal{P}_{\text{inf}})_u$ belongs to a geometrically linear convex system. The duality theory for such problems has been studied extensively (cf., e.g., Ekeland and Temam (1976)).

If there is a constraint in $\mathcal{U}_a$,

$$\mathcal{U}_a = \{ u \in \mathcal{U} | \ u(x) = 0 \ \forall x \in \partial\Omega, \ \int_\Omega u d\Omega = const. \} \tag{6.189}$$

say, then the Euler-Lagrange equation reads $A(u) = \mu$, with the constant Lagrange multiplier $\mu \in \mathbb{R}$ introduced for relaxing the integral constraint in $\mathcal{U}_a$. Since the curvature of the surface is a constant, the associated problem is also referred to as the H-surface problem, which has been studied extensively for the last thirty years. The multi-duality in this type of problem was discussed by Gao and Yang (1994).

In the obstacle problems, there is an inequality constraint in

$$\mathcal{U}_a = \{u \in \mathcal{U} | \ u(x) \geq \psi(x) \ \ \forall x \in \Omega, \ \ u(x) = 0 \ \ \text{on } \partial\Omega\}, \tag{6.190}$$

where the real-valued function ψ is a prescribed concave obstacle function. In this case, the primal problem $(\mathcal{P}_{\inf})_u$ is equivalent to the nonlinear complementarity problem

$$A(u) \leq 0 \ \ \mu \geq 0, \ \ \mu A(u) = 0 \ \ \text{in } \Omega. \tag{6.191}$$

The variational inequality in minimal surface type problems with a concave obstacle was studied, among many others, by Kinderlehrer (1973) and Nitsche (1989).

Generally speaking, although primal problems for non-parametric surfaces belong to a geometrically linear, convex system, the kinematically admissible space $\mathcal{U}_k \subset \mathcal{W}^{1,1}$ in these problems is not reflexive and the existence theorem (Theorem 5.1.6) does not apply. However, the dual problem possesses a unique solution. To make the problem general enough, we let $\mathcal{U}_a$ be a convex subset of $\mathcal{U}$. For a given external load $f \in \mathcal{U}^*$, the general external energy functional $F : \mathcal{U} \to \underline{\mathbb{R}}$ is defined by

$$F(u) := \int_\Omega f u d\Omega - \mathcal{I}_{\mathcal{U}_a}(u) = \begin{cases} \int_\Omega f u d\Omega & \text{if } u \in \mathcal{U}_a, \\ -\infty & \text{otherwise}, \end{cases} \tag{6.192}$$

where $\mathcal{I}_{\mathcal{U}_a}$ is the indicator function of the set $\mathcal{U}_a$. It is clear that the effective domain of F is $\text{dom}F(u) = \mathcal{U}_a$. Then, the total area $P(u)$ of the surface can be written in the general form

$$P(u) = \Phi(u, \Lambda u) = W(\Lambda u) - F(u), \tag{6.193}$$

where the functional Φ takes $\mathcal{U} \times \mathcal{E}$ into $\overline{\mathbb{R}}$. Thus, the primal problem $(\mathcal{P}_{\inf})_u$ is equivalent to the unconstrained problem

$$\inf_{u \in \mathcal{U}} \Phi(u, \Lambda u). \tag{6.194}$$

In this differential geometry problem, the conjugate variable $\boldsymbol{\sigma} \in \mathcal{E}^*$ of $\boldsymbol{\epsilon} \in \mathcal{E}$, defined by

$$\boldsymbol{\sigma} = DW(\boldsymbol{\epsilon}) = \boldsymbol{\epsilon}/\sqrt{1 + \boldsymbol{\epsilon} \cdot \boldsymbol{\epsilon}}, \tag{6.195}$$

is a surface tension vector. Since $W(\boldsymbol{\epsilon})$ is strictly convex on $\mathcal{E}$, the Legendre duality theorem holds on $\mathcal{E} \times \mathcal{E}^*$, and the Fenchel conjugate is

$$W^*(\boldsymbol{\sigma}) = \sup_{\boldsymbol{\epsilon} \in \mathcal{E}} \left\{ \int_\Omega [\boldsymbol{\sigma} \cdot \boldsymbol{\epsilon} - \sqrt{1 + |\boldsymbol{\epsilon}|^2}] d\Omega \right\} = \begin{cases} -\int_\Omega \sqrt{1 - |\boldsymbol{\sigma}|} d\Omega & \text{if } |\boldsymbol{\sigma}|^2 \leq 1, \\ +\infty & \text{if } |\boldsymbol{\sigma}| > 1. \end{cases}$$

Furthermore, we define the closed convex subset $\mathcal{E}_a^* \subset \mathcal{E}^*$ by

$$\mathcal{E}_a^* := \{\boldsymbol{\sigma} \in \mathcal{E}^* | \ |\boldsymbol{\sigma}(x)| \leq 1 \ \ \forall x \in \Omega\}, \tag{6.196}$$

and write the functional $W^* : \mathcal{E}_a^* \to \overrightarrow{\mathbb{R}}$ as

$$W^*(\sigma) = -\int_\Omega \sqrt{1-|\sigma|^2}\mathrm{d}\Omega + \mathcal{I}_{\mathcal{E}_a^*}(\sigma), \tag{6.197}$$

where $\mathcal{I}_{\mathcal{K}_1}$ is the indicator of the subset $\mathcal{E}_a^*$.

The Lagrangian $L : \mathcal{U}_a \times \mathcal{E}_a^* \to \mathbb{R}$ associated with $(\mathcal{P}_{\inf})_u$ is the right-saddle functional

$$L(u,\sigma) = \langle \Lambda u; \sigma\rangle - W^*(\sigma) - F(u) = \int_\Omega [(\nabla u)\cdot\sigma + \sqrt{1-|\sigma|^2} - fu]\mathrm{d}\Omega.$$

By the Lagrange duality, the dual functional is obtained as

$$P^d(\sigma) = \inf_{u\in\mathcal{U}_a} L(u,\sigma) = F^*(\bar{\Lambda}^*\sigma) - W^*(\sigma),$$

where $F^* : \mathcal{U}^* \to \overleftarrow{\mathbb{R}}$ is the Fenchel conjugate of $F(u)$, i.e.

$$F^*(\bar{\Lambda}^*\sigma) = \inf_{u\in\mathcal{U}_a} \{\langle \Lambda u, \sigma\rangle - F(u)\}.$$

For minimal non-parametric surface problems,

$$F^*(\bar{\Lambda}^*\sigma) = \begin{cases} \oint_{\partial\Omega} \bar{u}\sigma\cdot\mathbf{N}\mathrm{d}\Gamma & \text{if } \Lambda^*\sigma = 0 \text{ in } \Omega, \\ -\infty & \text{otherwise.}\end{cases}$$

For H-surface problems, if $f = 0$ and $\mathcal{U}_a$ is defined by (6.189), then

$$F^*(\bar{\Lambda}^*\sigma) = \begin{cases} 0 & \text{if } \Lambda^*\sigma = \mu \in \mathbb{R} \text{ in } \Omega, \\ -\infty & \text{otherwise.}\end{cases}$$

where μ is a Lagrange multiplier for the integral constraint in $\mathcal{U}_a$.

For the obstacle problem,

$$F^*(\bar{\Lambda}^*\sigma) = \begin{cases} -\int_\Omega \psi(\nabla\cdot\sigma + f)\mathrm{d}\Omega & \text{if } \Lambda^*\sigma \geq f \text{ in } \Omega, \\ -\infty & \text{otherwise.}\end{cases}$$

Let $\mathcal{U}_a^* = \operatorname{dom} F^*$ be the effective domain of F^*. On the statically admissible space

$$\mathcal{S}_s = \{\sigma \in \mathcal{E}_a^* |\ \bar{\Lambda}^*\sigma \in \mathcal{U}_a^*\},$$

the dual functional $P^d : \mathcal{S}_s \to \mathbb{R}$ is finite. For example, for minimal non-parametric surface problem, P^d is given by

$$P^d(\sigma) = \oint_{\partial\Omega} \bar{u}\sigma\cdot\mathbf{N}\mathrm{d}\Gamma + \int_\Omega \sqrt{1-|\sigma|^2}\mathrm{d}\Omega. \tag{6.198}$$

The problem dual to $(\mathcal{P}_{\inf})_u$ is

$$(\mathcal{P}_{\sup}^d)_\sigma : \ P^d(\sigma) = F^*(\bar{\Lambda}^*\sigma) - W^*(\sigma) \to \sup \ \forall \sigma \in \mathcal{S}_s. \tag{6.199}$$

Theorem 6.5.2 *For a given $\bar{u}$ such that $\mathcal{U}_a$ is defined by (6.186), the strong duality condition*

$$\inf_{u \in \mathcal{U}_k} P(u) = \sup_{\sigma \in \mathcal{S}_s} P^d(\sigma) \tag{6.200}$$

holds. Furthermore, the primal problem $(\mathcal{P}_{\inf})_u$ may or may not possess a solution, whereas the dual problem $(\mathcal{P}^d_{\sup})_\sigma$ possesses a unique solution $\sigma \in \mathcal{S}_s$. In the case that $(\mathcal{P}_{\inf})_u$ has a solution $\bar{u}$, the dual solution $\bar{\sigma}$ is an interior point of $\mathcal{S}_s$, i.e. $|\bar{\sigma}(x)| < 1$ for all $x \in \Omega$, and the extremality condition

$$\nabla \bar{u}(x) = -\frac{\bar{\sigma}(x)}{\sqrt{1 - |\bar{\sigma}(x)|^2}}$$

holds almost everywhere in Ω.

This theorem can be proved by using the right saddle Lagrange duality theory. Detailed studies of the regularity and minimization sequences were conducted by Ekeland and Temam (1976).

6.6 Applications and Commentary

6.6.1 3-D Elastic Cylindrical Tube

Let us first consider the deformation of a thick-walled right circular cylindrical tube composed of hyperelastic material into another right circular cylinder due to internal and external pressures. The tube is also subject to a fixed axial load. Such problems have been studied extensively in the literature (see, e.g., Ogden, 1984).

The undeformed configuration occupies the region defined, in cylindrical polar coordinates, by

$$\Omega = \{(R, \Theta, Z) \in \mathbb{R}^3 \,|\, 0 < A \leq R \leq B, \ 0 \leq \Theta \leq 2\pi, \ 0 \leq Z \leq L\}, \tag{6.201}$$

where A and B are the inner and outer radii of the undeformed cylinder, and L is its length. The deformation $\chi \in \mathbb{R}^3$ is a vector-valued function from Ω to another right cylinder and is given by

$$\chi = (r(R), \ \Theta, \ \epsilon_3 Z)^T : \Omega \to \mathbb{R}^3, \tag{6.202}$$

where ϵ_3 is a constant, the deformation $r(R)$ is assumed to be $C^1(A, B)$. Referred to cylindrical coordinates the deformation gradient and right Cauchy-Green strain tensor are

$$\mathbf{A} = \begin{pmatrix} r' & 0 & 0 \\ 0 & \frac{r}{R} & 0 \\ 0 & 0 & \epsilon_3 \end{pmatrix}, \quad \mathbf{C} = \mathbf{A}^T\mathbf{A} = \begin{pmatrix} (r')^2 & 0 & 0 \\ 0 & \frac{r^2}{R^2} & 0 \\ 0 & 0 & \epsilon_3^2 \end{pmatrix} = \Lambda(\chi). \tag{6.203}$$

Here, a prime denotes the derivative with respect to R. The principal stretches $\lambda_i = \sqrt{C_i}$ are given by

$$\lambda_1 = \sqrt{C_1} = |r'|, \ \lambda_2 = \sqrt{C_2} = |\frac{r}{R}|, \ \lambda_3 = \sqrt{C_3} = |\epsilon_3|. \tag{6.204}$$

For a homogeneous, isotropic hyperelastic material, the strain-energy function is symmetrical in its arguments, that is

$$\tilde{U}(\lambda_1, \lambda_2, \lambda_3) = \tilde{U}(\lambda_2, \lambda_1, \lambda_3) = \tilde{U}(\lambda_1, \lambda_3, \lambda_2). \tag{6.205}$$

By objectivity, there exists a stored-energy density $\bar{U}(\mathbf{C})$ such that

$$\tilde{U}(\lambda_1, \lambda_2, \lambda_3) = \bar{U}(C_1, C_2, C_3). \tag{6.206}$$

Similarly, $\bar{U}(\mathbf{C})$ is also symmetrical in its arguments. The canonical conjugate stress $\mathbf{T} = \mathbf{C}^*$ is the second Piola-Kirchhoff type stress tensor

$$\mathbf{T} = \begin{pmatrix} T_1 & 0 & 0 \\ 0 & T_2 & 0 \\ 0 & 0 & T_3 \end{pmatrix}, \quad T_i = \frac{\partial \bar{U}}{\partial C_i}. \tag{6.207}$$

We write simply $\mathbf{C} = \{C_i\}$, and $\mathbf{T} = \{T_i\}$. The first Piola-Kirchhoff stress is then

$$\mathbf{S} = \{S_i\} = 2\{\lambda_i T_i\} = 2(r' T_1, r T_2/R, \epsilon_3 T_3). \tag{6.208}$$

In the absence of body forces, the only non-trivial equilibrium equation in terms of the nominal stress tensor $\mathbf{S}$ is

$$S_1' + \frac{S_1 - S_2}{R} = 0 \quad \text{in } \Omega. \tag{6.209}$$

On the inner, outer and end surfaces, the traction boundary conditions are

$$S_1 = p_a \quad \text{on } R = A, \quad S_1 = p_b \text{ on } R = B, \quad S_3 = p_z \quad \text{on } Z = L.$$

It is easy to show that the statically admissible space $\mathcal{S}_s$ in this problem is

$$\mathcal{S}_a = \{(S_1, S_2, S_3)| \ S_1 = \frac{1}{R}\int_A^R S_2 dR + p_a, \ \ S_1|_{R=B} = p_b \ \ S_3|_{Z=L} = p_z\}. \tag{6.210}$$

Since $\mathbf{C} = \Lambda(\chi)$ is a quadratic function of the deformation χ with $a = 2, \beta = 0$, the gap functional (6.107) in this case is

$$G(\chi, \mathbf{T}) = \int_\Omega (r'^2 T_1 + r^2 T_2/R^2 + \epsilon^2 T_3) R dR d\Theta dZ. \tag{6.211}$$

For a statically admissible field $\mathbf{S} \in \mathcal{S}_s$, the pure complementary energy Π^d can be written as

$$\Pi^d(\mathbf{T}) = F_\Lambda^d(\mathbf{T}) - \int_\Omega \bar{U}^*(T_1, T_2, T_3) R dR d\Theta dZ, \tag{6.212}$$

where the dual functional $F_\Lambda^d(\mathbf{T})$ is given as

$$F_\Lambda^d(\mathbf{T}) = -\int_\Omega \frac{1}{4}(S_1^2/T_1 + S_2^2/T_2 + S_3^2/T_3) R dR d\Theta dZ. \tag{6.213}$$

For an unconstrained (i.e. compressible) material, the complementary energy density is simply

$$\bar{U}^*(T_1, T_2, T_3) = \sup_{C_1, C_2, C_3 > 0} \left\{ \sum_{i=1}^{3} C_i T_i - \bar{U}(C_1, C_2, C_3) \right\}. \tag{6.214}$$

The dual canonical Euler-Lagrange tensor equation (6.138) can be written as the three coupled equations

$$4T_i^2 \frac{\partial \bar{U}^*}{\partial T_i} = S_i^2, \quad i = 1, 2, 3. \tag{6.215}$$

For the given $\mathbf{S} = (S_1(S_2), S_2, S_3) \in \mathcal{S}_s$, the analytic solution χ is

$$r = \int_A^R \frac{S_1}{2T_1} dR + A = \frac{S_2 R}{2T_2}, \quad \epsilon_3 = \frac{S_3}{2T_3}. \tag{6.216}$$

For a given material such that $\bar{U}(\mathbf{C})$ is a canonical function, $\bar{U}^*(\mathbf{T})$ is well defined. For an incompressible material, the complementary energy density is obtained by the constrained optimization problem

$$\bar{U}^*(T_1, T_2, T_3) = \sup_{C_1, C_2, C_3 > 0} \left\{ \sum_{i=1}^{3} C_i T_i - \bar{U}(C_1, C_2, C_3) \right\} \quad s.t. \;\; C_1 C_2 C_3 = 1.$$

6.6.2 Deformation Theory with Internal Variables

In general finite deformation problems, the stored strain energy density U usually depends not only on the deformation gradient $\mathbf{A}$ but also on some *internal variables, or parameters* $\boldsymbol{\mu}$, where $\boldsymbol{\mu}$ could be represented in scalar, vector or tensor forms. By the principle of material frame-indifference, $U : \mathcal{A}(\Omega) \to \mathbb{R}$ should satisfy the condition

$$U(\mathbf{A}, \boldsymbol{\mu}) = U(\mathbf{QA}, \boldsymbol{\mu}) = \bar{U}(\mathbf{E}, \boldsymbol{\mu}) \;\; \forall \mathbf{A} \in \mathcal{A}, \;\; \mathbf{Q} \in \mathcal{M}^+_{ort}, \;\; \mathbf{E} \in \mathcal{E}. \tag{6.217}$$

Finite deformation problems with internal parameters appear frequently in phase transitions and hysteresis, shape memory alloys, elastoplasticity with residual stress, damage mechanics, post-buckling of large deformed structures, shape optimization and control, etc. Here, we consider only a special case, i.e., the internal parameter $\beta \in \mathbb{R}$ is regarded as fixed (inactive) at a constant initial value, β_0 say, until some threshold value of the deformation is reached, and thereafter, β depends on the deformation history. For example, in the deformation theory of elastoplasticity with strain hardening effects, such a threshold could be modeled as a hyper-surface $Y : \mathcal{M}_{sym} \to \mathbb{R}$ in strain space. Thus,

$$Y(\mathbf{E}, \beta) = 0. \tag{6.218}$$

For Hencky's plastic material in $\mathbb{R}^3$, if we let $\mathbf{E} = \mathbf{E}^{(1)}$ be the Green strain tensor, then $Y(\mathbf{E})$ is the plastic yield function

$$Y(\mathbf{E}, \beta) = \sqrt{\operatorname{tr}(\mathbf{E}^d \mathbf{E}^d)} - \beta = |\mathbf{E}^d| - \beta \tag{6.219}$$

in strain space, where $\mathbf{E}^d = \mathbf{E} - \frac{1}{3}\mathrm{tr}\mathbf{E}$ denotes the deviator of $\mathbf{E}$. The stored-energy $\bar{U}$ is a nonsmooth function of $\mathbf{E}$, given by

$$\bar{U}(\mathbf{E},\beta) = \begin{cases} \frac{1}{2}\kappa_0(\mathrm{tr}\mathbf{E})^2 + \mu_0|\mathbf{E}^d|^2 = \frac{1}{2}\mathbf{EHE} & \text{if } |\mathbf{E}^d| \le \beta, \\ \frac{1}{2}\kappa_0(\mathrm{tr}\mathbf{E})^2 + \frac{1}{4\mu_0}\beta^{*2} + \beta^*(|\mathbf{E}^d|^2 - \beta) & \text{if } |\mathbf{E}^d| \ge \beta, \end{cases} \tag{6.220}$$

where κ_0, μ_0 and β^* are material constants and $\mathbf{H}$ is the generalized Hooke's tensor. For perfectly plastic deformation, the internal parameter $\beta = \beta_0 = \beta_0^*/(2\mu_0)$ is a constant. However, if the material possesses harderning effects, β and β^* will depend on the deformation history. It is easy to prove that $\bar{U}(\mathbf{E},\beta)$ is convex, piecewisely Gâteaux differentiable, and the second Piola-Kirchhoff stress is given by

$$\mathbf{T} = D\bar{U}(\mathbf{E},\beta) = \begin{cases} \kappa_0(\mathrm{tr}\mathbf{E})\mathbf{I} + 2\mu_0\mathbf{E}^d & \text{if } |\mathbf{E}^d| - \beta \le 0, \\ \kappa_0(\mathrm{tr}\mathbf{E})\mathbf{I} + \beta^*\mathbf{E}^d/|\mathbf{E}^d| & \text{if } |\mathbf{E}^d| - \beta \ge 0. \end{cases} \tag{6.221}$$

The Fenchel-conjugate of $\bar{U}$ has the simple form

$$\begin{aligned} \bar{U}^*(\mathbf{T},\beta) &= \sup_{\mathbf{E}\in\mathcal{E}}\{\mathbf{T}:\mathbf{E} - \bar{U}(\mathbf{E},\beta)\} \\ &= \begin{cases} \frac{1}{18\kappa_0}(\mathrm{tr}\mathbf{T})^2 + \frac{1}{4\mu_0}|\mathbf{T}^d|^2 = \frac{1}{2}\mathbf{T}\mathbf{H}^{-1}\mathbf{T} & \text{if} |\mathbf{T}^d| \le \beta^*, \\ +\infty & \text{otherwise}. \end{cases} \end{aligned} \tag{6.222}$$

The yield surface in stress space is then

$$Y^*(\mathbf{T},\beta^*) = |\mathbf{T}^d| - \beta^*. \tag{6.223}$$

The inverse constitutive law can be written as

$$\mathbf{E} = D\bar{U}^*(\mathbf{T},\beta^*) = \begin{cases} \mathbf{H}^{-1}\mathbf{T} & \text{if } |\mathbf{T}^d| - \beta^* < 0, \\ \mathbf{H}^{-1}\mathbf{T} + \mu\frac{\partial Y^*(\mathbf{T},\beta^*)}{\partial \mathbf{T}} & \text{if } |\mathbf{T}^d| - \beta^* = 0,\ \mu > 0, \\ \emptyset & \text{if } |\mathbf{T}^d| - \beta^* > 0. \end{cases} \tag{6.224}$$

If we let $\mathbf{E} = \mathbf{E}^e + \mathbf{E}^p$, and $\mathbf{E}^e = \mathbf{H}^{-1}\mathbf{T}$ be the elastic strain, then the plastic strain $\mathbf{E}^p$ satisfies the well-known plastic flow law

$$\mathbf{E}^p = \mu\frac{\partial Y^*(\mathbf{T},\beta^*)}{\partial \mathbf{T}} = \mu\frac{\mathbf{T}^d}{|\mathbf{T}^d|}, \tag{6.225}$$

where the Lagrange multiplier $\mu \ge 0$ for the yield condition $Y^*(\mathbf{T},\beta^*) \le 0$ is the so-called *plastic flow factor*. It has to satisfy the *KKT constraint qualification condition*

$$\mu \ge 0,\ \ Y^*(\mathbf{T},\beta^*) \le 0,\ \ \mu Y^*(\mathbf{T},\beta^*) = 0. \tag{6.226}$$

The admissible space $\mathcal{T}_a \subset \mathcal{T}$, defined by

$$\mathcal{T}_a = \{\mathbf{T} \in \mathcal{T} |\ |\mathbf{T}^d(\mathbf{X})| \le \beta^*\ \ \forall \mathbf{X} \in \Omega\} \tag{6.227}$$

is the so-called *plastic admissible yield space*. For a given statically admissible nominal stress field $\mathbf{S} \in \mathcal{S}_s$, the dual canonical variational problem (Π^d_{sta}) for Hencky plasticity is then nonconvex variational problem

$$\Pi^d(\mathbf{T}) = \int_\Omega [\text{tr}\mathbf{S} - \frac{1}{2}\text{tr}(\mathbf{S}^T\mathbf{T}^{-1}\mathbf{S} + \mathbf{T}) - \frac{1}{2}\mathbf{T}\mathbf{H}^{-1}\mathbf{T}]d\Omega \quad \to \text{sta} \ \forall \mathbf{T} \in \mathcal{T}_s \quad (6.228)$$

with inequality constraints in $\mathcal{T}_s = \{\mathbf{T} \in \mathcal{T}_a | \ \det \mathbf{T} \neq 0 \quad \text{in } \Omega\}$. The dual canonical Euler-Lagrange equation in this case is then

$$\mathbf{T}\left[2\left(\mathbf{H}^{-1}\mathbf{T} + \mu\frac{\mathbf{T}^d}{|\mathbf{T}^d|}\right) + \mathbf{I}\right]\mathbf{T} = \mathbf{S}\mathbf{S}^T, \quad (6.229)$$

subjected to the KKT condition (6.226). Equation (6.145) then gives the analytical solution of the variational problem for Hencky plasticity.

6.6.3 Historical Remarks and Commentary

In contrast to one-dimensional problems, nonlinear variational problems in multi-dimensional systems are much more difficult and many basic questions remain unresolved. Most of these questions have been illuminated by a fertile interaction in the last 20 years with problems arising from solid mechanics and materials science (Ball, 1998). For example, most alloys undergo solid phase transformations involving a change of shape at one or more critical temperatures. The calculus of variations can give insight into why such phase transformations typically lead to patterns of fine microstructure, in the simplest case consisting of a laminate of parallel thin layers in which the spatial gradient of the deformation alternates between two values. In post-buckling analysis of large deformation thin-walled structures, the free energy has more than one potential well. The multiple local minimizers of a given variational problem lead to possible post-buckling states of the deformed structure. Generally speaking, the direct approaches for solving nonconvex, nonsmooth multi-dimensional primal variational problems are very difficult. Based on the classical Fenchel-Rockafellar duality theory, relaxation methods can only be used to find the "generalized" global minimizer. Because of the nonconvexity, there exists a duality gap between the real solution and any relaxed generalized solution. Estimation of the duality gap in nonconvex optimization was discussed by many mathematicians (cf., e.g., Aubin and Ekeland, 1976). The direct analysis of these nonlinear variational problem has posed many fundamental open questions concerning quasiconvexity, relaxation, and gradient Young measures (cf. Ciarlet, 1988; Ball, 1998). The earliest applications of Fenchel-Rockafellar duality theory in nonlinear elasticity theory are probably provided by Stumpf (1975, 1976, 1978). Dual extremum theorems and global error bounds were established for the solution of monotone hyperelastic boundary-value problems. These results were applied into convex large deformation structures (see, Stumpf, 1979).

The complementary energy principle in geometrically nonlinear nonconvex variational systems was first proposed by Hellinger in 1914. Since the boundary conditions were clarified by Reissner (1953, see also 1998), the dual variational principles and methods in finite deformation mechanics have been studied extensively by many mechanicians (see the books by Washizu, 1967; Oden and Reddy, 1983; Mason, 1980; Ogden, 1984 and many more). It is known that the Hellinger-Reissner principle involves both the Kirchhoff stress and the displacement, so it is not considered as a pure complementary energy principle. Though this principle has many important consequences in large deformation theory and computational mechanics, the extremality property of the generalized Hellinger-Reissner energy has been an open problem for more than forty years, and this raised many arguments in finite deformation theory and nonconvex mechanics (cf., e.g., Koiter, 1976; Nemat-Nasser, 1977; Atluri, 1980; Lee and Shield, 1980; Pian and Tong, 1980 and Bufler, 1983). The Levinson principle involves only the first Piola-Kirchhoff stress, but this principle is not valid unless the constitutive relation is invertible. It is known (see Ogden, 1977) that in finite deformation theory, the stored energy is usually nonconvex in the deformation gradient but convex in the right stretch tensor $\mathbf{U}$ for most reasonable hyperelastic materials, thus, the de Veubeke Principle is always true. But this principle involves both the first Piola-Kirchhoff stress $\mathbf{S}$ and the rotation tensor $\mathbf{R}$. Moreover, since $\mathbf{S} \in \mathcal{A}$ is not symmetric, it is very difficult to use this principle.

Ogden's regularity theorem reveals both the physical properties of finitely deformed materials and the geometrical structure intrinsic in Legendre transformation of the stored-energy. This nice result is of fundamental importance in understanding bifurcation phenomena in finite deformation boundary-value problems, in the formulation of complementary variational principles and in the understanding of the duality solutions proposed in this chapter. Ogden's results may play a potentially important role in solving general fully nonlinear boundary-value problems arising in hyperelasticity. Since Ogden's paper (Ogden, 1977) was written mainly for mechanicians and physicist, this important result did not receive the attention by pure mathematicians that it deserved.

Hill's conjugate strain-stress measures play an important role in finite deformation theory. Comprehensive studies and surveys on the strain measures and conjugate stress-strain tensor functions were given recently by Zheng (1994), Stumpf and Hoppe (1997) and Le and Stumpf (1997). For most of reasonable engineering materials, the stored-energy density is a canonical function (either strictly convex or concave) on certain Hill-Seth strain space. Thus, the canonical dual transformation method can be used to develop suitable complementary variational formulations.

The canonical dual transformation method and the associated duality theory were originally studied by Gao and Strang (1989a) for finite deformation problems in which the constitutive law is nonsmooth and the geometrical mapping Λ is a generally nonlinear partial differential operator. By using the canonical decomposition $\Lambda = \Lambda_t + \Lambda_c$, a complementary gap functional was discovered, by means of which the duality gap existing in classical Fenchel-Rockafellar duality theory was recovered. They proved that if, for the quadratic operator Λ with convex canonical energy, this gap functional is positive then, the canonical Lagrangian is a saddle functional, and they proposed the unified complementary energy principle (Theorem 6.3.1). In applications to elastoplasticity, Gao and

Strang (1989b) proved that the existence of the plastic limit load depends on the sign of the gap functional, i.e. when the gap functional is positive the Hencky plastic structure is sure to have a limit load. The primal and dual variational approaches lead to upper and lower bounds respectively on the safety factor. On the other hand, if this gap functional is negative, the structure may buckle elastically before the loading factor reaches its plastic limit. Unfortunately, Gao and Strang did not continue this work on the extremality of the canonical Lagrangian for when the gap functional is negative. Recently, when the author was studying post-buckling phenomena arising in the large deformation elastic beams, he discovered, first graphically and then theoretically, that when the gap functional is negative, the canonical Lagrangian is a super-critical point functional. Thus, the triality theory was proposed in mechanics and nonsmooth, nonconvex variations (Gao, 1997, 1998). By the sequential canonical transformation discussed in Chapter 4, the results presented in this chapter can certainly be generalized to general problems with multi-wells, as well as finitely deformed dynamical systems. These are very rich fields and duality theory will surely play an important role.

Since the dual canonical Euler-Lagrange equation is a nonlinear tensor equation, numerical methods will generally be needed for complicated problems in multi-dimensional systems. The triality theorem can be used to develop potentially powerful numerical algorithms. For any given statically admissible nominal stress field $\mathbf{S} \in \mathcal{S}_s$, the duality theorem ensures that the solutions for the associated canonical stress lead to alternative dual bounding of the cost functional. However, in order to find the primal solution, the key step is to find a rotation tensor $\mathbf{R}$ satisfying the compatibility condition. Though there are many open problems left, the canonical dual transformation method provides a new, powerful approach to solve general fully nonlinear, nonsmooth variational boundary-value problems.

Finally, the author has to write some words on the complementary gap functional which was proposed in Gao and Strang's paper in 1989. Actually, this gap functional was first announced at the 17th IUTAM Congress, August 26, 1988, in Grenoble, France. A similar result appeared soon after in a paper by J. J. Telega (1989) (submitted on February 2, 1989, and accepted on February 20, 1989), wherein the Gao and Strang work was mentioned in its French abstract. Thereafter, in a series publications (see e.g., Galka and Telega (1992) and the references cited therein), as well as in a review article by Telega (*Mathematics Review*, MR94d:73029, 1994), it was claimed that the complementary gap functional was first discovered by himself and his co-workers. Unfortunately, in this work the Green strain tensor $E_{ij} = \frac{1}{2}[u_{i,j} + u_{j,i} + u_{k,i}u_{k,j}]$ was decomposed into a (linear) symmetric part $\Lambda_1 \mathbf{u} = e_{ij} = u_{(i,j)}$ and another linear part $\Lambda_2 \mathbf{u} = \nabla \mathbf{u}$, and without definition, $\Lambda \mathbf{u} = (\Lambda_1 \mathbf{u}, \Lambda_2 \mathbf{u})$ was set equal to $(\mathbf{e}(\mathbf{u}), \nabla \mathbf{u})$. By introduction of the second Piola-Kirchhoff stress (in the notation $\mathbf{S} \in \mathbb{M}^3_s$) and a new tensor $\mathbf{T} \in \mathbb{M}^3$ (without definition), it was claimed that the complementary potential of $W(\mathbf{E})$ is (equation (22) of the paper)

$$g^*(\mathbf{S},\mathbf{T}) = \sup_{p\in\mathbb{M}^3_s, q\in\mathbb{M}^3} \left\{ S_{ij}p_{ij} + T_{ij}q_{ij} - W(p_{ij} + \frac{1}{2}q_{ki}q_{kj}) \right\} \quad (6.230)$$

$$= W^*(\mathbf{S}) + g_1^*(\mathbf{S},\mathbf{T}), \quad (6.231)$$

where $W^*(\mathbf{S})$ is the Fenchel-conjugate of $W(\mathbf{E})$, and

$$g_1^*(\mathbf{S},\mathbf{T}) = \sup\{T_{ij}q_{ij} - \frac{1}{2}S_{ij}q_{ki}q_{kj})|\ q \in \mathbb{M}^3\}.$$

It is clear that the function g_1^* is identical to $+\infty$ if S_{ij} is strictly negative-definite. However, the "extremality condition" $\mathbf{T} = \mathbf{S}(\nabla\mathbf{u})$ was defined through

$$\tilde{g}_d^*(\mathbf{S},\nabla\mathbf{u}) := g_1^*(\mathbf{S},\mathbf{T} = \mathbf{S}\nabla\mathbf{u}) = \frac{1}{2}S_{ij}u_{k,i}u_{k,j},$$

and was called the gap function. Since $p_{ij} = \frac{1}{2}(u_{i,j} + u_{j,i})$ is not independent of $q_{ij} = u_{i,j}$, it appears that when the supremum problem in (6.230) was split into two different problems in (6.231), an error such as $\sup(\sum(\cdots)) = \sum(\sup(\cdots))$ occurred. Unfortunately, this method has been repeated in a series of publications and so-called "truly complementary energy principles" (cf., Galka and Telega (1992)) have been produced. Since the extremality condition $\mathbf{T} = \mathbf{S}(\nabla\mathbf{u})$ used in the definition of $\tilde{g}_d^*$ is true only when $\mathbf{S}$ is non-negative-definite almost everywhere in the domain, his gap function is well defined only for this special case. By the triality theorem, we known that if the second Piola-Kirchhoff stress is negative-definite, the total potential is nonconvex. In this case, the *duality gap* for all the so-called "truly complementary energy principles" proposed by Telega and his co-workers is $+\infty$. Moreover, by Lemma 6.3.1 (proposed in Gao (1992)) we know that when the second Piola-Kirchhoff stress is positive-definite, the total potential is convex. In this case, the unified complementary energy principle Theorem 6.3.1 shows that all the well-known complementary energy principles are equivalent to the Levinson principle, which has the simplest formulation (see Gao (1992)). Interested readers are recommended to read the recent paper by Bojarski (1999) for further discussion.

The main concern in all scientific work must be the human being himself. This, one should never forget among all those diagrams and equations.

—Albert Einstein

The principal creator of three-dimensional elasticity is Cauchy (1789-1857). ⋯ It is fair to say that much of the algebra of vectors, matrices, and tensors grew out of Cauchy's work on the strain, local rotation, and stress in elastic bodies.

—Clifford Ambrose Truesdell (1983)

7 APPLICATIONS, OPEN PROBLEMS AND CONCLUDING REMARKS

By the seventh day God had finished the work he had been doing; so on the seventh day he rested from all his work.

—Genesis, 2.2

The greatest joy of a thinking man is to have explored the explorable and just to admire the unexplorable.

—Johann Wolfgang von Goethe

If one does not sometimes think the illogical, one will never discover new ideas in science.

—Max Planck, 1945

Truth is much too complicated to allow anything but approximations.

—John von Neumann

(a) Direct method (b) Dual method

Figure 7.1. Duality in solving large deformation problem.

In this last chapter, we shall make use of the techniques of duality for various problems arising from plastic limit analysis, contact problems of elastoplastic beams and the large deformation structures. All these problems can be formulated in terms of the canonical potential $\Pi(u) = \Phi(u, \Lambda(u))$. However, some problems, which do not have such a formulation, are listed for the future research topics.

7.1 Constitutive Nonlinearity: Plastic Limit Analysis

7.1.1 Constitutive Law and Super-Potentials

Consider a rigid-perfectly plastic material in $\mathbb{R}^3$ with domain $\Omega \subset \mathbb{R}^3$ and boundary $\Gamma = \partial\Omega = \Gamma_f \cup \Gamma_u$. Let $\mathcal{E}$ and $\mathcal{S} = \mathcal{E}^*$ be the admissible strain and stress space, respectively, defined by

$$\begin{aligned} \mathcal{E} &= \{\epsilon \in L^{\beta}(\Omega; \mathbb{R}^{3\times 3}) |\ \epsilon = \{\epsilon_{ij}\} = \epsilon^T\}, \\ \mathcal{S} &= \{\sigma \in L^{\beta^*}(\Omega; \mathbb{R}^{3\times 3}) |\ \sigma = \{\sigma_{ij}\} = \sigma^T\}, \end{aligned}$$

where β, β^* are dual numbers. The standard bilinear form $\langle *; * \rangle$ puts $\mathcal{E}$ and $\mathcal{S}$ in duality.

For the *Levy-Mises material*, the classical setting of the constitutive law has three elements:

$$\begin{array}{lll} \text{(a) incompressibility condition,} & \operatorname{tr} \epsilon = 0; & \\ \text{(b) yield condition,} & Y^c(\sigma) \leq 0; & \quad (7.1) \\ \text{(c) plastic flow law,} & \epsilon = \mu \dfrac{\partial Y^c(\sigma)}{\partial \sigma}. & \end{array}$$

In (7.1), Y^c is the plastic yield function, defined by

$$Y^c(\sigma) = |\sigma^d| - \sigma_b = \sqrt{\sigma^d_{ij}\sigma^d_{ij}} - \sigma_b, \tag{7.2}$$

where $\sigma^d = \sigma - \frac{1}{3}\operatorname{tr}\sigma$ is the *stress deviator*, $\sigma_b > 0$ is a material constant, $|*|$ is the Euclidean norm in $\mathbb{R}^{3\times 3}$, and $\mu \geq 0$ is the *plastic flow factor*. Let

$$\mathcal{E}_a = \{\epsilon \in \mathcal{E} \mid \operatorname{tr} \epsilon = 0 \quad \text{in } \Omega\}, \tag{7.3}$$

$$\mathcal{S}_a = \{\sigma \in \mathcal{S} \mid Y^c(\sigma) \leq 0 \quad \text{in } \Omega\}. \tag{7.4}$$

The *super-potential* associated with this Levy-Mises material is defined as

$$W(\epsilon) = \int_{\Omega} \sigma_b |\epsilon^d| d\Omega + \mathcal{I}_{\mathcal{E}_a}(\epsilon) = \sigma_b \|\epsilon\| + \mathcal{I}_{\mathcal{E}_a}(\epsilon), \tag{7.5}$$

where $\| * \|$ denotes the norm in $\mathcal{E}$, and $\mathcal{I}_{\mathcal{E}_a} : \mathcal{E} \to \overline{\mathbb{R}}$ is the indicator functional of the set $\mathcal{E}_a$, i.e.

$$\mathcal{I}_{\mathcal{E}_a}(\epsilon) = \begin{cases} 0 & \text{if } \epsilon \in \mathcal{E}_a, \\ +\infty & \text{if } \epsilon \notin \mathcal{E}_a. \end{cases} \tag{7.6}$$

It is clear that $W : \mathcal{E} \to \overrightarrow{\mathbb{R}}$ is convex, lower semicontinuous, so that the constitutive law for rigid-perfect plasticity can be written in the standard form

$$\sigma \in \partial W(\epsilon) = \begin{cases} \frac{\sigma_b}{|\epsilon^d|}\epsilon^d + p\mathbf{I} & \text{if } \epsilon \in \mathcal{E}_a, \\ \emptyset & \text{if } \epsilon \notin \mathcal{E}_a, \end{cases} \tag{7.7}$$

p being an undetermined parameter, and $\mathbf{I}$ the unit tensor in $\mathcal{E}$.

From the Fenchel transformation, the conjugate super-potential of $W(\epsilon)$ is computed as

$$\begin{aligned} W^*(\sigma) &= \sup_{\epsilon \in \mathcal{E}} \{ \langle \epsilon; \sigma \rangle - W(\epsilon) \} = \sup_{\epsilon \in \mathcal{E}_a} \{ \langle \epsilon; \sigma \rangle - \sigma_b \|\epsilon\| \} \\ &= \sup_{\epsilon \in \mathcal{E}_a} \int_\Omega |\epsilon^d|(|\sigma^d| - \sigma_b) \mathrm{d}\Omega = \begin{cases} 0 & \text{if } Y^c(\sigma) \le 0 \text{ in } \Omega, \\ +\infty & \text{otherwise .} \end{cases} \end{aligned}$$

Therefore, the conjugate super-potential for rigid-perfect plasticity is just the indicator functional of the convex set $\mathcal{S}_a$, i.e. $W^*(\sigma) = \mathcal{I}_{\mathcal{S}_a}(\sigma)$. Thus, the inverse form of the constitutive law can be written as

$$\epsilon \in \partial W^*(\sigma) = \begin{cases} \mu \frac{\partial Y^c(\sigma)}{\partial \sigma} & \text{if } Y^c(\sigma) = 0, \mu \ge 0, \\ \{0\} & \text{if } Y^c(\sigma) < 0, \\ \emptyset & \text{if } Y^c(\sigma) > 0. \end{cases} \tag{7.8}$$

Since $W : \mathcal{E} \to \overrightarrow{\mathbb{R}}$ is convex, l.s.c., the Fenchel duality relation

$$\sigma \in \partial W(\epsilon) \;\Leftrightarrow\; \epsilon \in \partial W^*(\sigma) \;\Leftrightarrow\; W(\epsilon) + W^*(\sigma) = \langle \epsilon; \sigma \rangle$$

holds on $\mathcal{E} \times \mathcal{E}^*$.

Let $\mathcal{U} = \mathcal{L}^2(\Omega; \mathbb{R}^3)$ be the configuration space, in duality with $\mathcal{U}^* = \mathcal{L}^2(\Omega; \mathbb{R}^3)$ by the standard bilinear form $\langle *, * \rangle : \mathcal{U} \times \mathcal{U}^* \to \mathbb{R}$. Let the geometric mapping $\Lambda : \mathcal{U} \to \mathcal{E}$ be a linear symmetrical operator

$$\Lambda \mathbf{u} = \nabla_s \mathbf{u} = \frac{1}{2}[\nabla \mathbf{u} + (\nabla \mathbf{u})^T].$$

For a given body force field $\bar{\mathbf{b}}$ in the domain Ω and the unit surface traction $\bar{\mathbf{f}}$ on Γ_f, the boundary-value problem of limit analysis is to find the safety factor $\varrho_c > 0$ and field functions $(\mathbf{u}, \epsilon, \sigma)$ such that

$$(\mathcal{P}_{\text{bv}}): \begin{cases} \epsilon = \Lambda \mathbf{u} = \nabla_s \mathbf{u} \text{ in } \Omega, & \mathbf{u} = 0 \text{ on } \Gamma_u, \\ \sigma \in \partial W(\epsilon) \text{ or } \epsilon \in \partial W^*(\sigma) & \text{in } \Omega, \\ \bar{\mathbf{b}} = \Lambda^* \sigma = -\text{div}\sigma \text{ in } \Omega, & \mathbf{n} \cdot \sigma = \varrho_c \bar{\mathbf{f}} \text{ on } \Gamma_f. \end{cases} \tag{7.9}$$

Introducing the admissible configuration space $\mathcal{U}_a$

$$\mathcal{U}_a = \left\{ \mathbf{u} \in \mathcal{U} \mid \mathbf{u}(x) = 0 \text{ on } \Gamma_u, \int_{\Gamma_f} \mathbf{u} \cdot \bar{\mathbf{f}} \mathrm{d}\Gamma = 1 \right\}. \tag{7.10}$$

Then, on $\mathcal{U} \times \mathcal{S}$, the Gauss-Green formula reads

$$\langle \Lambda \mathbf{u}; \boldsymbol{\sigma} \rangle = \int_\Omega (\nabla_s \mathbf{u}) : \boldsymbol{\sigma} d\Omega = \int_{\Gamma_f} \mathbf{u} \cdot \boldsymbol{\sigma} \cdot \mathbf{n} d\Gamma - \int_\Omega \mathbf{u} \cdot \operatorname{div} \boldsymbol{\sigma} d\Omega = \langle \mathbf{u}, \bar{\Lambda}^* \boldsymbol{\sigma} \rangle,$$

where $\bar{\Lambda}^* : \mathcal{S} \to \mathcal{U}^*$ is defined by

$$\bar{\Lambda}^* \boldsymbol{\sigma} = \begin{cases} -\nabla \cdot \boldsymbol{\sigma} & \text{in } \Omega, \\ \mathbf{n} \cdot \boldsymbol{\sigma} & \text{on } \Gamma \, . \end{cases} \tag{7.11}$$

The external energy $F : \mathcal{U} \to \overleftarrow{\mathbb{R}}$ is defined by

$$F(\mathbf{u}) = \langle \mathbf{u}, \bar{\mathbf{b}} \rangle - \mathcal{I}_{\mathcal{U}_a}(\mathbf{u}). \tag{7.12}$$

On the kinematically admissible space $\mathcal{U}_k = \{\mathbf{u} \in \mathcal{U}_a | \ \nabla_s \mathbf{u} \in \mathcal{E}_a\}$, the total potential $P(\mathbf{u}) = \varrho^\sharp(\mathbf{u})$ is the convex, Gâteaux differentiable functional

$$\varrho^\sharp(\mathbf{u}) = W(\Lambda \mathbf{u}) - F(\mathbf{u}) = \int_\Omega \sigma_b |(\nabla_s \mathbf{u})^d| d\Omega - \int_\Omega \mathbf{u} \cdot \bar{\mathbf{b}} d\Omega. \tag{7.13}$$

Thus, the primal problem is a convex variational problem

$$(\mathcal{P}_{\text{inf}}) : \quad \varrho^\sharp(\mathbf{u}) \to \inf \ \ \forall \mathbf{u} \in \mathcal{U}_k,$$

which is similar to the minimal surface type of problem discussed before in Chapter 6. Since $\mathcal{U}_k$ is a bounded convex subset, it is not difficult to prove that, if $\mathcal{U}_k$ is not empty, the minimum value of $\varrho^\sharp$ is the limit load factor, i.e.

$$\varrho_c = \varrho^\sharp(\bar{\mathbf{u}}) = \inf \varrho^\sharp(\mathbf{u}) \ \ \forall \mathbf{u} \in \mathcal{U}_k.$$

Thus, by solving the primal problem, we only have an upper bound for the safety factor ϱ_c, i.e.

$$\varrho_c \le \varrho^\sharp(\mathbf{u}) \ \ \forall \mathbf{u} \in \mathcal{U}_k.$$

To find a lower bound for ϱ_c, we need to establish dual formulations.

7.1.2 Complementary Bounding Theorems

The Lagrangian $L : \mathcal{U} \times \mathcal{S} \to \bar{\mathbb{R}}$ associated with $(\mathcal{P}_{\text{inf}})$ is

$$L(\mathbf{u}, \boldsymbol{\sigma}) = \langle \nabla_s \mathbf{u}; \boldsymbol{\sigma} \rangle - \mathcal{I}_{\mathcal{S}_a}(\boldsymbol{\sigma}) - \langle \mathbf{u}, \bar{\mathbf{b}} \rangle + \mathcal{I}_{\mathcal{U}_a}(\mathbf{u}). \tag{7.14}$$

This is a classical saddle functional. For a fixed $\boldsymbol{\sigma} \in \mathcal{S}$, the dual functional is given by

$$P^d(\boldsymbol{\sigma}) = \inf_{\mathbf{u} \in \mathcal{U}} L(\mathbf{u}, \boldsymbol{\sigma}) = F^*(\bar{\Lambda}^* \boldsymbol{\sigma}) - \mathcal{I}_{\mathcal{S}_a}(\boldsymbol{\sigma}),$$

where

$$F^*(\bar{\Lambda}^* \boldsymbol{\sigma}) = \inf_{\mathbf{u} \in \mathcal{U}_a} \{ \langle \mathbf{u}, \bar{\Lambda}^* \boldsymbol{\sigma} \rangle - \langle \mathbf{u}, \bar{\mathbf{b}} \rangle \}.$$

By introducing the Lagrange multiplier $\varrho^b > 0$, the integral constraint in $\mathcal{U}_a$ can be relaxed and we have

$$\begin{aligned} F^*(\bar{\Lambda}^*\boldsymbol{\sigma}) &= \sup_{\varrho^b \geq 0} \inf_{\mathbf{u}\in\mathcal{U}} \left\{ \langle \mathbf{u}, \bar{\Lambda}^*\boldsymbol{\sigma}\rangle - \langle \mathbf{u}, \bar{\mathbf{b}}\rangle - \varrho^b \left(\int_{\Gamma_f} \mathbf{u}\cdot\bar{\mathbf{f}} d\Gamma - 1 \right) \right\} \\ &= \sup_{\varrho^b \geq 0} \{ \varrho^b - \mathcal{I}_{\mathcal{U}_a^*}(\boldsymbol{\sigma}) \}, \end{aligned}$$

where $\mathcal{U}_a^*$ is a subset of $\mathcal{U}^*$ defined by

$$\mathcal{U}_a^* = \{\mathbf{u}^* \in \mathcal{U}^* |\ \mathbf{u}^* = \bar{\mathbf{b}} \text{ in } \Omega,\ \mathbf{u}^* = \varrho^b \bar{\mathbf{f}} \text{ on } \Gamma_f\}.$$

Thus, the statically admissible space $\mathcal{S}_s = \{\boldsymbol{\sigma} \in \mathcal{S}_a |\ \bar{\Lambda}^*\boldsymbol{\sigma} \in \mathcal{U}_a^*\}$ in this problem is a convex subset of $\mathcal{S}$, defined by

$$\mathcal{S}_s = \{\boldsymbol{\sigma} \in \mathcal{S} |\ Y^c(\boldsymbol{\sigma}) \leq 0,\ -\nabla\cdot\boldsymbol{\sigma} = \bar{\mathbf{b}} \text{ in } \Omega,\ \mathbf{n}\cdot\boldsymbol{\sigma} = \varrho^b \bar{\mathbf{f}} \text{ on } \Gamma_f\}.$$

Clearly, for any given $\boldsymbol{\sigma} \in \mathcal{S}_s$ and associated $\varrho^b(\boldsymbol{\sigma}) > 0$, the dual functional is simply the statically admissible loading factor, i.e.

$$P^d(\boldsymbol{\sigma}) = \varrho^b(\boldsymbol{\sigma}) \geq 0 \quad \forall \boldsymbol{\sigma} \in \mathcal{S}_s. \tag{7.15}$$

On the product space $\mathcal{U}_a \times \mathcal{S}_a$, $L(\mathbf{u}, \boldsymbol{\sigma})$ is finite and can be written as

$$\varrho^\natural(\mathbf{u}, \boldsymbol{\sigma}) = \langle \Lambda\mathbf{u}, \boldsymbol{\sigma}\rangle - \langle \mathbf{u}, \bar{\mathbf{b}}\rangle = \int_\Omega [(\nabla_s \mathbf{u}) : \boldsymbol{\sigma} - \mathbf{u}\cdot\bar{\mathbf{b}}] d\Omega. \tag{7.16}$$

From the saddle minimax theorem, it is easy to obtain the following result.

Theorem 7.1.1 (Complementary Bounding Theorem)
(T1) For any given $(\mathbf{u}, \boldsymbol{\sigma}) \in \mathcal{K}_a \times \mathcal{S}_s$, the safety factor ϱ_c of the limit state is bounded by $\varrho^\sharp(\mathbf{u})$ and $\varrho^b(\boldsymbol{\sigma})$, i.e.

$$\varrho^\sharp(\mathbf{u}) \geq \varrho_c \geq \varrho^b(\boldsymbol{\sigma}) \ \forall (\mathbf{u}, \boldsymbol{\sigma}) \in \mathcal{U}_k \times \mathcal{S}_s. \tag{7.17}$$

(T2) The safety factor ϱ_c is the saddle value of $\varrho^\natural(\mathbf{u}, \boldsymbol{\sigma})$ in the form

$$\inf_{\mathbf{u}\in\mathcal{U}_k} \sup_{\boldsymbol{\sigma}\in\mathcal{S}_a} \varrho^\natural(\mathbf{u}, \boldsymbol{\sigma}) = \varrho_c = \sup_{\boldsymbol{\sigma}\in\mathcal{S}_s} \inf_{\mathbf{u}\in\mathcal{U}_a} \varrho^\natural(\mathbf{u}, \boldsymbol{\sigma}). \tag{7.18}$$

(T3) For a given $\bar{\mathbf{f}} \in \mathcal{U}^*$, if $\mathcal{U}_k$ and $\mathcal{S}_s$ are not empty, then

$$\inf_{\mathbf{u}\in\mathcal{U}_k} \varrho^\sharp(\mathbf{u}) = \varrho_c = \sup_{\boldsymbol{\sigma}\in\mathcal{S}_s} \varrho^b(\boldsymbol{\sigma}). \tag{7.19}$$

7.1.3 Penalty-Duality Methods and Lower Bound Theorems

We now wish to relax the inequality constraint in the lower bound problem based upon construction of the complementary super-potential $W^*(\boldsymbol{\sigma}) = \mathcal{I}_{\mathcal{S}_a}(\boldsymbol{\sigma})$. Three methods will be discussed, by means of which some useful bounding theorems can be developed.

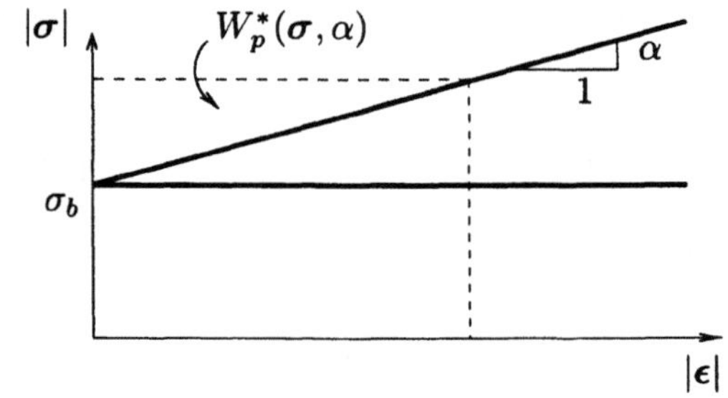

Figure 7.2. Rigid plastic model with linear hardening.

I. *Penalty-Construction.* A direct construction of the complementary super-potential $W^*(\boldsymbol{\sigma})$ is given by the well-known *exterior penalty method.* We define

$$W_p^*(\boldsymbol{\sigma},\alpha) = \int_\Omega \frac{1}{2\alpha}\,[Y^c(\boldsymbol{\sigma})]_+^2 \mathrm{d}\Omega, \tag{7.20}$$

where $\alpha > 0$ is a penalty factor and $[Y^c]_+ = \max\{0, Y^c\}$. It is clear that, for any given $\boldsymbol{\sigma} \in \mathcal{S}$, we have

$$\mathcal{I}_{\mathcal{S}_a}(\boldsymbol{\sigma}) = \sup_{\alpha>0} W_p^*(\boldsymbol{\sigma},\alpha). \tag{7.21}$$

Thus, a penalty-type lower bound for the safety factor is given as

$$\begin{aligned} \varrho_c &= \lim_{\alpha\to+0} \sup_{\boldsymbol{\sigma}\in\mathcal{S}} \{\, \varrho^b(\boldsymbol{\sigma}) - W_p^*(\boldsymbol{\sigma},\alpha)\}, \\ s.t. &\quad -\nabla\cdot\boldsymbol{\sigma} = \bar{\mathbf{b}} \ \text{ in } \Omega,\ \ \boldsymbol{\sigma}\cdot\mathbf{n} = \varrho^b\bar{\mathbf{f}} \ \text{ on } \Gamma_f. \end{aligned}$$

The physical significance of the exterior penalty method in constrained optimization was discussed by Gao (1986, 1988). To see the physical meaning of this penalty-type super-potential approach in limit analysis, let us consider a rigid-plastic model with linear hardening, as shown in Fig. 7.2. For a given hardening parameter $\alpha > 0$, the plastic complementary energy is just $W_p^*(\boldsymbol{\sigma},\alpha)$. As the hardening parameter $\alpha \to 0$, the hardening material approaches the perfectly plastic media, and the complementary energy $W_p^*(\boldsymbol{\sigma},\alpha)$ approaches the complementary super-potential $W^*(\boldsymbol{\sigma}) = \mathcal{I}_{\mathcal{S}_a}(\boldsymbol{\sigma})$, i.e.

$$W^*(\boldsymbol{\sigma}) = \lim_{\alpha\to 0^+} W_p^*(\boldsymbol{\sigma},\alpha)\ \ \forall \boldsymbol{\sigma} \in \mathcal{S}.$$

One of the main intrinsic disadvantages of the penalty method is that the problem becomes ill-conditioned when the penalty factor becomes sufficiently small. Also, the convergence rate for this method is rather slow.

II. *Dual Construction.* The Lagrange multiplier method is a duality-type construction of the W^*, with

$$W_d^*(\boldsymbol{\sigma},\mu) = \int_\Omega \mu[Y^c(\boldsymbol{\sigma})]_+ \mathrm{d}\Omega, \tag{7.22}$$

where the parameter $\mu \geq 0$ is a Lagrange multiplier for the inequality constraint $Y^c(\boldsymbol{\sigma}) \leq 0$. For any given $\boldsymbol{\sigma} \in \mathcal{S}$, we have

$$\mathcal{I}_{\mathcal{S}_a}(\boldsymbol{\sigma}) = \sup_{\mu \geq 0} W_d^*(\boldsymbol{\sigma}, \mu). \tag{7.23}$$

Let $\varrho_d : \mathcal{U} \times \mathcal{S} \to \mathbb{R}$ be defined by

$$\varrho_d(\mathbf{u}, \boldsymbol{\sigma}, \mu) = \frac{\int_\Omega [\,(\nabla \mathbf{u}) : \boldsymbol{\sigma} - \ \mu[Y^c(\boldsymbol{\sigma})]_+ - \bar{\mathbf{b}} \cdot \mathbf{u}] d\Omega}{\int_{\Gamma_t} \mathbf{u} \cdot \bar{\mathbf{f}} d\Gamma}. \tag{7.24}$$

Then, the duality-type generalized variational principle for the safety factor ϱ_c is obtained as

$$\varrho_c = \inf_{\mathbf{u} \in \mathcal{U}} \sup_{\mu \geq 0} \sup_{\boldsymbol{\sigma} \in \mathcal{S}} \varrho_d(\mathbf{u}, \boldsymbol{\sigma}, \mu). \tag{7.25}$$

Since the yield condition is a nonlinear inequality constraint, the critical point $(\bar{\mathbf{u}}, \bar{\boldsymbol{\sigma}}, \bar{\mu})$ of ϱ_d is subject to the KKT conditions

$$\bar{\mu} \geq 0, \ \ Y^c(\bar{\boldsymbol{\sigma}}) \leq 0, \ \ \bar{\mu} Y^c(\bar{\boldsymbol{\sigma}}) = 0 \ \text{ in } \Omega. \tag{7.26}$$

The formulation for the admissible Lagrange multiplier μ was discussed extensively by mechanicians about thirty years ago (cf., e.g., Qiang and Zhong, 1963; Mura and Lee, 1963; Wang *et al*, 1965 and Hsue, 1975). By the mathematical theory of duality, we know that the Lagrange multiplier is a dual variable of the function $Y^c(\boldsymbol{\sigma})$. For the von Mises material, from the criticality condition $D\varrho_d(\bar{\mathbf{u}}, \bar{\boldsymbol{\sigma}}, \bar{\mu}) = 0$ and the KKT condition, the correct form for the Lagrange multiplier $\bar{\mu}$ is simply the Legendre conjugate of $|\boldsymbol{\sigma}^d| \in \mathcal{S}$ (cf., Gao, 1983, 1988a), i.e.

$$\bar{\mu}(\bar{\mathbf{u}}) = Y(\bar{\boldsymbol{\epsilon}}) = |\boldsymbol{\epsilon}^d(\bar{\mathbf{u}})|.$$

Its physically meaning is the plastic flow factor.

To find the solution of (7.25) subject to the KKT conditions is usually difficult. This is one of the main disadvantages of the pure Lagrange multiplier method for inequality constrained problems. But, by the duality-type generalized variational principle, some useful bounding theorems can be developed. For the purpose of engineering applications, we give two very simple bounding theorems proposed originally by Gao (1983).

First, we let

$$\varrho_d^\natural(\mathbf{u}, \boldsymbol{\sigma}) = \int_\Omega [\,(\nabla \mathbf{u}) : \boldsymbol{\sigma} - \mu(\mathbf{u}) Y^c(\boldsymbol{\sigma}) - \bar{\mathbf{b}} \cdot \mathbf{u}] d\Omega. \tag{7.27}$$

Theorem 7.1.2 *For any given independent variables* $(\mathbf{u}, \boldsymbol{\sigma}) \in \mathcal{U}_a \times \mathcal{S}$, *the inequalities*

$$\varrho^\flat(\boldsymbol{\sigma}) \leq \varrho_d^\natural(\mathbf{u}, \boldsymbol{\sigma}) \ \ \forall\, \boldsymbol{\sigma} \in \mathcal{S}_s, \tag{7.28}$$

$$\varrho^\sharp(\mathbf{u}) \ \geq \varrho_d^\natural(\mathbf{u}, \boldsymbol{\sigma}) \ \ \forall\, \mathbf{u} \in \mathcal{U}_k \tag{7.29}$$

hold.

Proof. Since $Y^c(\boldsymbol{\sigma}) \leq 0 \; \forall \boldsymbol{\sigma} \in \mathcal{S}_s$ and $\mu(\mathbf{u}) \geq 0 \, \forall \mathbf{u} \in \mathcal{U}_a$, we have

$$\varrho_d^\natural(\mathbf{u}, \boldsymbol{\sigma}) - \varrho^\flat(\boldsymbol{\sigma}) = -\int_\Omega \mu(\mathbf{u}) Y^c(\boldsymbol{\sigma}) \mathrm{d}\Omega \geq 0 \;\; \forall (\mathbf{u}, \boldsymbol{\sigma}) \in \mathcal{U}_a \times \mathcal{S}_s.$$

On the other hand, for any given $(\mathbf{u}, \boldsymbol{\sigma}) \in \mathcal{U}_k \times \mathcal{S}$, we obtain

$$\varrho^\sharp(\mathbf{u}) - \varrho_d^\natural(\mathbf{u}, \boldsymbol{\sigma}) = \int_\Omega [|\boldsymbol{\sigma}^d||\boldsymbol{\epsilon}^d(\mathbf{u})| - \boldsymbol{\sigma}^d : \boldsymbol{\epsilon}^d(\mathbf{u}) \,] \mathrm{d}\Omega \geq 0 \tag{7.30}$$

by taking the Cauchy-Schwarz inequality into account. □

Let $\mathcal{S}_e \subset \mathcal{S}$ be an equilibrium admissible space defined by

$$\mathcal{S}_e = \{\boldsymbol{\sigma} \in \mathcal{S} | \;\; -\nabla \cdot \boldsymbol{\sigma} = \bar{\mathbf{b}} \;\text{ in } \Omega, \;\; \mathbf{n} \cdot \boldsymbol{\sigma} = \bar{\mathbf{f}} \;\text{ on } \Gamma_f\}.$$

Then, on $\mathcal{S}_e$, a simple lower bound theorem can be proposed.

Theorem 7.1.3 *For an arbitrarily given statically admissible field $\boldsymbol{\sigma} \in \mathcal{S}_e$ and associated $\varrho^\flat > 0$ on Γ_f, if there exists at least one point $x \in \Omega$ such that $[Y^c(\boldsymbol{\sigma}(x))]_+ \neq 0$, then*

$$\varrho_c \geq \frac{\sigma_b \varrho^\flat(\boldsymbol{\sigma})}{\max_{x \in \Omega}\{|\boldsymbol{\sigma}^d(x)|\}}. \tag{7.31}$$

Proof. By the duality-type generalized variational principle (7.25), we have

$$\begin{aligned}
\varrho_c &= \inf_{\mathbf{u}\in\mathcal{U}} \sup_{\boldsymbol{\sigma}\in\mathcal{S}} \sup_{\mu\geq 0} \varrho_d(\mathbf{u}, \boldsymbol{\sigma}, \mu) \\
&= \sup_{\mu\geq 0} \sup_{\boldsymbol{\sigma}\in\mathcal{S}_e} \left\{ \varrho^\flat(\boldsymbol{\sigma}) - \int_\Omega \mu [Y^c(\boldsymbol{\sigma})]_+ \mathrm{d}\Omega \right\} \\
&\geq \sup_{\mu\geq 0} \left\{ \varrho^\flat(\boldsymbol{\sigma}) - \int_\Omega \mu [Y^c(\boldsymbol{\sigma})]_+ \mathrm{d}\Omega \right\} \quad \forall \boldsymbol{\sigma} \in \mathcal{S}_e.
\end{aligned}$$

By assumption, there exists at least one point $x \in \Omega$ at which $[Y^c(\boldsymbol{\sigma}(x)]_+ = Y^c(\boldsymbol{\sigma}(x)) = |\boldsymbol{\sigma}^d(x)| - \sigma_b > 0$. Thus,

$$\begin{aligned}
\varrho_c &\geq \sup_{\mu\geq 0} \left\{ \varrho^\flat(\boldsymbol{\sigma}) - [\max_{x\in\Omega} Y^c(\boldsymbol{\sigma}(x))] \int_\Omega \mu \mathrm{d}\Omega \right\} \\
&= \varrho^\flat(\boldsymbol{\sigma}) + (\sigma_b - \max_{x\in\Omega} |\boldsymbol{\sigma}^d(x)|) \inf_{\mu\geq 0} \int_\Omega \mu \mathrm{d}\Omega.
\end{aligned} \tag{7.32}$$

From the fact that $\mu = |\boldsymbol{\epsilon}^d(\mathbf{u})|$, the primal variational principle leads to

$$\inf_{\mu\geq 0} \int_\Omega \sigma_b \mu \mathrm{d}\Omega = \inf_{\mathbf{u}\in\mathcal{U}_a} \int_\Omega \sigma_b \mu(\mathbf{u}) \mathrm{d}\Omega = \inf_{\mathbf{u}\in\mathcal{U}_a} \int_\Omega \sigma_b |\boldsymbol{\epsilon}^d(\mathbf{u})| \mathrm{d}\Omega = \varrho_c.$$

Substituting this into (7.32) we obtain the lower bound theorem (7.31). □

III. *Penalty-Duality Construction.* Combining both the penalty and the duality-type generalized variational approach, we obtain the *penalty-duality type construction* of the indicator $\mathcal{I}_{\mathcal{S}_a}(\boldsymbol{\sigma})$. Thus,

$$W^*_{pd}(\boldsymbol{\sigma}, \mu, \alpha) = \begin{cases} \int_\Omega [\mu Y^c(\boldsymbol{\sigma}) + \frac{1}{2\alpha}(Y^c(\boldsymbol{\sigma}))^2] d\Omega & \text{if } \mu\alpha + Y^c(\boldsymbol{\sigma}) > 0 \\ -\int_\Omega \frac{1}{2}\alpha\mu^2 d\Omega & \text{otherwise} \end{cases}$$
$$= \int_\Omega \frac{\alpha}{2}\left\{[\mu + \frac{1}{\alpha}Y^c(\boldsymbol{\sigma})]_+^2 - \mu^2\right\} d\Omega. \tag{7.33}$$

Clearly, for any given $\boldsymbol{\sigma} \in \mathcal{S}$,

$$\mathcal{I}_{\mathcal{S}_a}(\boldsymbol{\sigma}) = \sup_{\mu\geq 0}\sup_{\alpha>0} W^*_{pd}(\boldsymbol{\sigma}, \mu, \alpha). \tag{7.34}$$

In mathematical programming, this penalty-duality type construction is the so-called *augmented Lagrangian method.* Thus, the augmented Lagrangian for limit analysis can be defined by

$$\varrho_{pd}(\mathbf{u}, \boldsymbol{\sigma}, \mu; \alpha) = \int_\Omega [(\nabla \mathbf{u}) : \boldsymbol{\sigma} - \bar{\mathbf{b}} \cdot \mathbf{u}] d\Omega - W^*_{pd}(\boldsymbol{\sigma}, \mu, \alpha). \tag{7.35}$$

Theorem 7.1.4 *There exists an $\alpha_o > 0$ such that for any given $\alpha \in (0, \alpha_o]$, the safety factor ϱ_c is the stationary value of the variational problem*

$$\varrho_c = \inf_{\mathbf{u}\in\mathcal{U}_a} \sup_{\boldsymbol{\sigma}\in\mathcal{S}} \inf_{\mu\geq 0} \varrho_{pd}(\mathbf{u}, \boldsymbol{\sigma}, \mu; \alpha). \tag{7.36}$$

This theorem shows that, for a fixed penalty parameter $\alpha \in (0, \alpha_o]$, the safety factor can be obtained by modifying the Lagrange multiplier $\mu \geq 0$. Thus, an efficient algorithm for solving limit analysis problems can be suggested.

Penalty-Duality Algorithm

(i) Given penalty-duality parameters $\alpha_k > 0$, $\mu_k \geq 0$, determine $\boldsymbol{\sigma}_k$, $\mathbf{u}_k$, ϱ_k by solving

$$\varrho_k = \inf_{\mathbf{u}\in U} \sup_{\boldsymbol{\sigma}\in\Sigma} \varrho_{pd}(\mathbf{u}, \boldsymbol{\sigma}, \lambda_n; \alpha_n). \tag{7.37}$$

(ii) For a given $\omega > 0$, if $|\alpha_k\mu_k + Y^c(\boldsymbol{\sigma}_k)| \leq \omega$ then $(\mathbf{u}_k, \boldsymbol{\sigma}_k)$ is the desired solution. Otherwise, go to the next step.

(iii) Modify the penalty-duality parameter (α_k, μ_k) by

$$\alpha_{k+1} = \begin{cases} \gamma\alpha_k & \text{if } |Y^c(\boldsymbol{\sigma}_k)| \geq \theta|Y^c(\boldsymbol{\sigma}_{k-1})|, \\ \alpha_k & \text{otherwise;} \end{cases} \tag{7.38}$$

$$\mu_{k+1} = \left[\mu_k + \frac{1}{\alpha_k}Y^c(\boldsymbol{\sigma}_k)\right]_+ \quad \text{in } \Omega; \tag{7.39}$$

let $k = k + 1$; go to (i).

The parameters γ, θ in this algorithm are positive constants. Suggested values are $\gamma \in [0.1 \ , \ 0.25\ ]$, $\theta \in [\ 0.1 \ , \ 0.5\ ]$. The convergence rate of this algorithm is controlled mainly by the penalty factor α_k. The smaller α_k, the faster the convergence rate. But the disadvantages of pure penalty function method may appear in this penalty-duality type algorithm if a sufficiently small penalty factor is used. Based on this algorithm, a primal-dual finite element programming method has been developed by Gao (1988b) for solving general plastic limit analysis problems. Related work on mixed finite element methods for plastic limit analysis were discussed by Belytschko and Hodge (1970); Strang *et al*, (1979, 1980, 1988); Casciano and Cascini (1982); Christiansen and Larsen (1983).

7.2 Contact Problems of Extended Elastoplastic Beam Theory

As a typical example of bi-complementarity problems in geometrically linear systems, we will use the extended elastic-perfectly plastic beam model developed recently in (Gao, 1998) to illustrate the duality theory in variational inequalities. Large deformation problems will be discussed in the next section.

7.2.1 Extended Beam Model and Contact Problem

Let us consider an elastic-perfectly plastic beam, possibly in contact with a rigid obstacle which is described by a strictly concave function $\psi(x)$ (see Fig. 7.3). Suppose that the beam in the x-y plane is the rectangle

$$\Omega = \{(x, y) \in \mathbb{R}^2 | 0 \le x \le \ell, \ -h \le y \le h\}.$$

The beam is subjected to a given distributed load $\bar{\mathbf{p}} = (\bar{q}^+(x), -\bar{p}(x))^T$ on its top surface $y = h$. On the bottom surface, the beam is subjected to a frictional shear force $q^-(x)$, which depends on a friction law. Let $u(x, y)$ denote the horizontal displacement of the material point $(x, y) \in \Omega$ and $w(x)$ describe the displacement of the middle axis, which coincides with the x-axis in the equilibrium state. Thus, the general displacement space $\mathcal{U}$ is a vector space in $\mathbb{R}^2$, given by

$$\mathcal{U} := \left\{ \mathbf{u} = \begin{pmatrix} u(x,y) \\ w(x) \end{pmatrix} \mid u(x,y) \in \mathcal{L}^2(\Omega), w(x) \in \mathcal{L}^2(0,\ell) \right\}.$$

Ignoring body forces on the beam, the source space $\mathcal{U}^*$ is defined by

$$\mathcal{U}^* := \left\{ \mathbf{p} = \begin{pmatrix} q(x) \\ p(x) \end{pmatrix} \mid q(x), \ \ p(x) \in \mathcal{L}^2(0,\ell) \right\}.$$

The bilinear form $\langle *, * \rangle : \mathcal{U} \times \mathcal{U}^* \to \mathbb{R}$, defined by

$$\langle \mathbf{u}, \mathbf{p} \rangle = \int_0^\ell q^+(x) u(x,h) \ dx + \int_0^\ell q^-(x) u(x,-h) \ dx + \int_0^\ell p(x) w(x) \ dx$$

puts $\mathcal{U}$ and $\mathcal{U}^*$ into duality. For infinitesimal deformations of the beam, the general strain vector is defined by

$$\boldsymbol{\epsilon} = \begin{pmatrix} \epsilon \\ \gamma \end{pmatrix} = \Lambda \mathbf{u} = \begin{pmatrix} \frac{\partial}{\partial x} & 0 \\ \frac{\partial}{\partial y} & \frac{\partial}{\partial x} \end{pmatrix} \begin{pmatrix} u(x,y) \\ w(x) \end{pmatrix} = \begin{pmatrix} \frac{\partial u}{\partial x} \\ \frac{\partial u}{\partial y} + \frac{\partial w}{\partial x} \end{pmatrix}, \tag{7.40}$$

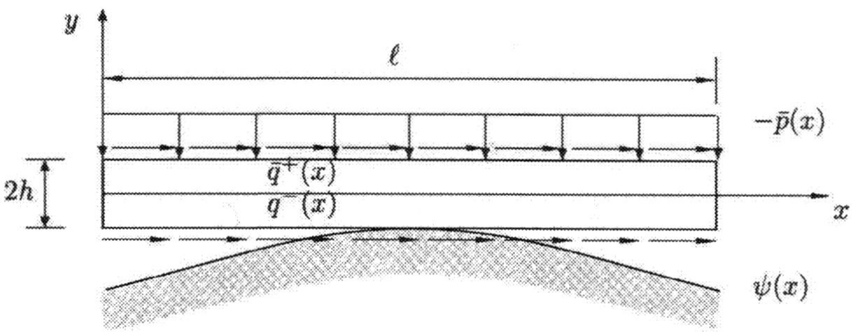

Figure 7.3. Frictional contact problem of elastoplastic beam

where the geometrical operator Λ is a linear differential operator. The *general strain space* $\mathcal{E}$ and its dual space $\mathcal{E}^* = \mathcal{S}$, the *general stress space*, are defined by

$$\mathcal{E} := \left\{ \boldsymbol{\epsilon} = \begin{pmatrix} \epsilon(x,y) \\ \gamma(x,y) \end{pmatrix} \mid \epsilon(x,y), \gamma(x,y) \in \mathcal{L}^2(\Omega) \right\}.$$

$$\mathcal{S} := \left\{ \boldsymbol{\sigma} = \begin{pmatrix} \sigma(x,y) \\ \tau(x,y) \end{pmatrix} \mid \sigma(x,y), \tau(x,y) \in \mathcal{L}^2(\Omega) \right\}.$$

The bilinear form $\langle *; * \rangle : \mathcal{E} \times \mathcal{S} \to \mathbb{R}$, i.e.

$$\langle \boldsymbol{\epsilon}; \boldsymbol{\sigma} \rangle = \int_\Omega (\sigma\epsilon + \tau\gamma) \mathrm{d}\Omega,$$

embodies the duality relationship between $\mathcal{E}$ and $\mathcal{S}$.

For any given $\mathbf{u} = (u, w)^T$ such that $\boldsymbol{\epsilon} = \Lambda \mathbf{u}$, the Gauss-Green formula leads to

$$\begin{aligned} \langle \Lambda \mathbf{u}; \boldsymbol{\sigma} \rangle &= \langle \mathbf{u}, \bar{\Lambda}^* \boldsymbol{\sigma} \rangle = -\int_\Omega \tau_{,x} w \mathrm{d}\Omega - \int_\Omega (\sigma_{,x} + \tau_{,y}) u \mathrm{d}\Omega \\ &+ \int_{-h}^{h} \sigma n_x u dy|_{x=0,\ell} + \int_{-h}^{h} \tau n_x w dy|_{x=0,\ell} + \int_0^\ell \tau n_y u dx|_{y=\pm h}. \end{aligned}$$

Thus, the adjoint operator $\bar{\Lambda}^* : \mathcal{S} \to \mathcal{U}^*$ is defined by

$$\bar{\Lambda}^* \boldsymbol{\sigma} = \begin{cases} \begin{pmatrix} -\frac{\partial}{\partial x} & -\frac{\partial}{\partial y} \\ 0 & -\frac{\partial}{\partial x} \end{pmatrix} \begin{pmatrix} \sigma \\ \tau \end{pmatrix} & \text{in } \Omega, \\ \sigma n_x & \text{at } x = 0, \ell \text{ in } x\text{-direction}, \\ \tau n_x & \text{at } x = 0, \ell \text{ in } y\text{-direction}, \\ \tau n_y & \text{at } y = \pm h. \end{cases}$$

The feasible set $\mathcal{U}_a \subset \mathcal{U}$ is a subspace incorporating the *essential* boundary conditions. For example, if the beam is clamped at both ends, this space is given as

$$\mathcal{U}_a = \{(u,w)^T \in \mathcal{U} |\ u = w = 0 \text{ at } x = 0, \ell\}.$$

Then, for any given $\mathbf{u} \in \mathcal{U}_a$, the virtual work principle

$$\langle \Lambda \mathbf{u}; \sigma \rangle = \langle \mathbf{u}, \bar{\Lambda}^* \sigma \rangle = \langle \mathbf{u}, \mathbf{p} \rangle$$

gives the equilibrium conditions

$$\bar{\Lambda}^* \sigma = \mathbf{p} \quad \Rightarrow \quad \begin{cases} \sigma_{,x} + \tau_{,y} = 0 \quad \forall (x,y) \in \Omega, \\ -\int_{-h}^{h} \tau_{,x} dy = p(x) \quad \forall x \in (0,\ell), \\ \tau(x, \pm h) = \pm q^{\pm}(x) \quad \forall x \in (0,\ell) \end{cases} \tag{7.41}$$

for this extended beam model.

For a linear elastic beam, the constitutive equation is given by Hooke's law. Thus,

$$\sigma = C\epsilon, \quad \begin{pmatrix} \sigma \\ \tau \end{pmatrix} = \begin{pmatrix} E & 0 \\ 0 & G \end{pmatrix} \begin{pmatrix} \epsilon \\ \gamma \end{pmatrix}, \tag{7.42}$$

where E, G are positive elastic constants. The fundamental equation $\Lambda^* C \Lambda \mathbf{u} = \mathbf{p}$ for this linear elastic beam model is the linear partial differential system

$$\begin{array}{ll} u_{,xx} + \beta u_{,yy} = 0 & \text{in } \Omega, \\ 2hw_{,xx}(x) + u_{,x}(x,h) - u_{,x}(x,-h) + p(x)/G = 0 & \forall x \in (0,\ell), \\ u_{,y}(x, \pm h) + w_{,x}(x) = \pm q^{\pm}(x)/G & \forall x \in (0,\ell), \end{array} \tag{7.43}$$

where $\beta = G/E$. For certain given boundary conditions, this linear system possesses a unique solution.

7.2.2 Elastoplastic Contact Problem and Bi-Complementarity

We assume that the shape $\psi(x)$ of the obstacle is a strictly concave function. For the given external load $\bar{\mathbf{p}}$, the external complementarity condition in the vertical direction is the well-known *Signorini condition*

$$\begin{array}{c} w(x) - \psi(x) \geq 0, \ \ \bar{p}(x) - p(x) \leq 0, \\ (w(x) - \psi(x))(\bar{p}(x) - p(x)) = 0 \end{array} \quad \forall x \in (0,\ell). \tag{7.44}$$

The operator $B(\mathbf{u}) = w - \psi$ is an affine operator, which depends only on w, and we denote it simply by $B(w)$. Its complementary operator $B^c(\mathbf{p}) = B(p) = -p + \bar{p}$ depends on the vertical force $p(x)$ only. Since the contact region $\ell_c \subset (0,\ell)$, defined by

$$\ell_c := \{x \in (0,\ell) |\ w(x) = \psi(x) \ \ \forall x \in (0,\ell)\},$$

and its boundary are unknown until the problem is solved, this problem is also referred as a free boundary-value problem.

For a lubricated contact problem, the lubrication implies that $q^-(x) = 0$. However, in the case of *Coulomb's law of dry friction*, we have

$$|q^-(x)| < -\nu\bar{p} \quad \Rightarrow \quad u(x,-h) = 0 \ \ \forall x \in \ell_c,$$
$$|q^-(x)| = -\nu\bar{p} \quad \Rightarrow \quad u(x,-h) = -\lambda^* q^-(x) \ \ \forall x \in \ell_c \ \text{ for some } \lambda^* \geq 0.$$

Here, ν is a coefficient of friction and $\lambda^* \geq 0$ is a physical parameter. In this case, the external complementarity condition in the x-direction is

$$|u(x,-h)| \geq 0, \ \ |q^-(x)| + \nu\bar{p}(x) \leq 0, \ \ u(x,-h)[|q^-(x)| + \nu\bar{p}(x)] = 0. \tag{7.45}$$

For general contact problems with non-classical friction laws, a super-potential $j : \mathcal{U} \to \overrightarrow{\mathbb{R}}$ and its Fenchel conjugate $j^* : \mathcal{U}^* \to \overrightarrow{\mathbb{R}}$ can be introduced such that the frictional complementarity condition on the bottom surface $y = -h$ in the x-direction can generally be written in one of the equivalent forms

$$-q^- \in \partial^- j(u) \ \ \Leftrightarrow \ \ u \in \partial^- j^*(-q^-) \ \ \Leftrightarrow \ \ j(u) + j^*(-q^-) = \langle u, -q^- \rangle.$$

For Coulomb's law of dry friction, this super-potential can be written as

$$j(u(x,-h)) = \sup_{|q^-| \leq -\nu\bar{p}} \int_0^\ell -q^-(x) u(x,-h) \, dx,$$

and the frictional law on the boundary $y = -h$ can be expressed as

$$\tau(x,-h) \in \partial j(u(x,-h)), \ \ \forall x \in (0,\ell).$$

When a certain combination of stresses in the beam results in strains exceeding the limit of elastic behavior, plastic deformation begins to take place. Let $\Omega_e \subset \Omega$ be the elastic zone and $\Omega_p \subset \Omega$ be the plastic zone such that $\Omega_e \cup \Omega_p = \Omega$, $\Omega_e \cap \Omega_p = \emptyset$. In the elastic zone Ω_e, the constitutive relation between ϵ and σ is controlled by Hooke's law (7.43). However, in the plastic zone Ω_p, the strain vector $\boldsymbol{\epsilon}$ can be decomposed into two parts as $\boldsymbol{\epsilon} = \boldsymbol{\epsilon}^e + \boldsymbol{\epsilon}^p$, i.e. the elastic strain $\boldsymbol{\epsilon}^e$, which is defined by the inverse of Hooke's law, and the plastic strain $\boldsymbol{\epsilon}^p$, which is given by *Hencky's plastic constitutive law*

$$\begin{aligned} \boldsymbol{\epsilon}^p &= \mu \frac{\partial Y^c(\boldsymbol{\sigma})}{\partial \boldsymbol{\sigma}} \quad \text{in } \Omega_p, \\ s.t. \ \ & \mu \geq 0, \ \ Y^c(\boldsymbol{\sigma}) \leq 0, \ \ \mu Y^c(\boldsymbol{\sigma}) = 0 \quad \text{in } \Omega. \end{aligned} \tag{7.46}$$

Here, μ is a plastic flow factor, $Y^c(\boldsymbol{\sigma})$ is the plastic yield function

$$Y^c(\boldsymbol{\sigma}) = |\boldsymbol{\sigma}|_\alpha - \sigma_b = \sqrt{\sigma^2 + \alpha\tau^2} - \sigma_b, \tag{7.47}$$

σ_b is a material constant and $\alpha > 0$ is a parameter. For the von Mises material, $\alpha = 3$, for the Tresca material, $\alpha = 4$ (cf., e.g., Chakrabarty (1987)), and $|\boldsymbol{\sigma}|_\alpha = \sqrt{\sigma^2 + \alpha\tau^2}$ is the effective stress, which is a norm for the stress vector

σ. For proportional loading problem, i.e. during plastic deformation, the strain vectors ϵ^e and ϵ^p remain in the same direction and it is easy to find that

$$\mu = Y(\epsilon) = |\epsilon|_{1/\alpha} - \epsilon_b = \sqrt{\epsilon^2 + \frac{1}{\alpha}\gamma^2} - \epsilon_b, \tag{7.48}$$

where $|\epsilon|_{1/\alpha} = \sqrt{\epsilon^2 + \frac{1}{\alpha}\gamma^2}$ is a norm for the strain vector and $\epsilon_b = \sigma_b/E$. By using the step function $\delta(Y^c)$, defined by

$$\delta(Y^c) = \begin{cases} 1 & \text{if } Y^c = 0, \\ 0 & \text{if } Y^c < 0, \end{cases}$$

the constitutive relation can be written as

$$\epsilon = \mathbf{C}^{-1}\sigma + Y\frac{\partial Y^c(\sigma)}{\partial \sigma}\delta(Y^c(\sigma)) \tag{7.49}$$

subject to the internal complementarity condition

$$Y(\epsilon) \geq 0, \quad Y^c(\sigma) \leq 0, \quad Y(\epsilon)Y^c(\sigma) = 0. \tag{7.50}$$

The elastic region Ω_e and the plastic region Ω_p are determined respectively by

$$\Omega_e := \{\ (x,y) \in \Omega |\ Y(\epsilon(x,y)) < 0 \quad \text{or} \quad Y^c(\sigma(x,y)) < 0\ \forall (x,y) \in \Omega\},$$

$$\Omega_p := \{\ (x,y) \in \Omega |\ Y^c(\sigma(x,y)) = 0 \quad \text{or} \quad Y(\epsilon(x,y)) \geq 0\ \forall (x,y) \in \Omega\}.$$

Thus, the interface of the elastic-plastic region $\Gamma_{ep} = \Omega_e \cap \Omega_p$ is defined by

$$\Gamma_{ep} := \{(x,y) \in \Omega |\ Y(\epsilon(x,y))Y^c(\sigma(x,y)) = 0\}.$$

The contact problem for this extended elastoplastic beam theory is then proposed as follows.

Problem 7.2.1 For a given obstacle function $\psi(x)$ and external load $\bar{p}(x)$, $\bar{q}^+(x)$, find the displacement field $\mathbf{u}(x,y) = (u(x,y), w(x))^T \in \mathcal{U}_a$ and the stress field $\sigma = (\sigma(x,y), \tau(x,y))^T$ such that the following equations are satisfied.

1. Geometrical equation :

$$\epsilon = \Lambda \mathbf{u}, \quad \text{or} \quad \begin{cases} \epsilon = u_{,x}, \\ \gamma = u_{,y} + w_{,x}. \end{cases}$$

2. Constitutive equation :

$$\epsilon = \mathbf{C}^{-1}\sigma + Y\frac{\partial Y^c(\sigma)}{\partial \sigma}\delta(Y^c(\sigma)), \quad \text{or} \quad \begin{cases} \epsilon = E^{-1}\sigma + Y\frac{\partial Y^c(\sigma)}{\partial \sigma}\delta(Y^c(\sigma)), \\ \gamma = G^{-1}\tau + Y\frac{\partial Y^c(\sigma)}{\partial \tau}\delta(Y^c(\sigma)). \end{cases}$$

3. Equilibrium conditions :

$$\Lambda^*\sigma = \mathbf{p}, \quad \text{or} \quad \begin{cases} \int_{-h}^{h} -\tau_{,x} \mathrm{d}y = p(x) \quad \forall x \in (0,\ell), \\ \sigma_{,x} + \tau_{,y} = 0 \quad \forall (x,y) \in \Omega, \end{cases}$$

$$\tau(x,h) = \bar{q}^{+}(x), \quad \tau(x,-h) \in \partial j(u(x,-h)) \quad \forall x \in (0,\ell).$$

4. External complementarity condition :

$$w(x) - \psi(x) \geq 0, \ \bar{p}(x) - p(x) \leq 0, \ (w - \psi)(\bar{p} - p) = 0.$$

5. Internal complementarity condition :

$$Y(\epsilon) \geq 0, \ Y^c(\sigma) \leq 0, \ Y(\epsilon)Y^c(\sigma) = 0.$$

♠

This is a constitutively nonlinear bi-complementarity problem. In the case of the frictional contact problem, there are two external complementarity conditions on the contact surface. Direct approaches for solving this problem are very difficult. We need to find its dual formulations.

7.2.3 Dual Variational Inequality and Applications

Let $\mathcal{S}_a$, defined by

$$\mathcal{S}_a = \{\sigma \in \mathcal{S} | \ Y^c(\sigma(x,y)) \leq 0 \ \forall (x,y) \in \Omega\}, \tag{7.51}$$

denote a convex subset of the stress space $\mathcal{S}$. The plastic super-potential $W_p^*(\sigma)$ is defined by

$$W_p^*(\sigma) = \begin{cases} 0 & \text{if } \sigma \in \mathcal{S}_a, \\ +\infty & \text{otherwise,} \end{cases} \tag{7.52}$$

which is the indicator function of the convex set $\mathcal{S}_a$. Its subdifferential is a point to set map

$$\partial^- W_p^*(\sigma) = \begin{cases} Y\frac{\partial Y^c(\sigma)}{\partial \sigma} & \text{if } Y^c(\sigma) = 0, Y \geq 0, \\ \{0\} & \text{if } Y^c(\sigma) < 0, \\ \emptyset & \text{if } Y^c(\sigma) > 0. \end{cases} \tag{7.53}$$

Hence, with this subdifferentiation notation, the plastic flow law (7.46) can be written in the form

$$\epsilon^p \in \partial^- W_p^*(\sigma). \tag{7.54}$$

Using the Legendre-Fenchel transformation, the conjugate function of W_p^* is given as

$$W_p(\epsilon) = \sup_{\sigma \in \mathcal{S}_a} \langle \sigma, \epsilon^p \rangle = \begin{cases} 0 & \text{in } \Omega_e, \\ \int_\Omega \sigma_b(\sqrt{\epsilon^2 + \frac{1}{\alpha}\gamma^2} - \epsilon_b) d\Omega & \text{in } \Omega_p, \end{cases}$$

which is the support function of the convex set $\mathcal{S}_a$. The elastic potential is the quadratic functional

$$W_e(\epsilon) = \frac{1}{2}\int_\Omega [\epsilon^T \mathbf{C} \epsilon \delta^c(Y(\epsilon)) + E\epsilon_b^2 \delta(Y(\epsilon))] d\Omega, \tag{7.55}$$

where $\delta^c = 1 - \delta$ is the complementary function of the step function δ. The total internal super-potential in this problem is the convex functional

$$W(\epsilon) = W_e(\epsilon) + W_p(\epsilon).$$

The constitutive equation and the internal complementarity condition can now be written in the subdifferential form

$$\sigma \in \partial^- W(\epsilon). \tag{7.56}$$

The external potential energy $F : \mathcal{U} \to \overleftarrow{\mathbb{R}}$ is defined by

$$\begin{aligned} F(\mathbf{u}) &= \langle \mathbf{u}, \bar{\mathbf{p}} \rangle - \mathcal{I}_{\mathcal{U}_a}(\mathbf{u}) - j(\mathbf{u}) \\ &= \begin{cases} \int_0^\ell [\bar{p}w + \bar{q}^+ u(x,h)] \, \mathrm{d}x - j(u(x,-h)) & \text{if } \mathbf{u} \in \mathcal{U}_a, \\ -\infty & \text{otherwise.} \end{cases} \end{aligned} \tag{7.57}$$

On the kinematically admissible space

$$\mathcal{U}_k = \{\mathbf{u} = (u,w)^T \in \mathcal{U}_a | \ \Lambda \mathbf{u} \in \mathcal{L}(\Omega; \mathbb{R}^2), \quad w(x) \geq \psi(x) \ \ \forall x \in (0,\ell)\},$$

the total potential energy $P : \mathcal{U}_k \to \overrightarrow{\mathbb{R}}$ has the complicated form

$$\begin{aligned} P(\mathbf{u}) &= W(\Lambda \mathbf{u}) - F(\mathbf{u}) \\ &= \int_\Omega \frac{1}{2}[(\Lambda \mathbf{u})^T C (\Lambda \mathbf{u}) \delta^c(Y(\Lambda \mathbf{u})) + (E\epsilon_b^2 + \sigma_b Y(\Lambda \mathbf{u}))\delta(Y(\Lambda \mathbf{u}))] \mathrm{d}\Omega \\ &- \int_0^\ell [\bar{p}w + \bar{q}^+ u(x,h)] \, \mathrm{d}x + j(u(x,-h)). \end{aligned} \tag{7.58}$$

The primal problem associated with this bi-complementarity problem is to find $\bar{\mathbf{u}} = (\bar{u}, \bar{w})^T \in \mathcal{U}_k$. Thus,

$$(\mathcal{P}_{\text{inf}}) : \quad P(\bar{\mathbf{u}}) = \inf P(\mathbf{u}) \ \ \forall \mathbf{u} = (u,w)^T \in \mathcal{U}_k. \tag{7.59}$$

The associated variational inequality problem $(\mathcal{P}_{\text{vi}})_1'$ is

$$\begin{aligned} &\langle \bar{\Lambda}^* \partial W(\Lambda \bar{\mathbf{u}}), \mathbf{u} - \bar{\mathbf{u}} \rangle + j(u(x,-h)) \\ \geq &\int_0^\ell [\bar{p}(w - \bar{w}) + \bar{q}^+(u(x,h) - \bar{u}(x,h))] \, \mathrm{d}x \\ &+ j(\bar{u}(x,-h)) \quad \forall \mathbf{u} = (u,w)^T \in \mathcal{U}_k. \end{aligned} \tag{7.60}$$

Clearly, direct approaches for solving this constitutively nonlinear primal variational inequality problem are very difficult. Now let us consider the dual approach.

The dual feasible set $\mathcal{S}_s \subset \mathcal{S}_a$ in this problem is

$$\mathcal{S}_s = \left\{ \boldsymbol{\sigma} = \begin{pmatrix} \sigma \\ \tau \end{pmatrix} \in \mathcal{S}_a \,\middle|\, \begin{array}{l} \sigma_{,x} + \tau_{,y} = 0 \ \text{ in } \Omega, \ \ \tau(x,h) = \bar{q}^+(x) \ \ \forall x \in (0,\ell), \\ \int_{-h}^h \tau_{,x} \mathrm{d}y + \bar{p}(x) \leq 0 \ \ \forall x \in (0,\ell) \end{array} \right\}.$$

Using the Legendre-Fenchel transformation, the external complementary energy $F^* : \mathcal{S} \to \overleftarrow{\mathbb{R}}$ can be obtained as

$$F^*(\Lambda^* \boldsymbol{\sigma}) = \begin{cases} \int_0^\ell -\psi(x)[\int_{-h}^h \tau_{,x} \mathrm{d}y + \bar{p}(x)] \, \mathrm{d}x - j^*(-\tau) & \text{if } \boldsymbol{\sigma} = (\sigma,\tau)^T \in \mathcal{S}_s \\ -\infty & \text{otherwise.} \end{cases} \tag{7.61}$$

The internal complementary energy $W^* : \mathcal{S} \to \overline{\mathbb{R}}$ is the convex, l.s.c. functional

$$W^*(\boldsymbol{\sigma}) = W_e^*(\boldsymbol{\sigma}) + W_p^*(\boldsymbol{\sigma}) = \begin{cases} \int_\Omega \frac{1}{2}\boldsymbol{\sigma}^T C^{-1}\boldsymbol{\sigma} \mathrm{d}\Omega & \text{if } \boldsymbol{\sigma} \in \mathcal{S}_a, \\ +\infty & \text{otherwise.} \end{cases} \tag{7.62}$$

On the dual feasible space $\mathcal{S}_s$, the total complementary energy functional $P^d(\boldsymbol{\sigma})$ can be simply written as

$$P^d(\sigma,\tau) = \int_0^\ell -\psi(x)[\bar{p}(x) + \int_{-h}^h \tau_{,x}\mathrm{d}y]\, \mathrm{d}x - \int_\Omega \frac{1}{2}[E^{-1}\sigma^2 + G^{-1}\tau^2]\mathrm{d}\Omega - j^*(-\tau). \tag{7.63}$$

Thus, the dual variational problem for this nonlinear bi-complementarity problem is to find $\bar{\mathbf{s}} = (\bar{\sigma},\bar{\tau})^T$, i.e.

$$(\mathcal{P}^d_{\text{sup}}): \quad P^d(\bar{\sigma},\bar{\tau}) \geq P^d(\sigma,\tau) \quad \forall (\sigma,\tau)^T \in \mathcal{S}_s. \tag{7.64}$$

The dual variational inequality associated with this problem has the simple form

$$\begin{aligned} (\mathcal{P}^d_{\text{vi}})'_1: \quad & \int_\Omega [E^{-1}\bar{\sigma}(\sigma - \bar{\sigma}) + G^{-1}\bar{\tau}(\tau - \bar{\tau})]\mathrm{d}\Omega + j^*(-\tau) \\ \geq & \int_\Omega \psi(x)(\bar{\tau} - \tau)_{,x}\mathrm{d}\Omega + j^*(-\bar{\tau}), \quad \forall (\sigma,\tau)^T \in \mathcal{S}_s. \end{aligned} \tag{7.65}$$

For the lubricated contact problem, the dual problem $(\mathcal{P}^d_{\text{sup}})$ is a quadratic programming problem on the closed convex subset $\mathcal{S}_s$, and the dual variational inequality $(\mathcal{P}^d_{\text{vi}})'_1$ is given simply by

$$(\mathcal{P}^d_{\text{vi}})'_1: \quad \int_\Omega [\beta\bar{\sigma}(\sigma - \bar{\sigma}) + \bar{\tau}(\tau - \bar{\tau})]\mathrm{d}\Omega \geq \int_\Omega G\psi(x)(\bar{\tau} - \tau)_{,x}\mathrm{d}\Omega \quad \forall (\sigma,\tau)^T \in \mathcal{S}_s. \tag{7.66}$$

Since $W^* : \mathcal{S} \to \mathbb{R}$ is a strictly convex functional, the solution of the dual problem is unique. In numerical analysis, dual approaches provide lower bounds for the solutions. Moreover, using the linear equilibrium constraint in the statically admissible space $\mathcal{S}_s$, the degrees of freedom in nonlinear iteration can be reduced via the mixed finite element method (see Gao, 1988b).

Example 7.2.1 (Rigid-Perfectly Plastic Beam and Limit Analysis)
As an application of duality theory in limit analysis, we will illustrate the dual approach to the safety factor of the extended beam theory without an obstacle. We assume that the external load $\bar{\mathbf{p}}$ is proportional to a positive scalar $\nu_c > 0$ such that $\bar{\mathbf{p}}(x) = \varrho\mathbf{p}_o(x)$, where $\mathbf{p}_o = (0,0)^T$ in Ω, $\mathbf{p}_o = (q_o(x), p_o(x))^T$ on $y = h$ is a given unit loading vector. The purpose of plastic limit analysis is to find the critical loading factor ϱ_c such that the structure collapses under the limit load $\varrho_c\mathbf{p}_o$. The elastic deformation is ignored and the material is treated as rigid-perfect plastic. In this case, Problem 7.2.1 is reduced as the follows.

For the given data $\mathbf{p}_o(x)$, find $\mathbf{u} = (u, w)^T \in \mathcal{U}_a$ and $\varrho_c > 0$ such that

1. Geometrical equation

$$\boldsymbol{\epsilon} = \Lambda \mathbf{u}, \quad \text{or} \quad \begin{pmatrix} \epsilon \\ \gamma \end{pmatrix} = \begin{pmatrix} \frac{\partial}{\partial x} & 0 \\ \frac{\partial}{\partial y} & \frac{\partial}{\partial x} \end{pmatrix} \begin{pmatrix} u(x,y) \\ w(x) \end{pmatrix};$$

2. Constitutive equation

$$\boldsymbol{\epsilon} = Y \frac{\partial Y^c(\boldsymbol{\sigma})}{\partial \boldsymbol{\sigma}} \delta(Y^c(\boldsymbol{\sigma})), \quad \text{or} \quad \begin{cases} \epsilon = Y \frac{\partial Y^c(\boldsymbol{\sigma})}{\partial \sigma} \delta(Y^c(\boldsymbol{\sigma})), \\ \gamma = Y \frac{\partial Y^c(\boldsymbol{\sigma})}{\partial \tau} \delta(Y^c(\boldsymbol{\sigma})); \end{cases}$$

3. Equilibrium condition

$$\Lambda^* \boldsymbol{\sigma} = \varrho_c \mathbf{p}_o, \quad \text{or} \quad \begin{cases} \int_{-h}^{h} -\frac{\partial \tau}{\partial x} \mathrm{d}y = \varrho_c p_o(x) \ \ \forall x \in (0, \ell), \\ \frac{\partial}{\partial x} \sigma + \frac{\partial}{\partial y} \tau = 0 \forall (x,y) \in \Omega, \\ \tau(x,h) = \varrho_c q_o(x), \quad \tau(x,-h) = 0 \forall x \in (0,\ell); \end{cases}$$

4. Complementarity conditions

$$Y(\boldsymbol{\epsilon}) \geq 0, \quad Y^c(\boldsymbol{\sigma}) \leq 0, \quad Y(\boldsymbol{\epsilon}) Y^c(\boldsymbol{\sigma}) = 0.$$

In terms of $\mathbf{u} = (u, w)^T$, the total potential energy P can be written as

$$P(u,w) = \int_\Omega \sigma_b \sqrt{u_{,x}^2 + \frac{1}{\alpha}(u_{,y} + w_{,x})^2} \mathrm{d}\Omega - \varrho_c \int_0^\ell (q_o u(x,h) + p_o w) \, \mathrm{d}x.$$

Let $\mathcal{U}_k \subset \mathcal{U}_a$ be the kinematically admissible space, defined by

$$\mathcal{U}_k = \left\{ \mathbf{u} = (u,w)^T \in \mathcal{U}_a \,\middle|\, \int_0^\ell (q_o u(x,h) + p_o w) \, \mathrm{d}x = 1 \right\},$$

and

$$\varrho^\sharp(u,w) = \int_\Omega \sigma_b \sqrt{u_{,x}^2 + \frac{1}{\alpha}(u_{,y} + w_{,x})^2} \mathrm{d}\Omega.$$

Then, as the primal variational inequality problem, we have the upper bound theorem

$$\varrho_c \leq \varrho^\sharp(\mathbf{u}) \quad \forall \mathbf{u} \in \mathcal{U}_k. \tag{7.67}$$

For the given unit external load $\mathbf{p}_o$, the associated equilibrium admissible stress field $\mathcal{S}_e^o$ is defined by

$$\mathcal{S}_e^o = \left\{ \boldsymbol{\sigma} = \begin{pmatrix} \sigma \\ \tau \end{pmatrix} \,\middle|\, \begin{array}{l} -\int_{-h}^{h} \tau_{,x} \mathrm{d}y = p_o(x), \quad \tau(x,h) = q_o(x), \quad \forall x \in (0,\ell), \\ \sigma_{,x} + \tau_{,y} = 0 \ \ \forall (x,y) \in \Omega, \quad \tau(x,-h) = 0 \ \ \forall x \in (0,\ell) \end{array} \right\}.$$

Let

$$\boldsymbol{\sigma} = \varrho^\flat \boldsymbol{\sigma}_o, \quad \text{with} \quad \boldsymbol{\sigma}_o = (\sigma_o, \ \tau_o)^T \in \mathcal{S}_e^o. \tag{7.68}$$

Then, by Theorem 7.1.3, we have

$$\varrho_c \geq \frac{\sigma_b}{\max_{(x,y)\in\Omega} \sqrt{\sigma_o^2(x,y) + \alpha \tau_o^2(x,y)}} \quad \forall \boldsymbol{\sigma}_o \in \mathcal{S}_e^o. \tag{7.69}$$

If the beam is simply supported at both ends and subjected to the external force $p_o(x) = \sin x\pi/\ell$ on $y = h$. The associated unit equilibrium admissible field is given by

$$\sigma_o(x,y) = \frac{\ell^2}{8h^2}\sin\frac{x\pi}{\ell}\sin\frac{y\pi}{2h}, \quad \tau_o(x,y) = \frac{\ell}{4h}\cos\frac{x\pi}{\ell}\cos\frac{y\pi}{2h}, \tag{7.70}$$

and this gives a lower bound

$$\varrho^- = \frac{8h^2\sigma_b}{\ell^2 \max_{(x,y)\in(0,\ell)\times(-h,h)} \sqrt{\sin^2\frac{x\pi}{\ell}\sin^2\frac{y\pi}{2h} + \frac{4h^2\alpha}{\ell^2}\cos^2\frac{x\pi}{\ell}\cos^2\frac{y\pi}{2h}}}.$$

It is easy to show that, if $4h^2\alpha/\ell^2 < 1$, then the solution for the problem

$$\max_{(x,y)\in(0,\ell)\times(-h,h)} \sqrt{\sin^2\frac{x\pi}{\ell}\sin^2\frac{y\pi}{2h} + \frac{4h^2\alpha}{\ell^2}\cos^2\frac{x\pi}{\ell}\cos^2\frac{y\pi}{2h}}$$

is $(x,y) = (\ell/2, \pm h)$. In this case, the lower bound theorem gives

$$\varrho^- = 8h^2\sigma_b/\ell^2.$$

For the same stress field, the lower bound from the classical beam model is $\pi^2h^2\sigma_b/\ell^2$, which is bigger than ϱ^- since the extended beam model is weaker than the classical one. ♡

7.3 Geometrical Nonlinearity: von Kármán Plate

We now turn our attention to the applications of duality to large deformation structural mechanics. We shall use the well-known von Kármán plate model for illustration. The von Kármán equations have many different versions. This section is based upon its refinement in the work of Ciarlet (1980). Related expositions are given by Ciarlet and Rabier (1980), Davet (1986) and Ciarlet (1990).

7.3.1 Large Deformation Thin Plate and Primal Problem

Consider a plate Ω in 3-dimensional space $\mathbb{R}^3$. The points of the plate are referred to a fixed, right-handed rectangular Cartesian coordinate system $0x_1x_2x_3$. The middle plane of the undeformed plate, assumed to have a constant thickness, coincides with the $0x_1x_2$-plane. The material points $x = (x_1, x_2, 0)$ of the undeformed plate constitute an open, bounded, connected subset $\Omega \subset \mathbb{R}^2$ with a Lipschitz boundary $\partial\Omega$. We denote by $u = (u_1, u_2)$ the horizontal and by ξ the vertical displacement of points $x \in \Omega$. If the plate is subjected to a distributed transverse load $f(x)$, the tri-canonical forms for the von Kármán plate model mentioned in (3.36) are as follows

$$
\begin{array}{ll}
\text{1.Geometric equations} & \xi = \tfrac{1}{2}[\nabla u + (\nabla u)^T] + \tfrac{1}{2}(\nabla\chi)\otimes(\nabla\chi) \quad \text{in } \Omega, \\
 & \kappa = -\nabla\nabla\chi \quad \text{in } \Omega. \\
\text{2.Constitutive equations} & \varsigma = C\xi, \quad M = D\kappa \quad \text{in } \Omega. \\
\text{3.Equilibrium equations} & \nabla\cdot(\nabla\cdot M) + h\nabla\cdot(\varsigma\nabla\chi) = f \quad \text{in } \Omega, \\
 & \nabla\cdot\varsigma = 0 \quad \text{in } \Omega.
\end{array}
\tag{7.71}
$$

Here, h is the plate thickness, $\varsigma = \{\varsigma_{\alpha\beta}\}$ (resp. $\xi = \{\xi_{\alpha\beta}\}$) denotes the stress (resp. strain) tensor in the plane of the plate, $M = \{M_{\alpha\beta}\}$ (resp. $\kappa = \{\kappa_{\alpha\beta}\}$) is the stress couple (resp. curvature) tensor, and $C = \{C_{\alpha\beta\gamma\delta}\}$ and $D = \{D_{\alpha\beta\gamma\delta}\}$ are the corresponding elasticity tensors. For an isotropic plate, C has the form

$$C_{\alpha\beta\lambda\mu} = \frac{E}{2(1-\nu^2)}((1-\nu)(\delta_{\alpha\lambda}\delta_{\beta\mu} + \delta_{\alpha\mu}\delta_{\beta\lambda}) + 2\nu\delta_{\alpha\beta}\delta_{\lambda\mu}).$$

The tensor D is assumed to be a positive-definite fourth-order tensor. It is given by

$$D_{\alpha\beta\gamma\delta} = K[\frac{1}{2}(1-\nu)(\delta_{\alpha\lambda}\delta_{\beta\mu} + \delta_{\alpha\mu}\delta_{\beta\lambda}) + \nu\delta_{\alpha\beta}\delta_{\lambda\mu}], \quad 0 \le \nu < \frac{1}{2},$$

where $K = Eh^3/12(1-\nu^2)$ is the bending rigidity of the plate, E is the modulus of elasticity and ν Poisson's ratio. Throughout this section Greek indices take values 1 and 2.

In buckling analysis, the buckling is caused by compressive load distribution on $\partial\Omega$, which is written

$$\varsigma\cdot n = \lambda t \quad \text{on } \partial\Omega, \tag{7.72}$$

where $t = \{t_1(x), t_2(x)\}, x \in \partial\Omega$, is a self-equilibrating compressive load distribution on $\partial\Omega$, i.e.

$$\int_{\partial\Omega} t_\alpha(x)\mathrm{d}\Gamma = 0, \qquad \int_{\partial\Omega}(t_1x_2 - t_2x_1)\mathrm{d}\Gamma = 0,$$

and λ is a real number measuring the magnitude of the load distribution. Suppose the plate is clamped on the boundary Γ_c and simply supported on Γ_s ($\Gamma_c \cup \Gamma_s = \Gamma = \partial\Omega$). Then, the displacement boundary conditions are

$$\chi(x) = 0 \quad \text{on } \Gamma, \quad u(x) = 0, \quad \frac{\partial\chi}{\partial n} = 0 \quad \text{on } \Gamma_c. \tag{7.73}$$

Following the traditional notations, we now introduce the two bilinear forms

$$
\begin{aligned}
a(\chi, w) &= \int_\Omega M_{\alpha\beta}(\chi)\kappa_{\alpha\beta}(w)\mathrm{d}\Omega \\
&= K\int_\Omega [(1-\nu)(\nabla\nabla\chi)(\nabla\nabla w) + \nu\Delta\chi\Delta w]\mathrm{d}\Omega,
\end{aligned}
\tag{7.74}
$$

$$b(\xi,\xi) = \int_\Omega h\varsigma_{\alpha\beta}(\xi)\xi_{\alpha\beta}\mathrm{d}\Omega = \int_\Omega h\xi_{\alpha\beta}C_{\alpha\beta\gamma\theta}\xi_{\gamma\theta}\mathrm{d}\Omega. \tag{7.75}$$

Let

$$\phi(\chi, w) := (\nabla\chi) \otimes (\nabla w) = \{\chi_{,\alpha} w_{,\beta}\}, \qquad \phi(\chi) := \phi(\chi, \chi),$$

$$\epsilon(u) := \frac{1}{2}[\nabla u + (\nabla u)^T],$$

and

$$\mathcal{U}_\chi = \{\chi \in \mathcal{H}^2(\Omega) |\ \chi = 0 \text{ on } \Gamma,\ \frac{\partial \chi}{\partial n} = 0 \text{ on } \Gamma_c\},$$

$$\mathcal{U}_u = \{u \in \mathcal{H}^1(\Omega; \mathbb{R}^2) |\ u(x) = 0 \text{ on } \Gamma_c\}$$

Thus, the primal variational problem for the von Kármán plate model can be proposed in its classical form.

Problem 7.3.1 For a given external load $p(x)$ and $\lambda \in \mathbb{R}$, find $(\bar{u}, \bar{\chi}) \in \mathcal{U}_u \times \mathcal{U}_\chi$ such that

$$a(\bar{\chi}, \chi - \bar{\chi}) + b(\xi(\bar{u}, \bar{\chi}), \phi(\bar{\chi}, \chi - \bar{\chi})) = \int_\Omega p(\chi - \bar{\chi}) d\Omega \quad \forall \chi \in \mathcal{U}_\chi, \tag{7.76}$$

$$b(\xi(\bar{u}, \bar{\chi}), \epsilon(u - \bar{u})) = \lambda \int_\Gamma ht(u - \bar{u}) d\Gamma \quad \forall u \in \mathcal{U}_u. \tag{7.77}$$

♠

Let $P : \mathcal{U}_u \times \mathcal{U}_\chi \to \mathbb{R}$ be the total potential energy

$$P(u, \chi) = \frac{1}{2} a(\chi, \chi) + \frac{1}{2} b(\xi(u, \chi), \xi(u, \chi)) - \int_\Omega p\chi d\Omega - \int_\Gamma \lambda ut d\Gamma. \tag{7.78}$$

It is easy to prove that Problem 7.3.1 is equivalent to

$$\delta P(\bar{u}, \bar{\chi}; u - \bar{u}, \chi - \bar{\chi}) = 0 \ \ \forall (u, \chi) \in \mathcal{U}_u \times \mathcal{U}_\chi.$$

Proposition 7.3.1 *Any deformed state defined by* $(\bar{u}, \bar{\chi})$ *has strictly negative potential energy, i.e.* $P(\bar{u}, \bar{\chi}) < 0$.

Proof. For any given deformed state $(\bar{u}, \bar{\chi})$, we have

$$\begin{aligned} P(\bar{u}, \bar{\chi}) \ &= \ \frac{1}{2}[a(\bar{\chi}, \bar{\chi}) + hb(\xi(\bar{u}, \bar{\chi}), \phi(\bar{\chi}))] + \frac{1}{2} b(\xi(\bar{u}, \bar{\chi}), \epsilon(\bar{u})) \\ &\quad -\frac{1}{4} hb(\xi(\bar{u}, \bar{\chi}), \phi(\bar{\chi})) - \int_\Omega p\bar{\chi} d\Omega - \lambda \int_\Gamma ut d\Gamma. \end{aligned}$$

Since $(\bar{u}, \bar{\chi})$ satisfies the variational equations (7.76) and (7.77) then, for $\chi \in \mathcal{U}_\chi$ and $u \in \mathcal{U}_u$, we may set $\chi = 2\bar{\chi}, u = 2\bar{u}$ to obtain

$$P(\bar{u}, \bar{\chi}) = -\frac{1}{2} b(\xi(\bar{u}, \bar{\chi}), \epsilon(\bar{u})) - \frac{1}{4} hb(\xi(\bar{u}, \bar{\chi}), \phi(\bar{\chi})) = -\frac{1}{2} hb(\xi(\bar{u}, \bar{\chi}), \xi(\bar{u}, \bar{\chi})).$$

Since $b(\xi, \xi)$ is strictly positive for any deformed state, we have $P(\bar{u}, \bar{\chi}) < 0$. □

7.3.2 Canonical Dual Transformation

To formulate the dual problem the key step is to choose the right dual variables and to introduce suitable geometrical operators. The general displacement space for this large deformation plate is defined by

$$\mathcal{U} := \{\mathbf{u} = \{u, \chi\} \in \mathbb{R}^3 \mid u \in \mathcal{L}^2(\Omega; \mathbb{R}^2), \chi \in \mathcal{L}^2(\Omega)\}. \tag{7.79}$$

The general strain space is:

$$\mathcal{E} := \{\mathbf{E} = \{\xi, \kappa\} \mid \xi = \{\xi_{\alpha\beta}\}, \ \ \kappa = \{\kappa_{\alpha\beta}\} \in \mathcal{M}_{sym}(\Omega; \mathbb{R}^{2\times 2})\},$$

where $\mathcal{M}_{sym}$ denotes the second-order symmetric tensor space, which was defined in Chapter 6. For a given displacement $\mathbf{u} = (u, \chi) \in \mathcal{U}$, the deformation of the plate is described by the nonlinear differential operator $\Lambda : \mathcal{U} \to \mathcal{E}$, defined by

$$\Lambda(\mathbf{u}) = \begin{pmatrix} \Lambda_\epsilon & \frac{1}{2}\Lambda_\phi(\chi) \\ 0 & \Lambda_\kappa \end{pmatrix} \begin{pmatrix} u \\ \chi \end{pmatrix} = \begin{pmatrix} \Lambda_\epsilon u + \frac{1}{2}\Lambda_\phi(\chi)\chi \\ \Lambda_\kappa \chi \end{pmatrix}, \tag{7.80}$$

in which

$$\Lambda_\epsilon = \nabla_s = \frac{1}{2}(\nabla + \nabla^t), \quad \Lambda_\phi(\chi) = (\nabla\chi) \otimes \nabla, \quad \Lambda_\kappa = -\nabla\nabla.$$

Then, the strain-displacement relation $\mathbf{E} = \Lambda(\mathbf{u})$ may be written in the matrix form

$$\mathbf{E} = \begin{pmatrix} \xi \\ \kappa \end{pmatrix} = \begin{pmatrix} \frac{1}{2}[\nabla u + (\nabla u)^T + (\nabla\chi) \otimes (\nabla\chi)] \\ -\nabla\nabla\chi \end{pmatrix} = \Lambda(\mathbf{u}).$$

Since Λ is a quadratic operator, $\mathbf{E}$ is actually a Cauchy-Green type strain tensor. By the canonical decomposition, Λ can be split into two parts, to give

$$\Lambda(\mathbf{u}) = \Lambda_t(\chi)\mathbf{u} + \Lambda_c(\chi)\mathbf{u},$$

where $\Lambda_t : \mathcal{U} \to \mathcal{E}$ is the Gâteaux derivative of $\mathbf{E}(\mathbf{u})$ at $\mathbf{u}$, and $\Lambda_c = \Lambda - \Lambda_t$ is the complementary operator, each depending only on the vertical displacement χ. Thus,

$$\Lambda_t(\chi) = \begin{pmatrix} \Lambda_\epsilon & \Lambda_\phi(\chi) \\ 0 & \Lambda_\kappa \end{pmatrix}, \quad \Lambda_c(\chi) = \begin{pmatrix} 0 & -\frac{1}{2}\Lambda_\phi(\chi) \\ 0 & 0 \end{pmatrix}.$$

Let $U(\mathbf{E})$ be the stored energy per unit volume, i.e.

$$U(\mathbf{E}) = \frac{1}{2}\mathbf{E}^T\mathbf{H}\mathbf{E} = \frac{h}{2}C_{\alpha\beta\lambda\mu}\xi_{\alpha\beta}\xi_{\lambda\mu} + \frac{1}{2}D_{\alpha\beta\lambda\mu}\kappa_{\alpha\beta}\kappa_{\lambda\mu}, \tag{7.81}$$

where $\mathbf{H} = \begin{pmatrix} hC & 0 \\ 0 & D \end{pmatrix}$ is the material matrix. For any given $\mathbf{u} = \{u, \chi\}$, the internal potential energy of the system is

$$W(\Lambda(\mathbf{u})) = \int_\Omega U(\Lambda(\mathbf{u}))d\Omega = \frac{1}{2}a(\chi, \chi) + \frac{1}{2}b(\xi(u, \chi), \xi(u, \chi)).$$

The conjugate variable of the general strain $\mathbf{E} \in \mathcal{E}$ is the general stress $\mathbf{T} = \{h\varsigma, M\}$, which is given by the linear constitutive equation

$$\mathbf{T} = DW(\mathbf{E}) = \frac{\partial U(\mathbf{E})}{\partial \mathbf{E}} = \mathbf{HE} \;\Rightarrow \left\{ \begin{array}{l} \varsigma_{\alpha\beta} = C_{\alpha\beta\lambda\mu}\xi_{\lambda\mu}, \\ M_{\alpha\beta} = D_{\alpha\beta\lambda\mu}\kappa_{\lambda\mu}. \end{array} \right. \tag{7.82}$$

Let $\mathcal{T}$ be the general stress space defined by

$$\mathcal{T} = \{\mathbf{T} = \{h\varsigma, M\} |\ \varsigma = \{\varsigma_{\alpha\beta}\},\ \ M = \{M_{\alpha\beta}\} \in \mathcal{M}_{sym}(\Omega; \mathbb{R}^{2\times 2}\}. \tag{7.83}$$

The standard bilinear form $\langle \mathbf{E}; \mathbf{T} \rangle : \mathcal{E} \times \mathcal{T} \to \mathbb{R}$ is defined by

$$\langle \mathbf{E}; \mathbf{T} \rangle = \langle h\xi; \varsigma \rangle + \langle \kappa; M \rangle = \int_\Omega [h\varsigma_{\alpha\beta}\xi_{\alpha\beta} + M_{\alpha\beta}\kappa_{\alpha\beta}] \mathrm{d}\Omega. \tag{7.84}$$

Let $\mathcal{U}^*$ be the general force space

$$\mathcal{U}^* = \{\mathbf{t} = (t, p) |\ t = \{t_1, t_2\} \in \mathcal{L}^2(\bar{\Omega}; \mathbb{R}^2),\ \ p \in \mathcal{L}^2(\bar{\Omega})\ \} \tag{7.85}$$

The standard bilinear form $\langle \mathbf{u}, \mathbf{t} \rangle : \mathcal{U} \times \mathcal{U}^* \to \mathbb{R}$ is defined by

$$\langle \mathbf{u}, \mathbf{t} \rangle := \int_\Omega p\chi \mathrm{d}\Omega + \int_\Gamma htu \mathrm{d}\Gamma.$$

For any given $\mathbf{u} = \{u, \chi\}$ such that $\mathbf{E} = \Lambda(\mathbf{u})$, the canonical decomposition of Λ leads to

$$\langle \Lambda(\mathbf{u}); \mathbf{T} \rangle = \langle \Lambda_t(\chi)\mathbf{u}; \mathbf{T} \rangle + \langle \Lambda_c(\mathbf{u}); \mathbf{T} \rangle = \langle \mathbf{u}, \Lambda_t^*(\chi)\mathbf{T} \rangle - G(\chi, \varsigma), \tag{7.86}$$

where Λ_t^* is the adjoint of Λ_t, given by

$$\Lambda_t^*(\chi) = \left(\begin{array}{cc} \Lambda_\epsilon^* & 0 \\ \Lambda_\phi^*(\chi) & \Lambda_\kappa^* \end{array} \right), \tag{7.87}$$

in which, Λ_ϵ^*, Λ_ϕ^* and Λ_κ^* are, respectively, the adjoint operators of Λ_ϵ, Λ_ϕ and Λ_κ, such that

$$\Lambda_\epsilon^* \varsigma = \left\{ \begin{array}{ll} -\nabla \cdot \varsigma & \text{in } \Omega, \\ n \cdot \varsigma & \text{on } \Gamma, \end{array} \right. \qquad \Lambda_\phi^*(\chi)\varsigma = \left\{ \begin{array}{ll} -\nabla \cdot (\varsigma \nabla \chi) & \text{in } \Omega, \\ n \cdot (\varsigma \nabla \chi) & \text{on } \Gamma, \end{array} \right.$$

$$\Lambda_\kappa^* M = \left\{ \begin{array}{ll} -\nabla \cdot (\nabla \cdot M) & \text{in } \Omega, \\ (\nabla \cdot M) \cdot n + \frac{\partial}{\partial s}(M_{\alpha\beta} n_\alpha s_\beta) & \text{on } \Gamma, \\ (n \cdot M \cdot n) \frac{\partial}{\partial n} & \text{on } \Gamma, \end{array} \right.$$

where $s = \{s_\alpha\} \in \mathbb{R}^2$ is the unit vector tangent to the boundary Γ. In this operator notations, the bilinear form $\langle \mathbf{u}, \Lambda_t^*(\chi)\mathbf{T} \rangle$ can be written as

$$\begin{aligned} \langle \mathbf{u}, \Lambda_t^*(\chi)\mathbf{T} \rangle &= -\int_\Omega [\nabla \cdot (\varsigma \nabla \chi) + \nabla \cdot (\nabla \cdot M)] \chi \mathrm{d}\Omega \\ &- \int_\Omega (\nabla \cdot \varsigma) u \mathrm{d}\Omega + \int_\Gamma n \cdot \varsigma \cdot u \mathrm{d}\Gamma + \int_\Gamma (n \cdot M \cdot n) \frac{\partial \chi}{\partial n} \mathrm{d}\Gamma \\ &+ \int_\Gamma [n \cdot (\varsigma \nabla \chi) + n \cdot (\nabla \cdot M) + \frac{\partial}{\partial s}(n_\alpha M_{\alpha\beta} s_\beta)] \chi \mathrm{d}\Gamma. \end{aligned}$$

The complementary gap function $G(\mathbf{u},\mathbf{T})$ in this plate model depends on the vertical displacement χ and the plane stress ς. Thus,

$$G(\mathbf{u},\mathbf{T}) = \langle -\Lambda_c(\chi)\mathbf{u};\mathbf{T}\rangle = \langle -\frac{1}{2}\Lambda_\phi(\chi)\chi;\varsigma\rangle = h\int_\Omega \frac{1}{2}(\nabla\chi)\varsigma(\nabla\chi)\mathrm{d}\Omega, \qquad (7.88)$$

and for any given $\mathbf{T} = \{h\varsigma, M\}$, $\mathbf{u} = \{u,\chi\}$, we can write simply $G(\mathbf{u},\mathbf{T}) = G(\chi,\varsigma)$.

For a given external force

$$\bar{\mathbf{t}}(x) = \{\lambda\bar{t}(x), p(x)\} = \begin{cases} \{(0,0),p(x)\} & \forall x\in\Omega, \\ \{\lambda(\bar{t}_1,\bar{t}_2),0\} & \forall x\in\Gamma, \end{cases}$$

and admissible displacement $\mathbf{u}\in\mathcal{U}_a$, the virtual work principle

$$\langle\delta\mathbf{E}(\bar{u};\mathbf{u});\mathbf{T}\rangle = \langle\Lambda_t(\bar{\chi})\mathbf{u};\mathbf{T}\rangle = \langle\mathbf{u},\Lambda_t^*(\bar{\chi})\mathbf{T}\rangle = \langle\mathbf{u},\bar{\mathbf{t}}\rangle \;\; \forall\mathbf{u}\in\mathcal{U}_a$$

leads to the equilibrium conditions

$$\Lambda_t^*(\bar{\chi})\mathbf{T} = \lambda\bar{\mathbf{t}} \;\Rightarrow\; \begin{cases} \Lambda_\epsilon^*\varsigma = \bar{t} & \text{in } \Omega\cup\Gamma, \\ \Lambda_\phi^*(\bar{\chi})\varsigma + \Lambda_\kappa^* M = p & \text{in } \Omega\cup\Gamma. \end{cases} \qquad (7.89)$$

Hence, the fundamental equation for the von Kármán plate has the standard abstract form

$$\Lambda_t^*(\bar{\chi})\mathbf{H}\Lambda(\bar{\mathbf{u}}) = \bar{\mathbf{t}} \quad \text{in } \Omega\cup\Gamma, \qquad (7.90)$$

and this belongs to a geometrically nonlinear but physically linear canonical system.

7.3.3 Complementary Variational Principles

Since the stored energy

$$W(\xi,\kappa) = \int_\Omega \frac{1}{2}D_{\alpha\beta\lambda\mu}\kappa_{\alpha\beta}\kappa_{\lambda\mu}\mathrm{d}\Omega + \int_\Omega \frac{h}{2}C_{\alpha\beta\lambda\mu}\epsilon_{\alpha\beta}\epsilon_{\lambda\mu}\mathrm{d}\Omega$$

for the von Kármán plate is a quadratic functional, its Legendre conjugate is simply given by

$$\begin{aligned} W^*(\varsigma,M) &= \sup_{(\xi,\kappa)\in\mathcal{E}} \{\langle\xi;\varsigma\rangle + \langle\kappa;M\rangle - W(\xi,\kappa)\} \\ &= \int_\Omega \frac{1}{2}hC^{-1}_{\alpha\beta\gamma\delta}\varsigma_{\alpha\beta}\varsigma_{\gamma\delta}\mathrm{d}\Omega + \int_\Omega \frac{1}{2}D^{-1}_{\alpha\beta\gamma\delta}M_{\alpha\beta}M_{\gamma\delta}\mathrm{d}\Omega. \end{aligned}$$

For a given external force $\bar{\mathbf{t}}\in\mathcal{U}^*$, the external energy F is a linear functional $F(\mathbf{u}) = \langle\mathbf{u},\bar{\mathbf{t}}\rangle$. Thus, the canonical Lagrangian associated with the primal problem takes the standard form

$$\begin{aligned} L(\mathbf{u},\mathbf{T}) &= \langle\Lambda(\mathbf{u});\mathbf{T}\rangle - W^*(\mathbf{T}) - F(\mathbf{u}) \\ &= \int_\Omega [h(u_{\alpha,\beta} + \frac{1}{2}\chi_{,\alpha}\chi_{,\beta})\varsigma_{\alpha\beta} + \chi_{,\alpha\beta}M_{\alpha\beta}]\mathrm{d}\Omega - W^*(\mathbf{T}) - F(\mathbf{u}). \end{aligned}$$

Let $\mathcal{T}_\varsigma$ be defined by

$$\mathcal{T}_\varsigma = \{\varsigma \in \mathcal{M}_{sym} | \ \Lambda_\epsilon^* \varsigma = \lambda \bar{t} \quad \text{in } \Omega \cup \Gamma\}.$$

Thus, on $\mathcal{U}_a \times \mathcal{T}_\varsigma \times \mathcal{M}_{sym}$, the Lagrangian L can be written as

$$L(\chi, \varsigma, M) = \int_\Omega [M_{\alpha\beta}\chi_{,\alpha\beta} + \frac{1}{2} h \varsigma_{\alpha\beta}\chi_{,\alpha}\chi_{,\beta} - p\chi] \mathrm{d}\Omega - W^*(\varsigma, M). \tag{7.91}$$

Since

$$\sup_{M \in \mathcal{M}_{sym}} \{\langle \Lambda_\kappa \chi; M\rangle - \int_\Omega \frac{1}{2} \mathrm{tr}[M D^{-1} M] \mathrm{d}\Omega\} = \frac{1}{2} \int_\Omega \mathrm{tr}[\kappa(\chi) D \kappa(\chi)] \mathrm{d}\Omega = \frac{1}{2} a(\chi, \chi),$$

one can define

$$\begin{aligned} L_M(\chi, \varsigma) &= \sup_{M \in \mathcal{M}_{sym}} L(\chi, \varsigma, M) = \frac{1}{2} a(\chi, \chi) - \frac{1}{2} b^*(\varsigma, \varsigma) + G(\chi, \varsigma) - \langle \chi, p\rangle \\ &= \frac{1}{2} \int_\Omega \mathrm{tr}[\kappa(\chi) D \kappa(\chi) - h\varsigma C^{-1} \varsigma] \mathrm{d}\Omega + \int_\Omega [\frac{1}{2} (\nabla \chi) \varsigma (\nabla \chi) - p\chi] \mathrm{d}\Omega. \end{aligned}$$

By the Poincaré inequality, we have

$$a(\chi, \chi) \geq \int_\Omega h \lambda_c |\nabla \chi|^2 \mathrm{d}\Omega \ \ \forall \chi \in \mathcal{H}_0^1(\Omega),$$

where $\lambda_c > 0$ is the pre-buckling load, which is the first eigenvalue of the bilinear form $a(\chi, \chi)$. Let

$$L_\lambda(\chi, \varsigma) = \int_\Omega \frac{1}{2} h[(\varsigma_{\alpha\beta} - \lambda_c \delta_{\alpha\beta}) \chi_{,\alpha}\chi_{,\beta} - C^{-1}_{\alpha\beta\gamma\theta} \varsigma_{\alpha\beta}\varsigma_{\gamma\theta}] \mathrm{d}\Omega - \int_\Omega p\chi \mathrm{d}\Omega. \tag{7.92}$$

We have

$$L_M(\chi, \varsigma) \geq L_\lambda(\chi, \varsigma) \ \ \forall (\chi, \varsigma) \in \mathcal{U}_\chi \times \mathcal{T}_\varsigma.$$

Thus, by the triality theorem proposed in finite deformation theory, we have the following result.

Theorem 7.3.1 *For any given external load $\bar{\mathbf{t}} = \{\lambda \bar{t}, \bar{p}\} \in \mathcal{U}^*$ such that $(\bar{\chi}, \bar{\varsigma})$ is a critical point of $L_\lambda(\chi, \varsigma)$ on $\mathcal{U}_\chi \times \mathcal{T}_\varsigma$, then if $\{\bar{\varsigma}_{\alpha\beta}(x) - \lambda_c \delta_{\alpha\beta}\}$ is positive-definite in Ω,*

$$\inf_{\chi \in \mathcal{U}_\chi} \sup_{\varsigma \in \mathcal{T}_\varsigma} L_\lambda(\chi, \varsigma) = L_\lambda(\bar{\chi}, \bar{\varsigma}) = \sup_{\varsigma \in \mathcal{T}_\varsigma} \inf_{\chi \in \mathcal{U}_\chi} L_\lambda(\chi, \varsigma). \tag{7.93}$$

If $\{\bar{\varsigma}_{\alpha\beta}(x) - \lambda_c \delta_{\alpha\beta}\}$ is negative-definite in Ω, then either

$$\inf_{\chi \in \mathcal{U}_\chi} \sup_{\varsigma \in \mathcal{T}_\varsigma} L_\lambda(\chi, \varsigma) = L_\lambda(\bar{\chi}, \bar{\varsigma}) = \inf_{\varsigma \in \mathcal{T}_\varsigma} \sup_{\chi \in \mathcal{U}_\chi} L_\lambda(\chi, \varsigma) \tag{7.94}$$

or

$$\sup_{\chi \in \mathcal{U}_\chi} \sup_{\varsigma \in \mathcal{T}_\varsigma} L_\lambda(\chi, \varsigma) = L_\lambda(\bar{\chi}, \bar{\varsigma}) = \sup_{\varsigma \in \mathcal{T}_\varsigma} \sup_{\chi \in \mathcal{U}_\chi} L_\lambda(\chi, \varsigma). \tag{7.95}$$

7.4 Large Deformation Beam Theory

As we mentioned before, in one-dimensional problems the von Kármán nonlinear plate model becomes a linear "beam" model. This section will show the reason for this paradox. In order to study the dynamic post-buckling phenomena, two geometrically nonlinear beam models are presented.

7.4.1 Extended Beam Model and the Reason for von Kármán's Paradox

Consider an elastic beam whose domain in the reference configuration space is a rectangle $\Omega = \{(x,y) \in \mathbb{R}^2 |\ 0 \le x \le \ell,\ -h \le y \le h\}$. For dynamical problems, we assume that the transverse displacement of the elastic axis is a function of x and time t only, while the horizontal displacement depends on x, y and t. Thus, the displacement of a material point $(x,y) \in \Omega$ is the vector

$$\mathbf{u} = \begin{pmatrix} u(x,y,t) \\ \chi(x,y,t) \end{pmatrix} = \begin{pmatrix} u_0(x,t) - y\theta(x,t) + v(x,y,t) \\ w(x,t) \end{pmatrix}, \tag{7.96}$$

where $u_0(x,t)$ is the horizontal displacement of the middle axis $y = 0$, θ is the *bending angle*, given by $\theta = \tan^{-1}(w_{,x}/(1 + u_{0,x}))$, measured in the counter-clockwise direction, and $v(x,y,t)$ is the shear displacement. The *shear angle* is $\gamma = \tan^{-1} v_{,y}(x,y,t)$, measured in the positive direction (see Fig. 7.4). The *net rotation angle* is therefore given by $\psi = \theta - \gamma$.

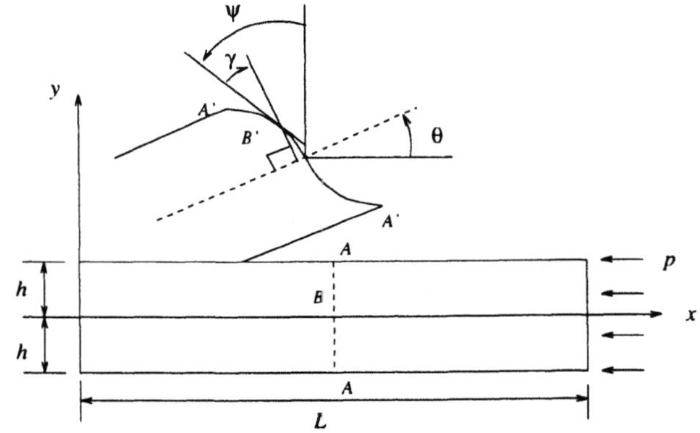

Figure 7.4. Geometry of the extended beam model

For moderately large beam deflection problems, we may make the assumptions

$$\begin{cases} h/\ell \sim w(x,t) \in O(1), & u_0 \sim v \sim w_{,x} \in O(\epsilon), \\ u_{0,x} \sim v_{,x} \sim v_{,y} \sim w_{,xx} \in O(\epsilon^2), \end{cases} \tag{7.97}$$

where the notation $\sim$ stands for "same order of magnitude". For the Taylor expansion, we have $\theta = \tan^{-1}[w_{,x}/(1 + u_{0,x})] = w_{,x} + O(\epsilon^3)$. Neglecting terms higher than $O(\epsilon^2)$, we approximate the net rotation angle by

$$\begin{aligned}\psi(x,y,t) &= -u_{,y}(x,y,t) = -[v(x,y,t) - y\theta(x,t)]_{,y} \\ &= w_{,x}(x,t) - v_{,y}(x,y,t).\end{aligned}$$

If $v(x,y,t)$ is a linear function of y, i.e. plane sections remain plane after deformation, then the shear angle γ depends on (x,t) only. This is the Timoshenko beam theory.

For the displacement $\mathbf{u}$ given by equation (7.96) the Green-St. Venant train tensor is defined by $\mathbf{E} = \frac{1}{2}[\nabla\mathbf{u} + (\nabla\mathbf{u})^T + (\nabla\mathbf{u})^T(\nabla\mathbf{u})]$. Neglecting terms higher than $O(\epsilon^2)$, and using the engineering strain tensor notations $\epsilon_x = \xi_{xx}, \epsilon_y = \xi_{yy}, \gamma = 2\xi_{xy}$, we then obtain

$$\begin{pmatrix} \epsilon_x \\ \epsilon_y \\ \gamma \end{pmatrix} = \begin{pmatrix} u_{0,x} - yw_{,xx} + v_{,x} + \frac{1}{2}w_{,x}^2 \\ \frac{1}{2}w_{,x}^2 \\ v_{,y} \end{pmatrix}, \tag{7.98}$$

where we simply denote $(w_{,x})^2$ by $w_{,x}^2$, etc. This geometrical equation shows clearly that both the horizontal strain ϵ_x and the thickness directional strain ϵ_y depend on the same quadratic nonlinear term $w_{,x}^2$. However, in the von Kármán theory, this quadratic term is considered in the plane deformation, while the strain in the thickness direction is completely ignored. For plate problems, the solution of the homogeneous in-plane equation $\nabla \cdot \varsigma = 0$ may not be a constant, and the nonlinear terms are remained. However, in one-dimensional problems, the constant axial stress ς leads to a linear equilibrium equation in the transverse direction, and this leads to the von Kármán paradox.

The stress vector in this plane deformation problem is given by the elastic constitutive equation

$$\begin{pmatrix} \sigma_x \\ \sigma_y \\ \tau \end{pmatrix} = \frac{E}{1-\nu^2} \begin{pmatrix} 1 & \nu & 0 \\ \nu & 1 & 0 \\ 0 & 0 & \beta \end{pmatrix} \begin{pmatrix} \epsilon_x \\ \epsilon_y \\ \gamma \end{pmatrix}, \tag{7.99}$$

where $\beta = (1-\nu)/2$.

We denote by $\rho(x,y)$ the mass density per unit area in the x,y plane, so that the kinetic energy at time $t \in (0,t_1)$ becomes

$$T(u,w) = \frac{1}{2}\int_\Omega \rho(x,y)[w_{,t}^2 + u_{,t}^2]\mathrm{d}\Omega. \tag{7.100}$$

We assume that the density is an even function of y. Then, with

$$\rho_h(x) = \int_{-h}^{h} \rho(x,y)\mathrm{d}y, \quad I_\rho(x) = \int_{-h}^{h} \rho(x,y)y^2\mathrm{d}y$$

and the deformation model represented in eqn (7.96), equation (7.100) becomes

$$\begin{aligned}T(u_0,v,w) &= \frac{1}{2}\int_0^\ell [\rho_h(x)(w_{,t}^2 + u_{0,t}^2) + I_\rho(x)w_{,xt}^2]\,\mathrm{d}x \\ &\quad + \frac{1}{2}\int_0^\ell \int_{-h}^{h} \rho(x,y)[-2yw_{,xt}(u_{0,t} + v_{,t}) + v_{,t}^2]\mathrm{d}y\,\mathrm{d}x.\end{aligned}$$

Suppose that the beam is subjected to a laterally distributed load $f(x,t)$ and a horizontal load with magnitude p at the end $x = \ell$ in the negative x-direction. From the virtual work principle, we then obtain the variational equation

$$\int_0^{t_1}\int_\Omega [\sigma_x\delta\epsilon_x + \sigma_y\delta\epsilon_y + \tau\delta\gamma]\mathrm{d}\Omega\mathrm{d}t - \int_0^{t_1}\int_0^\ell f(x,t)\delta w\,\mathrm{d}x\mathrm{d}t$$

$$= \int_0^{t_1} -p\delta u_0(\ell,t)]\mathrm{d}t + \int_0^{t_1}\delta T(u_0,v,w)\mathrm{d}t. \tag{7.101}$$

Substituting the geometrical relations (7.98) into (7.101), we see that if the horizontal velocities $u_{0,t}$, $v_{,t}$ and accelerations $u_{0,tt}$, $v_{,tt}$ can be ignored, we obtain the Euler-Lagrange equations

$$\sigma_{x,x} + \tau_{,y} = -\rho y w_{,ttx}, \quad \int_{-h}^{h}\sigma_{x,x}dy = 0,$$

$$\int_{-h}^{h}[y\sigma_{x,xx} + \sigma_{y,x}w_{,x} + (\sigma_x+\sigma_y)w_{,xx}]dy + f = \rho_h w_{,tt} - (I_\rho w_{,ttx})_{,x},$$

and the boundary variational equations

$$\int_{-h}^{h}\sigma_x|_{x=0}\delta u dy = 0, \quad \int_{-h}^{h}\sigma_x|_{x=\ell}\delta u_0 dy + p\delta u_0(\ell) = 0,$$

$$\int_{-h}^{h}\sigma_x|_{x=0,\ell}\delta v dy = 0, \quad \int_{-h}^{h}y\sigma_x|_{x=0,\ell}\delta w_{,x}dy = 0,$$

$$\int_{-h}^{h}(y\sigma_{x,x} + (\sigma_x+\sigma_y)w_{,x})|_{x=0,\ell}\delta w dy = 0, \quad \int_{-h}^{h}\tau\delta v|_{y=\pm h}dx = 0.$$

If the beam is simply supported at the ends $x = 0, x = L$, and if E, G and ρ are constants, then the governing equations of motion are

$$u_{0,xx} + (1+\nu)w_{,x}w_{,xx} = 0, \tag{7.102}$$

$$v_{,xx} + \beta v_{,yy} = y w_{,xxx} - \bar{\rho} y w_{,ttx}, \tag{7.103}$$

$$I w_{,xxxx} - 2h(1+\nu)(u_{0,x} + (2-\nu)w_{,x}^2)w_{,xx} - \int_{-h}^{h} y v_{,xxx}\mathrm{d}y$$

$$= 2h\bar{f} - \bar{\rho}[2hw_{,tt} - Iw_{,ttxx}], \tag{7.104}$$

where $\bar{\rho} = (1-\nu^2)\rho/E$, $\bar{f} = (1-\nu^2)f/(2hE)$. Integration of (7.102) gives

$$u_{0,x} = -\frac{1+\nu}{2}w_{,x}^2 - \frac{\mu}{2h(1+\nu)}, \text{ with } \mu = \frac{(1+\nu)(1-\nu^2)}{E}p.$$

Substituting this into (7.102-7.104) we eventually obtain the dynamic equations for the nonlinear beam model in the form

$$v_{,xx} + \beta v_{,yy} = y w_{,xxx} - \bar{\rho} y w_{,ttx} \quad \text{in } \Omega \times [0,+\infty) \tag{7.105}$$

$$I w_{,xxxx} - a w_{,x}^2 w_{,xx} + \mu w_{,xx} - \int_{-h}^{h} y v_{,xxx}\mathrm{d}y$$

$$= 2h(\bar{f} - \bar{\rho} w_{,tt}) + I\bar{\rho} w_{,ttxx} \;\; \forall (x,t) \in (0,\ell) \times [0,+\infty), \tag{7.106}$$

where $a = 3h(1-\nu^2)$ is a positive parameter. Since the shear displacement v is assumed to be an odd function of y, this fourth-order partial differential system can only be used to handle some special external shear loads. Ignoring the shear effects, this beam model has the simple formulation

$$\bar{\rho}w_{,tt} = \frac{1}{2h}(aw_{,x}^2 - \mu)w_{,xx} - \frac{h^2}{3}w_{,xxxx} + \bar{f}. \tag{7.107}$$

The total potential energy of this model is a double-well energy.

7.4.2 Extended Second-Order Dynamic Beam Model

In order to obtain a more general beam model, we use directly the axial displacement $u(x,y)$ as the unknown function instead of shear deformation $v(x,y,t)$. The Green-St. Venant train tensor in this case is

$$\mathbf{E} = \begin{pmatrix} u_{,x} + \frac{1}{2}(u_{,x}^2 + w_{,x}^2) & \frac{1}{2}(u_{,y} + w_{,x} + u_{,x}u_{,y}) \\ \frac{1}{2}(u_{,y} + w_{,x} + u_{,x}u_{,y}) & \frac{1}{2}u_{,y}^2 \end{pmatrix}.$$

Since $u(x,y) = u_0(x) - y\theta(x) + v(x,y)$, then by the assumption (7.97), we have

$$u_{,y}^2 = (-\theta + v_{,y})^2 = w_{,x}^2 + O(\epsilon^3).$$

Neglecting terms higher than $O(\epsilon^2)$ we then have

$$\begin{pmatrix} \epsilon_x \\ \epsilon_y \\ \gamma \end{pmatrix} = \begin{pmatrix} u_{,x} + \frac{1}{2}w_{,x}^2 \\ \frac{1}{2}w_{,x}^2 \\ w_{,x} + u_{,y} \end{pmatrix}. \tag{7.108}$$

Suppose that the beam is subjected to the external loading system

$$\mathbf{f}(x,t) = \begin{cases} \begin{pmatrix} q^+(x,t) \\ f(x,t) \end{pmatrix} \text{ at } y = h, & \begin{pmatrix} q^-(x,t) \\ 0 \end{pmatrix} \text{ at } y = -h, \\ \begin{pmatrix} -p \\ 0 \end{pmatrix} \text{ at } x = \ell. \end{cases} \tag{7.109}$$

The virtual work principle then leads to the dynamical equations for this beam model, namely,

$$\begin{aligned} &\textstyle\int_{-h}^{h}[(\sigma_x + \sigma_y)w_{,x} + \tau]_{,x}\mathrm{d}y + f(x) = \rho_h w_{,tt}, \\ &\sigma_{x,x} + \tau_{,y} = \rho u_{,tt}, \\ &\tau(x, \pm h, t) = \pm q^{\pm}(x,t), \end{aligned} \tag{7.110}$$

and the boundary variational conditions

$$\int_{-h}^{h} \sigma_x(0,y,t)\delta u(0,y,t)dy = 0,$$

$$\int_{-h}^{h} (\sigma_x(\ell,y,t) + p)\delta u(\ell,y,t)\mathrm{d}y = 0,$$

$$\int_{-h}^{h} [(\sigma_x + \sigma_y)w_{,x} + \tau]|_{x=0,\ell}\delta w\mathrm{d}y = 0.$$

For an elastic beam, substituting the constitutive relation into (7.110), for any $(x, y, t) \in \Omega \times [0, +\infty)$, we obtain

$$u_{,xx} + \beta u_{,yy} = -(1+\nu) w_{,x} w_{,xx} + \bar{\rho} u_{,tt}, \tag{7.111}$$

$$(3(1+\nu)w_{,x}^2 + \beta) w_{,xx} + \frac{1}{2h}\int_{-h}^{h} ((1+\nu)u_{,x}w_{,x} + \beta u_{,y})_{,x} \mathrm{d}y + \bar{f} = \bar{\rho} w_{,tt}, \tag{7.112}$$

$$u_{,y}(x, \pm h, t) = -w_{,x} \pm \bar{q}^{\pm}(x, t), \tag{7.113}$$

where $\bar{q}^{\pm} = \frac{2(1+\nu)}{E} q^{\pm}$. This beam model is a coupled second-order nonlinear partial differential system, which can be used to handle general large deformation problems with arbitrary external loads (7.109).

In the dynamic buckling analysis, we assume that the beam is clamped at $x = 0$ and simply supported at $x = \ell$. The boundary conditions can be given as

$$w(0) = 0, \quad w(\ell) = 0, \quad u(0, y) = 0,$$

$$u_{,x}(\ell, y) + \frac{1+\nu}{2} w_{,x}^2(\ell) = -\bar{p} \quad \forall y \in [-h, h],$$

where $\bar{p} = (1-\nu^2)p/E$, if the beam is subjected to a compressive load p applied in the negative direction of the x-axis at $x = \ell$. The distributed load $f(x, t)$ on the top of the beam is prescribed. When the combination of the external loads reaches a certain critical value, the beam may buckle. We assume that the compressive load p at $x = \ell$ changes neither its direction nor its magnitude during the deformation (i.e., p is a *dead load*), then the acceleration $u_{,tt}$ in equation (7.111) can be ignored. Since this beam model is linear in the displacement $u(x, y, t)$, we set $u(x, y, t) = -\bar{p}x + \bar{u}(x, y, t)$, and the governing equations for this beam model can be written as

$$\bar{u}_{,xx} + \beta \bar{u}_{,yy} = -(1+\nu) w_{,x} w_{,xx}, \tag{7.114}$$

$$[(1+\nu)(3w_{,x}^2 - \bar{p}) + \beta] w_{,xx} + \frac{1}{2h}\int_{-h}^{h} ((1+\nu)\bar{u}_{,x}w_{,x} + \beta \bar{u}_{,y})_{,x} \mathrm{d}y + \bar{f} = \bar{\rho} w_{,tt}, \tag{7.115}$$

$$\bar{u}_{,y}(x, \pm h, t) = -w_{,x} \pm q^{\pm}(x, t),$$

$$\bar{u}(0, y, t) = 0, \quad \bar{u}_{,x}(\ell, y, t) = -\frac{1+\nu}{2} w_{,x}^2.$$

For a given $w(x, t)$, equation (7.114) is a nonhomogeneous linear partial differential equation, the solution of which is

$$\bar{u}(x, y, t) = -\int \frac{1+\nu}{2} w_{,x}^2 \, \mathrm{d}x + v(x, y, t), \tag{7.116}$$

where $v(x,y,t)$ is a homogeneous solution of the boundary-value problem

$$\begin{aligned} v_{,xx} + \beta v_{,yy} &= 0, \\ v_y(x, \pm h, t) &= -w_{,x}(x,t) \pm q^{\pm}(x,t), \\ v(0,y,t) = 0, \quad & v_{,x}(\ell, y, t) = 0. \end{aligned}$$

For a given $w(x,t)$, this problem has a unique solution. Substituting $\bar{u}_{0,x}(x,y,t) = -\frac{1+\nu}{2} w_{,x}^2 + v_{,x}(x,y,t)$ into (7.115), we obtain

$$(\frac{3\alpha^2}{2} w_{,x}^2 + \beta - \lambda\alpha) w_{,xx} + \frac{\beta}{2h}[v_{,x}(x,h,t) - v_{,x}(x,-h,t)] + \bar{f} = \bar{\rho} w_{,tt} \quad \forall x \in (0,\ell), \tag{7.117}$$

where $\alpha = \sqrt{1-\nu^2}$, $\lambda = (1+\nu)\alpha p/E$. Thus, the dynamic buckling problem for large deformation thick beam theory can be proposed as follows.

Problem 7.4.1 For the given loads $\bar{f}(x,t)$ and $\lambda \geq 0$, find the shear deformation $v(x,y,t)$ and the deflection $w(x,t)$ satisfying the governing equations

$$\begin{aligned} & v_{,xx} + \beta v_{,yy} = 0, \\ & (\frac{3\alpha^2}{2} w_{,x}^2 + \beta - \lambda\alpha) w_{,xx} + \frac{\beta}{2h}[v_{,x}(x,h,t) - v_{,x}(x,-h,t)] + \bar{f} = \bar{\rho} w_{,tt}, \\ & v_y(x,\pm h,t) + w_{,x}(x,t) = \pm q^{\pm}(x,t) \end{aligned}$$

in the spacetime $\Omega \times (0, +\infty)$, with boundary and initial conditions

$$\begin{aligned} & v(0,y,t) = 0, \quad v_{,x}(\ell,y,t) = 0 \ , \\ & w(0,t) = 0, \quad w(\ell,t) = 0, \\ & v(x,y,0) = 0, \quad w(x,0) = w_{,t}(x,0) = 0. \end{aligned}$$

♠

This is a coupled, nonlinear second-order partial differential dynamic problem. If either $q^{\pm} = 0$ or $q^{+}(x) = -q^{-}(x)$, the formulation for the solution v is

$$v(x,y,t) = \sum_n c_n(w_n,t) \sin(\lambda_n x) \sinh(\lambda_n y/\sqrt{\beta}),$$

where $\lambda_n = \pi(2n+1)/2L$, w_n is the coefficient of Fourier sine series of $w(x,t)$. In this case, the shear deformation is an odd function of y, i.e. $v(x,-y,t) = -v(x,y,t)$.

If the shear effects are ignored, this problem reduces to the one-dimensional nonlinear dynamic equation

$$\bar{\rho} w_{,tt} = \frac{3}{2}\alpha^2 w_{,x}^2 w_{,xx} - \lambda\alpha w_{,xx} + \bar{f}. \tag{7.118}$$

This is a Duffing type equation. For a given periodic external load $\bar{f}(x,t)$ and $\lambda > 0$, this post-buckling system may have chaotic solutions.

7.4.3 Complementary Energy Variational Principles

We will discuss the primal and dual variational approaches to the static post-buckling problem of the beam model proposed in Problem 7.4.1. Let $\mathcal{U} = C^1(\Omega; \mathbb{R}^2)$ be a displacement space. An element $\mathbf{u} = (v, w)^t \in \mathcal{U}$ is a continuous, differentiable vector in $\mathbb{R}^2$ with domain Ω. For the clamped/simply supported beam subjected to a transverse load $\bar{f}$ and the buckling load at $x = \ell$, the admissible space $\mathcal{U}_a \subset \mathcal{U}$ is defined as

$$\mathcal{U}_a := \left\{ \mathbf{u} = \begin{pmatrix} v \\ w \end{pmatrix} \in \mathcal{U} \;\middle|\; \begin{array}{l} v(x,-y) = -v(x,y), \;\; w(0) = w(\ell) = 0, \\ v(0,y) = v_{,x}(\ell, y) = 0 \;\; \forall y \in [-h, h] \end{array} \right\}.$$

The boundary-value problem for the static post-buckling analysis can now be proposed.

Problem 7.4.2 For the given external loads $\bar{f}$ and $\lambda > 0$, find $(v(x,y), w(x))^T \in \mathcal{U}_a$ such that

$$\begin{aligned} & v_{,xx} + \beta v_{,yy} = 0 \;\; \forall (x,y) \in \Omega, \\ & (\frac{3\alpha^2}{2} w_{,x}^2 + \beta - \lambda\alpha) w_{,xx} + \frac{\beta}{h} v_{,x}(x,h) + \bar{f} = 0 \quad \forall x \in (0, \ell), \\ & v_y(x, \pm h) = -w_{,x}(x) \;\; \forall x \in (0, \ell). \end{aligned}$$

♠

The total potential energy for this buckling problem is the nonconvex functional

$$P(v,w) = \frac{1}{2}\int_\Omega [(v_{,x} + \frac{1}{2}\alpha w_{,x}^2 - \lambda)^2 + \beta(v_{,y} + w_{,x})^2] d\Omega - \int_0^\ell \bar{f} w \, dx. \quad (7.119)$$

If the shear effect is ignored, then P can be written as

$$P(w) = \int_0^\ell [\frac{1}{2}(\frac{1}{2}\alpha w_{,x}^2 - \lambda)^2 - \bar{f}(x) w(x)] \, dx. \quad (7.120)$$

This is the same as Ericksen's double-well energy for the extension of an elastic bar! For a given load $\bar{f}$, $P(w)$ may have two minimizers corresponding two possible locally stable states.

Since $P(v, w)$ is a strictly convex functional of v then, for each w, the problem $\inf_v P(v,w)$ has at most one solution $\bar{v}$. But $P(v,w)$ is not convex in w, and the solution for $\inf_w P(v,w)$ is not unique. Direct methods for solving this nonconvex minimization problem are very difficult.

In order to establish a simple complementary energy variational principle, we need to find the suitable dual variables and the conjugate functional of P. Instead of the engineering strain vector, we let $\boldsymbol{\xi} = (\xi, \gamma)^T$ be the generalized strain vector, defined by the geometrical equation

$$\boldsymbol{\xi} = \begin{pmatrix} \xi \\ \gamma \end{pmatrix} = \Lambda(\mathbf{u}) = \begin{pmatrix} \frac{\partial}{\partial x} & \frac{1}{2}\alpha w_{,x}\frac{\partial}{\partial x} \\ \frac{\partial}{\partial y} & \frac{\partial}{\partial x} \end{pmatrix} \begin{pmatrix} v(x,y) \\ w(x) \end{pmatrix}, \quad (7.121)$$

where $\Lambda(\mathbf{u})$ is quadratic operator. Hence, $\boldsymbol{\xi}$ is a Green-St. Venant type strain vector. In terms of the strain vector $\boldsymbol{\xi}$, the stored energy $\bar{W}$ is the strictly convex (quadratic) functional

$$\bar{W}(\boldsymbol{\xi}) = \int_\Omega \frac{1}{2}[(\xi - \lambda)^2 + \beta\gamma^2]\mathrm{d}\Omega.$$

The conjugate stress vector $\boldsymbol{\varsigma}$ depends linearly on $\boldsymbol{\xi}$. Thus,

$$\boldsymbol{\varsigma} = \begin{pmatrix} \varsigma \\ \tau \end{pmatrix} = \frac{\partial \bar{W}(\boldsymbol{\xi})}{\partial \boldsymbol{\xi}} = \begin{pmatrix} 1 & 0 \\ 0 & \beta \end{pmatrix} \begin{pmatrix} \xi - \lambda \\ \gamma \end{pmatrix}. \tag{7.122}$$

From the Legendre-Fenchel transformation, the complementary energy is uniquely defined by

$$\begin{aligned} \bar{W}^*(\boldsymbol{\varsigma}) &= \sup_{\boldsymbol{\xi}}\{\langle \boldsymbol{\xi}; \boldsymbol{\varsigma}\rangle - \bar{W}(\boldsymbol{\xi})\} = \sup_{\xi,\gamma}\{\int_\Omega (\varsigma\xi + \tau\gamma)\mathrm{d}\Omega - \bar{W}(\xi,\gamma)\} \\ &= \int_\Omega [\frac{1}{2}\varsigma^2 + \lambda\varsigma + \frac{1}{2\beta}\tau^2]\mathrm{d}\Omega. \end{aligned} \tag{7.123}$$

Let $F : \mathcal{U}_a \to \mathbb{R}$ be the external potential defined by $F(\mathbf{u}) = \int_0^\ell \bar{f} w \, \mathrm{d}x$. The total potential P can be written in the canonical form

$$\Pi(\mathbf{u}) = \bar{W}(\Lambda(\mathbf{u})) - F(\mathbf{u}).$$

If we use the real strain vector $\boldsymbol{\xi} = (\epsilon_u, \epsilon_y, \gamma)^t$ defined in (7.108), then the dual problem will have three variational arguments, namely, σ_x, σ_y and τ. This problem shows that in the so-called canonical dual transformation method, the choice of the geometrical operator Λ is not unique. A good choice will lead to a simple formulation. By the canonical decomposition of the geometrical operator $\Lambda(\mathbf{u}) = \Lambda_t(\mathbf{u})\mathbf{u}) + \Lambda_c(\mathbf{u})$, we have

$$\Lambda_t(\mathbf{u}) = \Lambda_t(w) = \begin{pmatrix} \frac{\partial}{\partial x} & \alpha w_{,x}\frac{\partial}{\partial x} \\ \frac{\partial}{\partial y} & \frac{\partial}{\partial x} \end{pmatrix}, \quad \Lambda_c(\mathbf{u}) = \begin{pmatrix} 0 & -\frac{1}{2}\alpha w_{,x}^2 \\ 0 & 0 \end{pmatrix}. \tag{7.124}$$

We let $\bar{\mathbf{f}} \in \mathbb{R}^2$ be the external load

$$\bar{\mathbf{f}} = \left\{ \begin{array}{llll} 0 & \text{in } \Omega, & 0 & \text{at } y = \pm h \\ \bar{f}(x) & \text{in } (0,\ell), & 0 & \text{at } x = L \end{array} \right\}, \tag{7.125}$$

and, with the standard bilinear forms $\langle \boldsymbol{\xi}; \boldsymbol{\varsigma}\rangle = \int_\Omega (\sigma e + \tau\gamma)\mathrm{d}\Omega$, $\langle \mathbf{u}, \bar{\mathbf{f}}\rangle = \int_0^\ell \bar{f} w \, \mathrm{d}x$, the virtual work principle

$$\langle \Lambda_t(\mathbf{u})\delta\mathbf{u}; \boldsymbol{\varsigma}\rangle = \langle \delta\mathbf{u}, \Lambda_t^*(\mathbf{u})\boldsymbol{\varsigma}\rangle = \langle \delta\mathbf{u}, \bar{\mathbf{f}}\rangle \ \ \forall \mathbf{u} \in \mathcal{U}_a \tag{7.126}$$

leads to the equilibrium equations

$$\Lambda_t^*(\mathbf{u})\boldsymbol{\varsigma} = \bar{\mathbf{f}} \ \Rightarrow \ \left\{ \begin{array}{ll} -\int_{-h}^h (\tau + \alpha w_{,x}\varsigma)_{,x}\mathrm{d}y = \bar{f} & \forall x \in (0,\ell), \\ -(\varsigma_{,x} + \tau_{,y}) = 0 & \forall (x,y) \in \Omega, \\ \int_{-h}^h (\tau + \alpha w_{,x}\varsigma)\mathrm{d}y = 0 & \text{at } x = \ell, \\ \tau(x,y) = 0 & \text{at } y = \pm h. \end{array} \right. \tag{7.127}$$

The complementary gap functional associated with the operator Λ is now a quadratic function of w, given by

$$G(w,\varsigma) = \int_\Omega \varsigma \cdot (-\Lambda_c(\mathbf{u}))d\Omega = \int_\Omega \frac{1}{2}\alpha\varsigma w_{,x}^2 d\Omega. \tag{7.128}$$

The canonical Lagrangian associated with the potential variational problem is

$$\begin{aligned} L(\mathbf{u},\varsigma) &= \langle \Lambda(\mathbf{u});\varsigma\rangle - W^*(\varsigma) - F(\mathbf{u}) \\ &= \int_\Omega [\varsigma(\frac{1}{2}aw_{,x}^2 + v_{,x}) + \tau(w_{,x} + v_{,y})]d\Omega \\ &- \int_\Omega (\frac{1}{2}\varsigma^2 + \lambda\varsigma + \frac{1}{2\beta}\tau^2)\, d\Omega - \int_0^\ell \bar{f}w\, dx. \end{aligned} \tag{7.129}$$

For the quadratic operator Λ, the pure complementary energy is obtained as

$$\Pi^d(\varsigma) = \text{sta}\{L(\mathbf{u},\varsigma)|\ \forall \mathbf{u} \in \mathcal{U}_a\} = F_\Lambda^*(\Lambda^*\varsigma) - W^*(\varsigma),$$

where $F_\Lambda^*(\Lambda^*\varsigma)$ is a generalized conjugate function of $F(\mathbf{u})$ defined by

$$F_\Lambda^*(\Lambda^*\varsigma) = \text{sta}\,\{\langle\Lambda(\mathbf{u});\varsigma\rangle - F(\mathbf{u})|\ \forall \mathbf{u} \in \mathcal{U}_a\}. \tag{7.130}$$

Let $\mathcal{S}_s \subset \mathcal{S}$ be the statically admissible space

$$\mathcal{S}_s = \left\{ \left(\begin{array}{c} \varsigma \\ \tau \end{array} \right) \left| \begin{array}{ll} \varsigma_{,x} + \tau_{,y} = 0\ \forall (x,y) \in \Omega, & \int_{-h}^h \varsigma dy \neq 0\ \forall x \in (0,\ell), \\ \tau(x,\pm h) = 0\ \forall x \in (0,\ell), & \int_{-h}^h \tau dy = \int_{-h}^h \varsigma dy = 0 \text{ at } x = \ell \end{array} \right. \right\}.$$

For any given $\varsigma \in \mathcal{S}_s$, the solution of the stationary problem (7.130) is unique and given by

$$\bar{w}_{,x} = \frac{g(x) - \int_{-h}^h \tau dy}{\alpha \int_{-h}^h \varsigma dy}, \tag{7.131}$$

where

$$g(x) = \int_x^\ell \bar{f}(s)\, ds. \tag{7.132}$$

On substituting of (7.131) into (7.130), the pure complementary energy functional Π^d for this nonlinear beam is then obtained as

$$\Pi^d(\mathbf{s}) = -\int_\Omega [\frac{1}{2}\varsigma^2 + \lambda\varsigma + \frac{1}{2\beta}\tau^2]d\Omega - \int_0^\ell \frac{[g(x) - \bar{\tau}(x)]^2}{2\alpha\bar{\varsigma}(x)}\, dx, \tag{7.133}$$

where

$$\bar{\tau}(x) = \int_{-h}^h \tau dy, \quad \bar{\varsigma}(x) = \int_{-h}^h \varsigma dy.$$

Theorem 7.4.1 *Suppose that $(\bar{\mathbf{u}}(= \bar{v}, \bar{w}), \bar{\varsigma}(= (\bar{\varsigma}, \bar{\tau}))$ is a critical point of L.*

(T1) $\bar{\mathbf{u}}$ and $\bar{\varsigma}$ are respectively the critical points of Π and Π^d, and the complementarity condition

$$P(\bar{\mathbf{u}}) = L(\bar{\mathbf{u}}, \bar{\mathbf{s}}) = \Pi^d(\bar{\mathbf{s}}) \tag{7.134}$$

holds.

(T2) $(\bar{\mathbf{u}}, \bar{\varsigma})$ *is a saddle point of L if and only if* $\int_{-h}^{h} \bar{\varsigma}(x,y)dy \geq 0 \ \forall x \in (0,\ell)$. *In this case,* $(\bar{\mathbf{u}}, \bar{\varsigma})$ *is a stable post-buckling state.*

(T3) $(\bar{\mathbf{u}}, \bar{\varsigma})$ *is a super-critical point of L if and only if* $\int_{-h}^{h} \bar{\varsigma}(x,y)dy \leq 0 \ \forall x \in (0,\ell)$. *In this case,* $(\bar{\mathbf{u}}, \bar{\varsigma})$ *is either a locally stable post-buckling state or an unstable post-buckling state.*

7.5 Optimal Shape Designs and Eigenvalue Problems

Structural optimization is a quest for excellence in design, and optimization methods are the means by which to achieve it. The primary problem often faced by designers of structural systems is to determine the shape of the structure. The goal in optimal shape design is to develop economical and reliable structures and systems under different environmental conditions and technological constraints. Let $\mathcal{O}$ be a family of admissible configurations $\Omega \subset \mathbb{R}^n$. With any $\Omega \in \mathcal{O}$ we associate an admissible deformation space $\mathcal{U}_k(\Omega)$ of state variables $u(x)$, defined in Ω. Let $\Phi(u, \Lambda(u), \Omega)$, with $\Omega \in \mathcal{O}$, $u \in \mathcal{U}_k(\Omega)$, be an objective (cost) functional. Then the ***abstract optimal shape design problem*** can be stated as follows.

Problem 7.5.1 Find $\Omega_o \in \mathcal{O}$ and the deformation state $\bar{u} \in \mathcal{U}_k$, such that

$$(\mathcal{P}_{\text{inf}}): \quad \Phi(\bar{u}, \Lambda(\bar{u}), \Omega_o) = \inf_{\Omega \in \mathcal{O}} \inf_{u \in \mathcal{U}_k} \Phi(u(\Omega), \Lambda(u(\Omega)), \Omega)\}. \tag{7.135}$$

♠

For geometrical linear systems, in which Λ is a linear operator, optimal shape design problems have been studied extensively by engineers during the past thirty years (cf., e.g., Haug, Choi and Komkov, 1986; Lasiecka and Sokolowski, 1991; Haftka and Gürdal, 1992; Sokolowski and Zolesio, 1992; Mróz, 1994). Recently, there has been increased mathematical interest in studying optimal shape design problems where the state equations are governed by variational inequalities, and in numerical approximations, sensitivity analysis and related areas (see, for example, Haslinger and Neittaanmäki, 1988).

However, we often, in practice, meet with design problems where the structures undergo large deformations. Moreover, for some engineering structures, such as composite materials, structures interact with actuators for example, where the cost functional could be nonsmooth (cf., e.g., Kohn and Strang, 1986). In these cases, direct methods for solving nonsmooth optimal shape design problems with large deformations are extremely difficult. This motivates the search for dual approaches and related numerical methods. Since both the configuration of a structure and the deformed state are unknown, to find the dual formulations and related extremality theorems for this coupled nonlinear system is, in general, very difficult. In this section, we propose some open problems arising in optimal shape designs as well as some unpublished partial results.

7.5.1 Optimal Shape Design of Beam Theory

The simplest nontrivial optimal shape design problem in structural mechanics is probably the elastic beam with a prescribed total volume. First, let us consider a pure bending beam with a given length $\ell > 0$, with its thickness $h(x)$ being a design variable such that the total volume of the beam is a given constant, say V_0. The admissible configuration space $\mathcal{O}$ is defined as

$$\mathcal{O} = \left\{ h(x) \in \mathcal{C}(\mathbb{R}) \mid \int_0^\ell h(x) \, dx = V_0 \right\}. \tag{7.136}$$

If the beam is subjected to a given distributed transverse load $f(x)$ on this top surface $y = h(x)$, then the total external energy is a functional of both the design variable $h(x)$ and the deflection $w(x)$ of the central x-axis. Thus,

$$F(w, h) = \int_0^\ell w(x) f(x) \sqrt{1 + h_{,x}^2} \, dx. \tag{7.137}$$

For the Euler-Bernoulli beam model, the stored internal energy is

$$W(\Lambda w, h) = \int_0^\ell \int_{-h}^h \frac{1}{2} E(-y w_{,xx})^2 dy \, dx = \int_0^\ell \frac{1}{3} E h^3(x) w_{,xx}^2 \, dx. \tag{7.138}$$

The geometrical operator Λ is a second-order ordinary differential operator and $E > 0$ is a material constant. The primal problem for this optimal shape design can now be proposed as follows.

Problem 7.5.2 For a given distributed load $f(x)$, find the deflection $\bar{w}$ of the beam and the optimal thickness $\bar{h}(x)$ such that

$$P(\bar{w}, \bar{h}) = \inf_{h \in \mathcal{O}} \inf_{w \in \mathcal{U}_k} \{W(\Lambda w, h) - F(w, h)\}, \tag{7.139}$$

where $\mathcal{U}_k$ is a given kinematically admissible space in which the essential boundary conditions are prescribed. ♠

The associated Euler-Lagrange equations form the coupled nonlinear differential system

$$\frac{2}{3} E \left(h^3 w_{,xx} \right)_{,xx} + f \sqrt{1 + h_{,x}^2} = 0, \tag{7.140}$$

$$E h^2 w_{,xx}^2 + \left(\frac{f w h_{,x}}{\sqrt{1 + h_{,x}^2}} \right)_{,x} = \lambda, \tag{7.141}$$

where $\lambda \in \mathbb{R}$ is a Lagrange multiplier for the integral constraint in $\mathcal{U}_k$. If w vanishes at both ends, multiplication by h on both sides of (7.141) followed by integration enables the Lagrange multiplier to be determined as

$$\lambda = \frac{1}{V_0} \int_0^\ell \left(E h^3 w_{,xx}^2 - \frac{f w h_{,x}^2}{\sqrt{1 + h_{,x}^2}} \right) dx. \tag{7.142}$$

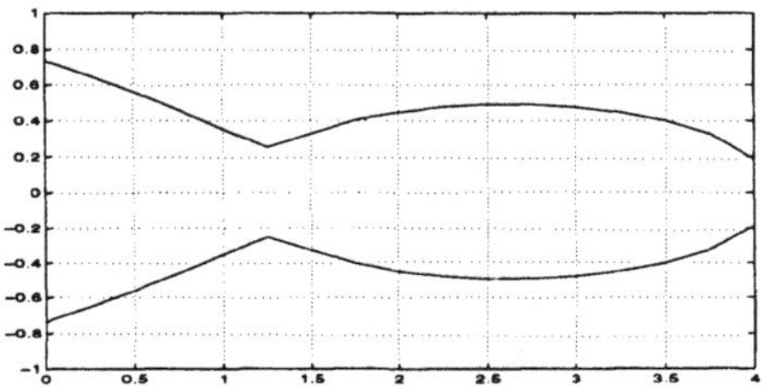

Figure 7.5. Optimal shape of the Euler-Bernoulli beam (clamped at $x = 0$ and simply supported at $x = \ell$)

This problem was originally studied by Gao (1996). For a fixed $h(x)$, the cost functional P is convex in the state variable w. The dual functional $P^d(M, h)$ is then easily obtained via the classical Lagrange duality theory. However, since P is nonconvex in the design variable h, the canonical dual transformation method developed in this book cannot be used directly for solving this type of variational problem. By using a so-called *evolutionary method*, this static problem can be converted into the initial-value problem

$$\begin{aligned} w_{,t} &= \tfrac{2}{3}E\left(h^3 w_{,xx}\right)_{,xx} + f\sqrt{1+h_{,x}^2} \;\; \forall (x,t) \in (0,\ell)\times(0,\infty), \\ h_{,t} &= Eh^2 w_{,xx}^2 + \left(\tfrac{fwh_{,x}}{\sqrt{1+h_{,x}^2}}\right)_{,x} - \lambda \;\; \forall (x,t) \in (0,\ell)\times(0,\infty). \end{aligned} \tag{7.143}$$

If the evolutionary solution $(w_{,t}, h_{,t}) \to (0,0)$ as the time $t \to \infty$, then the steady state $(w(x), h(x))$ satisfies the Euler-Lagrange equations and, hence, is a critical point of $P(w,h)$. It was discovered by this method that for a given uniformly distributed load f, the numerical solutions $\bar{h}(x)$ of the optimal shape for the Euler-Bernoulli beam model subjected differential boundary conditions are nonsmooth functions (see Fig. 7.5). Fig. 7.6 shows the shape evolution process for a beam which is clamped at its both ends.

Mathematically speaking, these nonsmooth "optimal shapes" are reasonable for infinitesimal pure bending beam models. However, the nonsmooth shape will yield stress concentrations, and the mathematical results are not then applicable to engineering structures. Thus, large deformation and shear effects have to be considered in optimal shape design problems. The extended beam model discussed in the previous section can be used for practical purposes. In this case, the stored energy W in Problem 7.5.2 is replaced by

$$\begin{aligned} \bar{W}(\Lambda(v,w),h) &= \int_0^\ell \int_0^{h(x)} [(v_{,x} + \frac{1}{2}\alpha w_{,x}^2 - \lambda)^2 + \beta(v_{,y} + w_{,x})^2] dy\, dx \\ &\quad - \int_0^\ell f(x) w \sqrt{1 + h_{,x}^2}\, dx. \end{aligned}$$

The geometrical operator Λ is now nonlinear. For fixed h, the total complementary energy Π^d can also be formulated in terms of the stresses and h. However,

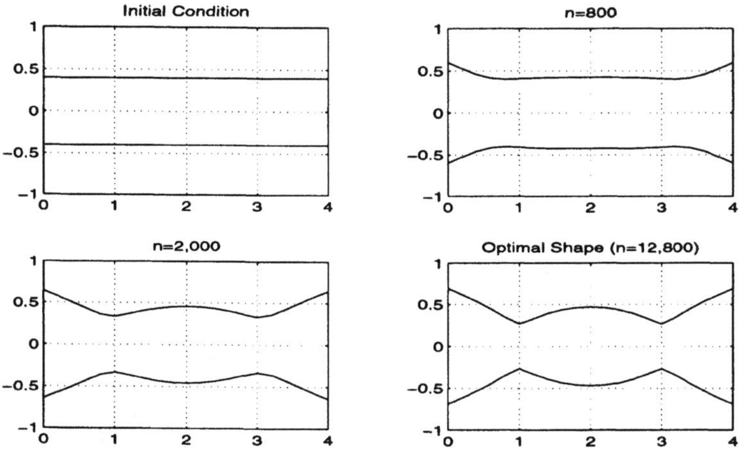

Figure 7.6. Evolution for the shape of the clamped beam.

dual formulation in terms of the dual design variable h^* and the extremality principles are not available as yet.

7.5.2 Eigenvalue Problems on Extremum Surfaces

Let $\Omega \subset \mathbb{R}^2$ be a given simply-connected domain with boundary $\Gamma = \partial\Omega$. Recall that a regular parametric surface $S \in \mathbb{R}^3$ is a $\mathcal{C}^2$-mapping

$$\chi : \Omega \to S \subset \mathbb{R}^3, \quad \chi(\mathbf{x}) = \{\chi^i(x_\alpha)\} i = 1,2,3, \ \alpha = 1,2,$$

such that rank $\nabla\chi(\mathbf{x}) = 2 \ \forall \mathbf{x} \in \Omega$. The induced Riemannian metric tensor (or the right Cauchy-Green strain tensor) $\mathbf{C} = \{C_{\alpha\beta} = \chi_{,\alpha} \cdot \chi_{,\beta}\} = \mathbf{C}^T$ is positive-definite and its inverse is denoted by $\mathbf{C}^{-1} = \{C^{\alpha\beta}\}$. Thus, the eigenvalue problem on the given surface S is to find an eigenfunction $\bar{u} : S \to \mathbb{R}$ such that

$$\lambda(\bar{u}) = \inf_{u \in \mathcal{H}_0^1(S)} \frac{\int_S |\nabla u|^2 \, dS}{\int_S u^2 \, dS} = \inf_{u \in \mathcal{U}_k} \int_\Omega |\nabla u|^2 \sqrt{\det \mathbf{C}} d\Omega, \tag{7.144}$$

where ∇ denotes *covariant differentiation* on the surface S, and

$$|\nabla u|^2 = C^{\alpha\beta} u_{,\alpha} u_{,\beta}, \ \ \alpha, \beta = 1, 2.$$

The kinematically admissible space $\mathcal{U}_k$ is defined by

$$\mathcal{U}_k = \left\{ u \in \mathcal{H}^1(S) | \ \int_S u^2 \, dS = 1, \ \ u = 0 \ \text{ on } \partial S \right\},$$

in which ∂S represents the boundary of the parametric surface S in $\mathbb{R}^3$.

Eigenvalue problems on a given surface have been studied extensively by both mathematicians and engineers for many years. For a primal problem, any direct method leads to an upper bound approach for the first eigenvalue. Duality methods for eigenvalues have also been discussed (cf., e.g., Auchmuty, 1986). Since the kinematically admissible space $\mathcal{U}_k$ is a nonconvex set, the duality theory for the first eigenvalue λ is still an open problem, although some effective lower bound estimates have been suggested (cf., e.g., Lichnerowicz, 1958; Li and Yau, 1980),

Motivated by the optimal shape design of thin-walled shells, Gao and Yau (1990) proposed the eigenvalue problem on an *extremum surface.*

Problem 7.5.3 For a given space curve $\chi_0 : \partial\Omega \subset \mathbb{R}^2 \to \partial S \subset \mathbb{R}^3$, find both the parametric surface $\bar{\chi} : \Omega \to S \subset \mathbb{R}^3$ and the eigenfunction $\bar{u} : S \to \mathbb{R}$ such that

$$\lambda(\bar{u}, \bar{\chi}) = \inf_{\chi \in \mathcal{X}_k} \inf_{u \in \mathcal{U}_k} \int_{S(\chi)} |\nabla u|^2 \, \mathrm{d}S, \tag{7.145}$$

where

$$\mathcal{X}_k = \{\chi \in \mathcal{W}^{1,1}(\Omega; \mathbb{R}^3) | \ \operatorname{rank} \chi(\mathbf{x}) = 2 \ \ \forall \mathbf{x} \in \Omega, \ \ \chi(\mathbf{x}) = \chi_0 \ \ \forall \mathbf{x} \in \partial S\}$$

is the kinematically admissible configuration space. ♠

This is a typical optimal shape design problem. It is not difficult to prove that the criticality condition for this nonconvex optimization problem leads to the coupled nonlinear boundary-value problem

$$\Delta_{\mathcal{M}} u - \lambda u = 0 \quad \text{on } S, \tag{7.146}$$
$$(|\nabla u|^2 - \lambda u^2) H(\chi) - B(\nabla u, \nabla u) = 0 \quad \text{on } S, \tag{7.147}$$
$$u(\chi(\mathbf{x})) = 0, \ \ \chi(\mathbf{x}) = \chi_0 \ \ \forall \mathbf{x} \in \partial\Omega, \tag{7.148}$$

where $\Delta_{\mathcal{M}} = \frac{1}{\sqrt{I_3}} \frac{\partial}{\partial x_\beta} (\sqrt{I_3} C^{\alpha\beta} \frac{\partial}{\partial x_\alpha})$ is the Laplace-Beltrami operator, $I_3(\mathbf{C}) = \det \mathbf{C}$, $H(\chi)$ is the mean curvature of the surface,

$$H = \frac{1}{2} C^{\alpha\beta} b_{\alpha\beta}, \quad B(\nabla u, \nabla u) = b_{\alpha\beta} C^{\alpha\gamma} C^{\beta\eta} u_{,\gamma} u_{,\eta},$$

and

$$b_{\alpha\beta} = \chi^i_{,\alpha\beta} N_i$$

is the *second fundamental form* of the surface χ, with $\mathbf{N} = \{N_i\}$ being the unit vector normal to the surface.

To solve this nonlinear coupled problem is very difficult. Similarly to the optimal shape design for elastic beam, an evolutionary method and associated algorithm have been developed by Gao and Yau (1991), and the interesting "non-smoothness" phenomenon was also found in this eigenvalue problem, i.e. when the total area of the surface is less than a certain value, there exist singularities in the steady solutions for both the eigenfunction $\bar{u}$ and the optimal surface χ (see Fig. 7.7).

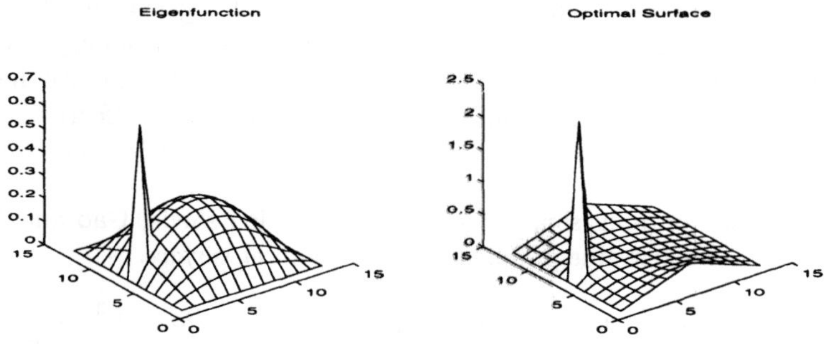

Figure 7.7. Singularity in the eigenvalue problem on an extremum surface

7.5.3 Singularities for Coupled Eigenvalue Problems on Extremum Surfaces

In order to study the singularities in this eigenvalue problem, we introduce a coupling factor $c \in \mathbb{R}^+$ instead of changing the domain for the surface. Thus, the primal problem can be modified as

$$\inf_{\chi \in \mathcal{X}_k} \inf_{u \in \mathcal{U}_k} \left\{ \int_{S(\chi)} [1 + c\frac{1}{2}|\nabla u|^2] \, \mathrm{d}S \right\}. \tag{7.149}$$

Clearly, if $c = 0$, this coupled eigenvalue problem reduces to a minimal surface problem.

On the space-time domain $\Omega \times (0, \infty)$ the associated evolutionary equations are

$$\chi_{,t} = (1 + c|\nabla u|^2 - c\lambda u^2) H(\chi) - cB(\nabla u, \nabla u) \quad \text{in } \Omega \times (0, +\infty) \tag{7.150}$$

$$u_{,t} = \Delta_{\mathcal{M}} u + \lambda u, \quad \text{in } \Omega \times (0, +\infty), \tag{7.151}$$

$$\chi(x, 0) = \chi_0(x), \quad u(x, 0) = u_0(x) \;\; \forall (x, t) \in \Omega \times \{0\},$$

$$\chi(x, t) = \chi_0, \quad u(x, t) = 0 \;\; \forall (x, t) \in \partial\Omega \times (0, +\infty).$$

From the condition $\int_{S(\chi)} u^2 \, \mathrm{d}S = 1$, we have

$$\frac{d}{dt} \int_{S(\chi)} u^2 \, \mathrm{d}S = 2 \int_{S(\chi)} u u_{,t} \, \mathrm{d}S + \int_{S(\chi)} u^2 \mathrm{div}_{\mathcal{M}} \chi \chi_{,t} \, \mathrm{d}S = 0$$

and by using the Gauss-Green formula we obtain

$$\int_{S(\chi)} u u_{,t} \, \mathrm{d}S - \int_{S(\chi)} (u u_{,\alpha} \chi_{,\alpha} + u^2 H(\chi)) \chi_{,t} \, \mathrm{d}S = 0. \tag{7.152}$$

Substituting equations (7.150) and (7.151) into (7.152) and solving for λ, we obtain, for each given coupling factor $c > 0$, the "evolution eigenvalue"

$$\lambda(t) = \frac{\int_{S(\chi)} \left[|\nabla u|^2 + E(u, u_{,\alpha}, \chi) \left((1 + c|\nabla u|^2) H(\chi) - cB(\nabla u, \nabla u) \right) \right] \, \mathrm{d}S}{\int_{S(\chi)} [u^2 + cu^2 H(\chi) E(u, u_{,\alpha}, \chi)] \, \mathrm{d}S}, \tag{7.153}$$

in which

$$E(u, u_{,\alpha}, \chi) = u u_{,\alpha} \chi_{,\alpha} + u^2 H(\chi). \tag{7.154}$$

For a non-parametric hypersurface defined by

$$\chi = \{x, y, f(x,y)\}, \quad f(x,y) : \Omega \to \mathbb{R}, \tag{7.155}$$

the induced metric on S is

$$\{C_{\alpha\beta}\} = \begin{pmatrix} 1 + f_{,x}^2 & f_{,x} f_{,y} \\ f_{,x} f_{,y} & 1 + f_{,y}^2 \end{pmatrix}. \tag{7.156}$$

Its inverse $C^{\alpha\beta}$ is

$$\{C^{\alpha\beta}\} = \frac{1}{I_3} \begin{pmatrix} 1 + f_{,y}^2 & -f_{,x} f_{,y} \\ -f_{,x} f_{,y} & 1 + f_{,x}^2 \end{pmatrix}, \tag{7.157}$$

where I_3 is the determinant of $C_{\alpha\beta}$ defined by

$$I_3(\mathbf{C}) = 1 + f_{,x}^2 + f_{,y}^2. \tag{7.158}$$

The unit vector normal to the surface is

$$\mathbf{N} = (-f_{,x}, -f_{,y}, 1)/\sqrt{I_3}. \tag{7.159}$$

For this non-parametric hypersurface, assuming that both functions f and u are functions of $(x, y, t) \in \Omega \times (0, \infty)$, the evolution equations can be written as

$$f_{,t} = [(1 + cC^{\alpha\beta} u_{,\alpha} u_{,\beta} - c\lambda u^2) H(f) - cB(u_{,\alpha}, u_{,\beta})]/\sqrt{I_3}, \tag{7.160}$$

$$u_{,t} = C^{\alpha\beta} u_{,\alpha\beta} + (\frac{1}{2} C^{\alpha\beta} C^{\gamma\theta} - C^{\alpha\gamma} C^{\beta\theta}) C_{\gamma\theta,\beta} u_{,\alpha} + \lambda u, \tag{7.161}$$

$$f(x, 0) = f_0(x), \quad u(x, 0) = u_0(x) \ \forall (x, t) \in \Omega \times \{0\}, \tag{7.162}$$

$$f(x, t) = \bar{f}(x), \quad u(x, t) = 0 \ \forall (x, t) \in \partial\Omega \times (0, +\infty), \tag{7.163}$$

in which the second fundamental form and the mean curvature are given as

$$\begin{aligned} b_{\alpha\beta}(f) &= \chi^i_{,\alpha\beta} N_i = f_{,\alpha\beta}/\sqrt{G}, \\ H(f) &= \frac{1}{2} C^{\alpha\beta} b_{\alpha\beta}(f) = \frac{1}{I_3^{3/2}} [(1 + f_{,y}^2) f_{,xx} + (1 + f_{,x}^2) f_{,yy} - 2 f_{,x} f_{,y} f_{,xy}]. \end{aligned}$$

To solve this *coupled nonlinear parabolic initial-boundary-value problem* by a finite difference method, for each given $\lambda(t)$, we write equations (7.160-7.161) as

$$\begin{aligned} f_{,t} &= F_f(f, u, f_{,\alpha}, u_{,\alpha}, f_{,\alpha\beta}, \lambda, c) \\ u_{,t} &= F_u(f, u, f_{,\alpha}, u_{,\alpha}, f_{,\alpha\beta}, \lambda, c). \end{aligned}$$

To discretize this initial-boundary-value problem, we use a regular three-dimensional net, generated from a regular net in the region Ω, with mesh size h, by translating it by multiples of j in the time direction. Then, for a given $\lambda(t_j)$, we seek approximations $f^j_{\mu,\nu}$, $u^j_{\mu,\nu}$ of the solutions $f(x,y,t)$, $u(x,y,t)$ at the grid point (x_μ, y_ν, t_j). Here, we use the standard finite difference formulations

$$
\begin{aligned}
u_{,x}(x_\mu, y_\nu, t_j) &= \frac{u^j_{\mu+1,\nu} - u^j_{\mu-1,\nu}}{2h}, \quad u_{,y}(x_\mu, y_\nu, t_j) = \frac{u^j_{\mu,\nu+1} - u^j_{\mu,\nu-1}}{2h}, \\
u_{,xx}(x_\mu, y_\nu, t_j) &= \frac{u^j_{\mu+1,\nu} - 2u^j_{\mu,\nu} + u^j_{\mu-1,\nu}}{h^2}, \\
u_{,yy}(x_\mu, y_\nu, t_j) &= \frac{u^j_{\mu,\nu+1} - 2u^j_{\mu,\nu} + u^j_{\mu,\nu-1}}{h^2}, \\
u_{,xy}(x_\mu, y_\nu, t_j) &= \frac{u^j_{\mu+1,\nu+1} - u^j_{\mu-1,\nu+1} - u^j_{\mu+1,\nu-1} + u^j_{\mu-1,\nu-1}}{h^2} \\
&= u_{,yx}(x_\mu, y_\nu, t_j), \\
u_{,t} &= \frac{u^{j+1}_{\mu,\nu} - u^j_{\mu,\nu}}{k},
\end{aligned}
$$

with similar expressions for f. Then, by the explicit method, at each regular grid point (x_μ, y_ν, t_j), we have

$$
f^{j+1}_{\mu,\nu} = f^j_{\mu,\nu} + kF_f(f^j_{\mu,\nu}, u^j_{\mu,\nu}, \ldots, \lambda(t_j), c), \tag{7.164}
$$

$$
u^{j+1}_{\mu,\nu} = u^j_{\mu,\nu} + kF_u(f^j_{\mu,\nu}, u^j_{\mu,\nu}, \ldots, \lambda(t_j), c). \tag{7.165}
$$

For a given decreasing coupling factor $c \to 0$, Fig. 7.8 shows that the position of the singular point moves over the surface. This interesting phenomenon has been studied by Gao and Yau since 1991. Theoretically speaking, the existence of the singularity in this coupled nonlinear system seems to be spurious, and may be produced by the numerical method or algorithm. However, similar results also appear in the shape optimization of beam models, where the nonsmooth shapes are physically reasonable for a pure bending beam. To obtain deeper insight into this new phenomenon, duality theory and the associated lower bound approaches may play a key role.

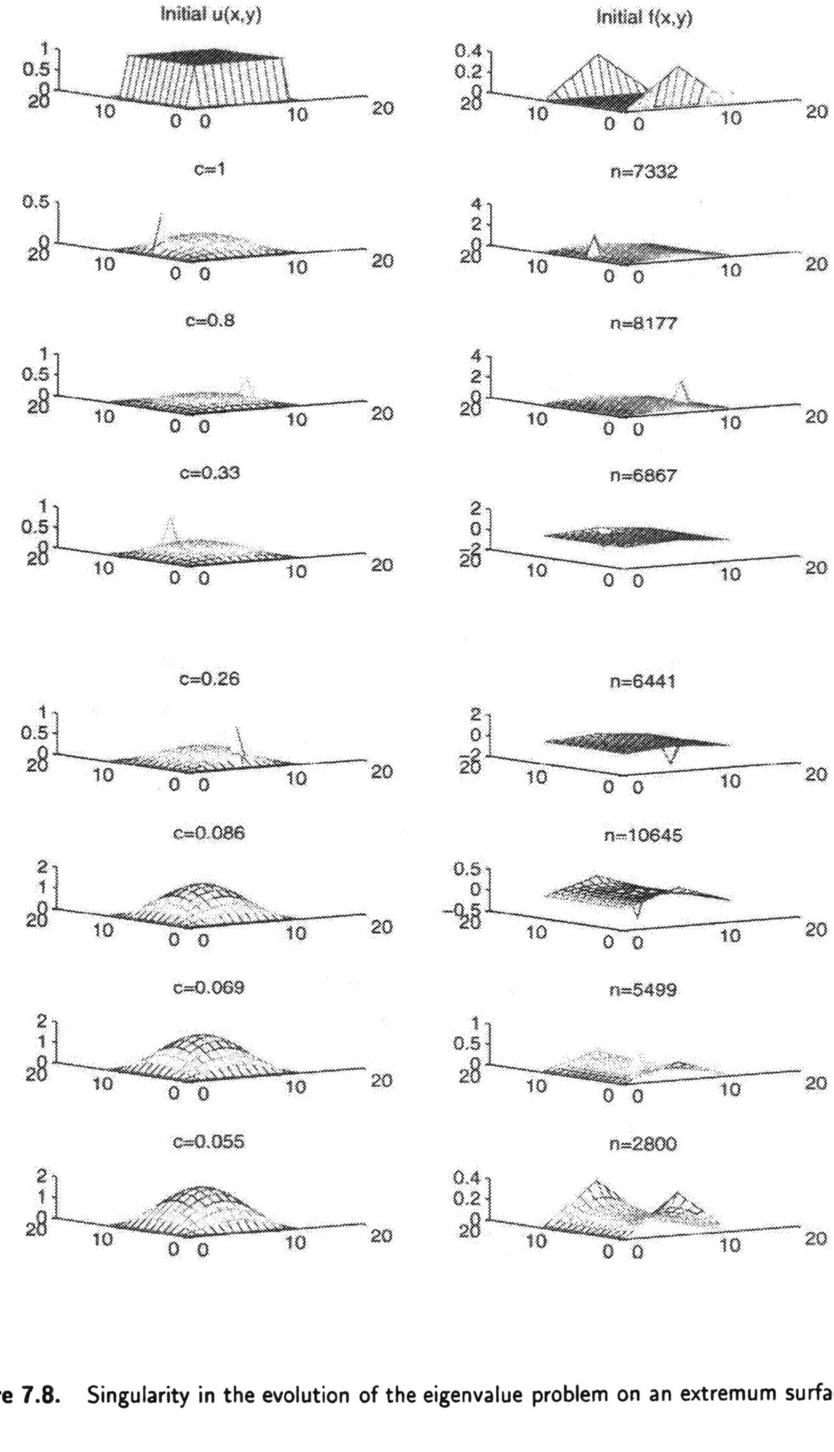

Figure 7.8. Singularity in the evolution of the eigenvalue problem on an extremum surface

7.6 Miscellaneous Open Problems

In many nonlinear systems, the governing equations are not Euler-Lagrange equations for variational problems. But, by using suitable transformations and bilinear forms, the "supersymmetry" can be found and the general framework still works for these systems. We list some well-known nonlinear problems for which the canonical dual transformation method cannot be applied directly to formulate the associated dual problems. New techniques and methods are needed in order to develop suitable dual formulations. This may be a key step towards fully understanding and solving these very difficult problems.

7.6.1 Potential Korteweg-deVries Equation

The *Korteweg-deVries Equation*

$$u_{,t} = au_{,xxx} + buu_{,x}, \quad a, b > 0, \tag{7.166}$$

is a well-known equation in nonlinear systems, but it is not the Euler-Lagrange equation for any variational problem. However, by introducing a potential function $\phi = \phi(x,t)$ and replacing u by $\phi_{,x}$, we form the *potential KdV equation*

$$\phi_{,tx} = a\phi_{,xxxx} + b\phi_{,x}\phi_{,xx}. \tag{7.167}$$

It is easy to check that this equation is the Euler-Lagrange equation for the variational problem

$$P(\phi) = \int_{I_t} \frac{1}{2}\left[a\phi_{,xx}^2 - \frac{1}{3}b\phi_{,x}^3 + \phi_{,t}\phi_{,x}\right] dxdt \ \to \text{sta}, \tag{7.168}$$

where $I_t = \mathbb{R}\times(0,\infty)$ is a spacetime domain. This is a nonconvex functional. To set this problem in our general framework, we introduce the generalized linear strain vector $\boldsymbol{\epsilon} = \Lambda\phi$, defined by

$$\boldsymbol{\epsilon} = \begin{pmatrix} \kappa \\ \epsilon \\ v \end{pmatrix} = \begin{pmatrix} \partial_x^2 \\ \partial_x \\ \partial_t \end{pmatrix} \phi = \Lambda\phi, \tag{7.169}$$

which also defines Λ, and let the constitutive matrix be

$$C(\epsilon) = \begin{pmatrix} a & 0 & 0 \\ 0 & -\frac{b}{3}\epsilon & \frac{1}{2} \\ 0 & \frac{1}{2} & 0 \end{pmatrix}. \tag{7.170}$$

The stored-energy density

$$V(\boldsymbol{\epsilon}) = \frac{1}{2}\boldsymbol{\epsilon}^T C(\epsilon)\boldsymbol{\epsilon} = \frac{1}{2}a\kappa^2 - \frac{1}{6}b\epsilon^3 + \frac{1}{2}v\epsilon \tag{7.171}$$

is then a nonconvex function. The conjugate variable $\boldsymbol{\sigma}$ of $\boldsymbol{\epsilon}$ depends nonlinearly on $\boldsymbol{\epsilon}$, and is given by

$$\boldsymbol{\sigma} = (M,\ \sigma,\ p)^T = DV(\boldsymbol{\epsilon}) = (a\kappa,\ -\frac{1}{2}b\epsilon^2 + \frac{1}{2}v,\ \frac{1}{2}\epsilon)^T. \tag{7.172}$$

The formal adjoint Λ^* associated with the standard bilinear form is the second-order linear partial differential operator

$$\Lambda^* = \left(\frac{\partial^2}{\partial x^2}, \; -\frac{\partial}{\partial x}, \; -\frac{\partial}{\partial t} \right),$$

and the equilibrium equation is

$$\Lambda^* \boldsymbol{\sigma} = M_{,xx} - \sigma_{,x} - p_{,t} = 0.$$

The potential KdV equation can then be written

$$A(\phi) = \Lambda^* C(\Lambda(\phi))\Lambda(\phi) = 0$$

in a framework. The Legendre conjugate of V is also a nonconvex function, namely

$$V^c(\boldsymbol{\sigma}) = \frac{1}{2a} M^2 + 2\sigma p + \frac{4}{3} b p^3. \tag{7.173}$$

Since the constitutive operator C is not a monotone operator, the system $\mathbb{S} = \{\langle \phi, \phi^* \rangle, \langle \boldsymbol{\epsilon}; \boldsymbol{\sigma} \rangle; \Lambda, C\}$ associated with the geometrically linear operator defined by (7.169) is a potential equilibrium system, but not a canonical system. The extremality conditions for this system are unknown.

7.6.2 Potential Boussinesq Equation

The *Boussinesq equation*

$$u_{,tt} + u u_{,xx} + u_{,x}^2 + u_{,xxxx} = 0 \tag{7.174}$$

is a well-known soliton equation, and arises as a model equation for the unidirectional propagation of solitary waves in shallow water. By a method similar to that used for the KdV equation, with u replaced by ψ_{xx}, the potential Boussinesq equation can be written as

$$\psi_{,ttxx} + \frac{1}{2}\partial_x^2 (\psi_{,xx})^2 + u_{,xxxxxx} = 0. \tag{7.175}$$

The associated variational problem is

$$P(\psi) = \int_{I_t} \frac{1}{2} \left[\psi_{,tx}^2 + \frac{1}{3}\psi_{,xx}^3 - \psi_{,xxx}^2 \right] dx dt \; \to \text{sta}. \tag{7.176}$$

Analogously to (7.169) and (7.170), we set

$$\boldsymbol{\epsilon} = \begin{pmatrix} \kappa \\ \gamma \\ \omega \end{pmatrix} = \Lambda \psi = \begin{pmatrix} \partial_x^2 \\ \partial_x^3 \\ \partial_{xt}^2 \end{pmatrix} \psi, \quad C(\kappa) = \begin{pmatrix} \frac{1}{2}\kappa & 0 & 0 \\ 0 & -1 & 0 \\ 0 & 0 & 1 \end{pmatrix}.$$

The stored-energy density

$$V(\boldsymbol{\epsilon}) = \frac{1}{2}\boldsymbol{\epsilon}^T C \boldsymbol{\epsilon} = \frac{1}{6}\kappa^3 - \frac{1}{2}\gamma^2 + \frac{1}{2}\omega^2$$

is then a nonconvex function. The conjugate variable $\boldsymbol{\sigma}$ of $\boldsymbol{\epsilon}$ is also nonlinearly dependent on $\boldsymbol{\epsilon}$. Thus,

$$\boldsymbol{\sigma} = (M,\ \sigma,\ \omega)^T = DV(\boldsymbol{\epsilon}) = (\frac{1}{2}\kappa^2,\ -\gamma,\ \omega)^T.$$

The formal adjoint Λ^* in this case is

$$\Lambda^* = \left(\frac{\partial^2}{\partial x^2},\ -\frac{\partial^3}{\partial x^3},\ \frac{\partial^2}{\partial x \partial t}\right),$$

and the equilibrium equation is

$$\Lambda^* \boldsymbol{\sigma} = M_{,xx} - \tau_{,xxx} + \omega_{,tx} = 0.$$

The fundamental equation for the potential Boussinesq problem reads

$$A(\psi) = \Lambda^* C(\Lambda(\psi))\Lambda(\psi) = 0.$$

Although A is a potential operator, the system generated by Λ and C is not a canonical system since C is not monotone. The Legendre conjugate

$$V^c(\boldsymbol{\sigma}) = \pm\frac{3-\sqrt{2}}{3}M^{3/2} - \frac{1}{2}\tau^2 + \frac{1}{2}\omega^2 \tag{7.177}$$

of V in this system is not unique. This leads to difficulty in understanding the physical meaning of this system.

7.6.3 Invariant Nonlinear Problems

Let $\Omega \subset \mathbb{R}^3$ be an open, bounded domain with boundary, and $u : \Omega \to \mathbb{R}$ a twice differentiable real-valued function. Then, the second derivative of u is the symmetric second-order tensor field

$$H(u) = \{u_{,ij}\} = \{\frac{\partial^2 u}{\partial x_i \partial x_j}\} = \{u_{,ji}\} \in \mathbb{R}^{3\times 3}. \tag{7.178}$$

An *eigenvalue*, λ say, of H is determined by the *characteristic equation*

$$\det(H - \lambda \mathbf{I}) = \det(u_{,ij} - \lambda\delta_{ij}) = 0.$$

Expansion of this determinant leads to the cubic equation

$$\lambda^3 - I_1(H)\lambda^2 + I_2(H)\lambda - I_3(H) = 0, \tag{7.179}$$

where

$$\begin{aligned} I_1(H) &= \mathrm{tr}H, \\ I_2(H) &= \det H(\mathrm{tr}H^{-1}) = \tfrac{1}{2}\left[(\mathrm{tr}H)^2 - \mathrm{tr}(H^2)\right], \\ I_3(H) &= \det H = \tfrac{1}{6}\left[(\mathrm{tr}H)^3 - 3(\mathrm{tr}H)\mathrm{tr}(H^2) + 2\mathrm{tr}H^3\right]. \end{aligned} \tag{7.180}$$

These are called the *principal invariants of H*. In terms of the eigenvalues $\lambda_1, \lambda_2, \lambda_3$ of H, (7.180) may be written

$$\begin{aligned} I_1(H) &= \lambda_1 + \lambda_2 + \lambda_3, \\ I_2(H) &= \lambda_1\lambda_2 + \lambda_2\lambda_3 + \lambda_3\lambda_1, \\ I_3(H) &= \lambda_1\lambda_2\lambda_3. \end{aligned} \tag{7.181}$$

For a given source $f : \Omega \to \mathbb{R}$, the first invariant of $H(u) = \{u_{,ij}\}$ gives the linear partial differential equation

$$I_1(u_{,ij}) = \Delta u = u_{,xx} + u_{,yy} + u_{,zz} = f(x,y,z). \tag{7.182}$$

This is the well-known Poisson's equation. Let Λ = grad, Λ^* = $-$div. Then, this equation can be written as $\Lambda^*\Lambda u + f = 0$. The associated potential is the quadratic functional

$$P(u) = \int_\Omega [|\nabla u|^2 + fu] d\Omega. \tag{7.183}$$

The second invariant leads to the nonlinear equation

$$\begin{aligned} I_2(u_{,ij}) &= \frac{1}{2}[u_{,ii}u_{,ii} - u_{,ij}u_{,ij}] \\ &= u_{,xx}u_{,yy} + u_{,yy}u_{,zz} + u_{,zz}u_{,xx} - u_{,xy}^2 - u_{,yz}^2 - u_{,zx}^2 = f. \end{aligned} \tag{7.184}$$

Unfortunately, this important equation does not have a name.

The third invariant gives the well-known *Monge-Ampere equation*

$$I_3(u_{,ij}) = u_{,xx}u_{,yy}u_{,zz} + 2u_{,xy}u_{,yz}u_{,xz} - u_{,xx}u_{,yz}^2 - u_{,yy}u_{,xz}^2 - u_{,zz}u_{,xy}^2 = f(x,y,z). \tag{7.185}$$

In two-dimensional space $I_2 = I_3$ leads to the classical Monge-Ampere equation

$$u_{,xx}u_{,yy} - u_{,xy}^2 = f(x,y), \quad (x,y) \in \Omega \tag{7.186}$$

which arises in differential geometry. The associated variational functional

$$P_{CH}(u) = \int_\Omega \frac{1}{2}\left[u_x^2 u_{,yy} - u_{,x}u_{,y}u_{,xy} + u_{,y}^2 u_{,xx} + 3fu\right] d\Omega \tag{7.187}$$

can be found in Courant and Hilbert (1953). Unfortunately, this functional does not give any clue concerning the functional spaces in which this variational problem belongs, nor does it suggest the appropriate generalization for functions in n-dimensional problems.

Another functional, whose Euler-Lagrange equation is (7.186) was found by Bakelman (1994). This is

$$P_B(u) = \int_\Omega \left[u(u_{xx}u_{yy} - u_{xy}^2) - 3fu\right] d\Omega. \tag{7.188}$$

If $\partial\Omega$ is a C^2-curve and $u(x,y) \in C^2(\bar{\Omega})$ with $u(x,y) = 0 \;\; \forall (x,y) \in \partial\Omega$, then $P_{CH} = -P_B$.

In n-dimensional problems, if $\Omega \subset \mathbb{R}^n$ is an open convex bounded domain in Euclidean space $\mathbb{R}^n$ and $\{x_i\}$ $(i = 1, 2, \cdots, n)$ are Cartesian coordinates in $\mathbb{R}^n$, then the Bakelman functional can be generalized for functions of n variables to

$$P_B(u) = \int_\Omega [u(x) \det(u_{,ij}(x)) - (n+1) f(x) u(x)] \mathrm{d}\Omega. \tag{7.189}$$

The formal Euler equation for this functional is the general Monge-Ampere equation

$$\det(u_{,ij}) = f(x_1, x_2, \cdots, x_n), \quad u_{,ij} = \frac{\partial^2 u}{\partial x_i \partial x_j}. \tag{7.190}$$

Unfortunately, we have not yet found suitable Λ and C to enable this equation to be expressed in terms of our canonical system. But, if we let

$$\Lambda(u) = \det(u_{,ij}),$$

then its directional derivative at $\bar{u}$ in the direction u is

$$\delta\Lambda(\bar{u}; u) = \Lambda_t(\bar{u})u,$$

and it is easy to show that, with the standard bilinear form,

$$\Lambda_t^*(\bar{u})\bar{u} = \Lambda(\bar{u}).$$

7.7 Commentary and Concluding Remarks

It is now the time to finish the writing on this endless topic. This last section includes a commentary on the topics presented in this chapter and a concluding remark on the general duality principle.

7.7.1 The Seventh Commentary

Continuum mechanics has always been a rich source of and inspiration for mathematics. As pointed by Maugin (1992), theoretical work on plasticity goes back to the papers by Saint Venant (1871), L'evy (1871), Tresca (1872), Haar and von Kármán (1909) and von Mises (1913). The classical mathematical theory of plastic limit analysis were started from the landmark works of Hill's maximal dissipation principle (1948), the minimum principle of Greenberg (1949) and Hodge and Prager (1949, see the book by Prager and Hodge, 1968), as well as Symonds' collapse mechanisms (1951). The applications of mathematical programming methods to lower bound approaches for the safety factor in engineering structural limit analysis were introduced by Maier (1969, 1970).

Starting from the pioneering work of Moreau (1963) in a frictional contact problem, where the notions of the super-potential and subdifferential were originally introduced, the subject of non-smooth/non-convex analysis and optimization has experienced significant development during the last three decades. In mathematics, this subject, with its diverse applications in modern calculus of variations, potential theory, mathematical programming, nonlinear functional analysis and partial differential equations, non-convex dynamic systems and control theory, numerical analysis, economics and game theory, has become a well established and fruitful fundamental discipline. These fields, on the other hand, serve as the foundation for the modern theoretical mechanics and computational mechanics.

In solid mechanics, Halphen and Nguyen Quoc Son's work on the generalized standard material (1975) was an elegant application of the super-potential and its subdifferential in elastoplasticity with internal variables. The so-called functions of bounded variations were introduced by Suquet (1979) and Temam and Strang (1980). Since then, the modern mathematical theory of convex analysis has been used extensively in the study of elastoplasticity. Many papers and monographs have been published in this area (cf., e.g., Temam, 1981, 1983; Oden, 1986; Panagiotopoulos, 1985; Šilhavý, 1997). The mathematical duality theory of elastoplasticity and related topics were studied by, among many others, Temam and Strang (1980), Kohn and Temam (1983).

The terminology *Non-Smooth Mechanics* was formally proposed by Moreau, Panagiotopoulos and Strang (1988). Several monographs have documented the basic theory, methods and applications of non-smooth/non-convex mechanics and their role in engineering (cf., e.g., Panagiotopoulos, 1985; Dem'yanov *et al*, 1996; Mistakidis and Stavroulakis, 1998). This field involves a powerful combination of theoretical analysis in mathematical modelling of natural systems, nonlinear optimization and variational principles, numerical methods and algorithms, software development and scientific computing. The application of this field to engineering mechanics has proven to be an exciting and fruitful endeavor. The practical scope and utility of non-smooth/non-convex mechanics continues to grow. However, monographs and scientific papers published in this field are primarily addressed to applied mathematicians and theoretically-oriented researchers.

Mathematically speaking, variational problems arising from engineering structural limit analysis are belong to a minimal surface type of problem (Gao and Yang, 1995). Direct methods and algorithms can provide only upper bound approaches for the safety factor. However, in structural analysis, engineers prefer to know the lower bound of the collapse load. Duality theory in plastic limit analysis has been well studied for infinitesmall deformation (geometrically linear) problems (cf., e.g., Temam, 1983, 1985; Temam and Strang, 1980; Gao and Hwang, 1987; Cesari and Yang, 1991; Kamenjarzh, 1996). Effects of geometry changes on the collapse load of structures were discussed by Onat (1960) and Maier and Drucker (1973). Their results indicated that the geometry changes can work for or against the capacity of the strength of the structure. However, the basic features of the mathematical problems of determining the limit load in large deformation plasticity remained somewhat obscure.

As an application of the extended Lagrange duality theory in geometrically linear systems, the duality theory for plastic limit analysis was presented clearly in the first section of this chapter, and some useful bounding formulations were developed. These results can be used directly to solve engineering problems. Based on the penalty-duality variational principle, an effective mixed finite element algorithm and computer code have been developed for serious engineering applications (see Gao, 1988a,b). Compared with the so-called primal-dual interior point method in linear programming, this mixed finite element formulation and algorithm can be considered as a ***primal-dual exterior point method.*** The connection between plastic limit analysis in engineering structures and minimum surface type problems in differential geometry was revealed by Gao and Yang (1995), where the multi-duality in nonlinear variational problems was discovered. Mathematical theory for large deformation elastoplasticity was studied by Gao and Strang (1989b), they proved that, if the complementary gap functional possesses a negative sign, the large deformed Hencky structures may be buckled elastically before the plastic deformation taking place. Duality formulations and extremality conditions were studied for finite deformation dynamics (Gao, 1990), rate variational problems (Gao and Onat, 1990) and elastoplastic plates and shells (see Gao *et al*, 1990-96). The triality theory developed in this book can be easily applied to these problems.

The simplest contact problem was first studied by Signorini (1959) for a heavy, linearly elastic body resting on a rigid, frictionless horizontal plane. The essential difficulty of this problem is that the region of contact between the body and the plane is not known a priori. Fichera (1964) was the first to study existence and uniqueness for this problem, which is nonlinear because the position fields satisfying the governing equations are subject to a unilateral constraint that restricts their values to a half space (Antman, 1983).

Variational inequality problems for unilateral (frictionless or frictional) boundary conditions were among the first applications of inequality mechanics (cf. e.g., Duvaut and Lions, 1972; Hlaváček *et al*, 1982; Baiocchi and Capelo, 1984; Friedman, 1993; Pfeiffer and Glocker, 1996; Adly *et al*, 1998; Motreanu and Panagiotopoulos, 1999). Direct approaches and associated numerical methods have been used in various works. In view of the complexity of the inequality constraints and the complementarity relations, augmented Lagrangian methods or the so-called mixed penalty Lagrangian techniques have also been used for the treatment of unilateral frictional problems (see, for examples, Oden and Kikuchi, 1988; Glowinski and Le Tallec, 1989; Laursen and Oancea, 1994; Bille *et al*, 1995). The dual variational inequality was first studied by Mosco (1967). Instead of ϵ^*, he used the dual variable of the configuration u as the dual variational argument. This yields some essential difficulties in finding the inverse of a partial differential operator (see Gao, 1998).

Using of the extended elastoplastic beam model developed by the author (1996), enabled the bi-complementarity to be proposed for the first time. The extended Lagrange duality theory presented in Chapter 5 leads to a very simple dual variational inequality formulation.

Buckling phenomena in thin-walled structures can be expressed through the multiplicity of solutions of the boundary-value problem associated with nonlinear elastic theory. The full mathematical treatment of this boundary-value problem, without extra assumptions on the shape of the structures or symmetry of the buckled state, proved to be extremely difficult due to the nature of the nonlinearity of the partial differential equations involved. Since these equations can be derived from certain potential variational principles (see Koiter (1967) for example), post-buckling problems of nonlinear elastic structures can be converted into nonconvex parametric variational problems.

According to Antman (1977) the historical development of nonlinear plate and shell theories has not been as felicitous as that of rod theories up to the creation of Kirchhoff's theory. Kirchhoff (1824-1887) established a satisfactory linear theory of plates following the work of Navier (1785-1836). A popular nonlinear plate theory was finally developed by von Kármán in 1910. These equations are "derived" by introducing, in the standard procedure of mechanics, a number of geometric approximations that are roughly analogous to the replacement of $\sin\phi$ in the rod equation

$$\phi'' + \sin\phi = 0$$

by its cubic approximation $\phi - \phi^3/6$. Moreover, von Kármán's equations are based upon a certain linear stress-strain law. Thus, von Kármán's plate model belongs to a geometrically nonlinear but physically linear canonical system. The total potential for this model in buckling analysis is a "double-well" energy.

An important mathematical virtue of the von Kármán equations is that they are semilinear. Consequently, their analysis is much simpler than for other large deformation problems in nonlinear elasticity. Direct approaches have been studied extensively by both pure and applied mathematicians. "The number of papers about plates is infinitely large, and gets larger and larger" (Mathematical folklore).

The mathematical analysis for buckling of von Kármán plates has been the subject of many investigations by, among many others, Berger and Fife (1968), Do (1977), Duvaut and Lions (1974), Naumann (1975), and Francu (1979). Recently the von Kármán model has been used for feedback control in smart material systems (cf., e.g., Lasiecka, 1998).

However, it is not overemphasis to say that the von Kármán model is valid only for (very) thin plate subjected to only "moderately large" rotations. As was pointed by Antman (1995), it is very important to note that the formal validity of this model is predicated on very special scalings of the loads in the thickness variables. It is very easy for mathematical analysis of these equations to be conducted for parameters outside the range of physical validity. A virtue of Ciarlet's approach (1980) is that the von Kármán equations characterize the leading term of a formal asymptotic expansion of the solution for the full three-dimensional equations. From Ciarlet's version of the von Kármán equations, we can see clearly that in one-dimensional problems this well-known nonlinear plate model reduces to a linear system. This phenomenon was noted by the author when he was reading the well-known book by Washizu (1967). It is reasonable to believe that this fact was also known already by many other people (but not all). Unfortunately, this one-dimensional von Kármán model was still considered as a nonlinear beam model, and some very complicated variational methods have been used to find its approximate solutions (cf., e.g., Galka and Telega, 1995).

The study of nonlinear dynamic beam theory has a long history (see for example, Libai and Simmonds, 1988; Antman, 1995). Although many large deformation beam theories have been developed, engineers are still looking for some simple but useful models for general engineering problems (cf., e.g., Kapania and Raciti, 1989; Noor and Burton, 1989; Pai and Nayfeh, 1994). Motivated by engineering control problems of smart materials, some extended beam models were proposed by Gao and Russell based upon the general theory of elastoplasticity (Gao and Russell, 1996a). Applications in contact problem of an elastoplastic beam theory lead to the so-called bi-complementarity problems (Gao, 1995, 1998). The generalizations in large deformation problems were due to the recent works by Gao and Russell (1996b) for the forth-order finite deformed beam theory and by Gao (1999) for the general second-order dynamical beam theory. These large deformation models shows clearly the reason that leads to the von Kármán's paradox. Also, the most recent research on the turbulent solutions of the Navier-Stokes equations has shown that the von Kármán-Prandtl universal logarithmic law in the wall-region of wall-bounded turbulent shear flow is based on an assumption which, though plausible, in fact is not quite correct (see Barenblatt and Chorin, 1998).

Complementary energy principles and related mixed/hybrid finite element methods for the von Kármán plate as well as for general thin-walled structures were mainly developed by theoretical mechanicians and applied engineers. The canonical dual transformation and extremality principles with positive gap functions were proposed by the author for the von Kármán plate (Gao, 1990) and for elastoplastic plates (Gao, 1994). The canonical dual approaches for post-buckling of the von Kármán plate were studied by Gao (1995), wherein the so-called lower bound of buckling factor is actually a critical value of a canonical complementary energy. In post-buckling analysis, the loading factor depends upon the deformation. Many interesting phenomena in post-buckling of solids have been discovered by engineers (cf., e.g., Hill and Hutchinson, 1975; Nguyen, 1994; Librescu and Lin, 1996). Dual formulations and the associated lower bound approaches to the pre-buckling load (the first eigenvalue) are still open problems.

Based on the systematic developments of the canonical dual transformation method and the general duality principles, the tri-canonical formulations and the canonical Lagrangian for the von Kármán plate as well as the newly proposed nonlinear beam models are clearly presented in this chapter. Detailed mathematical study and further applications of the triality theory for these models is left for interested readers. The open problems proposed in this chapter does not means that they are very difficult. It is hope that the study of duality theory on these problems will lead to some new variational techniques and mathematical tools.

7.7.2 Concluding Remarks

Duality in natural systems is amazingly beautiful. Motivated mainly by finite deformation theory, the canonical dual transformation method and general duality principles in nonconvex nonsmooth systems have been developed heuristically throughout this book. The extremality principles show nice symmetric forms in geometrically linear systems, whereas the triality theory reveals the intrinsic harmony in fully nonlinear systems, for which the simple symmetry is broken.

Duality is a fundamental concept that underlies almost all natural phenomena. For any given problem, as long as the tri-canonical forms can be characterized, the canonical dual transformation method and dual extremality principles can be used to establish nice theoretical results and to develop powerful alternative numerical methods. Applications of the general tri-duality principles described in Chapter 4 have into three aspects:

(1) to check the validity and completeness of the existence theorems;
(2) to develop new (dual) theories and methods based upon the known ones;
(3) to predict the new systems and possible theories by the polar-duality principles and its sequential extensions.

Duality in natural systems is a very broad and rich field. To theoretical scientists and philosophical thinkers as well as great artists, duality has always been a common topic in their different fields. It is really "a splendid feeling to realize the unity of a complex of phenomena that by physical perception appear to be completely separated". Although the author has many other topics in mind and really wants to write down, he has to stop here and to take care of his own chaotic system. It is pleasing to see that more and more knowledgeable researchers and scientists are working in this wonderland and exploring the intrinsic beauty of the nature.

Truthful words are not beautiful; beautiful worlds are not truthful.
Good words are not persuasive; persuasive words are not good.
He who knows has no wide learning; he who has wide learning does not know.
The sage does not hoard.
Having bestowed all he has on others, he has yet more;
Having given all he has to others, he is richer still.
The Tao of heaven benefits and does not harm; the Tao of the sage is bountiful and does not contend.

—Lao Chi, *Tao De Chin*, LXXXI

The function of an expert is not to be more right than other people, but to be wrong for more sophisticated reasons.

—David Butler

Appendices

Appendix A
Duality in Linear Analysis

Mathematics contains much that will neither hurt one if one does not know it nor help one if one does know it.

—J.B. Mencken, 1715

A mathematician, like a painter of poet, is maker of patterns. If his patterns are more permanent than theirs, it is because they are made with ideas.

—Godfrey Harold Hardy

Mathematics loves symmetry about all.

—J. C. Maxwell

A.1 Linear Spaces and Duality

Abstract vector spaces and their dual spaces are very important in physics and engineering science and allow us to study a large range of problems within a common mathematical structure. In this section, we shall briefly discuss some basic concepts and results from functional analysis.

Let $\mathbb{F}$ be a scalar field. The elements $\alpha, \beta, \gamma, \dots \in \mathbb{F}$ are called *scalars*. Here $\mathbb{F}$ is taken to be either the real number field, $\mathbb{R}$, or the complex number field, $\mathbb{C}$. In this book, where we do not explicitly mention $\mathbb{F}$ in examples we shall be taking $\mathbb{F} = \mathbb{R}$. Suppose that $\mathcal{U}$ is a set of elements $u, v, w, \cdots$ over $\mathbb{F}$. An element $u \in \mathcal{U}$ is called a *vector*, and could be a Euclidean vector if $\mathcal{U} = \mathbb{R}^n$, a matrix if $\mathcal{U} = \mathbb{R}^{m\times n}$ or a continuous function on a closed interval $[a, b]$ if $\mathcal{U}$ is a space of continuous functions defined on $[a, b] \subset \mathbb{R}$, etc.

The *Cartesian product* of two non-empty sets $\mathcal{U}$ and $\mathcal{V}$ is denoted by $\mathcal{U} \times \mathcal{V}$ and represents the set of all ordered pairs (u, v) with $u \in \mathcal{U}$ and $v \in \mathcal{V}$. Thus,

$$\mathcal{U} \times \mathcal{V} := \{(u, v) |\ u \in \mathcal{U} \text{ and } v \in \mathcal{V}\}.$$

Note that, in general, $\mathcal{U} \times \mathcal{V} \neq \mathcal{V} \times \mathcal{U}$.

On $\mathcal{U}$ we define an additive binary operation $\oplus : \mathcal{U} \times \mathcal{U} \to U$ with following properties.

1. Closure: $u \oplus v \in \mathcal{U} \quad \forall u, v \in \mathcal{U}$.
2. Operation laws:

(a) Associative law: $(u \oplus v) \oplus w = u \oplus (v \oplus w)$,
(b) Commutative Law: $u \oplus v = v \oplus u$.

3. Identity and Inverse Elements:

(a) There exists a unique zero element $0 \in \mathcal{U}$ such that $u \oplus 0 = u \ \ \forall u \in \mathcal{U}$,
(b) For each $u \in \mathcal{U}$, there exists a unique inverse element $-u \in \mathcal{U}$ such that $u \oplus -u = 0$.

The operation $\oplus$ with these three properties is called *addition*. The pair $\{\mathcal{U}; \oplus\}$ forms an additive Abelian (i.e. commutative) group. On the scalar field $\mathbb{F}$, the traditional operations such as the scalar addition "+" and multiplication "·" are automatically defined. On $\{\mathcal{U}; \oplus\}$, if we introduce another closed mapping $* : \mathbb{F} \times \mathcal{U} \to \mathcal{U}$, which we call *scalar multiplication*, such that for each $u \in \mathcal{U}$, $\alpha * u \in \mathcal{U}$ $\forall \alpha \in \mathbb{F}$, then an abstract vector space can be defined as follows.

Definition A.1.1 (Vector Space) A triple $\{\mathcal{U}; \oplus, *\}$ is an abstract *linear vector space (or, simply, a vector space)* over the field $\mathbb{F}$ if for arbitrary $u, v, w \in \mathcal{U}$; $\alpha, \beta, \gamma \in \mathbb{F}$ and an identity element $e \in \mathbb{F}$, the conditions

1. $(\alpha\beta) * u = \alpha * (\beta * u)$,
2. $(\alpha + \beta) * u = (\alpha * u) \oplus (\beta * u)$, $\alpha * (u \oplus v) = (\alpha * u) \oplus (\alpha * v)$,
3. $e * u = u$,

hold. ◇

For convenience, vector addition $\oplus$ is denoted simply by the usual symbol $+$, and scalar multiplication $*$ will be denoted by the usual symbolism, i.e. $\alpha * u$ will be written as αu or $\alpha \cdot u$.

Example A.1.1 The n-dimensional Euclidean space $\mathcal{U} = \mathbb{R}^n$, with vector addition and scalar multiplication defined by

$$u + v = (u_1 + v_1, u_2 + v_2, \cdots, u_n + v_n) \ \ \forall u, v \in \mathbb{R}^n,$$

$$\alpha u = (\alpha u_1, \alpha u_2, \cdots, \alpha u_n) \ \ \forall \alpha \in \mathbb{R},$$

is a vector space over $\mathbb{R}$, but not over $\mathbb{C}$. The set of m by n matrices, denoted by $\mathbb{R}^{m\times n}$, together with matrix addition and scalar multiplication is a linear space. This is true because the sum of two m by n matrices is also a matrix in $\mathbb{R}^{m\times n}$. ♡

Example A.1.2 The function space $\mathcal{C}^n(a, b)$ is a vector space. Its elements, the n-times continuously differentiable functions defined on an open interval $I = (a, b) \subset \mathbb{R}$, form a linear space. The reason is that the sum of any two such functions is another such function. For $n = 0$, $\mathcal{C}^0(I)$ is simply written as $\mathcal{C}(I)$. ♡

Definition A.1.2 (Linear Subspace) If $\mathcal{U}$ is a vector space over $\mathbb{F}$, then a subset $\mathcal{U}_a \subset \mathcal{U}$ is a *linear subspace* if $u, v \in \mathcal{U}_a \Rightarrow \alpha u + \beta v \in \mathcal{U}_a$ $\forall \alpha, \beta \in \mathbb{F}$. ◇

A feature of linear subspaces is that they all contain the zero element. If we translate the origin of a linear subspace, we obtain an affine subset (sometimes called a *linear manifold* or a *linear variety*).

Definition A.1.3 (Affine subset) If $\mathcal{U}$ is a vector space over $\mathbb{F}$ and $\mathcal{U}_a$ is a linear subspace of $\mathcal{U}$, then, an *affine subset* $\mathcal{M}$ is defined as

$$\mathcal{M} := \{u \in \mathcal{U} | \ \ u = c + v \ \ \forall v \in \mathcal{U}_a\}$$

for a fixed element $c \in \mathcal{U}$. ◇

Example A.1.3 In $\mathbb{R}^n$ all lines and planes through the origin are subspaces, whereas lines and planes not passing through the origin are affine subsets.

Definition A.1.4 (Hyperplane) A *hyperplane* $\mathcal{P}$ of a vector space $\mathcal{U}$ is a maximal proper affine subset $\mathcal{M}$ of $\mathcal{U}$, i.e., in $\mathcal{M}$, the subspace $\mathcal{S}$ is maximal in the sense that any other subspace containing $\mathcal{S}$ is either $\mathcal{S}$ or $\mathcal{U}$. ◇

Example A.1.4 Hyperplanes in $\mathbb{R}^2$ are lines; in $\mathbb{R}^3$ hyperplanes are planes.

If the vector space $\mathcal{U}$ is spanned by a finite set of linearly independent vectors $u_1, u_2, \cdots, u_n$, then we say $\mathcal{U}$ has dimension n, i.e. $\dim \mathcal{U} = n$. If no such set of vectors exists, $\mathcal{U}$ is said to have infinite dimension. Finite-dimensional vector spaces are very special and are just like $\mathbb{R}^n$ or $\mathbb{C}^n$. However, there are infinite numbers of infinite-dimensional vector spaces, such as $C^n(a, b)$ for any $n \geq 0$.

Definition A.1.5 (Isomorphic Vector Spaces)

Two vector spaces $\mathcal{U}$ and $\mathcal{V}$ are isomorphic if there is a bijective map $\Lambda : \mathcal{U} \to \mathcal{V}$ such that

$$\Lambda(\alpha u + \beta v) = \alpha(\Lambda u) + \beta(\Lambda v) \ \ \forall u, v \in \mathcal{U}$$

holds for $\alpha, \beta \in \mathbb{F}$. ◇

A major result of linear algebra is that all finite-dimensional real vector spaces are isomorphic to $\mathbb{R}^n$ and all finite-dimensional complex vector spaces are isomorphic to $\mathbb{C}^n$.

In the n-dimensional Euclidean vector space $\mathbb{R}^n$, the norm of a vector, defined by

$$\|u\| := \left(\sum_{i=1}^{n} |u_i|^2 \right)^{\frac{1}{2}},$$

is the length of the vector $u \in \mathbb{R}^n$. A generalization of this concept can be introduced for abstract spaces.

Definition A.1.6 (Norm and Semi-Norm)

(D1) A *norm* is a non-negative set function $\| * \|: \mathcal{U} \to \mathbb{R}$ on a vector space $\mathcal{U}$ such that

(a) positivity: (1) $\| u \| \geq 0$; (2) $\| u \| = 0 \ \Leftrightarrow u = 0$;
(b) homogeneity: $\| \alpha u \| = |\alpha| \, \| u \| \quad \forall \alpha \in \mathbb{F}$;
(c) triangle inequality: $\| u + v \| \leq \| u \| + \| v \| \quad \forall u, v \in \mathcal{U}$.

(D2) A function is called a *seminorm* on a vector space $\mathcal{U}$ it it possesses all of these properties except (a1). ◇

Definition A.1.7 (Normed linear space) A *normed linear space* is a vector space $\mathcal{U}$ with a norm $\| * \|_{\mathcal{U}}$ defined on it, and it is denoted by $\{\mathcal{U}; \| * \|_{\mathcal{U}}\}$. ◇

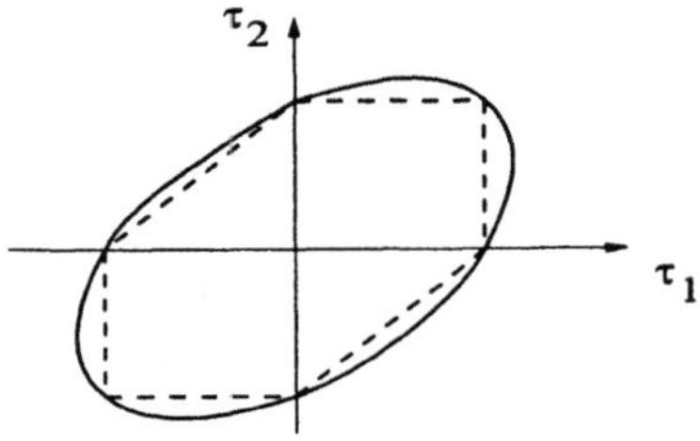

Figure A.1. Tresca (doted) and von Mises norms in $\mathbb{R}^2$

Usually, where there will be no confusion, we just write $\mathcal{U}$, and use $\| * \|$ for the norm in $\mathcal{U}$.

Example A.1.5 In $\mathbb{R}^n$, in addition to the Euclidean norm, there are several other possible norms, such as

$$\text{(a) the } \infty\text{-norm:} \qquad \|u\|_\infty := \max_{1\le i\le n}\{u_i\}; \tag{A.1}$$

$$\text{(b) the } p\text{-norms:} \qquad \|u\|_p := \left(\sum_{i=1}^n |u_i|^p\right)^{1/p} \quad 1 \le p < \infty. \tag{A.2}$$

In three-dimensional plastic deformation theory, if $\sigma = (\sigma_1, \sigma_2, \sigma_3) \in \mathbb{R}^3$ is a principal stress vector field, then $\tau = \{\tau_i = |\sigma_j - \sigma_k|, \quad i \ne j \ne k \ne i\}$ is a tangential stress vector. For a given material constant (elastic limit stress) $k_0 \in \mathbb{R}$, the criterion

$$\|\tau\|_\infty = \max\{\tau_1, \tau_2, \tau_3\} = k_0$$

defined by the ∞-norm is the so-called *Tresca yield condition*, which was experimentally obtained by H.E. Tresca (1872). Its mathematical formulation was given simultaneously by Barré de Saint-Venant in 1871. The criterion defined by the 2-norm

$$\|\tau\|_2 = \sqrt{\tau_1^2 + \tau_2^2 + \tau_3^2} = \sqrt{2}k_0$$

is the so-called *von Mises yield condition*, proposed by von Mises (1913) on a purely mathematical basis because of its obvious simple representation. However, it was shown in the well-known experiment by *Taylor* and *Quinney* (1931), in the case of metals, that the experimental results lie between these two criteria, although they are generally closer to the Mises criterion. The graphs of these two yield surfaces are shown in Fig. A.1

The triangle inequality for the Euclidean norm can be proved using the *Cauchy-Schwarz inequality*

$$\left(\sum_{i=1}^n |u_i v_i|\right) \le \left(\sum_{i=1}^n |u_i|^2\right)^{1/2} \left(\sum_{i=1}^n |v_i|^2\right)^{1/2}. \tag{A.3}$$

However, to prove that the p-norm is in fact a norm we need the *Minkowski inequality*

$$\left(\sum_{i=1}^{n} |u_i + v_i|^p\right)^{1/p} \leq \left(\sum_{i=1}^{n} |u_i|^p\right)^{1/p} + \left(\sum_{i=1}^{n} |v_i|^p\right)^{1/p}. \tag{A.4}$$

Both the Cauchy-Schwarz and Minkowski inequalities are valid for n finite or infinite. For $\mathbb{R}^n$, we use the notations

$$l_p^n = \{\mathbb{R}^n; \| * \|_p\}, \quad l_\infty^n = \{\mathbb{R}^n; \| * \|_\infty\}.$$

For an infinite-dimensional vector space $\mathcal{U}$, we introduce the notations

$$\begin{aligned} l_p &:= \{u \in \mathcal{U} | \ \|u\|_p < \infty\}, \\ l_\infty &:= \{u \in \mathcal{U} | \ \|u\|_\infty < \infty\} \end{aligned}$$

for the p-norm and the ∞-norm, respectively. Clearly, $l_p \subset l_\infty \ \forall p \geq 1$. A set $\mathcal{S}$ in a normed linear space $(\mathcal{U}; \| * \|_\mathcal{U})$ is bounded if $\sup_{u \in \mathcal{S}} \|u\|_\mathcal{U} < \infty$. Thus, l_p and l_∞ are bounded linear spaces.

Example A.1.6 Take $\mathcal{U} = \mathcal{C}(a, b)$ with the p-norm

$$\|u\|_p = \left(\int_a^b |u(x)|^p \, \mathrm{d}x\right)^{1/p}.$$

This is always finite for any finite a and b. The triangle inequality can be proved by using the integral form

$$\left(\int_a^b |u(x) + v(x)|^p \, \mathrm{d}x\right)^{1/p} \leq \left(\int_a^b |u(x)|^p \, \mathrm{d}x\right)^{1/p} + \left(\int_a^b |v(x)|^p \, \mathrm{d}x\right)^{1/p} \tag{A.5}$$

of Minkowski's inequality for $1 \leq p < \infty$. Hence, $\{\mathcal{C}(a, b); \| * \|_p\}$ is a l_p space.

Definition A.1.8 (Topologically Isomorphic Normed Linear Spaces) Let $\{\mathcal{U}, \| * \|_\mathcal{U}\}$ and $\{\mathcal{V}, \| * \|_\mathcal{V}\}$ be two normed linear spaces. Then, $\mathcal{U}$ and $\mathcal{V}$ are topologically isomorphic if there exists a linear bijective map $\Lambda : \mathcal{U} \to \mathcal{V}$ and positive constants α, β such that

$$\alpha \|u\|_\mathcal{U} \leq \|\Lambda u\|_\mathcal{V} \leq \beta \|u\|_\mathcal{U} \quad \forall u \in \mathcal{U}.$$

This topological isomorphism is denoted by $\mathcal{U} \cong \mathcal{V}$. ◊

Finite-dimensional spaces of the same dimension are topologically isomorphic.

Example A.1.7 Consider the two normed linear spaces

$$\begin{aligned} \mathcal{U} &= \{u \in \mathbb{R}^3 | \ u = a\mathbf{i} + b\mathbf{j} + c\mathbf{k} \ \forall a, b, c \in \mathbb{R}, \ \text{ with } \|u\| = \sqrt{a^2 + b^2 + c^2}\} \\ \mathcal{V} &= \{v(t) | \ v(t) = a + bt + \frac{1}{2}ct^2 \ \forall a, b, c, t \in \mathbb{R}, \ \text{ with } \|v\| = \max\{|a|, |b|, |c|\}\} \end{aligned}$$

of dimension three. Note that the map $\Lambda : \mathcal{U} \to \mathcal{V}$ given by

$$\Lambda(u) = a + bt + \frac{1}{2}ct^2$$

is linear and bijective; also, we have

$$\|\Lambda u\|_{\mathcal{V}} = \max\{a, b, c\} \leq \sqrt{a^2 + b^2 + c^2} = \|u\|_{\mathcal{U}}.$$

Since

$$\frac{1}{3}\|u\|_{\mathcal{U}} \leq \|\Lambda u\|_{\mathcal{V}} \leq \|u\|_{\mathcal{U}},$$

this implies that the two spaces are topologically isomorphic.

Definition A.1.9 (Isometrically Isomorphic Normed Linear Spaces)
Suppose that $(\mathcal{U}, \| * \|_{\mathcal{U}})$ and $(\mathcal{V}, \| * \|_{\mathcal{V}})$ are two topologically isomorphic normed linear spaces under the linear bijective map $\Lambda : \mathcal{U} \to \mathcal{V}$. If

$$\|\Lambda u\|_{\mathcal{V}} = \|u\|_{\mathcal{U}} \quad \forall u \in \mathcal{U},$$

then $\mathcal{U}$ and $\mathcal{V}$ are said to be *isometrically isomorphic*, and the map Λ is called an *isometry*. ◇

Any finite-dimensional space is a subspace of some higher-dimensional space, and may also be considered as a product space of $\mathbb{R}$ under a suitable norm.

Definition A.1.10 (Subspace of a normed linear space)
A subspace $\{\mathcal{U}_a; \| * \|_{\mathcal{U}}\}$ of a normed linear space $\{\mathcal{U}; \| * \|_{\mathcal{U}}\}$ is a subset $\mathcal{U}_a \subset \mathcal{U}$, together with the induced norm $\| * \|_{\mathcal{U}}$. ◇

Definition A.1.11 (Product of Normed Linear Spaces)
Let $\{\mathcal{U}; \| * \|_{\mathcal{U}}\}$ and $\{\mathcal{V}; \| * \|_{\mathcal{V}}\}$ be normed linear spaces over the same field $\mathbb{F}$. Then, under any of the following product norms

(1) $\|(u, v)\| = \|u\|_{\mathcal{U}} + \|v\|_{\mathcal{V}} \quad \forall u \in \mathcal{U}, v \in \mathcal{V},$
(2) $\|(u, v)\| = (\|u\|_{\mathcal{U}}^p + \|v\|_{\mathcal{V}}^p)^{1/p} \quad \forall u \in \mathcal{U}, v \in \mathcal{V}, \quad 1 \leq p < \infty,$
(3) $\|(u, v)\| = \max\{\|u\|_{\mathcal{U}}, \|v\|_{\mathcal{V}}\} \quad \forall u \in \mathcal{U}, v \in \mathcal{V},$

the product space $\mathcal{Z} = \mathcal{U} \times \mathcal{V} = \{(u, v) |\ u \in \mathcal{U},\ v \in \mathcal{V}\}$ is a normed linear space. ◇

In normed vector spaces we can study continuity and convergence. A sequence $\{u_n\}$ in a normed linear space $\{\mathcal{U}; \| * \|_{\mathcal{U}}\}$ converges to u_o if

$$\lim_{n \to \infty} \|u_n - u_o\|_{\mathcal{U}} = 0.$$

Continuity and convergence are closely related: a map $F : \mathcal{U} \to \mathcal{V}$ is continuous at u_o if and only if

$$\lim_{n \to \infty} F(u_n) = F(\lim_{n \to \infty} u_n)$$

for every sequence $\{u_n\}$ in $\mathcal{U}$ that converges to u_o.

A set $\mathcal{S}$ in a normed linear space $\mathcal{U}$ is *closed* if all convergent sequences in $\mathcal{S}$ have their limit points in $\mathcal{S}$. A set $\mathcal{S}$ is *open* if its algebraic complement $\mathcal{S}^c = \mathcal{U} - \mathcal{S}$ is closed. Alternatively, a set $\mathcal{S}$ is open if, for any point $u \in \mathcal{S}$, there exists an $\epsilon > 0$ such that the set $\{v| \, \|u - v\|_{\mathcal{U}} < \epsilon\}$ is wholly contained in $\mathcal{S}$. Of course, any set $\mathcal{S}$ is contained in a closed set, which is formed by adding all limit points of sequences in $\mathcal{S}$ to $\mathcal{S}$. This is the *closure* of $\mathcal{S}$, which is usually denoted as $\bar{\mathcal{S}}$. The concept of an open set is fundamental in functional analysis; it allows us to study the topological properties of spaces.

Definition A.1.12 (Boundness and Compactness)

(D1) A set $\mathcal{U}_a$ in a normed linear space $\{\mathcal{U}, \| * \|_{\mathcal{U}}\}$ is *bounded* if

$$\sup_{u \in \mathcal{U}_a} \|u\|_{\mathcal{U}} < \infty.$$

(D2) A set $\mathcal{U}_a$ is *sequentially compact* if every sequence in $\mathcal{U}_a$ contains a convergent subsequence with a limit point in $\mathcal{U}_a$. ◊

Compactness plays an important role in optimization because a continuous functional $F : \mathcal{U}_a \subset \mathcal{U} \to \mathbb{R}$ achieves its minimum (or maximum) if $\mathcal{U}_a$ is a compact subset of the normed linear space $\mathcal{U}$.

It is obvious that a compact set is closed and bounded. In finite-dimensional spaces, bounded sequences always have subsequences that converge to a point in the space. Unfortunately, this is not true for all infinite-dimensional spaces, so we need another definition.

Definition A.1.13 (Cauchy sequences)

A sequence $\{u_i\}$ of elements in a normed linear space $\mathcal{U}$ is called a *Cauchy sequence* if, for every $\epsilon > 0$, there exists a number $N(\epsilon)$ such that

$$\|u_n - u_m\|_{\mathcal{U}} < \epsilon, \quad \forall n, m > N.$$

◊

Of course, all convergent sequences are Cauchy sequences, but not all Cauchy sequences need converge to a point in the space. As this situation can be troublesome, we prefer to work with spaces where Cauchy sequences always have limits in the space. This desirable property is called *completeness* and for normed linear space with completeness we have a special name.

Definition A.1.14 (Completeness and Banach space)

(D1) A normed linear space $\mathcal{U}$ is said to be *complete* if every Cauchy sequence in $\mathcal{U}$ converges to a limit which is itself an element of $\mathcal{U}$.

(D2) A complete normed linear space is called a *Banach space*. ◊

The spaces l_p^n, l_∞^n, l_p and l_∞ are examples of Banach spaces, and $C(a, b)$ under the sup-norm is a Banach space. However, $C(a, b)$ is not complete under any p-norm satisfying $1 \leq p < \infty$.

Example A.1.8 Let $\mathcal{C}^k(\Omega)$ denote the spaces of real-valued functions of n variables defined on an open subset $\Omega \subset \mathbb{R}^n$ and which are continuously differentiable up to order k. Let $\alpha = (\alpha_1, \cdots, \alpha_n)$, where $\alpha_i \geq 0$ is an integer, and

$$|\alpha| = \sum_{i=1}^{n} \alpha_i.$$

Then, for $u(x) \in \mathcal{C}^k(\Omega)$, $x \in \Omega$, the derivatives

$$D^\alpha u = \frac{\partial^{|\alpha|} u}{\partial x_1^{\alpha_1} \cdots \partial x_n^{\alpha_n}} \quad |\alpha| \leq k$$

exist and are continuous, and, under the norm

$$\|u\| = \max_{0 \leq |\alpha| \leq k} \{\|D^\alpha u\|_\infty\}, \quad \text{where} \quad \|D^\alpha u\|_\infty := \sup_{x \in \Omega} |D^\alpha u(x)|,$$

$\mathcal{C}^k(\Omega)$ is a Banach space.

A.2 Bilinear Forms and Inner Product Spaces

Definition A.2.1 (Bilinear Form)

Let $\mathcal{U}$ be a real vector space. The *bilinear form* (or functional) $b(*,*) : \mathcal{U} \times \mathcal{U} \to \mathbb{R}$ is a bi-additive map with the following properties.

1. Positivity: $b(u,u) \geq 0 \ \forall u \in \mathcal{U}, \quad b(u,u) = 0$ iff $u = 0$;
2. Symmetry: $b(u,v) = b(v,u) \quad \forall u, v \in \mathcal{U}$;
3. Bi-linearity: $b(u, \alpha v + \beta w) = \alpha b(u,v) + \beta b(u,w) \ \forall u, v, w \in \mathcal{U}, \quad \forall \alpha, \beta \in \mathbb{R}$.

◇

Definition A.2.2 (Nondegenerate Bilinear Form)

A bilinear form $b(*,*)$ is said to be *nondegenerate* (or separable) if

$$b(u,v) = 0 \ \forall v \in \mathcal{U} \quad \Rightarrow \quad u = 0.$$

◇

With this definition, we can introduce inner product spaces.

Definition A.2.3 (Inner Product and Inner Product Space)

Suppose $\mathcal{U}$ is a real vector space.

(D1) An *inner product* $\langle *,* \rangle : \mathcal{U} \times \mathcal{U} \to \mathbb{R}$ on $\mathcal{U}$ is a nondegenerate bilinear form on $\mathcal{U}$.

(D2) The space $\mathcal{U}$ with the structure of the inner product $\langle *,* \rangle$ is called an *inner product space*, and is denoted by $\{\mathcal{U}, \langle *,* \rangle\}$. ◇

An inner product on a space $\{\mathcal{U}, \langle *,* \rangle\}$ automatically defines a norm, which is given by

$$\|u\|_{\mathcal{U}} = \sqrt{\langle u, u \rangle}.$$

This is called the norm induced by the inner product. However, not all normed spaces are inner product spaces.

Example A.2.1 On $\mathcal{C}(\Omega)$, the inner product can be defined as

$$\langle u, v\rangle = \int_\Omega u(x)v(x)\mathrm{d}\Omega.$$

With the norm induced by this inner product, $\mathcal{C}(\Omega)$ is a normed linear space, but is not complete. This space can be completed with respect to this induced norm by including all functions defined on $\Omega \subset \mathbb{R}^n$ for which the integral

$$\langle u, u\rangle = \int_\Omega [u(x)]^2\mathrm{d}\Omega$$

exists in the *Lebesgue* sense. Such functions are generally referred to as *square-integrable* functions, and the corresponding inner product space is denoted by $\mathcal{L}^2(\Omega)$. ♡

In general, for $1 \leq p < \infty$, the space

$$\mathcal{L}^p(\Omega; \mathbb{R}^n) := \{u : \Omega \to \mathbb{R}^n \mid \int_\Omega |u(x)|^p \mathrm{d}\Omega \text{ exists in the sense of Lebesgue}\}$$

is complete with respect to the norm

$$\|u\|_p = \left\{\int_\Omega |u(x)|^p \mathrm{d}\Omega\right\}^{1/p}. \tag{A.6}$$

Strictly speaking, $\|u\|_p$ is a seminorm since $\|u\|_p = 0$ only implies that $u = 0$ almost everywhere; in other words, u could differ from zero on a set of measure zero in Ω.

While the norm is connected with the length or magnitude of vector, the inner product is connected with the "angle" between two vectors. Two elements u and v of an inner product space $(\mathcal{U}, \langle *, *\rangle)$ are said to be ***orthogonal*** if $\langle u, v\rangle = 0$.

Example A.2.2 If $\mathcal{U} = C[0, L]$, the space of continuous functions defined on an interval $(0, L)$ of the real line. Then

$$\langle u, v\rangle = \int_0^L u(x)v(x)\ \mathrm{d}x \tag{A.7}$$

is an inner product on $\mathcal{U}$. Let $u(x) = \sin(\pi x/L)$, $v(x) = \cos(\pi x/L)$. Since $\int_0^L \sin(\pi x/L)\cos(\pi x/L)\ \mathrm{d}x = 0$, these functions are orthogonal. ♡

Theorem A.2.1 (Cauchy-Schwarz Inequality) *Let $\mathcal{U}$ be an inner product space with the norm $\| u \| = \langle u, u\rangle^{\frac{1}{2}}$ induced by the inner product. Then,*

$$|\langle u, v\rangle| \leq \sqrt{\langle u, u\rangle\langle v, v\rangle} = \|u\|\|v\| \quad \forall u, v \in \mathcal{U}. \tag{A.8}$$

Proof. The inequality is obviously true for $v = 0$. If $v \neq 0$, then for an arbitrary scalar $\lambda \in \mathbb{R}$,

$$\langle u - \lambda v, u - \lambda v\rangle \geq 0.$$

Hence

$$f(\lambda) = \lambda^2 \parallel v \parallel^2 - 2\lambda\langle u, v\rangle + \parallel u \parallel^2 \geq 0,$$

and the nonnegative property of $f(\lambda)$ implies that $f(\lambda) = 0$ can have at most one real root. Hence

$$\langle u, v\rangle^2 \leq \parallel u \parallel^2 \parallel v \parallel^2,$$

from which the assertion follows. ■

The Cauchy-Schwarz inequality is an important tool in many proofs in analysis. In real inner product spaces, we have other important geometrical features. For examples, the *parallelogram law*

$$\|u+v\|^2 + \|u-v\|^2 = 2\|u\|^2 + 2\|v\|^2 \quad \forall u, v \in \mathcal{U}$$

holds in such spaces. The *angle* θ between vectors u and v is defined by

$$\cos\theta = \frac{\langle u, v\rangle}{\|u\|\|v\|}. \tag{A.9}$$

A *sphere of radius r centered at u_0* is the set of vectors $u \in \mathcal{U}$ such that

$$\langle u - u_0, u - u_o\rangle = r^2. \tag{A.10}$$

By the parallelogram law, if u is orthogonal to v, then u and v satisfy the *Pythagorean theorem*

$$\|u+v\|^2 = \|u\|^2 + \|v\|^2.$$

Definition A.2.4 (Hilbert space)
A complete inner product space is called a *Hilbert space.* ◇

Every finite-dimensional inner product space is a Hilbert space since every finite-dimensional space is complete. Clearly, every Hilbert space is a Banach space. The converse, however, is not true. For example, for any $p \in [1, \infty)$, $\mathcal{L}^p(\Omega; \mathbb{R}^n)$ is a Banach space, but only $\mathcal{L}^2$ is a Hilbert space.

A.3 Linear functionals and Dual spaces

Let $\mathcal{U}$ be a vector space. A linear functional on $\mathcal{U}$ is a linear map $F : \mathcal{U} \to \mathbb{R}$, such that, for any given $u, v \in \mathcal{U}$, $F(\alpha u + \beta v) = \alpha F(u) + \beta F(v) \quad \forall \alpha, \beta \in \mathbb{F}$ holds.

Definition A.3.1 (Bounded Linear Functional and Norm)
(D1) A *linear functional* F on a normed space $\mathcal{U}$ is *bounded* if there is a constant $M > 0$ such that $|F(u)| \leq M\|u\| \quad \forall u \in \mathcal{U}$.
(D2) The smallest such constant M is called the *norm* of F, and is denoted by $\|F\|$, i.e.

$$\|F\| = \inf\{M \geq 0 | \ |F(u)| \leq M\|u\| \quad \forall u \in \mathcal{U}\}.$$

◇

Theorem A.3.1 *Let F be a linear function on a normed space $\mathcal{U}$. Then*

(T1) F is continuous at a single point implies that it is continuous through $\mathcal{U}$;

(T2) F is bounded if and only if F is continuous.

Example A.3.1 On the space $\mathcal{U} = \mathbb{R}^n$ with dot product, all linear functionals are of the form $F(x) = \langle y, x \rangle = \sum_{i=1}^{n} y_i x_i$ for fixed $y = (y_1, y_2, \cdots, y_n)$, and is continuous through $\mathbb{R}^n$.

Example A.3.2 Let $\mathcal{U} = C[a, b]$ and consider the linear functional

$$F(u) = \int_a^b u(x)\ \mathrm{d}x.$$

Since

$$|F(u)| \leq \max_{a \leq x \leq b} (b-a)|u(x)| = (b-a)\|u\|_{\mathcal{U}},$$

then $F : \mathcal{U} \to \mathbb{R}$ is continuous. If $u(x)$ is a constant c, we have $|F(u)| = |c|(b-a)$, and hence $\|F\| = |c|(b-a)$.

Example A.3.3 On the space $\mathcal{U} = \mathcal{L}^2[0,1]$, the functional $F(x) = \int_0^1 y(t)x(t)\mathrm{d}t$, for any given $y \in \mathcal{L}^2[0,1]$, is a continuous linear functional. Furthermore, it can be shown that all linear functionals on $\mathcal{L}^2$ are of this form.

The existence of non-trivial continuous linear functionals on general normed linear space is guaranteed by the following theorem.

Theorem A.3.2 (Hahn-Banach Theorem)

Every continuous linear functional $F : \mathcal{D} \subset \mathcal{U} \to \mathbb{F}$ defined on a linear subspace $\mathcal{D}$ of a normed linear space $\mathcal{U}$ can be extended to a continuous linear functional on the whole space $\mathcal{U}$ with preservation of norm.

This theorem has many different versions. A geometrical version is given in Chapter 5.

As we can see, the product space $\mathcal{U}^* \times \mathcal{U}$ and the inner product $\langle *, * \rangle : \mathcal{U}^* \times \mathcal{U} \to \mathbb{R}$ form an inner product space. For any given $u^* \in \mathcal{U}^*$, the inner product $\langle u^*, u \rangle$ defines a linear functional $u^*(u)$ on $\mathcal{U}$. All these linear functional on $\mathcal{U}$ form a linear space, i.e. $\mathcal{U}^*$, called the *algebraic dual* space of $\mathcal{U}$. This definition is independent of any topological structure on $\mathcal{U}$, such as might be induced by a norm on $\mathcal{U}$. The notion of dual spaces arises naturally in many areas of physics and plays an important role in the duality theory of variational principles. We are particular interested in the subspace of the algebraic dual space consisting of all bounded linear functionals on a normed space.

Definition A.3.2 (Dual or Conjugate Space and Norm)

The *topological dual or the conjugate space* of a normed linear space $\mathcal{U}$ is the normed linear space of all bounded linear functionals on $\mathcal{U}$, and is denoted by $\mathcal{U}^*$. The functional defined by

$$\|u^*\|_{\mathcal{U}^*} = \sup_{u \in \mathcal{U}} \{|u^*(u)| \mid \|u\|_{\mathcal{U}} = 1\}. \tag{A.11}$$

is called the *norm* of an $u^* \in \mathcal{U}^*$. ◇

The topological dual space could be written as $\mathcal{L}(\mathcal{U}, \mathbb{F})$, or $\mathcal{U}'$, but instead we choose the same symbol as the algebraic dual space for convenience. The dual space of $\mathcal{U}$ is always a Banach space even when $\mathcal{U}$ is not since $\mathbb{F}$ is a Banach space. In particular, the topological and algebraic duals of $\mathcal{U} = \mathbb{R}^n$ or $\mathcal{U} = \mathbb{C}^n$ are the same.

Since $u^*(u) : \mathcal{U} \to \mathbb{F}$ is a scalar-valued linear continuous functional (operator) then, by the definition of the norm of an operator (see Section B), the norm of an element in $\mathcal{U}^*$ can be expressed in the alternative ways

$$\begin{aligned} \|u^*\|_{\mathcal{U}^*} &= \sup_{u \in \mathcal{U}} \{|u^*(u)| \mid \|u\|_{\mathcal{U}} \leq 1\} \\ &= \inf\{M \geq 0 | \ |u^*(u)| \leq M\|u\| \quad \forall u \in \mathcal{U}\} \\ &= \sup_{u \in \mathcal{U}} \left\{ \frac{|u^*(u)|}{\|u\|_{\mathcal{U}}} \mid \forall u \neq 0 \right\}. \end{aligned}$$

Clearly, on the product space $\{\mathcal{U}^* \times \mathcal{U}; \langle *, * \rangle\}$, we always have

$$\langle u^*, u \rangle \leq \|u^*\|_{\mathcal{U}^*} \|u\|_{\mathcal{U}} \quad \forall u^* \in \mathcal{U}^*, \ u \in \mathcal{U}.$$

A vector $u^* \in \mathcal{U}^*$ is said to be *aligned* with a vector $u \in \mathcal{U}$ if $\langle u, u^* \rangle = \|u^*\|_{\mathcal{U}^*} \|u\|_{\mathcal{U}}$.

Example A.3.4 Let $\mathcal{U} = \mathcal{L}^p[a, b]$, $1 < p < \infty$ and $\mathcal{U}^* = \mathcal{L}^q[a, b]$, $1/p + 1/q = 1$. The condition for two functions $f(x) \in \mathcal{L}^p$, $f^*(x) \in \mathcal{L}^q$ to be *aligned* follows directly from the conditions for equality in the integral form of the *Hölder inequality*, namely

$$\int_a^b f(x) f^*(x) \, \mathrm{d}x = \left\{ \int_a^b |f(x)|^p \, \mathrm{d}x \right\}^{1/p} \left\{ \int_a^b |f^*(x)|^q \, \mathrm{d}x \right\}^{1/q}$$

if and only if $f(x) = \alpha[\mathrm{sgn} f^*(x)] |f^*(x)|^{q/p}$ almost everywhere for some constant α. ♡

To identify the dual spaces of $\mathcal{L}^p(\Omega; \mathbb{R}^n)$, we need the following *representation theorem*.

Theorem A.3.3 (Riesz Representation Theorem for $(\mathcal{L}^p(\Omega; \mathbb{R}^n))^*$)
Let $\Omega \subset \mathbb{R}^n$ be an open set and $1 \leq p < \infty$. For any linear continuous functional f on the space $\mathcal{L}^p(\Omega; \mathbb{R}^n)$, there exists a unique function $u^* \in \mathcal{L}^q(\Omega; \mathbb{R}^n)$, $1/p + 1/q = 1$, such that

$$f(u) = \langle f, u \rangle = \int_\Omega u^* u d\Omega \quad \forall u \in \mathcal{L}^p(\Omega; \mathbb{R}^n).$$

Moreover, $\|f\|_{(\mathcal{L}^p)^*} = \|u^*\|_{\mathcal{L}^q}$.

This theorem implies that there is an isometry between $(\mathcal{L}^p)^*$ and $\mathcal{L}^q$, i.e.

$$(\mathcal{L}^p(\Omega))^* \cong \mathcal{L}^q \quad \text{with } \frac{1}{p} + \frac{1}{q} = 1, \;\; 1 \leq p < \infty. \tag{A.12}$$

For the spaces l_p^n, we have a similar representation theorem, and so

$$(l_p^n)^* \cong l_q^n, \;\; \frac{1}{p} + \frac{1}{q} = 1, \;\; 1 \leq p < \infty. \tag{A.13}$$

To see this, we define the inner product

$$F_{u^*}(u) = u_1^* u_1 + \cdots + u_n^* u_n$$

for a fixed $u^* \in l_q^n$ with $u^* \neq 0$. Clearly $F_{u^*} : l_p^n \to \mathbb{R}$ is linear. To prove that it is also bounded, we need the Hölder inequality in $\mathbb{R}^n$, i.e.

$$\left|\sum_{i=1}^n u_i^* u_i\right| \leq \left(\sum_{i=1}^n |u_i^*|^q\right)^{1/q} \left(\sum_{i=1}^n |u_i|^p\right)^{1/p}. \tag{A.14}$$

Hence,

$$\begin{aligned} |F_{u^*}(u)| &= |u_1^* u_1 + \cdots + u_n^* u_n| \\ &\leq (|u_1^*|^q + \cdots + |u_n^*|^q)^{1/q}(|u_1|^p + \cdots + |u_n|^p)^{1/p} \\ &= \|u^*\|_q \|u\|_p. \end{aligned}$$

This shows that F_{u^*} is bounded and $\|F_{u^*}\|_{(l_p^n)^*} \leq \|u^*\|_q$. If we set

$$u_i = \begin{cases} |u_i^*|^q / u_i^* & \text{if } u_i^* \neq 0 \\ 0 & \text{if } u_i^* = 0 \end{cases}$$

then

$$\|u\|_p = \left(\sum_{i=1}^n |u_i^*|^{qp-p}\right)^{1/p}.$$

But $qp - p = p(q-1) = q$, and therefore

$$\|u\|_p = \left(\sum_{i=1}^n |u_i^*|^q\right)^{1/p}.$$

Since

$$|F_{u^*}(u)| = \sum_{i=1}^n |u_i^*|^q = \|u\|_p \left(\sum_{i=1}^n |u_i^*|^q\right)^{1-1/p} = \|u\|_p \|u^*\|_q,$$

it follows that

$$\|F_{u^*}\|_{\mathcal{U}^*} = \|u^*\|_q.$$

In fact, all continuous linear functionals on l_p^n have this form, so that $(l_p^n)^* \cong l_q^n$.

Similarly it can be shown that

$$(l_p)^* \cong l_q \quad \text{for } \frac{1}{p} + \frac{1}{q} = 1, \; 1 \leq p < \infty. \tag{A.15}$$

The above results have a special interpretation when $p = q = 2$, i.e.

$$(l_2^n)^* \cong l_2^n, \; l_2^* \cong l_2, \; (\mathcal{L}^2(\Omega;\mathbb{R}^n))^* \cong \mathcal{L}^2(\Omega;\mathbb{R}^n).$$

Such spaces are said to be *self-dual.*

The representation theorems do not include the case $p = \infty$. To identify the duals of $\mathcal{L}^\infty$, l_∞^n and l_∞, we need to consider so-called *functions of bounded variation.*

Since the dual of a normed linear space $\mathcal{U}$ is a Banach space $(\mathcal{U}^*, \; \|\cdot\|_{\mathcal{U}^*})$, it also makes sense to speak of the space of continuous linear functionals on $\mathcal{U}^*$.

Definition A.3.3 (Bi-dual Space)

The dual of the dual of a normed space $\mathcal{U}$ is a Banach space, denoted $\mathcal{U}^{**}$, called the *bi-dual* of $\mathcal{U}$, and we have $(\mathcal{U}^*)^* = \mathcal{U}^{**}$. ◊

Since the inner product $\langle u, u^* \rangle$ is a bilinear form over $\mathcal{U} \times \mathcal{U}^*$ that can be considered either as a family of linear forms on $\mathcal{U}$ depending on the parameter $u^* \in \mathcal{U}^*$, or as a family of linear forms over $\mathcal{U}^*$ depending on the parameter $u \in \mathcal{U}$, then, for each fixed element $u \in \mathcal{U}$, we may define a functional $F_u \in \mathcal{U}^{**}$ by means of

$$F_u(u^*) = u^{**}(u^*) = \langle u^*, u^{**} \rangle = \langle u, u^* \rangle, \quad u^* \in \mathcal{U}^*.$$

The norm of $F_u \in \mathcal{U}^{**}$ is defined in a natural way, namely,

$$\|F_u\|_{\mathcal{U}^{**}} = \sup_{u^* \in \mathcal{U}^*} \frac{|\langle u^*, u \rangle|}{\|u^*\|_{\mathcal{U}^*}}, \quad u \in \mathcal{U}, \; u^* \neq 0.$$

The correspondence $F_u(u^*) = \langle u^*, u \rangle$ describes a linear injective map of $\mathcal{U}$ into $\mathcal{U}^{**}$. It can be shown that the map $u \to F_u$ is an isometric isomorphism of $\mathcal{U}$ into $\mathcal{U}^{**}$ (i.e. $\|F_u\| = \|u\|$), and is called the *natural embedding* of $\mathcal{U}$ in $\mathcal{U}^{**}$. Thus, any Banach space $\mathcal{U}$ is isomorphic, in a natural way, to a subspace of its bi-dual $\mathcal{U}^{**}$. If this natural embedding is surjective, then we have the following definition.

Definition A.3.4 (Reflexive Space)

A normed linear space $\mathcal{U}$ is *reflexive* if its bi-dual $\mathcal{U}^{**}$ is isometrically isomorphic to $\mathcal{U}$ under the natural embedding $F_u : \mathcal{U} \to \mathcal{U}^{**}$, and we identify $\mathcal{U}$ with its bi-dual. Thus, $\mathcal{U} \cong \mathcal{U}^{**}$. ◊

Reflexive spaces enjoy a number of useful properties not found in other normed spaces. Here we list some of properties of reflexive spaces.

1. Any reflexive normed space is a Banach space.
2. A closed subspace of a reflexive Banach space is reflexive.
3. Any product of two reflexive spaces is reflexive.

4. A normed space is reflexive if and only if its dual is reflexive.

Example A.3.5 The l_p spaces of bounded infinite sequences of numbers, and the $\mathcal{L}^p$ spaces, $1 \leq p \leq \infty$, provide examples of reflexive and nonreflexive spaces. For $1 < p < \infty$, l_p and $\mathcal{L}^p$ are reflexive and

$$l_p^{**} \cong l_p, \quad (\mathcal{L}^p)^{**} \cong \mathcal{L}^p.$$

For $p = 1$, the spaces l_1 and l_∞ are not reflexive. But $l_1^* \cong l_\infty$; however, it is not true that $l_\infty^* \cong l_1$. The same applies for the spaces $\mathcal{L}^1$ and $\mathcal{L}^\infty$. ♡

In order to study the dual space of $C(a,b)$, we need the definition of a space of *bounded variation.*

Definition A.3.5 (Bounded Variation Space)
Let $x_i \in (a,b)$, $i = 0, 1, \cdots, n$ and consider the partition $a = x_0 < x_1 < \cdots < x_n < b$. A function $u : (a,b) \to \mathbb{R}^n$ is said to be of ***bounded variation*** if, for any partition of (a,b),

$$V(u) = \sum_{i=1}^{n} \|u(x_i) - u(x_{i-1})\|_{\mathbb{R}^n}$$

is finite. We write $\mathcal{B}_v(a,b)$ to denote the set of all functions of bounded variation on (a,b). ◊

If we define a subspace of $\mathcal{B}_v(a,b)$ by

$$\mathcal{B}_{va}(a,b) := \{u \in \mathcal{B}_v(a,b) |\ u(a) = 0,\ \ u(x^+) = \lim_{t \to x^+} u(t) = u(x),\ \ \forall x \in (a,b)\},$$

with $\|u\| = V(u)$, then the dual space of $C(a,b;\mathbb{R}^n)$ is $\mathcal{B}_{va}(a,b)$, the action of a linear functional being characterized by the *Stieltjes integral*

$$u^*(u) = \int_a^b u(x)\mathrm{d}u^*(x), \quad u \in C(a,b;\mathbb{R}^n), \ \ u^* \in \mathcal{B}_{va}(a,b).$$

Definition A.3.6 (Orthogonal Complement)
Let $\mathcal{K}$ be a subspace of a normed linear space $\mathcal{U}$. The *orthogonal complement* of $\mathcal{K}$, denoted $\mathcal{K}^\perp$, is defined by

$$\mathcal{K}^\perp := \{u^* \in \mathcal{U}^* |\ \langle u^*, u\rangle = 0 \ \ \forall u \in \mathcal{K}\}. \tag{A.16}$$

Similarly, the orthogonal complement of a subspace $\mathcal{M} \subset \mathcal{U}^*$ is the subset of $\mathcal{U}$ defined by

$$\mathcal{M}^\top := \{u \in \mathcal{U} |\ \langle u^*, u\rangle = 0 \ \ \forall u^* \in \mathcal{M}\}. \tag{A.17}$$

◊

It is easy to check that $\mathcal{K}^\perp$ is a closed subspace of $\mathcal{U}^*$ and $\mathcal{M}^\top$ is closed subspace of $\mathcal{U}$. Note that in defining the orthogonal complement of $\mathcal{M} \subset \mathcal{U}^*$ we refer back to the original space $\mathcal{U}$, and not to the bi-dual $\mathcal{U}^{**}$.

The set $\mathcal{M}^{\top}$ may be thought of as the intersection of $\mathcal{M}^{\perp}$ with $\mathcal{U}$, where $\mathcal{U}$ is considered to be embedded in $\mathcal{U}^{**}$ by the natural mapping. In particular, we have the following fundamental duality result.

Theorem A.3.4 *Let $\mathcal{K}$ be a subspace of a normed space $\mathcal{U}$. Then $\mathcal{K}$ is closed if and only if*

$$(\mathcal{K}^{\perp})^{\top} = \mathcal{K}.$$

The proof of this theorem can be found in Luenberger (1969, p. 118).

If $\mathcal{U} = \mathcal{H}$ is a Hilbert space and $\mathcal{M}$ is a closed subspace of $\mathcal{H}$, then $\mathcal{H}$ is the *direct sum* of $\mathcal{M}$ and $\mathcal{M}^{\perp}$, i.e., $\mathcal{H} = \mathcal{M} \oplus \mathcal{M}^{\perp}$.

Theorem A.3.5 (Projection Theorem) *Let $\mathcal{H}$ be a Hilbert space and let $\mathcal{M}$ be a closed subspace of $\mathcal{H}$. Then, every $u \in \mathcal{H}$ has a unique decomposition*

$$u = v + w \quad \text{with } v \in \mathcal{M}, \quad w \in \mathcal{M}^{\perp}.$$

Appendix B
Linear Operators and Adjointness

A study of operators and their adjoints is essential for a sophisticated approach to many equilibrium problems. This section only gives a brief introduction to operator theory with an emphasis on the concept of the adjoint of a linear operator which, being defined on the dual space, characterizes many aspects of duality theory.

B.1 Linear Operators

Let $\mathcal{U}$ and $\mathcal{E}$ be two linear spaces. An *operator* Λ is a mapping from $\mathcal{U}$ into $\mathcal{E}$ and we write $\Lambda : \mathcal{U} \to \mathcal{E}$. The fact that the operator Λ sends $u \in \mathcal{U}$ to $\epsilon = \Lambda(u) \in \mathcal{E}$ is denoted symbolically by $\epsilon = \Lambda(u)$, which is sometimes also called a transformation or function, etc. The special case in which the values of Λ are real numbers (i.e. $\mathcal{E} = \mathbb{R}$), is fundamentally important. Such operators are referred to as *functionals*, and much of what we are going to study in this book has to do with their properties. We allow for an operator being defined only on a subset $\mathcal{D} \subset \mathcal{U}$, called the *domain* of Λ, denoted as $\mathcal{D}(\Lambda)$. The set of elements given by $\Lambda(u)$ is called the *range* of Λ and is denoted by $\mathcal{R}(\Lambda)$, which is a subset of $\mathcal{E}$, i.e. $\mathcal{R}(\Lambda) \subset \mathcal{E}$.

Definition B.1.1 (Injective, Surjective and Bijective)

(D1) The mapping $\Lambda : \mathcal{D}(\Lambda) \subset \mathcal{U} \to \mathcal{R}(\Lambda)$ is said to be *into* $\mathcal{E}$ if $\mathcal{R}(\Lambda) \subset \mathcal{E}$.

(D2) The mapping is said to be *onto (or surjective)* $\mathcal{E}$ if $\mathcal{R}(\Lambda) = \mathcal{E}$.

(D3) The mapping is said to be *injective* (or one-to-one) if

$$u_1 \neq u_2 \quad \Rightarrow \quad \Lambda(u_1) \neq \Lambda(u_2) \quad \forall u_1, u_2 \in \mathcal{U}.$$

(D4) The mapping is said to be *bijective(or one-to-one and onto)* if it is both injective and surjective. ◊

If the mapping $\mathcal{U} \to \mathcal{E}$ is bijective, there exists an inverse operator Λ^{-1} : $\mathcal{E} \to \mathcal{U}$ with $\mathcal{D}(\Lambda^{-1}) = \mathcal{R}(\Lambda)$ and $\mathcal{R}(\Lambda^{-1}) = \mathcal{D}(\Lambda)$ that associates a unique $u = \Lambda^{-1}(v) \in \mathcal{R}(\Lambda^{-1})$ with each $v \in \mathcal{D}(\Lambda^{-1})$. An operator $\Lambda : \mathcal{U} \to \mathcal{U}$ is said to be *homogeneous* if for any given $\alpha \in \mathbb{R}$,

$$\Lambda(\alpha u) = \alpha\Lambda(u) \ \ \forall u \in \mathcal{U}, \tag{B.1}$$

and *additive* if

$$\Lambda(u_1 + u_2) = \Lambda(u_1) + \Lambda(u_2) \ \ \forall u_1, u_2 \in \mathcal{U}. \tag{B.2}$$

Definition B.1.2 (Linear Operator and its Nullspace)
(D1) An operator $\Lambda : \mathcal{U} \to \mathcal{E}$ is said to be linear if it is both homogeneous and additive.
(D2) The subset $\mathcal{N}$ of $\mathcal{U}$ defined by

$$\mathcal{N}(\Lambda) = \{u \in \mathcal{U} | \ \Lambda u = 0\} \tag{B.3}$$

is called the *nullspace* of the linear operator Λ. ◇

Remark B.1.1 It is a common convention that the element u acted upon by a linear operator Λ is denoted by Λu instead of $\Lambda(u)$. An operator Λ is invertible if and only if $\mathcal{N}(\Lambda) = 0$. Nullspace of a linear operator $\Lambda : \mathcal{U} \to \mathcal{E}$ is also called the *kernel* of Λ and is denoted by $\ker \Lambda$. □

Homogeneity and additivity are algebraic properties, i.e. they depend only on the algebraic structures of $\mathcal{U}$ and $\mathcal{E}$, vector addition and scalar multiplication. If $\mathcal{U}$ and $\mathcal{E}$ are Banach spaces, then they also have topological structure. We can also describe a variety of other properties for Λ; for example, continuity, compactness, boundness, etc.

Definition B.1.3 Let $\mathcal{U}$ and $\mathcal{E}$ be two normed spaces. Then, $\Lambda : \mathcal{U} \to \mathcal{E}$ is said to *bounded* if there is a constant M such that $\|\Lambda u\|_{\mathcal{E}} \leq M\|u\|_{\mathcal{U}}$ $\forall u \in \mathcal{U}$. The smallest such M which satisfies the above condition is denoted $\|\Lambda\|$ and called the *norm* of Λ. ◇

From this definition, the norm of Λ can be written as

$$\|\Lambda\| = \sup_{u \neq 0} \frac{\|\Lambda u\|}{\|u\|} = \sup_{\|u\|=1} \|\Lambda u\|.$$

Theorem B.1.1 *Let $\Lambda : \mathcal{U} \to \mathcal{E}$ be a linear operator, and let $\mathcal{U}$ and $\mathcal{E}$ be two normed spaces.*
(T1) Λ is continuous at every point in $\mathcal{U}$ if it is continuous at a single point.
(T2) Λ is bounded if and only if it is continuous.
(T3) If Λ^{-1} exists, then Λ^{-1} is linear.

Proof. We only prove the third statement. Suppose $\Lambda^{-1}\epsilon_1 = u_1$, $\Lambda^{-1}\epsilon_2 = u_2$, then

$$\Lambda u_1 = \epsilon_1, \ \ \Lambda u_2 = \epsilon_2,$$

and the linearity of Λ implies that $\Lambda(\alpha_1 u_1+\alpha_2 u_2) = \alpha_1\epsilon_1+\alpha_2\epsilon_2$. thus $\Lambda^{-1}(\alpha_1\epsilon_1+\alpha_2\epsilon_2) = \alpha_1\Lambda^{-1}\epsilon_1 + \alpha_2\Lambda^{-1}\epsilon_2$. ■

All of the bounded linear operators from the normed space $\mathcal{U}$ into the normed space $\mathcal{E}$ form a normed space, denoted by $\mathcal{L}(\mathcal{U},\mathcal{E})$. If $\mathcal{E}$ is complete, then $\mathcal{L}(\mathcal{U},\mathcal{E})$ is a Banach space.

Let Λ be a linear operator defined on its domain $\mathcal{D}(\Lambda) \subset \mathcal{U}$ with range $\mathcal{R}(\Lambda) \subset \mathcal{E}$. The *graph* of Λ is a subset of $\mathcal{U} \times \mathcal{E}$ defined by

$$\text{graph}\,\Lambda = \{(u,\epsilon) \in \mathcal{U} \times \mathcal{E} |\ \epsilon = \Lambda u \ \ \forall u \in \mathcal{D}(\Lambda)\}. \tag{B.4}$$

A linear operator Λ is said to be *closed* if its graph is a closed linear subspace of $\mathcal{U} \times \mathcal{E}$ (under the product topology).

Theorem B.1.2 *Let Λ be a linear operator from a Banach space $\mathcal{U}$ into a Banach space $\mathcal{E}$.*

(T1) **Open mapping theorem**: *If Λ is surjective, then the image of every open set $\mathcal{U}_a \subset \mathcal{U}$ under Λ is open in $\mathcal{E}$.*

(T2) **Bounded inverse theorem**: *If Λ is bijective, then the inverse map Λ^{-1} is continuous.*

(T3) **Closed graph theorem**: *If Λ is closed, then Λ is bounded.*

A simple corollary of the closed graph theorem is that if Λ is closed and invertible then Λ^{-1} is closed.

B.2 Adjoint Operators

Adjoint operators provide a convenient mechanism for describing the orthogonality and duality relations which permeate nearly every variational analysis. In this section, we assume that $\mathcal{U}$ and $\mathcal{E}$ are two normed linear spaces and that $\mathcal{U}^*$ and $\mathcal{E}^*$ are their dual spaces. On the product spaces $\mathcal{U} \times \mathcal{U}^*$ and $\mathcal{E} \times \mathcal{E}^*$, we define the bilinear forms

$$\langle *, * \rangle : \mathcal{U} \times \mathcal{U}^* \to \mathbb{R}, \qquad \langle *; * \rangle : \mathcal{E} \times \mathcal{E}^* \to \mathbb{R}. \tag{B.5}$$

Definition B.2.1 Let $\Lambda \in \mathcal{L}(\mathcal{U},\mathcal{E})$. The operator $\Lambda^* : \mathcal{E}^* \to \mathcal{U}^*$, defined by the equation

$$\langle u, \Lambda^*\epsilon^* \rangle = \langle \Lambda u; \epsilon^* \rangle \ \ \forall u \in \mathcal{D}(\Lambda),\ \epsilon^* \in \mathcal{D}(\Lambda^*), \tag{B.6}$$

is called the *adjoint* of Λ associated with the bilinear forms (B.5), or simply, the adjoint of Λ. ♢

If $\mathcal{U}$ is reflexive, then $\Lambda = \Lambda^{**}$ is the adjoint of Λ^*. Given a fixed $\epsilon^* \in \mathcal{E}^*$, and $\Lambda \in \mathcal{L}(\mathcal{U},\mathcal{E})$, the mapping $u^*(u) = \langle \Lambda u; \epsilon^* \rangle$ is a linear functional on $\mathcal{U}$. Since

$$|\langle \Lambda u; \epsilon^* \rangle| \le \|\epsilon^*\| \ \|\Lambda u\| \le \|\epsilon^*\| \ \|\Lambda\| \ \|u\|,$$

it follows that this functional is bounded and is therefore an element of $\mathcal{U}^*$. We may then state that $u^* = \Lambda^*\epsilon^*$.

In terms of operator (rather than bracket) notation, the adjoint satisfies the equation

$$\epsilon^*(\Lambda u) = (\Lambda^*\epsilon^*)(u) \ \ \forall u \in \mathcal{U},$$

so we may write

$$\epsilon^*\Lambda = \Lambda^*\epsilon^*,$$

where the left-hand side denotes the functional on $\mathcal{U}$ that is the composition of the operators Λ and ϵ^*, and the right-hand side is the functional obtained by operating on ϵ^* with Λ^*.

Theorem B.2.1 *If $\Lambda \in \mathcal{L}(\mathcal{U},\mathcal{E})$ then $\Lambda^* \in \mathcal{L}(\mathcal{E}^*,\mathcal{U}^*)$ and $\|\Lambda\| = \|\Lambda^*\|$.*

Proof. The proof of linearity is elementary and is left to the reader.

Since for any $\Lambda \in \mathcal{L}(\mathcal{U},\mathcal{E})$ and $\epsilon^* \in \mathcal{E}^*$,

$$|\langle u, \Lambda^*\epsilon^*\rangle| = |\langle \Lambda u; \epsilon^*\rangle| \leq \|\Lambda u\| \ \|\epsilon^*\| \leq \|\Lambda\| \ \|u\| \ \|\epsilon^*\|,$$

it follows that $\|\Lambda^*\epsilon^*\| \leq \|\Lambda\| \ \|\epsilon^*\|$, which implies that

$$\|\Lambda^*\| \leq \|\Lambda\|.$$

Thus Λ^* is bounded. Now let u_o be an nonzero element in $\mathcal{U}$. According to the Hahn-Banach theorem, there exists an element $\epsilon_o^* \in \mathcal{E}^*$, $\|\epsilon_o^*\| = 1$, such that $\langle \Lambda u_o, \epsilon_o^*\rangle = \|\Lambda u_o\|\|\epsilon_o^*\| = \|\Lambda u_o\|$. Hence,

$$\|\Lambda u_o\| = |\langle u_o, \Lambda^*\epsilon_o^*\rangle| \leq \|\Lambda^*\epsilon_o^*\| \ \|u_o\| \leq \|\Lambda^*\| \ \|u_o\|,$$

and we conclude that

$$\|\Lambda\| \leq \|\Lambda^*\|.$$

It now follows that $\|\Lambda\| = \|\Lambda^*\|$. ∎

The adjoint of a linear operator is obviously unique. Adjoints enjoy the following algebraic relations, which follow easily from the definition.

Proposition B.2.1 (Algebraic Properties of Adjoints)

1. *If Λ_1, $\Lambda_2 \in \mathcal{L}(\mathcal{U},\mathcal{E})$, $\alpha \in \mathbb{R}$, then $(\alpha\Lambda_1 + \Lambda_2)^* = \alpha\Lambda_1^* + \Lambda_2^*$.*
2. *If $\Lambda_1 \in \mathcal{L}(\mathcal{U},\mathcal{E})$, $\Lambda_2 \in \mathcal{L}(\mathcal{E},\mathcal{W})$, then $(\Lambda_2 \circ \Lambda_1)^* = \Lambda_1^* \circ \Lambda_2^*$.*
3. *If $\Lambda \in \mathcal{L}(\mathcal{U},\mathcal{E})$ and Λ has a bounded inverse, then $(\Lambda^{-1})^* = (\Lambda^*)^{-1}$.*

Proof. Properties (1) and (2) are trivial. To prove property (3), we let $\Lambda \in \mathcal{L}(\mathcal{U},\mathcal{E})$ have a bounded inverse Λ^{-1}. To show that Λ^* has an inverse, we must show that it is bijective. Let $\epsilon_1^* \neq \epsilon_2^* \in \mathcal{E}^*$; then

$$\langle u, \Lambda^*\epsilon_1^*\rangle - \langle u, \Lambda^*\epsilon_2^*\rangle = \langle \Lambda u; \epsilon_1^* - \epsilon_2^*\rangle \neq 0$$

for any $u \in \mathcal{U}$, and $u \neq 0$. Thus, $\Lambda^*\epsilon_1^* \neq \Lambda^*\epsilon_2^*$, and Λ^* is one-to-one. Now for any $u^* \in \mathcal{U}^*$ and any $u \in \mathcal{U}$ with $\Lambda u = \epsilon$, we have

$$\langle u, u^*\rangle = \langle \Lambda^{-1}\epsilon, u^*\rangle = \langle \epsilon; (\Lambda^{-1})^*u^*\rangle = \langle \Lambda u; (\Lambda^{-1})^*u^*\rangle = \langle u, \Lambda^*(\Lambda^{-1})^*u^*\rangle,$$

which shows that u^* is in $\mathcal{R}(\Lambda^*)$, and also that $(\Lambda^*)^{-1} = (\Lambda^{-1})^*$. ■

If $\mathcal{U}$ and $\mathcal{E}$ are two real Hilbert spaces, then they are their own duals and the operator Λ^* can be regarded as mapping $\mathcal{E}$ into $\mathcal{U}$. In Hilbert space we therefore have the additional property $\Lambda^{**} = \Lambda$.

Definition B.2.2 (Self-adjoint and Positive Definite Operator)
(D1) A bounded linear operator Λ mapping a real Hilbert space into itself is said to be *self-adjoint* if $\Lambda^* = \Lambda$.
(D2) A self-adjoint linear operator on a Hilbert space $\mathcal{H}$ is said to be *positive semidefinite* if $\langle u, \Lambda u\rangle \geq 0 \ \forall u \in \mathcal{H}$. It is said to be *positive definite* if the strict inequality holds. ◇

Example B.2.1 Let $\Lambda : \mathcal{C}[0,1] \to \mathcal{C}[0,1]$ be the integral operator

$$\Lambda u(x) = \int_0^1 K(x,y)u(y)\mathrm{d}y,$$

where $K(x,y)$ is a continuous *kernel function* defined on $[0,1] \times [0,1]$. With the inner product

$$\langle u, v\rangle = \int_0^1 u(x)v(x)\ \mathrm{d}x,$$

we have

$$\begin{aligned} \langle \Lambda u, v\rangle &= \int_0^1 \int_0^1 K(x,y)u(y)v(x)\mathrm{d}y\ \mathrm{d}x \\ &= \int_0^1 u(x) \int_0^1 K(y,x)v(y)\mathrm{d}y\ \mathrm{d}x = \langle u, \Lambda^* v\rangle. \end{aligned}$$

The adjoint of Λ is then given by

$$\Lambda^* v(x) = \int_0^1 K(y,x)v(y)\mathrm{d}y,$$

and is self-adjoint if the kernel is symmetric, i.e. $K(x,y) = K(y,x)$. ♡

Example B.2.2 Let $\mathcal{U} = \{u(t) \in \mathcal{C}^1[a,b] |\ u(a) = u(b) = 0\}$, with the standard bilinear form

$$\langle u, v\rangle := \int_a^b u(t)v(t)\mathrm{d}t. \tag{B.7}$$

Then $\mathcal{H}_1 = \{\mathcal{U}; \langle *, *\rangle\}$ is a Hilbert space. Let $\Lambda = \mathrm{d}/\mathrm{d}t$ be a differential operator on $\mathcal{U}$. Then

$$\langle v, \Lambda u\rangle = \int_a^b u'(t)v(t)\mathrm{d}t = \int_a^b -v'(t)u(t)\mathrm{d}t = \langle \Lambda^* v, u\rangle\ \ \forall u, v \in \mathcal{U}. \tag{B.8}$$

In this case, we have $\Lambda^* = -\mathrm{d}/\mathrm{d}t$. Further calculations show that, on $\mathcal{C}^2$, the second derivative $\mathrm{d}^2/\mathrm{d}t^2$ is a self-adjoint operator on $\mathcal{H}_1$. Usually, on the inner product space $\mathcal{H}_p = \{\mathcal{C}^p[a,b], \langle *, *\rangle\}$, $\Lambda^* = -\Lambda$ for odd derivative operators Λ, and $\Lambda^* = \Lambda$ for even derivative operators. ♡

B.3 Duality Relations for Range and Nullspace

Adjoints are extremely useful in our recurring task of translating between the geometric properties and the algebraic description of a given problem. The following theorems are of particular importance in the geometry of abstract vector spaces.

Theorem B.3.1 *Let $\mathcal{U}$ and $\mathcal{E}$ be Banach spaces and let $\Lambda \in \mathcal{L}(\mathcal{U}, \mathcal{E})$. Then,*

$$[\mathcal{R}(\Lambda)]^{\perp} = \mathcal{N}(\Lambda^*). \tag{B.9}$$

If $\mathcal{R}(\Lambda)$ is closed, we have the dual theorem

$$\mathcal{R}(\Lambda^*) = [\mathcal{N}(\Lambda)]^{\perp}. \tag{B.10}$$

Proof. Let $\epsilon^* \in \mathcal{N}(\Lambda^*)$ and $\epsilon \in [\mathcal{R}(\Lambda)]^{\perp}$. Then, $\epsilon = \Lambda v$ for some $v \in \mathcal{U}$. The fact that $\langle \epsilon; \epsilon^* \rangle = \langle \Lambda v; \epsilon^* \rangle = \langle v, \Lambda^* \epsilon^* \rangle = 0$ shows that $\mathcal{N}(\Lambda^*) \subset [\mathcal{R}(\Lambda)]^{\perp}$.

Now assume $\epsilon^* \in [\mathcal{R}(\Lambda)]^{\perp}$. For every $v \in \mathcal{U}$, $\langle \Lambda v, \epsilon^* \rangle = 0$. This implies $\langle v, \Lambda^* \epsilon^* \rangle = 0$ and hence that $[\mathcal{R}(\Lambda^*)]^{\perp} \subset \mathcal{N}(\Lambda^*)$. Hence, we must have $[\mathcal{R}(\Lambda)]^{\perp} = \mathcal{N}(\Lambda^*)$.

The proof of the dual theorem requires both the Banach inverse theorem and the Hahn-Banach theorem. See Luenberger (1969, p. 155) for details. ■

Recall the positive and negative conjugate cone defined in Chapter 5. The following theorem generalizes the result above.

Theorem B.3.2 *Let $\mathcal{U}$ and $\mathcal{E}$ be Banach spaces and let $\Lambda \in \mathcal{L}(\mathcal{U}, \mathcal{E})$. For any $\mathcal{K} \subset \mathcal{U}$,*

$$[\Lambda(\mathcal{K})]^{\oplus} = \Lambda^{*-1}(\mathcal{K}^{\oplus})\mathcal{N}(\Lambda^*), \quad [\Lambda(\mathcal{K})]^{\ominus} = \Lambda^{*-1}(\mathcal{K}^{\ominus})\mathcal{N}(\Lambda^*), \tag{B.11}$$

(where the inverse denotes the inverse range of $\mathcal{K}^{\oplus}$).

Proof. Assume $\epsilon^* \in [\Lambda(\mathcal{K})]^{\oplus}$ and $v \in \mathcal{K}$. Then, $\langle \Lambda v, \epsilon^* \rangle \geq 0$ and hence $\langle v, \Lambda^* \epsilon^* \rangle \geq 0$. Since v is arbitrary in $\mathcal{K}$, we have $\Lambda^* \epsilon^* \in \mathcal{K}^{\oplus}$, i.e. $\epsilon^* \in \Lambda^{*-1}(\mathcal{K}^{\oplus})$. The argument is reversible. ■

Remark B.3.1 If we put $\mathcal{S} = \mathcal{U}$, $\mathcal{K}^{\oplus} = \mathcal{K}^{\ominus} = \{0\}$, then the above result reduces to $[\mathcal{R}(\Lambda)]^{\perp} = \mathcal{N}(\Lambda^*)$. □

In many applications the range of the underlying operator is finite-dimensional, and hence satisfies the closure requirement. In other problems, however, this requirement is not satisfied and this generally leads to serious analytical difficulties. In Hilbert spaces we have additional useful relations between range and nullspace.

Theorem B.3.3 *Let Λ be a bounded linear operator acting between two real Hilbert spaces. Then,*

$$[\mathcal{R}(\Lambda)]^{\perp} = \mathcal{N}(\Lambda^*), \quad [\mathcal{R}(\Lambda^*)]^{\perp} = \mathcal{N}(\Lambda). \tag{B.12}$$

$$cl\mathcal{R}(\Lambda) = [\mathcal{N}(\Lambda^*)]^{\perp}, \quad cl\mathcal{R}(\Lambda^*) = [\mathcal{N}(\Lambda)]^{\perp}. \tag{B.13}$$

Proof. This theorem can be proved easily using the facts that $[\mathcal{R}(\Lambda)]^{\perp\perp} = [\mathcal{N}(\Lambda^*)]^{\perp}$ and $\Lambda^{**} = \Lambda$. ∎

For linear operators on finite-dimensional spaces, we have the following fundamental result

Theorem B.3.4 (Fundamental Theorem of Linear Algebra)
For any $\Lambda \in \mathcal{L}(\mathbb{R}^n, \mathbb{R}^m)$,

$$\dim \mathcal{R}(\Lambda^*) + \dim \mathcal{N}(\Lambda) = n, \tag{B.14}$$

$$\dim \mathcal{R}(\Lambda) + \dim \mathcal{N}(\Lambda^*) = m. \tag{B.15}$$

Example B.3.1 Let $\mathcal{U} = \mathbb{R}^n = \mathcal{U}^*$, $\mathcal{E} = \mathbb{R}^m = \mathcal{E}^*$, with the usual inner products

$$\langle u, u^* \rangle = \sum_{i=1}^{n} u_i u_i^*, \quad u, u^* \in \mathbb{R}^n,$$

$$\langle \epsilon; \epsilon^* \rangle = \sum_{j=1}^{m} \epsilon_j \epsilon_j^* \quad \epsilon, \epsilon^* \in \mathbb{R}^m,$$

respectively. If $\Lambda = (\lambda_{ij}) : \mathcal{U} \to \mathcal{E}$ is an $m \times n$ matrix, and hence satisfies the closure requirement, then

$$\langle \Lambda u; \epsilon^* \rangle = \sum_{j=1}^{m} \sum_{i=1}^{n} \lambda_{ji} u_i \epsilon_j^* = \sum_{i=1}^{n} u_i (\sum_{j=1}^{m} \lambda_{ji} \epsilon_j^*) = \langle u, \Lambda^* \epsilon^* \rangle.$$

Thus, $\Lambda^* = \Lambda^T$ is the adjoint of Λ. Since Λ consists of n column vectors $\boldsymbol{\lambda}_i, i = 1, 2, \cdots, n$, and since $\mathcal{R}(\Lambda)$ is the subspace of $\mathbb{R}^m$ spanned by these vectors, $[\mathcal{R}(\Lambda)]^{\perp}$ consists of those vectors in $\mathbb{R}^m$ that are orthogonal to each $\boldsymbol{\lambda}_i$.

On the other hand, the matrix $\Lambda^* = \Lambda^T$ has the $\boldsymbol{\lambda}_i$'s as its rows; hence the vectors in $\mathbb{R}^m$ orthogonal to $\boldsymbol{\lambda}_i$'s form the nullspace of Λ^*. Therefore, both $[\mathcal{R}(\Lambda)]^{\perp}$ and $\mathcal{N}(\Lambda^*)$ consist of all vectors orthogonal to each $\mathbf{a}_i$. Λ is a self-adjoint operator if and only if $m = n$ and Λ is symmetric, $\Lambda^T = \Lambda$. ♡

Appendix C
Nonlinear Operators

Let $\mathcal{U}$ and $\mathcal{U}^*$ be two Banach spaces such that the inner product $\langle *, * \rangle : \mathcal{U} \times \mathcal{U}^* \to \mathbb{R}$ places them in duality. In this section, we consider a special nonlinear operator $A : \mathcal{U} \to \mathcal{U}^*$ and the question of existence of solutions to nonlinear problems of the following type. Given $f \in \mathcal{U}^*$, find $u \in \mathcal{U}$ such that

$$\langle Au, v \rangle = \langle f, v \rangle \ \forall v \in \mathcal{U}. \tag{C.1}$$

Definition C.0.1 (Monotonicity, Boundedness and Coerciveness)
(D1) The operator $A : \mathcal{U} \to \mathcal{U}^*$ is said to be *monotone* on $\mathcal{U}$ if

$$\langle Au - Av, u - v\rangle \geq 0 \;\; \forall u, v \in \mathcal{D}(A),$$

and *strictly monotone* if this inequality is strict whenever $u \neq v$.
(D2) $A : \mathcal{U} \to \mathcal{U}^*$ is *bounded* if it maps bounded sets in $\mathcal{U}$ into bounded sets in $\mathcal{U}^*$.
(D3) $A : \mathcal{U} \to \mathcal{U}^*$ is *coercive* on $\mathcal{U}$ if and only if

$$\lim_{\|u\| \to \infty} \frac{\langle Au, u\rangle}{\|u\|} = +\infty.$$

◊

Definition C.0.2 (Hemi-, Demi- and Complete Continuity)
(D1) $A : \mathcal{U} \to \mathcal{U}^*$ is *hemicontinuous* at a point $u_o \in \mathcal{U}$ if

$$\phi(\theta) = \langle A(u_o + \theta u), v\rangle \;\; \forall \theta \in [0, 1]$$

is a continuous function of θ for every $u, v \in \mathcal{U}$.
(D2) $A : \mathcal{U} \to \mathcal{U}^*$ is *demicontinuous* at u, if and only if, for every sequence $\{u_n\} \in \mathcal{U}$ converging strongly to $u \in \mathcal{U}$, the sequence $\{A(u_n)\}$ converges weakly to $A(u)$ in $\mathcal{U}^*$.
(D3) $A : \mathcal{U} \to \mathcal{U}^*$ is *completely continuous* at u if, for every sequence $\{u_n\} \in \mathcal{U}$ converging strongly to $u \in \mathcal{U}$, the sequence $\{A(u_n)\}$ converges strongly to $A(u)$ in $\mathcal{U}^*$. ◊

It is clear that we have

Complete continuity $\Rightarrow$ Demicontinuity $\Rightarrow$ Hemicontinuity.

Lemma C.0.1 *Let $\mathcal{U}$ be a reflexive Banach space. If $A : \mathcal{U} \to \mathcal{U}^*$ is monotone, bounded, and hemicontinuous, then it is demicontinuous.*

Example C.0.2 Let $\mathcal{H}$ be a Hilbert space and $A : \mathcal{H} \to \mathcal{H}$ be a compact self-adjoint linear operator. Then, A is a monotone operator if all the eigenvalues of A are non-negative. ♡

Example C.0.3 Let $\mathcal{H}$ be a Hilbert space and C be a closed convex subset of $\mathcal{H}$. Let Pu denote the point of minimum distance of C from u, that is

$$Pu = \{\bar{u} \in C | \;\; \|\bar{u} - u\| = \inf_{v \in C} \|v - u\|\}.$$

Then, P is a monotone operator on $\mathcal{H}$. ♡

Example C.0.4 Let $\mathcal{U} = \{u \in C^2[0, 1] | \;\; u(0) = u(1) = 0\}$ with the inner product

$$\langle u, v\rangle = \int_0^1 uv \; dx.$$

Define a nonlinear operator $A : \mathcal{U} \to \mathcal{U}$ by

$$A(u) = -\frac{d^2u}{dx^2} + u^3.$$

Then, for any $u, v \in \mathcal{U}$,

$$\langle Av - Au, v - u\rangle = \int_0^1 (v-u)\left[-\frac{d^2(v-u)}{dx^2} + v^3 - u^3\right]\, dx$$

$$= \int_0^1 \left[\left(\frac{d(v-u)}{dx}\right)^2 + \frac{1}{2}(v-u)^2(v^2 + (v+u)^2 + u^2)\right]\, dx \geq 0$$

Hence, A is a monotone operator.

C.1 Operators on Finite-Dimensional Spaces

To better understand nonlinear problems on Banach spaces, we first focus on finite-dimensional problems. Let $\mathcal{U}$ be a finite-dimensional subspace of $\mathbb{R}^n$ and let $A : \mathcal{U} \to \mathcal{U}^*$. The problem

$$Au = f$$

can now be regarded as "n equations in n unknowns". Since the weak and strong topologies coincide for finite-dimensional spaces, the demicontinuity of $A : \mathcal{U} \to \mathcal{U}^*$ implies that A is continuous in $\mathcal{U}$. Our study is based mainly on the following result.

Theorem C.1.1 (Brouwer Fixed-Point Theorem)
Let C be a convex, compact, nonempty subset of $\mathbb{R}^n$, and suppose $A : C \to C$ is continuous. Then, A has a fixed point in C, i.e. there exists a $u \in C$ such that

$$A(u) = u.$$

Proof. For $n = 1$, the proof is easy to see graphically. In this case, the continuous function $A(u)$ maps an interval $C = [a, b]$ into itself, so the graph of $A(u)$ must lie in the box $[a, b] \times [a, b]$ in the (x, y)-plane. Since the function is continuous, the graph must cross the line $x = y$ somewhere. Any crossing gives us a fixed point. For the general case, we refer the reader to the literature (cf. Renardy and Rogers, 1993). ∎

Lemma C.1.1 *Let $f \in \mathbb{R}^n$ be given and let $A : \mathbb{R}^n \to \mathbb{R}^n$ be hemicontinuous. Then, if $\mathbf{u} \in \mathbb{R}^n$ is a solution of the variational inequality*

$$\langle A(v) - f, v - u\rangle \geq 0 \;\; \forall v \in \mathbb{R}^n, \tag{C.2}$$

it follows that u is also a solution of $A(u) = f$. Moreover, if A is monotone, then every solution of $A(u) = f$ is also a solution of the variational inequality.

Proof. Let $v = u + \theta w$ for $\theta > 0$. Then, the variational inequality gives

$$\langle A(u + \theta w) - f, w\rangle \geq 0 \;\; \forall w \in \mathbb{R}^n.$$

Since A is hemicontinuous then, for any $w \in \mathbb{R}^n$, letting $\theta \to 0^+$ gives

$$\langle A(u) - f, w\rangle \geq 0 \;\; \forall w \in \mathbb{R}^n.$$

Using $w = \pm(A(u) - f)$, it follows that $A(u) = f$.

Moreover, if A is monotone and $A(u) = f$, then, by the definition of monotonicity, we have

$$\langle A(v) - f, v - u\rangle = \langle A(v) - A(u), v - u\rangle \geq 0 \;\; \forall v \in \mathbb{R}^n.$$

Thus, u is also a solution of the variational inequality. ■

Based on the fixed-point theorem, we have the following results.

Theorem C.1.2 *Let $\mathcal{U}$ be a finite-dimensional normed linear space and $A : \mathcal{U} \to \mathcal{U}^*$ be a nonlinear mapping. For any $f \in \mathcal{U}^*$, let $\mathcal{K}$ be a solution set of $A(u) = f$.*

(T1) If A is demicontinuous and monotone, then the solution set $\mathcal{K}$ is convex and closed.

(T2) If A is demicontinuous and coercive, then the set $\mathcal{K}$ has at least one element.

(T3) If A is strictly monotone, then $\mathcal{K}$ has a unique element u such that $A(u) = f$.

Proof. For any fixed $v \in \mathbb{R}^n$, the set

$$\mathcal{K}_v = \{u \in \mathbb{R}^n \,|\, \langle A(v) - f, v - u\rangle \geq 0\}$$

is a closed half-space, and is therefore convex. By Lemma C.1.1 we know that the solution set of $A(u) = f$ should be

$$\mathcal{K} = \cap_{v \in \mathbb{R}^n} \mathcal{K}_v.$$

Hence, $\mathcal{K}$ must be closed and convex since the intersection of any collection of closed (resp. convex) sets is closed (resp. convex). The statements (T2) and (T3) in the theorem can be proved by using the Brouwer fixed point theorem. ■

Theorem C.1.3 *If $\mathcal{U}$ is finite dimensional and $A : \mathcal{U} \to \mathcal{U}^*$ is monotone, then A is bounded. Moreover, if A is hemicontinuous, then A is continuous.*

Proof. Suppose that the sequence $\{u_m\}$ converges strongly to u in $\mathcal{U}$ but that $A(u_m)$ is unbounded. Then,

$$u_m^* = \|Au_m\|_* \to +\infty \quad \text{as } m \to \infty.$$

By the monotonicity of A, we have

$$\frac{1}{u_m^*}\langle A(u_m) - A(u + \theta v), u_m - u - \theta v\rangle \geq 0 \;\; \forall u_m, v \in \mathcal{U}, \;\; \theta \in [0, 1].$$

For a fixed θ, we have $A(u+\theta v)/u_m^* \to 0$ as $m \to \infty$. Also $A(u_m)/u_m^*$ is bounded since $\|A(u_m)/u_m^*\|_* = 1$. Thus, $\langle A(u_m)/u_m^*, u_m - u\rangle \to 0$ as $m \to \infty$. Dividing by $\theta > 0$, we have

$$\lim_{m \to \infty} \inf\langle A(u_m)/u_m^*, -v\rangle \geq 0 \;\; \forall v \in \mathcal{U}.$$

Hence, $A(u_m)/u_m^* \to 0$ weakly in $\mathcal{U}^*$, and strongly as well since $\mathcal{U}$ is finite dimensional. But this contradicts the result $\|A(u_m)/u_m^*\|_* = 1$. Hence, A is bounded. Moreover, if A is also hemicontinuous, the continuity of A follows immediately from the previous theorem and the result proved above. ■

C.2 Monotone and Pseudo-Monotone Operators on Banach Spaces

Now we turn to the corresponding problems on infinite-dimensional Banach spaces. We assume that $\mathcal{U}$ and $\mathcal{U}^*$ are real, reflexive Banach spaces, and that the bilinear form $\langle *, * \rangle$ puts them in duality. Then, we can obtain the following analogous results.

Lemma C.2.1 *Let $f \in \mathcal{U}^*$ be given and let $A : \mathcal{U} \to \mathcal{U}^*$ be hemicontinuous. Then if $u \in \mathcal{U}$ is a solution of the variational inequality*

$$\langle A(v) - f, v - u \rangle \geq 0 \ \ \forall v \in \mathcal{U}, \tag{C.3}$$

u is also a solution of the equation $Au = f$.

Moreover, if A is monotone, then every solution of the equation $Au = f$ is also a solution of the variational inequality (C.3).

Theorem C.2.1 *Let $A : \mathcal{U} \to \mathcal{U}^*$ be (1) monotone, (2) hemicontinuous, (3) coercive and (4) bounded on $\mathcal{U}$. Then A is surjective, i.e., for any $f \in \mathcal{U}^*$, there exists at least one solution $u \in \mathcal{U}$ such that*

$$Au = f.$$

Moreover, if A is strictly monotone, then the solution is unique.

Definition C.2.1 Let $\mathcal{U}$ be a reflexive Banach space. An operator $A : \mathcal{U} \to \mathcal{U}^*$ is said to be *pseudo-monotone* if A is bounded and if, whenever $\{u_n\}$ convergent weakly to $\bar{u}$ (i.e. $u_n \rightharpoonup \bar{u}$, see Section 5.1) in $\mathcal{U}$ and

$$\limsup_{n\to\infty} \langle A(u_n), u_n - \bar{u} \rangle \leq 0, \tag{C.4}$$

it follows that

$$\liminf_{n\to\infty} \langle A(u_n), u_n - v \rangle \geq \langle A(\bar{u}), \bar{u} - v \rangle \ \ \forall v \in \mathcal{U}. \tag{C.5}$$

◊

The relationship between monotonicity and pseudo-monotonicity is given in the following statement.

Theorem C.2.2 *Let $\mathcal{U}$ be a reflexive Banach space and $A : \mathcal{U} \to \mathcal{U}^*$.*

(T1) If A is monotone and hemicontinuous, then A is pseudo-monotone.

(T2) If A is bounded and pseudo-monotone, then A is demicontinuous and, therefore, continuous if $\mathcal{U}$ is finite dimensional.

Theorem C.2.3 *Let $\mathcal{U}$ be a reflexive Banach space and suppose $A : \mathcal{U} \to \mathcal{U}^*$ is bounded, coercive, and pseudo-monotone. Then, for every $f \in \mathcal{U}^*$, there exists at least one solution $u \in \mathcal{U}$ of the equation $A(u) = f$, i.e. A is surjective.*

C.3 Potential Operators and Duality Mappings

We will now examine the relationship between the minimum problem

$$F(u) - \langle f, u\rangle \quad \to \min \quad \forall u \in \mathcal{U}$$

and the Euler-Lagrange equation

$$A(u) = DF(u) = f.$$

In particular, we will determine when the operator equations $Au = f$ can be written as a minimization problem, i.e. when a functional can be found such that $DP = A$.

Definition C.3.1 (Potential Operator)
Let $\mathcal{U}$ be a Banach space. The operator $A : \mathcal{U} \to \mathcal{U}^*$ is called a *potential operator* if and only if there exists a Gâteaux differentiable functional $P : \mathcal{U} \to R$ such that $A = DP$, P being called a potential of A. ◇

If A is hemicontinuous, then by the method given in (Zeidler, 1985), we can define the pseudo-potential P of A by

$$P_A(u) := \int_0^1 \langle A(\theta u), u\rangle d\theta. \tag{C.6}$$

The hemicontinuity of A guarantees the continuity of the integrand. We then have the following criteria for A to be a potential operator.

Proposition C.3.1 *If $A : \mathcal{U} \to \mathcal{U}^*$ is a hemicontinuous operator, then the following two assertions hold.*
(1) Integral criterion. A is a potential operator if and only if

$$P_A(u) - P_A(v) = \int_0^1 \langle A(v + \theta(u - v)), u - v\rangle d\theta \quad \forall u, v \in \mathcal{U}. \tag{C.7}$$

The pseudo-potential P_A is therefore a potential, and any other potential for A differs from P_A only by a constant.
(2) Derivative criterion. If DA exists on $\mathcal{U}$ as a Gâteaux derivative, with the condition

$$(\alpha, \theta) \mapsto \langle DA(w + \alpha u + \theta v)s, t\rangle \text{ is continuous on } [0, 1] \times [0, 1] \quad \forall u, v, w, s, t \in \mathcal{U}, \tag{C.8}$$

then A is a potential operator if and only if

$$\langle DA(u)v, w\rangle = \langle DA(u)w, v\rangle \quad \forall u, v, w \in \mathcal{U}, \tag{C.9}$$

i.e. DA is symmetric.

Proof. To show that the condition (C.7) is necessary, we let $A = DP$. For

$$\phi(\theta) := P(v + \theta(u - v)) \quad \forall \theta \in \mathbb{R},$$

we have

$$\phi'(\theta) = \langle A(v + \theta(u - v)), u - v\rangle.$$

Hence,

$$P(u) - P(v) = \phi(1) - \phi(0) = \int_0^1 \langle A(v + \theta(u - v)), u - v\rangle d\theta$$

and $P(u) = P(0) + P_A$.

Conversely, if the condition (C.7) is true, then by the definition of the Gâteaux variation, we have

$$\begin{aligned}\langle DP_A(v), w\rangle &= \lim_{\theta\to 0^+} \frac{P_A(v + \theta w) - P_A(v)}{\theta} \\ &= \lim_{\theta\to 0^+} \int_0^1 \langle A(v + \theta t w), w\rangle \mathrm{d}t = \langle A(v), w\rangle,\end{aligned}$$

i.e. $DP_A = A$. Passage to the limit and integration can be interchanged because the integrand is continuous since A is hemicontinuous.

Now we prove the derivative criterion. To show that C.9 is necessary, we let $W(\alpha, \theta) = P(w + \alpha u + \theta v)$ for all $\alpha, \theta \in \mathbb{R}$. Then,

$$W_{\alpha\theta} = \langle DA(w + \alpha u + \theta v)v, u\rangle,$$

$$W_{\theta\alpha} = \langle DA(w + \alpha u + \theta v)u, v\rangle.$$

By (C.8), the derivatives are continuous and we have $W_{\alpha\theta}(0,0) = W_{\theta\alpha}(0,0)$, which is (C.9).

Conversely, by letting

$$U(\alpha, \theta) := \langle A(\alpha v + \theta u), v\rangle, \quad V(\alpha, \theta) := \langle A(\alpha v + \theta u), u\rangle,$$

then (C.9) implies that $U_\theta = V_\alpha$ for all $\alpha, \theta \in \mathbb{R}$. Therefore, for the classical line integral,

$$\oint U(\alpha, \theta)\mathrm{d}\alpha + V(\alpha, \theta)\mathrm{d}\theta = 0$$

holds. In particular, if we choose the triangle shown in Fig. C.1, then we obtain (C.7), i.e. A is a potential operator by the sufficiency condition of the integral criterion. ∎

Example C.3.1 Let $\mathcal{U} = \mathbb{R}^3 = \mathcal{U}^*$ and let $u = (x, y, z)$ be a vector. If a Gâteaux differentiable three-dimensional force field $A = (a(u), b(u), c(u))$ is a potential operator, then there exists a real-valued function $P : \mathbb{R}^3 \to \mathbb{R}$ such that

$$A = \mathrm{grad} P \;\Rightarrow\; (a, b, c) = (P_x, P_y, P_z).$$

Since

$$DA = \begin{pmatrix} a_x & a_y & a_z \\ b_x & b_y & b_z \\ c_x & c_y & c_z \end{pmatrix}$$

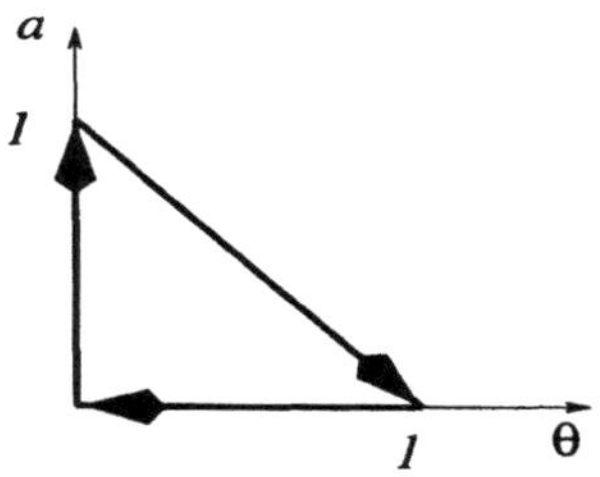

Figure C.1. Proof of Proposition

it can easily be shown that the condition (C.9) is equivalent to curl$A = 0$, i.e. $a_y = b_x, a_z = c_x, b_z = c_y$. Furthermore, it can easily be verified that the condition (C.7) simply means that $\oint Adu = 0$ in the sense of a classical line integral when one integrates around a triangle. Moreover, $P_A(u) = \int_0^u Adv$, where the integration is along the segment from 0 to u. ♡

Proposition C.3.2 *Let $A : \mathcal{U} \to \mathcal{U}^*$ be an operator on the real reflexive Banach space $\mathcal{U}$.*

Part I. If A is a monotone potential operator, then A is demicontinuous.

Part II. If the demicontinuous Gâteaux derivative DA exists on $\mathcal{U}$, then the following two statements are equivalent.

(1) A is a monotone potential operator.

(2) $\langle DA(u)v, v\rangle \geq 0 \ \ \forall u, v \in \mathcal{U}$.

Part III. The following assertions are equivalent.

(1) A is a monotone potential operator.

(2) A is monotone, hemicontinuous and (C.7) holds.

(3) A is a potential operator, i.e. $A = DP$, and P is convex and lower semicontinuous on $\mathcal{U}$.

Example C.3.2 Let $A = -\sum_{i,j=1}^{3} a_{ij} \frac{\partial^2}{\partial x_i \partial x_j} : \mathcal{U} \to \mathcal{U}^*$ be a continuous linear operator on the real Banach space $\mathcal{U} = \{u \in C^2(\Omega) | \ \ u(x) = 0 \ \ \forall x \in \partial\Omega\}$, $a_{ij} \in \mathbb{R} \ \ \forall i, j = 1, 2, 3$. We set $Bu = Au - f$ for all $u \in \mathcal{U}$ and fixed $f \in \mathcal{U}^*$. Then, $DB(u) = A \ \ \forall u \in \mathcal{U}$. Hence, from (C.9), we know that B is a potential operator if and only if A is symmetric. If this is the case, then B has the potential

$$\begin{aligned} P_B(u) &= \int_0^1 \langle A(\theta u) - f, u\rangle d\theta = \frac{1}{2}\langle Au, u\rangle - \langle f, u\rangle \\ &= \int_\Omega -\frac{1}{2}\sum_{i,j=1}^{3} a_{ij} u_{,ij} u d\Omega - \int_\Omega u f d\Omega = \int_\Omega \frac{1}{2}\sum_{i,j=1}^{3} a_{ij} u_{,i} u_{,j} d\Omega - \int_\Omega u f d\Omega. \end{aligned}$$

Thus, from the results of this section, we have the equivalent relations:

(1) A is a monotone potential operator $\Leftrightarrow$

(2) A is positive and symmetric (i.e. $a_{ij} = a_{ji}$) $\Leftrightarrow$

(3) P_B is convex and lower semicontinuous. ♡

Example C.3.3 In one-dimensional dissipative systems, the linear differential operator (2.46)

$$A = -\left(\frac{d}{dt} + \alpha\right)\left(m\frac{d}{dt}\right), \quad \alpha \in \mathbb{R}$$

is a symmetric the operator under the weighted bilinear forms

$$\langle v, w \rangle = \int_I e^{\alpha t} v(t) w(t) dt$$

since

$$\langle Av, w \rangle = \langle Aw, v \rangle, \quad \forall v, w \in \mathcal{H}_0^2(I).$$

The potential

$$P(u) = \frac{1}{2}\langle Au, u \rangle = \frac{1}{2}\langle m\Lambda u, \Lambda u \rangle = \frac{1}{2}\int_I e^{\alpha t} m(\frac{d}{dt}u)^2 dt.$$

is simply a quadratic form. ♡

Definition C.3.2 (Duality Mapping)
Let $\mathcal{U}$ be a Banach space. A map $A : \mathcal{U} \to \mathcal{U}^*$ is said to be a *duality mapping* if, for any $u \in \mathcal{U}$, (1) $\langle Au, u \rangle = \|Au\| \, \|u\|$, and (2) $\|Au\| = \|u\|$. ◊

A duality mapping can be constructed in any normed linear space in the following ways.

1. By the Hahn-Banach theorem, for any $u \in \mathcal{U}$, there exists at least one bounded linear functional $u_u^* = \langle u^*, u \rangle \in \mathcal{U}^*$ such that $\|u^*\| = 1$ and $\langle u^*, u \rangle = \|u\|$. Taking one such functional u_u^* and setting $Au = \|u\|u_u^*$ and $A(-u) = -\|u\|u_u^*$, we obtain $\|Au\| = \|u\|$ and $\langle Au, u \rangle = \|Au\| \; \|u\|$.
2. Let $F(u) = \frac{1}{2}\|u\|^2$. Its subdifferential $\partial F(u) : \mathcal{U} \to \mathcal{U}^*$ is a duality mapping.

Proposition C.3.3 $A(u) = \partial F(u)$ *consists of exactly all* $u^* \in \mathcal{U}^*$ *such that*

$$\langle u^*, u \rangle = \|u^*\| \; \|u\|, \quad \|u^*\| = \|u\|,$$

and $A(u)$ *is nonempty and convex as well as weak* $*$ *compact.*

Proof. This follows directly from results in Section 5.1.4. ■

Theorem C.3.1 *Let* $A : \mathcal{U} \to \mathcal{U}^*$ *be a duality map.*

(T1) A is multi-valued.

(T2) A is single-valued if $\mathcal{U}^*$ *is strictly convex.*

(T3) A is monotone and coercive. If $\mathcal{U}$ *is strictly convex, then A is strictly monotone.*

Proof. By definition, to each $u \in \mathcal{U}$, the duality mapping $A : \mathcal{U} \to \mathcal{U}^*$ is a set

$$\mathcal{U}_u^* = \{u^* \in \mathcal{U}^* | \langle u^*, u \rangle = \|u\|^2, \|u^*\| = \|u\|\}.$$

Thus, A is multi-valued. It is obvious that

$$\mathcal{U}_u^* \subset \mathcal{U}_v^* = \{v^* \in \mathcal{U}^* | \; \langle v^*, u \rangle = \|u\|^2, \; \|v^*\| \leq \|u\|\}.$$

Since, for any $v_1^*, v_2^* \in \mathcal{U}_v^*$, $\theta \in [0,1]$,

$$\langle v^*, u\rangle = \langle \theta v_1^* + (1-\theta)v_2^*, u\rangle = \|u\|^2,$$

$$\|v^*\| = \|\theta v_1^* + (1-\theta)v_2^*\| \le \|u\|,$$

we have $v^* = \theta v_1^* + (1-\theta)v_2^* \in \mathcal{U}_v^*$, and hence $\mathcal{U}_v^*$ is convex. By the definition of $\mathcal{U}_v^*$, we have $v^* = 0$ if $u = 0$; if $u \neq 0$, it follows from $\|u\|^2 = \langle v^*, u\rangle \le \|v^*\|\|u\|$ that $\|u\| \le \|v^*\|$, i.e. $\mathcal{U}_u^* = \mathcal{U}_v^*$. Hence, $\mathcal{U}_u^*$ is convex. Since $\mathcal{U}_u^* \subset \mathcal{S}_r^* = \{u^* \in \mathcal{U}^* | \ \|u^*\| = \|u\| = r\}$ for some r, it follows that, if $\mathcal{U}^*$ is strictly convex, then $\mathcal{U}_u^*$ consists of only one point and thus any duality mapping is therefore single-valued.

By definition, a duality mapping A is coercive because $\langle Au, u\rangle/\|u\| = \|u\|$; therefore,

$$\langle Au, u\rangle/\|u\| \to +\infty \text{ as } \|u\| \to \infty.$$

Suppose that there exists a pair of points $u, v \in \mathcal{U}$ such that

$$\langle Au - Av, u - v\rangle < 0.$$

Then, we have

$$\langle Au, u\rangle + \langle Av, v\rangle - \langle Au, v\rangle - \langle Av, u\rangle < 0.$$

Since $|\langle Av, u\rangle| \le \|Av\|\|u\|$ and $|\langle Au, v\rangle| \le \|Au\|\|v\|$, it follows from condition (T1) in the definition of the duality mapping that

$$\begin{aligned} 0 &> \|Au\| \ \|u\| + \|Av\|\|v\| - \|Av\|\|u\| - \|Au\|\|v\| \\ &= (\|Au\| - \|Av\|)(\|u\| - \|v\|). \end{aligned}$$

By condition (T2) in the definition of the duality mapping, we have

$$(\|Au\| - \|Av\|)(\|u\| - \|v\|) = (\|u\| - \|v\|)^2 \ge 0. \tag{C.10}$$

This contradiction shows that A is monotone.

Now we assume that $\mathcal{U}$ is strictly convex. Suppose that $\langle Au - Av, u - v\rangle = 0$. By the inequality (C.10), we have $\|u\| = \|v\|$. If $u \neq v$, it follows that $\langle Au, v\rangle < \|Au\|\|v\|$ since $\mathcal{U}$ is strictly convex. Similarly, $\langle Av, u\rangle < \|Av\|\|u\|$. Hence

$$\begin{aligned} 0 &= \langle Au - Av, u - v\rangle \\ &= \|Au\|\|u\| + \|Av\|\|v\| - \langle Au, v\rangle - \langle Av, u\rangle \\ &> (\|Au\| - \|Av\|)(\|u\| - \|v\|) \ge 0. \end{aligned}$$

This is a contradiction and hence it follows that A is strictly monotone. ■

References

Abeyaratne, R. and Knowles, J.K. (1989). Equilibrium shocks in plane deformations of incompressible elastic materials, *J. Elasticity*, **22**, 63-80.

Adly, S., Goeleven, D. and Motreanu, D. (1998). Unilateral Hamiltonian systems: a survey on the inequality approach. *Adv. Nonlinear Var. Inequal.* **1**, 11-26.

Alexander, A.E. and Johnson, P. (1949). *Colloid Science*, Oxford University Press.

Alibert, J.-J. and Dacorogna, B. (1992). An example of a quasiconvex function that is not polyconvex in two dimensions, *Arch. Rational Mech. Anal.*, **117**, 155-166.

Allaire, G. and Lods, V. (1999). Minimizers for a double-well problem with affine boundary conditions. *Proc. Roy. Soc. Edinburgh Sect.* **A 129**, 439-466.

Allen, G. (1977). Variational inequalities, complementarity problems, and duality theorems, *J. Math. Analy. Appl.*, **58**, 1-10.

Ambrosetti, A. and Rabinowitz, P.H. (1973). Dual variational methods in critical point theory and applications, *J. Funct. Anal.*, **14**, 349-381.

Antman, S. (1977). Birfurcation problems for nonlinearly elastic structures, in *Sym. Appl. of Bifurcation Theory*, Rabinowitz, P. (ed.), Academic Press, 73-125.

Antman, S. (1995). *Nonlinear Problems of Elasticity*, Springer-Verlag.

Arnold, V.I. (11978). *Mathematical Methods of Classical Mechanics*, Springer.

Arthurs, A.M. (1980). *Complementary Variational Principles*, Clarendon Press, Oxford.

Atai, A.A. and Steigmann, D. J. (1998). Coupled deformations of elastic curves and surfaces, *Internat. J. Solids Structures*, **35**, 1915-1952.

Atluri, S.N. (1980). On some new general and complementary energy theorems for the rate problems in finite strain, classical elasticity, *J Struct Mech*, **8**, 61-92

Aubert, G. (1987). On a counterexample of a rank 1 convex function which is not polyconvex in that case $N = 2$, *Proc. Royal Soc. Edinburgh*, **106 A**, 237-240.

Aubin, J.P. (1979). *Applied Functional Analysis*. Wiley Interscience, New York.

Aubin, J.P. (1993). *Optima and Equilibria, An Introduction to Nonlinear Analysis*. Springer-Verlag: Berlin, New York.

Aubin, J.P. and Ekeland, I. (1976). Estimates of the duality gap in nonconvex optimization. *Math. Oper. Res.***1**, no. 3, 225-245.

Aubin, J.P. and Ekeland, I. (1984). *Applied nonlinear analysis.* Wiley: New York.

Auchmuty, G. (1983). Duality for non-convex variational principles, *J. Diff. Eqns* **50**, 80-145.

Auchmuty, G. (1986). Dual variational principles for eigenvalue problems, *Proceedings of Symposia in Pure Math.*, **45**, Part 1, 55-71.

Auchmuty, G. (1994). Orthogonal decompositions and bases for three-dimensional vector fields, *Numer. Funct. Anal. Optimiz.* **15**, 455-488.

Auchmuty, G. (1997). Min-max problems for non-potential operator equations, *Contemporary Mathematics*, **209**, 19-28.

Baiocchi, C. and Capelo, A. (1984). *Variational and Quasivariational Inequalities, Applications to Free-Boundary Problems*, John Wiley and Sons, New York.

Bakelman, I.J. (1994). *Convex Analysis and Nonlinear Geometric Elliptic Equations*, Springer.

Ball, J.M. (1977). Convexity conditions and existence theorems in nonlinear elasticity, *Arch. Rat. Mech. Anal.* **63**, 337-403

Ball, J.M. (1998). The calculus of variations and materials science, *Qualt. Appl. Math.* **56**, 719-740

Barenblatt, G.I. and Chorin, A.J. (1998). Turbulence: an old challenge and new perspectives, *Meccanica*, **33**, 445-468.

Beardmore, R.E. and Song, Y.H. (1998). Differential-algebraic equaitons: A tutorial review, *Int. J. Bifurcation and Chaos*, **8**(7), 1399-1411.

Beatty, M.F. (1987). Topics in finite elasticity: hyperelasticity of rubber, elastomers, and biological tissues-with examples, *Appl. Mech. Reviews*, **40**, 1699-1734.

Belytschko, T. and Hodge, P.G. Jr. (1970). Plane stress limit analysis by finite element, *J. Eng. Mech. Div., ASCE,* **96**, 931-944.

Berger, M.S. and P.C. Fife (1968). On von Karman's equations and the buckling of a thin elastic plate, II. Plate with general edge conditions, *Comm. Pure Appl. Math.* **XXI**, 227-241.

Bille, J.P., Cescotto, S., Habraken, A.M. and Charlier, R. (1995). Numerical approach for contact using an augmented Lagrangian method. In Raous, M., Jean, M. and Moreau, J.J., editors, *Contact Mechanics*, Plenum Press: New York-London, pp. 243-246.

Blackorby, C., Primont, D. and Russell, R.R. (1978). *Duality, separability, and functional structure : theory and economic applications*, North-Holland: New York.

Blanchard, P. and Brüning, E. (1992). *Variational Methods in Mathematical Physics, A Unified Approach*, Springer-Verlag.

Blatz, P.J. and Ko, W.L. (1962). Application of finite elasticity to the deformation of rubbery materials, *Trans. Soc. Rheol.*, **6**, 223-251.

Bluman, G.W. and Kumei, S. (1989). *Symmetries and Differential Equations*, Springer Verlag.

Bojarski, J.L. (1999). Discussion on rectification to the paper of Doc. dr J.J. Telega, *Nonlinear Analysis*, **36**, 369-371.

Brenan, K.E., Campbell, S.L. and Petzold, L.R. (1996). *Numerical solution of Initial-Value Problems in Differential-Algebraic Equations*, SIAM, Philadelphia.

Brokate, M. and Sprekels, J. (1996). *Hysteresis and Phase Transitions*, Springer-Verlag, New York.

Bufler, H. (1983). On the work theorems for finite and incremental elastic deformations with discontinuous fields: a unified treatment of different versions, *Comput Meth Appl Mech Eng* **36**, 95-124.

Bui, H.D. (1992). On the variational boundary integral equations in elastodynamics with the use of conjugate functions, *J. Elasticity*, **28**, 247-256.

Bunch, B. (1989). *Reality's Mirror: Exploring the Mathematics of Symmetry*, Wiley, New York.

Candelas, P., de la Ossa, X.C., Green, P.S. and Parkes, L. (1991). A pair of Calabi-Yau manifolds as an exactly soluble superconformal theory. *Nuclear Phys. B* **359**, 21-74.

Capurso, M. (1970). On the incremental solution of elasto-plastic continue in the range of large displacements, *Meccanica*, **5**.

Carr, J., Gurtin, M.E. and Slemrod, M. (1984). Structured phase transitions on a finite interval, *Arch. Rat. Mech. Anal.* **86**, 317-351.

Casciaro, R. and Cascini, A. (1982). A mixed formulation and mixed finite elements for limit analysis, *Int. J. Solids and Struct.*, **19**, 169-184.

Cesari, L. and Yang, W.H. (1991). Serrin integrals and second order problems of plasticity, *Proc. Royal Soc. Edinburgh*, **177A**, 193-207.

Chadwick, P. and Haddon, E.W. (1972). Inflation-extension and reversion of a tube of incompressible isotropic elastic material, *J. Inst. Math. Appl.* **10**, 258-278.

Chakrabarty, J. (1987). *Theory of Plasticity*, McGraw-Hill Book Company.

Chang, K.C. (1986). *Critical point Theory and Its Applications*, Science and Technology Press, Shanghai (in Chinese).

Chang, K.C. (1993). *Infinite dimensional Morse theory and multiple solution problems*. Progr. Nonlin. Diff. Eq. Appl. **6**. Birkhäuser, Boston.

Chow, S.-N. and Hale, J.K. (1982). *Methods of bifurcation theory,* Grundlehren **251**, Springer, New York.

Christiansen, E. and Larsen, S. (1983). Computations in limit analysis for plastic plates, *Int. J. Nem. Meth. Eng.,* **19**, 169-184.

Ciarlet, P.G. (1980). A justification of the von Kármán equations, *Arch. Rational Mech. Anal.* **73**, 349-389.

Ciarlet, P.G. (1988). *Mathematical Elasticity. Vol. 1: Three-dimensional Elasticity*, North-Holland, Amsterdam, New York.

Ciarlet, P.G. and Geymonat, G. (1982). Sur les lois de comportement en élasticité nonlinéaire compressible, *C. R. Acad. Sci. Paris, Sér. A.*

Clarke, F.H. (1979). A classical variational principle for periodic Hamiltonian trajectories, *Proc. A.M.S.* **76**, 186-189.

Clarke, F.H. (1983). *Optimization and Nonsmooth Analysis*, John Wiley, New York.

Conley, C.C. (1978). *Isolated Invariant Sets and the Morse Index*, CBMS Regional Conf. Ser. in Math., **38**, Amer. Math. Soc., Providence, R.I.

Constantinescu, C. (1980). *Duality in Measure Theory*, Springer-Verlag: Berlin, New York.

Cornes, R. (1992). *Duality and Modern Economics*, Cambridge University Press: Cambridge, New York.

Cottle, R.W. (1964). *Nonlinear programs with positively bounded Jacobians*, PhD Thesis, Department of Mathematics, University of California, Berkeley.

Cottle, R.W., Giannessi, F. and J.-L. Lions, J.-L. (ed.) (1980). *Variational Inequalities and Complementarity Problems : Theory and Applications*, J. Wiley, New York.

Cottle, R.W., Pang, J.S. and Stone, R.E. (1992). *The linear Complementarity Problems*, Academic Press, New York.

Courant, R. and Hilbert, D. (1953). *Methods of Mathematical Physics,* Vol. I. Interscience, New York.

Dacorogna, D. (1989). *Direct Methods in the Calculus of Variations.* Springer-Verlag.

Dem'yanov, V.F., Stavroulakis, G.E., Polyakova, L.N. and Panagiotopoulos, P.D. (1996). *Quasidifferentiability and Nonsmooth Modelling in Mechanics, Engineering and Economics.* Kluwer Academic Publishers: Dordrecht.

Dierkes, U., Hildebrandt, S., Küster, A. and Wohlrab, O. (1992). *Minimal Surfaces I*, Springer-Verlag.

Do, C. (1977). Bifurcation theory of thin elastic plates subjected to unilateral conditions, *J. Math. Anal. Appl.*, **60**, 435-448.

Du, D.Z. and Pardalos, P.M. (1995). *Minimax and Applications*, Kluwer Academic Publishers: Dordrecht/Boston/London.

Duvaut, G. and Lions, J.L. (1976). *Inequalities in Mechanics and Physics*, Springer-Verlag, Berlin.

Einstein, A. (1934). *Essays in Science,* Philosophical Library.

Ekeland, I. (1977). Legendre duality in nonconvex optimization and calculus of variations, *SIAM J. Control and Optimization*, **15**, 905-934.

Ekeland, I. (1983). Dual variational methods in nonconvex optimization and differential equations. *Mathematical theories of optimization* (Genova, 1981), 108-120, Lecture Notes in Math., 979, Springer, Berlin-New York.

Ekeland, I. (1990). *Convex Methods and Hamiltonian Mechanics*, Springer-Verlag.

Ekeland, I. and Temam, R. (1976). *Convex Analysis and Variational Problems*, North-Holland Pub. Co. New York, Amsterdam.

Elliott, P.D.T.A. (1997). *Duality in Analytic Number Theory*, Cambridge University Press: New York.

Ericksen, J.L. (1975). Equilibrium of bars, *J. Elasticity* **5**, 191-202.

Ericksen, J.L. (1986). Stable equilibrium configurations of elastic crystals, *Arch. Rat. Mech. Anal.* **94**, 1-14.

Falk, F. (1990). Elastic phase transitions and nonconvex energy functions, in *Free Boundary Problems: Theory and Applications I.* (K.H. Hoffmann and J. Sprekels, eds.). Longman: London.

Fenchel, W. (1949). On conjugate convex functions, *Canad. J. Math.,* **1**, 73-77.

Feynman, R.P., Leighton, R.B. and Sands, M. (1964). *The Feynman Lectures in Physics,* Vol. 2. Addison-Wesley, Reading, Massachusetts.

Fichera, G. (1972). Existence theorems in elasticity-boundary value problems of elasticity with unilateral constraints, in *Encyclopedia of Physics*, (S. Flügger ed.), Vol. VI a/2, Mechanics of Solids(II), C. Truesdell, ed. Springer Verlag, p. 347-427.

Francu, J. (1979). On the Signorini problem for von Karman equations, the case of angular domain, *Apl. Matematiky*, **24**, 355-371.

Friedman, A. (1993). *Variational Principles and free-boundary problems.* Springer-Verlag: New York.

Fried, E. and Gurtin, M.E. (1996). Semi-quadratic variational problems for multiphase equilibria, *Quartl Appl. Math.* **54** (1), 73-84.

Galka, A. and Telega, J. J. (1992). The complementary energy principle as a dual problem for a specific model of geometrically nonlinear elastic shells with an independent rotation vector: general results. *European J. Mech. A Solids*, **11**, no. 2, 245-270.

Galka, A. and Telega, J.J. (1995). Duality and the complementary energy principle for a class of nonlinear structures, Part II. Anomalous dual variational principles for compressed elastic beams, *Arch. Mech.*, **47** (4) , 699-724.

Gao, D.Y. (1983). Extended bounding theorems of limit analysis *Appl. Math. Mech.*, **4**, 571-584.

Gao, D.Y. (1986a). *Complementarity Principles in Nonsmooth Elastoplastic Systems and Pan-penalty Finite Element Methods*, Ph.D. Thesis, Tsinghua University.

Gao, D.Y. (1986b). Tao of complementary-duality: on the general duality principles in thinking systems, *J. Hefei University* (Sociology edition), **1**, 13-22.

Gao, D.Y. (1988a). On the complementary bounding theorems for limit analysis, *Int. J of Solids Structures*, **24**, 545-556.

Gao, D.Y. (1988b). Panpenalty finite element programming for plastic limit analysis, *Compute & Structures,* **28**, 749-755.

Gao, D.Y. (1990). Dynamically loaded rigid-plastic analysis under large deformation, *Quart. Appl. Math.* **48**, 731-739.

Gao, D.Y. (1992). Global extremum criteria for finite elasticity, *ZAMP* **43**, 924-937.

Gao, D.Y. (1994). Stability and extremum principles for post yield analysis of finite plasticity, *Acta Mech Sinica* **10**, 311-325.

Gao, D.Y. (1995a). Duality theory in nonlinear buckling analysis for von Karman equations, *Studies in Applied Mathematic*, **94**, 423-444.

Gao, D.Y. (1995b). Limit analysis of plastic shells subjected to large deformations, *Eur. J. Mech. A/Solids*, **14**(3), 459-472.

Gao, D.Y. (1995c). Complementarity and duality: a unified framework in nonlinear equilibrium problems, *ICAM Report, Department of Mathematics, Virginia Tech.*, 95-12-2, 1-20.

Gao, D.Y. (1996a). Post-yield analysis for rigid-plastic plates with large deflections, *Eur. J. Mech., A/Solids,* **15**, 599-615.

Gao, D.Y. (1996b). Tao of the complementarity-duality: on the duality principles in natural science, *Selected Tsinghua's Ph.D. Theses on Philosophy*, Tsinghua University Press, 12-25

Gao, D.Y. (1996c). Singularity and a new phenomenon in optimal shape design of beam theory, *AMS-IMS-SIAM Summer Research Conference on Optimization Methods in Partial Differential Equations*, June 16-20, 1996, Mount Holyoke, MA.

Gao, D.Y. (1997). Dual extremum principles in finite deformation theory with applications in post-buckling analysis of nonlinear beam model, *Appl. Mech. Reviews, ASME* **50**, S64-S71.

Gao, D.Y. (1998a). Duality, triality and complementary extremum principles in nonconvex parametric variational problems with applications, *IMA J. Appl. Math.*, **61**, 199-235.

Gao, D.Y. (1998b). Minimax and triality theory in nonsmooth variational problems, in *Reformulation: Nonsmooth, Piecewise Smooth, Semismooth and Smoothing Methods*, Ed. Fukushima and L. Qi, Kluwer Academic Pub., Boston, p. 161-179.

Gao, D.Y. (1998c). Bi-complementarity and duality: A framework in nonlinear equilibria with applications to contact problem of elastoplastic beam theory, *J. Math. Analy. Appl.* **221**, 672-697.

Gao, D.Y. (1999a). Duality in nonconvex finite deformation theory: A survey and unified approach, to appear in *From Convexity to Nonconvexity, A Volume dedicated to the memory of Professor Gaetano Fichera*, (R. Gilbert, P.D. Panagiotopoulos and P. Pardalos eds). Kluwer Academic Publishers: Dordrecht.

Gao, D.Y. (1999b). Duality-Mathematics, in *Encyclopedia of Electrical and Electronics Engineering,* John G. Webster ed., John Wiely & Sons, Inc., **6**, 68-77.

Gao, D.Y. (2000a). Finite deformation beam models and triality theory in dynamical post-buckling analysis, *Int. J. Non-Linear Mechanics*, **35**, 103-131.

Gao, D.Y. (2000b). Analytic solutions and triality theory for nonconvex and nonsmooth variational problems with applications, to appear in *Nonlinear Analysis.*

Gao, D.Y. and Cheung, Y.K. (1989). On the extremum complementary energy principles for nonlinear elastic shells, *Int. J. Solids Struct.* **25**, 683-693.

Gao, D.Y. and Hwang, K.C. (1987). Complementary principles in elastic perfectly plastic media, *Scientia Sinica A*, **31**, 1447-1453.

Gao, D.Y. and Onat, E.T. (1990). Rate variational extremum principles for finite elastoplasticity, *Appl. Math. Mech.*, **11**(7), 659-667.

Gao, D.Y. and Russell, D.L. (1996a). An extended beam theory for smart materials applications, Part I. Extended beam theory, duality theory and finite element simulations, *Appl. Math. Optimization,* **34** (3), 279-298.

Gao, D.Y. and Russell, D.L. (1996b). Finite deformation extended beam theory and nonlinear buckling analysis, in *Contemporary Research in the Mathematics and Mechanics of Materials,* Ed by RC Batra and MF Beatty, CIMNE, Barcelona, Spain, 430-441.

Gao, D.Y. and Strang, G. (1989a). Geometric nonlinearity: Potential energy, complementary energy, and the gap function, *Quart. Appl. Math.* **47**, 487-504.

Gao, D.Y. and Strang, G. (1989b) Dual extremum principles in finite deformation elastoplasitc analysis, *Acta Applicandae Mathematicae.* **17**, 257-267.

Gao, D.Y. and Wirzbicki, T. (1989). Bounding theorem in finite plasticity with hardening effect, *Quart. Appl. Math.,* **47**, 395-403.

Gao, D.Y. and Yang, W.-H. (1995). Multi-duality in minimal surface type problems, *Studies in Appl. Math.*, **95**, 127-146.

Gao, D.Y. and Yau, S.-T. (1991). A new phenomenon in eigenvalue problem on extremum surface, Semina talks presented at the University of Michigan, Indiana University, and the *Int. Sym. Methods and Appl. of Analysis*, Hong Kong, Dec. 16-19, 1994.

Ghoussoub, N. (1993). *Duality and Perturbation Methods in Critical Point Theory,* Cambridge University Press, New York.

Giannessi, F. and Maier, G. (1991). Complementarity systems and optimization problems in structural engineering, *Eng. Optim.*, **18**, 43-66.

Glowinski, R., Lions, J.L. and Trémolières, R. (1981). *Numerical Analysis of Variational Inequalities*, North-Holland, Amsterdam, New York, Oxford.

Goldstine, H. H. (1980). *A History of the Calculus of Variations, From the 17th Through the 19th Century.* Springer-Verlag, New York.

Gruber, B. *et al* (editors). *Symmetries in Science I-VI,* Plenum, New York, 1980, 1986, 1989, 1990, 1991, 1993.

Gurtin, M.E. (1964). Variational principles for linear initial-value problems, *Q. Appl. Math.*, **22** (3), 252-256.

Haftka, R.T. and Gürdal, G. (1992). *Elements of Structural Optimization: Third Revised and Expanded Edition,* Kluwer Academic Publishers.

Hargittai, I. (1986). *Symmetry: Unifying Human Understanding,* Pergamon, Oxford.

Hargittai, I. (1989). *Symmetry 2: Unifying Human Understanding,* Pergamon, Oxford.

Haines, D.Q. and Wilson, W.D. (1979). Strain energy density of rubber-like materials, *J. Mech. Phys. Solids* **27**, 345-360

Hamilton, WR (1835). Second essay on a general method in dynamics, *Philosophical Transactions of the Royal Society of London,* Part I, 95-144.

Haslinger, J. and Neittaanmäki, P. (1988). *Finite Element Approximation for Optimal Shape Design. Theory and Applications.* John Wiley & Sons Ltd., Chichester.

Haslinger, J. and Panagiotopoulos, P.D. (1984). The reciprocal variational approach to the Signorini problem with friction, approximation results, *Proc. Roy Soc. Edinburgh Sect A*, **98**, pp. 365-383.

Haug, E.J., Choi, K.K. and Komkov, V. (1986). *Design Sensitivity Analysis of Structural Systems,* Academic Press, Orlando.

Hellinger, E. (1914). Die allgemeine Ansätze der Mechanik der Kontinua. Encyklopädie der Mathematischen Wissenschaften IV, 4, pp. 602-94.

Hill, R. (1948). A variational principle of maximum plastic work in classical theory of plasticity, *Quart. J. Mech. Appl. Math.*, **1**, 1-18.

Hill, R. (1950). *Mathematical Theory of Plasticity*, Oxford.

Hill, R. (1957). On uniqueness and stability in the theory of finite elastic strain, *J. Mech. Phys. Solids*, **5**, 229-241.

Hill, R. (1978). Aspects of invariance in solids mechanics, in *Adv. in Appl. Mech.* **18**, 1-75.

Hill, R. and Hutchinson, J.W. (1975). Bifurcation phenomena in the plane tension test, *J. Mech. Phys. Solids*, **23**(4-5), 239-264.

Hiriart-Urruty, J.B. and Lemaréchal, C. (1993). *Convex Analysis and Minimization Algorithms I*, Springer-Verlag.

Hlavacek, I., Haslinger, J., Necas, J. and Lovisek, J. (1982). *Solution of Variational Inequalities in Mechanics*, Springer-Verlag, New York, Berlin.

Hofmann, K.H. (1970). *The Duality of Compact Semigroups and C*-bigebras*, Lecture Notes in Mathematics, Vol. 129. Springer-Verlag: Berlin, New York.

Hoger, A. (1987). The stress conjugate to logarithmic strain,*Int. J. Solids Struct.*, **23**(12), 1645-1656.

Hoger, A. (1996). The elasticity tensor of a transversely isotropic hyperelastic material with residual stress, *J. Elasticity*, **42**, 115-132.

Horgan, C.O.(1996). Remarks on ellipticity for the generalized Blatz-Ko model for a compressible nonlinear elastic solid, *J. of Elasticity*, **42**, 165-176.

Horst, R. and Padalos, P.M. (1994). *Handbook in Global Optimization*, Kluwer Academic Publishers, Boston.

Hsue, D.W. (1975). Contributions on a set of theorems on limit analysis, *Chinese Science Bulletin*, **20**, 175-181 (in Chinese).

Hu, Hai-Chang (1955). On some variational principles in the theory of elasticity and the theory of plasticity, *Scientia Sinica* **4**, 33-54.

Huang, R.Q. (1990). Invariants of sets of linear varieties, *Proc. Natl. Acad. Sci. USA*, **87**, 4557-4560.

Huang, R.Q., Rota, G.-C. and Stein, J.A. (1989). Supersymmetric algebra, supersymmetric space, and invariant theory, in *Symmetry in Nature, A Volume in Honour of Luigi A. Radicati di Brozolo*, Tomo II, Sculola Normale Superiore, 407-432.

Isac, G. (1992). *Complementarity Problems.* Lecture Notes in Mathematics, Springer-Verlag, New York.

James, R.D. and Kinderlehrer, D. (1989). Theory of diffusionless phase transitions, in *Lecture Notes in Physics*, **344** (M. Rascle et al. eds). Springer-Verlag: Berlin, 147-176.

Jung, M. (1999). Duality theory for solutions to Volterra integral equations, *J. Math. Anal. Appl.*, **230**, 112-134.

Kaijser, S. and Pelletier, J.W. (1986). *Interpolation functors and duality*, Springer-Verlag: Berlin, New York.

Kamenjarzh, J.A. (1996), *Limit Analysis of Solid and Structures*, CRC Press, New York.

Kapania, R.K. and Raciti, S. (1989). Recent advances in analysis of laminated beams and plates, Part I: Shear effects and buckling; Part II: Vibrations and wave propagation, *AIAA J.* **27**, 923-946.

Karamardian, S. (1971). Generalized complementarity problems, *J. of Optimization Theory and Applications*, Vol. 6, **19**, 161-168.

Kaplan, W. (1999). *Maxima and Minima with Applications: Practical optimization and duality*, Wiley-Interscience Publication.

Kibble, T.W.B. (1997). Phase transitions and topological defects in the early universe, *Aust. J. Phys.* **50**, 697-722.

Kikuchi, N. and Oden, J.T. (1988). *Contact Problems in Elasticity: A Study of Variational Inequalities and Finite Element Methods*, SIAM, Philadelphia.

Kinderlehrer, D. (1971). Variational inequalities with lower dimensional obstacles, *Isr. J. Math.*, **10**, 339-348.

Kinderlehrer, D. and Stampacchia, G. (1980). *An Introduction to Variational Inequalities and Their Applications*, Academic Press, New York.

Klarbring, A., Mikelić, A. and Shillor, M (1990). Duality applied to contact problems with friction, *Appl. Math. Optim.*, **22**, pp. 211-226.

Klein Haneveld, W. K. (1986). *Duality in stochastic linear and dynamic programming*, Lecture Notes in Economics and Mathematical Systems, 274. Springer-Verlag: Berlin, New York.

Knowles, J.K. and Sternberg, E. (1978). On the failure of ellipticity and the emergence of discontinuous deformation gradients in plane finite elastoplastics, *J. Elasticity*, **8**, 329-379.

Knowles, J.K. and Sternberg, E. (1980). Discontinuous gradient near the tip of a crack in finite antiplane shear, *J. Elasticity* **10**, 81-110.

Koiter, W.T. (1973). On the principle of stationary complementary energy in the nonlinear theory of elasticity, *SIAM, J. Appl. Math.*, **25**, 424-434.

Koiter, W.T. (1976). On the complementary energy theorem in nonlinear elasticity theory, in *Trends in the Application of Pure Mathematics to Mechanics*, (G. Fichera ed.), Pitman.

Kohn, R. V. and Strang, G. (1986). Optimal design and relaxation of variational problems, I, II, III, *Communications on Pure and Applied Mathematics*, **XXXIX**, 113-137, 139-182, 353-377.

Kohn, R.V. (1991). The relaxation of a double-well energy, *Continuum Mech. Thermodyn.* **3**, 193-236.

Krasnoselskii, M.A. (1963). *Topological Methods in the Theory of Nonlinear Integral Equations.* Pergamon Press.

Kuyk, W. (1977). *Complementarity in mathematics: a first introduction to the foundations of mathematics and its history*, D. Reidel Pub. Co., Dordrecht, Holland, Boston, Hingham, Mass.

Lasiecka, I. (1998). Uniform stability of a full von Karman system with nonlinear boundary feedback, *SIAM J. Control Optim.*, **36**, 1376-1422.

Lasiecka, I. and Sokolowski, J. (1991). Sensitivity analysis of constrained optimal control problem for wave equation. *SIAM J. Control and Optimization*, **29**, 1128-1149.

Laursen, T.A. and Oancea, V.G. (1994). Automation and assessment of augmented Lagrangian algorithms for frictional contact problems, *ASME J. Appl. Mech.*, **61**, 956-963.

Lawvere, F.W. and Schanuel, S.H. (1997). *Conceptual Mathematics: A first introduction to categories*. Cambridge University Press.

Lazopoulos, K.A. (1995). Beam buckling as a coexistence of phase phenomenon, *Eur. J. Mech. A/Solids*, **14**, 589-604.

Lazopoulos, K.A. and Ogden, R.W. (1998). Nonlinear elasticity theory with discontinuous internal variables, *Math. and Mech. of Solids*, **3**, 29-51.

Le, K.C. and Stumpf, H. (1998). Strain measures, integrability condition and frame indifference in the theory of oriented media, *Int. J. Solids Structures*, **35**, 783-798.

Lee, R.C.H and Yung, S.P. (1997). Optimality conditions and duality for a nonlinear time-delay control problem, *Optimal Control Applications & Methods*, **18**, 327-340.

Lee, S.J. and Shield, R.T. (1980). Variational principles in finite elastics, *J. Appl. Math. Physics (ZAMP)*, **31**, 437-453.

Lee, S.J. and Shield, R.T. (1980). Applications of variational principles in finite elasticity, *J. Appl. Math. Physics (ZAMP)*, **31**, 454-472

Levinson, M.(1965). The complementary energy theorem in finite elasticity, *Trans. ASME, ser. E, J. Appl. Mech.*, **87**, 826-828.

Li, T.-Y. and Yorke, J. (1975). Period three implies chaos, *American Math. Monthly*, **82** (10), 985-992.

Lian, B.H., Liu, K. and Yau, S.-T. (1997). Mirror principle. I. *Asian J. Math.* **1**, 729-763.

Librescu, L. and Lin, Weiqing (1996). Postbuckling and vibration of shear deformable flat and curved panels on a non-linear elastic foundation, *Int. J. Non-Linear Mech.*, **32**, 211-225.

Logan, J.D. (1977). *Invariant Variational Principles*, Academic Press.

Lovísek, J. (1981). Duality in the obstacle and unilateral problem for the biharmonic operator, *Aplikace Matematiky*, Svazek 26, 291-303.

Luenberger, D.G. (1969). *Optimization by Vector Space Methods*, John Wiley & Sons, Inc., New York.

Luo, Z.Q., Pang, J.S. and Ralph, D. (1996). *Mathematical Programs with Equilibrium Constraints*, Cambridge University Press, Cambridge.

Lurie, K.A. and Cherkaev, A.V. (1988). On a certain variational problem of phase equilibrium, in *Material Instabilities in Continuum Mechanics*, (J.M. Ball, ed.), Oxford University Press, 257-268.

MacLane, S. (1971). *Categories for the working mathematician*, New York, Springer-Verlag.

Maier, G. (1969). Complementarity plastic work theorems in piecewise-linear elastoplasticity, *Int. J. Solids and Struct.*, **5**, 261-270.

Maier, G. (1970). A matrix structural theory of piecewise-linear plasticity with interacting yield planes, *Meccanica*, **5**, 55-66.

Maier, G. and Drucker, D.C. (1973). Effects of geometry changes on essential features of inelastic materials, *Engrg. Mech. Div. ASCE*, 819-834

Marsden, J. and Ratiu, T. (1995). *Introduction to Mechanics and Symmetry,* Springer.

Marsden, J.E. and Hughes, T.J.R. (1994). *Mathematical Foundations of Elasticity.* Dover Publications, Inc. New York.

Mason, J. (1980). *Variational, Incremental and Energy Methods in Solid Mechanics and Shell Theory,* Elsevier, Amsterdam, New York.

Matthies, H. and Strang, G. (1979). The solution of nonlinear finite element equations, *Inter. J. Num. Meth. in Eng.* **14**, 1613-1626.

Maugin, G.A. (1992). *The Thermomechanics of Plasticity and Fracture,* Cambridge University Press, Cambridge-New York.

Mawhin, J. and Willem, M. (1989). *Critical Point Theory and Hamiltonian Systems.* Appl. Math. Sci. **74**, Springer, New York.

Milnor, J. (1963). *Morse Theory,* Princeton Univ. Press, Princeton, N.J.

Morrey, C.B. (1952). Quasi-convexity and the lower semicontinuity of multiple integrals, *Pacific J. Math.* **2**, 25-53.

Moreau, J.J. (1968). La notion de sur-potentiel et les liaisons unilatérales en élastostatique, *C.R. Acad. Sc. Paris,* 267 A, 954-957.

Moreau, J.J., Panagiotopoulos, P.D. and Strang, G. (1988). Topics in nonsmooth mechanics. Birkhuser Verlag, Basel-Boston, MA.

Mosco, U. (1972). Dual variational inequalities, *J. Math. Analy. Appl.*, **40**, 202-206.

Motreanu, D. and Panagiotopoulos, P. D. (1999). *Minimax Theorems and Qualitative Properties of the Solutions of Hemivariational Inequalities.* Kluwer Academic Publishers, Dordrecht.

Mróz, Z. (1994). Variational methods in sensitivity analysis and optimal design, *Eur. J. Mech. A/Solids,* **13**, 115-147.

Mura, T. and Lee, S.L. (1963). Application of variational principle of limit analysis, *Quart. Appl. Math.,* **21**(3), 243-248.

Nakagami, Y. and Takesaki, M. (1979). *Duality for Crossed Products of von Neumann Algebras,* Springer-Verlag: Berlin, New York.

Nemat-Nasser, S. (1972). General variational principles in nonlinear and linear elasticity with applications, *Mechanics Today* **1**, 214-61.

Nitsche, J.C.C. (1989). *Lectures on Minimal Surfaces,* Cambridge University Press, Cambridge ; New York.

Nguyen, Q. S. (1994). 4Bifurcation and stability in dissipative media (plasticity, friction, fracture), *Appl. Mech. Rev.*, **47** (1) Part 1, 1-31.

Noble, B. and Sewell, M.J. (1972). On dual extremum principles in applied mathematics, *J. Inst. Math. Appl.*, **9**, 123-193.

Noor, A.K. and Burton, W.S. (1989). Assessment of shear deformation theories for multilayered composite plates, *Appl. Mech. Rev.*, **41**, ASME, 1-13.

Novak, B. and Tyson, J.J. (1993). Modeling the cell division cycle: M-phase trigger, oscillations, and size control, *J. Theor. Biol.*, **165**, 101-134.

Novozhilov, V.V. (1961). Theory of Elasticity, Pergamon Press, Oxford.

Naumann, J. (1975). On some unilateral boundary value problems for the von Karman equations. Part I, The coercive case, *Apl. Matematiky,* **20**, 96-125.

Oden, J.T. (1986). *Qualitative Methods in Nonlinear Mechanics,* Prentice-Hall: Englewood Cliffs.

Oden, J. T. and Lee, J. K. (1977). Dual-mixed hybrid finite element method for second-order elliptic problems. Mathematical aspects of finite element methods (Proc. Conf., Consiglio Naz. delle Ricerche (C.N.R.), Rome, 1975). In *Lecture Notes in Math.,* Vol. 606, Springer, Berlin, 275-291.

Oden, J.T. and Reddy, J.N. (1973). On dual complementary variational principles in mathematical physics, *Int. J. Engrg. Sci.* **12**, 1-29.

Oden, J.T. and Reddy, J.N. (1983). *Variational Methods in Theoretical Mechanics.* Springer-Verlag.

Ogden, R.W. (1972). Large deformation isotropic elasticity. On the correlation of theory and experiment for incompressible rubber-like solids, *Proc. Roy. Soc. London,* **A326**, 567-583

Ogden, R.W. (1976) Volume changes associated with the deformation of rubber-like solids, *J. Mech. Phys. Solids*, **24**, 323-338

Ogden, R.W. (1977). Inequalities associated with the inversion of elastic stress-deformation relations and their implications, *Math. Proc. Camb. Phil. Soc.* **81**, 313-324

Ogden, R.W. (1978). Extremum principles in non-linear elasticity and their application to composite - I, *Int. J. Solids Struc.* **14**, 265-282.

Ogden, R.W. (1984). *Non-linear elastic deformations*, Ellis Horwood: Chichester.

Ogden, R.W. and Gao, D.Y. (2000). On the complementary energy principles in nonconvex elasticity and applications, in preparation.

Ohtake, K., Oden, J. T. and Kikuchi, N. (1980). Analysis of certain unilateral problems in von Kármán plate theory by a penalty method. I. A variational principle with penalty. *Comput. Methods Appl. Mech. Engrg.*, **24**, no. 2, 187-213.

Olver, P.J. (1995). *Equivalence, Invariants, and Symmetry.* Cambridge University Press.

Onat, E.T. (1960). The influence of geometry changes on the load-deformation behavior of plastic solid, *Plasticity, Proc. 2nd US Naval Struct. Mech.*, Pergamon Press, Oxford. 225-238.

Oravas, G.A. and McLean, L. (1966). Historical development of energetical principles in elastomechanics, *Appl. Mech. Reviews*, **19**, 647-658 and 919-933.

Pai, P.F. and Nayfeh, A.H. (1994). A fully nonlinear theory of curved and twisted composite rotor blades accounting for warpings and three-dimensional stress effects, *Int. J. Solids and Structures,* **31** (9), 1309-1340.

Panagiotopoulos, P.D. (1985). *Inequality Problems in Mechanics and Applications.* Birkhäuser: Boston.

Panagiotopoulos, P.D. (1993). *Hemivariational Inequalities: Applications in Mechanics and Engineering.* Springer-Verlag.

Pfeiffer, F. and Glocker, C. (1996). *Multibody Dynamics with Unilateral Contacts*, John Wiley, New York.

Pian, T.H.H. and Tong, P. (1980). Reissner's principle in finite element formulations, in *Mechanics Today*, **5**, S. Nemat-Nasser (ed.), Pergamon Press, 377-395.

Pipkin, A.C. (1993). Relaxed energy densities for small deformations of membranes, *IMA J. Appl. Math.* **50**, 225-237.

Pitteri, M. and Zanzotto, G. (1999). *Continuum Models for Phase Transitions and Twinning in Crystals*, Chapamn and Hall.

Prager, W. and Hodge, P.G. Jr. (1968). *Theory of Perfect Plastic Solids*, Dover Publications, New York.

Qiang, L.X. and Zhong, W.X. (1963), On the limit analysis in solid mechanics and a generalized variational principle, *Acta Mechanica Sinica,* **6**, 287-303.

Rabinowitz, P.H. (1978). Some minimax theorems as applications to nonlinear partial differential equations, *Nonlinear Analysis: A collection of papers in honor of Erich Röthe*, Academic Press, New York, 161-177.

Rabinowitz, P.H. (1986). *Minimax Methods in Critical Point Theory with Applications to Differential Equations*, CBMS Regional Conf. Ser. in Math., **65**, AMS, Providence, R.I.

Rall, L.B. (1969). *Computational solution of Nonlinear Operator Equations*, Wiley, New York.

Reddy, B.D. (1992). Mixed variational inequalities arising in elastoplasticity, *Nonlinear Analysis, Theory, Methods & Applications,* **19**(11), 1071-1089.

Reissner, E (1953). On a variational theorem for finite elastic deformations, *J. Math. & Phys.* **32**, 129-153.

Reissner, E. (1996). *Selected works in applied mechanics and mathematics*, Jones and Bartlett Publishers: Boston.

Renardy, M. and Rogers, R. (1993). *An Introduction to Partial Differential Equations.* Springer-Verlag, New York.

Rockafellar, R.T. (1967). Duality and stability in extremum problems involving convex functions, *Pacific J. Math.* **21**, 167-187.

Rockafellar, R.T. (1970). *Convex analysis,* Princeton University Press.

Rockafellar, R.T. (1974). *Conjugate Duality and Optimization*, SIAM, Philadelphia.

Rockafellar, R.T. and Wets, R.J.B. (1997). *Variational analysis*, Springer: Berlin, New York.

Rogers, R.C. and Truskinovsky, L. (1997). Discretization and hysteresis, *Physica B* **233**, 370-375.

Rosakis, P. and Jiang, Q. (1993). Deformations with discontinuous gradients in plane elastostatics of compressible solids, *J. Elasticity*, **33**, 233-257.

Rosen, J. (1995). *Symmetry in science: An introduction to the general theory*, Springer.

Rowlinson, J.S. (1979). Translation of J. D. van der Waals' "The thermodynamic theory of capillarity under the hypothesis of a continuous variation of density". *J. Statist. Phys.* **20**, 197–244.

Serre, D. (1983). Formes quadratiques et calcul des variations, *J. Math. purres appl.*, **62**, 177-196

Sewell, M.J. (1987). *Maximum and Minimum Principles.* Cambridge University Press.

Šilhavý, M. (1997). *The Mechanics and Thermodynamics of Continuous Media*, Springer-Verlag, Berlin

Silverman, J.H. and Tate, J. (1992). *Rational Points on Elliptic Curves,* Springer-Verlag.

Sokolowski, J. and Zolesio, J.P. (1992). *Introduction to Shape Optimization, Shape Sensitivity Analysis.* Springer, New York.

Sprekels, J. and Tiba, D. (1999). A duality-type method for the design of beams. *Adv. Math. Sci. Appl.*, **9**, 89-102.

Stakgold, I. (1979). *Green's Functions and Boundary Value Problems*, John Wiley & Sons, New York.

Steigmann, D.J. and Ogden, R.W. (1997). Plane deformations of elastic solids with intrinsic boundary elasticity, *Proc. R. Soc. Lond. A* , **453**, 853-877.

Strang, G. (1979). A minimax problem in plasticity theory, *Functional Analysis Methods in Numerical Analysis,* M.Z. Nashed (ed.), Springer Lecture Notes **701**, 319-333.

Strang, G. (1979). A family of model problems in plasticity, in *Proc. Symp. Computing Methods in Applied Sciences,* R. Glowinski and J.L. Lions (ed.), Springer Lecture Notes **704**, 292-308.

Strang, G. (1986). *Introduction to Applied Mathematics* Wellesley-Cambridge Press, MA.

Strang, G. (1988). A framework for equilibrium equations, *SIAM Review*, **30**, 282-297.

Strang, G. and Fix, G. (1973). *An Analysis of the Finite Element Method,* Wellesley-Cambridge Press.

Strang, G., Matthies, H. and Christiansen, E. (1979). The saddle point of a differential program, *Energy Methods in Finite Element Analysis*, ed. by R. Glowinski, E. Rodin, and O.C. Zienkiewicz, John Wiley.

Strang, G. and Matthies, H. and Temam, R. (1980). Mathematical and computational methods in plasticity, *Proc. IUTAM Symp. on Variational Methods in the Mechanics of Solids,* S. Nemat- Nasser, ed., Pergamon, 20-28.

Stumpf, H. (1975). Dual extremum principles and error bounds in the theory of plates with large deflections, *Arch. Mech.*, **27**(3), 485-496.

Stumpf, H. (1976). The principle of complementary energy in non-linear plate theory, *J. Elasticity,* **6**(1), 1-10.

Stumpf, H. (1978). Dual extremum principles and error bounds in non-linear elasticity theory, *J. Elasticity,* **8** (4), 425-438.

Stumpf, H. (1979). The derivation of the dual extremum and complementary stationary principles in geometrical nonlinear shell theory, *Ing. Arch.* **48**, 221-237.

Stumpf, H. and Hoppe, U. (1997). The application of tensor algebra on manifolds to nonlinear continuum mechanics - invited survey article, *ZAMM,* **77** (5), 327-339.

Struwe, M. (1996). *Variational Methods: Applications to nonlinear partial differential equations and Hamiltonian systems* (Second edition). Springer, Berlin-New York.

Šverák, V. (1991). Quasiconvex functions with subquadratic growth, *Proc. Royal Soc. Edinburgh* **120**, 185-189

Šverák, V. (1992). New examples of quasiconvex functions, *Arch. Rational Mech. Anal.* **119**, 293-300

Synge, J.L. (1957). *The Hypercircle in Mathematical Physics*, Cambridge University Press.

Tabarrok, B. and Rimrott, F.P.J. (1994). *Variational methods and complementary formulations in dynamics.* Kluwer Academic Publishers: Dordrecht.

Taylor, G.I. and Quinney, H. (1931). The plastic distortion of metals, *Phil. Trans., Roy. Soc. London,* **A191**, 441-446.

Telega, J.J. (1989). On the complementary energy principle in nonlinear elasticity. I. von Kármán plates and three-dimensional solids, *C. R. Acad. Sci. Paris,* II Méc. Phys. Chim. Sci. Univers Sci. Terre **308**, no. 14, 1193-1198.

Temam, R. (1983). Dual variational principles in mechanics and physics. Semi-infinite programming and applications (Austin, Tex., 1981), 295-309, *Lecture Notes in Econom. and Math. Systems,* **215**, Springer, Berlin-New York.

Temam, R. (1985). *Mathematical problems in plasticity.* Gauthier-Villars.

Temam, R. and Strang, G. (1980). Duality and relaxation in variational problems of plasticity, *J. de Mecanique*, **19**, 493-527.

Terpstra, F.J. (1938). Die Darstellung biquadratischer Formen als Summen von Quadraten mit Anwendung auf die Variationsrechnung, *Math. Ann.* **116**, 166-180.

Toland, J.F. (1978). Duality in nonconvex optimization, *J. Math. Anal. Appl.* **66**, 399-415.

Toland, J.F. (1979). A duality principle for non-convex optimization and the calculus of variations, *Arch. Rational Mech. Anal.* **71,** 41-61.

Tonti, E. (1972). A mathematical model for physical theories, *Accad. Naz. dei Lincei*, Serie III, **LII**, I, 175-181; II, 350-356.

Tonti, E. (1972). On the mathematical structure of a large class of physical theories, *Accad. Naz. dei Lincei*, Serie VIII, **LII**, 49-56.

Trefftz, E. (1926). Ein Gegenstuck zum Ritzchen Verfahren, *Proc. of the 2nd International Congress of Applied Mechanics,* pp. 131-137, Zurich.

Truesdell, C. and Noll, W. (1965). The nonlinear field theories of mechanics, Handbuch der Physik, Vol. III/3. Springer-Verlag, Berlin.

Truesdell, C and Noll, W (1992). *The Non-linear Field Theories of Mechanics*, Second edition, Springer-Verlag.

Tsoutsinos, G.I. and Vinter, R.B. (1995). Duality theorems for convex problems with time delay, *J. Optim. Theory Appl.*, **87**, 167-195.

Uhlig, F. (1979). A recurring theorem about pairs of quadratic forms and extensions: a survey. *Lin. Alg. and its Appl.* **25** 219-237

van der Waals, J.D. (1895) (also see Rowlinson, 1979). *Arch. Neerl. Exactes Nat.*, **28**, 121-209.

van Hove, L. (1947). Surl'extension de la condition de Lagendre du calcul des variations aux intégrales multiples á plusieurs fonctions inconnues, *Proc. Koninkl. Ned. Akad. Wetenschap* **50**, 18-23

Vainberg, M.M. (1964). *Variational Methods for Study of Nonlinear Operator Equations*, Holden-Day, San Francisco.

von Kármán, T. (1910). Festigkeisprobleme im Maschinebau, in *Encyclopädie der Mathematischen Wissenschaften IV/4C*, 311-385.

Walk, M. (1989). *Theory of duality in mathematical programming*, Springer-Verlag: Wien, New York.

Wang, R. *et al* (1965). Discussion on the paper by Qiang, L.X. and Zhong, W.X. (1963), *Acta Mechanica Sinica,* **8**, 63-79.

Washizu, K. (1955) On the variational principles of elasticity and plasticity, Aeroelastic and Structures Research Laboratory, Technical Report 25-18, MIT, Cambridge.

Washizu, K. (1967) *Variational Methods in Elasticity and Plasticity*, Pergamon Press, New York

Wempner, G (1986). A general theory of shells and the complementary potentials, *J. Appl. Mech.* **53**, 881-885.

Weyl, H. (1952). *Symmetry*, Princeton University Press, Princeton.

Willem, M. (1982). Remarks on the dual least action principle, *Z. Anal. Anwendungen*, **1**, 85-90.

Wright, S.J.(1997). *Primal-Dual Interior-Point Methods.* SIAM. Philadelphia.

Wu, C.H. (1994). Stress-induced phase transformations in solids and the associated double-well potentials, *Int. J. Solids Struct.* **32**, 525-542.

Wu, X.M. (1984). *Approximation-Transforming Theory and Pansystem Methodology in Mathematics*, Science and Technology Press, Hunan (in Chinese).

Wu, X.M. (1987). Pansystems methodology and PSEA: Pedagogy, sociology, epistemology, aesthetics, *Busefal*, **31**, 157-161.

Yang, C.N. (1984). *A Brief History for Particle Physics*, Science and Technology Press, Shanghai (in Chinese).

Yang, W.H. (eds.) (1991). *Topics in Plasticity*, AM Press, Ann Arbor, Michigan.

Yang, W.H. (1991). On generalized Höder inequality, *Nonlinear Analysis, TMA*, **16**, 489-498.

Yau, S.-T. and Gao, D.Y. (1992). Obstacle problem for von Karman equations, *Adv. Appl. Math.* **13**, 123-141.

Yoshimura, H. (1995). *Modelling of Flexible Multibody Dynamics,* Doctoral Dissertation, Waseda University.

Yosida, K. (1966). *Functional Analysis.* Springer-Verlag, Berlin (6th edition, 1980).

Yuan, G.X.Z., Isac, G., Tan, K.K. and Yu, J. (1998). The study of minimax inequalities, abstract economics and applications to variational inequalities and Nash equilibria, *Acta Appl. Math.*, **54**, 135-166.

Zeidler, E. (1985). *Nonlinear Functional Analysis and its Applications: III Variational Methods and Optimization*, Springer-Verlag: New York-Berlin.

Zeidler, E. (1988). *Nonlinear Functional Analysis and its Applications: IV Applications to Mathematical Physics*, Springer.

Zheng, Q.-S. (1994). Theory of representations for tensor functions: A unified invariant approach to constitutive equations. *Appl. Mech. Rev.* **47**(11), 545-587.

Zubov, L.M., (1970). The stationary principle of complementary work in nonlinear theory of elasticity, *Prikl. Mat. Mech.*, **34**, pp. 228-232.

Index